Integrated Analytical Systems

This comprehensive and interdisciplinary series offers the most recent advances in all key aspects of development and applications of modern instrumentation for chemical and biological analysis on the microscale.

These key aspects will include (1) innovations in sample introduction through micro- and nano-fluidic designs, (2) new types and methods of fabrication of physical transducers and ion detectors, (3) materials for sensors that became available due to the breakthroughs in combinatorial materials science and nanotechnology, and (4) innovative data processing and mining methodologies that provide dramatically reduced rates of false alarms.

Clearly, a true multidisciplinary effort is required to meet objectives for a system with previously unavailable capabilities. This cross-discipline fertilization is driven by the expanding need for chemical and biological detection and monitoring and leads to the creation of instruments with new capabilities for new demanding applications. Indeed, instruments with more sensitivity are required today to analyze ultra-trace levels of environmental pollutants, pathogens in water, and low vapor pressure energetic materials in air. Sensor devices with faster response times are desired to monitor transient in-vivo events and bedside patients. More selective instruments are wanted to analyze specific proteins in vitro and analyze ambient urban or battlefield air. For these and many other applications, new features of modern microanalytical instrumentation are urgently needed. This book series is a primary source of both fundamental and practical information on both the current state of the art and future directions for microanalytical instrumentation technologies. This book series is addressed to the rapidly growing number of active practitioners and developers and those who are interested in starting research in this direction, directors of industrial and government research centers, laboratory supervisors and managers, students and lecturers.

For further volumes:
http://www.springer.com/series/7427

Ghenadii Korotcenkov

Handbook of Gas Sensor Materials

Properties, Advantages and Shortcomings for Applications Volume 2: New Trends and Technologies

Ghenadii Korotcenkov
Materials Science and Engineering
Gwangju Institute of Science and Technology
Gwangju, Korea, Republic of (South Korea)

ISSN 2196-4475 ISSN 2195-4483 (electronic)
ISBN 978-1-4939-4877-2 ISBN 978-1-4614-7388-6 (eBook)
DOI 10.1007/978-1-4614-7388-6
Springer New York Heidelberg Dordrecht London

Springer is part of Springer Science+Business Media (www.springer.com)

Preface

Sensing materials play a key role in the successful implementation of gas sensors, which, year by year, find wider application in various areas from environmental control to everyday monitoring of such activities as public safety, engine performance, medical therapeutics, and many more. Gas sensors can also be found in various industries such as chemical and petrochemical industries, food and drinks processing, semiconductor manufacturing, agriculture, fabrication industries, including the motor, ship, and aircraft industries, power generation, etc., where control and analysis of process gases are necessary. At present industrial processes increasingly involve the use and manufacture of highly dangerous substances, particularly toxic and combustible gases. Inevitably, occasional escapes of gas occur, creating a potential hazard to the industrial plant, its employees, and people living nearby. Gas sensors allow detection of toxic and combustible gases in atmosphere and, therefore, the use of these devices can prevent disastrous consequences for people.

However, the multidimensional nature of the interactions between function and composition, preparation method, and end-use conditions of sensing materials often make their rational design for real-world applications very challenging. Moreover, the world of sensing materials is very broad and practically all well-known materials could be used for the chemical sensors elaboration. Therefore, the selection of optimal sensing material for gas sensor is complicated and multivariate task. However, one should note that the number of published books describing the analysis of materials through their application in the field of gas sensors is very limited.

Moreover, most of them are devoted to analysis of one specific sensing material, for example, polymer or metal oxide. Therefore, it is very difficult to conduct a comparative analysis of various materials and to choose sensing material optimal for concrete application.

Taking this situlation into account, I decided to fill this gap. My main goal was to create a really useful encyclopedic handbook of gas sensor materials. The *Handbook of Gas Sensor Materials: Properties, Advantages and Shortcomings for Application* is the first book containing a comprehensive examination of materials suitable for gas sensor design. For convenience of practical use, the present *Handbook* is divided into two parts: Vol.1: *Conventional Approaches* and Vol.2: *New Trends in Materials and Technologies*. In these books one can find detailed analysis of conventional gas sensing materials such as metal oxides, polymers, metal films, and semiconductors. New trends in gas sensing materials include analysis of, among other materials, 1D metal oxide nanostructures, carbon nanotubes, fullerences, graphene, semiconductor quantum dots, and metal nanoparticles. The properties and applications of nanocomposites, photonic crystals, calixarenes-based compounds, ion conductors, ion liquids, metal-organic frameworks, porous semiconductors, ordered mesoporous materials, and zeolites are also discussed in the books. Close attention is given in these books to examining problems connected with stability and functionalizing of gas sensing materials. It is known that high stability is the main requirement for materials aimed for use as a gas sensor. The book chapters introduce analysis of general approaches to selection of sensing materials for gas sensor design. Auxiliary materials used in gas sensors such as substrates, catalysts, membranes, heaters, and electrodes are also discussed. Thus, in these two volumes, the reader can find comparative analyses of all materials acceptable for gas sensor design and can estimate their real advantages and shortcomings. This means that one can consider the present books as a selection guide of materials aimed for gas sensor manufacture. In addition, the books contain a large number of tables with information necessary for gas sensor design. The tables alone make these books very helpful and comfortable for the user. Hence, my belief that these books comprise an encyclopedic handbook of gas sensor materials, which answers many questions arising during selection of optimal sensor materials and promotes an understanding of the fundamentals of sensor functioning and development of the technological route of their fabrication for applications in various types of gas sensors.

These books will be of real interest to all materials scientists, especially to researchers working or planning to begin working in the field of gas sensing materials study and gas sensor design. The books will also be interesting for practicing engineers and project managers in industries and national laboratories who are interested in the development and fabrication of gas sensors for the sensor market. With many references to the vast resource of recently published literature on the subject, these books intend to serve as a significant source of valuable information, which will provide scientists and engineers with new insights for understanding and improving existing devices and for designing new materials for making better gas sensors.

I believe that these books will also be useful to university students, postdocs, and professors. The structure of the books offers the basis for courses in the field of materials science, chemical sensors, sensor technologies, chemical engineering, semiconductor devices, electronics, and environmental control. Graduate students could also find the books useful while conducting research and trying to understand the basics of gas sensor design and functioning. I hope very much that in these books all will find specific information that will be of interest and use in his/her area of scientific and professional interests.

Gwangju, South Korea Ghenadii Korotcenkov

Series Preface

In my career I've found that "thinking outside the box" works better if I know what's "inside the box."

Dave Grusin, composer and jazz musician

Different people think in different time frames: scientists think in decades, engineers think in years, and investors think in quarters.

Stan Williams, Director of Quantum Science Research, Hewlett Packard Laboratories

Everything can be made smaller, never mind physics;
Everything can be made more efficient, never mind thermodynamics;
Everything will be more expensive, never mind common sense.

Tomas Hirschfeld, pioneer of industrial spectroscopy

Integrated Analytical Systems

The field of analytical instrumentation systems is one of the most rapidly progressing areas of science and technology. This rapid development is facilitated by (1) the advances in numerous areas of research that collectively provide the impact on the design features and performance capabilities of new analytical instrumentation systems and by (2) the technological and market demands to solve practical measurement problems.

The book series *Integrated Analytical Systems* reflects the most recent advances in all key aspects of development and applications of modern instrumentation for chemical and biological analysis. These key development aspects include: (1) innovations in sample introduction through micro- and nano-fluidic designs, (2) new types and methods of fabrication of physical transducers and ion detectors, (3) materials for sensors that became available due to the breakthroughs in biology, combinatorial materials science and nanotechnology, (4) innovative data processing and mining methodologies that provide dramatically reduced rates of false alarms, and (5) new scenarios of applications of the developed systems.

A multidisciplinary effort is required to design and build instruments with previously unavailable capabilities for demanding new applications. Instruments with more sensitivity are required today to analyze ultra-trace levels of environmental pollutants, pathogens in water, and low vapor pressure energetic materials in air. Sensor systems with faster response times are desired to monitor transient in-vivo events and bedside patients. More selective instruments are sought to analyze specific proteins *in vitro* and analyze ambient urban or battlefield air. Distributed sensors for multiparameter measurements (often including not only chemical and biological but also physical measurements) are needed for surveillance over large terrestrial areas or for personal health monitoring as wearable sensor networks. For these and many other applications, new analytical instrumentation is urgently needed. This book series is intended to be a primary source on both fundamental and practical information of where analytical instrumentation technologies are now and where they are headed in the future.

Niskayuna, NY, USA

Radislav A. Potyrailo
GE Global Research

Acknowledgments

I would like to express my great gratitude to Gwangju Institute of Science and Technology, Gwangju, Korea, which invited me and gave me an opportunity to prepare this book for publication. This work was supported by Korean BK 21 Program, and by Basic Science Research Program through the National Research Foundation of Korea (NRF) funded by the Ministry of Education, Science and Technology (2012R1A1A2041564). I am also thankful to all my colleagues: Dr. Vladimir Brinzari (Moldova), Dr. Sang Do Han (Korea), Professors Beongki Cho (Korea), Johannes W. Schwank (USA), Albert Cornet (Spain), Joseph Stetter (USA), Joan R. Morante (Spain), Vladimir Matolin (Chezh Rep.), and Valeri P. Tolstoy (Russia). Our successful collaboration, lasting for a long time, was based on their important contributions to my work, such as valuable advises, important information on new phenomena, and participation in mutual experiments and discussions. I would also like to thank my wife Irina Korotcenkova for her patience, understanding, and support during my work on the manuscript.

Gwangju, South Korea
Ghenadii Korotcenkov

Contents

Part VI Technology and Sensing Material Selection

Contents of Volume I

Part I
Nanostructured Gas Sensing Materials

Chapter 1
Carbon-Based Nanostructures

As mentioned earlier in Volume 1, there are no ideal sensing materials which meet all requirements. That is why research is continually being conducted to search for new sensing materials with new properties which might be used in the development of gas sensors with new and unusual functional characteristics.

1.1 Carbon Black

Carbon black (CB) is one of numerous forms of carbon (see Table 1.1). CB is a material produced by the incomplete combustion of heavy petroleum products such as FCC (fluid catalytic cracking) tar, coal tar, and ethylene cracking tar, and a small amount comes from vegetable oil. Carbon black is a form of amorphous carbon that has a high surface-area-to-volume ratio. However, in spite of that fact, carbon black, due to specific conductivity and mechanical properties, is not being used as a sensing material in gas sensors. Only activated carbon, also called activated charcoal, activated coal or carbon activates, one can find in gas sensors where CB can be used as a filter. Activated carbon is a form of carbon that has been processed to make it extremely porous and thus to have a very large surface area available for either adsorption or chemical reactions. Due to its high degree of microporosity, just 1 g of activated carbon has a surface area in excess of 500 m^2.

Other possibility for carbon black to be integrated in gas sensors is connected with using composites, where another material provides the gas-sensing properties while carbon black plays the part of filler, characterized by high conductivity and high dispersion. The key carbon black properties useful for composites design are excellent dispersion, integrity of the carbon black structure or network, consistent particle size, specific resistance, structure, and purity. As a rule, carbon black is used mainly in polymer-based composites. The carbon black endows electrical conductivity to the films, whereas the different organic polymers such as poly(vinyl acetate) (PVAc), polyethylene (PE), poly(ethylene-*co*-vinyl acetate) (PEVA), and poly(4-vinylphenol) (PVP) are sources of chemical diversity between elements in the sensor array. In addition, polymers function as the insulating phase of the carbon black composites. The concentration of CB in composites is varied within the range 2–40 wt%. The conductivity of these materials and their response to compression or expansion can be explained using percolation theory (McLachlan et al. 1990). The compression of a composite prepared by mixing conducting and insulating particles leads to increased conductivity, and, conversely, expansion leads to decreased conductivity. This effect is especially strong in the composites with compositions around the percolation threshold; an extremely small volume change of the phase due to an extrinsic perturbation

G. Korotcenkov, *Handbook of Gas Sensor Materials*, Integrated Analytical Systems,
DOI 10.1007/978-1-4614-7388-6_1, © Springer Science+Business Media New York 2014

Table 1.1 The properties of carbon allotropes

	Carbon allotropes				
Parameter	Graphite	Diamond	Fullerene (C_{60})	Carbon nanotube	Graphene
Hybridized form	sp^2	sp^3	Mainly sp^2	Mainly sp^2	sp^2
Dimension	Three	Three	Zero	One	Two
Crystal system	Hexagonal	Octahedral	Tetragonal	Icosahedral	Hexagonal
Experimental specific surface area (m^2/g)	~10–20	20–160	80–90	~1,300	~1,500
Density (g/cm^3)	2.09–2.23	3.5–3.53	1.72	>1	>1
Optical properties	Uniaxial	Isotropic	Nonlinear optical response	Structure-dependent properties	97.7 % of optical transmittance
Thermal conductivity (W/m·K)	1,500–2,000[a], 5–10[c]	900–2,320	0.4	3,500	4,840–5,300
Hardness	High	Ultrahigh	High	High	Highest (single layer)
Tenacity	Flexible nonelastic	–	Elastic	Flexible elastic	Flexible elastic
Electronic properties	Electrical conductor	Insulator, semiconductor	Insulator	Metallic and semiconducting	Semimetal, zero-gap semiconductor
Electrical conductivity (S/cm)	Anisotropic, $2–3 \times 10^{4a}$, 6[b]		10^{-10}	Structure dependent	2,000

Source: Reprinted with permission from Wu et al. (2012). Copyright 2012 Elsevier
[a]a-axis direction
[b]c-axis direction

brings about the resistivity change of the composite (see Fig. 1.1). This means that the swelling of the polymer upon exposure to a vapor increases the resistance of the film, thereby providing an extraordinarily simple means for monitoring the presence of organic vapor such as toluene, benzene, ethyl acetate, methanol, ethanol, 2-propanol, hexane, chloroform, acetone, and tetrahydrofuran (THF) (Lonergan et al. 1996). Individual carbon black composites can also be explored as humidity sensors. Typical operating characteristics of CB–polymer composite-based gas sensors are shown in Fig. 1.1. It is seen that composites' resistance increases during organic vapor absorption and returns to the initial value when the vapor desorbs completely.

It should be noted that the percolation threshold strongly depends on both the parameters of CB used and the technology of composite preparation. In different articles the percolation threshold was observed at CB contents which were varied from <3 wt% (Chen et al. 2005) to 33 wt% (Lonergan et al. 1996). This means that reproducibility of sensor parameters designed on the base of CB–polymer composites is not high.

It was established that many factors can influence the response of electrical resistance of CB–polymer-based composite sensors against organic solvent vapors. For example, modification of carbon black surface by grafting polymerization (Chen and Tsubokawa 2000), crystallinity, and molecular weight of polymer matrix (Chen et al. 2002), and content and dispersivity of carbon black in the composites (Dong et al. 2003), are closely related to the response, reproducibility, and stability of the composites. It was found that the response habit of the composites is a function of temperature and vapor pressure (Matzger et al. 2000). The maximum responsivity of the composites decreases with decreasing vapor pressure at a given temperature. The slopes of the relationships change with solvent species as a result of different solubilities, which might help to construct sensors or sensor arrays capable of quantifying and discriminating vapors of interest using simple signal treatment. For the tests in saturated vapors, elevated temperatures usually increase the rate of response (see Fig. 1.2a).

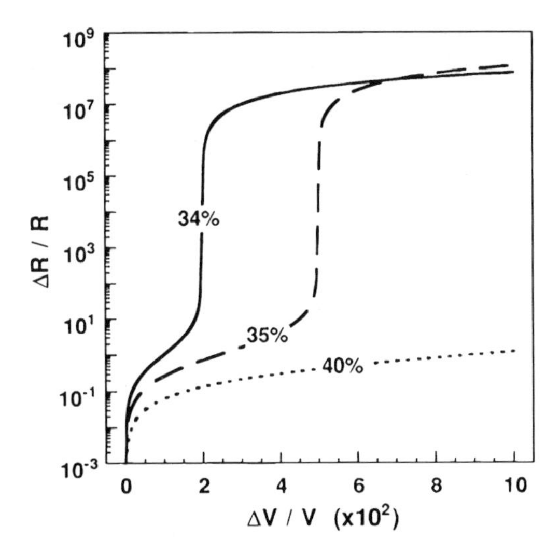

Fig. 1.1 Relative differential resistance change, $\Delta R/R$, predicted by percolation theory as a function of the relative volume change, $\Delta V/V$, of a carbon black–polymer composite upon swelling. The volume of carbon black is assumed to be unaffected by swelling, and the polymer matrix is assumed to have a conductivity 11 orders of magnitude lower than that of carbon black. The three separate lines are for composites with differing initial volume percentages of carbon black, as indicated. The percolation threshold for the system is at CB content = 0.33. The total volume change results in a change in the effective carbon black content. When, this value drops below the percolation threshold, a sharp increase in response is observed. Of course, the position of this sharp increase depends on the value of the percolation threshold (Reprinted with permission from Lonergan et al. 1996, Copyright 1996 American Chemical Society)

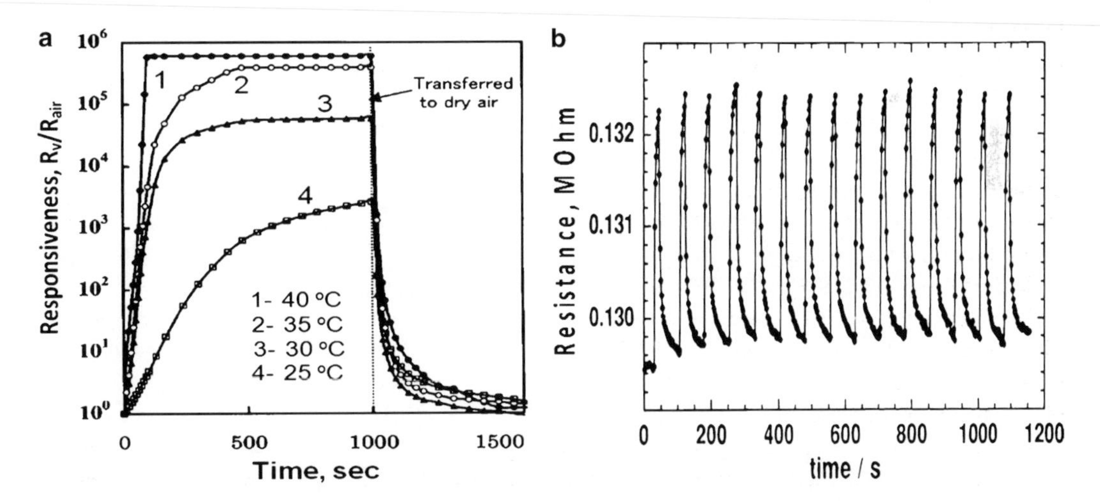

Fig. 1.2 (**a**) Effect of temperature on the responsiveness of the composite from low-density polyethylene (LDPE) and poly(ethylene-*block*-ethylene oxide) (PE-*b*-PEO)-grafted CB against the saturated cyclohexane vapor. The composites (CB/LDPE ~ 25 wt%) were exposed to the vapor with 1,000 s and then transferred to dry air (reprinted with permission from Chen and Tsubokawa (2000). Copyright 2000 Wiley). (**b**) Resistances, R, of carbon black composites of poly(4-vinylphenol) (PVP) upon 15 repeated exposures to methanol (at 1.5 ppt), respectively. The CB/PVP composite was fabricated from a 45 wt% carbon black mixture. Composite films were deposited onto glass slides. The exposure periods were for 15 s during which time the resistances increased as shown. These exposures were interlaced between recovery periods in which the resistances decreased. These traces demonstrate the good reproducibility and stability that can be achieved with carbon black composites (Reprinted with permission from Lonergan et al. 1996, Copyright 1996 American Chemical Society)

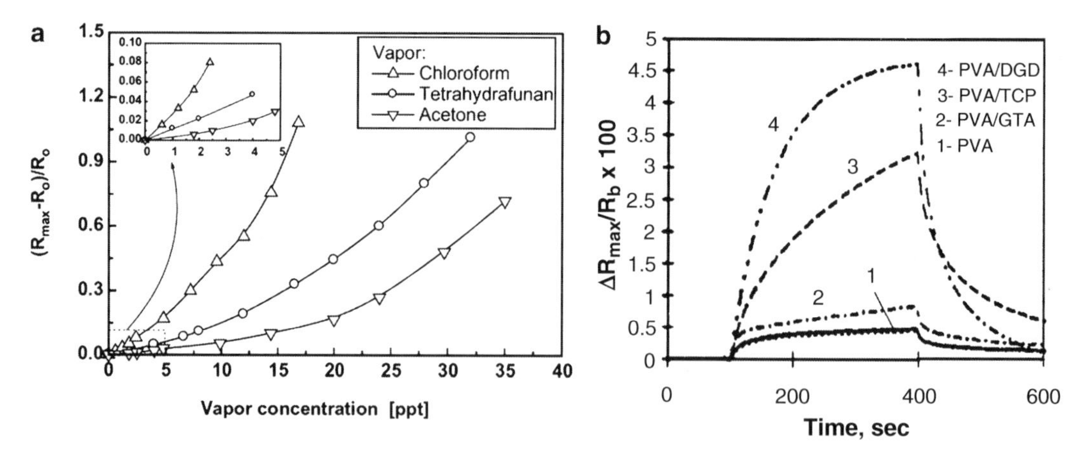

Fig. 1.3 (**a**) Vapor concentration dependences of the maximum electric resistance responses of carbon black/poly(methyl methacrylate) composites (CB/PMMA ~ 14 wt%) to the vapors of acetone, tetrahydrofuran, and chloroform. The solid regression lines are drawn according to equation $Q_e = kC_e^{1/n}$, where Q_e is an equilibrium adsorbate loading on the adsorbent, C_e is the equilibrium concentration of the adsorbate, and k and $1/n$ are constants, indicating adsorption capacity and intensity, respectively. The *inset* shows the situations at very low vapor concentrations. (Reprinted with permission from Dong et al. (2004). Copyright 2004 Elsevier). (**b**) Response curves for a poly(vinyl acetate)/carbon black composite detector compared to the response of plasticized PVA/carbon black detectors. The composites (CB/PVA ~ 20 wt%) were exposed to the acetone vapors (5 % of its vapor pressure at room temperature). The plasticizers, diethylene glycol dibenzoate (DGD), tricresyl phosphate (TCP), and glycerol triacetate (GTA), were present at 20 wt%. Results clearly indicate that the plasticizer had a distinct effect on the response properties of the base polymer used in the carbon black–polymer composite detectors (Reprinted with permission from Matzger et al. 2000, Copyright 2000 American Chemical Society)

However, at operating temperatures higher than 35–50 °C, a decrease of sensor response takes place (see Fig. 1.2a). It was established that as a rule the relationships between the electrical response and vapor concentration or partial pressure are nonlinear (see Fig. 1.3). According to Patel et al. (2000), the nonlinear relationship was once attributed to the interference from the content of CB, i.e., when the composites are near the percolation threshold, the next small addition of analyte causes a disproportionately large increase in electric resistance. This means that far from the percolation threshold the relationships between the electrical response and vapor concentration can be linear. However, Dong et al. (2004) believe that the vapor concentration dependence of the response expressed in terms of relative electric resistance variation of the composites should be described by the Freundlich isothermal adsorption model because solvent adsorption on the composites is the driving source for their resistance change. Therefore, the linear relationship between electric resistance response and vapor concentration at a relatively low concentration regime (Severin et al. 2000) is the first approximation of the absorption isotherm.

It was found that CB–polymer composite sensors have the following advantages. They are highly sensitive, inexpensive, easily controlled, and robust in many different environments. In addition they have simple fabrication processing and good compatibility with modern CMOS VLSI technology. As a result, CB–polymer-based sensors are promising for the design of various e-nose systems; they can be made to have diverse responses by choosing materials used as insulating polymers, additive plasticizers, and conductive carbon blacks, and by regulating the relative quantities of them, simply (see Fig. 1.3a). Moreover, these sensors provide the opportunity to fabricate very small size, low-power, and lightweight sensor arrays (Matzger et al. 2000; Kim et al. 2005; Xie et al. 2006).

There have been attempts to use metal oxide–CB composites for gas sensor design (Liou and Lin 2007). However, such an approach does not give any improvement in operating characteristics in comparison with conventional metal oxide or CB–polymer-based gas sensors.

Experiment has shown that for preparing polymer–CB composites, various methods can be used such as physical mixing of CB and the polymer matrix (Mather and Thomas 1997), in situ polymerization in the presence of CB (Dong et al. 2004), and ultrasonic mixing of CB and polymer powders (Ramos et al. 2005). However, it should be noted that generally, for most methods used, it is difficult to attain good dispersion of the CB into the polymer matrix which affects the percolation limit of the composite. This is due to the high surface energy, small particle size and strong agglomeration tendency of the CB, limited shear force of the mixer, and high viscosity of the polymer solution. To obtain CB which has high dispersibility, the CB is usually modified by coating the surface with an organic compound, such as an oligomer with a terminal active group (Chen and Tsubokawa 2000). Chen and Tsubokawa (2000) reported that this process is difficult because there are almost no active groups on the surface of conductive CB that can be used for a treatment reaction. Thus it is impossible to bind the organic compound directly onto the surface by a chemical reaction. Therefore, usually, a two-step modification process is used. In particular, Chen and Tsubokawa (2000) introduced carboxyl groups onto the CB surface through the trapping of 4-cyanopentanoic acid radicals, which came from the decomposition of 4,4′-azobis(4-cyanopentanoic acid) (ACPA). Then as a second step, poly(ethylene- *block*-ethylene oxide) (PE-*b*-PEO) was grafted onto the surface by direct condensation between terminal hydroxyl groups of PE-*b*-PEO and carboxyl groups on the CB surface in the presence of *N,N*′-dicyclohexylcarbodiimide (DCC), as a condensing agent. Chen and Tsubokawa (2000) established also that the responsibility of LDPE/CB composite with PE-*b*-PEO-grafted CB is more stable and reproducible than that from untreated CB. Arshak et al. (2005) found that the treatment of composites with surfactants such as Hypermer PS3 and Hypermer PS4 (Uniqema) also gives the improvement of gas-sensing characteristics. The percolation curves of surfactant-treated composites showed that the resistivity of the composite was increased due to better dispersion of the CB and also the prevention of the CB from reagglomerating after shear mixing. The TEM images confirmed that the surfactants significantly improved the level of dispersion of CB in the composites and prevented reagglomeration of the CB.

One can find a detailed analysis of the peculiarities of composites' application in gas sensor elaboration in Chaps. 12 and 13 (Vol. 2).

1.2 Fullerenes

Fullerenes are closed-cage carbon molecules containing pentagonal and hexagonal rings arranged in such a way that they have the formula C_{20+m}, with *m* being an integer number (Dresselhaus et al. 1996; Mauter and Elimelech 2008) (see Fig. 1.4). They are the fifth allotropic form of carbon, the others being

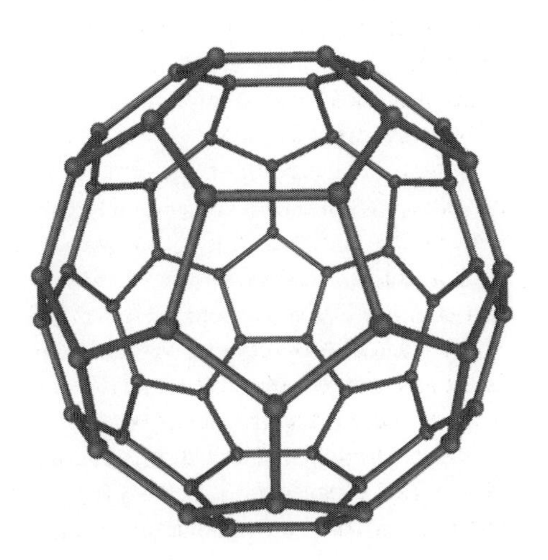

Fig. 1.4 Schematic view of fullerenes (Reprinted from http://commons.wikipedia.org)

Table 1.2 Physicochemical characteristics of carbon-based materials

Isomer	Fullerenes (C_{60})	Nanotubes	Graphite	Graphene	Diamond
Dimension	0D	1D	2D	2D	3D
Hybridization	sp^2-like	sp^2	sp^2	sp^2	sp^3
Density, g/cm^3	1.72	1.2–2.0	2.26	>1.0	3.515
Bond length, nm	0.14 (C=C) 0.146 (C–C)	0.144 (C=C)	0.142 (C=C) 0.144 (C–C)	0.142 (C=C)	0.154 (C–C)
Electronic properties	Semiconductor $E_g = 1.9$ eV	Metal or Semiconductor	Semimetal	Semimetal	Insulating $E_g = 5.47$ eV

Source: Data from Saito et al. (1998)

graphite, diamond, carbon nanotube (CNT), and graphene (see Table 1.2). Fullerenes comprise a wide range of isomers and homologous series, from the most studied C_{60} or C_{70} to the so-called higher fullerenes, C_{240}, C_{540}, and C_{720}. The first of these compounds was discovered in 1985 through spectrometric measurements on interstellar dust, and their structure was confirmed later in the laboratory (Kroto et al. 1985). Kroto, Smalley, and Curl received the Nobel Prize in 1996 for their work.

Physicists, chemists, and material scientists or engineers, among others, have found unusual potential in these new spherical carbon structures for use as superconductor materials, sources of new compounds, self-assembling nanostructures, and several optical devices (Dresselhaus et al. 1996). This initial attention led to an increasing number of investigations that revealed the special properties of fullerenes, some of which might lead to practical applications (Mauter and Elimelech 2008). Although a wide range of uses has been explored and several applications developed, fullerenes are not so far fulfilling their initial spectacular promise (Baena et al. 2002). Research on the application of fullerenes has proved to be slower than expected, but it must not be considered unsuccessful when one considers the great advances in the knowledge of the physical and chemical characteristics of fullerenes. Thanks to the additional information obtained during recent years, they have been found to be really useful in several fields, particularly in analytical chemistry.

It was established that a characteristic feature of fullerenes is their affinity for various organic molecules. Therefore, fullerene C_{60} with 60 π-electrons potentially can be used as a good adsorbent to adsorb and detect nonpolar and some polar organic molecules. However, fullerenes cannot adsord metal ions, anions, and most polar organic species. Taking into account the above-mentioned properties, fullerenes in analytical chemistry can be approached from two different points of view (Baena et al. 2002). The first sees fullerenes as analytes, which involves their determination in various samples such as biological tissues. The second sees fullerenes as analytical tools, including their use as chromatographic stationary phases, as electrochemical sensors based on their activity as electron mediators, and in the exploitation of their unique superficial characteristics as sorbent materials in continuous-flow systems.

Initially, to establish the analytical features of fullerenes as sensors, adsorption studies were carried out on organic molecules bound onto fullerenes. For this purpose, the adsorption of gases and organic vapors was studied with fullerene-coated devices sensitive enough to detect changes in mass or pressure related to the adsorption of gas molecules onto the fullerene layer. Such devices are surface acoustic wave- (SAW) and quartz microbalances-based (QMBs) gas sensors. Through these first investigations, the retention of certain monomeric gas molecules was demonstrated, and consideration was given to the possible use of C_{60} films as analytical sensors for volatile polar gases such as NH_3 (Synowczyk and Heinze 1993). Gas adsorption onto the fullerene film reduces the film resistance, resulting in a charge transfer to the electronic system. Sensitivity levels of a few milligrams per liter of NH_3 in air were achieved, but there were still some problems, such as the lack of selectivity vs. other gas vapors (which were also adsorbed, leading to the same electrical signal), response times of the order of seconds, the influence of humidity level on the calibration, or instability of the sensor

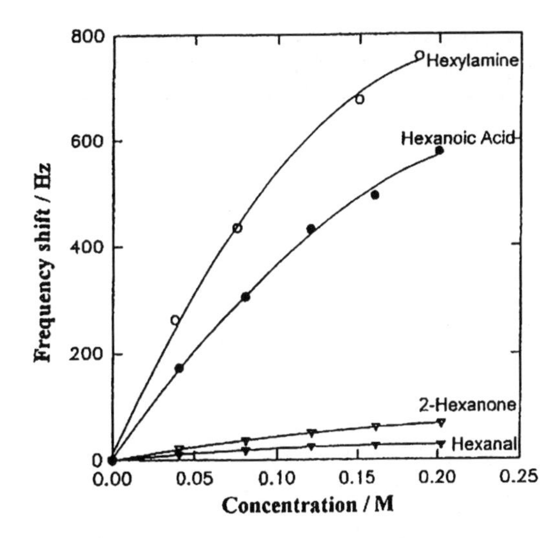

Fig. 1.5 Frequency responses of the C_{60}-coated PZ detector for polar organic molecules (Reprinted with permission from Shih et al. 2001, Copyright 2001 Elsevier)

Table 1.3 Detection limits of various organic vapors with the C_{60}-cryptand 22 coated SAW sensor

Organic vapors	Detection limit (mg/L)	Organic vapors	Detection limit (mg/L)
Methanol	0.80	Diethyl ether	3.60
Ethanol	0.70	Acetone	2.60
n-Propanol	0.48	Propionaldehyde	0.85
n-Butanol	0.25	Hexane	2.30
iso-Butanol	0.27	Hexene	0.80
tert-Butanol	0.70	–	–

Source: Reprinted with permission from Lin and Shih (2003). Copyright 2003 Elsevier

when exposed to air several times. Nevertheless, the use of fullerenes as modifiers was found to be a promising research topic in several fields, especially as coatings in QMBs and SAW sensors, since it is well known that the presence of fullerenes improves the electrochemical characteristics of the film or membrane by reducing the resistivity.

Various reusable and sensitive piezoelectric (PZ) quartz crystal microbalance (QCM) sensors have recently been developed to detect organic/inorganic vapors (see Fig. 1.5) and organic/inorganic biological species in solutions. Fullerene C_{60} and fullerene derivatives, among others, were synthesized and applied as coating materials on quartz crystals of QCM sensors (Chao and Shih 1998; Shih et al. 2001) and SAW-based sensors (Lin and Shih 2003). In particular, in sensors designed in (Shih et al. 2001; Lin and Shih 2003), C_{60}-cryptand 22 and C_{60}-dibenzo-16-crown-5-coated quartz crystals were used. Thus, chemisorption on C_{60} fullerene was observed for amines, diamines, dithiols, dienes, and alkynes, and only physical adsorption was found for carboxylic acids, aldehydes, alcohols, ketones, alkenes, and alkanes. This seems to imply that the nucleophilic addition to fullerene by polar electron-donor groups, as in amines and thiols, is easier than electrophilic addition. Furthermore, diamines and dithiols showed greater interactions than those for the monodentate form, behavior being attributed to the formation of stable cyclic compounds between fullerene and the bidentate ligand. It was established that SAW-based devices have higher sensitivity in comparison with QCM-based devices (Lin and Shih 2003). Detection limits of various organic vapors with the C_{60}-cryptand 22 coated SAW sensor are listed in Table 1.3.

The fullerene C_{60}-coated PZ crystal gas sensor was also set up and employed to study the interaction between C_{60} and some inorganic vapors, e.g., ozone, HCl, and HNO_3 (Shih et al. 2001). As shown in

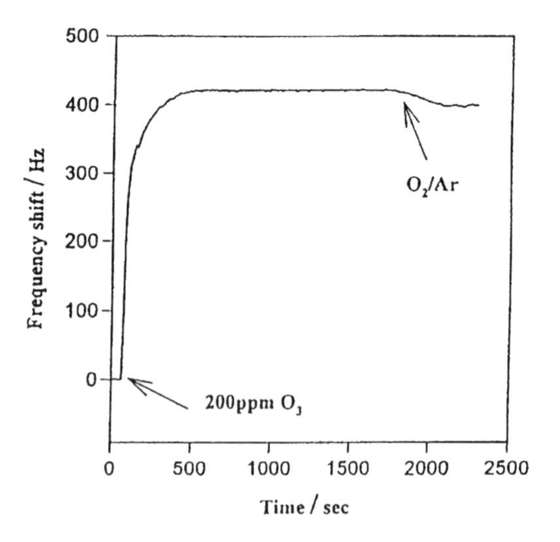

Fig. 1.6 Response of C_{60}-coated piezoelectric gas detector to ozone in the air (Reprinted with permission from Shih et al. 2001, Copyright 2001 Elsevier)

Fig. 1.6, frequency of the C_{60}-coated PZ crystal gas sensor shows increases after the adsorption of ozone molecules. The decomposition of fullerene C_{60} into small pieces after oxidation by ozone was reported by Taylor et al. (1991), which can lead to the decreased mass onto the PZ crystal and results in decrease in the frequency of the C_{60}-coated PZ crystal gas sensor. The oxidation of C_{60} can also be confirmed by the irreversible response (Fig. 1.6) after introducing the clean air. It was established that, after reacting with the ozone molecule, fullerene C_{60} exhibited greater adsorbing ability with organic species, e.g., propanol, than the original C_{60} molecule. This result may be attributed to the increased polarity of the fullerene molecule after reacting with the ozone molecule. Suzuki et al. (1991) also reported that the electrophilic property of fullerene C_{60} increased after the oxidation of C_{60} with ozone. The irreversible response and the change in IR peaks were also found in the study of the interaction between C_{60} and HCl or HNO_3 by using the C_{60}-coated PZ crystal gas sensor. The reactivity of HNO_3 toward the C_{60} molecule seems greater than that of HCl.

It was also established that mass-sensitive sensors coated by C_{60} film have high sensitivity to humidity (Radeva et al. 1997). This feature of fullerenes, of course, is a disadvantage for sensor application. Moreover, Sberveglieri et al. (1996) have found that humidity decreased the electrical response of the C_{60} film on hydrogen.

In independent studies, fullerene has been widely used as an electron mediator in electrodes, since the incorporation of C_{60} significantly reduces the electrical resistance of the coating membrane. By way of example, an iodide-sensitive sensor was reported (Wang et al. 1996) in which the bilayer lipid membrane supported on a copper wire—which acted as a modified electrode—also contained C_{60} fullerene. The resulting electrode was further used in a three-electrode system for the determination of iodide in solution, obtaining a detection limit of 10 nM.

Recently, the optical properties of C_{60} have also been applied to the development of a sensitive oxygen-sensing system based on the quenching of the photo-excited triplet state of fullerene molecules (Bouchtalla et al. 2002; Nagl et al. 2007; Baleizao et al. 2008). Although the amperometric oxygen electrode has been the most popular sensing system for this element, the instability of the electrode surface itself, and in the oxygen diffusion barrier, demands a practical alternative. Much attention has been given to optical sensing systems based on luminescence quenching of an indicator (organic dye, polyamide–hydrazide (PAH)-transition metal complex). The C_{60} and C_{70} fullerene can also be used as an indicator. It was found that fullerenes have strong thermally activated delayed fluorescence at elevated temperatures that is extremely oxygen sensitive (Baleizao et al. 2008). In addition, fullerenes can easily form thermally stable films with polymers, such as polystyrene (PS), and possess useful electronic and photochemical properties, such as a fairly long lifetime for the

Table 1.4 Comparison between the materials used so far for sensing of oxygen

Oxygen probe	Polymer	λ_{exc} (nm)[a]	Signal[b]	O_2 range (%)
Pt-TFPP	FIB	337	DT, TD	0–20
Ru-dpp	Sol–gel	470	DT, TD	0–100
Pt-TFPP lactone	FIB	390	DT, I, TD	0–20
Pt-TFPP	p-*t*BS-co-TFEM	465	I	0–20
Pd-TFPP in PSAN microbeads	Hydrogel for both particles	405	DT, FD	0–20
Pd-TFPP in PSAN microbeads	Hydrogel for both particles	470 (Ru)	DT, FD	0–100
Pt-TFPP in PSAN microbeads	Hydrogel for both particles	525 (Pd) 405	DT, TD	1–40
C_{70} in OS or EC film		470	DT, TD	0–0.005 (0–50 ppmv)

Source: Reprinted with permission from Baleizao et al. (2008). Copyright 2008 American Chemical Society
DT luminescence decay time, *I* luminescence intensity, *TD* time domain, *FD* frequency domain, *EC* ether ethyl cellulose, *OS* organosilica
[a]For both luminophores, except when mentioned otherwise
[b]Analytical signal

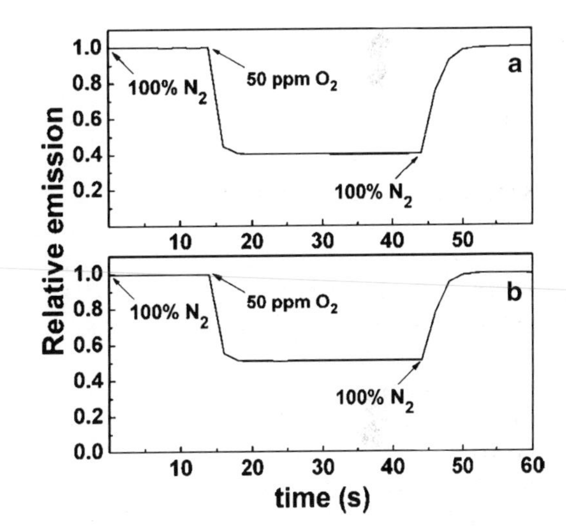

Fig. 1.7 Fluorescence intensity response time plots for (**a**) C_{70}/EC and (**b**) C_{70}/OS at 20 °C and for oxygen concentrations between 0 and 50 ppmv in nitrogen at atmospheric pressure (Reprinted with permission from Baleizao et al. 2008, Copyright 2008 American Chemical Society)

photo-excited triplet state (~100 μs). This lifetime is effectively quenched by oxygen and decreases with increasing oxygen concentration (Arbogast et al. 1991). Thus, by using time-resolved spectroscopy with laser-flash photolysis, a highly sensitive oxygen sensor can be obtained (see Table 1.4).

Baleizao et al. (2008) established that if the materials reported so far display operation temperatures between 0 and 70 °C, and 0 and 20 % or 0 and 100 % for oxygen concentrations (see Table 1.4), the fullerene-based sensor is specifically suited for the determination of trace amounts of oxygen and covers a very wide temperature range. Experiment has shown that the Ru(phen)3/PAN-C_{70}/EC and Ru(phen)3/PAN-C_{70}/OS sensing materials cover a temperature range between 0 and 120 °C and allow the measurement of oxygen concentrations between 0 and 50 ppmv with LODs in the ppbv range. The response time of the oxygen sensor within the concentration range used is less than a few seconds (see Fig. 1.7). The cross sensitivity of C_{70} to temperature is accounted for by means of the temperature sensor.

The unavailability and high cost of fullerenes have probably deterred their use in analytical chemistry. Many firms now supply fullerenes at reasonable prices. It is therefore optimistically forecasted

(Baena et al. 2002) that the advantages of fullerenes as sorbent materials, chromatographic stationary phases, and active microzones in sensors, based on their unique characteristics, will be consolidated and extended in the near future. One of the foreseeable trends is the use of synthetic fullerene derivatives that exhibit better properties than the original fullerenes. The introduction of radicals into the fullerene spheres can lead to an increase in the reversible sorption of organic molecules, as well as to direct retention and elution of metal traces by covalent binding of typical ligands such as EDTA and DDC. The unusual electrical properties of fullerenes can be fully exploited by progressively substituting the conventional carbon forms in building macro- and microelectrodes.

1.3 Carbon Nanotubes

CNTs were first fabricated in 1991 by Iijima (1991). Starting from this time, a great deal of effort has been devoted to the fundamental understanding of their properties and of their use in a wide range of applications such as electronics, catalysis, filters, and sensors, (Schnorr and Swager 2011). It was established that there are two types of CNT morphology (Saito et al. 1998; Dresselhaus et al. 1996; Varghese et al. 2001; Terrones et al. 2004). On the one hand, single-walled carbon nanotubes (SWCNTs) consist of a honeycomb network of carbon atoms and can be visualized as a cylinder rolled from a graphitic sheet. On the other hand, multi-walled carbon nanotubes (MWCNTs) are a coaxial assembly of graphitic cylinders generally separated by a plane space of graphite (Dresselhaus et al. 1996) (see Fig. 1.8). Each tubule in MWCNTs has a diameter ranging typically from 2 to 25 nm in size with 0.34 nm distance between sheets close to the interlayer spacing in the graphite. The diameter and the length of the SWCNTs typically vary from 0.5 to 3 nm and from 1 to 100 μm, respectively. CNTs have the tendency to aggregate, usually forming bundles that consist of tens to hundreds of nanotubes in parallel and in contact with each other. This effect is due to strong van der Waals interactions between the nanotubes. Synthesis methods for SWCNTs and MWCNTs include arc discharge, laser ablation, pyrolysis, chemical vapor deposition (CVD), and gas-phase catalytic growth (Terrones et al. 2004; Mamalis et al. 2004; Kuchibhatla et al. 2007; Zhang and Zhang 2009). However, till now these methods have not produced a monodisperse product with controlled physical and chemical properties.

It was established that this novel material shows extraordinary physical, mechanical, and chemical properties. Actually, CNTs have demonstrated very high carrier mobility in field-effect transistors, a very high electromigration threshold, a very high thermal conductivity, and exceptional mechanical properties. The electronic structure of SWCNTs can be either metallic or semiconducting, depending

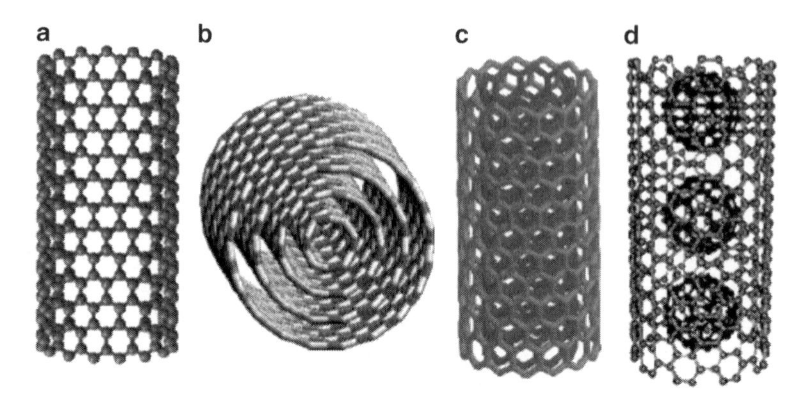

Fig. 1.8 Schematic diagrams of (**a**) a single-wall carbon nanotube (SWNT), (**b**) a multiwall carbon nanotube (MWNT), (**c**) a double-wall carbon nanotube (DWNT), and (**d**) a peapod nanotube consisting of an SWNT filled with fullerenes (e.g., C_{60}) (Reprinted with permission from Dresselhaus et al. 2003, Copyright 2003 Elsevier)

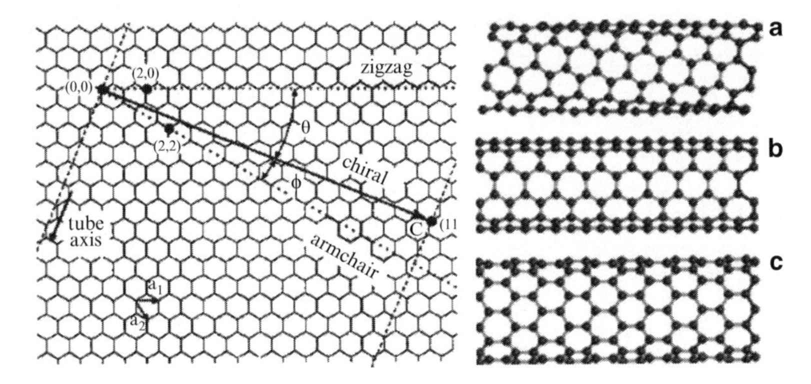

Fig. 1.9 The carbon lattice and the ways it can be rolled up to form a zigzag (**a**), an armchair (**b**), or a chiral (**c**) single-walled nanotube, depicted with its chiral angle (Adapted from Mamalis et al. 2004, Copyright 2004 Elsevier)

on their diameter, chirality, or helicity (symmetry of the two-dimensional carbon lattice) (see Fig. 1.9) (Dresselhaus et al. 1996, 2003; Odom et al. 1998; Varghese et al. 2001). Semiconducting SWCNTs are *p*-type semiconductors with holes as the main charge carriers. The bandgap of semiconducting SWCNTs is inversely related to their diameter and corresponds to ~0.8 eV for a tube with a diameter of 1 nm. It is important to note that theoretical calculations indicate that all armchair tubules have metallic electronic properties only (Saito et al. 1992). It is supposed (Valentini et al. 2003) that these diverse electronic properties of CNTs make it possible to develop nanoelectronic devices as metal/semiconductor heterojunctions by combining metallic and semiconducting nanotubes. A possible approach is the modification of different parts of a single nanotube to have different electronic properties using controlled mechanical or chemical processes (e.g., nanotube bending or gas molecule adsorption).

It should be noted that CNTs have the same developed surface as fullerenes, and therefore their applications should lie in the same general area, namely in the field analytical chemistry, in particular gas sensing (Mauter and Elimelech 2008). Moreover, CNTs seem to be more suitable for adsorption and detection of gases because small diameter and hollow structure makes them extremely sensitive to changes in their surroundings; all the atoms on a CNT are exposed to its environment, and the extremely small diameter forces electrical signals traveling along the tube to interact with even tiny defects on or near the tube. As a result, gas adsorption on CNTs is now the focus of intense experimental and theoretical studies.

Results of research carried out in this area have shown that CNTs may really find successful applications in the design of room-temperature adsorption/desorption type gas sensors such as SAW, QCM, and capacitance, where their peculiar structural features could be realized (Varghese et al. 2001; Kuchibhatla et al. 2007; Zhang et al. 2008; Schnorr and Swager 2011). In particular, Wei et al. (2003) designed QCM-based gas sensors with deposited CNTs bundles. This sensor detected CO, NO_2, H_2, and N_2 by detecting changes in oscillation frequency and was more effective at higher temperatures (200 °C). Moreover, it was established that such sensors can be extremely sensitive. For example, research conducted by Penza et al. (2004a, b) showed that, at room temperature, CNT-based SAW sensors were up to three to four orders of magnitude more sensitive than existing organic layer–coated SAW sensors. The mass sensitivity of CNT sensors can reach zeptograms (10^{-21} g). Therefore, they have a very low limit of detection, and 1 ppm of ethanol or toluene is easily sensed. Numerous studies have shown that SWNT-based sensors usually have better a performance compared to MWNT sensors while preparation of MWNT is easier.

It was established that selectivity to volatile organic compounds (VOCs) can be affected by the type of organic solvent used to disperse the CNTs as sensing materials onto QCM and SAW sensors

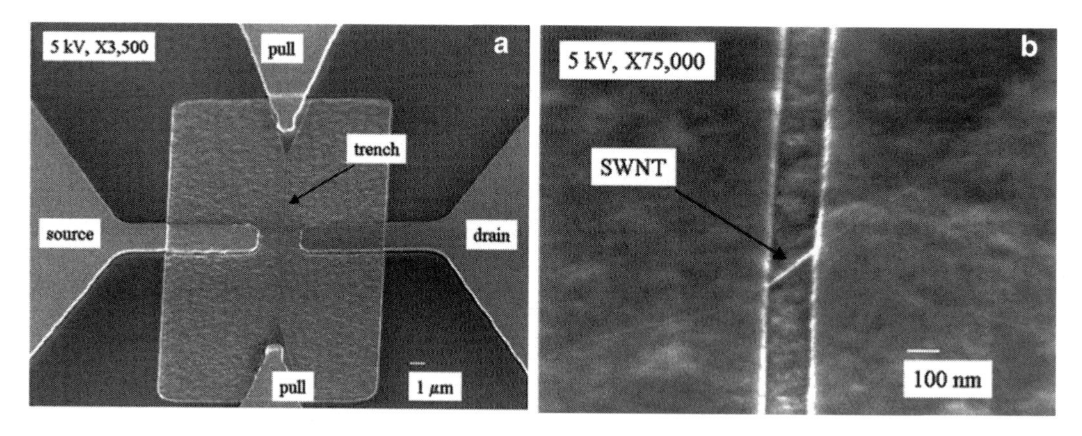

Fig. 1.10 (**a**) Low- and (**b**) high-magnification SEM image of a CNT-based FET. (Reprinted with permission from Kaul et al. 2006, Copyright 2006 American Chemical Society)

(Penza et al. 2004a, b). The interaction between the CNTs' surface and VOCs plays a main role in the sensing mechanism. Mass spectrometry measurements indicated that the interaction between the CNTs and solvents used becomes stronger with solvents that form hydrogen bonds (e.g., ethanol), suggesting a possible role for the chemisorbed oxygen on CNTs as chemical mediators between the CNTs and the dispersing agents. This means that the sensing effects are strongly dependent on the chemical affinity between the analytes to be detected and the solvent used. Also, CNTs form a net supporting adsorbed molecules, producing a sensing structure that is stable at room temperature.

CNT-based gas sensors have shown good electrical response as well. It was established that chemiresistors and chemical field-effect transistors are probably the most promising types of gas sensors based on CNTs (Kong et al. 2000; Bondavalli et al. 2009; Wang and Yeow 2009; Zhang and Zhang 2009; Hu, et al. 2010). Typical configuration of such sensors is shown in Fig. 1.10. It has to be pointed out that for this kind of sensor the research has essentially focused on SWCNTs, because MWCNTs are only metallic and therefore unsuitable to fabricate chemiresistors and transistors.

Many studies have shown that although CNTs are robust and inert structures, their electrical properties are extremely sensitive to the effects of charge transfer and chemical doping by various molecules. The electronic structures of target molecules near the semiconducting nanotubes cause measurable changes to the nanotubes' electrical conductivity (Zhang et al. 2008; Consales et al. 2008). In particular, Valentini et al. (2003, 2004a, b) and Kong et al. (2000) established that NO_2 exposure drastically decreases the electrical resistance of CNT-based sensors. NH_3, H_2O, C_6H_6, and ethanol exposure also increases the electrical resistance. Oxygen also strongly affected the electronic properties of CNTs (Collins et al. 2000). Further, the sensitivity achieved was pretty good, and removing the gas totally restored the initial resistance. The threshold of NO_2 detection in many sensors was smaller at 10 ppb. Such CNT behavior indicates that charge transfer due to the interaction of CNTs with adsorbates is an important mechanism in changing conductivity in the CNTs upon adsorption of NO_2, water vapor, NH_3, C_3H_6, and ethanol gases. In addition, CNT-based sensors demonstrated a faster response and a higher sensitivity than, for example, metal oxide sensors operated at room temperature (Valentini et al. 2003, 2004a, b). Table 1.5 summarizes sensing performance of selected CNT-based chemiresistors and ChemFETs.

However, it was established that CO and H_2 exposure does not affect the resistance of CNT-based sensors at room temperature (Kong et al. 2001; Sayago et al. 2005; Guo and Jayatissa 2008). Instead, they could operate for CO and H_2 gases only at elevated temperatures (Sayago et al. 2005; Guo and Jayatissa 2008). When noble-metal catalysts such as Pd or Pt functionalize the surface of CNTs, the CNTs could sense the CO and H_2 at room temperature. A Pd- and Pt-functionalized *p*-type

Table 1.5 Summary of selected sensing performance of CNT-based chemiresistors and ChemFETs

CNT type	Sensor configuration	Targeted analytes	Detection limit	Response time (s)	Reversibility
Single SWCNT	ChemFET	NO_2 NH_3	2 ppm 0.1 %	<600	Irreversible
SWCNTs	ChemFET	Alcoholic vapors	N/S	5–150	Reversible
SWCNTs	ChemFET	DMMP	<1 ppb	1,000	Reversible
SWCNTs	Chemiresistor	O_2	N/S	N/S	Reversible
MWCNTs	Chemiresistor	NO_2	5–10 ppb	~600 (165 °C)	Reversible (165 °C)
SWCNTs	Chemiresistor	NO_2 Nitrotoluene	44 ppb 262 ppb	600	Reversible
MWCNTs	Chemiresistor	NH_3	10 ppm	~100	Reversible
SWCNTs	Chemiresistor	$SOCl_2$, DMMP	100 ppm	~10	Irreversible
SWCNTs	Chemiresistor	O_3	6 ppb	<600	Reversible
SWCNTs	Chemiresistor	Methanol, acetone	N/S	~100	N/S
SWCNTs	Chemiresistor	H_2O	N/S	10–100	Reversible
Carboxylated SWCNT	Chemiresistor	CO	1 ppm	~100	Reversible

Source: Data from Zhang et al. (2008); Zhang and Zhang (2009)
SWNT single-walled carbon nanotubes, *MWNT* multiwalled carbon nanotubes

single-walled carbon nanotube (SWCNT) gave an increase in resistance when exposed to H_2 by dissociation of H_2 molecules into reactive H atoms (Kong et al. 2001; Kumar and Ramaprabhu 2006). Chemically treated multiwalled carbon nanotubes also showed a good H_2- and CO-sensing response at room temperature (Kim et al. 2011). For example, even without any conventional catalysts they could detect 1 ppm of CO gas at room temperature. Hybrid materials of CNT/SnO_2 were shown to have a good sensing performance for detection of reducing gases including H_2 at room temperature as well (Lu et al. 2009).

Thus, experiments have shown that nanosensors based on changes in electrical conductance of CNTs are highly sensitive. In addition it was established that sensor response is fast. For example, Kong et al. (2000) reported that the response time of the CNT-based devices to 200 ppm NO_2 was a few seconds, and the response (defined as the ratio between resistance after and before gas exposure) was approximately 100–1,000. The response time to approximately 1 % NH_3 was a few minutes with the response between 10 and 100 (Kong et al. 2000; Li et al. 2003). However, during the same experiments it was established that CNT-based sensors are also limited by factors such as their inability to identify analytes with low adsorption energies, poor diffusion kinetics, and poor charge transfer with CNTs (Modi et al. 2003). In addition, it was also found that CNT-based sensors operated at room temperatures have long recovery times. Some nanotube sensors need several hours to release the adsorbed analytes at room temperature before they can be reused. Moreover, in some cases these sensors have incomplete recovery. For example, Kong et al. (2000) succeeded in recovering the initial transistor characteristics after interaction with NO_2 only by heating the sample for 1 h at 200 °C in air or by exposing the sample to pure Ar (at room temperature) for around 12 h. The same results were presented by Zhang et al. (2006) and Li et al. (2003). For example, Li et al. (2003) reported that the recovery time of CNT-based chemiresistors was very long, on the order of 10 h. This effect was explained by high bonding energy between CNTs and NO_2. The strong bonding between NH_3 molecules and the CNTs causes the slow recovery of the CNT-based sensor as well (Nguyen and Huh 2006).

The explanations of observed gas-sensing effects in CNT-based devices are usually based on the analysis of adsorption/desorption phenomena taking place on the surface of CNTs (Peng and Cho 2000; Zhao et al. 2001, 2002; Bauschlicher and Ricca 2004). According to this approach, the resistance of CNT-based gas sensors is conditioned by the change of the CNT resistance caused by interaction with analyte. In particular, in many papers the interaction of NO_2 with the nanotube was interpreted as strictly connected to a bulk doping effect. Actually, NO_2 can be bound to a semiconducting nanotube

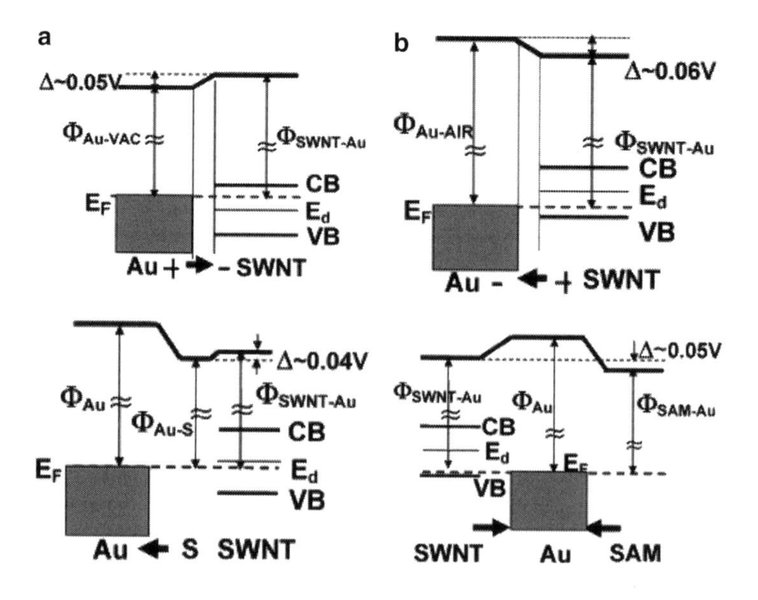

Fig. 1.11 Effect of oxygen on the Fermi level alignment: (**a**) Au/CNT contact in vacuum (*n*-type case), (**b**) Au/CNT contact in air (*p*-type case) (Reprinted with permission from Cui et al. 2003, Copyright 2003 American Chemical Society)

with a subsequent electron charge transfer from the tube to the adsorbed molecules (Peng and Cho 2000; Zhao et al. 2001, 2002; Bauschlicher and Ricca 2004). The hole carrier concentration in the nanotube increases and so does the conductance, with a consequent shift of the activation bias (V_{ON}) to a larger positive voltage. This effect has been assimilated to a sort of "molecular gating" of the CNT, due to the gas molecule adsorption. In contrast, NH_3 molecules, as has been demonstrated, have no binding affinity with semiconducting SWCNTs (Peng and Cho 2000; Zhao et al. 2001, 2002; Bauschlicher and Ricca 2004). Two possible reasons were proposed in order to explain the effect of the NH_3 molecules on the CNTFET channel: (1) the binding between NH_3 molecules and the hydroxyl groups on the SiO_2 substrate leading to a reduction of the negative charges on the oxide, equivalent to a positive electrostatic gating of the SWCNT and (2) the interaction of NH_3 with oxygen species adsorbed on SWCNT.

Thus, the above-mentioned approach attributes the key role in the gas sensing of CNT-based chemiresistors and FETs to the change in the properties of the nanotubes and not of the metal/SWCNT junctions. In fact, they assert that the gas molecules dope the nanotube and so change its conductance. However, it is necessary to take into account that there is other opinion on the mechanism of gas sensitivity of CNT-based devices. Several scientific teams have adopted a different point of view and have focused their studies on demonstrating that the metal/SWCNT junctions are the key players in the sensing mechanism (Leonard and Tersoff 2000; Cui et al. 2003; Auvray et al. 2005; Zhang et al. 2006; Bondavalli et al. 2009). For example, Leonard and Tersoff (2000) and Cui et al. (2003) have shown that the interaction of oxygen at the junction between the metal electrode and the SWCNT changes the metalwork function and also the Fermi level alignment. They assumed that the Fermi level at the contact is not pinned by "metal-induced gap states" (MIGs), as happens for contacts of most metals with normal semiconductors (Si, GaAs, etc.), but that it is controlled by the metalwork function. In the light of this analysis, they have concluded that oxygen raises the metal electrode (Au in this case) work function, thus permitting the switching of the electrical behavior of CNTs from *n*-type (in vacuum) to *p*-type in air (Fig. 1.11). According to Bondavalli et al. (2009), the main sensing mechanism in CNT-based sensors seems to be the modulation of the Schottky barrier height at the contacts, due to the buildup of interface dipoles that depend on the gas species, and also the chemical reactivity of the metal constituting the electrodes. Bondavalli et al. (2009) also believe that the "wetting" of the contact metal on the nanotubes is also a parameter to take into account in the sensing mechanism, since it can shape the interface, leading to the formation of a transition region of paramount importance.

Optical and fiber-optic sensors can be designed based on CNTs as well (Penza et al. 2004b; Barone et al. 2005; Cusano et al. 2006). In particular, Consales et al. (2008) demonstrated CNT-based

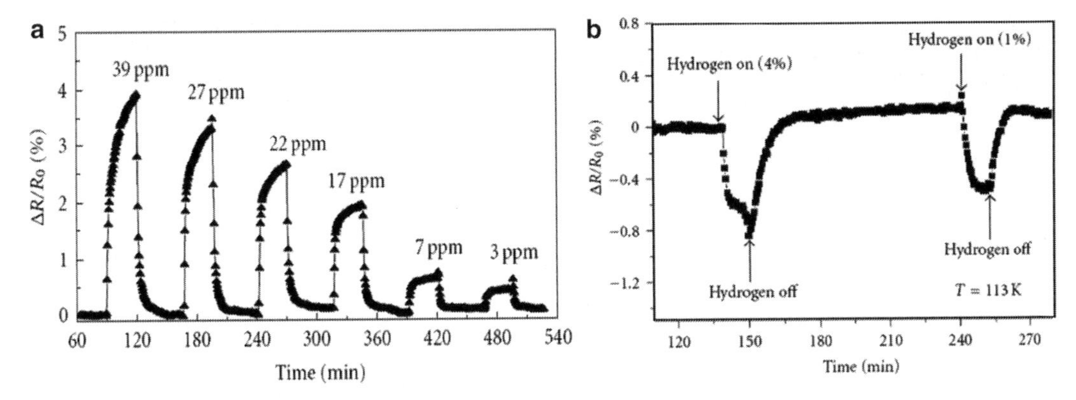

Fig. 1.12 Time responses of the CNT-based optochemical sensor, exposed to (**a**) four decreasing concentration pulses of toluene vapors, at room temperature, and (**b**) to decreasing concentration pulses of gaseous hydrogen (<5 %), at 113 K (Reprinted from Consales et al. 2008, Published by Hindawi Publishing Corporation)

fiber-optic optochemical nanosensors with reflectometric configuration. The adopted optical configuration was based on an extrinsic low-finesse Fabry–Perot interferometer. The principle of operation of this device was based on the measurement of the changes in the amount of power reflected at the fiber–film interface, occurring as a consequence of the changes in the optical (complex refractive index) and geometrical properties (thickness) of the sensitive elements, caused by the interaction of the sensing layers with the target analyte molecules present in the environment. The realized chemical sensors have been tested against VOCs and several gases in different conditions, including harsh environments at cryogenic temperatures suitable for space application (see Fig. 1.12b). In most of the investigated cases the fiber-optic chemosensors coated by SWCNTs demonstrated their strong potentiality as well as the ability of detecting environmental pollutants around or well below the ppm threshold. Typical operation characteristics are shown in Fig. 1.12a.

Research has shown that the ionization gas sensor (IGS) is another possible area of CNT application (Wang and Yeow 2009). This type of gas sensor was discussed in Chap. 19 (Vol. 1). It is known that, in the case of chemical gas sensors, it is difficult to detect gas molecules with low adsorption energy. In IGS there is no adsorption and chemical interaction between the device and target molecules. Therefore they are not limited to identifying gases with low adsorption energy and poor charge transfer with the sensing materials. Compared to standard gas sensors, the IGS is based on the ionization characteristics of the detected gases. The ionization of detected gas is caused by the collisions of molecules with accelerated electrons. However, the issues related to conventional IGSs are their bulky architectures, considerable high power consumption, and breakdown voltage, which is inefficient and risky in operation. It was established that the application of CNTs in IGS can considerably improve their characteristics. It is known that nanotubes are good electron emitters due to their sharp tip curvature and low electron escaping work function (De Heer et al. 1995). This means that the incorporation of CNTs can induce a large field enhancement factor and thereby intensively increase the electric field around the tips to initiate corona discharge at very low voltage (Hou et al. 2006). Therefore, the effects of gas adsorption on the field emission properties of CNTs and CNT-enhanced IGSs have attracted a great deal of research interest (Modi et al. 2003; Kim 2006).

A brief analysis of results obtained indicates that CNTs are really promising materials for gas sensor applications (Li et al. 2008; Kalcher et al. 2009; Bondavalli et al. 2009). However, similar to other sensing materials, CNTs have disadvantages as well. Technological difficulties related to sensor fabrication, bad reproducibility, slow response, and low selectivity are the main shortcomings of these devices (Fam et al. 2011). These shortcomings are subject to the following conditions (Bondavalli et al. 2009).

First, till now there has been no method which can fabricate only semiconducting SWCNTs. As a result, one cannot predict whether a SWCNT is metallic or semiconducting. In addition, it is known that to obtain CNTs of high purity and uniformity is one of the big issues that still impact the applications of CNTs as gas-sensing materials. The as-prepared CNTs usually contain a lot of impurities; some of them, such as amorphous carbon and fullerenes, can hardly be completely removed from the raw materials, and the purity is difficult quantify. Thus, the measured physical and chemical properties of CNTs are peculiar to different research groups. Moreover, the structures of the CNTs obtained always possess surface defects and are not identical in geometrical structure, which causes the actual mechanical strength, electrical and thermal conductivity, as well as other properties to lie far from the theoretical predictions. Recently, some promising results on controlled synthesis of nanotubes in terms of morphology and diameter have been reported. However, chirality of the nanotube is difficult to control. Several strategies have also been reported to increase homogeneity. In particular, Arnold et al. (2006) proposed to differentiate CNTs using selective chemistry, which would involve the use of surface functionalization and/or surfactants which will interact with the surface of the CNT with specific chiralities, thereby sorting them. However, the overall cost of synthesis of pure CNTs strongly increases with the complexity of the separation techniques adopted (Fam et al. 2011). In addition, the synthesis of pure and ideal CNTs is still challenging and costly. It is very difficult to grow defect-free nanotubes continuously to macroscopic length. The precise control over the growth or dispersion of CNTs on surfaces is another problem (Wang and Yeow 2009).

Second, it is quite laborious to identify the position of a single SWCNT on a sensor platform using standard methods. Proper manipulation techniques are required for applying a single tube or thin films of CNTs on substrates that do not allow direct growing methods. Various proposals exist for their incorporation into devices in single-tube or thin-film architectures (Bachtold et al. 2001; Consales et al. 2008). However, though understanding that the realization of homogeneous thin films of CNTs with a controllable thickness and tube size is an important basis for the future development of CNT-based devices for the sensor market, the development of reasonable technologies for separation and selection tubes with similar diameter and manipulation with nanotubes is still a task of great importance.

Third, considering that the CNTFET electrical characteristics are dependent on the individual SWCNT physical characteristics (bandgap in particular, which depends on diameter for semiconductor specimens), it is very difficult to obtain reproducible devices. Depending on the preparation technique and process, the property and behavior of the sensors can vary significantly, which is very crucial for devices aimed at the sensor market. Therefore, the ability to synthesize identical and reproducible CNTs with consistent properties is very important for the application of CNTs in all areas (Wang and Yeow 2009).

Tendency to deformation can also be considered as a disadvantage of CNTs. It was established that both the Young modulus and tensile strength of various CNTs are significantly more elevated than those for stainless steel and Kevlar. This means that CNTs tend to a permanent deformation under a strong tensile strain. Furthermore, due to the hollow structure and high aspect ratio of CNTs, it seems that their strength is limited under compression or bending stress. Experiments confirmed these conclusions. In particular, Ruoff et al. (1993) established that two near nanotubes can be deformed even by van der Waals forces, and Yu et al. (2000) and Palaci et al. (2005) have shown that individual CNTs are rather soft in their radial direction.

It should be noted that the above-mentioned disadvantages mainly relate to sensors based on single CNTs. In the case of sensors such as SAW, QCM, and capacitance and optical sensors based on using CNT networks (mats) or composites inclusive of CNTs, the disadvantages indicated are not so important for sensor operation. Due to the integral effect, there is no need to control parameters of individual CNTs in the devices indicated. There have also been attempts to design FET-based sensors using SWCNT mats as channels (see Fig. 1.13) (Snow et al. 2003; Star et al. 2006; Kumar et al. 2006; Chang et al. 2007). In particular, Star et al. (2006) designed NO sensors using assays of CNT transistors arranged in an array format. These CNT transistors were based on random network architectures that display relatively large tube-to-tube variations. In a network configuration, however, the difference is averaged and the device performance is defined by the mean properties of the CNT architecture.

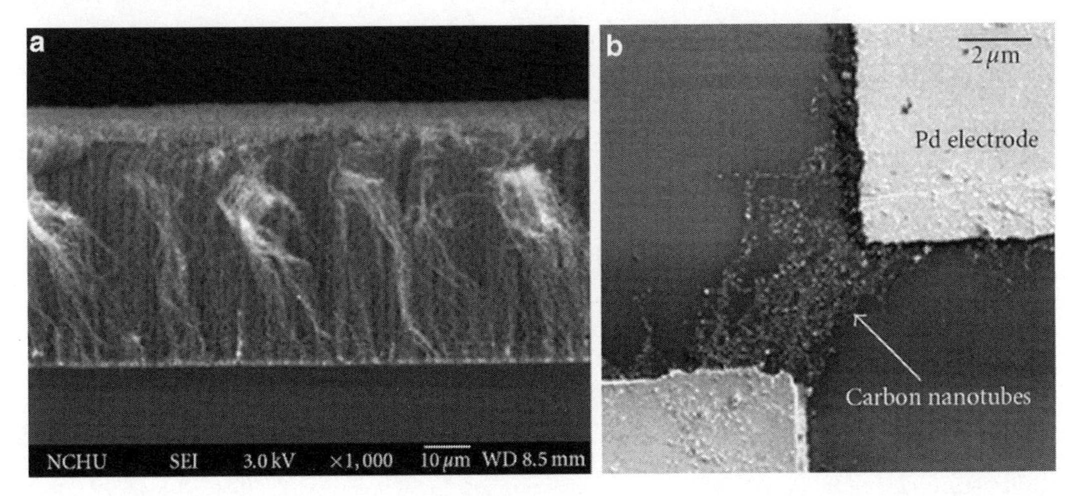

Fig. 1.13 SEM images of (**a**) vertically aligned CNT mat and (**b**) CNTs trapped in castellated microelectrode gaps of sensors by positive dielectrophoresis (DEP). DEP is the electrokinetic motion of dielectrically polarized materials in nonuniform electric fields and has been used to manipulate CNTs for separation, orientation, and positioning of CNTs ((**a**) Reprinted with permission from Huang et al. 2005, Copyright 2005 Elsevier. (**b**) Reprinted with permission from Suehiro et al. 2007, Copyright 2007 Elsevier)

This method employing SWCNT mats is very attractive, but we have to recognize that the theoretical modeling of this kind of sensor is challenging. In addition, it should be noted that, in this case, results are not as impressive as in the case of single SWCNT-based sensors.

As indicated before, slow response and recovery, caused by the nature of the gas adsorption and desorption processes to the nanotubes, is another disadvantage of CNT-based gas sensors. According to Valentini et al. (2003, 2004a, b), sensor reversibility was characterized by fast recovery at only 165 °C. However, it was demonstrated that, by integrating a microheater under the CNT sensing layer or short exposure to UV light, the response time of the sensor can be improved (Cho et al. 2005; Ueda et al. 2008). For example, Li et al. (2003) reported that by using ultraviolet (UV) light, the recovery time was shortened to about 10 min. The UV exposure decreases the desorption-energy barrier to ease the NO_2 desorption. We need to recognize, however, that though all these methods improved the recovery of the CNT-based sensors to some extent, the recovery time is still not satisfactory.

The low solubility of CNTs is other factor limiting CNT applications (Tasis et al. 2003). Several attempts have been made to overcome this limitation (Star et al. 2002; Li et al. 2005; Backes et al. 2009). It was found that pristine CNTs are essentially insoluble, especially in polar solvents such as water. This has enabled the use of solution processing techniques such as drop-casting, spin-casting, or spraying, which facilitate the fabrication of CNT-based devices. Moreover, treatments used for solubility improvement were often accompanied by strong changes in SWCNTs conductivity, which is not a permitted option for many applications. The development of novel methods that facilitate the processing of CNTs while having little impact on their electrical properties or providing the option to restore the conductivity in a subsequent step would therefore be desirable (Schnorr and Swager 2011).

According to Pumera (2009), there is also a problem connected with the features of CNTs synthesis. CNTs are typically grown from carbon-containing gas with the use of metallic catalytic nanoparticles. It is well documented that such nanoparticles remain in the CNTs even after extensive purification procedures, leading to two very significant problems (Pumera et al. 2007). It has been shown that such residual metallic impurities are electrochemically active even when intercalated within the CNTs and

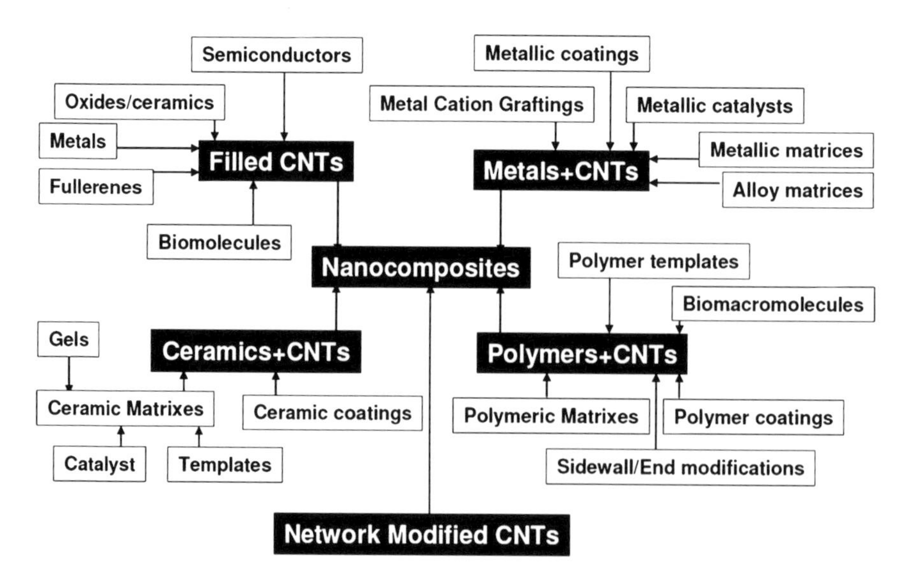

Fig. 1.14 Types of composites prepared using carbon nanotubes. Data from Zeng (2003)

that they can dominate the electrochemistry of CNTs (Liu et al. 2007). This is a significant problem for the construction of reliable sensors with reproducible parameters.

Interference from relative humidity at room temperature, low selectivity, and limited range of operating temperatures can be considered as disadvantages as well (Zhang and Zhang 2009). However, it should be noted that numerous research projects focusing on the improvement of the selectivity of CNT-based sensors were carried out and different routes to improve selectivity have been proposed. They are based on the diversification of metal electrodes, polymer functionalization, metal particle decoration of the SWCNTs, time desorption resolution, etc. (Bondavalli et al. 2009). Approaches used for CNT functionalization will be discussed in Chap. 25 (Vol. 2). Other interesting results connected with CNT preparation and study have also been reported in the literature (Dresselhaus et al. 1996; Harris 1999; Poulin et al. 2002; Valentini et al. 2003; Cantalini et al. 2003; Penza et al. 2004; Kuchibhatla et al. 2007; Mauter and Elimelech 2008; Zhang et al. 2008; Bondavalli et al. 2009; Zhang and Zhang 2009; Wang and Yeow 2009; Schnorr and Swager 2011).

It must be noted that CNTs are promising material for preparing various composites, which can also be used for gas sensor design. Types of composites formed using CNTs according to their chemical composition and structures are summarized in Fig. 1.14. Features of CNT/polymer composites, which have the most evident advantages for gas sensor application, will be discussed in Chap. 13 (Vol. 2). It was established that polymer/CNT composites combine the unique properties of nanotubes with the ease of processability of polymers. Moreover, for design multifunctional materials based on CNT/polymer composites, a very low fraction content of CNT is required.

1.4 Graphene

Graphene is another carbon-based nanomaterial which is promising for sensor applications (Kauffman and Star 2010; Ratinac et al. 2011). Geim and Novoselev received the Nobel Prize in 2010 for their work related to graphene. Graphene is a two-dimensional (2D), single layer of sp2 hybridized carbon that can be considered the “mother of all graphitic forms” of nanocarbon, including, as discussed earlier, 1D CNTs (Geim and Novoselov 2007; Allen et al. 2010). Graphene has two atoms per unit

Fig. 1.15 Schematic diagram of graphene structure. From (http://en.wikipedia.org/wiki/File:Graphen.jpg)

cell. In graphene, carbon atoms are arranged in planar and hexagonal form. Graphene is most easily visualized as an atomic-scale chicken wire made of carbon atoms and their bonds (see Fig. 1.15). The crystalline or "flake" form of graphite consists of many graphene sheets stacked together. Its honeycomb structure has important consequences for the charge carriers (Marchenko et al. 2011). The π and π^* bands of freestanding graphene form cones which touch each other in a single point signaling the presence of massless relativistic electrons (Morozov et al. 2008) giving rise to outstanding transport properties (Novoselov et al. 2007). A high mobility of charge carriers and the ability to modify the electronic properties by doping (Rossi and Sarma 2008), by deformation (Huertas-Hernando et al. 2006), or by interaction with different substrates (Ran et al. 2009) place graphene among the most promising materials for future electronic devices.

It has been reported that methods such as CVD, reactive ion etching, thermal decomposition of SiC, direct-current arc discharge with graphite rods in He atmospheres, chemical modification of graphite, and simple mechanical exfoliation, or "peeling off," of layers from highly oriented pyrolytic graphite can be used for the production of graphene (Allen et al. 2010; Choi et al. 2010; Singh et al. 2011). Epitaxial graphene growth on SiC substrates is also possible (Nomani et al. 2010). The mechanical exfoliation method is low cost, but the graphene produced is of poor quality with limited area. It is particularly difficult and time-consuming to obtain single-layer graphene in a large scale with this method. The graphene obtained by epitaxial growth showed poor uniformity and contained a multitude of domains. Currently, however, the most popular method for graphene production relies on the chemical modification of graphite using the Hummers method, which involves the oxidation of graphite in the presence of strong acids and oxidants (Park and Ruoff 2009). In this case, oxidized graphite is cleaved via rapid thermal expansion or ultrasonic dispersion, and subsequently the graphene oxide sheets were reduced to graphene. This method produces isolated, water-soluble graphite oxide (GO) sheets with many oxygen-containing defect sites (He et al. 1998). Graphite oxide can be transferred in graphene, also called reduced graphene oxide (RGO) or chemically converted graphene (CCG), by chemical reduction with aqueous hydrazine as a reducing agent (Stankovich et al. 2007). However, RGO does have limitations. A serious drawback of this method is that the oxidation process induces a variety of defects which would degrade the electronic properties of graphene. Therefore, at present most interest is in preparing graphene using the CVD method (Li et al. 2009). It was established that CVD graphene tends to be more atomically smooth, whereas RGO usually has many oxygen-containing defect groups (Bagri et al. 2010). CVD-grown graphene can also show several orders of magnitude

Table 1.6 Graphene-based gas sensors

Active material	Reduction method	Analyte	Measurement	Detection limits
RGO	Thermal	NO_2, NH_3	*I*	~100 ppm
RGO+Pd	Chemical	H_2	*R*	N/A
RGO	Chemical	NO_2, NH_3, DNT	*R*	~ppm
RGO	Thermal	NO_2	*I*	~ppm
RGO	Thermal, chemical	Water vapor	*R*	N/A
RGO (inkjet printer)	Chemical	NO_2, Cl_2	*R*	~ppm
Pt/RGO/SiC	Thermal	H_2	*I*	N/A
Pristine graphene	Micromechanical cleavage of graphite	NO_2, NH_3, H_2O and CO (vacuum)	*R*	<1 ppm
Pristine graphene	Micromechanical cleavage of graphite	CO_2	*G*	~ppm

Source: Data from Schedin et al. (2007); Singh et al. (2011); Yoon et al. (2011)
I current, *G* conductivity, *R* resistance, *DNT* dinitrotoluene

lower resistivity, as compared to RGO (Li et al. 2009). One can find a description of all the above-mentioned methods in a review by Singh et al. (2011).

Similar to CNTs – which can be functionalized with polymers, nanoparticles (NPs), or atomic dopants – different approaches toward graphene functionalization have also been reported (Xu et al. 2008; Kauffman and Star 2010; Allen et al. 2010; Singh et al. 2011; Vedala et al. 2011). In particular, recently, N substitutionally doped graphene was first synthesized by a CVD method with the presence of CH_4 and NH_3 (Wei et al. 2009). As doping is accompanied by the recombination of carbon atoms into graphene in the CVD process, dopant atoms can be substitutionally doped into the graphene lattice, which is hard to realize by other synthetic methods. The process of graphene doping will be discussed later in Chap. 25 (Vol. 2).

Taking into account the unique properties of graphene, it was assumed that graphene-based devices should be viable candidates for the development of low-temperature gas sensors. This assumption was based on the following facts. First, graphene's electronic properties are strongly affected by the adsorption of molecules (Lin and Avouris 2008), a prerequisite for design of any type of gas sensor. It was established that the adsorption of gas molecules from the surrounding atmosphere is accompanied by doping of the graphene layers with electrons or holes depending on the nature of the adsorbed gas. As a result, by monitoring changes in resistivity, one can sense minute concentrations of certain gases present in the environment. Second, the 2D structure of graphene constitutes an absolute maximum of the surface-area-to-volume ratio in a layered material, which is essential for high sensitivity. According to Pumera (2009), graphene has a theoretical surface area of 2,630 m^2/g, surpassing that of graphite (~10 m^2/g), and is double that of CNTs (1,315 m^2/g). Third, graphene has good long-term stability of parameters (Marchenko et al. 2011) and good compatibility with standard microelectronic technologies such as conventional lithographic processes (Berger et al. 2004; Shao et al. 2009). It should be noted that graphene is more suitable for device integration than are CNTs because the planar nanostructure of the former makes it advantageous for use in standard microfabrication techniques. In addition, the recent improvements made to graphene deposition methods have contributed to an increase in the applicability of graphene for device integration (Li et al. 2009; Reina et al. 2009).

Experiments carried out confirmed the statements made above. For most gas sensor applications, graphene synthesized by various methods was deposited on Si or Si/SiO_2 substrates, while electrical contacts were prepared with Au/Ti or other metals, which provided good adhesion and ohmic contact with graphene (Schedin et al. 2007; Sundaram et al. 2008; Fowler et al. 2009). In addition, it was found that the role of electrode electrical contacts in the sensing mechanism of graphene was minimal. The sensing mechanism was primarily attributed to charge transfer at the graphene surface (Fowler et al. 2009).

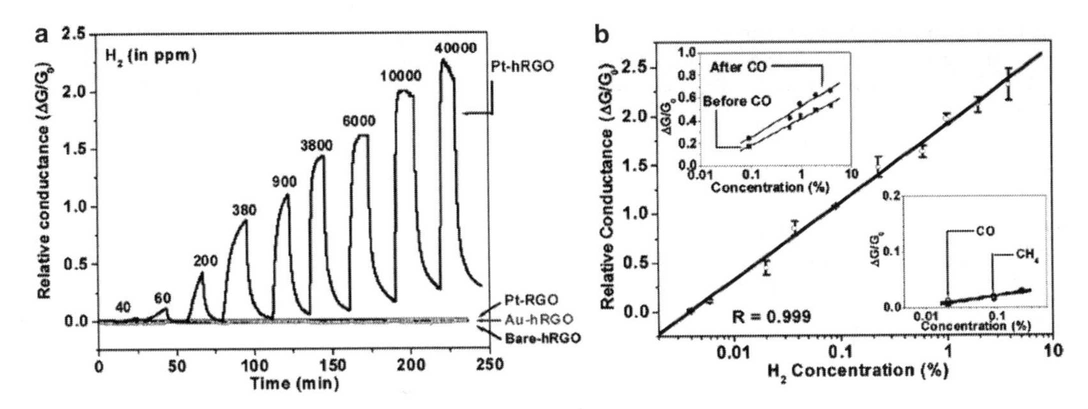

Fig. 1.16 Room temperature response of Pt–hRGO to hydrogen gas. (**a**) Relative conductance ($\Delta G/G_0$) vs. time curves for H_2 concentrations 40–40,000 ppm (in N_2) for bare–hRGO, Au–hRGO, Pt–hRGO, and Pt–reduced graphene oxide (RGO). After each H_2 exposure, the devices were allowed to recover in synthetic air. (**b**) Calibration curve of Pt–hRGO for response to H_2 gas; *left inset* shows the device response to H_2 before and after exposure to CO (0.25 % in N_2), and *right inset* shows the response to CO (0.05–0.25 % in N_2) and CH_4 (0.4–4 % in N_2). Holey RGO (hRGO) samples were prepared by enzymatic oxidation of grapheme oxide followed by chemical reduction using hydrazine hydrate. The flakes were decorated with platinum NPs by pulsed potentiostatic electrodeposition from aqueous solutions containing Pt^{4+} or Au^{3+} metallic ions (Reprinted with permission from Ratinac et al. 2011, Copyright 2011 Wiley)

It was established that, similar to CNTs, pristine graphene interacts with numerous gases with large binding energies such as NO_2, NH_3, CO, H_2, CO_2, and H_2O (see Table 1.6). Detection of various vapors like nonanol, octanoic acid, trimethylamine, acetone, HCN, dimethyl methylphosphonate, dinitrotoluene (DNT), iodine, ethanol, and hydrazine hydrate has also been reported (Schedin et al. 2007; Robinson et al. 2008; Fowler et al. 2009; Dan et al. 2009; Allen et al. 2010). It should be noted that, due to stability requirements, graphene oxide (GO) is usually used in gas sensors. RGO has superior conductivity compared to GO but inferior to pristine graphene (Singh et al. 2011). Typical operating characteristics of graphene-based sensors are shown in Fig. 1.16.

Experiment has shown that the detection limits of graphene-based sensors range between parts per billion (ppb) and parts per million (ppm) levels. For example, Nomani et al. (2010) reported about 10 ppb noise limited sensitivity to NO_2. This and even higher levels of sensitivity are sought for industrial, environmental, and military monitoring. However, Schedin et al. (2007) believe that, due to unique properties, graphene makes it possible to increase the sensitivity to its ultimate limit and detect individual dopants. They gave the following explanation of this statement. First, graphene is a strictly two-dimensional material and, as such, has its whole volume, i.e., all carbon atoms, exposed to surface adsorbates, which maximizes their effect. Second, graphene is highly conductive, exhibiting metallic conductivity and, hence, low Johnson noise even in the limit of no charge carriers, where a few extra electrons can cause notable relative changes in carrier concentration, *n*. The mobility of electrons in graphene can be more than 100,000 cm^2/Vs at room temperature, much higher than in other materials. For comparison, the mobility of electrons in silicon equaled ~1,400 cm^2/Vs. As a result, graphene has resistivity (~1.0 μΩ cm) which is about 35 % less than the resistivity of copper. Third, graphene has few crystal defects (Novoselov et al. 2005; Geim and Novoselov 2007), which ensures a low level of excess (1/*f*) noise caused by their thermal switching. Fourth, graphene allows four-probe measurements on a single-crystal device with electrical contacts that are ohmic and have low resistance (Schedin et al. 2007; Fowler et al. 2009). All of these features contribute to make a unique combination that maximizes the signal-to-noise ratio to a level sufficient for detecting changes in a local concentration by less than one electron charge, *e*, at room temperature.

However, one should take into account that it is nearly impossible to produce only single layers of graphene with current fabrication methods (Ratinac et al. 2011). Therefore, "graphene" can include anything from one to many layers, and the exact number of layers, *N*, critically affects properties, especially for low values of *N*. In particular, stacks with $N \geq 12$ or so tend to behave more like thin-film graphite than graphene (Partoens and Peeters 2006). This means that the importance of knowing what sort of "graphene" you are working with is great.

Graphene-based composite materials have been studied for gas sensors as well. For example, Pt/RGO/SiC-based devices were fabricated for hydrogen gas sensing (Shafiei et al. 2010). Experiments have shown that the flexible gas sensor can also be designed on the basis of RGO (Dua et al. 2010).

Kauffman and Star (2010) concluded that there are several hurdles to overcome before graphene can compete with CNTs as the preferred carbon nanostructure for sensing platforms. According to Kauffman and Star (2010), graphene-based sensor platforms in comparison with CNT or other nanowire (NW)-based sensors suffer from the following major disadvantages. First, the 2D nature of graphene inherently limits the sensor response. Second, graphene is a zero-bandgap semiconductor and behaves as a semimetallic material (Geim and Novoselov 2007). It was established that it is very difficult to turn the graphene into an electrically conductive "off-state" because thermal promotion of charge carriers produces nonzero electrical conductance at any applied gate voltage in FET structures. The absence of optical spectroscopy is another major limitation of graphene application as a sensor platform. Unlike SWNT, graphene does show UV-region absorbance (Liang et al. 2009). In addition, its luminescence is weak unless bandgaps are created through chemical oxidation or size reduction (Gokus et al. 2009). Kauffman and Star (2010) believe that atomic doping of graphene (Boukhvalov and Katsnelson 2008) or decorating graphene with NPs may improve sensor response and the development of spectroscopic techniques for graphene will undoubtedly serve to help further the field of graphene-based gas sensors.

It is also necessary to take into account that graphene's single layers are not completely flat; instead, the flexible sheets have a tendency to fold, buckle, and corrugate (Ratinac et al. 2011). This flexibility is related to the out-of-plane phonons (flexural vibrations) that occur in soft membranes, which means that freestanding graphene tends to crumple (Castro Neto et al. 2009). Thus, the larger-scale distortions like folds and "pleats" (Novoselov et al. 2004) are seen to be an unavoidable by-product of graphite-cleaving techniques. When working with a soft membrane such as graphene, invariably some of the individual layers will fold and buckle during the process of mechanical peeling and subsequent solution deposition onto the substrate.

Some researchers have recently observed that inclusion of lithographic (photo or e-beam) steps in the preparation of graphene can cause some negative effects on the sensing properties of graphene due to the presence of residual polymers on the graphene surface. In work by Dan et al. (2009), a cleaning process was demonstrated to remove the contamination on the sensor device structure, allowing the intrinsic chemical response of graphene-based sensors. The contamination layer was removed by a high temperature cleaning process in a reducing (H_2/Ar) atmosphere (Dan et al. 2009).

In addition, we need to take into account that, like CNT and other nanomaterials, the key challenge in synthesis and processing of bulk-quantity graphene sheets is aggregation (Singh et al. 2011). Unless well separated from each other, graphene tends to form irreversible agglomerates or even restack to form graphite through van der Waals interactions. The prevention of aggregation is essential for graphene sheets because most of their unique properties are only associated with individual sheets.

Bad selectivity of graphene-based sensors, which is typical of most of solid-state gas sensors designed, and long response and recovery times, especially at room temperatures, can also be considered as disadvantage of graphene-based sensors (Schedin et al. 2007; Pearce et al. 2011). In particular, Schedin et al. (2007) established that adsorbed molecules such as NO_2, NH_3, H_2O, and CO were strongly attached to the graphene devices at room temperature, and therefore the elimination of tested gas only led to small and slow changes in the film conductivity. They found that the initial undoped state could be recovered only by annealing at 150 °C or by short-time ultraviolet illumination, which should be considered as an alternative to thermal annealing (see Fig. 1.17). For more inert gases, such as CO_2, the situation is better. Yoon et al. (2011) found that, because of the weak interaction between CO_2 and graphene, the sensor response was rapid and reproducible. Even at room temperatures the

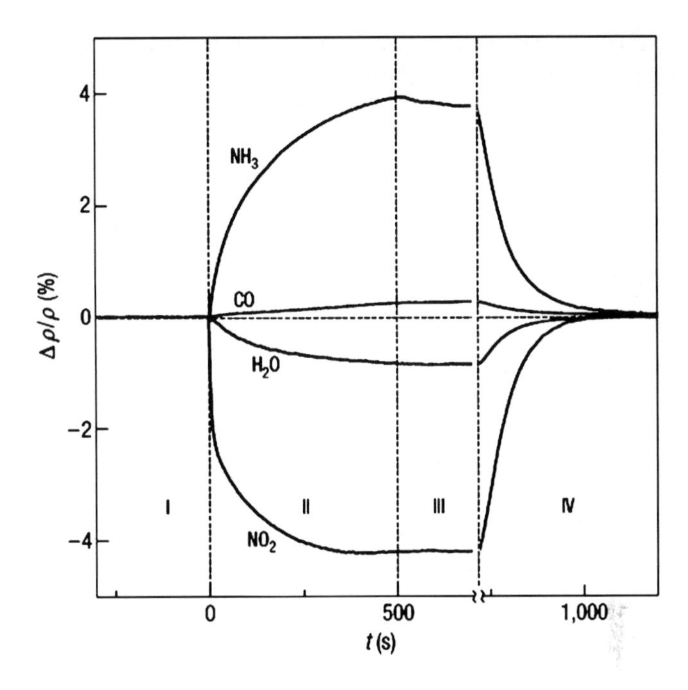

Fig. 1.17 Resistivity response of pristine graphene monocrystals to 1 ppm concentrations of different reducing and oxidizing gases. Regions: (*I*) response in vacuum before gas exposure, (*II*) exposure to 1 ppm of gases, (*III*) gas removed by vacuum, and (*IV*) gas desorption by annealing at 150 °C (Reprinted with permission from Schedin et al. 2007, Copyright 2007 Nature Publishing Group)

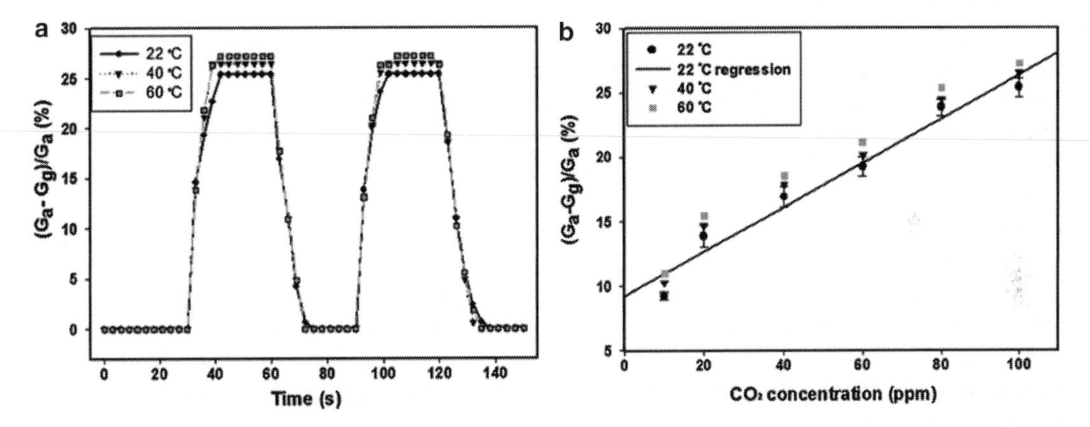

Fig. 1.18 (**a**) Time response of the graphene CO_2 gas sensor in the presence of 100 ppm CO_2, at different temperatures; (**b**) conductance changes at different concentrations of CO_2 (Reprinted with permission from Yoon et al. 2011, Copyright 2011 Elsevier)

response time of graphene-based sensor to CO_2 was less than 10 s (see Fig. 1.18). This result is interesting, because metal oxides such as SnO_2 and In_2O_3 do not show any sensitivity to CO_2.

1.5 Nanodiamond Particles

Nanodiamonds (NDs) are members of the diverse structural family of nanocarbons discussed in the present chapter. Therefore, it is anticipated that the attractive properties of NDs will be exploited in a similar manner to other carbon nanoparticles, in particular for gas sensor design. However, due to

Table 1.7 Physicochemical properties of NDs important for biomedical applications

Property	Characteristics
Structural	Small size of primary monocrystalline particles (~4–5 nm)
	Availability of variable sizes and narrow size fractions
	Different forms (i.e., particulate, coating/film, substrate)
	Large specific surface area (300–400 m^2/g)
	Low porosity/permeability of films
	High specific gravity (3.5 g/cm^2)
Chemical	Chemically resistant to degradation/corrosion, pH stability
	High chemical purity
	Possible sp^2 carbon shells
	Numerous oxygen-containing groups on surface
	Ease of surface functionalization (chemical, photochemical, mechanochemical, enzymatic, plasma- and laser-assisted methods)
	Radiation/ozone resistance
	Large number of unpaired electrons on the surface
Biological	High biocompatibility, low toxicity
	Readily bind bio-active substances (i.e., proteins, DNA) with retained functionality
	Solid phase carrier
Optical	Photoluminescence: non-photobleaching, nonblinking, originates from N-vacancy defects
	High refractive index, optical transparency
	Unique Raman spectral signal
Mechanical	High strength and hardness
	Fine abrasive
Electrochemical	Electrochemical plating with metals
	Redox behavior of DND
Thermal	Can withstand very high/low temperatures

Source: Data from Schrand et al. (2009)

technological difficulties related to synthesis of this material, nanodiamond particles are used mainly in bioapplications (Schrand et al. 2009). The benefits of using nanodiamonds in biomedical applications, including purification, sensing, imaging, and drug delivery, are based upon their desirable chemical, biological, and physical (optical, mechanical, electrical, thermal) properties (Table 1.7). In particular, it was established that diamond is a biocompatible material. In addition, nanodiamonds are unique among the class of carbon nanoparticles because of their intrinsic hydrophilic surface, which is one of the many reasons that these nanocarbon particles are envisioned for biomolecular applications. The surface of nanodiamond particles contains a complex array of surface groups, including carboxylic acids, esters, ethers, lactones, and amines. Therefore alterations in detonation nanodiamond surface groups can produce a high density of chemical functionalities.

References

Allen MJ, Tung VC, Kaner RB (2010) Honeycomb carbon: a review of graphene. Chem Rev 110:132–145

Arbogast JW, Darmanyan AP, Foote CS, Rubin Y, Diederich FN, Alvarez MM, Anz SJ, Whitten RL (1991) Photophysical properties of sixty atom carbon molecule (C_{60}). J Phys Chem 95:11–12

Arnold MS, Green AA, Hulvat JF, Stupp SI, Hersam MC (2006) Sorting carbon nanotubes by electronic structure using density differentiation. Nat Nanotechnol 1:60–65

Arshak K, Moore E, Cavanagh L, Harris J, McConigly B, Cunniffe C, Lyons G, Clifford S (2005) Determination of the electrical behaviour of surfactant treated polymer/carbon black composite gas sensors. Compos A 36:487–491

Auvray S, Borghetti J, Goffman MF, Filoramo A, Derycke V, Bourgoin JP (2005) Carbon nanotubes transistor optimization by chemical control of nanotubes-metal interface. Appl Phys Lett 84(25):5106

Bachtold A, Hadley P, Nakanishi T, Dekker C (2001) Logic circuits with carbon nanotube transistors. Science 294(5545):1317–1320

Backes C, Schmidt CD, Hauke F, Bottcher C, Hirsch A (2009) High population of individualized SWCNTs through the adsorption of water-soluble perylenes. J Am Chem Soc 131:2172–2184

Baena JR, Gallego M, Valcarcel M (2002) Fullerenes in the analytical sciences. Trends Anal Chem 21(3):187–198

Bagri A, Mattevi C, Acik M, Chabal YJ, Chhowalla M, Shenoy VB (2010) Structural evolution during the reduction of chemically derived graphene oxide. Nat Chem 2:581–587

Baleizao C, Nagl S, Schaferling M, Berberan-Santos MN, Wolfbeis OS (2008) Dual fluorescence sensor for trace oxygen and temperature with unmatched range and sensitivity. Anal Chem 80:6449–6457

Barone PW, Baik S, Heller DA, Strano MS (2005) Near-infrared optical sensors based on single-walled carbon nanotubes. Nat Mater 4(1):86–92

Bauschlicher CW, Ricca A (2004) Binding of NH_3 to graphite and to (9,0) carbon nanotubes. Phys Rev B 70:115409

Berger C, Song Z, Li T, Li X, Ogbazghi AY, Feng R, Dai Z, Marchenkov AN, Conrad EH, First PN, de Heer WA (2004) Ultrathin epitaxial graphite: 2D electron gas properties and a route toward graphene-based nanoelectronics. J Phys Chem B 108:19912–19916

Bondavalli P, Legagneux P, Pribat D (2009) Carbon nanotubes based transistors as gas sensors: state of the art and critical review. Sens Actuators B 140:304–318

Bouchtalla SL, Janot JM, Deronzier A, Moutet JC, Seta P (2002) (60)Fullerene immobilized in a thin functionalized polypyrrole film: basic principles for the elaboration of an oxygen sensor. Mater Sci Eng C 21:125–129

Boukhvalov DW, Katsnelson MI (2008) Tuning the gap in bilayer graphene using chemical functionalization: density functional calculations. Phys Rev B 78:085413

Cantalini C, Valentini L, Lozzi L, Armentano I, Kenny JM, Santucci S (2003) NO_2 gas sensitivity of carbon nanotubes obtained by plasma enhanced chemical vapor deposition. Sens Actuators B 93:333–337

Castro Neto AH, Guinea F, Peres NMR, Novoselov KS, Geim AK (2009) The electronic properties of graphene. Rev Mod Phys 81:109–162

Chang YW, Oh JS, Yoo SH, Choi HH, Yoo K-H (2007) Electrically refreshable carbon nanotube-based gas sensors. Nanotechnology 18:435504

Chao YC, Shih JS (1998) Adsorption study of organic molecules on fullerene with piezoelectric crystal detection system. Anal Chim Acta 374:39–46

Chen J, Tsubokawa N (2000) A novel gas sensor from polymer-grafted carbon black: responsiveness of electric resistance of conducting composite from LDPE and PE-b-PEO-grafted carbon black in various vapors. Polym Advan Technol 11:101–107

Chen J, Tsubokawa N, Maekawa Y, Yoshida M (2002) Vapor response properties of conducting composites prepared from crystalline oligomer-grafted carbon black. Carbon 40:1602–1605

Chen SG, Hu JW, Zhang MQ, Rong MZ (2005) Effects of temperature and vapor pressure on the gas sensing behavior of carbon black filled polyurethane composites. Sens Actuators B 105:187–193

Cho W-S, Moon S-I, Lee Y-D, Lee Y-H, Park J, Ju BK (2005) Multiwall carbon nanotube gas sensor fabricated using thermomechanical structure. IEEE Electr Device L 26(7):498–500

Choi W, Lahiri I, Seelaboyina R, Kang YS (2010) Synthesis of graphene and its applications: a review. Crit Rev Solid State 35(1):52–71

Collins PG, Bradley K, Ishigami M, Zettl A (2000) Extreme oxygen sensitivity of electronic properties of carbon nanotubes. Science 287(5459):1801–1804

Consales M, Cutolo A, Penza M, Aversa P, Giordano M, Cusano A (2008) Fiber optic chemical nanosensors based on engineered single-walled carbon nanotubes. J Sensor 2008:936074

Cui X, Freitag M, Martel R, Brus L, Avouris P (2003) Controlling energy-level alignment at carbon nanotubes/Au contact. Nano Lett 3(6):783–787

Cusano A, Pisco M, Consales M, Cutolo A, Giordano M, Penza M, Aversa P, Capodieci L, Campopiano S (2006) Novel optochemical sensors based on hollow fibers and single walled carbon nanotubes. IEEE Photonic Tech L 18(22): 2431–2433

Dan Y, Lu Y, Kybert NJ, Luo Z, Johnson ATC (2009) Intrinsic response of graphene vapor sensors. Nano Lett 9:1472–1475

De Heer WA, Chatelain A, Ugarte D (1995) A carbon nanotube field-emission electron source. Science 270(5239):1179–1180

Dong XM, Fu RW, Zhang MQ, Zhang B, Li JR, Rong MZ (2003) Vapor induced variation in electrical performance of carbon black/poly(methyl methacrylate) composites prepared by polymerization filling. Carbon 41:371–374

Dong XM, Fu RW, Zhang B, Rong MZ (2004) Electrical resistance response of carbon black filled amorphous polymer composite sensors to organic vapors at low concentrations. Carbon 42:2551–2559

Dresselhaus MS, Lin YM, Rabin O, Jorio A, Souza Filho AG, Pimenta MA, Saito R, Samsonidze G, Dresselhaus G (2003) Nanowires and nanotubes. Mater Sci Eng C 23:129–140

Dresselhaus MS, Dresselhaus G, Eklund PC (1996) *Science of fullerenes and carbon nanotubes*. Academic Press, San Diego, CA.

Dua V, Surwade SP, Ammu S, Agnihotra SR, Jain S, Roberts KE (2010) All-organic vapor sensor using inkjet-printed reduced graphene oxide. Angew Chem Int Ed 49:2154–2157

Fam DWH, Palaniappan A, Tok AIY, Liedberg B, Moochhala SM (2011) A review on technological aspects influencing commercialization of carbon nanotube sensors. Sens Actuators B 157:1–7

Fowler JD, Allen MJ, Tung VC, Yang Y, Kaner RB, Weiller BH (2009) Practical chemical sensors from chemically derived graphene. ACS Nano 3(2):301–306

Geim AD, Novoselov KS (2007) The rise of graphene. Nat Mater 6:183–191

Gokus T, Nair RR, Benetti A, Bohmler M, Lombardo A, Novoselov KS, Geim AK, Ferrari AC, Hartshuh A (2009) Making graphene luminescent by oxygen plasma treatment. ACS Nano 3:3963–3968

Guo K, Jayatissa AH (2008) Hydrogen sensing properties of multi-walled carbon nanotubes. Mater Sci Eng C 28:1556–1559

Harris PJE (1999) Carbon nanotubes and related structures. Cambridge University Press, Cambridge, UK

He H, Klinowski J, Forster M, Lerf A (1998) A new structural model for graphite oxide. Chem Phys Lett 287:53–56

Hou Z, Xu D, Cai B (2006) Ionization gas sensing in a microelectrode system with carbon nanotubes. Appl Phys Lett 89(21) 213502, 1–3

Hu PA, Zhang J, Li L, Wang Z, O'Neill W, Estrela P (2010) Carbon nanostructure-based field-effect transistors for label-free chemical/biological sensors. Sensors (Basel) 10(5):5133–5159

Huang CS, Huang BR, Jang YH, Tsai MS, Yeh CY (2005) Three-terminal CNTs gas sensor for N_2 detection. Diam Relat Mater 14(11–12):1872–1875

Huertas-Hernando D, Guinea F, Brataas A (2006) Spin-orbit coupling in curved graphene, fullerenes, nanotubes, and nanotube caps. Phys Rev B 74:155426

Iijima S (1991) Helical microtubules of graphitic carbon. Nature 354:56–58

Kalcher K, Svancara I, Buzuk M, Vytras K, Walcarius A (2009) Electrochemical sensors and biosensors based on heterogeneous carbon materials. Monatsh Chem 140:861–889

Kauffman DR, Star A (2010) Graphene versus carbon nanotubes for chemical sensor and fuel cell applications. Analyst 135(11):2790–2797

Kaul AB, Wong EW, Epp L, Hunt BD (2006) Electromechanical carbon nanotube switches for high-frequency applications. Nano Lett 6:942–947

Kim S (2006) CNT sensors for detecting gases with low adsorption energy by ionization. Sensor 6(5):503–513

Kim YS, Ha S-C, Yang Y, Kim YJ, Cho SM, Yang H, Kim YT (2005) Portable electronic nose system based on the carbon black–polymer composite sensor array. Sens Actuators B 108:285–291

Kim D, Pikhitsa PV, Yang H, Choi M (2011) Room temperature CO and H_2 sensing with carbon nanoparticles. Nanotechnology 22:485501

Kong J, Franklin NR, Zhou C, Chapline MG, Peng S, Cho K, Dai H (2000) Nanotube molecular wires as chemical sensors. Science 287:622–625

Kong J, Chapline MG, Dai H (2001) Functionalized carbon nanotubes for molecular hydrogen sensors. Adv Mater 13:1384–1386

Kroto HW, Heath JR, O'Brien SC, Curl RF, Smalley RE (1985) C_{60}: buckminsterfullerene. Nature 318:162–163

Kuchibhatla SVNT, Karakoti AS, Bera D, Seal S (2007) One dimensional nanostructured materials. Prog Mater Sci 52:699–913

Kumar MK, Ramaprabhu S (2006) Nanostructured Pt functionlized multiwalled carbon nanotube based hydrogen sensor. J Phys Chem B 110:11291–11298

Kumar S, Pimparkar N, Murthy JY, Alam AA (2006) Theory of transfer characteristics of nanotube networks transistors. Appl Phys Lett 88(12):123505

Leonard F, Tersoff J (2000) Role of Fermi-level pinning in nanotube Schottky diodes. Phys Rev Lett 84(20):4693

Li J, Lu Y, Ye Q, Cinke M, Han J, Meyyappan M (2003) Carbon nanotube sensors for gas and organic vapor detection. Nano Lett 3(7):929–933

Li H, Cheng F, Duft AM, Adronov A (2005) Functionalization of single-walled carbon nanotubes with well-defined polystyrene by "click" coupling. J Am Chem Soc 127:14518–14524

Li C, Thostenson ET, Chou TW (2008) Sensors and actuators based on carbon nanotubes and their composites: a review. Compos Sci Technol 68:1227–1249

Li X, Cai W, An J, Kim S, Nah J, Yang D, Piner R, Valemakanni A, Jung I, Tutuc E, Banerjee SK, Colombo L, Ruoff RS (2009) Large-area synthesis of high-quality and uniform graphene films on copper foils. Science 324:1312–1314

Liang Y, Wu D, Feng X, Mullen K (2009) Dispersion of graphene sheets in organic solvent supported by ionic interactions. Adv Mater 21:1679–1683

Lin YM, Avouris P (2008) Strong suppression of electrical noise in bilayer graphene nanodevices. Nano Lett 8:2119–2125

Lin H-B, Shih J-S (2003) Fullerene C_{60}-cryptand coated surface acoustic wave quartz crystal sensor for organic vapors. Sens Actuators B 92:243–254

Liou W-J, Lin N-M (2007) Nanohybrid TiO_2/carbon black sensor for NO_2 gas. China Particuology 5:225–229

Liu X, Gurel V, Morris D, Murray D, Zhitkovich A, Kane AB, Hurt RH (2007) Bioavailability of nickel in single-wall carbon nanotubes. Adv Mater 19:2790–2796

Lonergan MC, Severin EJ, Doleman BJ, Beaber SA, Grubbs RH, Lewis NS (1996) Array-based vapor sensing using chemically sensitive, carbon black-polymer resistors. Chem Mater 8:2298–2312

Lu G, Ocola LE, Chen J (2009) Room-temperature gas sensing based on electron transfer between discrete tin oxide nanocrystals and multiwalled carbon nanotubes. Adv Mater 21:2487–2491

Mamalis AG, Vogtländer LOG, Markopoulos A (2004) Nanotechnology and nanostructured materials: trends in carbon nanotubes. Precis Eng 28:16–30

Marchenko D, Varykhalov A, Rybkin A, Shikin AM, Rader O (2011) Atmospheric stability and doping protection of noble-metal intercalated graphene on Ni(111). Appl Phys Lett 98:122111

Mather PJ, Thomas KM (1997) Carbon black/high density polyethylene conducting composite materials. J Mater Sci 32:1711–1715

Matzger AJ, Lawrence CE, Grubbs RH, Lewis NS (2000) Combinatorial approaches to the synthesis of vapor detector arrays for use in an electronic nose. J Comb Chem 2:301–304

Mauter MS, Elimelech M (2008) Environmental applications of carbon-based nanomaterials. Environ Sci Technol 42(16):5843–5859

McLachlan DS, Blaszkiewicz M, Newnham RE (1990) Electrical resistivity of composites. J Am Ceram Soc 73(8):2187–2203

Modi A, Koratkar N, Lass E, Wei B, Ajayan PM (2003) Miniaturized gas ionization sensors using carbon nanotubes. Nature 424(6945):171–174

Morozov SV, Novoselov KS, Katsnelson MI, Schedin F, Elias DC, Jaszczak JA, Geim AK (2008) Giant intrinsic carrier mobilities in graphene and its bilayer. Phys Rev Lett 100:016602

Nagl S, Baleiz o C, Borisov SM, Schferling M, Berberan-Santos MN, Wolfbeis OS (2007) Optical sensing and imaging of trace oxygen with record response. Angew Chem Int Ed 46:2317–2319

Nguyen H-Q, Huh J-S (2006) Behavior of single-walled carbon nanotube-based gas sensors at various temperatures of treatment and operation. Sens Actuators B 117:426–430

Nomani MWK, Shishir R, Qazi M, Diwan D, Shields VB, Spencer MG, Tompa GS, Sbrockey NM, Koley G (2010) Highly sensitive and selective detection of NO_2 using epitaxial graphene on 6H-SiC. Sens Actuators B 150:301–307

Novoselov KS, Geim AK, Morozov SM, Jiang D, Zhang Y, Dubonos SV, Grigorieva IV, Firsov AA (2004) Electric field effect in atomically thin carbon films. Science 306(5696):666–669

Novoselov KS, Jiang D, Schedin F, Booth TJ, Khotkevich VV, Morozov SV, Geim AK (2005) Two-dimensional atomic crystals. Proc Natl Acad Sci USA 102:10451

Novoselov KS, Jiang Z, Zhang Y, Morozov SV, Stormer HL, Zeitler U, Maan JC, Boebinger GS, Kim P, Geim AK (2007) Room-temperature quantum hall effect in graphene. Science 315(5817):1379

Odom TW, Huang JL, Kim P, Lieber CM (1998) Atomic structure and electronic properties of single-walled carbon nanotubes. Nature 391(6662):62–64

Palaci I, Fedrigo S, Brune H, Klinke C, Chen M, Riedo E (2005) Radial elasticity of multiwalled carbon nanotubes. Phys Rev Lett 94(17):175502

Park S, Ruoff RS (2009) Chemical methods for the production of graphenes. Nat Nanotechnol 4:217–224

Partoens B, Peeters FM (2006) From graphene to graphite: electronic structure around the *K* point. Phys Rev B 74:075404

Patel SV, Jenkins MW, Hughes RC, Yelton WG, Ricco AJ (2000) Differentiation of chemical components in a binary solvent vapor mixture using carbon/polymer composite-based chemiresistors. Anal Chem 72:1532–1542

Pearce R, Iakimov T, Andersson M, Hultman L, Lloyd Spetz A, Yakimova R (2011) Epitaxially grown graphene based gas sensors for ultrasensitive NO_2 detection. Sens Actuators B 155:451–455

Peng S, Cho K (2000) Chemical control of nanotube electronics. Nanotechnology 11:57–60

Penza M, Antolini F, Vittori-Antisari M (2004a) Carbon nanotubes as SAW chemical sensors materials. Sens Actuators B 100:47–59

Penza M, Cassano G, Aversa P, Antolini F (2004b) Alcohol detection using carbon nanotubes acoustic and optical sensors. Appl Phys Lett 85(12):2379–2381

Poulin P, Vigolo B, Launois P (2002) Films and fibers of oriented single wall nanotubes. Carbon 40:1741–1749

Pumera M (2009) Electrochemistry of graphene: new horizons for sensing and energy storage. Chem Rec 9:211–223

Pumera M, Merkoci A, Alegret S (2007) Carbon nanotube detectors for microchip CE: Comparative study of single-wall and multiwall carbon nanotube, and graphite powder films on glassy carbon, gold, and platinum electrode surfaces. Electrophoresis 28(8):1274–1280

Radeva E, Georgiev V, Spassov L, Koprinarov N, St K (1997) Humidity adsorptive properties of thin fullerene layers studied by means of quartz micro-balance. Sens Actuators B 42:11–13

Ramos MV, Al-Jumaily A, Puli VS (2005) Conductive polymer-composite sensor for gas detection, In: Proceedings of 1st international conference on sensing technology, Palmerston North, New Zealand, 21–23 November 2005, p 213–216

Ran QS, Gao MZ, Guan XM, Wang Y, Yu ZP (2009) First-principles investigation on bonding formation and electronic structure of metal-graphene contacts. Appl Phys Lett 94:103511

Ratinac KR, Yang W, Gooding JJ, Thordarson P, Braet F (2011) Graphene and related materials in electrochemical sensing. Electroanal 23(4):803–826

Reina A, Jia X, Ho J, Nezich D, Son H, Bulovic V, Dresselhaus MS, Kong J (2009) Large area, few-layer graphene films on arbitrary substrates by chemical vapor deposition. Nano Lett 9:30–35

Robinson JT, Perkins FK, Snow ES, Wei Z, Sheehan PE (2008) Reduced graphene oxide molecular sensors. Nano Letters 8(10):3137–3140

Rossi E, Sarma SD (2008) Ground state of graphene in the presence of random charged impurities. Phys Rev Lett 101:166803

Ruoff RS, Tersoff J, Lorents DC, Subramoney S, Chan B (1993) Radial deformation of carbon nanotubes by van der Waals forces. Nature 364(6437):514–516

Saito R, Fujita M, Dresselhaus G, Dresselhaus MS (1992) Electronic structure of graphene tubules based on C_{60}. Phys Rev B Condens Matter 46:1804–1811

Saito R, Dresselhaus G, Dresselhaus MS (1998) Physical properties of carbon nanotubes. Imperial College, London

Sayago I, Terrado E, Lafuente E, Horrillo MC, Maser WK, Benito AM, Navarro R, Urriolabeitia EP, Martinez MT, Gutierrez J (2005) Hydrogen sensors based on carbon nanotubes thin films. Synth Met 148:15–19

Sberveglieri G, Faglia G, Perego C, Nelli P, Marks RN, Virgili T, Taliani C, Zamboni R (1996) Hydrogen and humidity sensing properties of C_{60} thin films. Synth Met 77:273–275

Schedin F, Geim AK, Morozov SV, Hill EW, Blake P, Katsnelson MI, Novoselov KS (2007) Detection of individual gas molecules adsorbed on graphene. Nat Mater 6:652–655

Schnorr JM, Swager TM (2011) Emerging applications of carbon nanotubes. Chem Mater 23:646–657

Schrand AM, Ciftan Hens SA, Shenderova OA (2009) Nanodiamond particles: properties and perspectives for bioapplications. Cr Rev Sol State 34:18–74

Severin EJ, Doleman BJ, Lewis NS (2000) An investigation of the concentration dependence and response to analyte mixtures of carbon black/insulating organic polymer composite vapor detectors. Anal Chem 72(4):658–668

Shafiei M, Spizzirri PG, Arsat R, Yu J, du Plessis J, Dubin S, Kaner RB, Kalantar-zadeh K, Wlodarski W (2010) Platinum/graphene nanosheet/SiC contacts and their application for hydrogen gas sensing. J Phys Chem C 114:13796–13801

Shao Q, Liu G, Teweldebrhan D, Balandin AA, Roumyantes S, Shur MS, Yan D (2009) Flicker noise in bilayer graphene transistors. IEEE Electr Device L 30:288–290

Shih J-S, Chao Y-C, Sung M-F, Gau G-J, Chiou C-S (2001) Piezoelectric crystal membrane chemical sensors based on fullerene C_{60}. Sens Actuators B 76:347–353

Singh V, Joung D, Zhai L, Das S, Khondaker SI, Seal S (2011) Graphene based materials: past, present and future. Prog Mater Sci 56:1178–1271

Snow ES, Novak JP, Campbell PM, Park D (2003) Random networks of carbon nanotubes as an electronic material. Appl Phys Lett 82(13):2145–2147

Stankovich S, Kikin DA, Piner RD, Kohlhaas KA, Kleinhammes A, Jia Y, Wu Y, Nguyen ST, Ruoff RS (2007) Synthesis of graphene-based nanosheets via chemical reduction of exfoliated graphite oxide. Carbon 45:1558–1565

Star A, Steuerman DW, Heath JR, Stoddart JF (2002) Starched carbon nanotubes. Angew Chem Int Ed 41:2508–2512

Star A, Joshi V, Skarupo S, Thomas D, Gabriel JCP (2006) Gas sensor array based on metal-decorated carbon nanotubes. J Phys Chem B 110:21014–21020

Suehiro J, Hidaka S, Yamane S, Imasaka K (2007) Fabrication of interfaces between carbon nanotubes and catalytic palladium using dielectrophoresis and its application to hydrogen gas sensor. Sens Actuators B 127:505–511

Sundaram RS, Navarro CG, Balasubramaniam K, Burghard M, Kern K (2008) Electrochemical modification of grapheme. Adv Mater 20:3050–3053

Suzuki T, Maruyama Y, Akasaka T, Ando W, Kobayashi K, Naggase S (1991) Redox properties of organofullerenes. J Am Chem Soc 11:1359–1363

Synowczyk AW, Heinze J (1993) Application of fullerenes as sensor materials. In: Kuzmany H, Mehring N, Fink J (eds) Electronic properties of fullerenes Springer series in solid state sciences, vol 117. Springer, Berlin, pp 73–77

Tasis D, Tagmatarchis N, Georgakilas V, Prato M (2003) Soluble carbon nanotubes. Chemistry 9:4000–4008

Taylor R, Parsons JP, Avent AG, Rannard SP, Dennis TJ, Hare JP, Kroto HW, Walton DRM (1991) Degradation of C_{60} by light. Nature 351:277

Terrones M, Jorio A, Endo M, Rao AM, Kim Y, Hayashi T, Terrones H, Charlier JC, Dresselhaus G, Dresselhaus MS (2004) New direction in nanotube science. Mater Today 2004 (October):30–45

Ueda T, Bhuiyan MMH, Norimatsu H, Katsuki S, Ikegami T, Mitsugi F (2008) Development of carbon nanotube based gas sensors for NO*x* gas detection working at low temperature. Physica E 40(7):2272–2277

Valentini L, Cantalini C, Lozzi L, Armentano I, Kenny JM, Santucci S (2003) Reversible oxidation effects on carbon nanotubes thin films for gas sensing applications. Mater Sci Eng C 23:523–529

Valentini L, Cantalini C, Armentano I, Kenny JM, Lozzi L, Santucci S (2004a) Highly sensitive and selective sensors based on carbon nanotubes thin films for molecular detection. Diam Relat Mater 13:1301–1305

Valentini L, Bavastrello V, Stura E, Armentano I, Nicolini C, Kenny JM (2004b) Sensors for inorganic vapor detection based on carbon nanotubes and poly(*o*-anisidine) nanocomposite material. Chem Phys Lett 383(5–6):617–622

Varghese OK, Kichambre PD, Gong D, Ong KG, Dickey EC, Grimes CA (2001) Gas sensing characteristics of multi-wall carbon nanotubes. Sens Actuators B 81:32–41

Vedala H, Sorescu DC, Kotchey GP, Star A (2011) Chemical sensitivity of graphene edges decorated with metal nanoparticles. Nano Lett 11:2342–2347

Wang Y, Yeow JTW (2009) A review of carbon nanotubes-based gas sensors. J Sensor 2009:493904

Wang LG, Wang X, Ottova AL, Tien HT (1996) Iodide sensitive sensor based on a supported bilayer lipid membrane containing a cluster form of carbon (fullerene C_{60}). Electroanal 8:1020–1022

Wei B-Y, Lin C-S, Lin H-M (2003) Examining the gas sensing behaviors of carbon nanotubes using a piezoelectric quartz crystal microbalance. Sensor Mater 15(4):177–190

Wei D, Liu Y, Wang Y, Zhang H, Huang L, Yu G (2009) Synthesis of N-doped graphene by chemical vapor deposition and its electrical properties. Nano Lett 9:1752–1758

Wu Z-S, Zhou G, Yin L-C, Ren W, Li F, Cheng H-C (2012) Graphene/metal oxide composite electrode materials for energy storage. Nano Energ 1:107–131

Xie H, Yang Q, Sun X, Yang J, Huang Y (2006) Gas sensor arrays based on polymer-carbon black to detect organic vapors at low concentration. Sens Actuators B 113:887–891

Xu C, Wang X, Zhu J (2008) Graphene–metal particle nanocomposites. J Phys Chem C 112:19841–19845

Yoon H-J, Jun D-H, Yang J-H, Zhou Z, Yang S-S, Cheng MM-C (2011) Carbon dioxide gas sensor using a graphene sheet. Sens Actuators B 157:310–313

Yu M-F, Kowalewski T, Ruoff RS (2000) Investigation of the radial deformability of individual carbon nanotubes under controlled indentation force. Phys Rev Lett 85(7):1456–1459

Zeng HC (2003) Carbon nanotube-based nanocomposites. In: Nalwa HS (ed) Handbook of organic hybrid materials and nanocomposites. American Scientific, Stevenson Ranch, pp 151–180

Zhang W-D, Zhang W-H (2009) Carbon nanotubes as active components for gas sensors. J Sensor 2009:160698

Zhang J, Boyd A, Tselev A, Paranjape M, Barbara P (2006) Mechanism of NO_2 detection in carbon nanotube field effect transistor chemical sensors. Appl Phys Lett 88:123112

Zhang T, Mubeen S, Myung NY, Deshusses MA (2008) Recent progress in carbon nanotube-based gas sensors. Nanotechnology 19:332001

Zhao J, Buldum A, Han J, Lu JP (2001) Gas molecules adsorption on carbon nanotubes. Mat Res Soc Proc 633:A13.48.1–A13.48.6

Zhao J, Buldum A, Han J, Lu JP (2002) Gas molecule adsorption in carbon nanotubes and nanotube bundles. Nanotechnology 13:195–200

Chapter 2
Nanofibers

2.1 Approaches to Nanofiber Preparation

Various methods can be used for preparing polymer nanofibers (see Table 2.1), including drawing, hard and soft template synthesis, phase separation, self-assembly, and electrospinning (Jayaraman et al. 2004; Liu and Zhang 2009; Long et al. 2011). Among these methods, electrospinning seems to be the simplest and most versatile technique capable of generating 1D nanostructures (mainly nanofibers). Electrospinning is the technique which uses a strong electric field to produce polymer nanofibers from polymer solution or polymer melt (see Fig. 2.1). If electrostatic forces overcome the surface tension of a solution, a charged jet is ejected and moves toward a grounded electrode. Generally, the electrospun fibers are deposited on a fixed collector in a 3D nonwoven membrane structure with a wide range of fiber diameter distribution from several nanometers to a few micrometers (see Fig. 2.2). More recently, aligned electrospun fibers were obtained by using a rotating or prepatterned collector (Theron et al. 2001; Kameoka et al. 2003).

One of the most important advantages of the electrospinning technique is that it is relatively easy and not expensive to produce large numbers of different kinds of nanofiber (Lu et al. 2009). Other advantages of the electrospinning technique are the ability to control the fiber diameters, the high surface-to-volume ratio, high aspect ratio, and pore size as nonwoven fabrics. Moreover, nanofiber composites can easily be made via the electrospinning technique with the only restriction being that the second phase needs to be soluble or well dispersed in the initial solution. The advantage of the facile formation of 1D composite nanomaterials by electrospinning affords the materials' multifunctional properties for various applications. Electrospinning has been used to convert a large variety of polymers into nanofibers and may be the only process that has the potential for mass production. To date, it is believed that more than 100 different polymers have been successfully electrospun into nanofibers by this technique. It should be noted that composite-based nanofibers can also be synthesized using electrospinning (Huang et al. 2003b). One can find a detailed description of this method and fibers prepared using this method in reviews prepared by Huang et al. (2003b), Ding et al. (2009), and Lu et al. (2009). During the process of electrospraying, the liquid drop elongates with increasing electric field. When the repulsive force induced by the charge distribution on the surface of the drop is balanced by the surface tension of the liquid, the liquid drop distorts into a conical shape. Once the repulsive force exceeds the surface tension, a jet of liquid ejects from the cone tip. Small droplets form as a result of the varicose breakup of the jet in the case of low-viscosity liquids. If this phenomenon is applied to polymer solutions, a solid fiber is generated instead of breaking up into individual drops for the electrostatic repulsions between the surface charges and the evaporation of solvent. Extensive research on electrospinning shows that parameters such as the fiber collectability uniformity of fibers,

G. Korotcenkov, *Handbook of Gas Sensor Materials*, Integrated Analytical Systems,
DOI 10.1007/978-1-4614-7388-6_2, © Springer Science+Business Media New York 2014

Table 2.1 Synthesis methods of conducting polymer nanotubes and nanofibers

Synthesis methods	Advantages	Disadvantages	Examples
Hard physical template method Porous membranes	Aligned arrays of tubes and wires with controllable length and diameter	A post-synthesis process is needed to remove the template	PANI; PPY; P3MT; PEDOT; PPV nanotubes/nanowires; CdS–PPY; Au–PEDOT–Au; Ni/PPV; MnO_2/PEDOT nanowires
Soft chemical template method Interfacial polymerization Other methods Dilute polymerization, template-free method, rapidly mixed reaction, reverse emulsion polymerization, ultrasonic shake-up, radiolytic synthesis	Simple self-assembly process without an external template	Relatively poor control of the uniformity of the morphology (shape, diameter); poorly or non-oriented 1D nanostructures	A variety of PANI and PPY micro-/nanostructures such as tubes, wires/fibers, hollow microspheres, nanowire/nanotube junctions and dendrites
Electrospinning	Mass fabrication of continuous polymer fibers	Usually in the form of nonwoven web; possible alignment	PANI/PEO; PPY/PEO; PANI; PPY; P3HT/PEO
Nanoimprint lithography or embossing	Rapid and low cost	A micromold is needed	PEDOT nanowires; 2D nanodots of semiconducting polymer and aligned CP arrays
Directed electrochemical nanowire assembly	Electrode–wire–electrode or electrode–wire–target growth of CP micro-/nanowires	Micro-/nanowires with knobby structures	PPY, PANI, and PEDOT nanowires
Other methods Dip-pen nanolithography method, molecular combing method, whisker method, strong magnetic field-assisted chemical synthesis, etc.			

Source: Reprinted with permission from Long et al. (2011). Copyright 2011 Elsevier

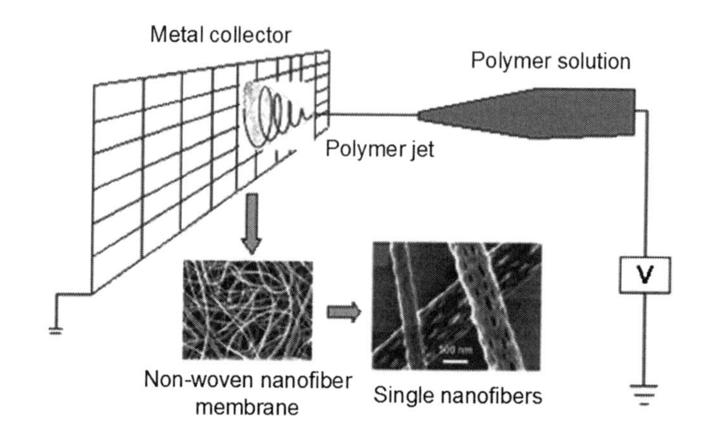

Fig. 2.1 Schematic diagram to show polymer nanofibers by electrospinning (Reprinted with permission from Huang et al. 2003b, Copyright 2003 Elsevier)

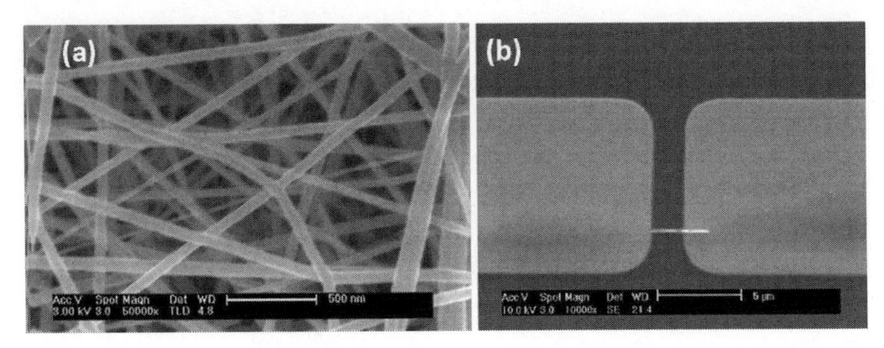

Fig. 2.2 A SEM image of (**a**) typical electrospun fibers and (**b**) single-polypyrrol (PPY) nanowire 200 nm wide and 3 μm long (Reprinted with permission (**a**) from Ding et al. 2009. Published by MDPI and (**b**) from Ramanathan et al. 2004, Copyright 2004 American Chemical Society)

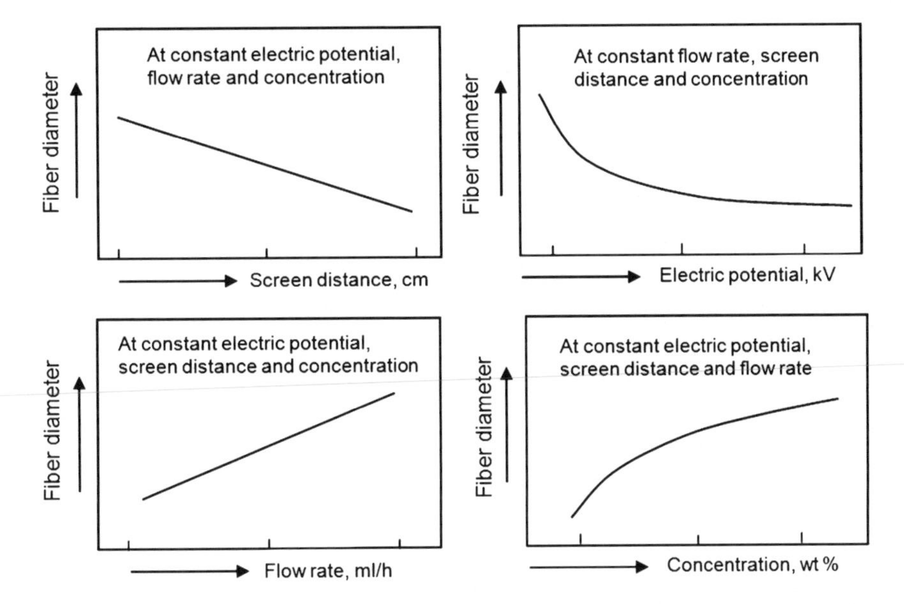

Fig. 2.3 Effect of process parameters on fiber diameter, produced by electrospinning

average fiber diameter, fiber diameter distribution, and fiber porosity are strongly affected by the solution properties (Fong et al. 1999; Ding et al. 2002) and processing parameters (McCann et al. 2006; Ding et al. 2006b). Correlations between the diameter of the fiber and parameters of the electrospinning process are shown in Fig. 2.3.

As shown by Huang and Kaner (2004), polymer nanofibers can also by synthesized by a general chemical route using interfacial polymerization at an aqueous/organic interface as well. This approach, in particular, was realized for PANI (Huang and Kaner 2004) and polypyrrole (PPy) (Huang et al. 2005). The nanofibers synthesized were 60–100 nm in diameter. In the template-free method reported, called the simplified template-free method (STFM), the polymer nanotubes were obtained via a self-assembly process (Liu and Zhang 2009). This process is schematically shown in Fig. 2.4. This figure illustrates the formation mechanism of the PANI nanotubes and nanofibers with or without surfactant. In the presence of a surfactant, micelles formed by anilinium cations and surfactant anions were regarded as templates in the formation of the nanostructures. As in the absence of a surfactant, on the other hand, micelles formed by anilinium cations were considered as templates. However, the size of PANI nanostructures was slightly affected by the addition of the surfactant during the polymerization.

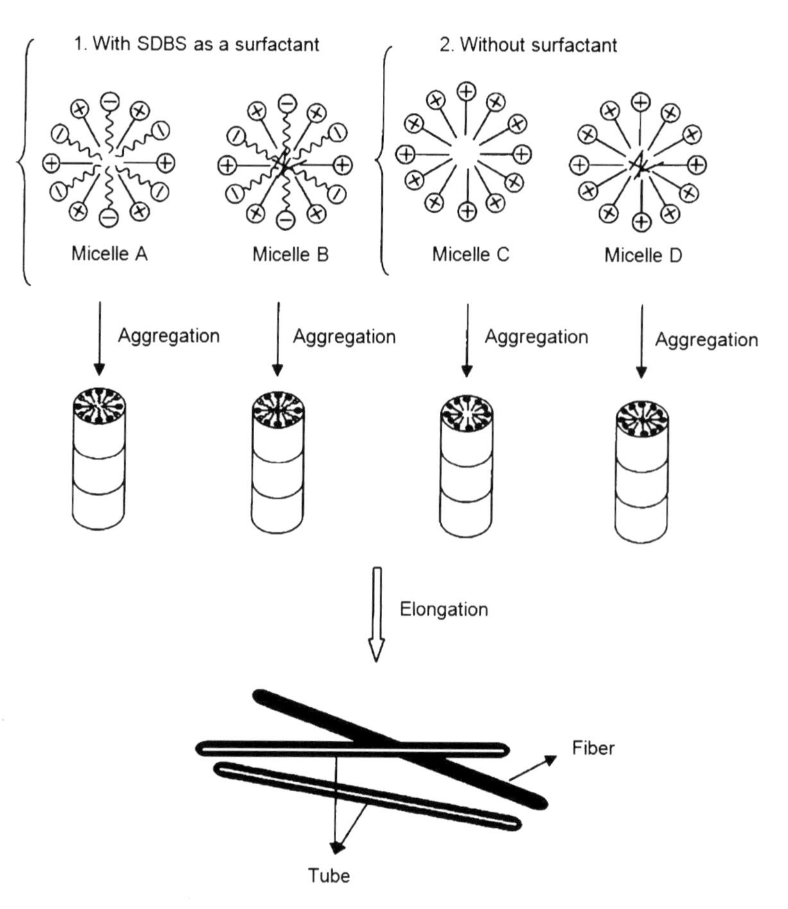

Fig. 2.4 Formation mechanism for PANI nanotubes and nanofibers synthesized by a self-assembly process (Reprinted with permission from Zhang et al. 2002, Copyright 2002 American Chemical Society)

Obviously, STFM is a facile and efficient approach to synthesize polymer, in particular PANI, nanostructures because it not only omits hard template and post-treatment of template removal but also simplifies reagents. However, the self-assembly mechanism of the conductive nanotubes of PANI by the STFM is not yet understood. It might be due to the formation of aniline dimer cation radicals which could act as effective surfactants to shape the polyaniline morphology.

Regarding the template method, we can say that template synthesis can be organized using "soft templates" such as surfactants, organic dopants, or polyelectrolytes that assist in the self-assembly of polyaniline nanostructures and "hard templates" such as porous membranes or zeolites, where the templated polymerization occurs in the 1D nanochannels (Long et al. 2011). Nanoporous anodic aluminum oxide (AAO) membranes are the most extensively used templates for nanofiber synthesis (see Fig. 2.5a). To synthesize nanofibers, materials have to be filled into the nanopores in some way. Electrochemistry is a powerful method for such applications and has been used to synthesize nanofibers consisting of various materials, including conducting polymers (Dan et al. 2007). This process is shown schematically in Fig. 2.5. However, it should be noted that this method cannot make one-by-one continuous nanofibers. In addition, the process requires many operations including phase separation which consists of dissolution, gelation, extraction using a different solvent, freezing, and drying, resulting in a nanoscale porous foam. Thus, template synthesis takes a relatively long period of time to transfer the solid polymer into the nanoporous foam. The self-assembly is a process in which

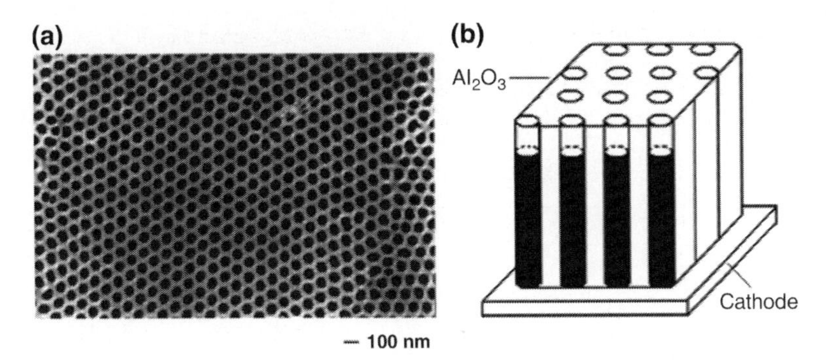

Fig. 2.5 (**a**) The SEM image of AAO membrane surface (Reprinted with permission from Jessensky et al. 1998, Copyright 1998 American Institute of Physics) and (**b**) AAO membrane used as a template to make nanofibers by electroplating

individual, preexisting components organize themselves into the desired patterns and functions. However, similar to the phase separation, the self-assembly is time-consuming when processing continuous polymer nanofibers.

2.2 Nanofiber-Based Gas Sensors

Of course, polymer nanofibers are not 1D structures in the classical understanding as are carbon nanotubes or metal oxide nanowires. Research has shown, however, that features of polymer nanofiber configuration and dimensional factors play a positive role during the design of polymer-based gas sensors. Polymer nanofibers, which usually have diameters in the range of 10–1,000 nm and a length from several micrometer up to centimeters and meters, posses many unique properties since these fibers have very large surface areas per unit mass and small pore sizes. In addition, polymer nanofibers permit easier addition of surface functionalities compared with conventional polymers. Innis and Wallace (2002) reported that nanodimensional conducting polymers can also exhibit unique properties such as greater conductivity and more rapid electrochemical switching speeds.

Experiment has shown that polymer nanofibers with controllable thickness, fine structures, diversity of materials, and large specific surface is an ideal candidate as sensing materials for gas sensors (Ding et al. 2009). So far, many attempts (listed in Table 2.2) have been carried out to prepare ultrasensitive gas sensors to detect vapors of NH_3, H_2S, CO, NO_2, O_2, CO_2, moisture, and VOCs (CH_3OH, C_2H_5OH, $C_5H_{10}C_{12}$, $C_6H_5CH_3$, C_4H_8O, etc.) with new and improved detection limits using nanofibrous membranes as sensing structures. For example, Huang et al. (2003a) and Virji et al. (2004) developed polyaniline nanofiber thin-film sensors and compared them to conventional polyaniline sensors. They found that the response of nanofiber-based sensors was higher and faster (see Fig. 2.6). Without any doubt, such a situation is conditioned by the high surface area and high porosity of nanofiber-based sensing materials. Liu et al. (2004) also created individual polyaniline/poly(ethylene oxide) nanowire sensors for detecting NH_3 at concentrations as low as 0.5 ppm with rapid response and recovery times.

In numerous research projects it was shown that, similar to other nanowire-based sensors discussed in previous chapters, either a fabric of nanofibers or a single nanofiber can be used for gas sensor design. The first design is easy to realize but efficient in promoting the sensitivity of chemiresistors (Zhang et al. 2004; Ma et al. 2006). The single nanowire is more difficult to use in sensors (Liu et al. 2005a, b). One of the possible approaches to resolving this problem was proposed by Dong et al. (2005a, b). They developed a new technology combined with nanoimprint lithography and a

Table 2.2 Characteristics of polymer nanofiber-based gas sensors

Type	Polymer		Fiber diameter	Analyte	Detection limit	References
Acoustic wave	PAA–PVA	N	100–400 nm	NH_3	50 ppm	Ding et al. (2004a)
	PAA	N	1–7 μm	NH_3	130 ppb	Ding et al. (2005)
	PEI–PVA	N	100–600 nm	H_2S	500 ppb	Ding et al. (2006a)
Resistive	PANI	N	40–80 nm	HCl, NH_3, ethanol	–	Alam et al. (2005)
	PANI	N	0.3–1.5 μm	Amines	100 ppm	Gao et al. (2008)
	PANI	N	50–80 nm	NH_3	1 ppm	Wang et al. (2006a)
	HCSA–PANI/PEO	S	100–500 nm	NH_3	500 ppb	Liu et al. (2004)
	PMMA–PANI	N	250–600 nm	$(C_2H_5)_3N$	20 ppm	Ji et al. (2008)
	HCSA–PANI	S	20–150 nm	Alcohols	No data	Pinto et al. (2008)
	Pd/PPy and PANI	N	75 nm–1 μm	H_2	1 nM	Im et al. (2006a)
	PDPA–PMMA	N	~400 nm	NH_3	1 ppm	Manesh et al. (2007)
	PPy	O	80–180 nm	NH_3	1 ppm	Wang et al. (2006a)
	HCSA–POT/PS	N	0.2–1.9 μm	H_2O	No data	Aussawasathien et al. (2008)
	CB–PECH, PEO, PIB, PVP	O	~3 μm	CH_3OH	1,000 ppm	Kessick and Tepper (2006)
				$C_5H_{10}Cl_2$	5 ppm	
				$C_6H_5CH_3$	250 ppm	
				C_2HCl_3	500 ppm	

N nonwoven, *S* single, *O* oriented, *PAA* polyacrylic acid, *PVA* polyvinyl alcohol, *PEI* polyethyleneimine, *HCSA* 10-camphorsulfonic acid, *PANI* polyaniline, *PEO* poly(ethylene oxide), *PDPA* polydiphenylamine, *PMMA* polymethyl methacrylate, *POT* poly-*o*-toluidine, *PS* polystyrene, *MWCNT* multiwalled carbon nanotube, *CB* carbon black, *PECH* polyepichlorohydrin, *PIB* polyisobutylene, *PVP* polyvinylpyrrolidone, *PAN* polyacrylonitrile, *VOC* volatile organic compound

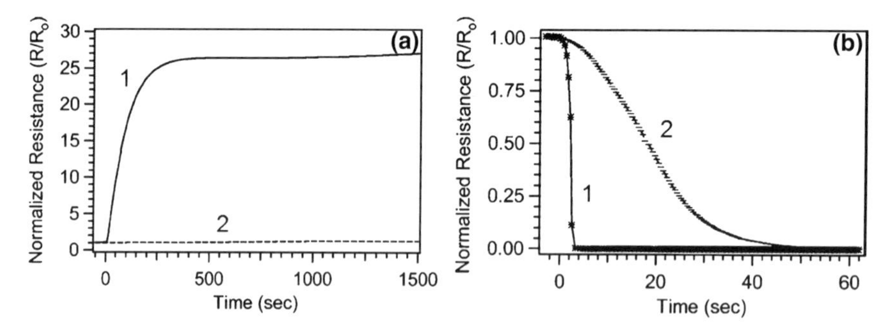

Fig. 2.6 Response of 0.3-μm nanofiber (1) and conventional polyaniline (2) thin films to (**a**) 3 ppm of hydrazine and (**b**) 100 ppm HCl (Reprinted with permission from Virji et al. 2004, Copyright 2004 American Chemical Society)

lift-off process to fabricate a PPY nanowire between microelectrodes. They also reported that the sensitivity and response time of a single nanowire are influenced by its diameter (Dong et al. 2005a). This means that the control of nanofiber diameter, similar to other nanowire-based gas sensors, is a required step in gas sensor fabrication for achieving acceptable reproducibility of sensor parameters.

It should be noted that the decrease of response and recovery times is a great advantage of nanofiber-based sensors, slow response and recovery being one of the most important disadvantages of conventional polymer gas sensors. The above-mentioned optimization of gas-sensing characteristics allowed Wang et al. (2004) to design a device which demonstrated real-time electronic sensing in the gas phase using an array of polyaniline nanoframework–electrode junctions. Wang et al. (2004) believe that this sensing device could be used for chemical sensing of HCl, NH_3, and ethanol vapor.

Several other examples of gas sensors based on polymer nanofibers are listed in Table 2.2. In particular, Ding et al. (2004a) used electrospun nanofibrous membranes on QCM electrodes as highly sensitive NH_3

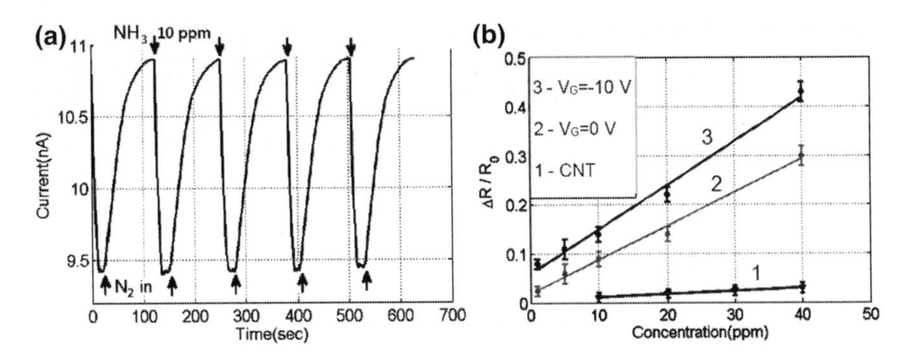

Fig. 2.7 (**a**) Room temperature repeatability test with 10 ppm NH_3 (V_G=−10 V). (**b**) Sensitivity compare of Pani fiber sensor with or without gate voltage and CNT sensor (Reprinted from Chen et al. 2011. Published by MDPI)

gas sensors. A series of composite fibers of polyacrylic acid (PAA) and poly(vinyl alcohol) (PVA) with diameters of 100–400 nm containing various weight percentages of PAA to PVA were deposited on QCM via electrospinning and characterized with regard to their morphology and sensitivity to NH_3. PAA, a weak anionic polyelectrolyte, interacts with NH_3 gas. Besides QCM, SAW, and resistive gas sensors, polymer nanofibers are used in the design of photoelectric (Yang et al. 2007a, b), cantilever, and optical gas sensors. For example, Luoh and Hahn (2006) first used electrospun nanocomposite fiber mats as optical sensors in conjunction with FTIR spectroscopy to detect CO_2 gas. The nanocomposite fiber mats were prepared by electrospinning polyacrylonitrile (PAN) solutions containing nanoparticles including iron oxide, antimony tin oxide, and zinc oxide with diameters ranging from 10 to 70 nm.

FET gas sensors based on individual nanofibers were designed as well (Pinto et al. 2008; Chen et al. 2011). The extracted sensitivities, $S=\Delta R/R_0$, of both the Pani nanofiber under two gate voltages (curves 2 and 3) and CNT sensor (curve 1) are shown in Fig. 2.7b. It is seen that (1) nanofibers have higher sensitivity in comparison with CNTs, (2) the sensitivity improved when the gate potential was increased from 0 V to −10 V, and (3) sensor response is reversible and reproducible (Fig. 2.7a). This improvement indicates that the single-PANI fiber FET sensor offers tunable sensitivity with field effect. Liu et al. (2005b) reported a single-nanofiber field-effect transistor from electrospun poly (3-hexylthiophene).

We note that for explanation of the gas-sensing effect in polymer nanofibers-based devices models designed for conventional polymer-based sensors can be used. For example, the sensing mechanism for chloroform molecule sensing depends on the fact that chloroform molecules are relatively small and can diffuse efficiently into the polymeric matrix, which expands the structure and decreases the conductivity of the film. Small alcohol molecules have a different response mechanism to polyaniline from those of halogenated solvents. They interact with the nitrogen atoms of polyaniline, leading to an expansion of the compact polymer chains into a linear form, thus decreasing the resistance of the film (Huang and Choi 2007). According to Huang et al. (2003a) and Virji et al. (2004), there are five different response mechanisms such as acid doping (HCl), base dedoping (NH_3), reduction (with N_2H_4), swelling (with $CHCl_3$), and polymer chain conformational changes (induced by CH_3OH), which can be used for polymer nanofiber-based gas sensors.

It should be noted that, besides polymer and polymer-based composite nanofibers (Lu et al. 2009), various inorganic nanofibers can be fabricated using various methods (Shao et al. 2002; Ding et al. 2003, 2004b; Raible et al. 2005; Luoh and Hahn 2006; Yang et al. 2007a, b). In particular, using template methods, nanofibers consisting of metals (Favier et al. 2001; Murray et al. 2004; Im et al. 2006b), semiconductors (Routkevitch et al. 1996), and metal oxides (Miao et al. 2002) were synthesized. Electrospinning also can be used for semiconductor and metal oxide nanofiber fabrication (Lim et al. 2010;

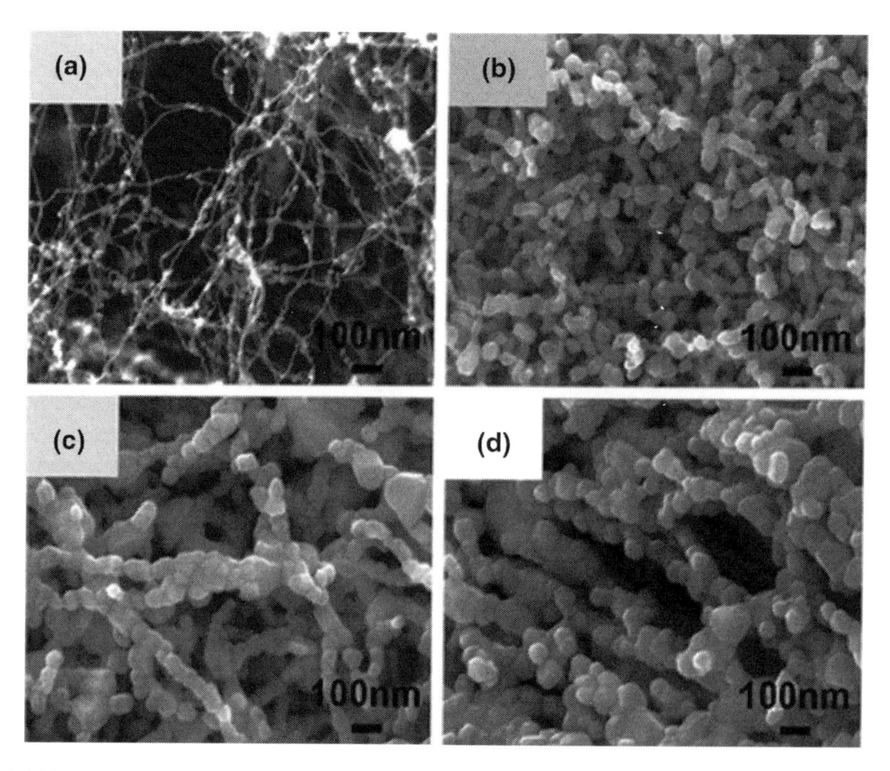

Fig. 2.8 (**a**) Morphology of the SWCNT template. (**b–d**) WO_3 nanowire morphology for W deposition times of 10 s, 20 s, and 60 s, respectively. The WO_3 structures were fabricated by oxidation at 700 °C in air for 2 h (Reprinted with permission from Vuong et al. 2012, Copyright 2012 Royal Society of Chemistry)

Park et al. 2010). For example, there are reports related to nanofibers of TiO_2 (Kim et al. 2006; Zhang et al. 2008; Landau et al. 2009), SnO_2 (Yang et al. 2007a; Wang et al. 2007; Zhang et al. 2008), WO_3 (Wang et al. 2006c), TiO_2: LiCl (Li et al. 2008), ZnO (Yang et al. 2007b), $SrTi_{0.8}Fe_{0.2}O_{3-}\delta$ (Sahner et al. 2007), etc. All indicated nanofibers were used for gas sensor design, including humidity sensors (Li et al. 2008) and optical gas sensors (Wang et al. 2002). For preparing metal oxide nanofibers, a hybrid solution, which is a mixture of the metal oxide sol precursor, polymer, and solvent, is normally used. In order to make the inorganic nanoparticles disperse effectively in a polymer, a surfactant is sometimes needed. It is necessary to take into account that sintering at elevated temperatures is usually required for preparing metal oxide fibers. This thermal treatment is necessary for both the transformation of hydroxides into oxides, and decomposition and removal of polymeric components used for electrospinning. Figure 2.8 shows WO_3 nanofibers prepared by deposition of W on an SWCNT template with subsequent annealing in an oxygen-containing atmosphere. The same images for In_2O_3-based nanofibers are shown in Fig. 2.9. As can be seen, metal oxides in nanofibers are polycrystalline. Experiments have shown that these materials are also acceptable for gas sensor design (see Table 2.3). Extremely high porosity is the main advantage of these sensors, which show very good operating characteristics (great and fast response) in comparison with sensors based on conventional materials. This unique morphology facilitates effective penetration of the surrounding gas into the porous ceramic layer, which is believed to be the main reason for the exceptionally high gas sensitivity of metal oxide gas sensors produced by this method (Kim et al. 2006). Unlike conventional screen printing methods that produce mesoporous granular layers with densely packed nanoparticles that give rise to poor gas transport, sensors produced by electrospinning display

Fig. 2.9 SEM images of (**a**) as prepared PVA/indium acetate composite nanofibers, (**b**) after annealing at T_{an} = 400 °C, (**c**) T_{an} = 500 °C, and (**d**) T_{an} = 600 °C (Reprinted with permission from Lim et al. 2010, Copyright 2010 Elsevier)

Table 2.3 Characteristics of several metal oxide nanofiber-based conductometric gas sensors

Material	Type	Fiber diameter (nm)	Gases tested	T_{oper} (°C)	Detection limit	References
TiO_2	N	200–500	NO_2	150–400	500 ppb	Kim et al. (2006)
TiO_2	N	120–850	CO, NO_2	300–400	50 ppb	Landau et al. (2009)
TiO_2	N	400–500	CO	200	<1 ppm	Park et al. (2010)
LiCl-TiO_2	N	150–260	H_2O	RT	11 %	Li et al. (2008)
SnO_2	S	700	H_2O	RT	N/A	Wang et al. (2007)
SnO_2	N	~100	C_2H_5OH	330	10 ppb	Zhang et al. (2008)
MWCNT/SnO_2	N	300–800	CO	RT	47 ppm	Yang et al. (2007a, b)
In_2O_3	N	~100	CO	300	~1 ppm	Lim et al. (2010)
WO_3	N	20–140	NH_3	350	50 ppm	Wang et al. (2006b)
WO_3	N	32–82	NO	300	30 ppb	Vuong et al. (2012)
V_2O_5	N	~10	NH_3	RT	30 ppb	Raible et al. (2005)
$SrTi_{0.8}Fe_{0.2}O_{3-}\delta$	N	~100	CH_3OH	400	5 ppm	Sahner et al. (2007)

N nonwoven, *S* single

a bimodal pore size distribution comprising both large and small pores that enhance gas transport and sensitivity in these layers (Kim et al. 2010). Gas-sensing characteristics of WO_3 fiber-based sensors are shown in Fig. 2.10.

To solve the problem of poor adhesion between fiber mats and the substrate, Kim et al. (2006) introduced an additional hot-pressing step after titania fiber deposition. Besides improving adhesion, this treatment was found to have an impact on the microstructure of the fibers as shown in Fig. 2.11. The as-spun metal oxide–polymer composite fibers exhibit a range of diameters from 200 to 500 nm (Fig. 2.11a). When calcined without hot-pressing to remove the organic vehicle, a bundle structure composed of sheaths of 200–500 nm diameters was obtained. In some cases, the outer sheaths were broken, revealing cores filled with ~10-nm-thick fibrils as shown in Fig. 2.11c. By introducing the

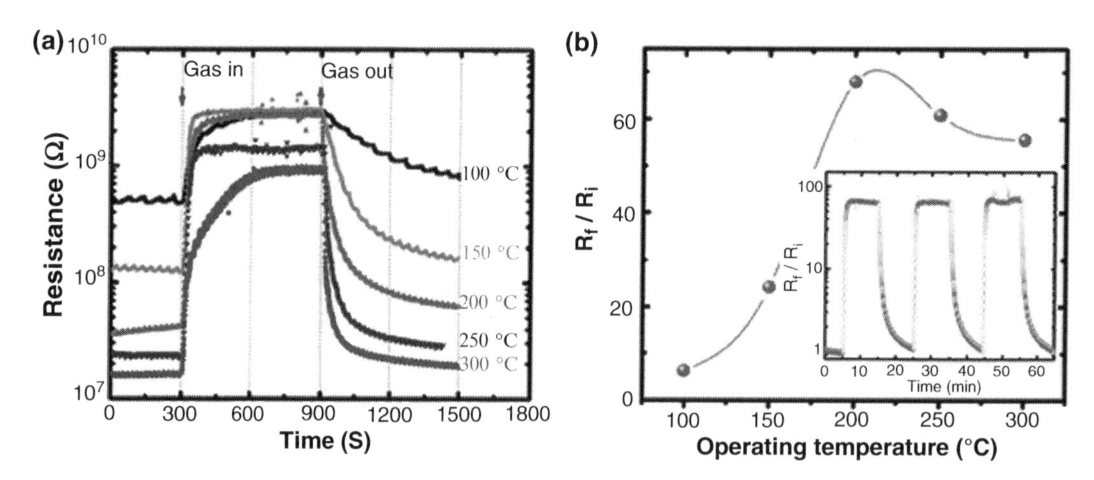

Fig. 2.10 Sensing properties of a thick WO_3 nanofibers with a crystallite size of 82 nm measured to NO gas concentrations diluted in dry air. (**a**) The response–recovery characteristics for 5 ppm NO gas and (**b**) the summary of responses (R_f/R_i) against the operation temperature. The *inset* shows three response cycles illustrating good reproducibility. The *curve* is shown as a guide (Reprinted with permission from Vuong et al. 2012, Copyright 2012 Royal Society of Chemistry)

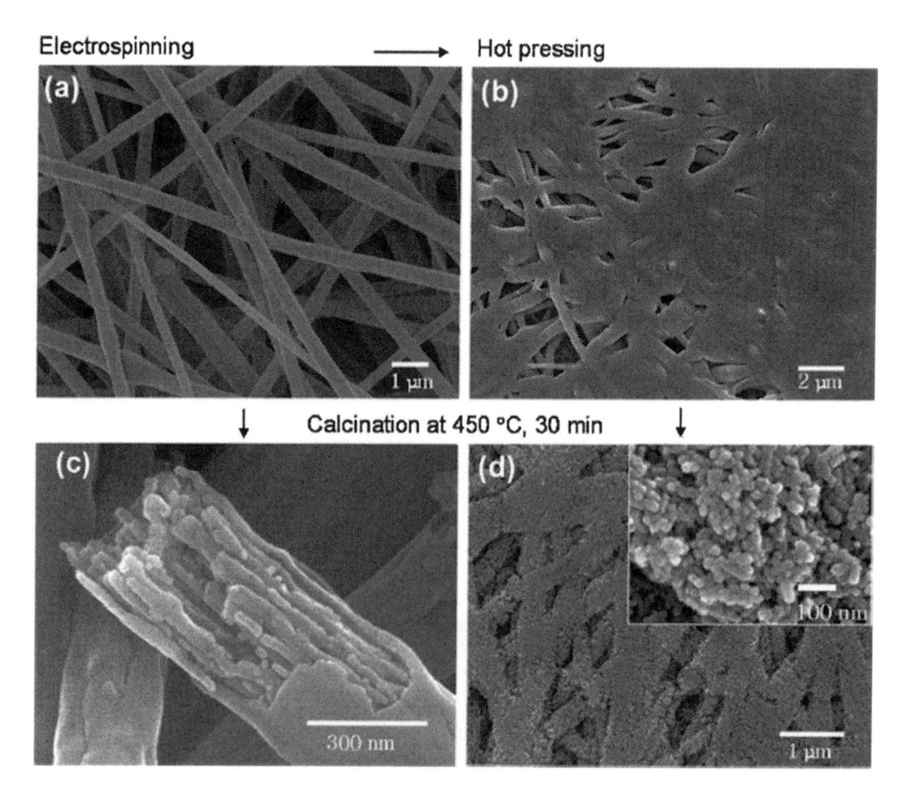

Fig. 2.11 Electrospinning and hot-pressing of metal oxide materials. (**a**) SEM image of the as-spun TiO_2/PVAc composite fibers fabricated by electrospinning from a DMF solution. (**b**) SEM image of TiO_2/PVAc composite fibers after hot-pressing at 120 °C for 10 min. (**c**) SEM image of unpressed TiO_2 nanofibers after calcination at 450 °C. (**d**) SEM image of hot-pressed TiO_2 nanofibers after calcinations at 450 °C (Reprinted with permission from Kim et al. 2006, Copyright 2006 American Chemical Society)

hot-pressing step prior to calcination, an interconnected morphology of the TiO_2/polymer composite fibers was obtained, as illustrated in Fig. 2.11b, due to the partial melting of the polymer vehicle. Subsequent calcination resulted in the exceptionally high surface area structures shown in Fig. 2.11d. The mechanical pressure applied during the hot-pressing served to break the outer sheaths, thereby exposing the fibrils and leading to an exceptionally high surface-to-volume ratio. The hot-pressed TiO_2 fiber sensors exhibited improved NO_x characteristics with detection limits down to the ppb range (Kim et al. 2006). The combined hot-pressing/electrospinning technique was successfully transferred to SnO_2 and SnO_2–TiO_2 composites (Kim et al. 2010; Sahner and Tuller 2010).

References

Alam MM, Wang J, Guo Y, Lee SP, Tseng H-R (2005) Electrolyte-gated transistors based on conducting polymer nanowires junction arrays. J Phys Chem B 109:12777–12784

Aussawasathien D, Sahasithiwat S, Menbangpung L (2008) Electrospun camphorsulfonic acid doped poly(o-toluidine)-polystyrene composite fibers: chemical vapor sensing. Synth Met 158:259–263

Chen D, Lei S, Chen Y (2011) A single polyaniline nanofiber field effect transistor and its gas sensing mechanisms. Sensors 11:6509–6516

Dan Y, Cao Y, Mallouk TE, Johnson AT, Evoy S (2007) Dielectrophoretically assembled polymer nanowires for gas sensing. Sens Actuators B Chem 125:55–79

Ding B, Kim H, Lee S, Lee D, Choi K (2002) Preparation and characterization of nanoscaled poly(vinyl alcohol) fibers via electrospinning. Fiber Polym 3:73–79

Ding B, Kim H, Kim C, Khil M, Park S (2003) Morphology and crystalline phase study of electrospun TiO_2-SiO_2 nanofibers. Nanotechnology 14:532–537

Ding B, Kim J, Miyazaki Y, Shiratori S (2004a) Electrospun nanofibrous membranes coated quartz crystal microbalance as gas sensor for NH_3 detection. Sens Actuators B Chem 101:373–380

Ding B, Kim C, Kim H, Seo M, Park S (2004b) Titanium dioxide nanofibers prepared by using electrospinning method. Fiber Polym 5:105–109

Ding B, Yamazaki M, Shiratori S (2005) Electrospun fibrous polyacrylic acid membrane-based gas sensors. Sens Actuators B Chem 106:477–483

Ding B, Kikuchi M, Li C, Shiratori S (2006a) Electrospun nanofibrous polyelectrolytes membranes as high sensitive coatings for QCM-based gas sensors. In: Dirote EV (ed) Nanotechnology at the leading edge, Nova Science Publishers. New York, USA, pp 1–28

Ding B, Li C, Miyauchi Y, Kuwaki O, Shiratori S (2006b) Formation of novel 2D polymer nanowebs via electrospinning. Nanotechnology 17:3685–3691

Ding B, Wang M, Yu J, Sun G (2009) Gas sensors based on electrospun nanofibers. Sensors 9:1609–1624

Dong B, Krutschke M, Zhang X, Chi LF, Fuchs H (2005a) Fabrication of polypyrrole wires between microelectrodes. Small 1:520–524

Dong B, Zhong DY, Chi LF, Fuchs H (2005b) Patterning of conducting polymers based on a random copolymer strategy: toward the facile fabrication of nanosensors exclusively based on polymers. Adv Mater 17:2736–2741

Favier F, Walter EC, Zach MP, Benter T, Penner RM (2001) Hydrogen sensors and switches from electrodeposited palladium mesowire arrays. Science 293:2227–2231

Fong H, Chun I, Reneker D (1999) Beaded nanofibers formed during electrospinning. Polymer 40:4585–4592

Gao Y, Gong J, Fan B, Su Z, Qu L (2008) Polyaniline nanotubes prepared using fiber mats membrane as the template and their gas-response behavior. J Phys Chem C 112:8215–8222

Huang Z-M, Zhang Y-Z, Kotaki M, Ramakrishna S (2003a) A review on polymer nanofibers by electrospinning and their applications in nanocomposites. Compos Sci Technol 63:2223–2253

Huang JX, Virji S, Weiller BH, Kaner RB (2003b) Polyaniline nanofibers: facile synthesis and chemical sensors. J Am Chem Soc 125:314–315

Huang J, Kaner RB (2004) Nanofiber formation in the chemical polymerization of aniline: a mechanistic study. Angew Chem Int Ed 43:5817–5821

Huang K, Wan MX, Long Y, Chen Z, Wei Y (2005) Multi-functional polypyrrole nanofibres via a functional dopant-introduced process. Synth Met 155:495–500

Huang X-J, Choi Y-K (2007) Chemical sensors based on nanostructured materials. Sens Actuators B Chem 122: 659–671

Im Y, Vasquez RP, Lee C, Myung N, Penner R, Yun M (2006a) Single metal and conducting polymer nanowire sensors for chemical and DNA detections. J Phys Conf Ser 38:61–64

Im Y, Lee C, Vasquez RP, Bangar MA, Myung NV, Menke EJ, Penner RM, Yun M (2006b) Investigation of a single Pd nanowire for use as a hydrogen sensor. Small 2:356–358

Innis PC, Wallace GG (2002) Inherently conducting polymeric nanostructures. J Nano Sci Nanotechnol 2:441–451

Jayaraman K, Kotaki M, Zhang Y, Mo X, Ramakrishna S (2004) Recent advances in polymer nanofibers. J Nanosci Nanotechnol 4(1–2):52–65

Jessensky O, Muller F, Gosele U (1998) Self-organized formation of hexagonal pore arrays in anodic alumina. Appl Phys Lett 72:1173–1175

Ji S, Li Y, Yang M (2008) Gas sensing properties of a composite composed of electrospun poly(methyl methacrylate) nanofibers and in situ polymerized polyaniline. Sens Actuators B Chem 133:644–649

Kameoka J, Orth R, Yang Y, Czaplewski D, Mathers R, Coates G, Craighead H (2003) A scanning tip electrospinning source for deposition of oriented nanofibers. Nanotechnology 14:1124–1129

Kessick R, Tepper G (2006) Electrospun polymer composite fiber arrays for the detection and identification of volatile organic compounds. Sens Actuators B Chem 117:205–210

Kim I, Rothschild A, Lee B, Kim D, Jo S, Tuller H (2006) Ultrasensitive chemiresistors based on electrospun TiO_2 nanofibers. Nano Lett 6:2009–2013

Kim I-D, Jeon E-K, Choi S-H, Choi D-K, Tuller HL (2010) Electrospun SnO_2 nanofiber mats with thermo-compression step for gas sensing applications. J Electroceram 25:159–167

Landau O, Rothschild A, Zussman E (2009) Processing-microstructure-properties correlation of ultrasensitive gas sensors produced by electrospinning. Chem Mater 21:9–11

Li Z, Zhang H, Zheng W, Wang W, Huang H, Wang C, MacDiarmid A, Wei Y (2008) Highly sensitive and stable humidity nanosensors based on LiCl doped TiO_2 electrospun nanofibers. J Am Chem Soc 130:5036–5037

Lim SK, Hwang SH, Chang D, Kim S (2010) Preparation of mesoporous In_2O_3 nanofibers by electrospinning and their application as a CO gas sensor. Sens Actuators B Chem 149:28–33

Liu HQ, Kameoka J, Czaplewski DA, Craighead HG (2004) Polymeric nanowire chemical sensor. Nano Lett 4:671–675

Liu XL, Ly J, Han S, Zhang DH, Requicha A, Thompson ME, Zhou CW (2005a) Synthesis and electronic properties of individual single-walled carbon nanotube/polypyrrole composite nanocables. Adv Mater 17:2727–2732

Liu HQ, Reccius CH, Craighead HG (2005b) Single electrospun regioregular poly(3-hexylthiophene) nanofiber field-effect transistor. Appl Phys Lett 87:253106-1–253106-3

Liu P, Zhang L (2009) Hollow nanostructured polyaniline: preparation, properties and applications. Crit Rev Solid State Mater Sci 34:75–87

Long Y-Z, Li M-M, Gub C, Wan M, Duvail J-L, Liue Z, Fan Z (2011) Recent advances in synthesis, physical properties and applications of conducting polymer nanotubes and nanofibers. Progr Polymer Sci 36:1415–1442

Lu X, Wang C, Wei Y (2009) One-dimensional composite nanomaterials: synthesis by electrospinning and their applications. Small 5(21):2349–2370

Luoh R, Hahn HT (2006) Electrospun nanocomposite fiber mats as gas sensors. Compos Sci Technol 66:2436–2441

Ma XF, Li G, Wang M, Cheng YN, Bai R, Chen HZ (2006) Preparation of a nanowire structured polyaniline composite and gas sensitivity studies. Chemistry 12:3254–3260

Manesh K, Gopalan A, Lee K, Santhosh P, Song K, Lee D (2007) Fabrication of functional nanofibrous ammonia sensor. IEEE Trans Nanotechnol 6:513–518

McCann J, Marquez M, Xia Y (2006) Highly porous fibers by electrospinning into a cryogenic liquid. J Am Chem Soc 128:1436–1437

Miao Z, Xu D, Ouyang J, Guo G, Zhao X, Tang Y (2002) Electrochemically induced sol–gel preparation of single-crystalline TiO_2 nanowires. Nano Lett 2:717–720

Murray BJ, Walter EC, Penner RM (2004) Amine vapor sensing with silver mesowires. Nano Lett 4:665–670

Park J-A, Moon J, Lee S-J, Kim SH, Zyung T, Chu HY (2010) Structure and CO gas sensing properties of electrospun TiO2 nanofibers. Mater Lett 64:255–257

Pinto N, Ramos I, Rojas R, Wang P, Johnson A Jr (2008) Electric response of isolated electrospun polyaniline nanofibers to vapors of aliphatic alcohols. Sens Actuators B Chem 129:621–627

Raible I, Burghard M, Schlecht U, Yasuda A, Vossever T (2005) V_2O_5 nanofibers: novel gas sensors with extremely high sensitivity and selectivity to amines. Sens Actuators B Chem 106:730–735

Ramanathan K, Bangar MA, Yun M, Chen W, Mulchandani A, Myung N (2004) Individually addressable conducting polymer nanowires array. Nano Lett 4:1237–1239

Routkevitch D, Bigioni T, Moskovits M, Xu J (1996) Electrochemical fabrication of CdS nanowire arrays in porous anodic aluminum oxide templates. J Phys Chem 100:14307

Sahner K, Gouma P, Moos R (2007) Electrodeposited and sol–gel precipitated p-type $SrTi_{1-x}Fe_xO_{3-}\delta$ semiconductors for gas sensing. Sensors 7:1871–1886

Sahner K, Tuller HL (2010) Novel deposition techniques for metal oxide: prospects for gas sensing. J Electroceram 24:177–199

Shao C, Kim H, Gong J, Lee D (2002) A novel method for making silica nanofibers by using electrospun fibers of polyvinyl alcohol/silica composite as precursor. Nanotechnology 13:635–637

Theron A, Zussman E, Yarin AL (2001) Electrostatic field-assisted alignment of electrospun nanofibers. Nanotechnology 12:384–390

Virji S, Huang JX, Kaner RB, Weiller BH (2004) Polyaniline nanofiber gas sensors: examination of response mechanisms. Nano Lett 4:491–496

Vuong NM, Jung H, Kim D, Kim H, Hong S-K (2012) Realization of an open space ensemble for nanowires: a strategy for the maximum response in resistive sensors. J Mater Chem 22:6716–6725

Wang X, Drew C, Lee S-H, Senecal KJ, Kumar J, Samuelson LA (2002) Electrospun nanofibrous membranes for highly sensitive optical sensors. Nano Lett 2:1273–1275

Wang J, Chan S, Carlson RR, Luo Y, Ge G, Ries RS, Heath JR, Tseng GR (2004) Electrochemically fabricated polyaniline nanoframework electrode junctions that function as resistive sensors. Nano Lett 4:1693–1697

Wang J, Bunimovich YL, Sui G, Savvas S, Wang J, Guo Y, Heath JR, Tseng H-R (2006a) Electrochemical fabrication of conducting polymer nanowires in an integrated micro fluidic system. Chem Commun 2006:3075–3077

Wang G, Ji Y, Huang X, Yang X, Gouma P, Dudley M (2006b) Fabrication and characterization of polycrystalline WO_3 nanofibers and their application for ammonia sensing. J Phys Chem B 110:23777–23782

Wang Y, Ramos I, Santiago-Aviles J (2007) Detection of moisture and methanol gas using a single electrospun tin oxide nanofiber. IEEE Sens J 7:1347–1348

Yang A, Tao X, Wang R (2007a) Room temperature gas sensing properties of SnO_2/multiwall-carbon nanotube composite nanofibers. Appl Phys Lett 91:133110

Yang M, Xie T, Peng L, Zhao Y, Wang D (2007b) Fabrication and photoelectric oxygen sensing characteristics of electrospun Co doped ZnO nanofibers. Appl Phys Lett 89:427–430

Zhang ZM, Wei ZX, Wan MX (2002) Nanostructures of polyaniline doped with inorganic acids. Macromolecules 35:5937–5942

Zhang XY, Goux WJ, Manohar SK (2004) Synthesis of polyaniline nanofibers by "nanofiber seeding". J Am Chem Soc 126:4502–4503

Zhang Y, He X, Li J, Miao Z, Huang F (2008) Fabrication and ethanol-sensing properties of micro gas sensor based on electrospun SnO_2 nanofibers. Sens Actuators B Chem 132:67–73

Chapter 3
Metal Oxide-Based Nanostructures

3.1 Metal Oxide One-Dimensional Nanomaterials

As research on carbon fullerenes and carbon nanotubes has shown, dimensionality is a very important factor in determining the properties of nanomaterials. Therefore, control of the size and shape of metal oxide crystallites is of great interest with regard to the application of such materials in gas sensors. To date, various kinds of 1D metal oxide structures have been synthesized as nanowires, nanotubes, nanospheres, nanorods, and nanobelts (Li et al. 2002; Jung et al. 2003; Varghese et al. 2003a, b; Kam et al. 2004; Guha et al. 2004; Kuchibhatla et al. 2007; Barth et al. 2010; Soldano et al. 2012) (see Fig. 3.1). Among the large numbers of MOX nanowires synthesized up to now, SnO_2, In_2O_3, and ZnO nanowires are usually considered to be the best candidates for developing gas sensors (see Fig. 3.2) due to their relatively low cost of production and well-known properties (Mathur et al. 2007; Huang and Choi 2007; Barth et al. 2010; Choi and Jang 2010). It was found that, due to the progress achieved in the synthesis of 1D metal oxide nanomaterials, at present it is possible to synthesize high-quality nanomaterials with the length of individual nanowires up to 10–500 μm (Li et al. 2002; Zhang et al. 2004a). Interest in 1D ZnO structures is encouraged by the easy synthesis of high-quality and single-crystalline 1D ZnO nanostructures. The synthesis of 1D nanostructures based on other sensitive metal oxides such as TiO_2 and WO_3 has been reported to be difficult compared to other oxides.

Among many strategies to synthesize oxide NWs such as physical evaporation, silica-assisted Fe-catalytic growth, thermal reaction between powders and active carbon, alumina-assisted catalytic growth, and carbon-assisted synthesis, bottom-up growth is conventionally considered to be the most cost effective way to produce NWs in large quantities (Choi et al. 2008; Barth et al. 2010). These synthesis techniques, such as vapor–liquid–solid (VLS) or vapor–solid (VS) growth, yield randomly oriented assemblies of NWs. Most studies to date have been carried out on individual NWs selected from these assemblies and wired with appropriate contacts for transport measurements (see Fig. 3.3). This has been done largely because the measurement of individual NWs may allow study of the fundamentals of the gas-sensing effect in these MOX nanostructures (Kolmakov and Moskovits 2004).

3.1.1 1D Structures in Gas Sensors

It was established that over a wide range of operating temperatures, electronic properties of 1D structures are strongly influenced by surface processes (Kolmakov and Moskovits 2004; Vuong et al. 2012). It was found that, depending on the surrounding atmosphere in many cases, the NW resistance

G. Korotcenkov, *Handbook of Gas Sensor Materials*, Integrated Analytical Systems,
DOI 10.1007/978-1-4614-7388-6_3,

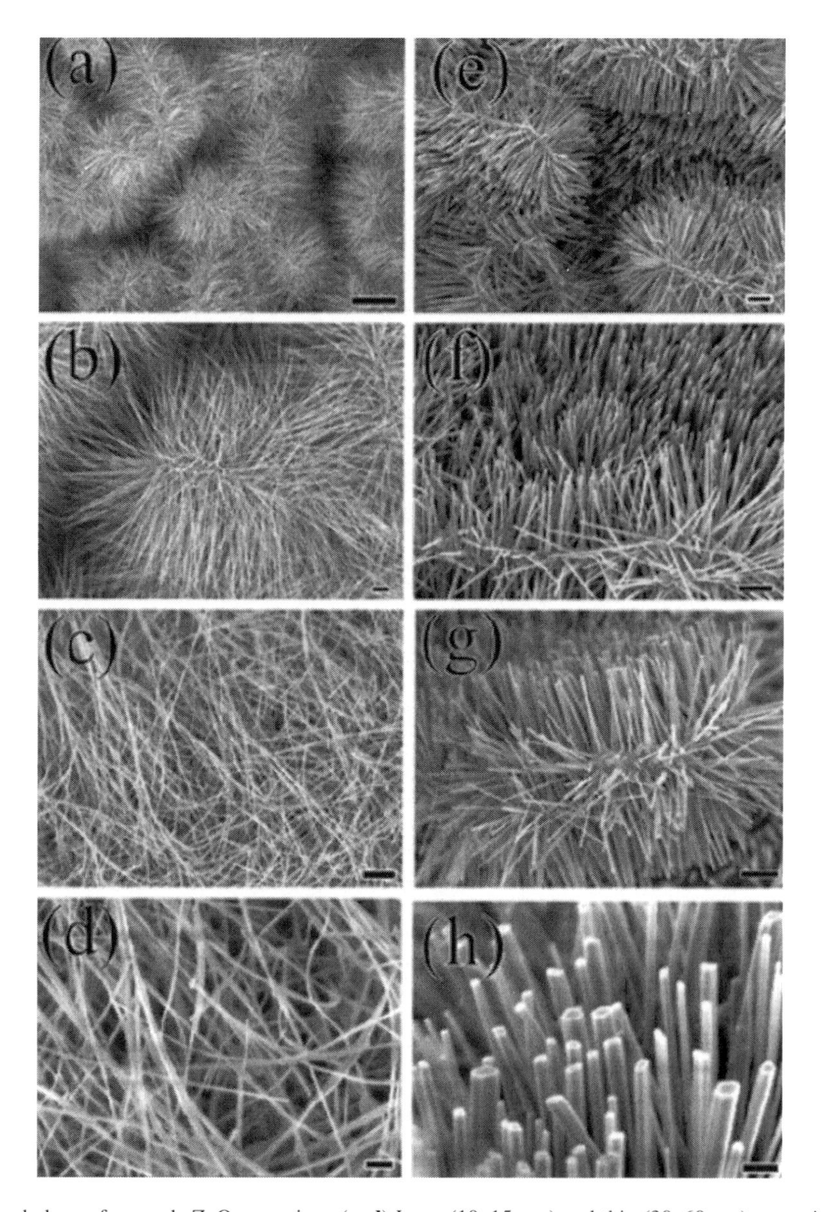

Fig. 3.1 Morphology of as-made ZnO nanowires. (**a**–**d**) Long (10–15 μm) and thin (30–60 nm) nanowires grown in the higher-temperature region of the furnace. (**e**–**h**) Short (1–2 μm) and thick (60–100 nm) nanowires grown in the lower-temperature region of the furnace. The *scale bar* for (**a**) is 10 μm; for (**b**, **c**, **e**–**g**) is 1 μm; and for (**d**, **h**) is 200 nm (Reprinted with permission from Banerjee et al. (2004), Copyright 2004 IOP)

can vary from almost a dielectric state to a highly conducting state entirely on the basis of the chemistry occurring at its surface. This means that 1D structures can show high gas-sensing performance (see, e.g., Fig. 3.4). Therefore, research has recently been carried out in many laboratories to synthesize and characterize NWs and to explore their integration in proof-of-concept gas sensors (Zhang et al. 2005; Comini et al. 2009; Hernandez-Ramirez et al. 2009; Choi and Jang 2010; Soldano et al. 2012). Some results of research related to gas sensitivity of 1D metal oxide nanostructures are presented in Tables 3.1 and 3.2.

It was established that the sensitivity of gas sensors based on NWs is very much dependent on the diameter of the NWs, an effect similar to grain-size influence on the response of sensors based on

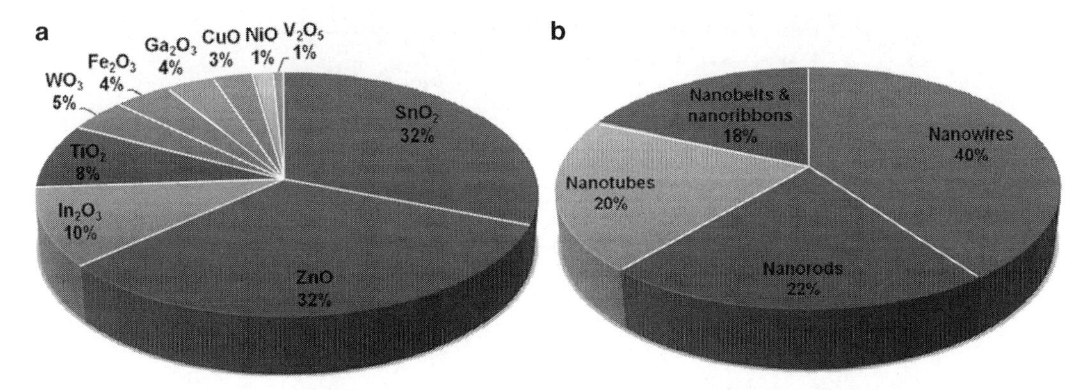

Fig. 3.2 (**a**) Top ten materials and (**b**) element forms of 1D metal oxide nanostructures used for gas sensor applications in publications in SCI Journals during 2002–2009 (Reprinted from Choi and Jang (2010), Published by MDPI)

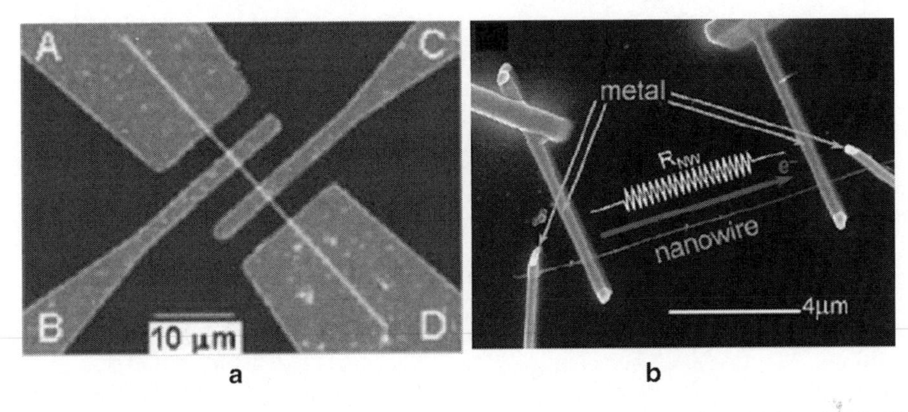

Fig. 3.3 SEM images of gas sensors based on single-metal oxide nanobelts and nanowires. (**a**) (Reprinted with permission from Fields et al. (2006), Copyright 2006 American Institute of Physics). (**b**) (Reprinted with permission from Hernandez-Ramirez et al. (2009), Copyright 2009 Royal Society of Chemistry)

polycrystalline metal oxides (see Fig. 3.5). For example, it was found that the Debye length (L_D) in most semiconducting oxide NWs is bigger than the NW radius, and, as a result, the gas sensitivity of such NWs is not high. As in the case of macroscopic semiconductors, the Debye length is a measure of the electronic "crosstalk" between the surface processes and bulk electronic structure. When a semiconductor oxide is moderately doped (10^{17}–10^{18} cm^{-3}), L_D falls into the 10–100-nm range. For nanostructures with radii r comparable to the L_D, this implies that surface band bending due to ionosorbed chemical or biological agents will comprise the entire volume of the nanostructure (Kolmakov 2008). Therefore the resultant electron depletion/accumulation due to adsorption/desorption events will drastically influence the conductivity of the nanostructure. Hernandez-Ramirez et al. (2007b) have shown that in real 1D SnO_2 structures, Debye length is ~20–30 nm, and therefore the performance of nanowire sensors can be significantly improved only when the diameter is smaller than 25 nm, that is comparable to the grain size in thin-film sensors. Research carried out by Li et al. (2003b, c) and Zhang et al. (2004b) confirmed this conclusion. They established that an extremely low detection limit was achieved for In_2O_3 and ZnO nanowires with 10 nm diameter only. In particular, such In_2O_3 nanowire sensors demonstrated record sensitivity to NO_x (up to five parts per billion (ppb) level) Zhang et al. (2004b).

However, the controllable growth and device fabrication of very thin metal oxide nanowires still remains an experimental challenge. Kolmakov and coworkers (Lilach et al. 2005) have found that

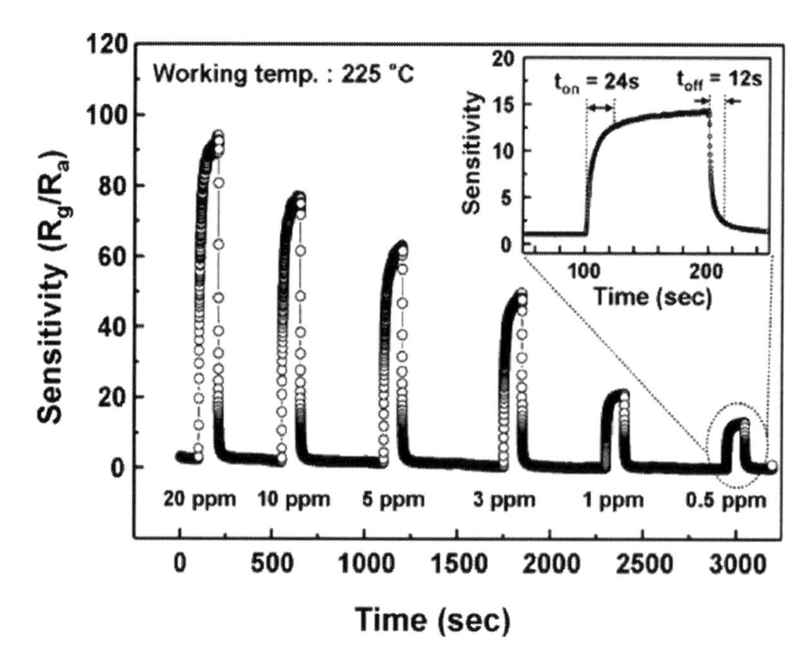

Fig. 3.4 Gas sensitivity curves of the network-structured ZnO-nanowire gas sensor under exposure to 20, 10, 5, 3, 1, and 0.5 ppm of NO_2 gas at the measurement temperature of 225 °C. The *inset* shows the response and recovery behavior of the gas sensor at 0.5 ppm (Reprinted with permission from Ahn et al. (2008), Copyright 2008 American Institute of Physics)

one of the most promising ways to address this challenge is to fabricate a single-crystal quasi-1D chemiresistor (can be mesoscopic) with one (or a few) very fine segments which adheres to the condition $r \sim L_D$. Since the narrow segment(s) will dominate the electron transport and sensing performance, such a device would have all the advantages of ultrathin single-crystal nanowires (see Fig. 3.5). Studies carried out by Dmitriev et al. (2007) have shown that the narrow segments serve as ideal "necks," as observed between particles in conventional polycrystalline thin-film gas sensors, but provide the significant advantages of greater morphological integrity and stability. Reports of the methodology of segmented oxide nanowire (SN) controllable growth via a programmable change in the local $SnxOy$ (x, y = 1, 2 ...) vapor supersaturation ratios in the vicinity of the wires during their vapor–solid growth have been published (Lilach et al. 2005).

Zhang et al. (2003) and Wan and Wang (2005) have shown that the doping can also be used for improvement of gas-sensing characteristics of metal oxide nanowires. In particular, Wan and Wang (2005) established that SnO_2 doping by Sb improves the kinetics of sensor response. They believed that the presence of Sb^{+5} in SnO_2 lattice accelerates the absorption of oxygen molecules and the formation of oxygen ions on the surface. Doping during growth can be very simple (Wallentina and Borgstrom 2011), because by modifying the composition of the precursor used for nanowire synthesis, the dopant may easily be introduced into the lattice of the growing nanowire, but fine control of the amount of dopant remains challenging. In addition, the formation of precipitates, the segregation, and nucleation of second phases must be carefully avoided during the doping. In single crystals, dopant atoms may produce defects and distortions in the lattice and, thanks to the confined dimension of nanowires, much more tensile stress can be managed (Fan et al. 2006; Hsu et al. 2006). Furthermore, unintentional defect formation may result in a decrease in the charge carrier mobility and a change in electrical and electronic properties, avoiding any structural and chemical stability deterioration.

Another approach to increase the sensitivity is to enhance molecule adsorption by modifying the surface (Kolmakov 2008). The surface of the nanowire sensors is always critical to their performance,

Table 3.1 Table of conductometric gas sensors based on 1D nanostructures of metal oxide classified by sensing oxide

Type of metal oxide	Gases tested	Multiple or single nanowire	Cross-section, diameter (nm)	Working temperature (°C)
SnO_2	CO, O_2	Single nanowire	60	200–300
SnO_2	CO, relative humidity	Single nanowire	25, 70	200–295
SnO_2	NH_3, CO	Single nanowire	100	60–300
SnO_2	NO_2	Single nanobelt	80–120/10–30	RT, UV activated
SnO_2–Pt/Ni	H_2, CO	Single nanowire	60	400
SnO_2	H_2, CO	Array of single nanowires	100	350
SnO_2	Ethanol, CO, NO_2	Mesh of nanobelts	200/20–40	200–400
SnO_2	Ozone	Mesh of nanobelts	100/20–40	400
SnO_2:Sb	Ethanol	Mesh of nanowire	40–100	300
ZnO	Ethanol	Bridging nanowires	80	300
ZnO	Ethanol	Mesh of nanowire	25	300
ZnO	Ethanol, H_2S, HCHO, LPG, gasoline, CO, ammonia	Nanowire, paste	40–80	330
ZnO	Methanol, NO/NO_2	Aligned nanowires	60–80	325
ZnO–Pt	H_2, ethanol	Nanowires, nanotubes, paste	60–100	150
ZnO–He+ implanted	H_2S	Single nanowire	400	RT
In_2O_3	Acetone, NO_2	Mesh of nanowires	100–500	200–500
In_2O_3	Ethanol, methanol, CH_4	Nanowires, paste	60–160	370
WO_3	CO, NO_2	Bundles of fibers	20–100	150
V_2O_5	Ethanol, methanol, H_2S, H_2, NH_3, CO, NO_x	Mesh of nanobelts	60–100/10–20	150–400
V_2O_5	NH_3, propanol, toluene, butylamine	Mesh of nanobelts	10/1.5	RT

Diameter is used for wires, while for nanostructure with rectangular cross-section, width/thickness are reported
Source: Reprinted with permission from Comini et al. (2009b). Copyright 2009 Springer

Table 3.2 Properties of gas sensors based on single 1D oxide nanostructure classified by tested gas

Target gas	Material (sensor type)	Detection limit (temp.)	Sensitivity (conc.)	Response time	Reference
NO_2	SnO_2 (nr-R)	2 ppm (25 °C)	7 (100 ppm)	~1 min	Law et al. (2002)
	SnO_2 (nw-R)	<0.1 ppm (25 °C)	1 (10 ppm)	~1 min	Prades et al. (2009b)
	In_2O_3 (nw-FET)	0.5 ppm (25 °C)	10^6 (100 ppm)	5 s	Li et al. (2003b)
	In_2O_3 (nw-FET)	0.02 ppm (25 °C)	0.8 (1 ppm)	15 min	Zhang et al. (2004b)
	ZnO (nw-R)	<0.1 ppn (225 °C)	100 (20 ppm)	24 s	Ahn et al. (2008)
H_2	SnO_2 (nw-FET)	<1 ppm (200 °C)	4 (1 ppm)	~50 s	Kolmakov et al. (2005)
	VO_2 (nw-R)	N/A (50 °C)	1,000 (100 %)	~10 min	Baik et al. (2009)
	$WO_{2.72}$ (nw-R)	<100 ppm (25 °C)	22 (1,000 ppm)	40 s	Rout et al. (2007)
CO	SnO_2 (nb-R)	5 ppm (400 °C)	7 (250 ppm)	30 s	Qian et al. (2006)
	SnO_2 (nw-FET)	100 ppm (25 °C)	15 (500 ppm)	~10 min	Kuang et al. (2008)
	ZnO (nw-R)	<50 ppm (275 °C)	3,200 (400 ppm)	~50 min	Wei et al. (2009)
	NiO (nw-R)	N/A (150 °C)	0.25 (800 ppm)	~2 h	Tresback and Padture (2008)
	CeO_2 (nw-R)	<10 ppm (25 °C)	2 (200 ppm)	~10 s	Liao et al. (2008)
H_2S	SnO_2 (nw-R)	<1 ppm (150 °C)	6×10^6 (50 ppm)	N/A	Kumar et al. (2009)
	ZnO (nw-R)	N/A (25 °C)	8 (300 ppm)	~50 s	Liao et al. (2007)
	In_2O_3 (nw-FET)	1 ppm (25 °C)	1 (20 ppm)	48 s	Zeng et al. (2009)
Ethanol	SnO_2 (nt-R)	N/A (400 °C)	20 (7.8 %)	~80 s	Liu and Liu (2005)
O_2	β-Ga_2O_3 (nw-R)	<50 ppm (25 °C)	20 (50 ppm)	1 s	Feng et al. (2006)

Sensitivity (R_{gas}/R_{air} or R_{air}/R_{gas}); *R* resistive; *nw* nanowire; *nt* nanotube; *nb* nanobelt; *nr* nanoribbon

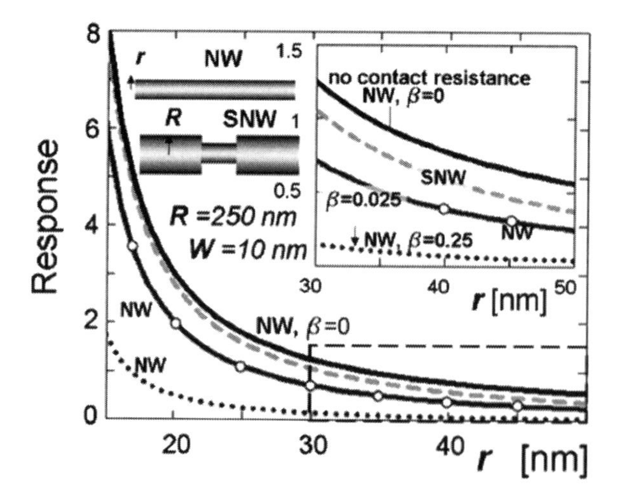

Fig. 3.5 Response of straight and segmented nanowires as a function of their radius at various contact resistances ($b=0$, 0.025, 0.25). For segmented nanowires (SNWs), the curve is drawn vs the radius of the smaller segment. The *solid curve* (*top*) corresponds to the nanowire with no contact resistance; the *dashed curve* corresponds to the SNW with thick segments of 500 nm diameter and $b=0.025$; the *solid curve* marked with *circles* corresponds to the SNW with $b=0.025$; the *dotted curve* corresponds to the SNW with $b=0.25$. The depletion width is ~10 nm at all cases (Reprinted with permission from Dmitriev et al. (2007), Copyright 2007 IOP)

and surface modification by chemical functionalization has been shown to enhance sensitivity significantly. As was shown before (Chap. 10, Vol. 1), addition of a small amount of noble metals over the MO_x surface such as Au, Pd, Pt, and Ag can speed up surface reactions and improve selectivity toward target gas species. For example, SnO_2 nanowires doped with Pd particles exhibit significantly improved sensitivity (Kolmakov et al. 2005) due to the formation of Schottky barrier junctions induced by Pd particles on the nanowire surface. This mechanism of surface functionalizing was discussed in Chap. 10 (Vol. 1).

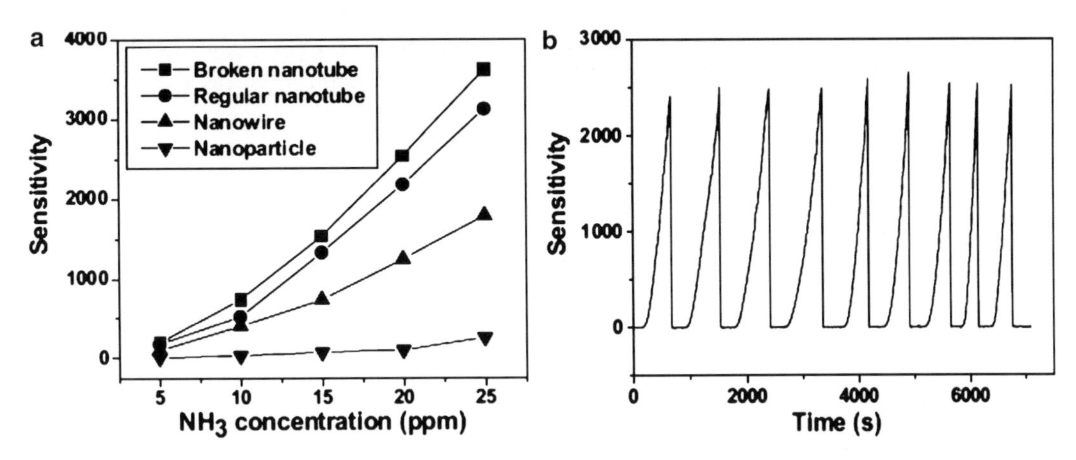

Fig. 3.6 (**a**) Sensitivity response vs ammonia concentration at room temperature for four types of gas sensors based on In_2O_3 nanostructures including broken In_2O_3 nanotubes, regular In_2O_3 nanotubes, In_2O_3 nanowires, and In_2O_3 nanoparticles. (**b**) Sensitivity response of the gas sensor based on broken In_2O_3 nanotubes vs the time for a concentration of NH_3 of 20 ppm at room temperature (Reprinted with permission from Du et al. (2007), Copyright 2007 Wiley)

Du et al. (2007) established that the morphology of nanostructures also influences sensor response. They investigated the response of an In_2O_3-based ammonia sensor fabricated using In_2O_3 nanostructures such as broken nanotubes, regular nanotubes, nanowires, and nanoparticles. It was found that broken In_2O_3 nanotubes showed a high response. Figure 3.6 shows the response of these In_2O_3 nanostructures. It is seen that porous (broken) nanotubes have better responses. Response and recovery times at room temperature were less than 20 s.

Most designers believe that the use of single-crystal or polycrystalline nanowires (NWs) as gas sensors has potential advantages compared to conventional thick-film and thin-film devices because of the intrinsic properties of NWs, such as well-defined geometry, single crystallinity, and high surface-to-volume ratio (Law et al. 2004). In addition, 1D metal oxide nanomaterials should be more thermodynamically stable in comparison to nanograins, promoting stable operation of chemical sensors at higher temperatures. In particular, single-nanowire sensors have no contribution of necks and grain boundaries to the device operation. In addition, Hernandez-Ramirez et al. (2009) showed that the absence of nooks and crannies in nanowire-based devices facilitates directs adsorption/desorption of gas molecules, improving the dynamic behavior of these prototypes to various gases. Therefore, it has been suggested that the use of NWs can help to improve stability and reduce temporal drift of the parameters of conductometric gas sensors (Hernandez-Ramirez et al. 2009; Sysoev et al. 2009).

Experiments have shown that NW-based gas sensors, because of their low energy consumption, are very promising for integration in portable systems. It has been demonstrated by several authors that self-heating may allow the development of gas sensors without external heaters (Strelcov et al. 2008; Prades et al. 2008; Kolmakov 2008). It was established that Joule self-heating of NWs operated under a passing electric current can be applied for NW-based gas sensors. Meier et al. (2007a, b) and Hernandez-Ramirez et al. (2007a, b) have shown that sensors based on individual NWs with self-heating can be fabricated with power consumption of only a few tens of microwatts, much less than that required for thin-film sensors mounted on microhotplates, which usually require milliwatts to work in continuous mode. Moreover, Prades et al. (2008) have demonstrated that the response of the sensors to 0.5 ppm NO_2 without a heater ($I_m = 10$ nA) was the absolute equivalent to that with a heater ($T = 175$ °C) (see Fig. 3.7). Of course, the indicated approach is not simple, because it requires improved quality of ohmic contacts and a special power supply with strong protection from external influence. Prades et al. (2009a, b) have also demonstrated the equivalence between thermal and room-temperature UV light-assisted responses of single-SnO_2 nanowire gas sensors. For instance, the response of the sensors to 0.5 ppm NO_2 at room temperature under UV light illumination was the

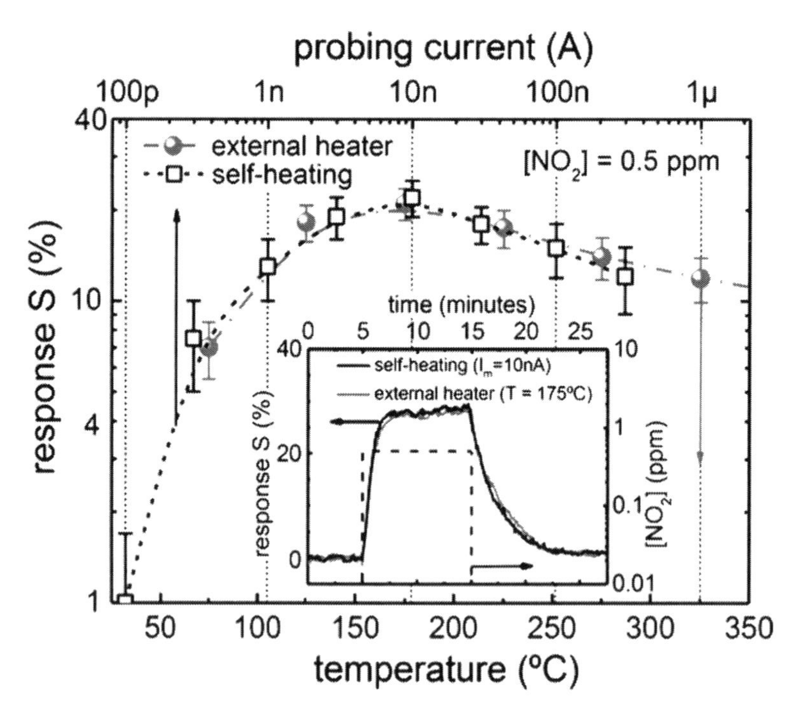

Fig. 3.7 Response S of the SnO_2 nanowires operated in self-heating mode and with external microheater to $[NO_2]=0.5$ ppm. The maximum response to this gas with and without heater ($I_m=10$ nA) is the absolute equivalent (*inset*) (Reprinted with permission from Prades et al. (2008), Copyright 2008 American Institute of Physics)

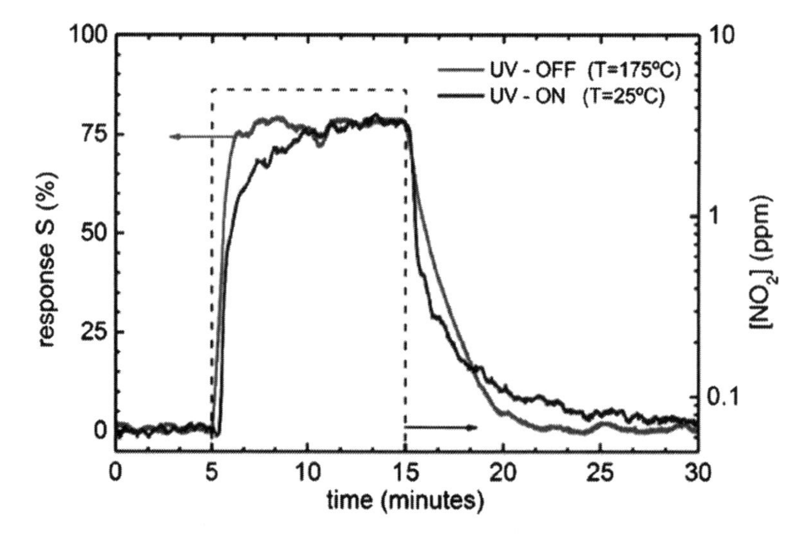

Fig. 3.8 Comparison of the response of a SnO_2 nanowire, operated at $T=175$ °C in dark conditions and at room temperature ($T=25$ °C) under UV illumination (Eph $=3.67\pm0.05$ eV, Φph $=30\times1022$ ph/m2s) to a pulse of 5 ppm NO_2 (Reprinted with permission from Prades et al. (2009a), Copyright 2009 Royal Society of Chemistry)

absolute equivalent to that operating at 175 °C in dark conditions (see Fig. 3.8). The experimental results revealed that nearly identical responses, similar to thermally activated sensor surfaces, could be achieved by choosing the optimal illumination conditions.

However, it should be noted that the controlled separation, manipulation, and characterization of NWs are not straightforward processes due to the intrinsic problems of working at the nanoscale.

Table 3.3 Summary of NW assembly technologies

NW assembly technologies	Advantages	Disadvantages
Flow-assisted alignment in microchannels	Parallel and crossed NW arrays can be assembled Compatible with both rigid and flexible substrates	Area for NW assembly is limited by the size of fluidic microchannels Difficult to achieve very high density of NW arrays NW suspension needs to be prepared first
Bubble-blown technique	Area for NW assembly is large Compatible with both rigid and flexible substrates	It is difficult to achieve high-density NW arrays NW suspension needs to be prepared first
Contact printing	Area for NW assembly is large High-density NW arrays can be achieved Parallel and crossed NW arrays can be assembled Direct transfer NW from growth substrate to receiver substrate Compatible with both rigid and flexible substrates NW assembly process is fast	Growth substrate needs to be planar The process works the best for long NWs
Differential roll printing	Area for NW assembly is large High-density NW arrays can be achieved Direct transfer NW from growth substrate to receiver substrate Compatible with both rigid and flexible substrates NW assembly process is fast	Growth substrate needs to be cylindrical The process works the best for long NWs
Langmuir–Blodgett technique	Area for NW assembly is large High-density NW arrays can be achieved Parallel and crossed NW arrays can be assembled Compatible with both rigid and flexible substrates	NWs typically need to be functionalized with surfactant The assembly process is slow and has to be carefully controlled NW suspension needs to be prepared first
Electric field-assisted orientation	NWs can be placed at specific location Compatible with both rigid and flexible substrates NW assembly process is fast	Patterned electrode arrays are needed Area for NW assembly is limited by the electrode patterning NW density is limited It works the best for conductive NWs NW suspension needs to be prepared first

Source: Reprinted with permission from Liu et al. (2012). Copyright 2012 American Chemical Society

These procedures are complex and require well-established methodologies which are not yet fully developed. Several promising techniques, however, have been reported to align or orientate NW assemblies with the help of microfluidics, electrostatic or magnetic fields, surface prepatterning, self-assembly, and templating. These architecture principles, which have made it possible to design prototype devices, have been comprehensively reviewed recently (Xia et al. 2003; Whang et al. 2003; Lieber 2003; Qi et al. 2003; Liu et al. 2012). The advantages and disadvantages of these approaches are briefly summarized in Table 3.3. All these fabrication routes, however, are still in a preliminary

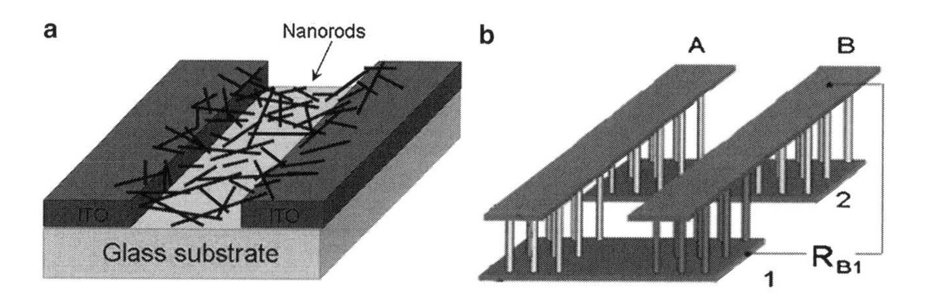

Fig. 3.9 (**a**) Schematic diagrams for multinanowire-based chemical sensors. (**b**) Sketch of planar nanowire array sensor composed of individually addressable sensing elements based on template synthesis (Adapted with permission from Kolmakov and Moskovits (2004), Copyright 2004 A Nonprofit Scientific Publisher)

stage of development. Thus, the use of NWs in real devices is still in a preliminary stage and needs a breakthrough in order to integrate them with low-cost industrial processes (Hernandez-Ramirez et al. 2009). However, taking into account the specificity of chemical sensors, one can expect that wide applications of 1D structures in sensor technology will come sooner than in standard microelectronics. Gas sensors can have very simple construction, which does not require a lot of connections as in integrated circuits. In addition, requirements for ohmic contacts in chemical sensors are not so strict. Gas sensors also permit considerably more range in the parameters of sensing materials.

However, research has shown that utilization of 1D structure arrays (see Fig. 3.9a) that allow the use of standard sensor technology for chemical sensor fabrication already provides an opportunity to realize in gas sensors the range of incontestable advantages of 1D nanostructures. As can be seen from results presented in Table 3.1, exactly such an approach was used in most gas sensors based on 1D structures. It is important that the observed sensitivity of 1D metal oxide sensors is significantly higher than the reported sensitivity of carbon nanotubes (Kong et al. 2000). It is assumed that the difference is related to the nature of the metal oxide surface, which can react readily with ambient species, compared to the inert sidewall of carbon nanotubes (Kong et al. 2000). Another approach involves growing parallel NWs or nanotubes in a perpendicular direction on substrates (Kolmakov and Moskovits 2004) (see Fig. 3.9b). Varghese and co-workers (Varghese and Grimes 2003; Varghese et al. 2009) used such an approach to develop devices from TiO_2 nanotubes. This technique is similar to thin-film production but promises higher accuracy in making the columnar nanoelements.

The approach proposed by Choi et al. (2008) and Ahn et al. (2008, 2009) is also interesting for gas sensor design (see Fig. 3.10). Figure 3.10b, d shows side- and top-view scanning electron microscope (SEM) images of ZnO nanowires grown on patterned electrodes. ZnO nanowires grown only on the patterned electrodes have many nanowire/nanowire junctions as seen in Fig. 3.10c. These junctions act as electrical conducting path for electrons. The device structure in this work is very simple and efficient compared with those adopted by previous researchers, because the electrical contacts to nanowires are self-assembled during the synthesis of nanowires. Ahn et al. (2008, 2009) asserted that this method of on-chip fabrication of nanowire-based gas sensors is scalable and reproducible. However, this statement raises doubts.

One of the challenges still to be resolved is that the reduced contact area between metal electrodes and NWs magnifies the contribution of the contact electrical properties and may hide the phenomena which take place on the surface of NWs (Nam et al. 2005; Hernandez-Ramirez et al. 2006, 2007a, b; Lin and Jian 2008). Comini et al. (2009b) have noted that the technical difficulties in producing reliable electrical contacts on one individual nanostructure in a controlled fabrication process at the nanoscale level has restricted the number of projects on this topic. Indeed the study on this topic leads to better comprehension of the electrical transport mechanisms which take place in these nanostructures.

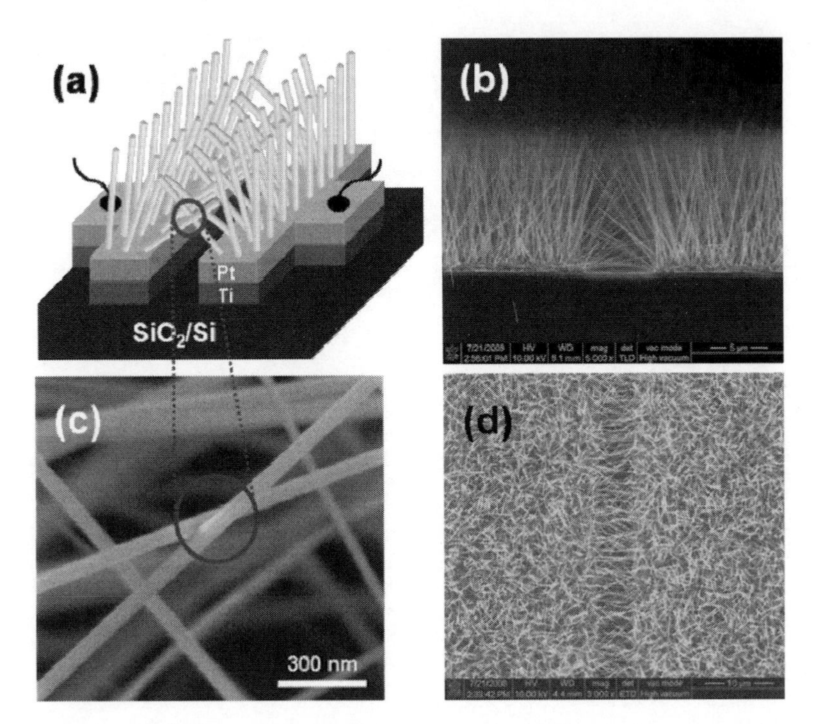

Fig. 3.10 (**a**) The schematic illustration of ZnO-nanowire air bridges over the SiO_2/Si substrate. (**b**) *Side-* and (**d**) *top-view* SEM images clearly show selective growth of ZnO nanowires on Ti/Pt electrode. (**c**) The junction between ZnO nanowires grown on both electrodes (Reprinted with permission from Ahn et al. (2009), Copyright 2009 Elsevier)

On another front, to accomplish electrical measurements of individual NWs, free of parasitic effects, and to develop competitive sensors, various fabrication and characterization strategies have been evaluated. For instance, low-current measurement protocols have been found to allow the devices to operate long-term without degradation of their performance (Hernandez-Ramirez et al. 2007a, b). Thus the present state of development of NW-based technologies has led to complete and well-controlled characterization of proof-of-concept devices, which were previously unattainable (Comini et al. 2009).

3.1.2 The Role of 1D Structures in the Understanding of Gas-Sensing Effects

From the extensive literature data available now in the field, it can be concluded that the most significant results related to employing NWs as gas sensor elements will be obtained by studying the fundamentals of their receptor function (Korotcenkov 2008). The random aggregation of nanoparticles in poly(nano)crystalline MOXs, as well as the scatter in their size, makes it difficult to study the gas transduction phenomena accurately at the nanoscale. The single crystallinity of the NWs, however, makes easier to interpret the experimental data (Fig. 3.11).

On the other hand, the geometry of individual single-crystal NWs promotes a detailed analysis of-the gas-surface interactions, because there are no necks and boundaries. The decrease in the number of parameters which control the sensor response of 1D structures should facilitate a better understanding of the nature of the observed effects. In addition, 1D metal oxide nanomaterials have excellent crystallinity and clear faceting with a fixed set of planes. It is expected that these nanomaterials will have less concentration of point defects and specific adsorption and catalytic properties, conditioned by a particular combination of crystallographic planes. In other words, semiconducting 1D metal

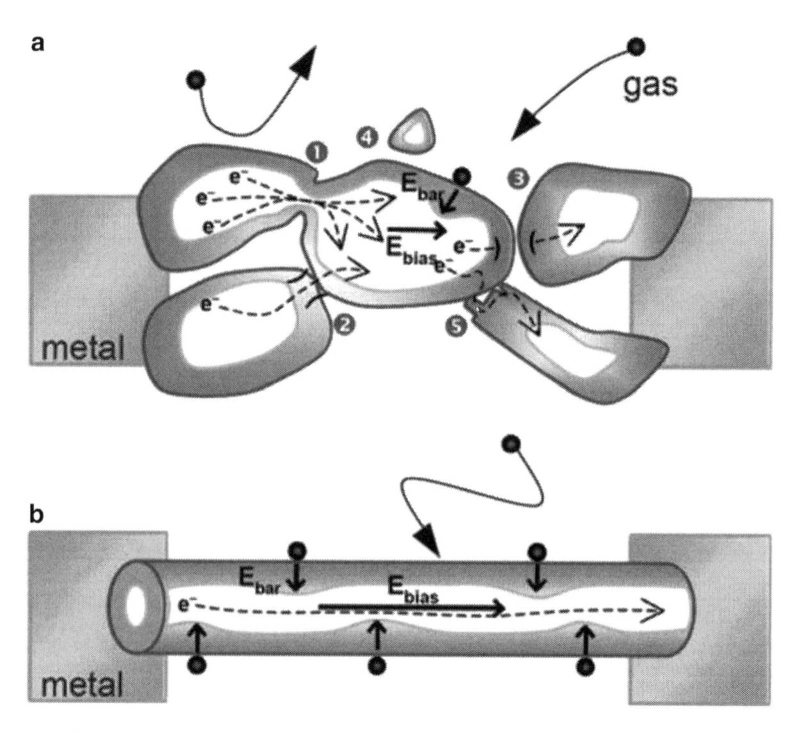

Fig. 3.11 Diagrams illustrating difference in gas-sensing effects in (**a**) polycrystalline material and (**b**) individual nanowires. One can see that any intergrain necks or boundaries are absent in 1D-based gas sensors. Moreover, E_{bar} and E_{bias} fields are always orthogonal and independent (Reprinted with permission from Hernandez-Ramirez et al. (2009), Copyright 2009 Royal Society of Chemistry)

oxide structures with well-defined geometry and perfect crystallinity could represent a perfect model-material family for systematic experimental study and theoretical understanding of the fundamentals of gas-sensing mechanisms in metal oxides. Theoretical simulation of gas-sensing effects in 1D-based devices becomes significantly simpler. In many cases, only two planes participate in gas-sensing effects in 1D structures. Therefore, in simulations, 1D structures should be considered as single crystals with limited sizes. This means that the NW studies provide a way to gain deeper insight into the chemical and electrical transduction mechanisms which control the gas-sensing performance of MOXs and may become an extremely helpful tool to further advance sensors in the future (Wang 2000). In particular, Hernandez-Ramirez et al. (2009) suggested that the use of NWs as a platform to develop novel gas sensors may provide the opportunity to find solutions to problems which remain partially unsolved under other conventional technologies, such as the lack of stability, poor selectivity toward different gas species, and power consumption. The assembly of single-crystal NWs as 3D or 2D networks or mats may be useful not only as a platform for practical sensors but also as a model to study the fundamentals of gas-sensing effects in polycrystalline samples (Go et al. 2009).

For example, Korotcenkov (2008) believes that the determination of crystallographic planes with optimal combinations of adsorption/desorption and catalytic parameters and the development of methods for metal oxide grains deposition with indicated faces should be considered as a main contemporary goal for thin-film technology as applied to metal oxide gas sensors. Research presented by Batzill et al. (2005), Korotcenkov et al. (2005c), Gurlo (2011), and Han et al. (2009) could be considered as the first step in this direction. A proper choice for crystallite deposition technology or synthesis with a necessary grain facet can help to increase sensitivity, improve selectivity, and decrease humidity effects in gas sensors. For example, in Golovanov et al. (2005b), using a Mulliken population analysis, it was shown that the chemisorption of OH groups on the (110) face is accompanied by the

Table 3.4 Crystallographic geometry of 1D oxide nanostructures

Nanostructures	Crystal structure	Growth direction	Top surface	Side surface
ZnO-belt	Wurtzite	(0001) or $(01\bar{1}0)$	$\pm(2\bar{1}\bar{1}0)$ or $\pm(2\bar{1}\bar{1}0)$	±(01–10) or ±(0001)
Ga_2O_3-belt	Monoclinic	(001) or (010)	±(100) or ±(100)	$\pm(010)$ or $\pm(10\bar{1})$
Ga_2O_3-sheet	Monoclinic	(101) (normal)	±(100)	±(010) $\pm(10-1)$ and $\pm(21\bar{2})$
t-SnO_2-belt	Rutile	(101)	$\pm(10\bar{1})$	±(010) and ±(10–1)
SnO_2-belt	Rutile	(100)	±(001)	±(010)
t-SnO_2-wire	Rutile	(101)	$\pm(10\bar{1})$	
SnO_2-belt				
Zigzag—initial	Rutile	(101)	±(010)	±(10–1) and ±(100)
Zigzag—final	Rutile	(101)	±(010)	±(100)
α-SnO_2-wire	Orthorhombic	(010)	±(100)	±(001)
SnO_2-diskette	Tetragonal	±(100) and ±(110)	±(001)	±(100) and ±(110)
SnO_2-ribbon	Rutile	(101)	$(10\bar{1})/(\bar{1}01)$	(010)/(0–10)
SnO_2-ribbon (sandwich)	Rutile/orthorombic	$(110)_o/(6\bar{5}3)_t$	$\pm(100)_o/\pm(231)_t$	$\pm(001)_o/\pm(10\bar{1})_t$

Source: Data from Dai et al. (2001, 2003), Ma et al. (2003), Huang et al. (2005), Korotcenkov (2008)

localization of negative charge to a greater extent than the chemisorption of OH groups at the (011) surface of SnO_2. This means that adsorption/desorption processes and surface reactions with water vary with different SnO_2 crystallographic planes. Understandably, the preparation of polycrystalline metal oxides with the necessary grain faceting is difficult to control. However, it is achievable for 1D sensors and should be a high-priority area of research. Crystallographic study of metal oxides nanobelts has shown that the planes and faceting in 1D structures depend on the parameters of synthesis. For example, in Liang et al. (2001), Li et al. (2003a, b), Dai et al. (2003), and Kong et al. (2003), it was reported that, depending on the synthesis route, there was a possibility to synthesize In_2O_3 nanobelts with (100), (120), (111), (110), and (001) growth directions. In_2O_3 nanobelts grown in (100) and (120) directions had the top and bottom surfaces being (001), while the (100) nanobelts had side surface of (010) and a rectangular cross-section (Kong et al. 2003). The (120) nanobelts had a parallelogram cross-section. In_2O_3 nanobelts grown in (111) and (110) directions had other set of planes; side and top surfaces were (100) planes (Liang et al. 2001; Zhang et al. 2004a). In the case of the (001) growth direction, In_2O_3 nanobelts were enclosed by (100) and (010) planes (Dai et al. 2003). The crystallographic geometry of other metal oxide 1D nanostructures is presented in Table 3.4. It is important to note that, for example, the SnO_2 nanostructures are not enclosed by (110) planes, which are the most stable crystallographic planes in the SnO_2 lattice. Thus, the results presented testify that we really have the ability to control faceting planes of metal oxide 1D structures. This means that in 1D-based sensors, we have additional opportunities for controlling the sensor's performance parameters. On the other hand, in 1D gas sensors the role of contacts increases because of their small area and therefore greater specific resistance.

With regard to the sensors base on nanowire arrays, the same regulations apply to them as to those on based on the usual nanoparticle films. Impedance spectroscopy studies have showed that the gas-sensing mechanism for sensors based on networked nanowire thin films involves changes in both the nanowire and the inter-nanowire boundary resistances (Deb et al. 2007). For gas sensor design, thin films containing nanowires in a highly networked fashion are promising. A study has shown that the sensors based on those films behave similarly to single-nanowire devices without much post-processing effort (Deb et al. 2007).

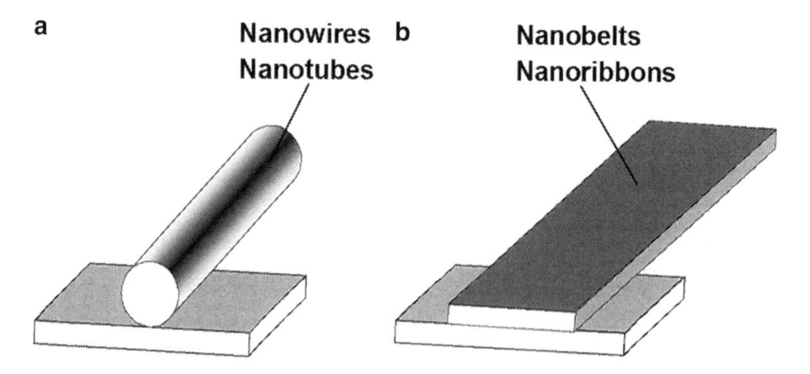

Fig. 3.12 Position of (**a**) nanowires or nanotubes and (**b**) nanobelts on the contact pad (Reprinted with permission from Korotcenkov (2008), Copyright 2008 Elsevier)

3.1.3 What Kind of 1D Structures Is Better for Gas Sensor Design?

It has been shown that there are many types of 1D structures which can be used for gas sensor design, including nanowires, nanobelts, nanotubes, nanoribbons, nanowhiskers, nanorings, and nanorods (Dai et al. 2002; Hu et al. 2002; Li et al. 2003a, b; Kam et al. 2004; Zhang et al. 2004a; Soldano et al. 2012). However, analyzing the opportunities for 1D structures of various types for their practical use in gas sensors, it is necessary to admit that nanobelts (nanoribbons) might be the most demanding. Nanobelts are thin and plain belt-type structures with rectangular cross-sections (see Fig. 3.12). At present nanobelts are available for practically all basic oxides used in chemical sensors and there is plenty of information about the synthesis of nanobelts based on SnO_2, In_2O_3, ZnO, Ga_2O_3, TiO_2, etc. (Li et al. 2000; Pan et al. 2001; Dai et al. 2002; Lee et al. 2002; Konga and Wang 2003; Varghese et al. 2003a, b; Jung et al. 2003; Kong et al. 2003; Guha et al. 2004). Synthesis of nanobelts can be accomplished using various methods (Liu et al. 2001; Wang et al. 2002; Li et al. 2002; Xu et al. 2002), creating good conditions for widening research into such nanosize materials.

In addition, nanobelts do not have the mechanical strength of nanotubes. Their crystallographic perfection is a very good attribute. Because there are zero defects in their structure, there is no problem, as there is with nanotubes, that defects may destroy quantum-level properties. The suitable geometry of nanobelts is also their important advantage for mass manufacturing. They have high structural homogeneity and long length. Typical nanobelts have widths of 30–300 nm, thicknesses of 10–15 nm, and lengths from several micrometers to hundreds, or even some thousands, of micrometers (Dai et al. 2001; Pan et al. 2001). Moreover, nanobelts are flexible and can therefore be curved over 180° without being damaged. This fact gives additional advantages to those materials in device design.

3.2 Mesoporous, Macroporous, and Hierarchical Metal Oxide Structures

Mesoporous, macroporous, and hierarchical metal oxide structures are other modern directions in sensing materials design. It was established that the ability to create macroporous objects from nano-scaled components may create new resources for optimization of gas sensor parameters. The pores in 3D MOX structures are developed in the submicrometer or nanodimensional domain. Therefore, these structures are frequently called *mesoporous*. These rather new materials with extremely high surface area offer a high degree of versatility in terms of structure and texture. The most successful approaches to developing mesoporous structures are based on the synthesis of pore-containing particles, including

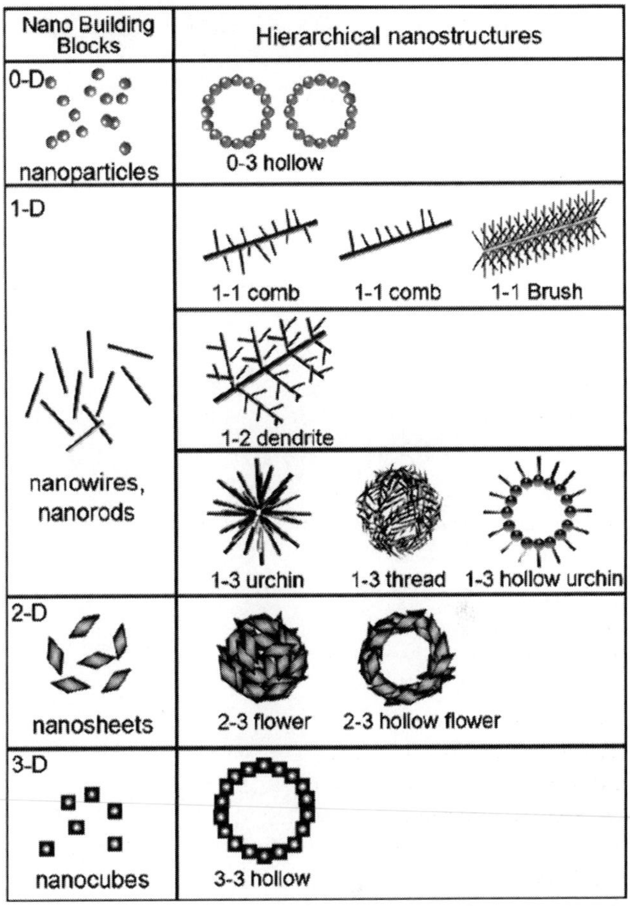

Nano Building Blocks	Hierarchical nanostructures
0-D nanoparticles	0-3 hollow
1-D nanowires, nanorods	1-1 comb, 1-1 comb, 1-1 Brush
	1-2 dendrite
	1-3 urchin, 1-3 thread, 1-3 hollow urchin
2-D nanosheets	2-3 flower, 2-3 hollow flower
3-D nanocubes	3-3 hollow

Fig. 3.13 Nomenclature of hierarchical structures according to the dimensions of the nano-building blocks (the former number) and of the consequent hierarchical structures (the latter number) (Reprinted with permission from Lee (2009), Copyright 2009 Elsevier)

pore core–solid shell nanostructures, via templating or via a facile wet-chemical approach combining with an annealing process (Lin et al. 2012).

The hierarchical nanostructures also have extremely high surface areas and have little tendency to agglomerate, which allows one to employ them as high-performance gas sensor materials. A "hierarchical structure" means the higher dimension of a micro- or nanostructure composed of many, low dimensional, nano-building blocks (Lee 2009). The various hierarchical structures can be classified according to the dimensions of nano-building blocks and the consequent hierarchical structures, referring to the dimensions, respectively, of the nano-building blocks and of the assembled hierarchical structures (see Fig. 3.13). For example, "1–3 urchin" means that 1D nanowires/nanorods are assembled into a 3D urchin-like spherical shape, and "2–3 flower" indicates that a 3D flowerlike hierarchical structure is assembled from many 2D nanosheets. Under this framework, the hollow spheres can be regarded as the assembly of 1D nanoparticles into the 3D hollow spherical shape. Thus, strictly speaking, the 0–3 hollow spheres should be regarded as one type of hierarchical structure.

Various methods have been considered for synthesizing such hierarchical hollow-particle structures, including spray-drying (Okuyama et al. 2006; Colombo et al. 2010), sol–gel (Hayashi et al. 2009), layer-by-layer (LbL) templating (Rothschild and Tuller 2006; Wang et al. 2008), electrodeposition (Wadea and Wegrowe 2005), vapor-phase impregnation (Yue and Zhou 2008), interface growth, and pulse laser deposition (Sanchez et al. 2008). However, the most promising technologies seem to be methods based on sol–gel, aerosol spray, and LbL deposition (Lee 2009).

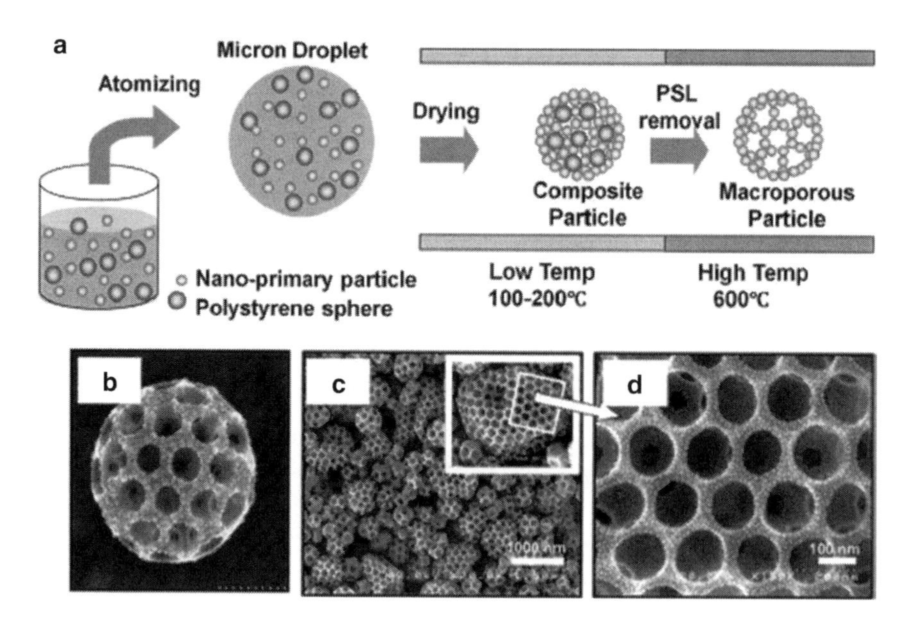

Fig. 3.14 (**a**) Experimental apparatus for developing particles containing ordered pores. (**b–d**) Photos of the particles (Reprinted with permission from Iskandar (2009), Copyright 2009 Elsevier)

The conventional LbL process starts with electrostatic adsorption of a charged species onto a substrate which is a priori charged with opposite sign. The species adsorption results in recharging of the substrate surface. Then the first layer is covered with the next one via deposition of an oppositely charged species and so on until the required film thickness is achieved. The LbL approach is versatile for assembling various materials—polymers, lipids, proteins, dye molecules, etc.—on a number of substrates via not only electrostatic interactions but also hydrogen bonding, hydrophobic interactions, covalent bonding, and complementary base pairing (Ariga et al. 2007). The properties of LbL films, such as composition and thickness, can be readily adjusted by simply varying the species adsorbed, the number of layers deposited, and the conditions employed during the assembly process. Removal of the templating substrate following LbL film formation can give rise to freestanding nanostructured materials with different morphologies and functions. Further details about this technique and its applications may be found elsewhere (Caruso et al. 2000; Ariga et al. 2007; Zhang et al. 2007; Wang et al. 2008).

One route to fabricating mesoporous hierarchical structures using an aerosol spray method, and photos of the structures produced, are given in Fig. 3.14.

Synthesis of mesoporous materials using liquid deposition techniques is shown in Fig. 3.15. This overall complex transformation can be seen simply as direct polycondensation of the inorganic precursors around the organic micelles (or mesophase), which freezes the liquid-crystal mesostructure.

Shimizu and coworkers (Hieda et al. 2008; Hayashi et al. 2009; Morio et al. 2009; Hyodo et al. 2010) have shown that, by using a modified sol–gel technique, pyrolysis or a physical vapor deposition process employing a polymethylmethacrylate (PMMA) microsphere film as a macropore template, macroporous (*mp*-) films of various materials promising for gas sensor application can be fabricated. They established that different kinds of gas sensors fabricated with the *mp*-semiconductor films (SnO_2 and In_2O_3 to detect H_2, NO_X, and H_2S), photoluminescence-type (SnO_2 mixed with Eu_2O_3 to detect H_2 and NO_2), and quartz crystal microbalance type ($BaCO_3$ to detect NO_2) showed stronger gas responses as well as fast response and recovery speeds in comparison with those fabricated with a conventional film and powders without macropores (Hayashi et al. 2009; Yuan et al. 2011; Hyodo and Shimizu 2011).

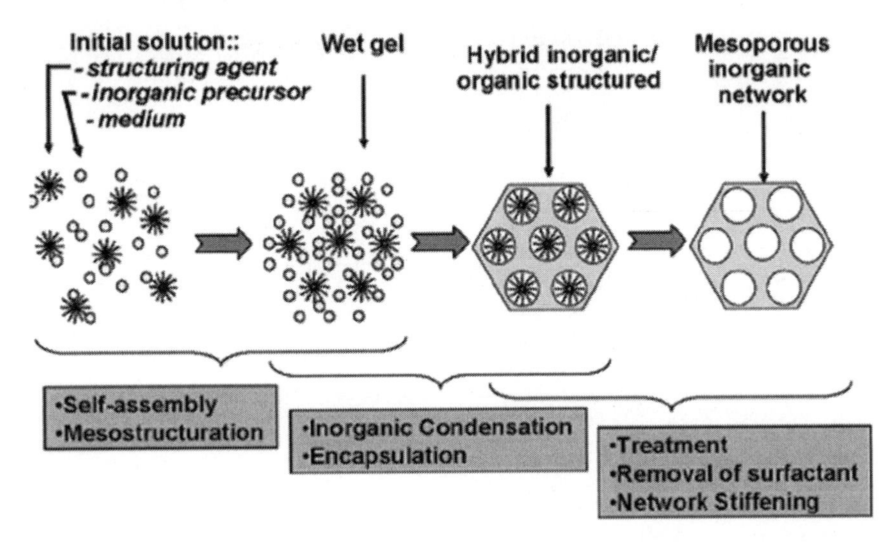

Fig. 3.15 Templating approach applied to make thin films via liquid deposition (Reprinted with permission from Sanchez et al. (2008), Copyright 2008 American Chemical Society)

Planar substrates containing defined pore structures (e.g., macroporous membranes) have also been widely employed as templates to develop mesoporous MOXs (Wang et al. 2008). Planar templates are typically fabricated using solid substrates such as quartz slides, Si wafers, and metal electrodes. The porous planar templates allow deposition of metals or MOXs with well-defined 3D morphologies. For example, membranes containing cylindrical pores provide the possibility to form (nano)tubes (hollow cylinders). The outer diameter, length, composition, and thickness of the tubes are controlled by the pore diameter, membrane thickness, the type of species deposited, and the number of layers assembled, respectively. The open ends and large surface area associated with tubes render them useful for delivery applications. In particular, they can be readily loaded with large quantities of gas species.

Spherical colloids can also be employed as templates. The deposition of MOX films onto the outer surface of colloidal particles, which are then chemically or thermally removed, further gives the possibility to design hollow spherical structures like small capsules (Fig. 3.16). This method permits good control over the properties of the hollow spheres (e.g., size, composition, thickness, permeability, function) by proper choice of the sacrificial colloids and the film components (Wang et al. 2008).

Chemical methods can also be applied for hierarchical structures synthesis (Lee 2009). For example, experiment has shown that the hollow precursor or oxide particles can be prepared either by the chemically induced self-assembly of surfactants into micelle configuration (Zhao et al. 2006) (see Fig. 3.17a) or by the polymerization of carbon spheres and subsequent encapsulation of metal hydroxide during the hydrothermal/solvothermal reaction (Yang et al. 2007). The core polymer parts are normally removed by heat treatment at elevated temperature (500–600 °C). Thus, hollow oxide structures can be used stably as gas detection materials at the sensing temperature of 200–400 °C without thermal degradation. Other approaches based on Ostwald ripening and the Kirkendall effect can also be used. These processes are illustrated in Fig. 3.17b, c. The Kirkendall effect is realized during the oxidation of dense and crystalline metal particles, when the outward diffusion of metal cations through the oxide shell layers is very rapid compared to the inward diffusion of oxygen to the metal core (Liu and Zeng 2004; Fan et al. 2007). Solid evacuation is the common aspect of Ostwald ripening and the Kirkendall effect. However, in principle, the shell layers developed by the Kirkendall effect are denser and less permeable than those by Ostwald ripening.

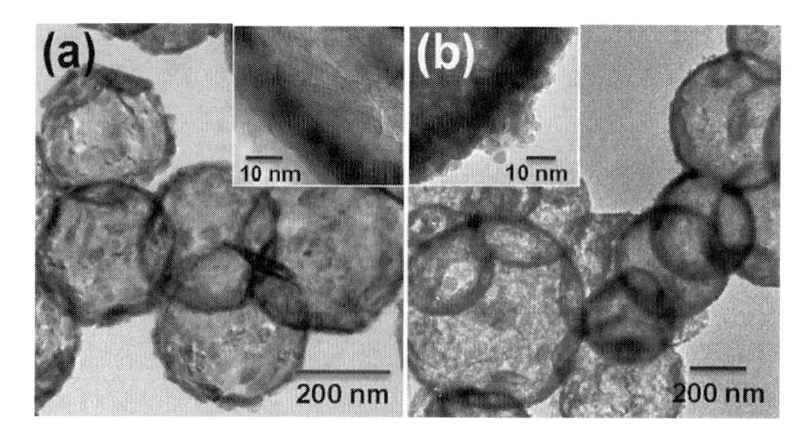

Fig. 3.16 (**a**) TiO_2 hollow spheres and (**b**) SnO_2 hollow spheres prepared by the encapsulation of Ti and Sn precursors on Ni spheres and the removal of core metal templates by dilute HCl aqueous solution after heat treatment at 400 °C (Reprinted with permission from Lee (2009), Copyright 2009 Elsevier)

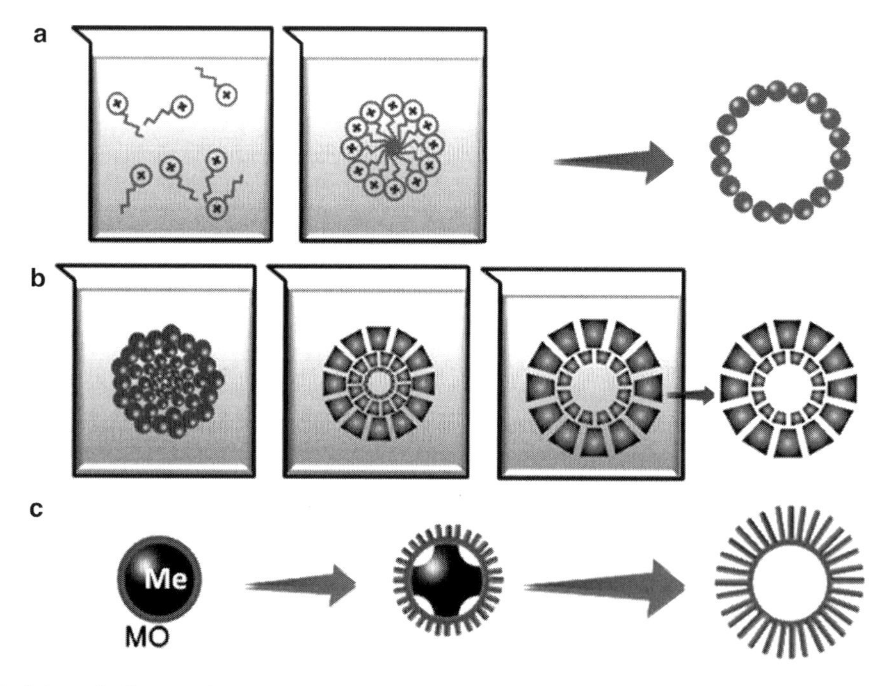

Fig. 3.17 Schematic diagrams for the preparation of hollow structures using the (**a**) self-assembled hydrothermal/solvothermal reaction, (**b**) Ostwald ripening of porous secondary particles, and (**c**) solid evacuation by the Kirkendall effect (Reprinted with permission from Lee (2009), Copyright 2009 Elsevier)

It should be noted that technologies for fabricating mesoporous and hierarchical nanostructures have been developed for all the basic MOXs (SnO_2, In_2O_3, TiO_2, WO_3, Fe_2O_3, etc.) utilized to develop conductometric gas sensors (Lee 2009). The gas-sensing performance of sensors based on mesoporous and hollow nanostructures is well reviewed elsewhere (Tiemann 2007; Lee 2009), and it is agreed that such structures are really attractive platforms for gas-sensing applications (Shimizu et al. 2004, 2005; Yue and Zhou 2008). Mesoporous and hollow structures have been reported to show very high gas-sensing response (Devi et al. 2002; Hyodo et al. 2003; Wagner et al. 2007) and fast response

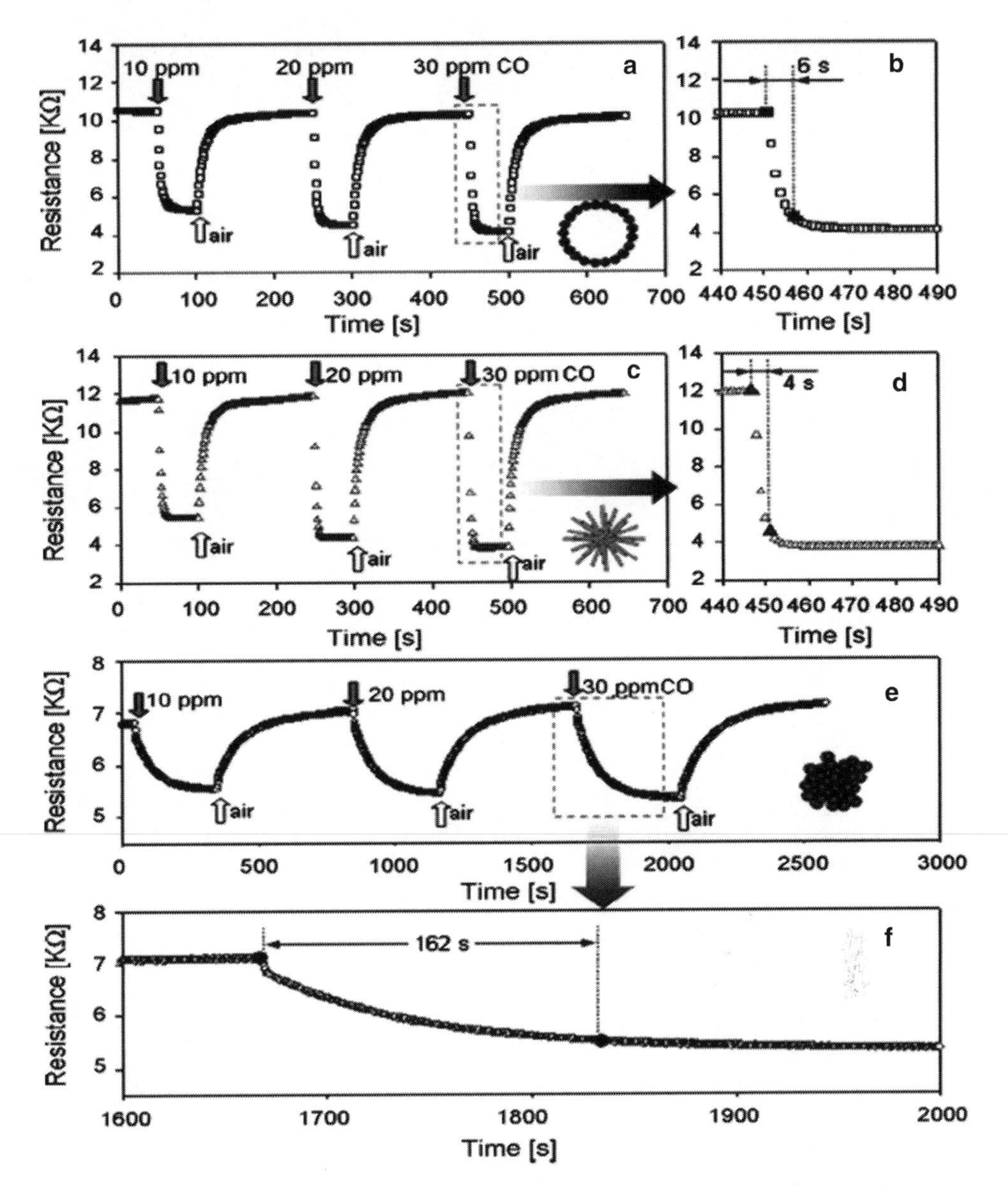

Fig. 3.18 Dynamic CO sensing transients of (**a**, **b**) the hollow In_2O_3 microspheres, (**c**, **d**) the urchin-like In_2O_3 microspheres, and (**e**, **f**) the agglomerated In_2O_3 powder (measurement at 400 °C) (Reprinted with permission from Choi et al. (2009b), Copyright 2009 Elsevier)

kinetics (Liu et al. 2007), which are attributed to their high surface area and well-defined porous architecture. A particularly large difference in kinetics of sensor response was observed in comparison with sensors fabricated using agglomerated powders (see Fig. 3.18). It was established that the hollow nanostructures follow the same basic trends as mentioned for the thin-film layers. When the shells are rather dense and thick, the gas-sensing reaction occurs only near the surface of the hollow spheres, and the inner parts of these spheres are inactive. However, if the shell is sufficiently thin, the primary particles in the entire hollow sphere are able to participate in gas-sensing reactions even when the shells are less permeable. In addition, the rate of sensor response of hollow spheres increases with the

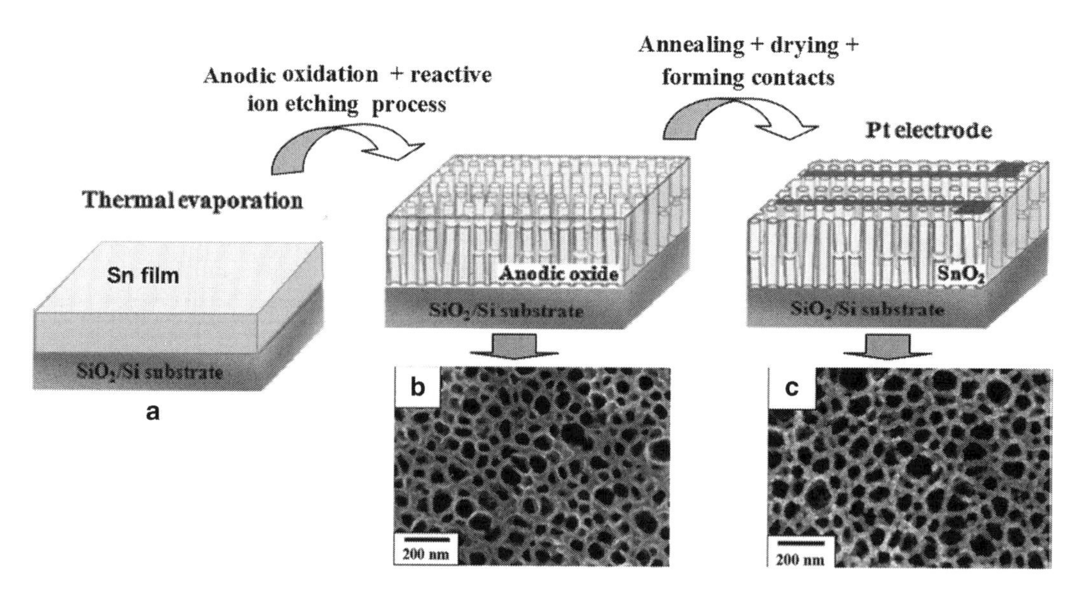

Fig. 3.19 (**a**) Schematic diagram of the experimental procedures of SnO_2-based sensor fabrication using Sn anodic etching with following oxidation and plain views of (**b**) as-etched and (**c**) as-annealed films (Reprinted with permission from Jeun and Hong (2010), Copyright 2010 Elsevier)

thinner shell configuration due to the faster gas diffusion. It has also been found that the sensor response and response kinetics of the mesoporous sensing materials, similarly to the conventional metal oxide matrix, can be further improved by surface modification (Hyodo et al. 2003) and doping by catalytic materials (Rossinyol et al. 2007a, b; He et al. 2009).

From our point of view, electrochemical etching of metal films with subsequent oxidation of fabricated porous structure is also a very promising approach to designing a mesoporous gas-sensing matrix. Such an approach was discussed with reference to TiO_2 and SnO_2 in Varghese et al. (2003b), Li et al. (2009b), Rani et al. (2010), and Jeun and Hong (2010). Anodic oxidation is the most commonly employed method for the synthesis of self-ordered porous semiconductor structures, which is a relatively simple, low-cost, and high-yield process. The process used by Jeun and Hong (2010) is shown schematically in Fig. 3.19. It was found that, after annealing at 700 °C for 1 h, the pore wall of Sn films changed into a granular SnO_2 structure due to sintering, but the microscopic features of the nanosized pores and vertical nanochannels remained the same, indicating a high thermal stability of the anodized films (see Fig. 3.19b, c).

It was established that the sensitivity of the above-mentioned mesoporous and hierarchical structures can be extremely high (Mor et al. 2004; Lin et al. 2012), and, similar to other gas-sensing materials, this sensitivity depends on pore size, wall thickness, and tube length (Varghese et al. 2003b). In particular, Varghese et al. (2003a, b) have shown that the crystallized nanoscale walls and intertubular connecting points play critical roles in determining the remarkable hydrogen sensitivities of the TiO_2 nanotube arrays. The TiO_2 nanotube sample showing the highest sensitivity had a wall thickness of ~13 nm. However, we need to recognize that, for commercialization of the results obtained, the continuation of research in this direction is required. Till now there have been technological problems such as the reproducibility, the forming of low-resistance contacts, and structure stabilization during and following annealing which need to be resolved. For example, Varghese et al. (2003a, b) have found that TiO_2 samples, prepared using anodization and consisting of well-defined nanotube arrays, were found to be stable up to temperatures around 580 °C (in oxygen ambient) even after crystallization of the tube walls. However, at higher temperatures, the crystallization of the titanium support disturbed the nanotube architecture, causing it to collapse and densify. When subjected to rapid annealing in oxygen at up to 950 °C, the structure did not collapse completely although a complete crystallization occurred; the tubes coalesced and formed a wormlike pattern. The anodization technique needs to be optimized as well (Li et al. 2009; Jeun and Hong 2010).

It should be noted that the problem of structural instability exists for all types of the above-mentioned structures independent of material used. Moreover, resolving this problem does not have a universal approach. Unfortunately, every material used for mesoporous, macroporous, and hollow structure fabrication requires a specific approach for resolving it. For example, Shimizu and coworkers (Hyodo et al. 2002) found that the most important key to the drastic improvement of thermal stability of mesoporous (*m*-) SnO_2 powders is to immerse them in a phosphoric acid aqueous solution before calcination and consequently loading phosphorus components on the surface of *m*-SnO_2 crystallites. Such treatment enabled the preparation of the *m*-SnO_2 powders with small crystallite size (2–3 nm in diameter) and large specific surface area (>300 m^2/g) even after calcination at 600 °C.

References

Ahn M-W, Park K-S, Heo J-H, Park J-G, Kim D-W, Choi KJ, Lee J-H, Hong S-H (2008) Gas sensing properties of defect-controlled ZnO-nanowire gas sensor. Appl Phys Lett 93:263103

Ahn M-W, Park K-S, Heo J-H, Kim D-W, Choi KJ, Park J-G (2009) On-chip fabrication of ZnO-nanowire gas sensor with high gas sensitivity. Sens Actuators B Chem 138:168–173

Ariga K, Hill JP, Ji Q (2007) Layer-by-layer assembly as a versatile bottom-up nanofabrication technique for exploratory research and realistic application. Phys Chem Chem Phys 9:2319–2340

Baik JM, Kim MH, Larson C, Yavuz CT, Stucky GD, Wodtke AM, Moskovits M (2009) Pd-sensitized single vanadium oxide nanowires: highly responsive hydrogen sensing based on the metal-insulator transition. Nano Lett 9:3980–3984

Banerjee D, Lao JY, Wang DZ, Huang JY, Steeves D, Kimball B, Ren ZF (2004) Synthesis and photoluminescence studies on ZnO nanowires. Nanotechnology 15:404–409

Barth S, Hernandez-Ramirez F, Holmes JD, Romano-Rodriguez A (2010) Synthesis and applications of one-dimensional semiconductors. Prog Mater Sci 55:563–627

Batzill M, Katsiev K, Burst JM, Diebold U, Chaka AM, Delley B (2005) Gas-phase-dependent properties of SnO_2 (110), (100), and (101) single-crystal surfaces: structure, composition, and electronic properties. Phys Rev B 72:165414

Caruso F, Trau D, Möhwald H, Renneberg R (2000) Enzyme encapsulation in Layer-by-Layer engineered multilayer capsules; Langmuir 16(4):1485–1488

Choi K-J, Jang HW (2010) One-dimensional oxide nanostructures as gas-sensing materials: review and issues. Sensors 10:4083–4099

Choi Y-J, Hwang I-S, Park J-G, Choi KJ, Park J-H, Lee J-H (2008) Novel fabrication of an SnO_2 nanowire gas sensor with high sensitivity. Nanotechnology 19:095508

Choi K, Kim HR, Lee JH (2009) Enhanced CO sensing characteristics of hierarchical and hollow In_2O_3 microspheres. Sens Actuators B Chem 138:497–503

Colombo P, Vakifahmetoglu C, Costacurta S (2010) Fabrication of ceramic components with hierarchical porosity. J Mater Sci 45:5425–5455

Comini E, Faglia G, Sberveglieri G (2009) Electrical-based gas sensing. In: Comini E, Faglia G, Sberveglieri G (eds) Solid state gas sensing. Springer, New York, pp 47–107

Dai ZR, Pan ZW, Wang ZL (2001) Ultra-long single crystalline nanoribbons of tin oxide. Solid State Comm 118:351–354

Dai ZR, Gole JL, Stout JD, Wang ZL (2002) Tin oxide nanowires, nanoribbons, and nanotubes. J Phys Chem B 106:1274–1279

Dai ZR, Pan ZW, Wang ZL (2003) Novel nanostructures of functional oxides synthesized by thermal evaporation. Adv Funct Mater 13(1):9–24

Deb B, Desai S, Sumanasekera GU, Sunkara MK (2007) Gas sensing behaviour of mat-like networked tungsten oxide nanowire thin films. Nanotechnology 18:285501

Devi GS, Hyodo T, Shimizu Y, Egashira M (2002) Synthesis of mesoporous TiO_2-based powders and their gas-sensing properties. Sens Actuators B Chem 87:122–129

Dmitriev S, Lilach Y, Button B, Moskovits M, Kolmakov A (2007) Nanoengineered chemiresistors: the interplay between electron transport and chemisorption properties of morphologically encoded SnO_2 nanowires. Nanotechnology 18:055707

Du N, Zhang H, Chen B, Ma X, Liu Z, Wu J, Yang D (2007) Porous indium oxide nanotubes: layer-by-layer assembly on carbon-nanotube templates and application for room-temperature NH_3 gas sensors. Adv Mater 19:1641–1645

Fan HJ, Fuhrmann B, Scholz R, Himcinschi C, Berger A, Leipner H, Dadgar A, Krost A, Christiansen S, Gosele U, Zacharias M (2006) Vapour-transport-deposition growth of ZnO nanostructures: switch between c-axial wires and a-axial belts by indium doping. Nanotechnology 17:S231–S8239

Fan HJ, Gosele Y, Zacharias M (2007) Formation of nanotubes and hollow nanoparticles based on Kirkendall and diffusion processes: a review. Small 3:1660–1671

Feng P, Xue YX, Liu YG, Wan Q, Wang TH (2006) Achieving fast oxygen response in individual β-Ga_2O_3 nanowires by ultraviolet illumination. Appl Phys Lett 89:112114

Fields LL, Zheng JP, Cheng Y, Xiong P (2006) Room temperature low-power hydrogen sensor based on a single tin dioxide nanobelt. Appl Phys Lett 88:263102

Go J, Sysoev VV, Kolmakov A, Pimparkar N, Alam MA (2009) A novel model for (percolating) nanonet chemical sensors for microarray-based E-nose applications. In: Proceedings of IEEE international electron devices meeting, Baltimore, MD, 7–9 December. Abstract 26.6(1–4) (doi: 10.1109/IEDM.2009.5424266)

Golovanov V, Pekna T, Kiv A, Litovchenko V, Korotcenkov G, Brinzari V, Cornet A, Morante J (2005) The influence of structural factors on sensitivity of SnO_2-based gas sensors to CO in humid atmosphere. Ukr Phys J 50(4):374–380

Guha P, Chakrabarti S, Chaudhuri S (2004) Synthesis of β-Ga_2O_3 nanowire from elemental Ga metal and its photoluminescence study. Physica E 23:81–85

Gurlo A (2011) Nanosensors: towards morphological control of gas sensing activity. SnO_2, In_2O_3, ZnO and WO_3 case studies. Nanoscale 3:154–165

Han X-G, He H-Z, Kuang Q, Zhou X, Zhang X-H, Xu T, Xie Z-X, Zheng L-S (2009) Controlling morphologies and tuning the related properties of nano/microstructured ZnO crystallites. J Phys Chem C 113:584–589

Hayashi M, Hyodo T, Shimizu Y, Egashira M (2009) Effects of microstructure of mesoporous SnO_2 powders on their H_2 sensing properties. Sens Actuators B Chem 141:465–470

He L, Jia Y, Meng F, Li M, Liu J (2009) Development of sensors based on CuO-doped SnO_2 hollow spheres for ppb level H_2S gas sensing. J Mater Sci 44:4326–4333

Hernandez-Ramirez F, Tarancon A, Casals O, Rodrıguez J, Romano-Rodriguez A, Morante JR, Barth S, Mathur S, Choi TY, Poulikakos D, Callegari V, Nellen PM (2006) Fabrication and electrical characterization of circuits based on individual tin oxide nanowires. Nanotechnology 17:5577–5583

Hernandez-Ramirez F, Prades JD, Tarancon A, Barth S, Casals O, Jimenez-Diaz R, Pellicer E, Rodrıguez J, Juli MA, Romano-Rodriguez A, Morante JR, Mathur S, Helwig A, Spannhake J, Mueller G (2007a) Portable microsensors based on individual SnO_2 nanowires. Nanotechnology 18:495501

Hernandez-Ramirez F, Tarancon A, Casals O, Arbiol J, Romano-Rodriguez A, Morante JR (2007b) High response and stability in CO and humidity measures using a single SnO_2 nanowire. Sens Actuators B Chem 121:3–17

Hernandez-Ramirez F, Prades JD, Jimenez-Diaz R, Fischer T, Romano-Rodriguez A, Mathur S, Morante JR (2009) On the role of individual metal oxide nanowires in the scaling down of chemical sensors. Phys Chem Chem Phys 11:7105–7110

Hieda K, Hyodo T, Shimizu Y, Egashira M (2008) Preparation of porous tin dioxide powder by ultrasonic spray pyrolysis and their application to sensor materials. Sens Actuators B Chem 133:144–150

Hsu CL, Lin YR, Chang SJ, Lu TH, Lin TS, Tsai SY, Chen IC (2006) Influence of the formation of the second phase in ZnO/Ga nanowire systems. J Electrochem Soc 153:G333–G336

Hu JQ, Ma XL, Shang NG, Xie ZY, Wong NB, Lee CS, Lee ST (2002) Large-scale rapid oxidation synthesis of SnO_2 nanoribbons. J Phys Chem B 106:3823–3826

Huang X-J, Choi Y-K (2007) Chemical sensors based on nanostructured materials. Sens Actuators B Chem 122:659–671

Huang L, Pu L, Shi Y, Zhang R, Gu B, Du Y, Wright S (2005) Controlled growth of well-faceted zigzag tin oxide mesostructures. Appl Phys Lett 87:163124

Hyodo T, Shimizu Y (2011) Microstructural design of gas-sensing materials by utilizing various templates. In: Proceedings of 4th GOSPEL workshop: gas sensors based on semiconducting metal oxides: basic understanding and applications, 6–7 June 2011, Tübingen

Hyodo T, Nishida N, Shimizu Y, Egashira M (2002) Preparation and gas-sensing properties of thermally stable mesoporous SnO_2. Sens Actuators B Chem 83:209–215

Hyodo T, Shimizu Y, Egashira M (2003) Gas-sensing properties of ordered mesoporous SnO_2 and effects of coating thereof. Sens Actuators B Chem 93:590–600

Hyodo T, Inoue H, Motomura H, Matsuo K, Hashishin T, Tamaki J, Shimizu Y, Egashira M (2010) NO_2 sensing properties of macroporous In_2O_3-based powders fabricated by utilizing ultrasonic spray pyrolysis employing polymethylmethacrylate microspheres as a template. Sens Actuators B Chem 151:265–273

Iskandar F (2009) Nanoparticle processing for optical applications—a review. Adv Powder Tech 20:283–292

Jeun J-H, Hong S-H (2010) CuO-loaded nano-porous SnO_2 films fabricated by anodic oxidation and RIE process and their gas sensing properties. Sens Actuators B Chem 151:1–7

Jung SW, Park WI, Yi GC, Kim M (2003) Fabrication and controlled magnetic properties of Ni/ZnO nanorod heterostructures. Adv Mater 15(15):1358–1361

Kam KC, Deepak FL, Cheetham AK, Rao CNR (2004) In_2O_3 nanowires, nanobouquets and nanotrees. Chem Phys Lett 397:329–334

Kolmakov A (2008) Some recent trends in fabrication, functionalisation and characterization of metal oxide nanowire gas sensors. Int J Nanotechnol 5:450–474

Kolmakov A, Moskovits M (2004) Chemical sensing and catalysis by one-dimensional metal oxide nanostructures. Annu Rev Mater Res 34:151–180

Kolmakov A, Klenov DO, Lilach Y, Stemmer S, Moskovits M (2005) Enhanced gas sensing by individual SnO_2 nanowires and nanobelts functionalized with Pd catalyst particles. Nano Lett 5:667–673

Kong J, Franklin NR, Zhou C, Chapline MG, Peng S, Cho K, Dai H (2000) Nanotube molecular wires as chemical sensors. Science 287:622–625

Kong XH, Sun XM, Li YD (2003) Synthesis of ZnO nanobelts by carbothermal reduction and their photoluminescence properties. Chem Lett 32:546–547

Konga YY, Wang ZL (2003) Structures of indium oxide nanobelts. Solid State Comm 128:1–4

Korotcenkov G (2008) The role of morphology and crystallographic structure of metal oxides in response of conductometric-type gas sensors. Mater Sci Eng R 61:1–39

Korotcenkov G, Cornet A, Rossinyol E, Arbiol J, Brinzari V, Blinov Y (2005) Faceting characterization of SnO_2 nanocrystals deposited by spray pyrolysis from $SnCl_4$-$5H_2O$ water solution. Thin Solid Films 471(1–2):310–319

Kuang Q, Lao CS, Li Z, Liu YZ, Xie ZX, Zheng LS, Wang ZL (2008) Enhancing the photon- and gas-sensing properties of a single SnO_2 nanowire based nanodevice by nanoparticle surface functionalization. J Phys Chem C 112:11539–11544

Kuchibhatla SVNT, Karakoti AS, Bera D, Seal S (2007) One dimensional nanostructured materials. Progr Mater Sci 52:699–913

Kumar V, Sen S, Muthe KP, Gaur NK, Gupta SK, Yakhmi JV (2009) Copper doped SnO_2 nanowires as highly sensitive H_2S gas sensor. Sens Actuators B Chem 138:587–590

Law M, Kind H, Messer B, Kim F, Yang PD (2002) Photochemical sensing of NO_2 with SnO_2 nanoribbon nanosensors at room temperature. Angew Chem Int Ed 41:2405–2408

Law M, Goldberger J, Yang P (2004) Semiconductor nanowires and nanotubes. Annu Rev Mater Res 34:83–122

Lee JH (2009) Gas sensors using hierarchical and hollow oxide nanostructures: overview. Sens Actuators B Chem 140:319–336

Lee JS, Park K, Nahm S, Kim SW, Kima S (2002) Ga_2O_3 nanomaterials synthesized from ball-milled GaN powders. Crystal Growth 244:287–295

Li JY, Qiao ZY, Chen XL, Chen L, Cao YG, He M, Li H, Cao ZM, Zhang Z (2000) Synthesis of β-Ga_2O_3 nanorods. J Alloys Compounds 306:300–302

Li ZJ, Li HJ, Chen XL, Li L, Xu YP, Li KZ (2002) β-Ga_2O_3 nanowires on unpatterned and patterned MgO single crystal substrates. J Alloys Comp 345:275–279

Li C, Zhang D, Han S, Liu TT, Zhou C (2003a) Diameter-controlled growth of single-crystalline In_2O_3 nanowires and their electronic properties. Adv Mater 15:143–146

Li C, Zhang D, Liu X, Han S, Tang T, Han J, Zhou C (2003b) In_2O_3 nanowires as chemical sensors. Appl Phys Lett 82:1613–1615

Li C, Zhang D, Lei B, Han S, Liu X, Zhou C (2003c) Surface treatment and doping dependence of In_2O_3 nanowires as ammonia sensors. J Phys Chem B 107:12451–12455

Li Y, Yu X, Yang Q (2009) Fabrication of TiO_2 nanotube thin films and their gas sensing properties. J. Sensors 2009:402174(1–19)

Liang C, Meng G, Lei Y, Phillip F, Zhang L (2001) Catalytic growth of semiconducting In_2O_3 nanofibers. Adv Mater 13:1330–1333

Liao L, Lu HB, Li JC, Liu C, Fu DJ, Liu YL (2007) The sensitivity of gas sensor based on single ZnO nanowire modulated by helium ion radiation. Appl Phys Lett 91:173110

Liao L, Mai HX, Yuan Q, Lu HB, Li JC, Liu C, Yan CH, Shen ZX, Yu T (2008) Single CeO_2 nanowire gas sensor supported with Pt nanocrystals: gas sensitivity, surface bond states, and chemical mechanism. J Phys Chem C 112:9061–9065

Lieber CM (2003) Nanoscale science and technology: building a big future from small things. MRS Bull 28:486–491

Lilach Y, Zhang JP, Moskovits M, Kolmakov A (2005) Encoding morphology in oxide nanostructures during their growth. Nano Lett 5:2019–2022

Lin Y-F, Jian W-B (2008) The impact of nanocontact on nanowire based nanoelectronics. Nano Lett 8:3146–3150

Lin Z, Song W, Yang H (2012) Highly sensitive gas sensor based on coral-like SnO_2 prepared with hydrothermal treatment. Sens Actuators B Chem Chem 173:22–27

Liu Y, Liu M (2005) Growth of aligned square-shaped SnO_2 tube arrays. Adv Mater 15(1):57–62
Liu B, Zeng HC (2004) Fabrication of ZnO "dandelions" via a modified Kirkendall process. J Am Ceram Soc 126:16744–16746
Liu Y, Zheng C, Wang W, Zhan Y, Wang G (2001) Production of SnO_2 nanorods by redox reaction. Crystal Growth 233:8–12
Liu Q, Zhang W-M, Cui Z-M, Zhang B, Wan L-J, Song W-G (2007) Aqueous route for mesoporous metal oxides using inorganic metal source and their applications. Micropor Mesopor Mater 100:233–240
Liu X, Long Y-Z, Liao L, Duan X, Fan Z (2012) Large-scale integration of semiconductor nanowires for high-performance flexible electronics. ACS Nano 6(3):1888–1900
Ma XL, Li Y, Zhu YL (2003) Growth mode of the SnO_2 nanobelts synthesized by rapid oxidation. Chem Phys Let 376:794–798
Mathur S, Ganesan R, Grobelsek I, Shen H, Ruegamer T, Barth S (2007) Plasma-assisted modulation of morphology and composition in tin oxide nanostructures for sensing applications. Adv Eng Mater 9:658–663
Meier DC, Semancik S, Button B, Strelcov E, Kolmakov A (2007a) Coupling nanowire chemiresistors with MEMS microhotplate gas sensing platforms. Appl Phys Lett 91:063118
Meier DC, Evju JK, Boger Z, Raman B, Benkstein KD, Martinez CJ, Montgomery CB, Semancik S (2007b) The potential for and challenges of detecting chemical hazards with temperature-programmed microsensors. Sens Actuators B Chem 121:282–294
Mor GK, Carvalho MA, Varghese OK, Pishko MV, Grimes CA (2004) A room-temperature TiO_2-nanotube hydrogen sensor able to self-clean photoactively from environmental contamination. J Mater Res 19:628–634
Morio M, Hyodo T, Shimizu Y, Egashira M (2009) Effect of macrostructural control of an auxiliary layer on the CO_2 sensing properties of NASICON-based gas sensors. Sens Actuators B Chem 139:563–569
Nam CY, Tham D, Fischer JE (2005) Disorder effects in focused-ion-beam-deposited Pt contacts on GaN nanowires. Nano Lett 5:2029–2033
Okuyama K, Abdullan M, Llenggoro IW, Iskandar F (2006) Preparation of functional nanostructured particles by spray drying. Adv Powder Technol 17:587–611
Pan ZW, Dai ZR, Wang ZL (2001) Nanobelts of semiconducting oxides. Science 291:1947–1949
Prades JD, Jimenez-Diaz R, Hernandez-Ramırez F, Barth S, Cirera A, Romano-Rodrıguez A, Mathur S, Morante JR (2008) Ultralow power consumption gas sensors based on self-heated individual nanowires. Appl Phys Lett 93:123110
Prades JD, Jimenez-Diaz R, Manzanares M, Hernandez-Ramirez F, Cirera A, Romano-Rodriguez A, Mathur S, Morante JR (2009a) A model for the response towards oxidizing gases of photoactivated sensors based on individual SnO_2 nanowires. Phys Chem Chem Phys 11:10881–10889
Prades JD, Jimenez-Diaz R, Hernandez-Ramirez F, Barth S, Cirera A, Romano-Rodriguez A, Mathur S, Morante JR (2009b) Equivalence between thermal and room temperature UV light-modulated responses of gas sensors based on individual SnO_2 nanowires. Sens Actuators B Chem 140:337–342
Qi P, Vermesh O, Grecu M, Javey A, Wang Q, Dai H (2003) Toward large arrays of multiplex functionalized carbon nanotube sensors for highly sensitive and selective molecular detection. Nano Lett 3:347–351
Qian LH, Wang K, Li Y, Fang HT, Lu QH, Ma XL (2006) CO sensor based on Au-decorated SnO_2 nanobelt. Mater Chem Phys 10:82–84
Rani S, Roy SC, Paulose M, Varghese OK, Mor GK, Kim S, Yoriya S, LaTempa TJ, Grimes CA (2010) Synthesis and applications of electrochemically self-assembled titania nanotube arrays. Phys Chem Chem Phys 12:2780–2800
Rossinyol E, Prim A, Pellicer E, Rodriguez J, Peir F, Cornet A, Morante JR, Tian B, Bo T, Zhao D (2007a) Mesostructured pure and copper-catalyzed tungsten oxide for NO_2 detection. Sens Actuators B Chem 126:18–23
Rossinyol E, Prim A, Pellicer E, Arbiol J, Hernandez-Ramirez F, Peir F, Cornet A, Morante JR, Solovyov LA, Tian B, Bo T, Zhao D (2007b) Synthesis and characterization of chromium-doped mesoporous tungsten oxide for gas sensing applications. Adv Funct Mater 17:1801–1806
Rothschild A, Tuller HL (2006) Gas sensors: new materials and processing approaches. J Electroceram 17:1005–1012
Rout CS, Kulkarni GU, Rao CNR (2007) Room temperature hydrogen and hydrocarbon sensors based on single nanowires of metal oxides. J Phys D 40:2777–2782
Sanchez C, Boissière C, Grosso D, Laberty C, Nicole L (2008) Design, synthesis, and properties of inorganic and hybrid thin films having periodically organized nanoporosity. Chem Mater 20:682–737
Shimizu Y, Hyodo T, Egashira M (2004) Mesoporous semiconducting oxides for gas sensor application. J Eur Ceram Soc 24:1389–1398
Shimizu Y, Jono A, Hyodo T, Egashira M (2005) Preparation of large mesoporous SnO_2 powders for gas sensor application. Sens Actuators B Chem 108:56–61
Soldano C, Comini E, Baratto C, Ferroni M, Faglia G, Sberveglieri G (2012) Metal oxides mono-dimensional nanostructures for gas sensing and light emission. J Am Ceram Soc 95(3):831–850

Strelcov E, Dmitriev S, Button B, Cothren J, Sysoev V, Kolmakov A (2008) Evidence of the self-heating effect on surface reactivity and gas sensing of metal oxide nanowire chemiresistors. Nanotechnology 19:355502

Sysoev VV, Schneider T, Goschnick J, Kiselev I, Habicht W, Hahn H, Strelcov E, Kolmakov A (2009) Percolating SnO_2 nanowire network as a stable gas sensor: direct comparison of long-term performance versus SnO_2 nanoparticle films. Sens Actuators B Chem 139:699–703

Tiemann M (2007) Porous metal oxides as gas sensors. Chem Eur J 13:8376–8388

Tresback JS, Padture NP (2008) Low-temperature gas sensing in individual metal-oxide-metal heterojunction nanowires. J Mater Res 23:2047–2052

Varghese OK, Grimes CA (2003) Metal oxide nanoarchitectures for environmental sensing. J Nanosci Nanotechnol 3:277–293

Varghese OK, Gong D, Paulose M, Grimes CA, Dickey EC (2003a) Crystallization and high-temperature structural stability of titanium oxide nanotube arrays. J Mater Res 18(1):156–165

Varghese OK, Gong D, Paulose M, Ong KG, Grimes CA (2003b) Hydrogen sensing using titania nanotubes. Sens Actuators B Chem 93:338–344

Varghese OK, Paulose M, Grimes CA (2009) Long vertically aligned titania nanotubes on transparent conducting oxide for highly efficient solar cells. Nat Nanotechnol 4:592–597

Vuong NM, Jung H, Kim D, Kim H, Hong S-K (2012) Realization of an open space ensemble for nanowires: a strategy for the maximum response in resistive sensors. J Mater Chem 22:6716–6725

Wadea TL, Wegrowe J-E (2005) Template synthesis of nanomaterials. Eur Phys J Appl Phys 29:3–22

Wagner T, Waitz T, Roggenbuck J, Froeba M, Kohl C-D, Tiemann M (2007) Ordered mesoporous ZnO for gas sensing. Thin Solid Films 515:8360–8363

Wallentina J, Borgstrom MT (2011) Doping of semiconductor nanowires. J Mater Res 26:2142–2156

Wan G, Wang TH (2005) Single-crystalline Sb-doped SnO_2 nanowires: synthesis and gas sensor application. Chem Commun 30:3841–3843

Wang ZL (2000) Characterizing the structure and properties of individual wire-like nanoentities. Adv Mater 12:1295–1298

Wang W, Xu C, Wang X, Liu Y, Zhan Y, Zheng C, Song F, Wang G (2002) Preparation of SnO_2 nanorods by annealing SnO_2 powder in NaCl flux. J Mater Chem 12:1922–1925

Wang Y, Angelatos AS, Caruso F (2008) Template synthesis of nanostructured materials via layer-by-layer assembly. Chem Mater 20:848–858

Wei TY, Yeh PH, Lu SY, Wang ZL (2009) Gigantic enhancement in sensitivity using Schottky contacted nanowire nanosensor. J Am Chem Soc 131:17690–17695

Whang D, Jin S, Wu Y, Lieber CM (2003) Large-scale hierarchical organization of nanowire arrays for integrated nanosystems. Nano Lett 3:1255–1259

Xia YN, Yang PD, Sun YG, Wu YY, Mayers B (2003) One-dimensional nanostructures: synthesis, characterization, and applications. Adv Mater 15:353–389

Xu C, Xu G, Liu Y, Zhao X, Guanghou WG (2002) Preparation and characterization of SnO_2 nanorods by thermal decomposition of SnC_2O_4 precursor. Scri Mater 46:789–794

Yang HX, Qian JF, Chen ZX, Ai XP, Cao YL (2007) Multilayered nanocrystalline SnO_2 hollow microspheres synthesized by chemically induced self-assembly in the hydrothermal environment. J Phys Chem 111:14067–14071

Yuan L, Hyodo T, Shimizu Y, Egashira M (2011) Preparation of mesoporous and/or macroporous SnO_2-based powders and their gas-sensing properties as thick film sensors. Sensors 11(2):1261–1276

Yue W, Zhou W (2008) Crystalline mesoporous metal oxide. Progr Nat Sci 18:1329–1338

Zeng ZM, Wang K, Zhang ZX, Chen JJ, Zhou WL (2009) The detection of H_2S at room by using individual indium oxide nanowire transistors. Nanotechnology 20:045503

Zhang D, Li C, Liu X, Han S, Tang T, Zhou C (2003) Doping dependent NH_3 sensing of indium oxide nanowires. Appl Phys Lett 83:1845–1847

Zhang Y, Ago H, Liu J, Yumura M, Uchida K, Ohshima S, Iijima S, Zhu J, Zhang X (2004a) The synthesis of In, In_2O_3 nanowires and In_2O_3 nanoparticles with shape-controlled. J Cryst Growth 264:363–368

Zhang DH, Liu ZQ, Li C, Tang T, Liu XL, Han S, Lei B, Zhou CW (2004b) Detection of NO_2 down to ppb levels using individual and multiple In_2O_3 nanowire devices. Nano Lett 4:1919–1924

Zhang Y, Kolmakov A, Libach Y, Moskovits M (2005) Electronic control of chemistry and catalysis at the surface of an individual tin oxide nanowire. J Phys Chem B 109:1923–1929

Zhang X, Chen H, Zhang HY (2007) Layer-by-layer assembly: from conventional to unconventional methods. Chem Commun 2007:1395–1405

Zhao Q, Gao Y, Bai X, Wu C, Xie Y (2006) Facile synthesis of SnO_2 hollow nanospheres and applications in gas sensors and electrocatalysts. Eur J Inorg Chem 2006(8):1643–1648

Chapter 4
Metal-Based Nanostructures

Through the study of nanoporous Pd films described in Chap. 4 (Vol. 1), it was demonstrated that the H_2 detection limit and response time could be improved in nanoporous structures with increased surface area and decreased distance for bulk diffusion. Taking into account mentioned above, one can conclude that films from metal nanoparticles and metal nanowires would be ideal structures for fast detection of low gas concentrations. Experiment has shown that this assumption is valid and noble metal nanoparticles can be successfully incorporated into gas sensors. The selection of noble metals such as Au and Pt for gas sensor fabrication is based on their chemical inertness (Dovgolevsky et al. 2009). It was established that, compared to conventional metal oxide chemiresistors, MNP-based devices have the advantage that they can be operated at room temperature or slightly above, which enables easy device integration and low-power operation (Joseph et al. 2008; Saha et al. 2012).

4.1 Metal Nanoparticles

4.1.1 Properties

Metal nanoparticles (MNPs) are discrete clusters of a finite number of atoms, generally in the range of 1–100 nm in size. For example, an Au NP 5.2 nm in diameter consists of 2,951 atoms. It is known that the surface area of nanocrystals increases markedly with the decrease in size (Rao et al. 2002). Thus, a small metal nanocrystal 1 nm in diameter will have 100 % of its atoms on the surface. A nanocrystal 10 nm in diameter, on the other hand, would have about 15 % of its atoms on the surface. A small nanocrystal with a higher surface area would therefore be expected to be more reactive (Rao et al. 2005). Furthermore, the qualitative change in the electronic structure arising due to quantum confinement in small nanocrystals will also bestow unusual adsorption and catalytic properties on these particles, totally different from those of the bulk metal (Daniel and Astruc 2004). For example, a low-temperature study of the interaction of elemental O_2 with Ag nanocrystals of various sizes has revealed the ability of smaller nanocrystals to dissociate dioxygen to atomic oxygen species (Rao et al. 1992). On bulk Ag, the adsorbed oxygen species at 80 K is predominantly in the form of O_2^-. The ability of Cu, Pd, Pt, and Ni nanoparticles to absorb CO at increased temperatures has been thoroughly investigated as well (Matolin et al. 1990). Carbon monoxide from a bulk Cu surface desorbs above 250 K. Small Cu particles, however, retain CO up to much higher temperatures (Santra et al. 1994). A similar observation has been made in the case of Pd particles (Gillet et al. 1986). The results obtained with Ni particles are more interesting. In addition to showing a trend similar to the above,

G. Korotcenkov, *Handbook of Gas Sensor Materials*, Integrated Analytical Systems,
DOI 10.1007/978-1-4614-7388-6_4, © Springer Science+Business Media New York 2014

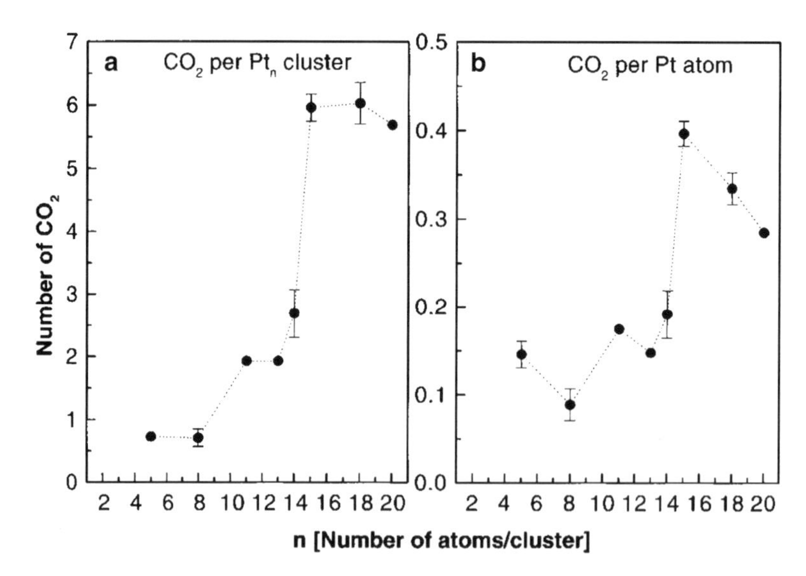

Fig. 4.1 (**a**) Total number of catalytically produced CO_2 molecules as a function of cluster size. (**b**) Total number of produced CO_2 molecules per atom as a function of cluster size. The CO_2 molecules produced by oxidation of CO are studied by means of temperature-programmed desorption mass spectrometry (Reprinted with permission from Heiz et al. 1999, Copyright 1999 American Chemical Society)

small Ni particles are also capable of dissociating CO to form carbidic species on the particle surface (Doering et al. 1982). Heiz and co-workers (1999) have studied the ability of a size-selected Pt cluster with nuclearity between 5 and 20 atoms to oxidize CO and found that the small Pt clusters are all catalytically active and exhibit a different temperature and activity profile, depending on the nuclearity (see Fig. 4.1).

It was established that as metal particles are reduced in size, the collective oscillation of electrons in the conduction band causes changes in the electrical, optical, and magnetic properties as well (Rao et al. 2005). For example, electron spectroscopic techniques such as UPS and XPS have shown that as the metal particle size decreases, the core-level binding energy of metals such as Au, Ag, Pd, Ni, and Cu increases sharply. This is shown in the case of Pd in Fig. 4.2, where the binding energy increases by over 1 eV at small size. The increase in the core-level binding energy in small particles is a manifestation of the size-induced metal–nonmetal transition in nanocrystals. According to experiments carried out, an electronic gap manifests itself for a nanoparticle with a diameter of 1–2 nm and possessing 300 ± 100 atoms. The variation in the binding energy is negligible at large particle size, since the binding energies are close to those of the bulk, macroscopic metals. Thus, experiment shows that the electronic properties of metal NPs are neither those of bulk metal nor those of molecular compounds. This means that metal nanoparticles (NPs) with indicated sizes effectively bridge between bulk materials and atomic/molecular structures (Zhang 1997; McConnell et al. 2000; Schmid and Corain 2003; El-Sayed 2004; Daniel and Astruc 2004; Van Dijk et al. 2006).

4.1.2 Synthesis

By now, metal NPs of Au, Ag, Pd, Pt, Cu, Co, and Ni can be easily synthesized, and many different types are commercially available. NPs can be synthesized either by physical methods such as vapor deposition and laser ablation or by chemical methods such as metal salt reduction or micelles (De Jongh 1994; Schmid 1994; Braunstein et al. 1999; Brust and Kiely 2002; Matolin et al. 1990; Masala and Seshadri

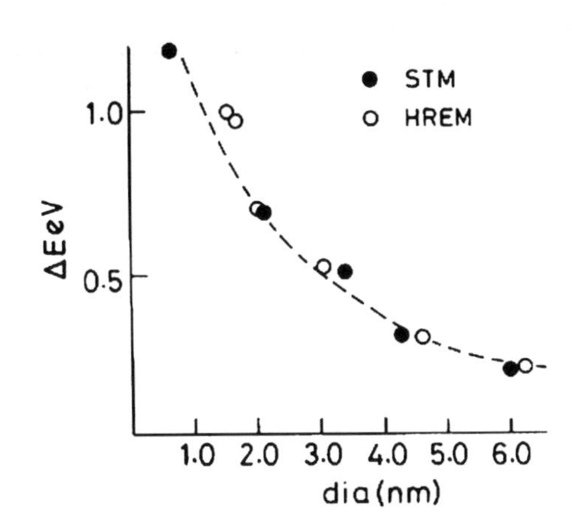

Fig. 4.2 Variation of the shift, ΔE, in the core-level binding energy (relative to the bulk metal value) of Pd with the nanoparticle diameter. The diameters were obtained from HREM and STM images (Reprinted with permission from Aiyer et al. 1994, Copyright 1994 Elsevier)

2004; Daniel and Astruc 2004; Burda et al. 2005; Rao et al. 2005; Matolin et al. 1990; Zabet-Khosousi and Dhirani 2008; Saha et al. 2012).

Chemical synthesis of sols of metals results in nanoparticles embedded in a layer of ligands or stabilizing agents, which prevent the aggregation of particles. The stabilizing agents employed include surfactants such as long-chain thiols or amines or polymeric ligands such as polyvinylpyrrolidone (PVP). Reduction of metal salts dissolved in the appropriate solvents produces small metal particles of varying size distributions (Turkevich et al. 1951; Schmid 1994; De Jongh 1994; Kulkarni et al. 2000). A variety of reducing agents have been employed for reduction. These include electrides, alcohols, glycols, metal borohydrides, and certain specialized reagents such as tetrakis(hydroxymethyl)phosphonium chloride. It should be noted that although the reaction temperature and reagent concentrations provide a rudimentary control of the main steps of synthesis such as nucleation, growth, and termination by the capping agent or ligand, it is often impossible to control them independently, and so the nanocrystals obtained usually exhibit a distribution in size. Typically, the distribution is log-normal with a standard deviation of 10 % (Rao et al. 2005). Given the fact that properties of the nanocrystals are size dependent, it is significant to be able to synthesize nanocrystals of precise dimensions with minimal size distributions. This can be accomplished to a limited extent by size-selective precipitation either by centrifugation or by use of a miscible solvent–nonsolvent liquid mixture to precipitate nanocrystals.

Successful nanocrystals synthesis has also been carried out employing soft templates such as the water pool in a reverse micelle, the interface of two phases. Reverse micellar methods have been successfully utilized in the preparation of Ag, Au, Co, Pt, and Co nanocrystals (Pileni 1993; Ahmadi et al. 1996). The synthesis of nanocrystals at the air–water interface as in Langmuir–Blodgett films or at a liquid–liquid interface is currently attracting wide attention (Shipway et al. 2000; Platt et al. 2002). It has been shown recently that films of metal nanocrystals can be prepared using a water–toluene interface (Rao et al. 2003a, b). Traditionally, clusters of controlled sizes have been generated by ablation of a metal target in vacuum followed by mass selection of the plume to yield cluster beams (Sattler et al. 1980; Milani and Iannotta 1999). Such cluster beams could be subject to in situ studies or be directed onto solid substrates. In order to obtain nanocrystals in solution, Harfenist et al. (1996, 1997) steered a mass-selected Ag cluster beam through a toluene solution of thiol and capped the vacuum-prepared particles.

Colloids of alloys have been made by the chemical reduction of the appropriate salt mixture in the solution phase. Thus, Ag–Pd and Cu–Pd colloids of varying composition have been prepared by alcohol reduction of mixtures of silver nitrate or copper oxide with palladium oxide (Vasan and Rao 1995). Fe–Pt alloy nanocrystals have been made by thermal decomposition of the Fe and Pt acetylacetonates in high-boiling organic solvents (Sun et al. 2000). Au–Ag alloy nanocrystals have been made by co-reduction of silver nitrate and chloroauric acid with sodium borohydride (Sandhyarani et al. 2000; He et al. 2002).

4.1.3 Gas Sensor Applications

According to Tisch and Haick (2010), there are several good reasons to exploit metal nanoparticles for gas-sensing applications. First, there is the possibility to synthesize MNPs of nearly any chemical composition one wishes. The core can be composed of either a single metal or an alloy of two or more metals. Second, there is the possibility to vary rather freely the particles' size and shape and, therefore, the surface-to-volume ratio. Using NPs with controlled shape allows one to tailor their properties with a greater versatility than can be achieved by controlling any other of their characteristic features. For sensing applications in particular, this allows deliberate control over the domination of the surface properties and, consequently, over the interaction "quality" with the analyte molecules (Dovgolevsky et al. 2009). Third, MNPs easily self-assemble into ordered macroscopic 1D, 2D, or 3D arrays with controllable porous properties. Assembly into ordered 3D structures grants control over the interparticle distance and makes it possible to obtain nearly uniform interparticle distances in the composite films. Finally, NPs can be successfully capped with a wide variety of molecular ligands, which allows precise control of both the chemical and the physical parameters of the NPs, including solubility, reactivity, and optical properties. It was found that ligands control the interaction between the NPs as well. For example, capping ligands plays an important role in the charge transport through NP solid films used as chemiresistor sensors. This means that by using molecules with specific functionalities (e.g., redox activity or biorecognition), those functionalities can be incorporated into the NP assembly. Generally, any mono- or bifunctional molecule that has a tendency to bind with NP core material can be used as a capping ligand or cross-linker, respectively. For example, thiol groups are most commonly used with Au or Ag NPs (Love et al. 2005), while amines and isocyanides are ligands used with Pd and Pt NPs (Masala and Seshadri 2004). For sensing applications these attributes imply that one can obtain MNPs with a wide variety of synergetic combinations of chemical and physical properties which, in turn, affect the sensitivity and the selectivity of the sensors. In addition, capping, the metal NPs with an organic monolayer can control the particles' size and protect and stabilize them against aggregation.

Experimental studies have shown that the above-mentioned features of metal nanoparticles really may be applied in sensor technology, and, by exploiting these nanoscale properties, a highly efficient gas sensor can be designed and fabricated (Shipway et al. 1999; Ahn et al. 2004; Raschke et al. 2004; Drake et al. 2007; Dovgolevsky et al. 2009; Saha et al. 2012). In particular, metal nanoparticles can be incorporated in optical hydrogen and VOCs sensors (Cioffi et al. 2002; Ahn et al. 2004; Filenko et al. 2005).

Generally speaking, MNP-based gas sensors can operate either on analyte-induced effects such as controlled assembly (i.e., aggregation) of MNPs or on swelling of MNP solid films, and/or altering their surface properties through MNP–analyte interaction via hydrogen bonding, π–π, host–guest, van der Waals, electrostatic, or charge transfer (Tisch and Haick (2010)). The indicated MNP–analyte interactions can affect the following parameters: (1) the individual NP's size; (2) the aggregate's size; (3) the interparticle distance; and (4) composition, periodicity, and thermal stability of the aggregates. These parameters alter the optical, mechanical, and electrical properties of the MNPs. This means that for design of NT-based gas sensor, various platforms can be used. In particular, in the literature one can find descriptions of SAW, QCM, conductometric, electrochemical, and optical NP-based gas sensors.

It should be noted that optical sensors based on surface plasmon resonance (SPR) are the most studied NT-based gas sensors. Mostly Au nanoparticles with sizes ranging from 2 to 300 nm are used for optical sensing applications. It was found that SPRs are excited neither in smaller NPs (quantum dimensions <1 or 2 nm), where electrons occupy discrete energy levels nor in bulk gold (Au), which has a continuous absorbance in the UV-visible spectral region. Ag and Cu nanoparticles can be applied as well. However, Ag, and especially Cu, tend to oxidize. The large density of free electrons in these metals yields highly intense SPRs. In fact, the light-absorption and light-scattering cross sections of NPs

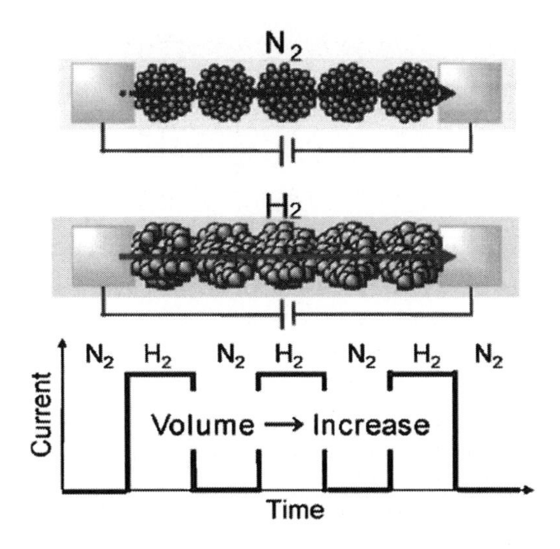

Fig. 4.3 Schematic illustration of the mechanism of electronic H_2 sensors for Pd–polyelectrolyte hybrid nanoparticles (Reprinted with permission from Dovgolevsky et al. 2009, Copyright 2009 Elsevier)

are several orders of magnitude higher than those of conventional highly colored dyes, for example, porphyrins and azo dyes (Kriebig and Vollmer 1995; Yguerabide and Yguerabide 1998). The SPR frequency depends strongly on the dielectric constant, i.e., on the refractive index, RI, of the surrounding medium, which makes metal NPs and their assemblies very attractive for optical sensing. For example, adsorption of chemical species at the NP's surface, or in between assembled NPs can generate the change in the dielectric constant, which will be accompanied by SPR shift. It should be noted that SPR-based gas sensors were designed for volatile organic compounds (VOCs). In particular, Cheng et al. (2007) demonstrated that sensors based on silver and gold nanoparticles and gold nanoshells had fast and reversible (approximately 8 s) response. The limit of detection for toluene was 5 ppm.

Regarding NT-based conductometric gas sensors, one can say that chemiresistors based on MNPs were first demonstrated in 1998 by Wohltjen and Snow (1998). They fabricated a chemiresistor by deposition of a thin film of octanethiol-coated AuNPs ($d \approx 2$ nm) onto an interdigitated microelectrode. A rapid decrease in the conductance due to film swelling was observed in the presence of toluene, tetrachloroethylene, 1-propanol, and water vapor with a detection limit of ~1 ppm.

A Pd hydrogen sensor, based on the swelling property (volume increase) of Pd, is another example of NP-based conductometric gas sensors (Favier et al. 2001; Kim et al. 2006). When hydrogen is present and dissolved into the Pd, the hydrogen-swollen Pd grains expand to "close" the nanoscale gaps in between. The grain chains are connected and provided a current pass that decreased the resistance of the Pd grain networks. These gaps are "reopened" when the Pd grains in Pd grain networks return to their equilibrium dimensions in the absence of hydrogen (see Fig. 4.3). For fabrication of such sensors, electrodeposition (Favier et al. 2001; Kim et al. 2006) and sputtering (Lith et al. 2007) are normally used. It was established that indicated sensors have fast response (~70–75 ms) and high sensitivity. However, experiment has shown that the control of the size of the nanogap is complex, and therefore the reproducibility of sensor parameters is low. For optimal sensor operation, the gap between Pd grains should be in the range 1–5 nm. In addition, this type of sensor can work efficiently only in the range of high hydrogen concentrations (1 % and above). All the factors mentioned above might limit its application.

QCM and SAW NT-based sensors can also operate at room temperatures. Yang and co-workers (2006) published a complementary study of the vapor-sensing properties of RT QCM gas sensors coated with films of spherical Au NPs functionalized with 2-naphthalenethiol, 2-benzothiazolethiol, and 4-methoxythiolphenol. Although gold is a poor catalyst in bulk form, nanometer-sized gold nanoparticles can exhibit excellent catalytic activity due to their relative high surface-area-to-volume ratio and their interface-dominated properties, which differ significantly from their bulk counterparts

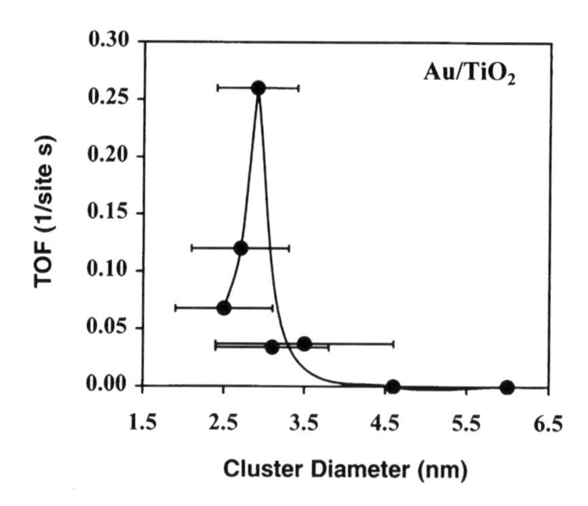

Fig. 4.4 CO oxidation turnover frequencies (TOFs) at 300 K as a function of the average size of the Au clusters supported on a high surface area TiO_2 support. The Au/TiO_2 catalysts were prepared by the deposition–precipitation method, and the average cluster diameters were measured by TEM. The *solid line* serves merely to guide the eye (Reprinted with permission from Valden et al. 1998, Copyright 1998 AAAS)

(Henglein 1989; Brus 1991). For example, Haruta (1997) demonstrated that gold nanoparticles (<5 nm) supported on oxides display high catalytic activity for the chemical and electrochemical oxidation of carbon monoxide and methanol (see Fig. 4.4). Valden et al. (1998) reported that the high catalytic activity of gold nanoparticles in catalyzing CO oxidation is related to the bandgap of a metallic-insulator transition for particles in the range of a few nanometers.

It was established that well-designed NP-based gas sensors can detect parts-per-billion levels of concentration and in some cases even reach sub-ppb sensitivity (Ahn et al. 2004). However, there is a trade-off between high sensitivity and reversibility of the sensor–analyte interaction. Reversible sensing signals with ppb sensitivity can be obtained if the capping ligands of the MNPs interact semipermanently with the analyte of interest, whereas sub-ppb sensitivity can only be achieved if a chemical reaction between analyte and capping ligand occurs, which leads to irreversible changes of the sensing material. This means that the main problem with metal nanoparticles is related to their low stability. They can aggregate because of the high surface-free energy and can be oxidized, contaminated by air, moisture, SO_2, and so on.

However, it was established that metallic nanoparticles can be stabilized in an organic medium using surface functionalization (Yu et al. 2003; Drake et al. 2007; Thanha and Green 2010; Saha et al. 2012). These materials are known as core–shell nanoparticles. The monolayer-protected core shell nanoparticles have a few nanometer metal cores with an organic compound (Pang et al. 2005). Different functional groups can be introduced either by the alkyl chain or in the chain terminal, changing the electrochemical properties of the system (Evans et al. 2000; Kang et al. 2001; Cai and Zellers 2002; Erathodiyil and Ying 2011). Chemistry applied for surface functionalization of nanoparticles is shown in Table 4.1. One of the approaches used for AuNPs functionalization is shown in Fig. 4.5. The presence of protected covering avoids aggregation and changes the dielectric constant of the medium surrounding the nanoparticles as well. For example, due to strong thiol–gold interaction, the thiol-protected AuNPs have superior stability, and they can be easily handled, characterized, and functionalized. The nanoparticles can be thoroughly dried and then redispersed in organic solvents without any aggregation or decomposition.

The sensitivity can be further enhanced by incorporating various polymer "linker" molecules. Different terminal groups incorporated into the linker molecules can be utilized to achieve enhanced sensitivity. For example, Chen and Lu (2010) have shown that the functionalization of Ag and Au nanoparticles by decanethiol, naphthalene thiol, and 2-mercaptobenzothiazole makes it possible to

Table 4.1 Functionalization chemistry of NPs

Ligand	Substrate	Ligand Attached to Substrate	Reaction
NP—SH	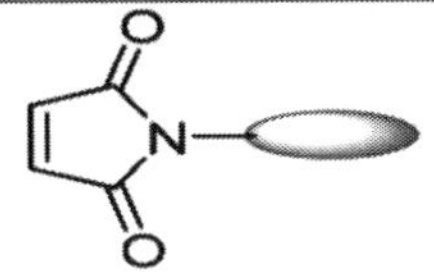 O, N—, O	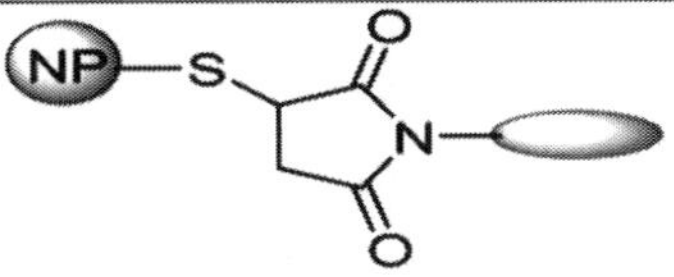 NP—S, O, N—, O	Michael Addition
$NP—NH_2$	 O	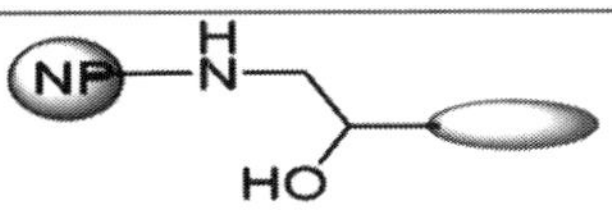 NP—N(H), HO	Epoxide Opening
$NP—NH_2$	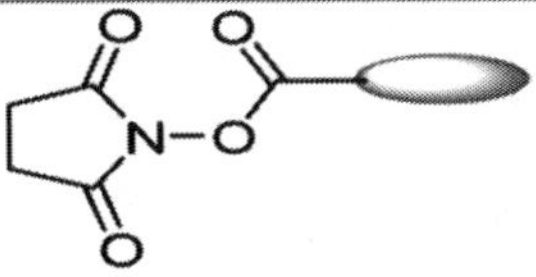 O O, N—O, O	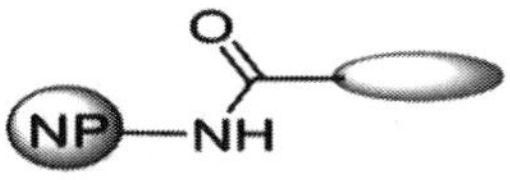 O, NP—NH	Amidation
$NP—NH_2$	 HOOC—	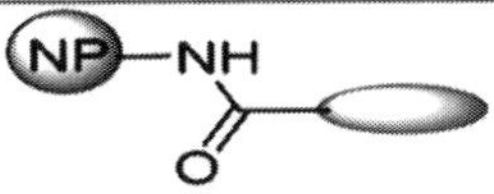 NP—NH, O	Amide Bond Formation
NP—COOH	 H_2N—	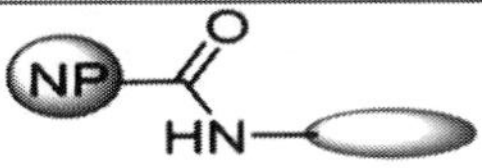 NP, O, HN—	Amide Bond Formation
NP—CHO	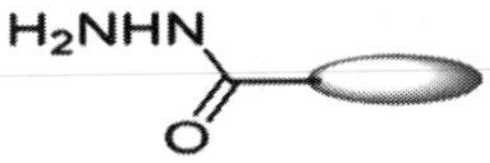 H_2NHN, O	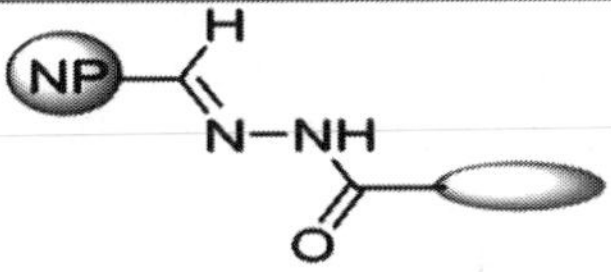 NP, H, N—NH, O	Imine Formation
NP—CHO	 H_2NO—	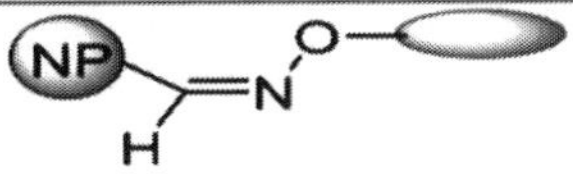 NP, O—, N, H	Imine Formation
$NP—N_3$		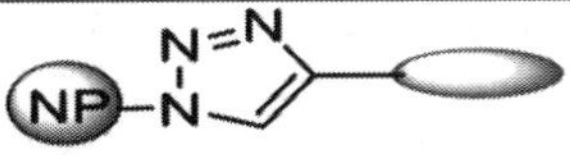 N=N, NP—N	Click Chemistry
$NP—NH_2$	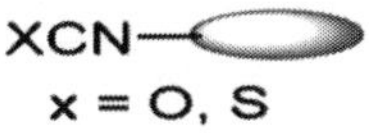 XCN— x = O, S	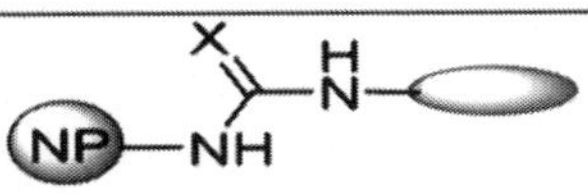 X, H, N—, NP—NH	Addition of Amine to Cyanates
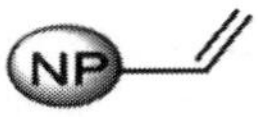 NP		 NP	Ring-Closing or Ring-Opening Metathesis
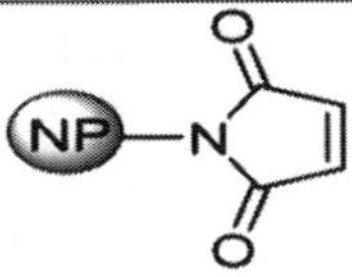 O, NP—N, O	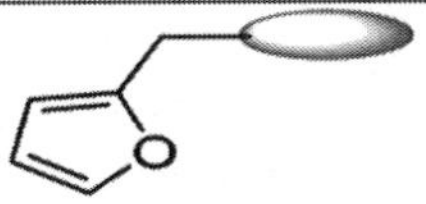 O	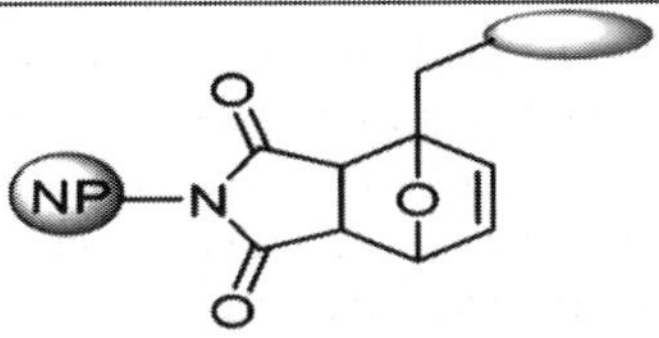 O, NP—N, O, O	Diels-Alder Reaction

Source: Reprinted with permission from Erathodiyil and Ying (2011). Copyright 2011 American Chemical Society

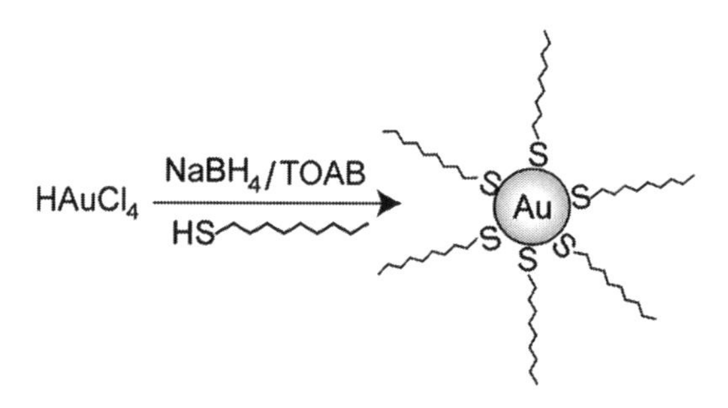

Fig. 4.5 Brust–Schiffrin method for two-phase synthesis of AuNPs by reduction of gold salts in presence of external thiol ligands (Reprinted with permission from Saha et al. 2012, Copyright 2012 American Chemical Society)

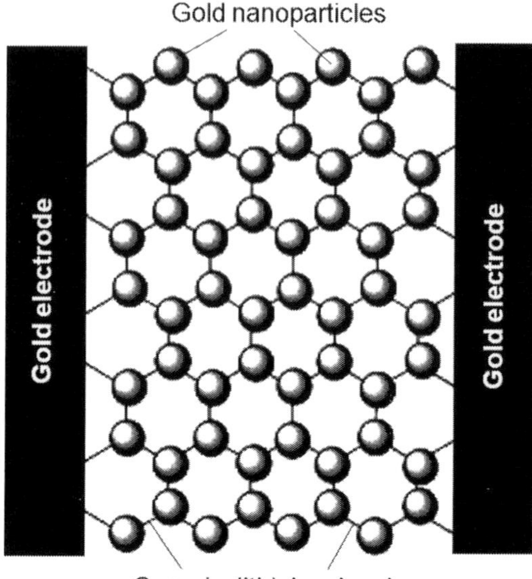

Fig. 4.6 Schematic illustration of the micro-gap electrodes which are connected with the network of organic dithiols and AuNPs. *Shaded circles* indicate AuNPs, and *lines* indicate organic dithiol molecules (Reprinted with permission from Ogawa et al. 2001, Copyright 2001 Elsevier)

increase the number of detected vapors. Joseph et al. (2003) have systematically investigated the sensing of toluene and tetrachloroethylene using films consisting of dodecylamine-stabilized AuNPs and dithiols with different chain lengths (C_6, C_9, C_{12}, C_{16}). At a given concentration of toluene it was observed that the resistance responses increase exponentially with increase of $-CH_2$ units. This effect was attributed to the augmentation of sorption sites with increasing ligand length.

The place exchange reactivity of these core–shell nanoparticles, reported by the Murray group (Templeton et al. 1998a, b; Hostetler et al. 1999), was one of the pivotal advances of these core–shell nanoparticles. The electron conduction between the metal cores (see Fig. 4.6) can be altered by the vapor sorption that causes the monolayer to swell and the dielectric properties to change (Ahn et al. 2004). This matrix swelling changes the distance between the metal nanoparticle cores, lowering the ability of electrons to conduct from core to core and resulting in increased resistivity. It should be noted that the distance between the nanoparticle cores determines the effectiveness of the polymer swelling and the dielectric changes of the matrix in the presence of the various gases. The other controlling factors in vapor sensing include the size and the length of the matrix or the linker molecule, the position of the sorption site, the nature of the matrix molecules, and the metal nanoparticle cores.

Experiment has shown that the electronic conduction of the VOC sensing metal nanoparticle films is also dependant on the metal core size itself and how the nanoparticles are linked to other polymer molecules (Leibowitz et al. 1999; Zheng et al. 2000; Han et al. 2001) (see Fig. 4.7). Brust et al. (1994) studied electronic transport properties, chemical composition, and the vapor sensitivity of 1,*n*-alkylenedithiol interlinked gold nanoparticles films as a function of alkylene chain length and

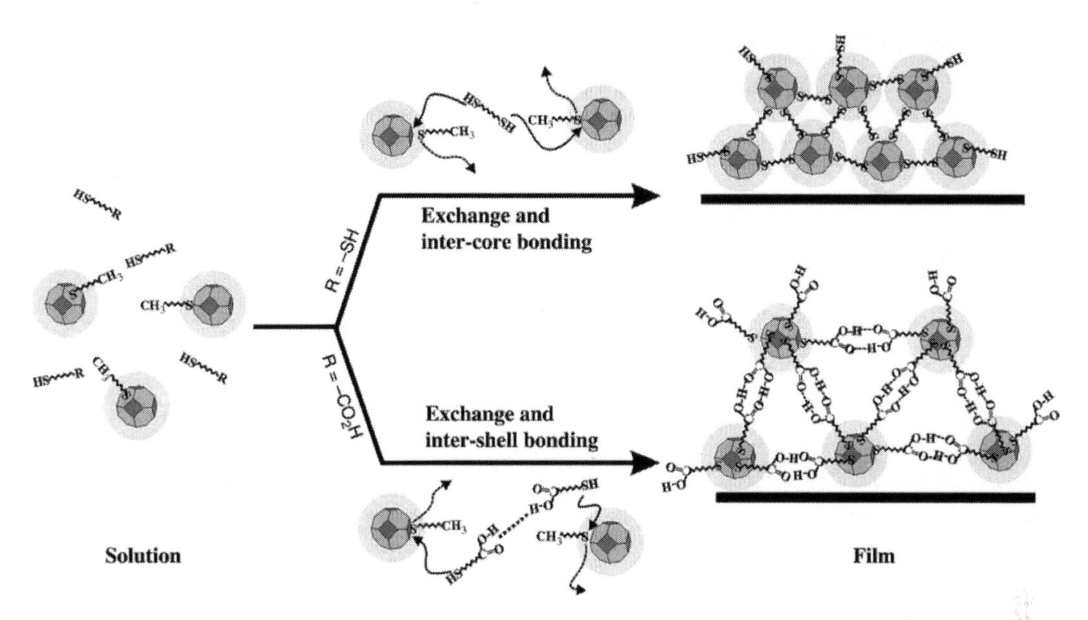

Fig. 4.7 Schematic illustrations of two types of interparticle molecular reactivities leading to the formation of nanoparticle thin films (Reprinted with permission from Han et al. 2001, Copyright 2001 Royal Society of Chemistry)

found that the response increased strongly with an increase in length of the linker molecules. The gold particles in these films were approximately 4 nm in size. It was supposed that the increase in the linker molecule length impacts the average tunneling distance between the neighboring particles. Evans et al. (2000) and Zhang et al. (2002) have shown that, in many cases, the conductivity of films of modified nanoparticles can be understood as being in accordance with an activated electron hopping model suggested by Neugebauer and Webb (1962). In this model, the electronic conductivity σ in the metallic nanoparticle film can be expressed as

$$\sigma \propto e^{-2\delta\beta} e^{-E_C/(kT)} \tag{4.1}$$

where δ is the core–core separation, β is a quantum mechanical tunneling factor, and E_C is the activation energy for charge hopping. If the other parameters remain constant, (4.1) shows that the conductivity will decay exponentially as the core–core separation δ is increased. Meanwhile, the activation energy E_C is essentially the Coulomb energy associated with the charging of a nanoparticle, which can be represented by

$$E_C \approx \frac{e^2}{4\pi\varepsilon_\gamma \varepsilon_0 r} \tag{4.2}$$

where ε_r is the relative permittivity of the dielectric medium surrounding the metallic cores and r is the radius of a nanoparticle ($r \leq \delta$). Equation (4.2) shows that increasing ε_r would result in a reduction of the energy barrier required for charge carrier formation and therefore will act to increase the conductivity. It is apparent from (4.1) and (4.2) that any process that changes either the core–core separation or the permittivity of the medium between the cores will lead to a change in conductivity. On this basis, the response of nanoparticle films to chemical vapors can be readily understood. Increasing the permittivity would give rise to an increase in σ, while increasing δ would decrease σ. In a later study it was found that the decrease in conductivity did not follow simple monoexponential decay with

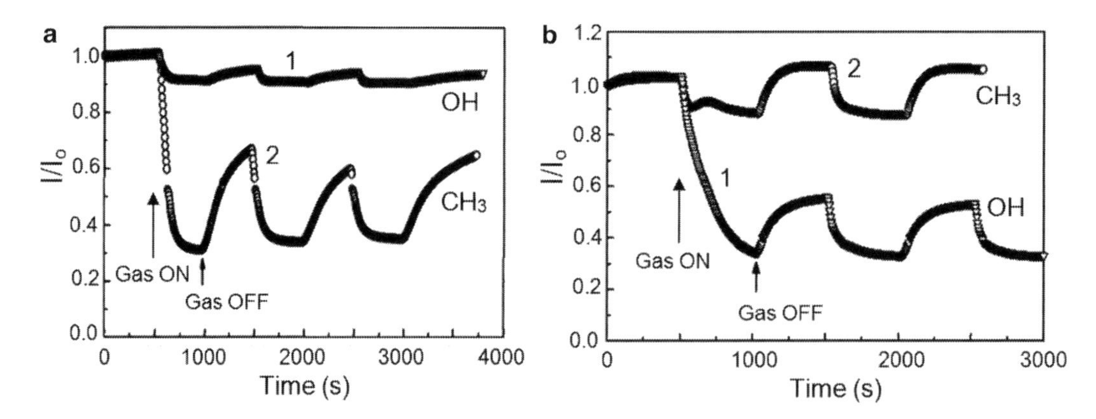

Fig. 4.8 Electrical responses of (1) OH- and (2) CH_3-functionalized nanoparticle films to (**a**) dichloromethane (DCM) vapor (about 90 ppt) and (**b**) methanol vapor (about 90 ppt) (Reprinted with permission from Zhang et al. 2002, Copyright 2002 IOP)

increasing linker chain length (Drake et al. 2007). This suggests that the charge transport may not be exclusively through tunneling in the backbone of the linker molecules. The tunneling distances may not scale linearly with increasing alkaline chain length.

Shen and co-workers (Evans et al. 2000; Zhang et al. 2002) have shown that aromatic functionalized AuNPs exhibited different sensory responses depending on the nature of the terminal functionality (OH, CH_3, NH_2, COOH) of aromatic thiols. In particular, Fig. 4.8 demonstrates how the terminal functional group of the ligand coating the nanoparticle can determine the electrical responses of thin nanoparticle films to different chemical vapors. It is seen that the CH_3-functionalized nanoparticles (4-methylbenzenethiol-functionalized gold nanoparticles) are more sensitive to vapors of nonpolar solvents like dichloromethane (DCM), while the OH-functionalized nanoparticles (4-mercaptophenol-functionalized gold nanoparticles) give more pronounced changes upon exposure to vapors of polar solvents, e.g., methanol. The different affinities of the various solvents to the nanoparticles are consistent with the solubility of the nanoparticles in the corresponding solvents. DCM is a good solvent for CH_3 nanoparticles but not for OH nanoparticles, while the OH nanoparticles are soluble in methanol but not in DCM. These results testify that, through surface functionalizing, we can influence on the selectivity of sensor response as well.

It should be noted that gas sensors can also be designed based on metal nanoparticles embedded into a polymer matrix. This approach to the design of gas-sensing materials will be discussed in Chap. 13 (Vol. 2). The embedding of nanoscopic metals into dielectric matrices represents a valid solution to the manipulation and stabilization problems. For example, Karakouz et al. (2008) developed SPR-based gas sensors in which evaporated gold island films were coated with the polymers polystyrene sulfonic acid (PSS) and polystyrene (PS). These polymers swell and/or shrink upon exposure to the various gases used (chloroform, water vapor, etc.), affecting the local refractive index and inducing SPR peak shifts (see Fig. 4.9). The limit of detection of this method was reported in terms of the vapor pressure of the gas: $0.05P_{sat}$. One of the major driving forces for using polymer matrices is to make the sensor chemically resistant to corrosive gases and VOCs. The nanoparticles are chemically susceptible to corrosive analytes and can adversely affect any unprotected sensing system. The chemical susceptibility of the nanoparticles and the type of polymer matrix used can cause issues for long-term reliability of the sensor (Drake et al. 2007). However, we must recognize that till now the problem of temporal stability of NT-based gas sensors has been of great importance.

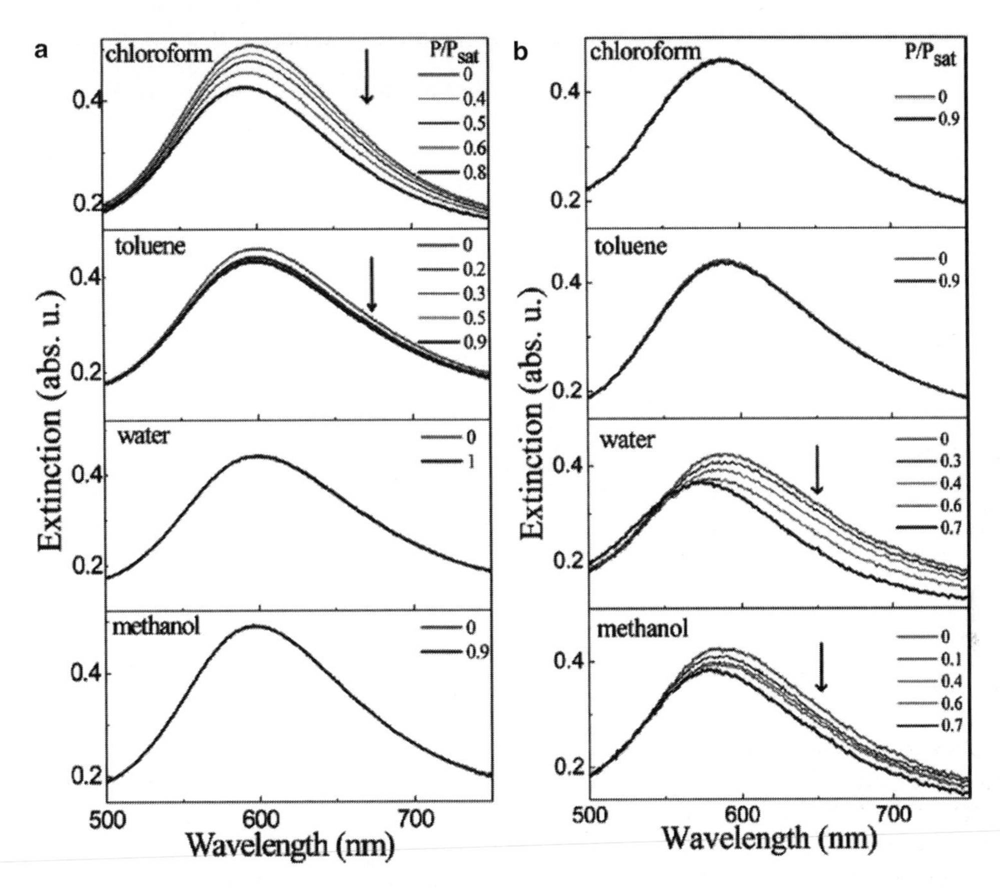

Fig. 4.9 UV–vis spectral response of the vapor sensor based upon 5-nm gold island films coated with (**a**) 55-nm polystyrene (PS) and (**b**) 85-nm polystyrene sulfonic acid, sodium salt (PSS). *Arrows* indicate the direction of change of the SP band intensity with increasing analyte pressure (P/P_{sat}) (Reprinted with permission from Karakouz et al. 2008, Copyright 2008 American Chemical Society)

Relatively low thermal stability is another major disadvantage of NT-based sensors (Tisch and Haick (2010)). At temperatures higher than about 80 °C, the analyte can affect the organic layer irreversibly. Due to the percolation mechanism of conductivity, the sorption-induced signal of chemiresistors strongly depends on the structure and NTs parameters. Therefore, it is difficult to achieve good batch-to-batch reproducibility of sensor parameters during their manufacture. It was found that MNT-based sensors also suffer from some interference by responding to chemical species that are structurally or chemically similar to the desired analyte (Tisch and Haick 2010). Utilizing different core types of the responsive MNPs could in principle overcome this limitation. That is so because different types of metallic cores can lead, via, for example, either pinholes in the capping monolayer and/or temporary exchange of *physically* adsorbed capping molecules, to distinctive interactions with otherwise similar analytes (Joseph et al. 2003, 2004). However, great success in this direction was not achieved, and research focused on selectivity improvement is required.

It should also be noted that, as a rule, the response of NT-based sensors, especially of conductometric types, is very low in comparison with conventional metal oxide-based gas sensors (Drake et al. 2007). In particular, Fig. 4.10 shows that appreciable response is observed at concentrations of VOCs exceeding 10,000 ppm.

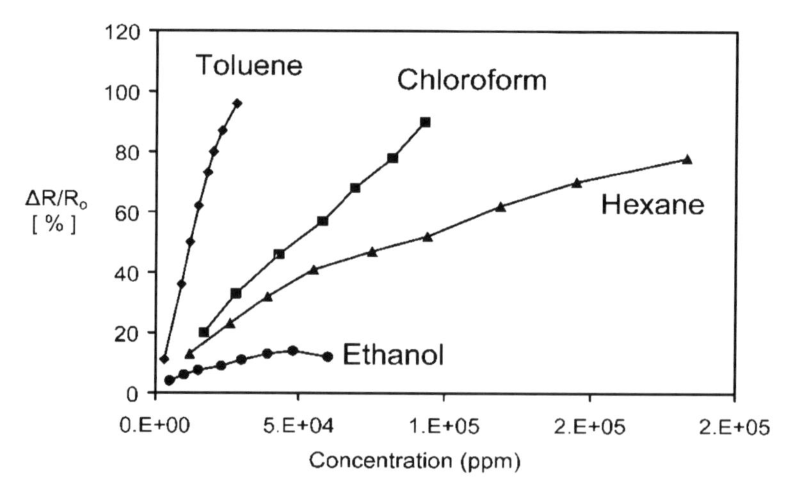

Fig. 4.10 Response of 12-(3-thienyl)dodecanethiol-protected gold nanoparticle film to various organic vapors as function of concentration: lines through data are included to guide general trend (Reprinted with permission from Ahn et al. 2004, Copyright 2004 American Chemical Society)

4.2 Metal Nanowires

At present many and various metal nanowires have been formed using different methods such as lithographic patterning, bottom-up growth, mainly through electrodeposition, ion milling, and grain structuring (Noh et al. 2011). In particular, in the literature one can find information about Cu, Mo, Ni, Pd, Pt, Au, etc., nanowires with diameters 50–900 nm. For example, Favier et al. (2001) and Walter et al. (2002a, b) produced different metal nanowires using the electrochemical step-edge decoration technique (ESED), i.e., electrodeposition from aqueous solutions onto step edges naturally present on highly oriented pyrolytic graphite surfaces. Depending essentially on the kinetics of the reduction reaction, two different ways of preparing metal nanowires have been developed. Metal cations with a slow transfer rate, such as molybdenum or cadmium, metal oxide wires (MoO_x and $Cd(OH)_2$, respectively) were electrodeposited from aqueous solutions. The resulting wires were converted into pure metal wires by reduction with H_2, or with H_2S, at high temperature. In contrast, from acidic solutions of palladium, gold, silver, and copper, no oxide formation is observed and nanowires are obtained as pure metals by electrodeposition. Freshly deposited metal nanowires are then detached from the graphite and transferred onto a glass substrate by means of cyanoacrylate. Another technique, which simplifies the previous one by avoiding the use of a template, is to manufacture a Pd nanowire array directly onto a crystalline silicon substrate. The choice of a silicon substrate opens the way to the direct integration of this kind of sensor device in microelectronics. Pd nanowire arrays were actually assembled directly onto a silicon chip by means of AC dielectrophoresis using a metal salt solution as a feed material (Cheng et al. 2005; La Ferrara et al. 2008). Experiment has shown that metal nanowires usually consist of agglomerated metal grains with “intergrain” nanogaps. As an example of this bottom-up approach, Pd nanowires can also be grown by electrodeposition into nanochannels of AAO templates from an aqueous solution of $PdCl_2$, using an Au cathode layer (Jeon et al. 2008). One can find the description of this technology in Schönenberger et al. (1997) and Yin et al. (2001). Pd nanowires prepared using different methods are shown in Fig. 4.11.

Di Francia et al. (2009) believe that the use of metal nanowires can contribute to improvement in operating characteristics of gas sensors, because metal nanowires have features such as ductility and chemical stability. Furthermore, when the diameter of these structures is in the nanorange, they could represent interesting transducers since the surface/volume ratio increases with the inverse of the wire

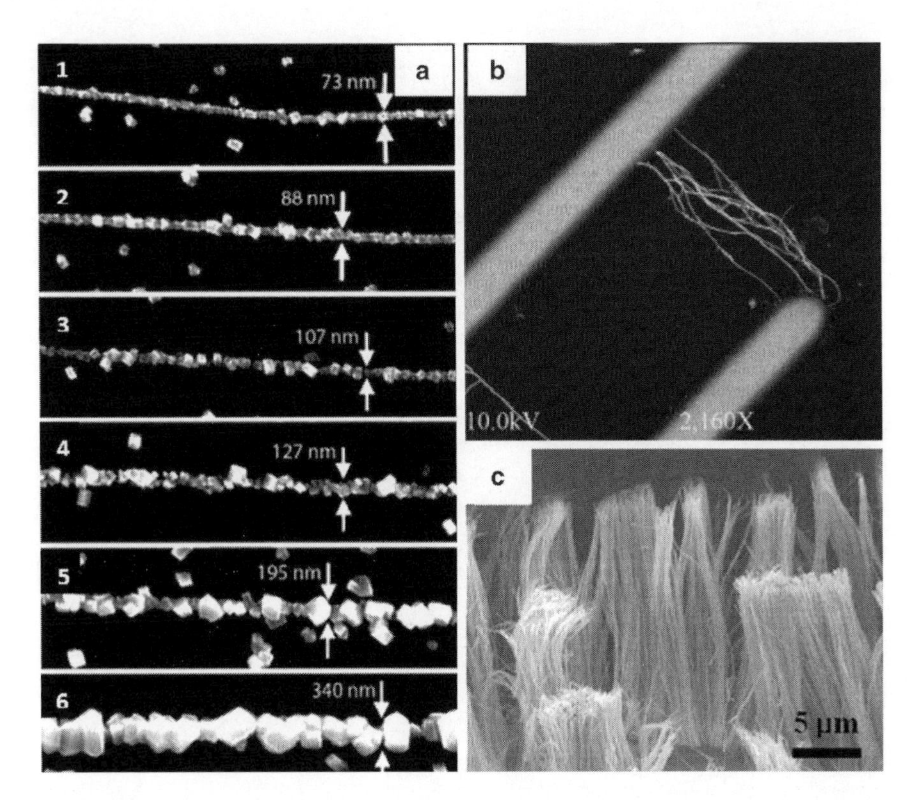

Fig. 4.11 SEM images of metal nanowires. (**a**) Copper nanowires. These nanowires were electrodeposited from the solution, using $E_{nucl}=-800\ mV_{SCE}$ and $E_{grow}=-5\ mV_{SCE}$. The growth times employed in each experiment were 1–120, 2–180, 3–300, 4–600, 5–900, and 6–2,700 s (Reprinted with permission from Walter et al. 2002c, Copyright 2002 American Chemical Society). (**b**) Pd nanowires assembled between electrodes under an applied AC signal of 10 V_{rms} and 300 kHz. The nanowires have an approximate diameter of 100 nm. Experiments involved palladium acetate dissolved in 10 mM HEPES buffer, pH=6.5. Gap spacing between electrodes 20 μm (Reprinted with permission from Cheng et al. 2005, Copyright 2005 American Chemical Society). (**c**) Arrays of Pd nanowires grown by electrodeposition into nanochannels of AAO templates; the Pd nanowires liberated from the electroplated AAO template by dissolution in a 2 % hydrofluoric acid solution (Reprinted with permission from Jeon et al. 2008, Copyright 2008 IOP)

diameter. The same conclusion was reached by Noh et al. (2011). As a consequence, it was demonstrated that a thin Pd nanowire could detect H_2 concentrations as low as 20 ppm. In addition, they established that the shortest response time obtained from the lithographically patterned Pd nanowires ($t=20$ nm) was $\tau_{0.4}\sim 3$ s at 1,000 ppm of H_2. This is much smaller than that of Pd thin films. However, we need to note that Pd and Ag nanowires only found application in gas sensor design (Favier et al. 2001; Walter et al. 2002a, b; Murray et al. 2004). Typical structures of gas sensors with individual metal nanowire are shown in Fig. 4.12.

It was established that the resistance of Pd nanowires, similar to Pd films and Pd nanoparticles, is sensitive to H_2. Many researchers believe that palladium is an ideal hydrogen sensor material due to its properties such as high sensitivity and selectivity to hydrogen gas, fast response, and operability at room temperature. Favier et al. (2001) believe that H_2 nanosensors based on Pd nanowires work (1) due to the swelling effect caused by a change in the Pd crystalline phase upon exposure to low concentrations of H_2 and/or (2) due to surface conversion into the more insulating PdH_2 phase after interaction with H_2. When exposed to hydrogen, the gas diffuses into the lattice and reacts with the metal, forming a metal hydride (PdH_x) and resulting in a volumetric wire expansion with a partial or total closure of the gaps. As a result, the swelling effect increases the electrical conductivity, while the

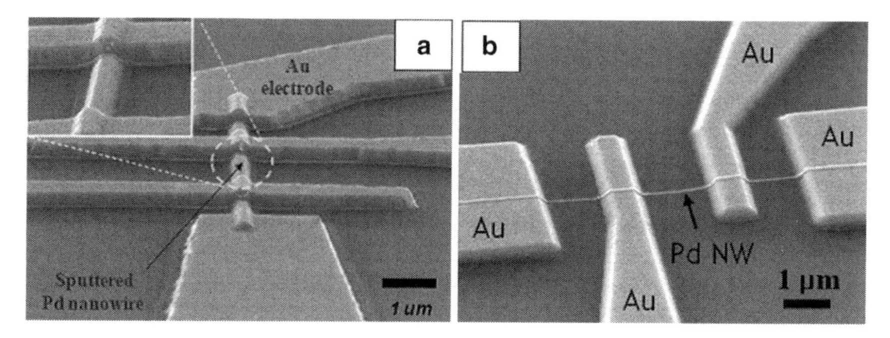

Fig. 4.12 SEM image of gas sensors with metal nanowires. (**a**) Lithographically patterned Pd nanowire with $t=100$ nm, $w=300$ nm, and $l=10$ µm. Four Ti/Au inner electrodes were patterned on the Pd nanowire. (**b**) Pd nanowire grown by electrodeposition into nanochannels of AAO templates: $d=20$ nm (Reprinted with permission from Jeon et al. 2008 and 2009, Copyright 2008 and 2009 IOP)

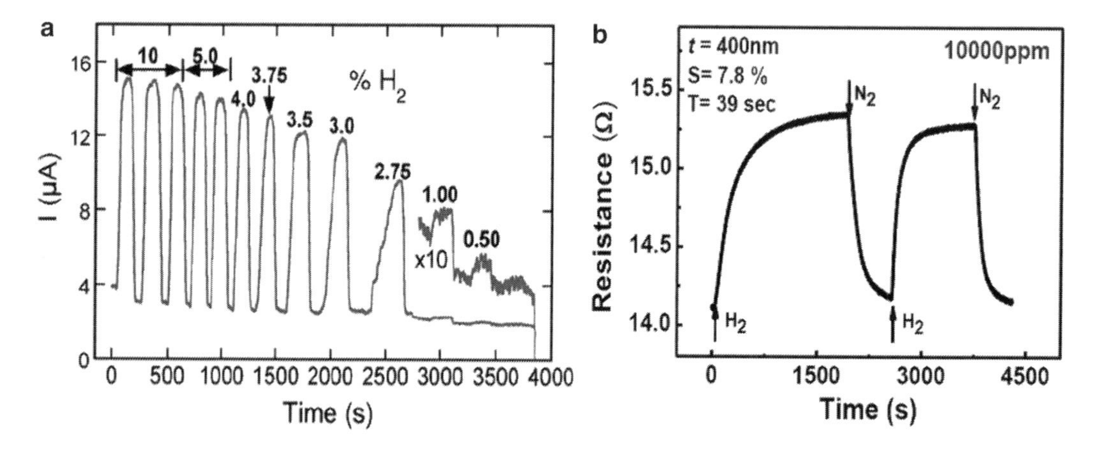

Fig. 4.13 The real-time electrical responses at room temperature of H_2 gas sensors based on Pd nanowires. (**a**) Grained structure of Pd nanowires (Reproduced with permission from Walter et al. 2002a, Copyright 2002 Elsevier). (**b**) Lithographically patterned Pd nanowires with $t=400$ nm (Reprinted with permission from Jeon et al. 2009, Copyright 2009 IOP)

formation of metal hydride increases the electrical resistance. The first effect is usually observed in grained structures, whereas the second effect takes place in dense structures, particularly in Pd nanowires formed by lithographic method (see Fig. 4.13). The indicated mechanisms were discussed in the previous section.

Favier and co-workers (Favier et al. 2001; Walter et al. 2002a, b) have shown that, due to the small (nano)gaps between Pd nanograins in nanowires, the response of H_2 sensors can be as fast as 20 ms, when devices are characterized at high H_2 concentrations (>8 %). It was established that the grain-structured Pd nanowires could detect H_2 concentrations as low as 2–5 ppm. However, Favier and co-workers (Favier et al. 2001; Walter et al. 2002a, b) found that grain swelling is not completely reversible. After H_2 removal, the grains come back to the initial volume but not to the initial position. La Ferrara et al. (2008) also found that response time depends on the nanowire diameter. It was shown that thinner nanowires (<90 nm wide) responded faster than the thicker ones, with up to 140 % current changes in the presence of 4 % H_2. The same effect was observed for lithographically patterned Pd nanowires. Jeon et al. (2009) have found that the response time remains almost constant below 100 nm, whereas it gradually decreases with decreasing thickness down to this value. Jeon et al. (2009) believe that the general decrease of response time with decreasing the thickness is due to the reduced hydrogen diffusion distance mentioned above. The steady response time along with the sensitivity trend in the thickness range smaller than 100 nm suggests that the clamping effect dominates the hydrogen absorption dynamics in this range. The shortest response time obtained from the lithographically patterned Pd nanowires was ~3 s at 1,000 ppm of H_2 for nanowires with $t=20$ nm. This is much smaller than that of Pd thin films.

It is important that other metal nanowires can also be used for gas sensor design. Murray et al. (2004) found that Ag nanowires, with diameters ranging from 150 to 950 nm and lengths up to 100 μm, upon exposure to ammonia vapor at room temperatures, showed an increase in electrical resistance (up to 10,000 %) that was fast (<5 s) and reversible. The same reversible behavior, although characterized by a slower response time (1 min), was recorded in the presence of liquid amine vapor, while an irreversible resistance increase was found when they were exposed to hydrogen sulfide. Conversely, carbon monoxide, oxygen, hydrocarbons, argon, and water caused no change in resistance for exposures up to 10 s. Tao et al. (2003) demonstrated that silver nanowire can also be applied for detection of 2,4-dinitrotoluene (2,4-DNT), the most common nitroaromatic compound for detecting buried landmines and other explosives by utilizing surface-enhanced Raman spectroscopy. Li et al. (2000) have shown that Cu nanowire arrays are also highly sensitive to gas surroundings. They demonstrated that the mechanism depended on the molecule–nanowire interactions and the conductance changes should be specific for different adsorptions. The specificity can be improved if the nanowires are pre-adsorbed with functionalized molecules because a specific interaction of a sample molecule with the functionalized molecules provides identity information about the sample molecule. However, experiments have shown that, as a rule, metal nanowires do not show sensitivity to gas surroundings and temporal stability is required for gas sensor development. Ag nanowires, due to irreversible changes in properties, will also hardly find any real application in spite of sensitivity to NH_3. Moreover, we have to say that only Pt nanowires have required chemical stability.

It should be noted that the advantages and disadvantages of metal nanowires application in gas sensors are similar to the advantages and disadvantages of metal nanoparticles discussed above. This means that bad reproducibility and low stability of sensor parameters are the main shortcomings of room temperature gas sensors based on metal nanowires. For example, many electrodeposited Pd nanowires can have different morphologies depending on the growth conditions and subsequent treatments. Hu et al. (2008) and Yang et al. (2009) have demonstrated that electrodeposited Pd nanowires with different morphologies led to sharp contrasts in their respective response behaviors. Pd nanowires fabricated by lithography techniques have the best reproducibility, and they could be good hydrogen sensors with fast response and low detection limit. However, they possibly face the risk of structural deformations at high H_2 concentrations because the nanowire body sticks to the substrate. This problem is characteristic for Pd films, which were discussed earlier in Chap. 4 (Vol. 1). However, Jeon et al. (2008) believe that this problem would be eliminated by using bottom-up grown Pd nanowires, which have no direct bonds with the substrate.

References

Ahmadi TS, Wang L, Henglein A, El-Sayed MA (1996) "Cubic" colloidal platinum nanoparticles. Chem Mater 8:1161–1163

Ahn H, Chandekar A, Kang B, Sung C, Whitten JE (2004) Electrical conductivity and vapor sensing properties of ω-(3-thienyl)alkanethiol-protected gold nanoparticle films. Chem Mater 16:3274–3278

Aiyer HN, Vijayakrishnan V, Subanna GN, Rao CNR (1994) Investigations of Pd clusters by the combined use of HREM, STM, high-energy spectroscopies and tunneling conductance measurements. Surf Sci 313:392–398

Braunstein P, Oro G, Raithby PR (eds) (1999) Metal clusters in chemistry. Wiley-VCH, Weinheim

Brus L (1991) Quantum crystallites and nonlinear optics. Appl Phys A 53:465–474

Brust M, Kiely CJ (2002) Some recent advances in nanostructure preparation from gold and silver particles: a short topical review. Colloids Surf A 202:175–186

Brust M, Walker M, Bethell D, Schiffrin DJ, Whyman R (1994) Synthesis of thiol-derivatised gold nanoparticles in a two-phase liquid-liquid system. J Chem Soc Chem Commun 7:801–802

Burda C, Chen X, Narayanan R, El-Sayed MA (2005) Chemistry and properties of nanocrystals of different shapes. Chem Rev 105:1025–1102

Cai QY, Zellers ET (2002) Dual-chemiresistor GC detector employing monolayer-protected metal nanocluster interfaces. Anal Chem 74:3533–3539

Chen KJ, Lu CJ (2010) A vapor sensor array using multiple localized surface plasmon resonance bands in a single UV-vis spectrum. Talanta 81:1670–1675

Cheng C, Gonela RK, Gu Q, Haynie DT (2005) Self-assembly of metallic nanowires from aqueous solution. Nano Lett 5(1):175–178

Cheng CS, Chen YQ, Lu CJ (2007) Organic vapour sensing using localized surface plasmon resonance spectrum of metallic nanoparticles self assemble monolayer. Talanta 73:358–365

Cioffi N, Losito I, Torsi L, Farella I, Valentini A, Sabbatini L, Zambonin PG, Bleve-Zacheo T (2002) Analysis of the surface chemical composition and morphological structure of vapor-sensing gold–fluoropolymer nanocomposites. Chem Mater 14:804–811

Daniel M-C, Astruc D (2004) Gold nanoparticles: assembly, supramolecular chemistry, quantum-size-related properties, and applications toward biology, catalysis, and nanotechnology. Chem Rev 104:293–346

De Jongh LJ (ed) (1994) Physics and chemistry of metal cluster compounds. Kluwer, Dordrecht

Di Francia G, Alfano B, La Ferrara V (2009) Conductometric gas nanosensors. J Sensors 659275(18)

Doering DL, Dickinson JT, Poppa H (1982) UHV studies of the interaction of CO with small supported metal particles, Ni mica. J Catal 73:91–103

Dovgolevsky E, Tisch U, Haick H (2009) Chemically sensitive resistors based on monolayer-capped cubic nanoparticles: towards configurable nanoporous sensors. Small 5:1158–1161

Drake C, Deshpande S, Bera D, Seal S (2007) Metallic nanostructured materials based sensors. Int Mater Rev 52(5):289–317

El-Sayed MA (2004) Small is different: shape-, size-, and composition-dependent properties of some colloidal semiconductor nanocrystals. Acc Chem Res 37:326–333

Erathodiyil N, Ying JY (2011) Functionalization of inorganic nanoparticles for bioimaging applications. Accounts Chem Res 44(10):925–935

Evans SD, Johnson SR, Cheng YL, Shen T (2000) Vapour sensing using hybrid organic–inorganic nanostructured materials. J Mater Chem 10:183–186

Favier F, Walter EC, Zach MP, Benter T, Penner RM (2001) Hydrogen sensors and switches from electrodeposited palladium mesowire arrays. Science 293(5538):2227–2231

Filenko D, Gotszalk T, Kazantseva Z, Rabinovych O, Koshets I, Shirshov Y (2005) Chemical gas sensors based on calixarene-coated discontinuous gold films. Sens Actuators B 111–112:264–270

Gillet E, Channakhone S, Matolin V, Gillet M (1986) Chemisorptional behaviour of Pd small supported particles depending on size and structure: TDS, SSIMS and TEM investigation. Surf Sci 152/153:603–614

Han L, Maye MM, Leibowitz FL, Ly NK, Zhong CJ (2001) Quartz-crystal microbalance and spectrophotometric assessments of inter-core and inter-shell reactivities in nanoparticle thin film formation and growth. J Mater Chem 11:1259–1264

Harfenist SA, Wang ZL, Alvarez MM, Vezmar I, Whetten RL (1996) Highly oriented molecular Ag nanocrystal arrays. J Phys Chem 100(3):13904–13910

Harfenist SA, Wang ZL, Whetten RL, Vezmar I, Alvarez MM (1997) Three-dimensional hexagonal close-packed superlattice of passivated Ag nanocrystals. Adv Mater 9:817–822

Haruta M (1997) Size- and support-dependency in the catalysis of gold. Catal Today 36:153–166

He ST, Xie SS, Yao JN, Gao HJ, Pang SJ (2002) Self-assembled two-dimensional superlattice of Au–Ag alloy nanocrystals. Appl Phys Lett 81:150–152

Heiz U, Sanchez A, Abbet S, Schneider W-D (1999) Catalytic oxidation of carbon monoxide on monodispersed platinum clusters: each atom counts. J Am Chem Soc 121:3214–3217

Henglein A (1989) Small-particle research: physicochemical properties of extremely small colloidal metal and semiconductor particles. Chem Rev 89:1861–1873

Hostetler MJ, Templeton AC, Murray RW (1999) Dynamics of place-exchange reactions on monolayer-protected gold cluster molecules. Langmuir 15:3782–3789

Hu Y, Perello D, Mushtaq U, Yun M (2008) A single palladium nanowire via electrophoresis deposition used as an ultrasensitive hydrogen sensor. IEEE Trans Nanotechnol 7:693–699

Jeon KJ, Jeun M, Lee E, Lee JM, Lee KI, von Allmen P, Lee W (2008) Finite size effect on hydrogen gas sensing performance in single Pd nanowires. Nanotechnology 19:4955011–4955016

Jeon KJ, Lee JM, Lee E, Lee W (2009) Individual Pd nanowire hydrogen sensors fabricated by electron-beam lithography. Nanotechnology 20:1355021–1355025

Joseph Y, Besnard I, Rosenberger M, Guse B, Nothofer H-G, Wessels JM, Wild U, Knop-Gericke A, Su D, Schogl R, Yasuda A, Vossmeyer T (2003) Self-assembled gold nanoparticle/alkanethiol films: preparation, electron microscopy, XPS-analysis, charge transport, and vapor-sensing properties. J Phys Chem B 107:7406–7413

Joseph Y, Guse B, Yasuda A, Vossmeyer T (2004) Chemiresistor coatings from Pt- and Au-nanoparticle/nonanedithiol films: sensitivity to gases and solvent vapors. Sens Actuators B 98:188–195

Joseph Y, Guse B, Vossmeyer T, Yasudaa A (2008) Gold nanoparticle/organic networks as chemiresistor coatings: the effect of film morphology on vapor sensitivity. J Phys Chem C 112:12507–12514

Kang JF, Ulman A, Liao S, Jordan R, Yang G, Liu GY (2001) Self-assembled rigid monolayers of 4 -substituted-4-mercaptobiphenyls on gold and silver surfaces. Langmuir 17:95–106

Karakouz T, Vaskevich A, Rubinstein I (2008) Polymer-coated gold island films as localized plasmon transducers for gas sensing. J Phys Chem B 112:14530–14538

Kim KT, Sim SJ, Cho SM (2006) Hydrogen gas sensor using Pd nanowires electro-deposited into anodized alumina template. IEEE Sens J 6:509–513

Kriebig U, Vollmer M (eds) (1995) Optical properties of metal clusters. Springer, Berlin

Kulkarni GU, Thomas PJ, Rao CNR (2000) Nanocrystals: synthesis and mesoscalar organization. In: Contescu CI, Putyera K (eds) Dekker encyclopedia of nanoscience and nanotechnology. CRC Press, New York, pp 2676–2696

La Ferrara V, Alfano B, Massera E, Di Francia G (2008) Palladium nanowires assembly by dielectrophoresis investigated as hydrogen sensors. IEEE Trans Nanotechnol 7(6):776–781

Leibowitz FL, Zheng WX, Maye MM, Zhong CJ (1999) Structures and properties of nanoparticle thin films formed via a one-step exchange–cross-linking–precipitation route. Anal Chem 71:5076–5083

Li CZ, He HX, Bogozi A, Bunch JS, Tao NJ (2000) Molecular detection based on conductance quantization of nanowires. Appl Phys Lett 76:1333–1335

Lith JV, Lassesson A, Brown SA, Schulze M, Partridge JG, Ayesh A (2007) A hydrogen sensor based on tunneling between palladium clusters. Appl Phys Lett 91:181910

Love JC, Estroff LA, Kriebel JK, Nuzzo RG, Whitesides GM (2005) Self-assembled monolayers of thiolates on metals as a form of nanotechnology. Chem Rev 105:1103–1169

Masala O, Seshadri R (2004) Synthesis routes for large volumes of nanoparticles. Annu Rev Mater Res 34:41–81

Matolin V, Gillet E, Reed NM, Vickerman JC (1990) CO oxidation over small Pd particle model catalysts. A static secondary ion mass spectrometry study. J Chem Soc Faraday Trans 86:2749–2755

McConnell WP, Novak JP, Brousseau LCI, Fuierer RR, Tenent RC, Feldheim DL (2000) Electronic and optical properties of chemically modified metal nanoparticles and molecularly bridged nanoparticle arrays. J Phys Chem B 104:8925–8930

Milani P, Iannotta S (1999) Cluster beam synthesis of nanostructured materials. Springer, Berlin

Murray BJ, Walter EC, Penner RM (2004) Amine vapor sensing with silver mesowires. Nano Lett 4(4):665–670

Neugebauer CA, Webb MB (1962) Electrical conduction mechanism in ultrathin, evaporated metal films. J Appl Phys 33:74–82

Noh J-S, Lee JM, Lee W (2011) Low-dimensional palladium nanostructures for fast and reliable hydrogen gas detection. Sensors 11:825–851

Ogawa T, Kobayashi K, Masuda G, Takase T, Maeda S (2001) Electronic conductive characteristics of devices fabricated with 1,10-decanedithiol and gold nanoparticles between 1-m electrode gaps. Thin Solid Films 393:374–378

Pang P, Guo Z, Cai Q (2005) Humidity effect on the monolayer-protected gold nanoparticles coated chemiresistor sensor for VOCs analysis. Talanta 65:1343–1348

Pileni MP (1993) Reverse micelles as microreactors. J Phys Chem 97:6961–6973

Platt M, Dryfe RAW, Roberts EPL (2002) Controlled deposition of nanoparticles at the liquid-liquid interface. Chem Commun 2002:2324–2325

Rao CNR, Vijayakrishnan V, Santra AK, Prins MWJ (1992) Dependence of the reactivity of Ag and Ni clusters deposited on solid substrates on the cluster size. Angew Chem Int Ed 31:1062–1064

Rao CNR, Kulkarni GU, Thomas PJ, Edward PP (2002) Size-dependent chemistry: properties of nanocrystals. Chem - A Eur J 8(1):28–35

Rao CNR, Kulkarni GU, Thomas PJ, Agrawal VV, Saravanan P (2003a) Films of metal nanocrystals formed at aqueous—organic interfaces. J Phys Chem B 107:7391–7395

Rao CNR, Kulkarni GU, Thomas PJ, Agrawal VV, Gautam UK, Ghosh M (2003b) Nanocrystals of metals, semiconductors and oxides: novel synthesis and applications. Curr Sci 85:1041–1045

Rao CNR, Kulkarni GU, Thomas PJ (2005) Physical and chemical; properties of nano-sized metal particles. In: Nicolais L, Carotenuto G (eds) Metal-polymer nanocomposites. Wiley, Hoboken, New Jersey, pp 1–36

Raschke G, Brogl S, Susha S, Rogach AL, Klar TA, Feldman J, Fieres B, Petkov N, Bein T, Nichtl A, Kuzinger K (2004) Gold nanoshells improve single nanoparticle molecular sensors. Nano Lett 4:1853–1857

Saha K, Agasti SS, Kim C, Li X, Rotello VM (2012) Gold nanoparticles in chemical and biological sensing. Chem Rev 112:2739–2779

Sandhyarani N, Reshmi MR, Unnikrishnan R, Vidyasagar K, Ma S, Antony MP, Selvam GP, Visalakshi V, Chandrakumar N, Pandian K, Tao YT, Pradeep T (2000) Monolayer protected cluster superlattices: structural, spectroscopic, calorimetric and conductivity studies. Chem Mater 12:104–113

Santra AK, Ghosh S, Rao CNR (1994) Dependence of the strength of interaction of carbon monoxide with transition metal clusters on the cluster size. Langmuir 10:3937–3939

Sattler K, Mhlback J, Recknagel E (1980) Generation of metal clusters containing from 2 to 500 atoms. Phys Rev Lett 45:821–824

Schmid G (ed) (1994) Clusters and colloids: from theory to applications. VCH, Weinheim

Schmid G, Corain B (2003) Nanoparticulated gold: syntheses, structures, electronics, and reactivities. Eur J Inorg Chem 17:3081–3098

Schönenberger C, van der Zande BMI, Fokkink LGJ, Henny M, Schmid C, Krüger M, Bachtold A, Huber R, Birk H, Staufer U (1997) Template synthesis of nanowires in porous polycarbonate membranes: electrochemistry and morphology. J Phys Chem B 101:5497–5505

Shipway AN, Lahav M, Blonder R, Willner I (1999) Bis-bipyridinium cyclophane receptor- Au nanoparticle superstructures for electrochemical sensing applications. Chem Mater 11:13–15

Shipway AN, Katz E, Willner I (2000) Nanoparticle arrays on surfaces for electronic, optical, and sensor applications. Chemphyschem 1:18–52

Sun S, Murray CB, Weller D, Folks L, Maser A (2000) Monodisperse FePt nanoparticles and ferromagnetic FePt nanocrystal superlattices. Science 287:1989–1992

Tao A, Kim F, Hess C, Goldberger J, He RR, Sun YG, Xia YN, Yang PD (2003) Langmuir–Blodgett silver nanowire monolayers for molecular sensing using surface enhanced Raman spectroscopy. Nano Lett 3:1229–1233

Templeton AC, Hostetler MJ, Kraft CT, Murray RW (1998a) Reactivity of monolayer-protected gold cluster molecules: steric effects. J Am Chem Soc 120:1906–1911

Templeton AC, Hostetler MJ, Warmoth EK, Chen S, Hartshorn CW, Krishnamurthy VM, Forbes MDE, Murray RW (1998b) Gateway reactions to diverse, polyfunctional monolayer-protected gold clusters. J Am Chem Soc 120:4845–4849

Thanha NTK, Green LAW (2010) Functionalisation of nanoparticles for biomedical applications. Nano Today 5:213–230

Tisch U, Haick H (2010) Sensors based on monolayer-capped metallic nanoparticles. In: Korotcenkov G (ed) Chemical sensors: fundamentals of sensing materials. Vol. 2. Nanostructured materials. Momentum, New York, pp 141–202

Turkevich J, Stevenson PC, Hillier J (1951) A study of the nucleation and growth processes in the synthesis of colloidal gold. Discuss Faraday Soc 11:55–56

Valden M, Lai X, Goodman DW (1998) Onset of catalytic activity of gold clusters on titania with the appearance of nonmetallic properties. Science 281:1647–1650

Van Dijk MA, Tchebotareva AL, Orrit M, Lippitz M, Berciaud S, Lasne D, Cognet L, Lounis B (2006) Absorption and scattering microscopy of single metal nanoparticles. Phys Chem Chem Phys 8:3486–3495

Vasan HN, Rao CNR (1995) Nanoscale Ag–Pd and Cu–Pd alloys. J Mater Chem 5:1755–1757

Walter EC, Ng K, Zach MP, Penner RM, Favier F (2002a) Electronic devices from electrodeposited metal nanowires. Microelectron Eng 61–62:555–561

Walter EC, Favier F, Penner RM (2002b) Palladium mesowire arrays for fast hydrogen sensors and hydrogen-actuated switches. Anal Chem 74(7):1546–1553

Walter EC, Murray B, Favier F, Kaltenpoth G, Grunze M, Penner RM (2002c) Noble and coinage metal nanowires by electrochemical step edge decoration. J Phys Chem B 106:11407

Wohltjen H, Snow AW (1998) Colloidal metal-insulator-metal ensemble chemiresistor sensor. Anal Chem 70:2856–2859

Yang C-Y, Li C-L, Lu C-J (2006) A vapor selectivity study of microsensor arrays employing various functionalized ligand protected gold nanoclusters. Anal Chim Acta 565:17–26

Yang F, Taggart DK, Penner RM (2009) Fast, sensitive hydrogen gas detection using single palladium nanowires that resist fracture. Nano Lett 9:2177–2182

Yguerabide J, Yguerabide EE (1998) Light-scattering sub microscopic particles as highly fluorescent analogs and their use as tracer labels in clinical and biological applications. Anal Biochem 262:137–156

Yin AJ, Li J, Jian W, Bennett AJ, Xu JM (2001) Fabrication of highly ordered metallic nanowire arrays by electrodeposition. Appl Phys Lett 79:1039–1041

Yu A, Liang Z, Cho J, Caruso F (2003) Nanostructured electrochemical sensor based on dense gold nanoparticle films. Nano Lett 3:1203–1207

Zabet-Khosousi A, Dhirani A (2008) Charge transport in nanoparticle assemblies. Chem Rev 108:4072–4124

Zhang JZ (1997) Ultrafast studies of electron dynamics in semiconductor and metal colloidal nanoparticles: effects of size and surface. Acc Chem Res 30:423–429

Zhang HL, Evans SD, Henderson JR, Miles RE, Shen TH (2002) Vapour sensing using surface functionalized gold nanoparticles. Nanotechnology 13:439–444

Zheng WX, Maye MM, Leibowitz FL, Zhong CJ (2000) Imparting biomimetic ion-gating recognition properties to electrodes with a hydrogen-bonding structured core–shell nanoparticle network. Anal Chem 72:2190–2199

Chapter 5
Semiconductor Nanostructures

5.1 Quantum Dots

5.1.1 General Consideration

Semiconducting nanocrystals, otherwise known as quantum dots (QDs), were first discovered in the early 1980s (Ekimov et al. 1985). Since then, interest in QDs as alternatives to traditional organic dyes has increased dramatically (Costa-Fernandez 2006; Jorge et al. 2007; Callan et al. 2007; Smith and Nie 2010). Typically, QDs are colloidal nanocrystalline particles, roughly spherical, with particle diameters typically ranging from 1 to 12 nm. At such small sizes (close to or smaller than the dimensions of the exciton Bohr radius within the corresponding bulk material), these nanostructured materials behave differently from bulk solids because of quantum-confinement effects (Alivisatos 1996).

As a rule, QDs consist of elements of Groups II and VI (Gaponik et al. 2002; Smith and Nie 2010). Much of the research in quantum dots has been concentrated on cadmium selenide quantum dots due to their intense light emission properties. However, other II–VI compounds, such as CdS, CdTe, PbS, PbSe, or ZnS, and III–V compounds, such as InP, InAs, or InGaAs, are used as well. A major drawback of the cadmium selenide dots is their toxicity. Zinc sulfide quantum dots could be much less toxic than cadmium selenide. However, ZnS does not possess the level of light emission intensity of cadmium selenide. Recent research has shown that this shortcoming can be overcome partially by doping the zinc sulfide quantum dots with an appropriate metal ion such as manganese and lanthanides (Mohanta et al. 2003).

At present there are a variety of methods which can be used for synthesis of nanomaterials with specific size and shape, including QDs. However, Guo and Wang (2011) believe that diverse wet-chemical and electrochemical approaches have advantages due to their simplicity and rapidity during preparing high-quality nanomaterials with required morphologies. A general scheme shown in Fig. 5.1 demonstrates how to use simple wet-chemical and electrochemical methods to make nanomaterials.

It was established that quantum dots, in contrast to organic dyes, have broad absorption spectra, higher quantum yields, better chemical and photoluminescence stability, reduced photobleaching, and narrow emission spectra without red tailing (Jaiswal and Simon 2004). Moreover, QDs have the size-dependent nature of the emission wavelength (see Fig. 5.2), which is related to the three-dimensional quantum confinement of their charge carriers. The smaller the dot, the greater the blue shift observed relative to the typical *Eg* of the bulk semiconductor. This means that by controlling the growth of the nanocrystal, the emission wavelength can be tailored. In fact, QDs can be made to emit luminescence from the ultraviolet to the near-infrared spectral region. In particular, depending on the particles size, the emission of CdSe quantum dots can be continuously tuned from 465 to 640 nm, corresponding to a size ranging from 1.9 to 6.7 nm (diameter), respectively. For CdTe the emission is observed in the

G. Korotcenkov, *Handbook of Gas Sensor Materials*, Integrated Analytical Systems,
DOI 10.1007/978-1-4614-7388-6_5, © Springer Science+Business Media New York 2014

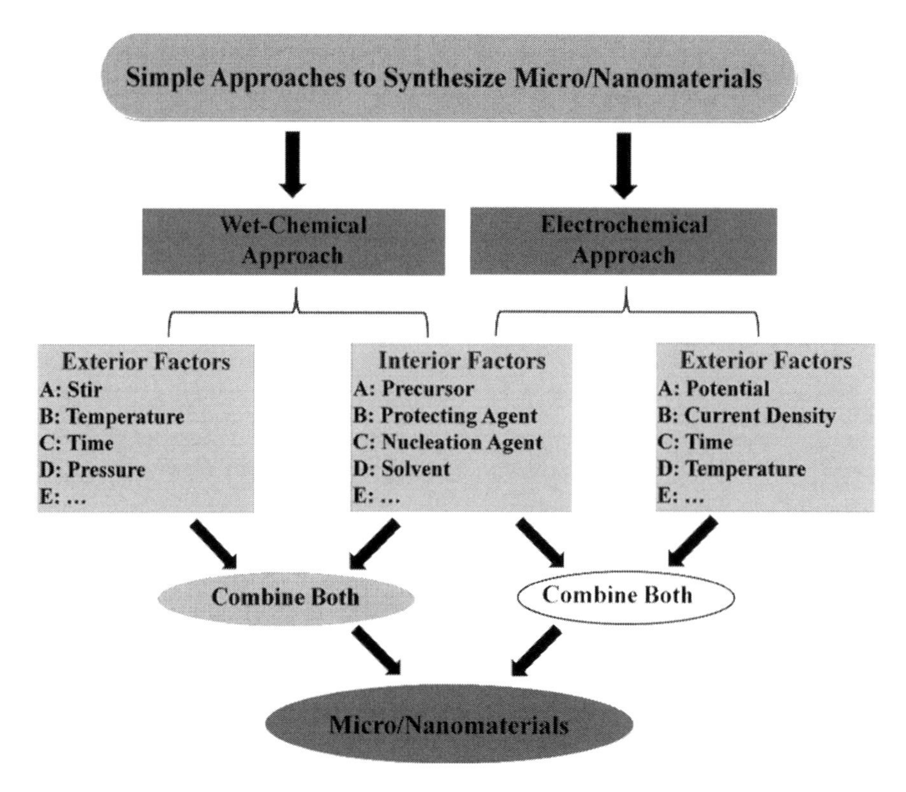

Fig. 5.1 Schematic procedure for simple synthesis of micro/nanomaterials (MNMs) (Reprinted with permission from Guo and Wang 2011, Copyright 2011 American Chemical Society)

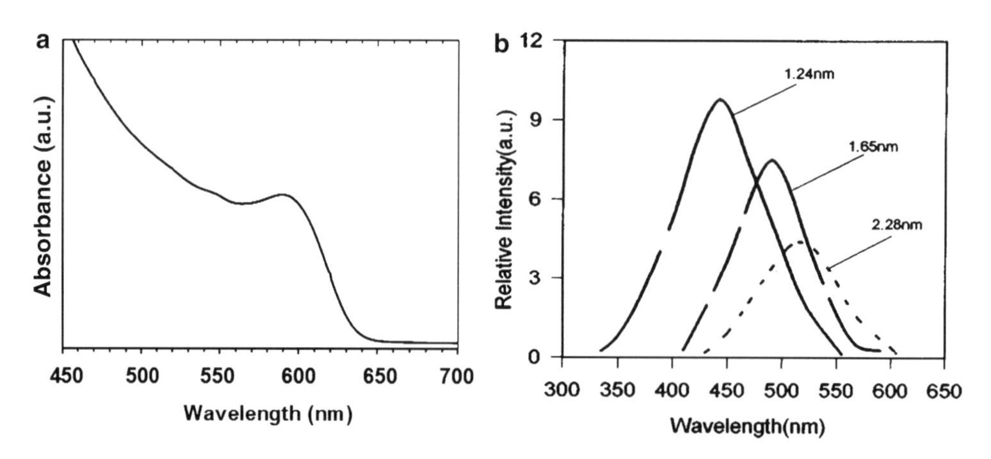

Fig. 5.2 (**a**) Typical absorption spectrum of CdSe nanocrystal QDs (Data from Jorge et al. 2007. Published by MDPI. Open access.). (**b**) Photoluminescence spectra of ZnS nanoparticles and ZnS nanoclusters in zeolite Y (λ_{exc} = 280 nm) (Reprinted with permission from Chen et al. 1997, Copyright 1997 American Institute of Physics)

red range from 600 to 725 nm. Higher-energy emission is possible with CdS and ZnZ (350–550 nm). Infrared emission is also available (780–2,000 nm) using indium arsenide (InAs) or lead selenide/sulfide (PbSe/PbS) nanocrystals (Jorge et al. 2007). Due to their broad absorption spectra (see Fig. 5.2), all QDs, independent of size, can be excited with a single wavelength, i.e., by the same optical source.

In addition, fundamental studies have revealed that luminescence of QDs is very sensitive to their surface states. It was established that emission processes in QDs result from electron–hole recombination

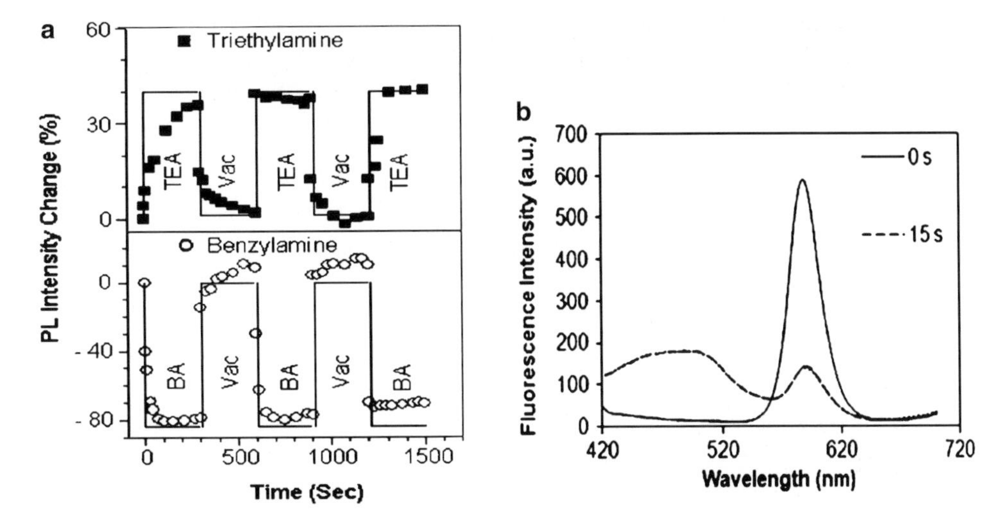

Fig. 5.3 (**a**) Dynamic response of the PL intensity with an altered atmosphere between vacuum and amine gases. (*Top*) *TEA* triethylamine. (*Bottom*) *BZA* benzylamine. ($\lambda_{excitation}$ = 514.5 nm) (Reprinted with permission from Nazzal et al. 2003, Copyright 2003. American Chemical Society). (**b**) Fluorescent spectra of CdSe/CdS QDs in dichloromethane after the injection of 100 ppm H_2S ($\lambda_{excitation}$ = 365 nm) (Reprinted with permission from Xu et al. 2010, Copyright 2010 Elsevier)

and are strongly dependent on the competition between such radiative processes and non-radiative recombination mechanisms. Non-radiative processes occur mainly at defects located at the nanocrystal surface. In this context, the large surface/volume ratio of QDs allows one to obtain enhanced quantum yields by control of their surface chemistry and passivation of surface defects (Jorge et al. 2007). This means that eventual chemical or physical interactions between a given chemical species with the surface of the nanoparticles would result in changes of the QD surface charges and would affect the QD photoluminescence emission very significantly (Chen and Rosenzweig 2002). It is clear that, due to indicated properties, quantum dots should have gained enormous attention for sensing applications. It is important that all the above-mentioned features of QDs are highly attractive for both optical fiber and planar platforms applications. Experiment has confirmed this prediction, and, in spite of a relatively recent field of research, a number of sensing applications have recently appeared. It was established that while most dyes present severe photodegradation when illuminated by energetic radiation, quantum dots have demonstrated to be photostable in most situations. Although photobleaching has been reported in bare dots, nanocrystals with an adequate protective shell are known to remain extremely bright even after several hours of exposure to moderate to high levels of UV radiation. On the other hand, the luminescence emission of common dyes can vanish completely after a few minutes (Jorge et al. 2007). It is necessary to note that the increased photostability of core–shell QDs is a key feature important for sensor applications.

5.1.2 Gas Sensor Applications of Quantum Dots

At present, QDs are mainly used for cation detection and biosensing in solution-sensing assays (Costa-Fernandez 2006; Callan et al. 2007). However, it was established that the surrounding gas can also influence the fluorescence properties of QDs embedded in gas permeable polymer. In particular, Nazzal et al. (2003) found that the PL properties of the CdSe nanocrystals, stabilized by trioctylphosphine oxide (TOPO) and incorporated in poly(methyl methacrylate) (PMMA) polymer films, can respond to the environment in a reversible and species-specific fashion (see Fig. 5.3). However, reversible response was observed in an oxygen-free atmosphere only. In oxygen, the CdSe was slowly oxidized, and therefore QDs have shorter lifetimes in air compared with those in an atmosphere of nitrogen.

Table 5.1 QD-based gas sensors

QDs	Matrix	Gas tested	References
CdSe	Polymethylmethacrylate	Triethylamine; benzylamine	Nazzal et al. (2003)
		Polar and nonpolar vapors	Potyrailo and Leach (2006)
		Aromatic hydrocarbons (xylenes, toluene)	Vassiltsova et al. (2007)
CdSe/CdS	Poly(dimethylsiloxane)	H_2S	Xu et al. (2010)
CdSe/ZnS	Sol–gel silica matrix + amphiphilic polymer	Ozone	Saren et al. (2011)
		VOCs	Hasani et al. (2010)
CdTe	Poly(dimethyldiallylammonium chloride)	Formaldehyde	Ma et al. (2009)
		VOCs	Norhayati et al. (2010)
CdTe/CdS	Sol–gel silica matrix + $PtF_{20}TPP$	Oxygen	Wang et al. (2008)
InP	Uncapped epitaxial InP QDs grown on InGaP layer	Methanol	De Angelis et al. (2012)

As reported by van Sark et al. (2001), the PL emission peak in an oxygen atmosphere irreversibly shifted by about 30 nm in total.

Potyrailo and Leach (2006) used the same approach. They evaluated the response of the luminescence from CdSe nanoparticles incorporated into a polymer film to the exposure of polar and nonpolar vapors. It was found that different size nanocrystals provided individual photoluminescence response patterns when the sensor film was exposed to methanol and toluene. Ma et al. (2009) have shown that under optimal conditions the PL intensity of CdTe QDs incorporated in poly(dimethyldiallylammonium chloride) decreased linearly with the increase of formaldehyde concentration in the range 5–500 ppb. The detection limit for formaldehyde was 1 ppb. Other examples of QD-based gas sensors are presented in Table 5.1.

The results presented are indicative of the potential that QDs have for use as gas sensors. However, we need to recognize that till now a clear understanding of sensing mechanism is absent. Some authors explain the observed quenching effect by reversible change in the dielectric environment of QDs, which affects the optoelectronic properties of the entrapped QDs (Potyrailo and Leach 2006). A second group assumes that these changes take place due to reversible changes in the physical integrity of the sensing material and its corresponding wetting properties when exposed to the different gaseous analytes (Vassiltsova et al. 2007). A third group believes that the PL intensity quenching of QDs is due to adsorption of electron acceptors on the surface of QDs following electron transfer between QDs and adsorbed electron acceptors (Burda et al. 1999).

As seen in Table 5.1, gas sensors were mainly based on CdSe and CdTe QDs. As a rule, QDs used in gas sensors have core–shell structures embedded in polymer matrix (see Fig. 5.4). The reasons for surface capping are mainly to prevent aggregation of the QDs caused by steric hindrance or charge and to passivate dangling bonds at the surface. Surface passivation involves coating the core QD with a substance that has a larger bandgap such as ZnS or CdS. It was found that overcoating the nanocrystal core (CdSe, CdTe) with an outer shell of a higher bandgap semiconductor (ZnS, CdS) is a successful strategy to produce materials with high quantum efficiencies (in the 50 % range). Besides the increased brightness, core–shell systems provide increased photostability and chemical resistance (Hines and Guyot-Sionnest 1996; Dabbousi et al. 1997). More recently, highly luminescent CdSe core nanocrystals capped with a multi-shell layer (CdS and ZnS) have been reported, displaying quantum yields in a 70–85 % range (Xie et al. 2005).

Modified thiols can also be used to cap the core and also provide an opportunity for structural modifications (Rogash et al. 1999). One can find in a review prepared by Chaudhuri and Paria (2012) a more detailed discussion of core–shell nanoparticles including their synthesis and properties.

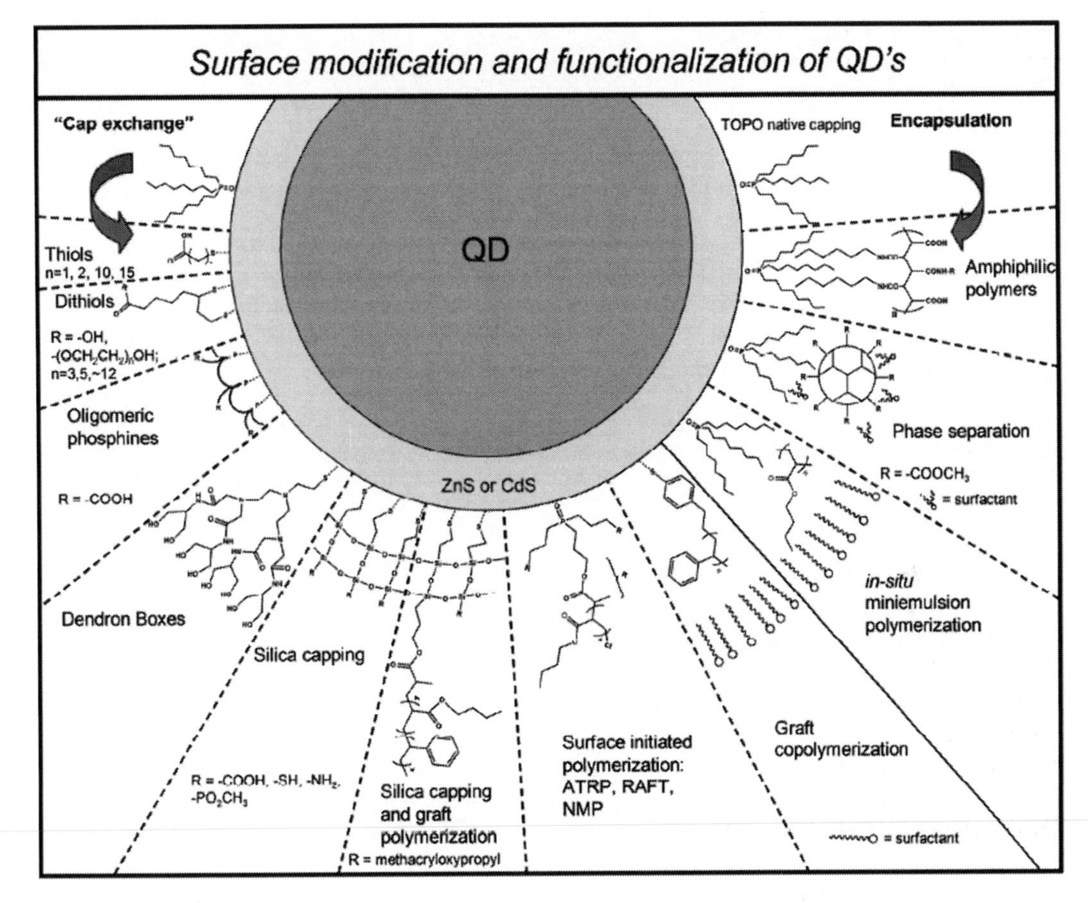

Fig. 5.4 Scheme illustrating some of the methods for chemical surface modification of QDs (Reprinted from Jorge et al. 2007, Published by MDPI)

Different techniques like molecular beam epitaxy (MBE), radio-frequency (RF) sputtering, and liquid phase epitaxy (LPE) can be used to synthesize semiconductor quantum dots, but the chemical route has been found to be the most attractive method (Mohanta et al. 2003). The chemical route of QDs synthesis is usually based on colloidal chemistry techniques which are very often associated with molecular precursor chemistry. For these methods, the semiconductor nanoparticles are homogeneously generated in a coordinating solvent or in the presence of a chemical stabilizer. Synthetic preparation of QDs normally involves the high temperature addition of a source of metal of Group IIB, for example, $CdMe_2$ or CdO, to a chalcogenic element (S, Se, or Te) in a strongly coordinating solvent such as trioctylphosphine (TOP) or TOPO (Green 2002). The reaction time, temperature, and metal to chalcogenide ratio can be varied to control the size of the nanocrystal and hence its spectral properties. Because QDs produced in this way have their surfaces capped with organic ligands, they are compatible with further (bio)chemical surface modification necessary for various applications, including gas sensors. It is well known that, in order to tailor the physical properties of QDs to their desired use, i.e., as sensors, the surface ligands (TOP/TOPO) should be exchanged with ligands of suitable functionality. Surface ligand displacement normally occurs by heating a solution of the desired ligand with the core QD/core–shell QD. The ligand usually bears a pendant thiol group for surface attachment although amines and alcohols have also been used (Jorge et al. 2007). Figure 5.4 summarizes some of the innovative chemical strategies that have been used to produce organically

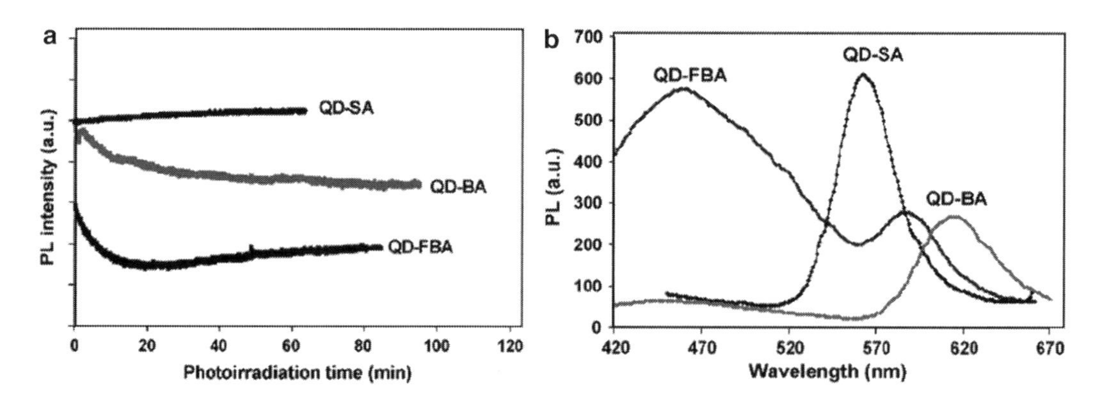

Fig. 5.5 (**a**) Peak PL vs time for the CdSe QD stabilized by TOPO and stearic acid (*QD–SA*), benzoic acid (*QD–BA*), and pentafluorobenzoic acid (*QD–FBA*) as monitored during the N_2 purged phototreatment process ($\lambda_{excitation}$ = 350 nm). (**b**) PL spectra taken after photoirradiation under N_2 for the *QD–SA*, *QD–BA*, and *QD–FBA* films (Reprinted with permission from Vassiltsova et al. 2007, Copyright 2007 Elsevier)

capped and/or polymer encapsulated QDs. One can find a detailed description of the approaches used for QDs functionalizing in reviews by Selvan et al. (2010) and Thanh and Green (2010).

Thus, the results presented testify that the use of QDs is a promising approach to gas sensor design, because via coating the QDs' surfaces with suitable ligands, which have a strong effect on luminescent response of QDs to specific chemical species (see Fig. 5.3), we make it possible to create selective gas sensors. However, we need to take into account that these sensors are more complicated in design and exploitation, and more expensive. For sensor functioning, an additional UV source is needed. High selectivity was also not achieved.

In addition, parameters of the indicated sensors depend strongly on the size of the nanodots, but the QDs synthesized have a broad particle size distribution, and there are difficulties with both particle size separation and homogeneous distribution of QDs inside the polymer matrix (Jorge et al. 2007). Moreover, QDs typically show a problem of agglomeration of the primary particles, losing their advantage as nanoparticles. This means that reproducibility of sensor parameters can be poor.

We also know that thermal and long-term instability of the properties of nanoparticles with sizes ~2–4 nm is a big problem as well. In spite of better characteristics in comparison with organic dyes, till now QDs do not have photostability required for applications in real devices for sensor market (Jorge et al. 2007). Much research has been carried out regarding QDs' photostability, and apparently contradictory results are often reported (Zhelev et al. 2004; Korsunska et al. 2005). It was found that the photostability of nanocrystals depends on their surface coatings (bare dots, core shell, or other) and on the surrounding environment (solution or solid matrix) (see Fig. 5.5). As a result, QD-based sensors have significantly shorter emission lifetimes in comparison with the lifetime of conventional, e.g., metal oxide, gas sensors. Moreover, the presence of UV irradiation creates an additional source of gas sensor instability caused by polymer degradation (see Chaps. 18 and 19 (Vol. 2)). We also need to note that the recovery process in QD-based gas sensors is usually slow. For example, Xu et al. (2010) found that the fluorescence of CdSe/CdS QDs recovers completely after interaction with H_2S in the open air for 2 h. Strong dependence on temperature is another problem (Pugh-Thomas et al. 2011). The change of the temperature and the influence of analyte can have the same effect. This means that temperature stabilization is required.

Thus, analysis carried out in this area testifies that there are still a number of practical issues which need to be improved, and much research needs to be done to develop an understanding of the underlying sensing mechanisms along with methods for achieving the desired sensitivity, selectivity, and reliability of QD-based gas sensors.

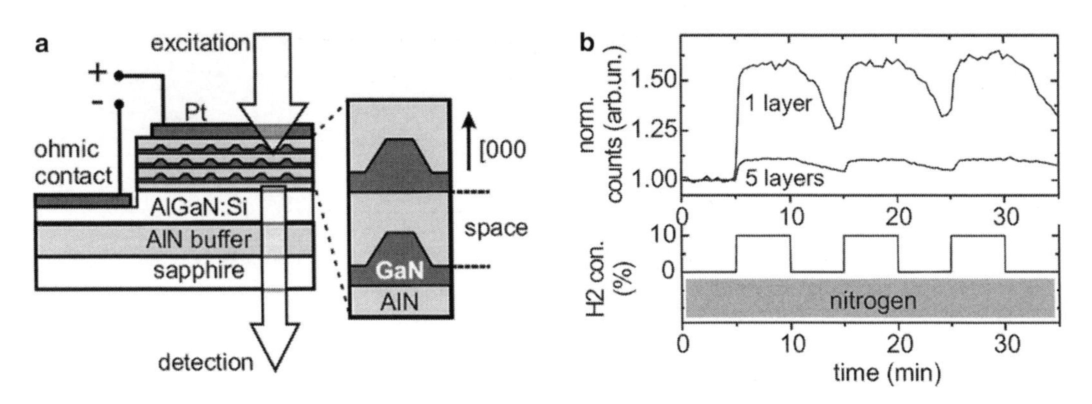

Fig. 5.6 (**a**) Schematic of the sample structure and the optical path for the PL measurements and (**b**) response of the PL intensity to introduction of 10 % hydrogen in a nitrogen atmosphere for samples with one and five QD layers. The PL intensity was recorded at a fixed detection wavelength near the emission maximum (due to a different size of the QDs, the maximum of the PL emission is located at an energy of 3.9 eV for the single QD layer) (Reprinted with permission from Weidemann et al. 2009, Copyright 2009 American Institute of Physics)

Another approach to the design of QD-based gas sensors was proposed by Weidemann et al. (2009). For these purposes, they used GaN/AlN QD superlattices (SL) grown on optically transparent AlN-on-sapphire substrates by plasma-assisted MBE and capped with a catalytic Pt-contact, which interact with gas, in particular hydrogen (Fig. 5.6a). Weidemann et al. (2009) established that, due to chemically induced changes of the surface potential at the Pt/AlN interface, caused by adsorption of dissociated hydrogen, the shift in the emission energy and a change in the photoluminescence (PL) intensity take place in the structures designed (see Fig. 5.6b). Of course, the tested structures do not possess several of the disadvantages of QDs discussed above, such as aggregation and so on. However, the sensitivity of proposed structures for the present is low in comparison with conventional hydrogen sensors, but the cost is too high for real applications.

One can find in reviews prepared by Alivisatos (1996), Green (2002), Costa-Fernandez (2006), Jorge et al. (2007), and Callan et al. (2007) additional information about quantum dots, their synthesis, and applications.

5.2 Semiconductor Nanowires

In other chapters it was shown that conventional semiconductors do not have the stability required for operation at high temperature in harsh environment (see Chap. 5 (Vol. 1) and Chap. 18 (Vol. 2)). However, in spite of such a situation, semiconductor 1D structures such as 1D nanowires continue to be of interest to gas sensor designers. Of course, this interest is conditioned first of all by technological resources of modern microelectronics and the possibilities of gas sensor integration with electronic devices. At present one can find research related to the design of 1D nanowire gas sensors based on Si (McAlpine et al. 2005, 2007; Skucha et al. 2010; Demami et al. 2012; Sadeghian and Islam 2011), GaN (Dobrokhotov et al. 2006; Chen et al. 2009; Wright et al. 2010; Aluri et al. 2011), InN (Wright et al. 2010), ZnS (Chen et al. 2008; Wang et al. 2012), InAs (Offermans et al. 2010; Dedigama et al. 2012), etc. (Xu et al. 2009). However, most of these studies are devoted to silicon. First, silicon is the basis of modern electronics. Therefore, they are easy to fabricate with existing silicon fabrication techniques that reduce the cost of the sensor designed, improve reproducibility, and ensure integrability with conventional CMOS devices used for signal processing and analysis. Second, native silicon oxide has the best capsulation properties. As a result, there is the possibility to operate Si-based

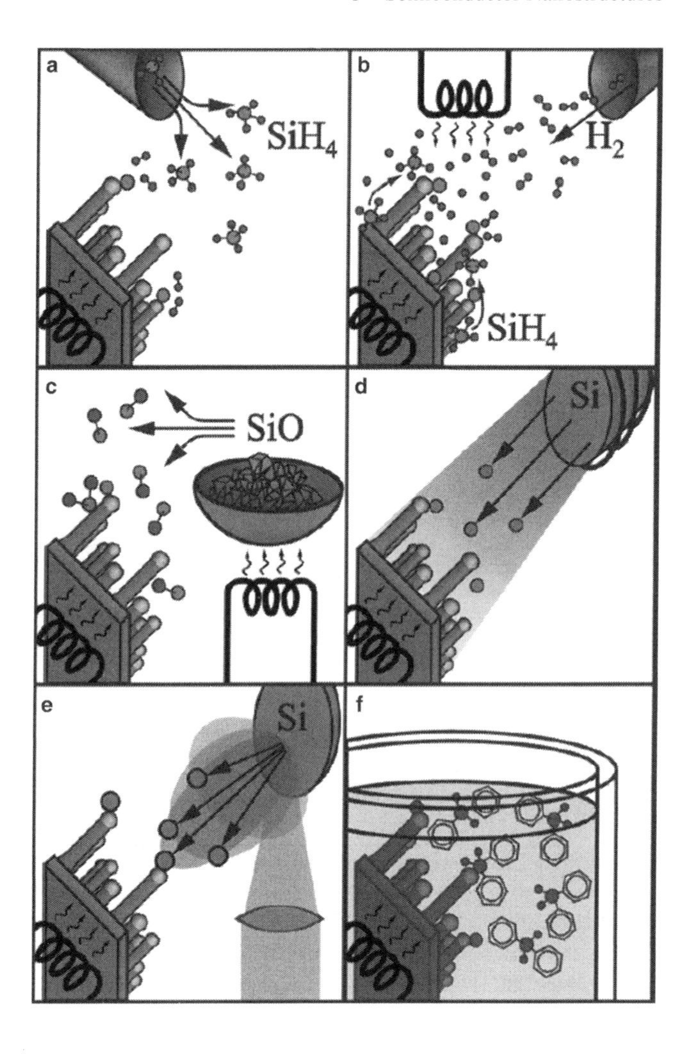

Fig. 5.7 Schematics of experimental setups for silicon nanowire growth: (**a**) CVD, (**b**) annealing in reactive atmosphere, (**c**) evaporation of SiO, (**d**) MBE, (**e**) laser ablation, and (**f**) solution-based growth (Reprinted with permission from Schmidt et al. 2009, Copyright 2009 Wiley)

devices at room temperature over a long time. In addition, the surface state density at the SiO_2/Si interface is very low. As a result, the pinning of surface Fermi level is absent. As a result, SiNW-based gas sensors usually have better gas-sensing characteristics in comparison with gas sensors based on other semiconductors. The possibility to control the concentration of free charge carriers is other merit of Si nanowires. It was established that SiNWs can be doped (Cui et al. 2000; Schmidt et al. 2009) *n*-type or *p*-type, in a similar manner to bulk silicon (see next section).

5.2.1 Synthesis of Semiconductor Nanowires

Experiments have shown that for fabrication of Si nanowires (SiNWs) various methods can be used including chemical vapor deposition (CVD), annealing in reactive atmosphere, evaporation of SiO, MBE, laser ablation, and solution-based techniques (Teo and Sun 2007; Schmidt et al. 2009, 2010; Bandaru and Pichanusakorn 2010). Illustrations of the above-mentioned methods are presented in Fig. 5.7. SiNWs prepared using various technologies are shown in Fig. 5.8. As a rule, the growth of SiNWs takes place according to vapor–liquid–solid (VLS) mechanism. The name VLS mechanism

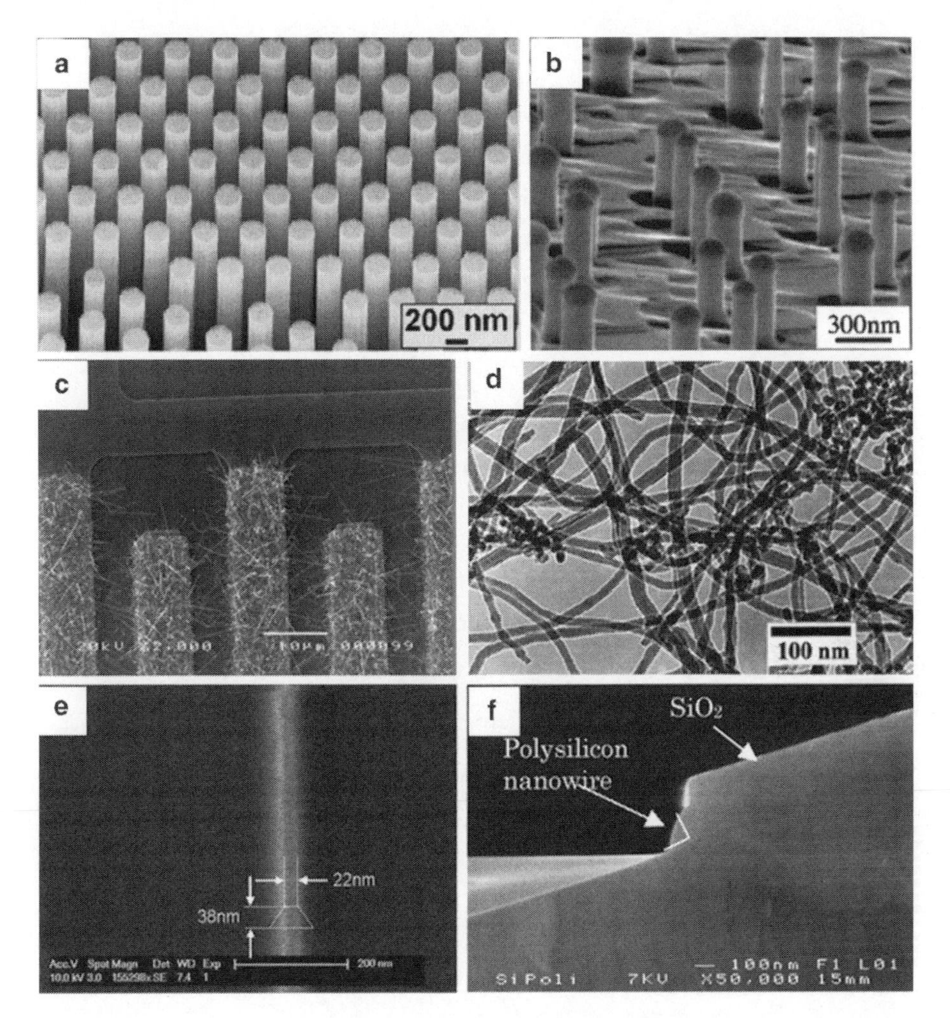

Fig. 5.8 SEM images of Si nanowires prepared using different methods. (**a**) Aligned, hexagonally ordered arrays of SiNWs fabricated using photolithography and etching. The nanowires are approximately 200 nm in diameter and 4 μm in length (Reprinted with permission from Field et al. 2011, Copyright 2011 American Chemical Society). (**b**) Au-catalyzed SiNWs on Si (111) grown by MBE (Reprinted with permission from Zakharov et al. 2006, Copyright 2006 Elsevier). (**c**) SiNWs synthesized by VLS method (Reprinted with permission from Demami et al. 2011, Copyright 2011 Elsevier). (**d**) SINWs synthesized by laser ablation using Ar (5 % H_2) carrying gas (Reprinted with permission from Zhang et al. 1999, Copyright 1999 American Institute of Physics). (**e**) SiNWs with top widths of 22 nm fabricated by trilayer nanoimprint and wet etching (Reprinted with permission from Gao et al. 2010, Copyright 2010 Elsevier). (**f**) Polysilicon NWs (100-nm radius curvature) fabricated using the sidewall spacer method (Reprinted with permission from Demami et al. 2011, Copyright 2011 Elsevier)

refers to the fact that silicon from the vapor passes through a liquid droplet and finally ends up as a solid. The appearance of VLS mechanism is based on two observations: that the addition of certain metal impurities is an essential prerequisite for growth of silicon wires in experiments and that small globules of the impurity are located at the tip of the wire during growth (see Fig. 5.8b). Nanowires of other semiconductors can be grown using the same approaches (Law et al. 2004).

There are many top-down fabrication methods which also make it possible to fabricate SiNWs (Teo and Sun 2007; Schmidt et al. 2009; Bandaru and Pichanusakorn 2010). Due to processing-related differences, one should distinguish between the fabrication of horizontal nanowires, that is, nanowires lying in the substrate plane (see Fig. 5.8e, f), on the one hand, and the fabrication of vertical nanowires, that is, nanowires oriented more or less perpendicular to the substrate, on the other (see Fig. 5.8a). Horizontal silicon nanowires are mostly fabricated from either silicon-on-insulator (SOI) wafers or bulk silicon wafers using a sequence of lithography and etching steps, often employing electron-beam lithography and reactive-ion etching. The interested reader is referred to the excellent articles of Singh et al. (2008) and Suk et al. (2005, 2008) and the references therein. In most cases, horizontal nanowire processing is finalized by an oxidation step, which also serves to reduce the silicon nanowire diameter. In this way, diameters well below 10 nm have been achieved in the past (Suk et al. 2008; Singh et al. 2008). Experiment has shown that no further thinning of the nanowires is necessary for gas sensor design.

Using standard silicon technology, vertical silicon nanowires can also be produced (Teo and Sun 2007; Schmidt et al. 2009). Reactive-ion etching is often used to etch vertical silicon nanowires out of a silicon wafer. The diameter of the nanowires is defined by a lithography step preceded by reactive-ion etching. A variety of different methods of nano-structuring, such as electron-beam lithography (Liu et al. 1993), nanosphere lithography (Hsu et al. 2008), nanoimprint lithography (Morton et al. 2008), or block copolymers (Zschech et al. 2007), have been employed for this purpose. As an alternative to reactive-ion etching, the so-called metal-assisted etching of silicon attracted some attention recently. In this approach, Si is wet-chemically etched, with the Si dissolution reaction being catalyzed by the presence of a noble metal that is added as a salt to the etching solution (Peng et al. 2002; Hochbaum et al. 2008). Alternatively, a continuous but perforated noble-metal film can be used. During etching, this perforated metal film will etch down into the silicon producing vertical silicon nanowires at the locations of the holes in the metal film (Huang et al. 2007).

SiNWs, fabricated through lithographic patterning or chemical synthesis, can be further reduced in diameter through a self-limiting oxidation process. For example, Liu et al. (1994) have shown that with an initial Si pillar diameter of ~30 nm, 6-nm Si nanowires can be obtained using this approach.

The doping of SiNWs can be achieved in two ways: in situ doping (doping during the growth process) and postdoping (doping after growth through an implantation and annealing process) (Yu et al. 2000; Teo and Sun 2007). For example, laser-assisted VLS growth was used to introduce either boron or phosphorus dopants during the vapor-phase growth of SiNWs (Cui et al. 2000; Cui and Lieber 2001; Ma et al. 2001; Tang et al. 2001, 2002). Gaseous compounds such as diborane, trimethylboron, and phosphine have also been used as dopants in a gas-phase VLS-CVD approach in which silane (SiH_4) or silicon tetrachloride, etc., is used as Si source (Cui et al. 2003; Zheng et al. 2004). Compared to solid dopants, the doping level can easily be controlled by tuning the concentration (flow rate) of the gas dopant(s) in the gas-phase silicon source. Measurements made on individual boron-doped and phosphorus-doped SiNWs showed that these materials behave as *p*-type and *n*-type materials, respectively (Cui et al. 2000, 2003). In other studies, metals such as Zn, Au (Yu et al. 2000; Chung et al. 2000), and Li (Zhou et al. 1999) have been doped into Si nanowires in order to change the electronic transport properties and morphology of the Si nanowires.

The doping process can also be carried out after Si nanowire growth. In one example, *n*-type Bi-doped SiNWs were fabricated by allowing bismuth vapor to diffuse into the SiNWs after the growth of SiNWs in a sealed evacuated (10^{-6} to 10^{-7} Torr) quartz tube at 1,000 °C (Byon et al. 2005). The doping concentrations depend on the oxide sheath thickness. Postdoping has advantages such as facilitating an appropriate choice of the vapor-phase dopants and selective patterning of the doped SiNWs.

5.2.2 Gas-Sensing Properties of Si Nanowires

Using SiNWs fabricated by various methods, resistive gas sensors (Gao et al. 2010; Field et al. 2011), FET (McAlpine et al. 2007; Paska et al. 2011), Schottky type gas sensors (Skucha et al. 2010), and field-ionized gas sensors (Sadeghian and Islam 2011) have been designed. Examples of such devices

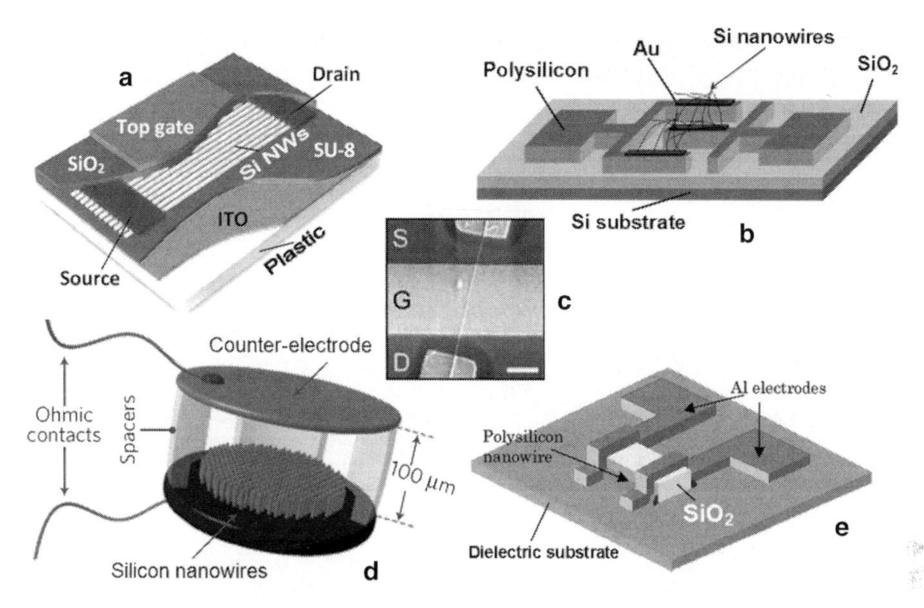

Fig. 5.9 Diagrams illustrating the construction of SiNW-based gas sensors. (**a**) FET-type gas sensor with ordered nanowire array on flexible substrate (Reprinted with permission from McAlpine et al. 2007, Copyright 2007 Nature Publishing Group). (**b**) Resistive SiNW-based sensors with SiNWs synthesized by VLS method with Au catalyst (Reprinted with permission from Demami et al. 2011, Copyright 2011 Elsevier). (**c**) SEM image of the 20-nm SiNW FET transistor on a mylar substrate. *Scale bar* is 1 μm. SiNWs were synthesized using a gold nanocluster-catalyzed VLS growth process (Reprinted with permission from McAlpine et al. 2005, Copyright 2005 IEEE). (**d**) SiNW-based device used to measure gas ionization. Nanowires were planted at the anode. d_{gap} = 100 μm is the spacing between anode and cathode (Reprinted with permission from Sadeghian and Islam 2011, Copyright 2011 Nature Publishing Group). (**e**) Polysilicon (10-μm length) NW-based resistive gas sensor fabricated using spacer method (Reprinted with permission from Demami et al. 2011, Copyright 2011 Elsevier)

Table 5.2 Parameters of SiNW-based gas sensors operated at room temperature

Type	Technology	Diameter/length	Target gas	Threshold limit	Res. time	References
R (A)	Vertically aligned array (chem. etching)	200 nm/4 μm	NH_3 NO_2	<0.5 ppm <0.25 ppm	1–3 min 5–8 min	Field et al. (2011)
FET (OA)	Super lattice NW pattern transfer technique	18 nm	NO_2	~20 ppb	1–4 min	McAlpine et al. (2007)
Schottky (Pd) (S)	NW contact printing technology. NWs are grown using VLS process	30 nm	H_2	<5 ppm	30–1,000 s	Skucha et al. (2010)

R resistive, *FET* field-effect transistor, *S* single nanowire, *A* nanowire ordered array, *OA* nanowire ordered array

are shown in Fig. 5.9. As seen in Fig. 5.9, SiNW sensors can be designed using single nanowires (McAlpine et al. 2005; Demami et al. 2012), nanowire arrays (Demami et al. 2012), and ordered nanowire arrays (McAlpine et al. 2007; Field et al. 2011).

It was found that response of SiNW-based gas sensors obeys the regularities established for nano-structured metal oxide-based sensors. The decrease of the width (diameter) or the thickness of the wires and the concentration of free charge carriers was accompanied by an increase of sensor response (McAlpine et al. 2007; Wan et al. 2009; Gao et al. 2010). Parameters of the best SiNW-based gas sensors are summarized in Table 5.2. All these sensors operate at room temperature. Operating characteristics of several devices from Table 5.2 are shown in Fig. 5.10.

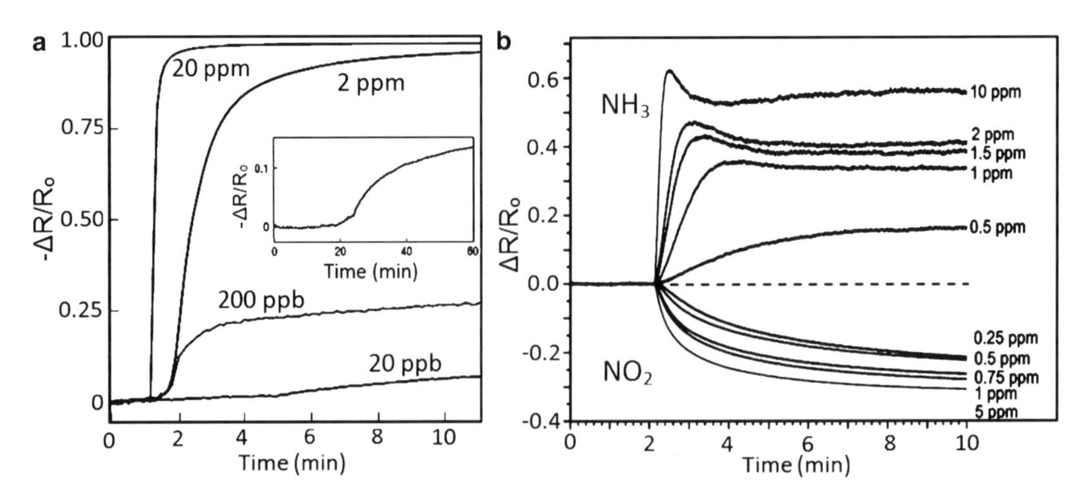

Fig. 5.10 (**a**) Response of FET-based sensor shown in Figure 5.9c to NO_2 diluted in N_2. *Inset*: an extended response of the sensor to 20 ppb NO_2 (Reprinted with permission from McAlpine et al. 2007, Copyright 2007 Nature Publishing Group). (**b**) Response to NH_3 and NO_2 at various concentrations for an ordered, vertically aligned silicon nanowire-based resistive gas sensor with a porous electrode (Reprinted with permission from Field et al. 2011, Copyright 2011 American Chemical Society)

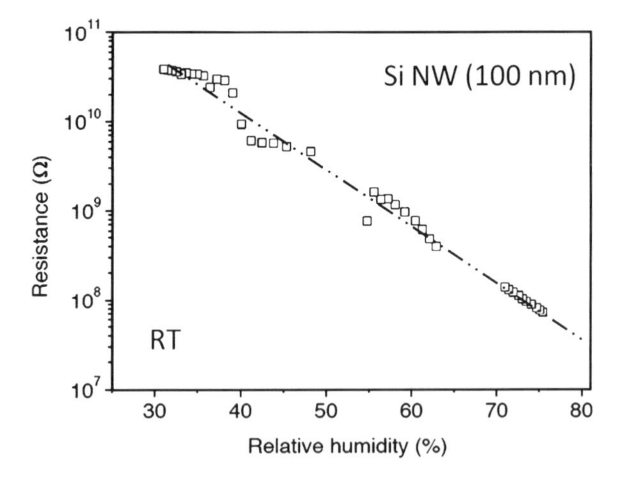

Fig. 5.11 Resistance variations vs relative humidity of the SiNW-based resistor. SiNWs (100 nm diameter) were synthesized using VLS method (Reprinted with permission from Demami et al. 2010, Copyright 2010 Elsevier)

Results presented in Table 5.2 and in Fig. 5.10 demonstrate that SiNW-based sensors with optimal geometry of silicon nanowires have high sensitivity. However, response and especially recovery times are long enough (Zhou et al. 2003). For example, Field et al. (2011) have found that the sensor needed at least 1 h of clean air exposure to desorb the analyte partially from the nanowire surfaces and return to a stable, flat baseline at 40 °C. Zhou et al. (2003) reported that the typical recovery times for SiNW-resistors nanowires after interaction with NH_3 and water vapor were 5 and 0.5 h, respectively. Because of irreversible adsorption of analytes on the nanowires, Field et al. (2011) also observed that the baseline never fully recovered to its original, preexposure resistance but reached a new equilibrium resistance and over time the sensor lost sensitivity. The incomplete desorption of analyte from the nanowire surface during exposure limited the number of exposures and prevented replicate measurements for each concentration of ammonia or nitrogen dioxide. The recovery and lifetime can probably be improved with a higher operating temperature since adsorption/desorption is temperature dependent but is a trade-off with sensitivity and requires additional optimization (Field et al. 2011). The increase of operating temperatures will also be accompanied by both the worsening of stability and the intensification of temporal drift. In addition, it is necessary to take into account that, due to operation at room temperatures, the effect of humidity should be strong (see Fig. 5.11).

Strong sensitivity to polar analytes in the gas phase (N_2O, NO, CO, etc.) and low sensitivity to nonpolar analytes (nonpolar VOCs) are, of course, disadvantages of SiO_2-coated SiNW-based devices (Paska et al. 2011). For example, SiNW FETs exhibited detection limits down to 20 ppb NO_2, but the same devices responded weakly to 1,000 ppm hexane (McAlpine et al. 2007). However, research has shown that the low sensitivity toward nonpolar analytes can in principle be improved by modifying the SiNW surface with appropriate organic receptors. In particular, it was found that oxide-coated SiNW FETs modified with alkanesilanes, aldehyde-silanes, or amino-silanes showed an improved response and sensitivity when exposed to hexane at 1,000 ppm (Tisch and Haick 2010). Paska et al. (2011) established that attaching dense hydrophobic organic hexyltrichlorosilane (HTS) monolayers that passivate most of the SiO_2/SiNW surface trap states in a FET device could also serve as a successful and simple strategy to achieve and maintain high sensitivity toward nonpolar VOCs. However, it must be admitted that these improvements in parameters of SiNW-based sensors are still far away from those required for successful applications. According to Paska et al. (2011), such situations occur due to (1) the weak adsorption of nonpolar VOCs in molecule-free sites, (2) the lack of suitable nonpolar organic functionalities that can be attached to the SiNWs, (3) the high density of surface states at the SiO_2/Si interface, and (4) the high density of trap states at the air/SiO_2 interface. Several studies have shown that removing the oxide and attaching an organic monolayer directly to the Si surface, through chemical (e.g., Si–C) bonds, increases the transconductance values and allows the formation of SiNW FETs with higher on–off ratios, as compared to SiO_2/SiNW FETs (Haick et al. 2006; Nolan et al. 2007; Bashouti et al. 2009a; Haight et al. 2009; Tisch and Haick 2010; De Smet et al. 2011). Nevertheless, there are practical and technical limitations to these approaches, including, but not confined to, the poor oxidation resistance (Bashouti et al. 2008, 2009a, b) and the complexity of the functionalization procedure, especially when the SiNWs are already integrated in a device platform. This means that, due to stability worsening, the application of this approach (removing oxide layer) in devices for the sensor market is not realistic.

References

Alivisatos AP (1996) Semiconductor clusters, nanocrystals, and quantum dots. Science 271:933–937

Aluri GS, Motayed A, Davydov AV, Oleshko VP, Bertness KA, Sanford NA, Rao MV (2011) Highly selective GaN-nanowire/TiO_2-nanocluster hybrid sensors for detection of benzene and related environment pollutants. Nanotechnology 22:295503

Bandaru PR, Pichanusakorn P (2010) An outline of the synthesis and properties of silicon nanowires. Semicond Sci Technol 25:024003

Bashouti MY, Stelzner T, Berger A, Christiansen S, Haick H (2008) Chemical passivation of silicon nanowires with C1-C6 alkyl chains through covalent Si-C bonds. J Phys Chem C 112:19168–19172

Bashouti MY, Tung RT, Haick H (2009a) Tuning electrical properties of Si nanowire field effect transistors by molecular engineering. Small 5:2761–2769

Bashouti MY, Stelzner T, Berger A, Christiansen S, Haick H (2009b) Covalent attachment of alkyl functionality to 50 nm silicon nanowires through a chlorination/alkylation process. J Phys Chem C 113:14823–14828

Burda C, Green TC, Link S, El-Sayed MA (1999) Electron shuttling across the interface of CdSe nanoparticles monitored by femtosecond laser spectroscopy. J Phys Chem B 103:1783–1788

Byon K, Tham D, Fischer JE, Johnson AT (2005) Synthesis and postgrowth doping of silicon nanowires. Appl Phys Lett 87:193104

Callan JF, De Silva AP, Mulrooney RC, McCaughan B (2007) Luminescent sensing with quantum dots. J Incl Phenom Macrocycl Chem 58:257–262

Chaudhuri RD, Paria S (2012) Core/shell nanoparticles: classes, properties, synthesis mechanisms, characterization, and applications. Chem Rev 112:2373–2433

Chen Y, Rosenzweig Z (2002) Luminescent CdS quantum dots as selective ion probes. Anal Chem 74:5132–5138

Chen W, Wang Z, Lin Z, Lin L, Efros AL, Rosen M (1997) Absorption and luminescence of the surface states in ZnS nanoparticles. J Appl Phys 82:3111–3115

Chen Z-G, Zou J, Liu G, Lu HF, Li F, Lu GQ, Cheng HM (2008) Silicon-induced oriented ZnS nanobelts for hydrogen sensitivity. Nanotechnology 19:055710

Chen RS, Lu CY, Chen KH, Chen LC (2009) Molecule-modulated photoconductivity and gain-amplified selective gas sensing in polar GaN nanowires. Appl Phys Lett 95:233119

Chung SW, Yu JY, Heath JR (2000) Silicon nanowire devices. Appl Phys Lett 76:2068–2070

Costa-Fernandez JM (2006) Optical sensors based on luminescent quantum dots. Anal Bioanal Chem 384:37–40

Cui Y, Lieber CM (2001) Functional nanoscale electronic devices assembled using silicon nanowire building blocks. Science 291:851–853

Cui Y, Duan X, Hu J, Lieber CM (2000) Doping and electrical transport in silicon nanowires. J Phys Chem B 104:5213–5216

Cui Y, Zhong ZH, Wang DL, Wang WU, Lieber CM (2003) High performance silicon nanowire field effect transistors. Nano Lett 3:149–152

Dabbousi BO, Rodriguez-Viejo J, Mikulec FV, Heine JR, Mattoussi H, Ober R, Jensen KF, Bawendi MG (1997) (CdSe) ZnS core-shell quantum dots: synthesis and characterization of a size series of highly luminescent nanocrystallites. J Phys Chem B 101(46):9463–9475

De Angelis R, Casalboni M, Hatami F, Ugur A, Masselink WT, Prosposito P (2012) Vapour sensing properties of InP quantum dot luminescence. Sens Actuators B 162:149–152

De Smet LCPM, Ullien D, Mescher M, Sudhölter EJR (2011) Organic surface modification of silicon nanowire-based sensor devices. In: Hashim A (ed) Nanowires—implementations and applications. InTech, Manhattan, pp 267–288

Dedigama A, Angelo M, Torrione P, Kim T-H, Wolter S, Lampert W, Atewologun A, Edirisoorya M, Collins L, Kuech TF, Losurdo M, Bruno G, Brown A (2012) Hemin-functionalized InAs-based high sensitivity room temperature NO gas sensors. J Phys Chem C 116:826–833

Demami F, Ni L, Rogel R, Salaun AC, Pichon L (2010) Silicon nanowires synthesis for chemical sensor applications. Procedia Eng 5:351–354

Demami F, Ni L, Rogel R, Salaun AC, Pichon L (2012) Silicon nanowires based resistors as gas sensors. Sens. Actuators B Chem 170:158–162

Dobrokhotov V, McIlroy DN, Norton MG, Abuzir A, Yeh WJ, Stevenson I, Pouy R, Bochenek J, Cartwright M, Wang L, Dawson J, Beaux M, Berven C (2006) Principles and mechanisms of gas sensing by GaN-nanowires functionalized with gold nanoparticles. J Appl Phys 99:104302

Ekimov AI, Efros AL, Onushchenko AA (1985) Quantum size effect in semiconductor microcrystals. Solid State Commun 56:921–924

Field CR, In HJ, Begue NJ, Pehrsson PE (2011) Vapor detection performance of vertically aligned, ordered arrays of silicon nanowires with a porous electrode. Anal Chem 83:4724–4728

Gao C, Deng S-R, Wana J, Lu B-R, Liu R, Huq E, Qu X-P, Chen Y (2010) 22 nm silicon nanowire gas sensor fabricated by trilayer nanoimprint and wet etching. Microelectron Eng 87:927–930

Gaponik N, Talapin DV, Rogach AL, Hoppe K, Shevchenko EV, Kornowski A, Eychmüller A, Weller H (2002) Thiol-capping of CdTe nanocrystals: an alternative to organometallic synthetic routes. J Phys Chem B 106: 7177–7185

Green M (2002) Solution routes to III–V semiconductor quantum dots. Curr Opin Solid State Mater Sci 6:355–363

Guo S, Wang E (2011) Functional micro/nanostructures: simple synthesis and application in sensors, fuel cells, and gene delivery. Acc Chem Res 44(7):491–500

Haick H, Hurley PT, Hochbaum AI, Yang P, Lewis NS (2006) Electrical characteristics and chemical stability of non-oxidized, methyl-terminated silicon nanowires. J Am Chem Soc 128:8990–8991

Haight R, Sekaric L, Afzali A, Newns D (2009) Controlling the electronic properties of silicon nanowires with functional molecular groups. Nano Lett 9:3165–3170

Hasani M, Coto Garcia AM, Costa-Fernandez JM, Sanz-Medel A (2010) Sol–gels doped with polymer-coated ZnS/CdSe quantum dots for the detection of organic vapors. Sens Actuators B 144:198–202

Hines MA, Guyot-Sionnest P (1996) Synthesis and characterization of strongly luminescing ZnS-capped CdSe nanocrystals. J Phys Chem 100:468–471

Hochbaum AI, Chen R, Delgado D, Liang W, Garnett EC, Najarian M, Majumdar A, Yang P (2008) Enhanced thermoelectric performance of rough silicon nanowires. Nature 451:163–167

Hsu C-M, Connor ST, Tang MX, Cui Y (2008) Wafer-scale silicon nanopillars and nanocones by Langmuir–Blodgett assembly and etching. Appl Phys Lett 93:133109

Huang Z, Fang H, Zhu J (2007) Fabrication of silicon nanowire arrays with controlled diameter, length, and density. Adv Mater 19:744–748

Jaiswal JK, Simon SM (2004) Potentials and pitfalls of fluorescent quantum dots for biological imaging. Trends Cell Biol 14:497–504

Jorge P, Martins MA, Trindade T, Santos JL, Farahi F (2007) Optical fiber sensing using quantum dots. Sensors 7:3489–3534

Korsunska NE, Dybiec M, Zhukov L, Ostapenko S, Zhukov T (2005) Reversible and non-reversible photo-enhanced luminescence in CdSe/ZnS quantum dots. Semicond Sci Technol 20(8):876–881

Law M, Goldberger J, Yang P (2004) Semiconductor nanowires and nanotubes. Annu Rev Mater Res 34:83–122

Liu HI, Biegelsen DK, Johnson NM, Ponce FA, Peace RFW (1993) Self-limiting oxidation of Si nanowires. J Vac Sci Technol B 11(6):2532–2537

Liu HI, Biegelsen DK, Ponce FA, Johnson NM, Pease RFW (1994) Self limiting oxidation for fabricating sub 5 nm silicon nanowires. Appl Phys Lett 64:1383–1385

Ma DDD, Lee CS, Lee ST (2001) Scanning tunneling microscopic study of boron-doped silicon nanowires. Appl Phys Lett 79:2468–2470

Ma Q, Cui H, Su X (2009) Highly sensitive gaseous formaldehyde sensor with CdTe quantum dots multilayer films. Biosens Bioelectron 25:839–844

McAlpine MC, Friedman RS, Lieber CM (2005) High-performance nanowire electronics and photonics and nanoscale patterning on flexible plastic substrates. Proc IEEE 93(7):1357–1363

McAlpine MC, Ahmad H, Wang D, Heath JR (2007) Highly ordered nanowire array on plastic substrates for ultrasensitive flexible chemical sensors. Nat Mater 6:379–384

Mohanta D, Nath SS, Mishara NC, Choudhury A (2003) Irradiation induced gain growth and surface emission enhancement of ZnS:Mn/PVOH semiconductor nano particles by Cl^{+9} ion impact. Bull Mater Sci 26:289–294

Morton KJ, Nieberg G, Bai S, Chou SY (2008) Wafer-scale patterning of sub-40 nm diameter and high aspect ratio (>50:1) silicon pillar arrays by nanoimprint and etching. Nanotechnology 19:345301

Nazzal AY, Qu L, Peng X, Min XM (2003) Photoactivated CdSe nanocrystals as nanosensors for gases. Nano Lett 3(6):819–822

Nolan M, O'Callaghan S, Fagas G, Greer JC, Frauenheim T (2007) Silicon nanowire band gap modification. Nano Lett 34:34–38

Norhayati AB, Aidhia R, Akrajas AU, Muhamad MS, Yahaya M (2010) Fluorescence gas sensor using CdTe quantum dots film to detect volatile organic compounds. Mater Sci Forum 663–665:276–279

Offermans P, Crego-Calama M, Brongersma SH (2010) Gas detection with vertical InAs nanowire arrays. Nano Lett 10:2412–2415

Paska Y, Stelzner T, Christiansen S, Haick H (2011) Enhanced sensing of nonpolar volatile organic compounds by silicon nanowire field effect transistors. ACS Nano 5(7):5620–5626

Peng K-Q, Yan Y-J, Gao S-P, Zhu J (2002) Synthesis of large-area silicon nanowire arrays via self-assembling nanoelectrochemistry. Adv Mater 14:1164–1167

Potyrailo RA, Leach AM (2006) Selective gas nanosensors with multisize CdSe nanocrystal/polymer composite films and dynamic pattern recognition. Appl Phys Lett 88(13):134110

Pugh-Thomas D, Walsh BM, Gupta MC (2011) CdSe(ZnS) nanocomposite luminescent high temperature sensor. Nanotechnology 22:185503

Rogash AL, Kornowski A, Gao M, Eychmuller A, Weller H (1999) Synthesis and characterization of a size series of extremely small thiol-stabilized CdSe nanocrystals. J Phys Chem B 103:3065–3069

Sadeghian RB, Islam MS (2011) Ultralow-voltage field-ionization discharge on whiskered silicon nanowires for gas-sensing applications. Nat Mater 10:135–140

Saren AA, Kuznetsov SN, Kuznetsov AS, Gurtov VA (2011) Excitonic chemiluminescence in Si and CdSe nanocrystals induced by their interaction with ozone. Chemphyschem 12(4):846–853

Schmidt V, Wittemann JV, Senz S, Gosele U (2009) Silicon nanowires: a review on aspects of their growth and their electrical properties. Adv Mater 21:2681–2702

Schmidt V, Wittemann JV, Gosele U (2010) Growth, thermodynamics, and electrical properties of silicon nanowires. Chem Rev 110:361–388

Selvan ST, Tan TTY, Dong Kee Yi DK, Jana NR (2010) Functional and multifunctional nanoparticles for bioimaging and biosensing. Langmuir 26(14):11631–11641

Singh N, Buddharaju KD, Manhas SK, Agarwal A, Rustagi SC, Lo GC, Balasubramanian N, Kwong D-L (2008) Si, SiGe nanowire devices by top–down technology and their applications. IEEE Trans Electron Devices 55:3107–3118

Skucha K, Fan Z, Jeon K, Javey A, Boser B (2010) Palladium/silicon nanowire Schottky barrier-based hydrogen sensors. Sens Actuators B 145:232–238

Smith AM, Nie S (2010) Semiconductor nanocrystals: structure, properties, and band gap engineering. Acc Chem Res 43(2):190–200

Suk SD, Lee S-Y, Kim S-M, Yoon E-J, Kim M-S, Li M, Oh CW, Yeo KH, Kim SH, Shin D-S, Lee K-H, Park HS, Han JN, Park CJ, Park J-B, Kim D-W, Park D, Ryu B-I (2005) High performance 5 nm radius twin silicon nanowire MOSFET (TSNWFET): fabrication on bulk Si wafer, characteristics, and reliability. In: Proceedings of IEEE international electron devices meeting, 5–7 Dec 2005. IEDM Technology Digest, IEEE, Washington, DC, pp. 717–720.

Suk SD, Yeo KH, Cho KH, Li M, Yeoh YY, Lee S-Y, Kim SM, Yoon EJ, Kim MS, Oh CW, Kim SH, Kim D-W, Park D (2008) High-performance twin silicon nanowire MOSFET (TSNWFET) on bulk Si wafer. IEEE Trans Nanotechnol 7:181–184

Tang YH, Sun XH, Au FCK, Liao LS, Peng HY, Lee CS, Lee ST, Sham TK (2001) Microstructure and field-emission characteristics of boron-doped Si nanoparticle chains. Appl Phys Lett 79:1673–1675

Tang YH, Sham TK, Jurgensen A, Hu YF, Lee CS, Lee ST (2002) Phosphorus-doped silicon nanowires studied by near edge x-ray absorption fine structure spectroscopy. Appl Phys Lett 80:3709–3711

Teo BK, Sun XH (2007) Silicon-based low-dimensional nanomaterials and nanodevices. Chem Rev 107:1454–1532

Thanh NTK, Green LAW (2010) Functionalisation of nanoparticles for biomedical applications. Nano Today 5:213–230

Tisch U, Haick H (2010) Nanomaterials for cross-reactive sensor arrays. MRS Bull 35:797–803

Van Sark WGJHM, Frederix PLTM, Van den Heuvel DJ, Gerritsen HC, Bol AA, van Lingen JNJ, de Donega C, Meijerink A (2001) Photooxidation and photobleaching of single CdSe/ZnS quantum dots probed by room-temperature time-resolved spectroscopy. J Phys Chem B 105:8281–8284

Vassiltsova OV, Zhao Z, Petrukhina MA, Carpenter MA (2007) Surface-functionalized CdSe quantum dots for the detection of hydrocarbons. Sens Actuators B 123:522–529

Wan J, Deng SR, Yang R, Shu Z, Lu BR, Xie SQ, Chen Y, Huq E, Liu R, Qu XP (2009) Silicon nanowire sensor for gas detection fabricated by nanoimprint on SU8/SiO_2/PMMA trilayer. Microelectron Eng 86:1238–1242

Wang XU, Chen X, Xie Z-X, Wang X-R (2008) Reversible optical sensor strip for oxygen. Angew Chem Int Ed 47:7450–7453

Wang X, Xie Z, Huang H, Liu Z, Chen D, Shen G (2012) Gas sensors, thermistor and photodetector based on ZnS nanowires. J Mater Chem 22:6845–6850

Weidemann O, Kandaswamy PK, Monroy E, Jegert G, Stutzmann M, Eickhoff M (2009) GaN quantum dots as optical transducers for chemical sensors. Appl Phys Lett 94:113108

Wright JS, Lim W, Norton DP, Pearton SJ, Ren F, Johnson JL, Ural A (2010) Nitride and oxide semiconductor nano-structured hydrogen gas sensors. Semicond Sci Technol 25:024002

Xie R, Kolb U, Li J, Basché T, Mews A (2005) Synthesis and characterization of highly luminescent CdSe-Core CdS/$Zn_{0.5}Cd_{0.5}S$/ZnS multishell nanocrystals. J Am Chem Soc 127:7480–7488

Xu J, Zhang W, Yang Z, Ding S, Zeng C, Chen L, Wang Q, Yang S (2009) Large-scale synthesis of long crystalline $Cu_{2-x}Se$ nanowire bundles by water-evaporation-induced self-assembly and their application in gas sensing. Adv Funct Mater 19:1759–1766

Xu H, Wu J, Chen C-H, Zhang L, Yang K-L (2010) Detecting hydrogen sulfide by using transparent polymer with embedded CdSe/CdS quantum dots. Sens Actuators B 143:535–538

Yu JY, Chung SW, Heath JR (2000) Silicon nanowires: preparation, device fabrication, and transport properties. J Phys Chem B 104:11864–11870

Zakharov ND, Werner P, Gerth G, Schubert L, Sokolov L, Gosele U (2006) Growth phenomena of Si and Si/Ge nanowires on Si (1 1 1) by molecular beam epitaxy. J Cryst Growth 290:6–10

Zhang YF, Tang YH, Peng HY, Wang N, Lee CS, Bello I, Lee ST (1999) Diameter modification of silicon nanowires by ambient gas. Appl Phys Lett 75:1842–1844

Zhelev Z, Jose R, Nagase T, Ohba H, Bakalova R, Ishikawa M, Baba Y (2004) Enhancement of the photoluminescence of CdSe quantum dots during long-term UV-irradiation: privilege or fault in life science research? J Photochem Photobiol B 75(1–2):99–105

Zheng GF, Lu W, Jin S, Lieber CM (2004) Synthesis and fabrication of high-performance *n*-type silicon nanowire transistors. Adv Mater 16:1890–1893

Zhou GW, Li H, Sun HP, Yu DP, Wang YQ, Huang XJ, Chen LQ, Zhang Z (1999) Controlled Li doping of Si nanowires by electrochemical insertion method. Appl Phys Lett 75:2447–2449

Zhou XT, Hu JQ, Li CP, Ma DDD, Lee CS, Lee ST (2003) Silicon nanowires as chemical sensors. Chem Phys Lett 369:220–224

Zschech D, Kim DH, Milenin AP, Scholz R, Hillebrand R, Hawker CJ, Russell TP, Steinhart M, Gosele U (2007) Ordered arrays of <100>-oriented silicon nanorods by CMOS-compatible block copolymer lithography. Nano Lett 7:1516–1520

Part II
Other Trends in Design of Gas Sensor Materials

Chapter 6
Photonic Crystals

6.1 Photonic Crystals in Gas Sensors

The application of photonic crystals (PhCs) is a new direction in gas sensor design (Lambrecht et al. 2007; Sünner et al. 2008; Srivastava et al. 2011; Zhao et al. 2011). The first concept of PhCs was proposed by Yablonovitch (1987) and John (1987) in 1987. Essentially, PhCs contain regularly repeating internal regions of high and low dielectric constant. In particular, as shown in Fig. 6.1, PhCs can consist of periodic arrangements of dielectric materials with different refractive indexes, which can be divided into 1D PhCs, 2D PhCs, and 3D PhCs according to the structures.

Compared with 1D PhCs (e.g., gratings) and 3D PhCs (e.g., opals), 2D PhCs are easier to fabricate and cover a wider research value. In general, there are two kinds of 2D PhCs: one is the air-hole type PhC and the other is the pillar-column type PhC. The fabrication of 2D air-hole type PhCs with triangular lattice structure is simple, and it has the largest PhC bandgap (Jamois et al. 2002). In particular, these structures can consist of a slab of material (such as silicon) which can be patterned using techniques borrowed from the semiconductor industry. Such chips offer the potential to combine photonic processing with electronic processing on a single chip. Therefore, 2D PhCs attract widespread interest and are the most frequently used structures (Jamois et al. 2002; Beiu and Beiu 2008; García-Rupérez et al. 2010).

Imprinting in polymers is one of the approaches widely used for fabrication of 2D PhC sensors (Boersma et al. 2011). Photonic crystal fiber (see below) is another approach to design such 2D structures. The fabrication of these fibers using microstructured polymeric materials (MPOF) was also reported by Van Eijkelenborg et al. (2003). For 3D PhCs, various techniques have been used including photolithography and etching techniques similar to those used for integrated circuits. Some of these techniques are already commercially available. To circumvent nanotechnological methods with their complex machinery, alternative approaches have been followed to grow PhCs as self-assembled structures from colloidal crystals. An example of a PhC prepared using this approach is shown in Fig. 6.2. Mass-scale 3D PhC films and fibers can now be produced using a shear-assembly technique which stacks 200–300-nm colloidal polymer spheres into perfect films of fcc lattice.

Research has shown that PhCs possess unique properties such as a photonic bandgap (PBG) (Cheng et al. 2009), photonic localization (Maloshtan and Kilin 2007), slow light devices based on PhCs (Adachi et al. 2010), and so on (Zhao et al. 2011). It was established that the propagation of photons in PhCs is similar to the propagation of electrons in conducting crystals. In analogy with the electronic structure, the PhC presents a periodic potential to a photon propagating through it, resulting in the photonic band. In other words, the incident light whose wavelength lies within the PBG cannot propagate through the PhC region, and the transmission spectrum exhibits a wide bandgap. It can be concluded that the constituent of the crystals and the geometry of the lattice are two key factors for

G. Korotcenkov, *Handbook of Gas Sensor Materials*, Integrated Analytical Systems,
DOI 10.1007/978-1-4614-7388-6_6, © Springer Science+Business Media New York 2014

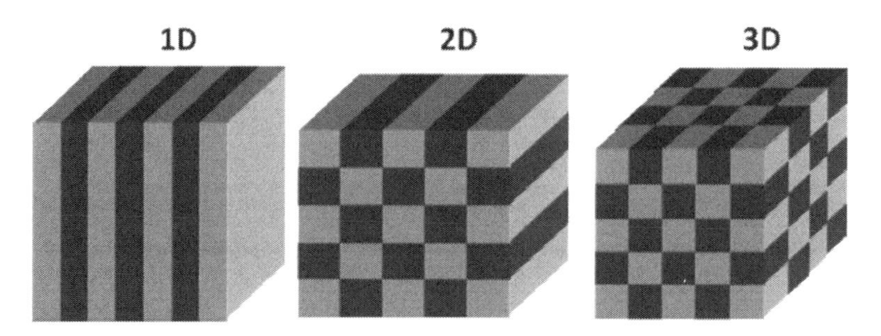

Fig. 6.1 Schematic illustration of 1, 2, and 3D PhCs (Reprinted with permission from Zhao et al. (2011). Copyright 2011 Elsevier)

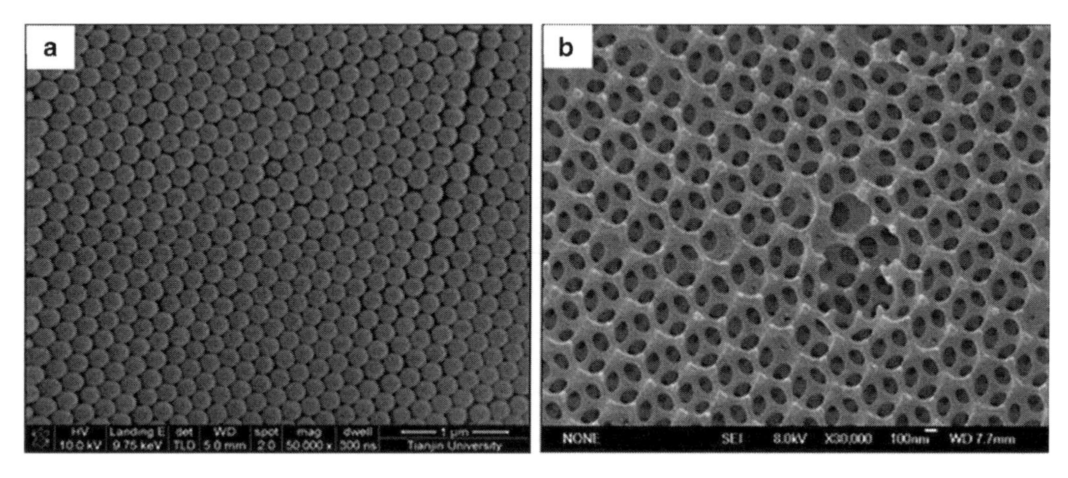

Fig. 6.2 SEM images of (**a**) the PhC template and (**b**) the fabricated polyacrylamide-based PhCs (PAM-imprinted PhCs). *Scale bars*: (**a**) 1 mm, (**b**) 100 nm (Reprinted with permission from Yuan et al. (2012). Copyright 2012 Wiley)

the PBG. Thus, PhCs with PBGs could prevent light from propagating in certain directions with specified frequencies (Cheng et al. 2009). Nowadays, many methods for the numerical simulation of the electromagnetic field propagation in PhCs have been developed, such as plane wave expansion (PWE) (Sakoda 2001), finite-difference time-domain algorithm (FDTD) (Kosmidou et al. 2005), finite element method (FEM) (Fujisawa and Koshiba 2006), and transfer-matrix method (TMM) (Pendry 1994). Experimental and theoretical studies have shown that the position and shape for PBG will be changed by any perturbation of the parameter to be measured; thus, it can be used in a sensor if this corresponding relationship is set up (Xiao et al. 2007). However, it is necessary to take into account that since the basic physical phenomenon is based on diffraction, the periodicity of the PhC structure has to be of the same length scale as half the wavelength of the electromagnetic waves, i.e., ~350 nm (blue) to 700 nm (red) for PhCs operating in the visible part of the spectrum—the repeating regions of high and low dielectric constants have to be of this dimension. This makes the fabrication of optical PhCs cumbersome and complex.

Another feature of the PhC is photonic localization (Maloshtan and Kilin 2007). By introducing certain defects into PhCs, their existing periodicity or symmetry will be destroyed. When this happens, a rather narrow defective state occurs within the PBG, and corresponding photons would be confined to defect. In other words, by locally breaking the period into PhC, one can introduce a photonic defect mode within the bandgap and, as a result, the transmission spectrum has a relatively sharp transmission peak (see Fig. 6.3b). For example, a PhC microcavity can be formed by removing a

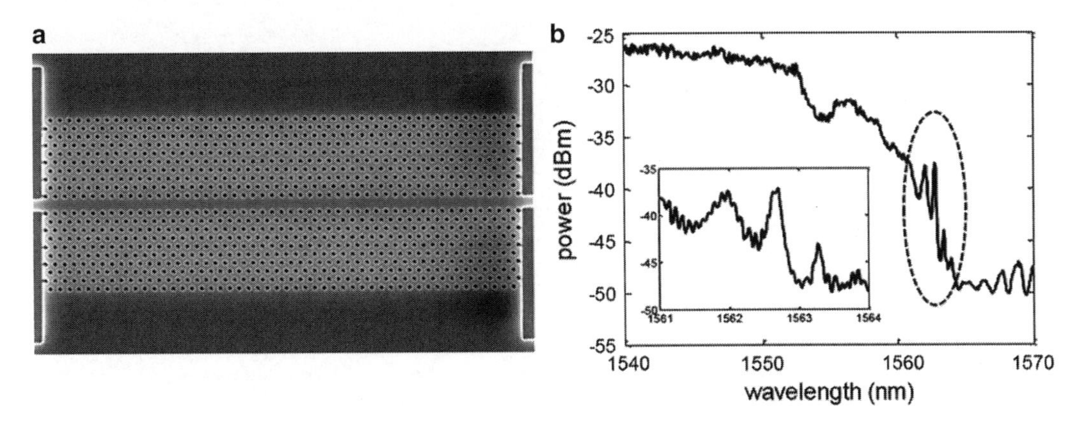

Fig. 6.3 (**a**) SEM image of the fabricated SOI (silicon-on-insulator) planar photonic crystal waveguide, where a row of holes is removed in the Γ–K direction (W1-type). (**b**) Spectrum of the photonic crystal waveguides (PCW) in the region of the band edge when having deionized water (DIW) as upper cladding. Transmission fringes at the band edge are marked with *dashed red line* and enlarged in the *inset* (Reprinted with permission from García-Rupérez et al. 2010. Copyright 2010 The Optical Society of America)

single hole, thus forming an energy well for photons similar to that for electrons in a quantum wire structure (Painter et al. 1999). The spectral position of the center of the transmission peak is highly sensitive to changes in the local environmental conditions. If gas molecules become bound to the defect, the local environmental condition, such as the refractive index, changes. Hence, it can be used as the sensing transduction signal. The PhC line defect is analogous to a waveguide—the light can only travel through it along the defect direction (see Fig. 6.3a). In another case, light will be localized on the defect plane; thus the photonic plane defect is just like a perfect mirror. Properties of the defective state would be changed if a certain analyte is infiltrated into the defect position, so the refractive index or other parameters of analyte can be obtained by observing the variation of resonant wavelength (Wang et al. 2008a). Photons can also be localized vertically by total internal reflection (TIR) at the air-slab interface. The combination of Bragg reflection from the 2D PhC and TIR from the low-index cladding (air) results in a three-dimensionally confined optical mode. Because the light confinement provided by the PBG is very strong, and because it is easily possible to adjust the defect mode wavelength across the PBG by finely tuning the structural parameters, PhC sensors based on photonic localization have received huge attention.

The possibility of reducing in PhCs the velocity of light whose wavelength matches the absorption peak of analyte is another important property of PhCs. When the PhC slow light structure is introduced into the sample cell, the interaction of light and matter will be enhanced, so the absorption coefficient can be greatly increased, and this technology offers a potential for the realization of small and high sensitivity sensors (Lambrecht et al. 2007).

Apart from the above-mentioned sensing properties of PhC, some other properties of PhCs can also be used for sensing applications. The self-collimation (SC) effect of PhCs can be applied to an interferometer, whose transmission spectra can reflect the surrounding parameters (Wang et al. 2008b). The SC effect can steer a light beam with almost no diffraction in PhCs (Kosaka et al. 1999). If rubber material such as PMMA is used to make a colloidal PhC, the structural color will vary according to different analytes, so the measuring parameters can be obtained with the naked eye. The mechanism of such influence is shown in Fig. 6.4. Moreover, the surface electromagnetic waves of 1D PhCs are strongly sensitive to surface modifications, which can also be used for monitoring analytes (Descrovi et al. 2007).

PhCs can be based on various materials (mostly dielectric), including semiconductors (Si, GaAs, Ge, SiC, etc.), polymers (PMMA, polyacrylamide (PAM), poly(2-hydroxyethyl methacrylate) (PHEMA), etc.), and metal oxides (SiO_2, TiO_2, Al_2O_3, WO_3).

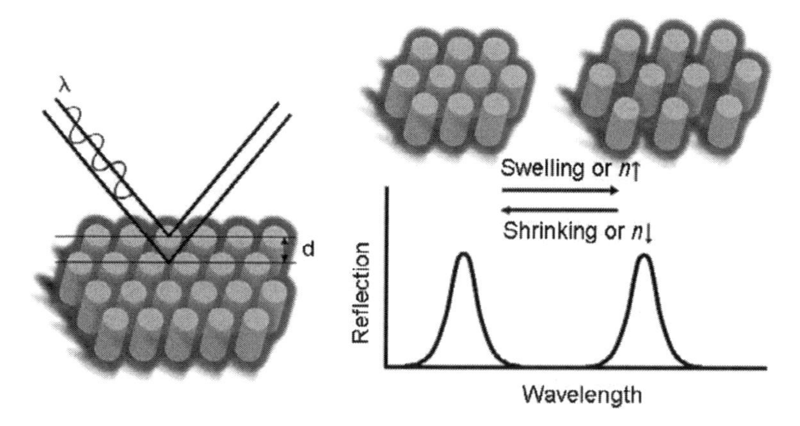

Fig. 6.4 Principle of a PhC gas sensor: the wavelength of the reflected light changes upon swelling/shrinking or when the refractive index changes (Reprinted with permission from Boersma et al. (2011). Copyright 2011 Elsevier)

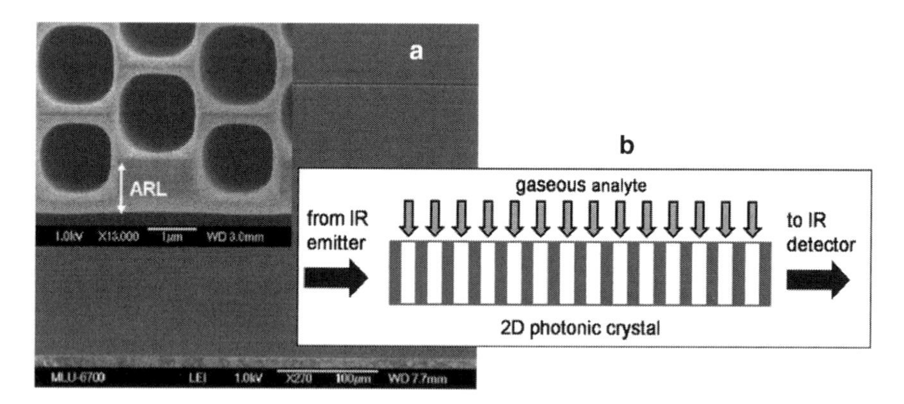

Fig. 6.5 (**a**) SEM picture of a complete 0.25-mm PhC gas cell device. The *inset* shows the coupling layer (ARL) at the interface with the PhC. 2D PhCs were prepared by photoelectrochemical etching of n-type silicon in HF-based solution. (**b**) Schematic diagram of a typical PhC gas sensor. The light impinging from the left side is absorbed by the gas molecules inside the PhC. Due to the reduced group velocity, the interaction path is effectively reduced (Reprinted with permission from Pergande et al. (2011). Copyright 2011 American Institute of Physics)

Experiment has shown that PhCs can be applied in practically all types of optical gas sensors, including PhC gas sensors based on spectroscopic absorption (Lambrecht et al. 2007; Jensen et al. 2008; Pergande et al. 2011), resonance properties (Wang et al. 2007; Sünner et al. 2008; Awad et al. 2010), surface electromagnetic waves (Colodrero et al. 2008; Hidalgo et al. 2009; Hidalgo et al. 2010), and self-collimation effects (Wang et al. 2008b). These sensors can detect a wide range of gases (O_2, CO_2, NO_2, VOCs, N_2, etc.), including the humidity. Pergande et al. (2011) believe that the use of PhCs makes it possible to obtain compact, robust, and low-cost spectroscopic gas sensors. The replacement of the interaction volume in a conventional sensor by a PhC is shown in Fig. 6.5.

However, it should be noted that most research in the area of PhCs is related to the study of PhC fibers (PCFs) (Hoo et al. 2003; Fini 2004; Li et al. 2007; Frazao et al. 2008; Skorobogatiy 2009). PCFs are two-dimensionally periodic PhCs fabricated using fiber draw techniques developed for communications fiber. The PCF has a periodic dielectric structure whose periodicity is of the order of a wavelength, giving rise to the PBG. In addition, PBG fibers show promise as a viable technology for the mass production of highly integrated and intelligent sensors in a single manufacturing step (Skorobogatiy 2009). In particular, PCFs are commercially available.

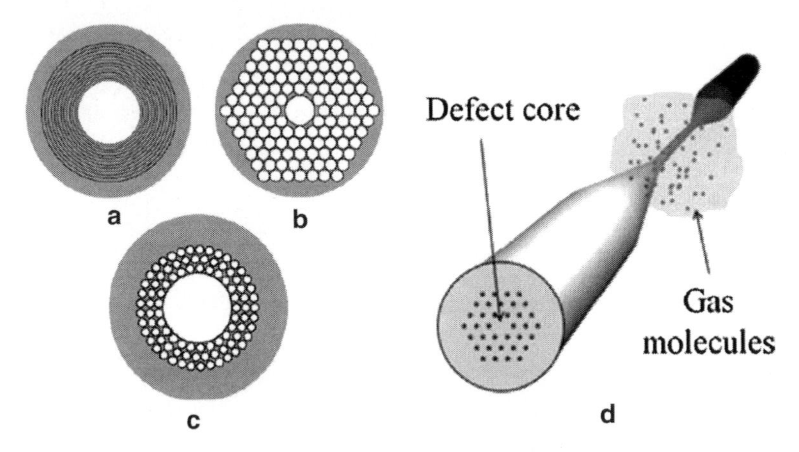

Fig. 6.6 Various types of hollow-core photonic bandgap fibers: (**a**) Bragg fiber featuring large hollow core surrounded by a periodic sequence of high- and low-refractive index layers; (**b**) photonic crystal fiber featuring a small hollow core surrounded by a periodic array of large air holes; (**c**) microstructured fiber featuring a medium-sized hollow core surrounded by several rings of small air holes separated by nanosize bridges (Reprinted from Skorobogatiy (2009). Published by Hindawi Publishing Corporation). (**d**) Photonic crystal fiber micro-taper structure for use in gas sensing (Reprinted with permission from Lee et al. (2009). Copyright 2009 Elsevier)

According to the cross-sectional distribution of the dielectric function, PCFs can be categorized as follows: PBG fibers (PCFs that utilize the PBG and the defect mode), holey fibers (PCFs with air holes along the axis of light propagation), hole-assisted fibers (PCFs consisting of a conventional higher-index core with air holes), and Bragg fibers (PBG fibers with concentric rings of different refractive index). PBG fibers can contain periodic sequences of micrometer-sized layers of different materials (Fig. 6.6a), periodically arranged micrometer-sized air voids (Fig. 6.6b), or rings of holes separated by nanosupports (Fig. 6.6c). PBG fibers are currently available in silica glass, polymer, and specialty soft glass implementations. An example of a PCF sensor with air holes is also shown in Fig. 6.6d.

It is found that PCFs have many unique characteristics that differ remarkably from those of conventional fibers, such as single-mode propagation over a wide range of wavelengths (Kerbage et al. 2000), sensitive structure manageable dispersion properties (Saitoh and Koshiba 2005), high birefringence (Lou et al. 2004; Antkowiak et al. 2005; Alam et al. 2005), and extra-strong nonlinear effects (Hu et al. 2004). Thus, due to specific properties, PCF essentially enables a substantial increase in design flexibility, making possible new or improved sensing solutions relative to the situation where the choice of components and devices was limited to the standard optical fiber technology. In particular, gas can be used to fill the air holes of PCFs because of the structure of PCFs, which offers the possibility of studying gas sensing by the use of PCFs (Hoo et al. 2002, 2003; Fini 2004; Ritari et al. 2004). This research has several advantages; for example, only a submicroliter of a gas sample is needed to achieve the longer interaction length. In comparison with conventional fibers, PCFs do not need to be stripped of the cladding and the coating, which makes them durable. Moreover, research on gas sensing that uses solid-core hole fiber with random hole distributions in the cladding can be carried out by the evanescent field absorption mechanism (Pickrell et al. 2004). As result, the core hole in the case of PBG PCF brings qualitatively better performances when compared with sensing solutions implemented with the standard fiber (Frazao et al. 2008).

6.2 Problems in the Sensing Application of PhCs

In spite of the obvious merits of PhCs for gas sensor design, we should note that there are disadvantages of PhCs which can limit their application. For example, Zhao et al. (2011) emphasized the following problems.

6.2.1 Problems on the Fabrication of Photonic Crystal

There are a lot of parameters in PhC, such as dielectric constant, dielectric rods, the material of background, and the structure of the crystal lattice. All these parameters affect the transmission characteristics of incident light. For example, Pergande et al. (2011) suggested on the basis of numerical estimation that the overall transmission of bulk PhC was limited by fluctuations of the pore diameter. An attenuation in the transmission of 15 dB/mm would be caused by 1 % pore radius fluctuation. By using perturbation theory, it was estimated that for a transmission above 90 % of a device 1 mm in length, the pore positional variation has to be below 0.3 % and the pore diameter fluctuations have to be below 0.5 %. However, because the lattice size is typically in the micron scale, a lot of difficult micro electric and mechanical processes are absolutely necessary to fabricate PhCs with high performance. This will remain a major challenge in the future. So far, various kinds of fabrication methods have been put forward, which have their own advantages and disadvantages. Precise mechanical processing technology (Özbay et al. 1995) is recognized as the most stable and reliable method to fabricate PhCs, but exceedingly complicated and very expensive, and it failed in some types of smaller-scale, adulterate, and defective PhCs. Lithography technology (Bogaerts et al. 2002) is suitable for making high-quality 3D-ordered PhCs, as hyperfine structures and defects can be easily introduced into 3D PhCs using this technology, but this method is time-consuming and quite expensive; thus, it is only suitable for small-scale fabrication of samples. As compared with the above methods, there exists a simple and economical method based on the self-assembly of microparticles in a colloidal suspension (Takagi et al. 2004), but there still exists the problem of how to introduce few and controlled defects. In addition, there are still many other methods that have been put forward. However, all these methods are in their infancy. With the development of PhC technology and the advance of manufacturing technology, more and better methods will arrive on the scene, which will open up a new breakthrough in the study of PhC sensors.

6.2.2 Problems with Coupling Losses

The transmission light of PCW (photonic crystal waveguides) is the Bloch wave, while the standard single-mode fiber guides light by the mechanism of TIR; different spatial patterns would cause impedance matching. Furthermore, the width of line defect PCW is typically less than 1 μm, while the diameter of the fiber core is 2–8 μm; the resulting mode field distribution would also lead to impedance mismatching. All of these impedance mismatchings would bring coupling losses. Meanwhile, a lower group velocity corresponding to a higher refractive index makes the in-and-out coupling of radiation difficult. For the structure of PCW in Delphine et al. (2008), the transmission efficiency of direct coupling is only 18 %. To avoid the influence of coupling loss on the output of sensors, many coupling structures and techniques, such as interface resonant mode (Barclay et al. 2004), J-coupler structure (Prather et al. 2002), and adiabatic coupling (Mekis and Joannopoulos 2001; Sanchis et al. 2002; Delphine et al. 2008), have been proposed and applied by many researchers. Interface resonant mode (Barclay et al. 2004) can achieve a very high transmission in principle, but the useful bandwidth is limited by the resonance width. The J-coupler mode (Prather et al. 2002) has a transmission of greater than 90 %, but it generally introduced radiation loss, which lowers the coupling efficiency. Adiabatic coupling (Mekis and Joannopoulos 2001; Sanchis et al. 2002; Delphine et al. 2008) predicts near 100 % transmission over a large frequency range provided the change is slow; conversely, it requires very small changes over relatively large length scales, making manufacturing tolerances very strict. Above all, the coupling loss was improved to a certain extent but is not yet able to be avoided completely.

6.2.3 Problems with Signal Detection

While PhC sensors show many unique advantages, there still exist a great many problems in signal detection. Most PhC sensors monitor and control the test samples by real time measuring of the transmittance or reflectance in a broad spectra range (Hasek et al. 2007), which needs a high-resolution optical spectrometer or related instruments and can be easily affected by environment factors, thus restricting the application development of PhC sensors. In addition, the disturbance of pressure (Descrovi et al. 2007; Sünner et al. 2008), temperature (Wild et al. 2004; Sünner et al. 2008), and moisture (Zhang et al. 2009) would also influence the output characteristics of PhC sensors. This would cause cross sensitivity. Some suitable means of signal detection are particularly necessary to overcome these questions.

References

Adachi J, Ishikura N, Sasaki H, Baba T (2010) Wide range tuning of slow light pulse in SOI photonic crystal coupled waveguide via folded chirping. IEEE J Sel Top Quantum Electron 16(1):192–199

Alam MS, Saitoh K, Koshiba M (2005) High group birefringence in air–core photonic bandgap fibers. Opt Lett 30:824–826

Antkowiak M, Kotynski R, Nasilowski T, Lesiak P, Wojcik J, Urbanczyk W, Berghmans F, Thienpont H (2005) Phase and group modal birefringence of triple-defect photonic crystal fibres. J Optic Pure Appl Optic 7:763–766

Awad H, Hasan I, Mnaymneh K, Majid S, Hall TJ, Mnaymneh K, Andonovic I (2010) Wireless enabled multi gas sensor system based on photonic crystals. In: Berghmans F, Mignani AG, van Hoof CA (eds) Optical sensing and detection. Proc. SPIE 7726, 77260K, Brussels

Barclay PE, Srinivasan K, Borselli M, Painter O (2004) Efficient input and output fiber coupling to a photonic crystal waveguide. Opt Lett 29(7):697–699

Beiu RM, Beiu V (2008) Fiber optical mechanical sensor based on a triangular-lattice photonic crystal. In: Proceedings of IEEE Photonics Global, Singapore, Vols. 1–2, pp. 183–186

Boersma A, Bourghoorn M, Saalmik M (2011) Imprinted photonic crystal chemical sensors. Procedia Eng 25:27–30

Bogaerts W, Wiaux V, Taillaert D, Beckx S, Luyssaert B, Bienstman P, Baets R (2002) Fabrication of photonic crystals in silicon-on-insulator using 248-nm deep UV lithography. IEEE J Sel Top Quantum Electron 8(4):928–934

Cheng SC, Wu JN, Yang TJ, Hsieh W-F (2009) Effect of atomic position on the spontaneous emission of a three-level atom in a coherent photonic-band-gap reservoir. Phys Rev A 79(1):013801

Colodrero S, Ocana M, Gonzalez AR, Miguez H (2008) Response of nanoparticle-based one-dimensional photonic crystal to ambient vapor pressure. Langmuir 24(16):9135–9139

Delphine MM, Eric C, Damien B, Guillaume M, Laurent V (2008) Ultracompact tapers for light coupling into two-dimensional slab photonic-crystal waveguides in the slow light regime. Opt Eng 47(1):014602

Descrovi E, Frascella F, Sciacca B, Geobaldo F, Dominici L, Michelotti F (2007) Coupling of surface waves in highly defined one-dimensional porous silicon photonic crystals for gas sensing applications. Appl Phys Lett 91(24):241109

Fini JM (2004) Microstructure fibres for optical sensing in gases and liquids. Meas Sci Technol 15:1120–1128

Frazao O, Santos JL, Araujo FM, Ferreira LA (2008) Optical sensing with photonic crystal fibers. Laser Photon Rev 2(6):449–459

Fujisawa T, Koshiba M (2006) Analysis of photonic crystal waveguide gratings with coupling-mode theory and finite-element method. Appl Opt 45(17):4114–4121

García-Rupérez J, Toccafondo V, Banuls MJ, Castelló JG, Griol A, Peransi-Llopis S, Maquieira A (2010) Label-free antibody detection using band edge fringes in SOI planar photonic crystal waveguides in the slow light regime. Opt Express 18(23):24276–24286

Hasek T, Kurt H, Citrin DS, Koch M (2007) A fluid sensor based on a sub-terahertz photonic crystal waveguide. In: Adibi A, Lin S-Y, Scherer A (eds) Photonic crystal materials and devices. Proceedings of SPIE 6480, 64801I, San Jose, CA

Hidalgo N, Calvo ME, Miguez H (2009) Mesostructured thin films as responsive optical coatings of photonic crystals. Small 5(20):2309–2315

Hidalgo N, Calvo ME, Colodrero S, Miguez H (2010) Porous one-dimensional photonic crystal coatings for gas detection. IEEE Sensors J 10(7):1206–1212

Hoo YL, Jin W, Ho HL, Wang DN, Windeler RS (2002) Evanescent-wave gas sensing using microstructure fiber. Opt Eng 41:8–9

Hoo YL, Jin W, Shi C, Ho HL, Wang DN, Ruan SC (2003) Design and modeling of a photonic crystal fiber gas sensor. Appl Opt 42:3509–3515

Hu M, Wang C-Y, Li Y, Chai L, Kondrat'ev YN, Sibilia C, Zheltikov AM (2004) An anti-Stokes-shifted doublet of guided modes in a photonic-crystal fiber selectively generated and controlled with orthogonal polarizations of the pump field. Appl Phys B Laser Optic 79:805–809

Jamois C, Wehrspohn R, Schilling J, Muller F, Hillebrand R, Hergert W (2002) Silicon-based PhC slabs: two concepts. IEEE J Sel Top Quantum Electron 38(7):805–810

Jensen KH, Alam MN, Scherer B, Lambrecht A, Mortensen NA (2008) Slow-light enhanced light-matter interactions with applications to gas sensing. Opt Commun 281(21):5335–5339

John S (1987) Strong localization of photons in certain disordered physics dielectric superlattices. Phys Rev Lett 58(23):2486–2489

Kerbage C, Eggleton B, Westbrook P, Windeler R (2000) Experimental and scalar beam propagation analysis of an air–silica microstructure fiber. Opt Express 7:113–122

Kosaka H, Kawashima A, Tomita A, Notomi M, Tamamura T, Sato T, Kawakami S (1999) Self-collimating phenomena in photonic crystals. Appl Phys Lett 74:1212–1214

Kosmidou EP, Kriezis EE, Tsiboukis TD (2005) FDTD analysis of photonic crystal defect layers filled with liquid crystals. Opt Quantum Electron 37(1):149–160

Lambrecht A, Hartwig S, Schweizer SL, Wehrspohn RB (2007) Miniature infrared gas sensors using photonic crystals. Proc SPIE 6480:64800D

Lee B, Roh S, Park J (2009) Current status of micro- and nano-structured optical fiber sensors. Opt Fiber Technol 15:209–221

Li S-G, Liu S-Y, Song Z-Y, Han Y, Cheng T-L, Zhou G-Y, Hou L-T (2007) Study of the sensitivity of gas sensing by use of index-guiding photonic crystal fibers. Appl Optics 46(22):5183–5188

Lou S-Q, Wang Z, Ren G-B, Jian S-S (2004) Propagation properties of an index guiding high birefringence fibre. Chin Phys 13:1493–1499

Maloshtan AS, Kilin SY (2007) Dynamic control of light localization in photonic crystals. Opt Spectroscopy 103(3):354–359

Mekis A, Joannopoulos JD (2001) Tapered couplers for efficient interfacing between dielectric and photonic crystal waveguides. J Lightwave Technol 19(6):861–865

Özbay E, Tuttle G, Sigalas M, Soukoulis CM, Ho KM (1995) Defect structure in layer-by-layer photonic band gap crystal. Phys Lett B 51:13961–13965

Painter O, Lee RK, Scherer A, Yariv A, O'Brien JD, Dapkus PD, Kim I (1999) Two-dimensional photonic band-gap defect mode laser. Science 284(5421):1819–1821

Pendry JB (1994) Photonic band structures. J Mod Opt 41(2):202–229

Pergande D, Geppert TM, Rhein AV, Schweizer SL, Wehrspohn RB, Moretton S, Lambrecht A (2011) Miniature infrared gas sensors using photonic crystals. J Appl Phys 109(8):083117

Pickrell G, Peng W, Wang A (2004) Random-hole optical fiber evanescent-wave gas sensing. Opt Lett 29:1476–1478

Prather DW, Murakowski J, Shi S, Venkataraman S, Sharkawy A, Chen C, Pustai D (2002) High-efficiency coupling structure for a single-line-defect photonic-crystal waveguide. Opt Lett 27(18):1601–1603

Ritari T, Tuominen J, Ludvigsen H, Petersen JC, Sorensen T, Hansen TP, Simonsen HR (2004) Gas sensing using air-guiding photonic bandgap fibers. Opt Express 12:4080–4087

Saitoh K, Koshiba M (2005) Empirical relations for simple design of photonic crystal fibers. Opt Express 13:267–274

Sakoda K (2001) Optical properties of photonic crystals. Springer, Berlin

Sanchis P, Marti J, Blasco J, Martinez A, Garcia A (2002) Mode matching technique for highly efficient coupling between dielectric waveguides and planar photonic crystal circuits. Opt Express 10(24):1391–1397

Skorobogatiy M (2009) Microstructured and photonic bandgap fibers for applications in the resonant bio- and chemical sensors. J Sensors 2009:524237

Srivastava T, Das R, Jha R (2011) Highly accurate and sensitive surface plasmon resonance sensor based on channel photonic crystal waveguides. Sens Actuators B Chem 157:246–252

Sünner T, Stichel T, Kwon SH, Schlereth TW, Höfling S, Kamp M, Forchel A (2008) Photonic crystal cavity based gas sensor. Appl Phys Lett 92(26):261112

Takagi K, Seno K, Kawasaki A (2004) Fabrication of a three-dimensional terahertz photonic crystal using monosized spherical particles. Appl Phys Lett 85(17):3681–3683

Van Eijkelenborg MA, Argyros A, Barton G, Bassett IM, Fellew MG, Henry G, Issa NA, Large MCJ, Manos S, Padden W, Poladian L, Zagari J (2003) Recent progress in microstructured polymer optical fiber fabrication and characterization. Opt Fiber Technol 9:199–209

Wang S-W, Chen X, Lu W, Li M, Wang H (2007) Fractal independently tunable multichannel filters. Appl Phys Lett 90:211113

Wang XL, Xu ZF, Lu NG, Zhu J, Jin F (2008a) Ultracompact refractive index sensor based on microcavity in the sandwiched photonic crystal waveguide structure. Opt Commun 281:1725–1731

Wang ZY, Han K, Shen XP (2008b) Subminiature gas sensor based on the photonic crystals. In: Proceedings of IEEE nanoelectronics conference, INEC 2008, 24–27 March, Pudong, Shanghai, pp. 303–306

Wild B, Ferrini R, Houdré R (2004) Temperature tuning of the optical properties of planar photonic crystal microcavities. Appl Phys Lett 84(6):846–848

Xiao SS, Pedersen J, Mortensen NA (2007) Liquid-infiltrated photonic crystals for lab-on-a-chip applications. Proc SPIE 6645:66451L

Yablonovitch E (1987) Inhibited spontaneous emission in solid-state physics and electronics. Phys Rev Lett 58(20):2059–2062

Yuan Y, Li Z, Liu Y, Gao J, Pan Z, Liu Y (2012) Hydrogel photonic sensor for the detection of 3-pyridinecarboxamide. Chem Eur J 18:303–309

Zhang WG, Yan J, Wang G, Li H-X, Zhang G-S (2009) A natural humidity sensitive two dimensional tunable photonic band gap material and its optic properties. J Inorg Mater 24(1):57–60

Zhao Y, Zhang Y-N, Wang Q (2011) Research advances of photonic crystal gas and liquid sensors. Sens Actuators B Chem 160:1288–1297

Chapter 7
Ionic Liquids in Gas Sensors

No commercial gas sensors use ionic liquids (ILs), but the research into their use is gaining momentum rapidly as of this writing (Welton 1999; Buzzeo et al. 2004d; Silvester and Compton 2006; Ahmad 2009; Sun and Armstrong 2010; Silvester 2011; Singh et al. 2012). Room-temperature ionic liquids (RTILs) are a unique class of compounds containing organic cations and anions, which melt at or close to room temperature, and thus are known as room-temperature molten salts (Seddon 1997; Demus et al. 1998). The simplest explanation that can be given for this circumstance focuses on the difficulty of finding efficient packing modes for the more complex and size-mismatched ions characteristic of IL and inorganic salt hydrates. The big difference in the size of a bulky organic cation and a small organic or inorganic anion does not allow packing of lattice, which happens in many inorganic salts; instead, the ions are disorganized. The result of this is that some of these salts remain liquid at the room temperature. The solvent properties of ILs such as melting point, dielectric constant, viscosity, polarity, and water miscibility can be tailored by combining different cations with suitable anions (Demus et al. 1998; Wasserschied and Welton 2003). The influence of chloride, water, and organic solvents on the physical properties of ILs has been investigated by Seddon et al. (2000).

The cations in RTILs are generally organic compounds with asymmetrically substituted N-containing cation that are bulky in nature with varying heteroatom functionality paired with charged diffused anion. The class of cations explored till now includes 1-allyl-3-methylimidazolium, *N*-alkylpyridinium, *N*-methylalkyl pyrrolidinium, pyrazolium, and tetraalkyl ammonium types; more importantly, phosphonium salts are also finding greater utility. Regarding anions acceptable for synthesis of ILs, it was found that a wide range of organic and inorganic species can be used for these purposes. In particular, anions ranging from simple halides, which generally inflect high melting points, to inorganic anions such as tetrafluoroborate, $[BF_4]^-$, and hexafluorophosphate, $[PF_6]^-$, and to large organic anions like bistriflimide, $[(CF_3SO_2)_2N]^-$, triflate, $[CF_3SO_3]^-$, or tosylate may be employed (Ahmad 2009). Commonly used cations and anions are shown in Fig. 7.1. Typical abbreviations used for ILs are listed in Table 7.1. It was established that the change of anion dramatically affects the chemical behavior and stability of the ionic liquid; the change of cation has a profound effect on the physical properties, such as melting point, viscosity, and density (Bowlas et al. 1996). There are also many interesting uses of ionic liquids with simple non-halogenated organic anions such as formate, alkylsulfate, alkylphosphate, or glycolate. The melting point of 1-butyl-3-methylimidazolium tetrafluoroborate with an imidazole skeleton is about −80 °C, and it is a colorless liquid with high viscosity at room temperature (Earle and Seddon 2000; Ahmad 2009).

Ionic liquids possess some unique properties (Seddon 1997; Demus et al. 1998; Welton 1999; Earle and Seddon 2000; Pinkert et al. 2009; Silvester 2011; Singh et al. 2012) and can be classified as a special category of nonaqueous electrolytes as well as serve as a gas-permeable membrane or a solvent. It was found that a diverse range of organic, inorganic, and organometallic compounds are

G. Korotcenkov, *Handbook of Gas Sensor Materials*, Integrated Analytical Systems,
DOI 10.1007/978-1-4614-7388-6_7, © Springer Science+Business Media New York 2014

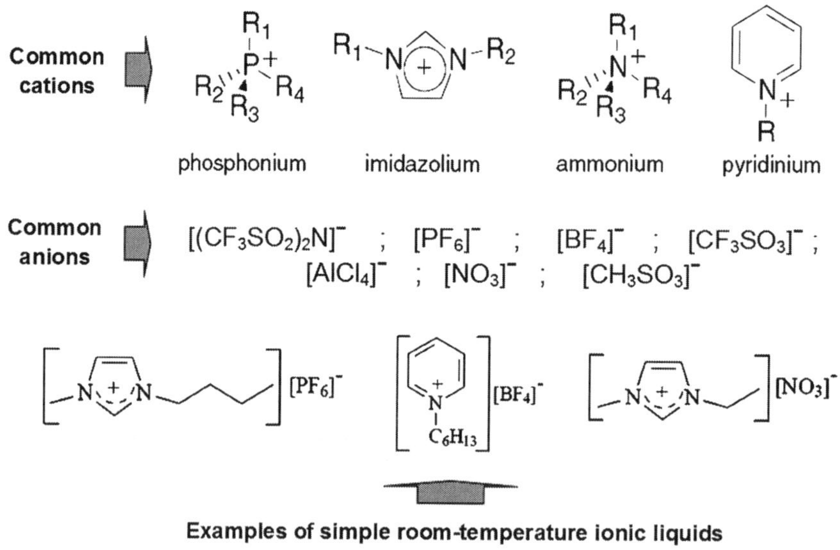

Common cations: phosphonium, imidazolium, ammonium, pyridinium

Common anions: $[(CF_3SO_2)_2N]^-$; $[PF_6]^-$; $[BF_4]^-$; $[CF_3SO_3]^-$; $[AlCl_4]^-$; $[NO_3]^-$; $[CH_3SO_3]^-$

Examples of simple room-temperature ionic liquids

Fig. 7.1 Commonly used cations and anions for ionic liquids

Table 7.1 List of some abbreviations used for ILs

Abbreviation	Name
[BEIm]	1-Butyl-3-ethylimidazolium
[BMIm]BF_4	1-Butyl-3-methylimidazolium tetrafluoroborate
[BMIm]PF_6	1-Butyl-3-methylimidazolium hexafluorophosphate
[BMIm]Tf_2N/NTf_2	1-Butyl-3-methylimidazolium bis(trifluoromethylsulfonyl)imide
[BMP]Tf_2N/NTf_2	1-Butyl-1-methylpyrrolidinium bis(trifluoromethylsulfonyl)imide
[BMMIm]Tf_2N/NTf_2	1-Butyl-2,3-dimethylimidazolium bis(trifluoromethylsulfonyl)imide
[BMIm]TfO	1-Butyl-3-methylimidazolium trifluoromethanesulfonate
[EMIm]BF_4	1-Ethyl-3-methylimidazoliumtetrafluoroborate
[EMIm]Cl	1-Ethyl-3-methylimidazolium chloride
[EMIm]PF_6	1-Ethyl-3-methylimidazolium hexafluorophosphate
[OMIm]Tf_2N/NTf_2	1-Octyl-3-methylimidazolium bis(trifluoromethylsulfonyl)imide
[PMP]	1-Propyl-1-methylpyrrolidinium
TFA	Trifluoroacetate
Tf_2N/NTf_2	Bis(trifluoromethylsulfonyl)imide
TfO	Trifluoromethanesulfonate
BF_4	Tetrafluoroborate
PF_6	Hexafluorophosphate

Source: Data from Singh et al. (2012). Published by Hindawi Publishing Corporation

soluble in ILs. Thus, ILs are nonaqueous polar alternatives for phase transfer processes. The solubility of gases such as O_2, benzene, nitrous oxide, ethylene, ethane, and carbon monoxide is also good, which makes them attractive solvent systems for catalytic hydrogenations, carbonylations, hydroformylation, and aerobic oxidations. ILs can be liquid over a range of 300 °C. This wide liquid range is a distinct advantage over traditional solvent system that has a much narrower liquid range; for example, water has a liquid range of 100 °C or toluene 206 °C. Moreover, ILs are immiscible with many organic solvents. In addition, the solvent properties of ILs can be tuned for a specific application by varying the anion–cation combinations. The high boiling point and thermal stability can combine the

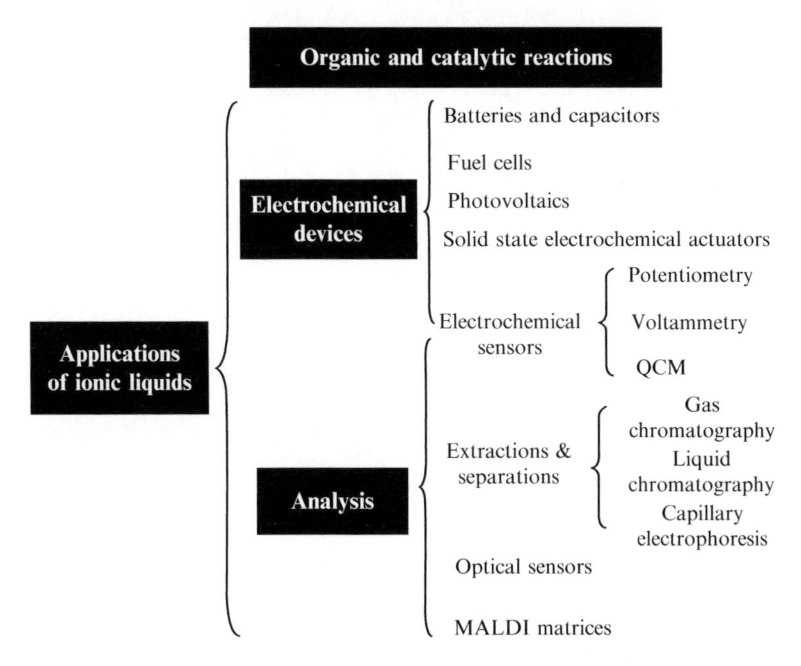

Fig. 7.2 Applications of ILs (Reprinted with permission from Wei and Ivaska (2008). Copyright 2008 Elsevier)

benefits of both solid and liquid electrolytes. Some of the immediately obvious benefits of ILs include the following. ILs have high ion conductivity, wide potential windows (up to 5.5 V) (potential region without significant background current), high heat capacity, and good chemical and electrochemical stability, and they have been explored as media in electrochemical devices including supercapacitors, fuel cells, lithium batteries, photovoltaic cells, electrochemical mechanical actuators, electroplating, electrochemical sensors, and other analytical applications (Buzzeo et al. 2004a; Wei and Ivaska 2008; Silvester 2011; Dossi et al. 2012). For example, ILs can improve separation of complex mixtures of both polar and nonpolar compounds when used either as stationary phase or as additives in gas–liquid chromatography (Anderson and Armstrong 2003), liquid chromatography (Peng et al. 2007), and capillary electrophoresis (Qi et al. 2006). They are also used in optical sensors (Fletcher et al. 2002; Oter et al. 2006a, b, 2008) and to enhance the analytical performance of the matrix-assisted laser desorption ionization mass spectrometry (MALDI-MS) (Mank et al. 2004). Figure 7.2 briefly lists the general applications of ILs. The use of ILs in different applications is determined by their intrinsic properties.

ILs have negligible vapor pressures, so there is no “drying out” of the electrolytes, thus reducing hazards associated with flash points and flammability (Baker et al. 2005; Anastas 2007). The low volatility of ILs has been demonstrated in gas-separation membranes for separation of SO_2 and CO_2 (Jiang et al. 2007). The SO_2 selectivity of separations using IL membranes has been shown to be 9–19 times that of CO_2.

ILs possess high thermal stability and oxidative stability, which allows regeneration and decontamination of the sensor electrolyte as well as enabling operation at elevated temperatures, thus increasing the rate of mass transfer and hence signals (Yu et al. 2005; Jin et al. 2006). It was established that ILs can be thermally stable up to temperatures of 450 °C with decomposition temperatures around 300–500 °C. The thermal stability of ILs is limited by the strength of their heteroatom-carbon and their heteroatom-hydrogen bonds, respectively (Mantz and Trulove 2003). The thermal decomposition temperature decreases as the anion hydrophobicity increases. Halide anions reduce the thermal stability of ILs, with decomposition occurring at least 373 K below corresponding ILs with nonhalide

Table 7.2 Electrochemistry of gases in ionic liquids or nonaqueous solvents

Gas	Electrolyte	Electrode	Mechanism
O_2	ILs-BmimHFP; dmbimHFP; $EmimBF_4$; $BmimPF_6$	Au, Pt, and glassy carbon electrodes	$O_2 + e \rightarrow O_2^-$
SO_2	ILs-$BmiBF_4$,	Pt electrode	$SO_2 + O_2 + 2e \rightarrow SO_4^{2-}$
NH_3	ILs-$EmimNTF_2$	Glass carbon electrode	QH_2 (hydroquinone) $+ 2NH_3 \rightarrow 2NH_4^+ + Q^{2-}$ or $4NH_3 \rightarrow 3NH_4^+ + +½N_2 + 3e$
NO_2	ILs-$C_2mimNTF_2$	Pt electrode	$NO_2 \rightarrow NO_2^+ + e$
CH_4	Nonaqueous 2 M $NaClO_4$ in γ-butyrolactone	Pt black electrode	$CH_4 \rightarrow CH_3(ad) + H^+ + e$

BmimHFP and $BmimPF_6$: 1-*n*-butyl-3-methylimidazolium hexafluorophosphate; $BmiBF_4$: 1-butyl-3-methylimidazolium tetrafluoroborate; $BmimPF_6$: 1-*n*-butyl-3-methylimidazolium hexafluorophosphate; $C_2mimNTF_2$ and $EmimNTF_2$: 1-ethyl-3-methylimidazolium bis[(trifluoromethyl)sulfonyl]imide; $EmimBF_4$: polyethylene-supported 1-ethyl-3-methylimidazolium tetrafluoroborate; $EmimNTF_2$: 1-ethyl-3-methylimidazolium bis(trifluoromethylsulfonyl)imide
Source: Data from Stetter et al. (2011)

anions. Relative anion stabilities have been suggested by Huddleston et al. (2001) as [BMIm] $PF_6^- > NTf_2^- \approx BF_4^- >$ halides. This means that the ILs [BMIm][PF_6], [BMIm][NTf_2], and [BMIm][BF_4] have decomposition temperature higher than the corresponding halide IL [BMIm][I]. The trend of thermal stability with respect to cation species appears to go as follows: phosphonium > imidazolium > tetraalkyl ammonium pyrrolidinium (Kroon et al. 2007).

ILs suppress conventional solvation and solvolysis phenomena and provide a medium able to dissolve a vast range of molecules up to very high concentrations (Welton 1999; Kou et al. 2006);

ILs are excellent solvents that can support many types of solvent–solute interactions (Welton 1999). They also offer other advantages such as decontamination, product recovery, and recyclable properties. Other pertinent properties include high intrinsic ionic conductivity, nonflammability, wide electrochemical stable window, broad liquid range and excellent heat transfer properties, and most importantly their hydrophobic nature (Ahmad 2009). The hydrophobic nature of ILs and the possibility to form various composites with polymers (Scott et al. 2002) are also important features for sensor applications (Wang et al. 2004). These composite materials can be used as conductive materials, semi-permeable membranes, and electrodes.

The use of ILs as electrolytes can also eliminate the need for a membrane and added supporting electrolytes, which are needed in conventional "Clark"-type gas sensors. The negligible vapor pressure and high thermal stability make the gas sensors based on ILs promising in more extreme operating conditions, such as high temperatures and pressures (Buzzeo et al. 2004a). Unlike the oxide electrolytes in solid-state gas sensors, which operate at temperatures of several hundred degrees, the high ionic conductivity at room temperature allows ILs to be excellent electrolytes for the fabrication of quasi-solid-state electrochemical gas sensors working at ambient temperatures. As shown in Table 7.2, electrochemical oxidation of NH_3 (Buzzeo et al. 2004b, 2004d), NO_2 (Broder et al. 2007), SO_2 and H_2 (Silvester et al. 2008a; O'Mahony et al. 2008), and Cl_2 (Huang et al. 2008), and electrochemical reduction of O_2 in ILs (Wang et al. 2004) have been reported.

As a result, it was established that ILs can be used in the development of stable electrochemical sensors for gaseous analytes such as O_2 (AlNashef et al. 2002; Wang et al. 2004; Buzzeo et al. 2003, 2004a; Wang et al. 2011), CO_2 (AlNashef et al. 2002; Buzzeo et al. 2004c), SO_2 (Cai et al. 2001), NH_3 (Giovanelli et al. 2004), NO_2 (Broder et al. 2007), ethylene (Zevenbergen et al. 2011), and vapors of ethanol and organic solvents (Lee and Chou 2004; Seyama et al. 2006). A promising property of ILs in electrochemical gas sensor development is that the physicochemical properties of ILs depend on the structure and size of both their cations and anions, which can easily be tuned by controlled organic synthesis (Silvester 2011). For example, in IL-based gas sensors, one can find ion liquids such as

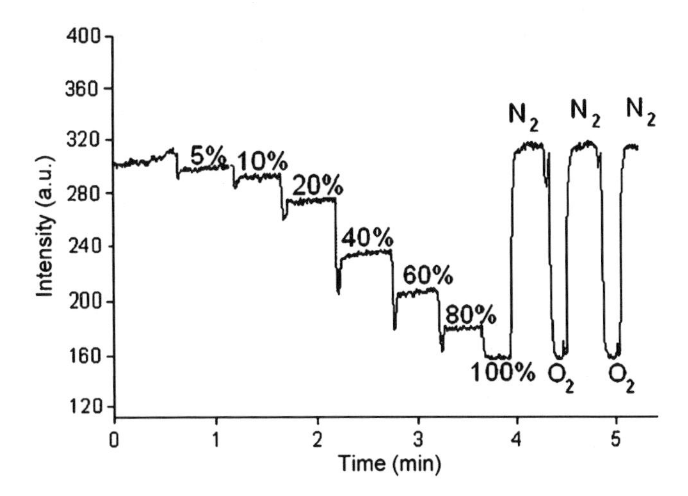

Fig. 7.3 Response kinetics of optical (fluorescence) sensors based on ruthenium complex incorporated in IL-modified sol–gel matrix (Reprinted with permission from Oter et al. (2009). Copyright 2009 Elsevier)

[EMIm][BF_4] (O_2), [BMIm][PF_6] (Cl_2, O_2), [EMIm][Tf_2N] (NH_3, NO_2, DMF), [C_2MIm][BF_4] (SO_2), [BMIm][BF_4] (VOCs), and [C_4C_1Im][PF_6] (VOCs) (Singh et al. 2012). There are potentially many more useful ionic liquids; for example, at least a million binary ILs, and 10^{18} ternary ILs, are potentially possible. Therefore, it is likely that highly sensitive and selective determination of gaseous analytes can be realized with optimized ILs. For example, conductive polymers are often regarded as polyions after they are doped. A recent study by Yu et al. (2008) shows that PANI in its doped state is a positively charged polymer and the negatively charged anions of the ionic liquid butylmethylimidazolium camphorsulfonate (BMICS) can be strongly absorbed on the PANI polymer backbone as counter ions. The electrostatic interactions and the van der Waals interactions between the IL and the charged conductive polymer template not only help increase the wettability of IL film electrolytes but also increase the selectivity by forming IL-PANI composite porous structures.

It is important to note that the application of ILs makes it possible to achieve very close contact between the electrode material and IL, thus allowing analytes from gaseous samples to undergo electron transfer just as they reach the working-electrode-material/IL interphase, without involving any analyte diffusion and/or dissolution step. Concomitantly, the IL medium available in close contact with the electrode material can ensure the transfer of charged species from the working electrode to the counter electrode. Thus, IL-based membrane-free amperometric gas sensors with fast response can be designed (Dossi et al. 2012).

Experiment has shown that ILs can be incorporated in optical sensors. For example, Oter et al. (2006b) reported an optical CO_2 sensor using the ILs ([MBIM][BF_4] or [MBIM][Br]) as the matrix with 8-hydroxypyrene-1,3,6-trisulfonic acid trisodium salt (HPTS). The detection of CO_2 was based on the fluorescence signal change of HPTS when pairing to CO_2. The same group then reported that ionic liquid modification of an ethyl cellulose matrix extended the detection range to 0–100 % pCO_2 (Oter et al. 2008). Recently, Oter et al. (2009) proposed to use the dye, tris(2,2′-bipyridyl)ruthenium(II) chloride, incorporated in ionic liquid ([EMIM]BF_4)-modified sol–gel matrices for oxygen-sensing purposes (see Fig. 7.3). A hybrid electrochemical–colorimetric-sensing platform for detecting explosives was developed by Forzani et al. (2009). The product of the electrochemical reaction was detected by a colorimetric device. A thin layer of [BMIM][PF6] played an important role in this platform: The IL coating selectively preconcentrated explosives and quickly transported them to the electrodes; it also facilitated the formation of reduction products.

It was established that IL-based chemically sensitive field-effect transistors and quartz crystal microbalance sensors can be developed as well (see, e.g., Fig. 7.4). For example, chemically sensitive field-effect transistors based on a composite of camphorsulfonic acid (CSA)-doped polyaniline (PAN) and an IL of 1-butyl-3-methylimidazolium bis(trifluoromethanesulfonyl)-imide, BMI(Tf_2N) could

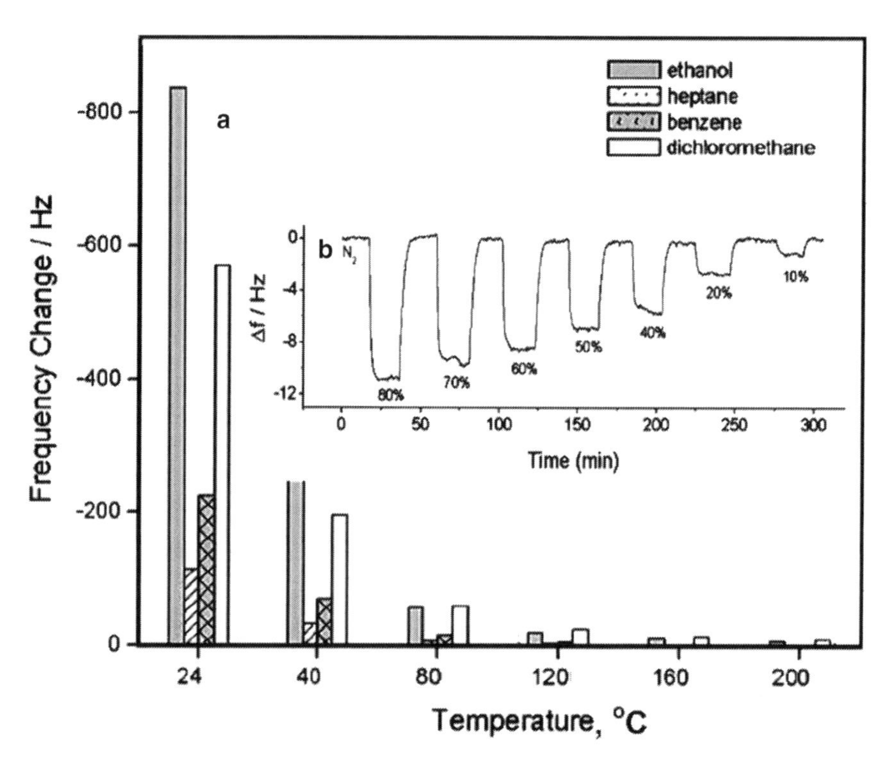

Fig. 7.4 (**a**) Frequency change of the IL/QCM sensor exposed to ethanol, heptane, benzene, and dichloromethane at various temperatures. The same concentration was used for the vapors at all temperatures. (**b**) Sensorgram of benzene at 120 °C from various concentrations (Adapted with permission from Yu et al. (2005). Copyright 2005 Royal Society of Chemistry)

detect ammonia gas in the range from 0.5 to 694 ppm in air. Quartz crystal microbalance sensors based on ionic liquid phosphonium dodecylbenzenesulfonate (i.e., $P_{6,6,6,14}$DBS) were sensitive toward various polar and nonpolar organic vapors (Yu et al. 2008). The sensor had linear, fast, and reversible response at temperatures up to 200 °C. However, we need to recognize that maximum sensor response was observed at T_{oper} = 24 °C and the increase of operating temperature was accompanied by considerable decrease of sensor response. In addition, the sensitivity was not high.

It was found that IL-based QCM sensors are also promising materials for detection of organic vapors (Liang et al. 2002; Goubaidoulline et al. 2005; Jin et al. 2006). Liang et al. (2002) showed that changes in viscosity of the IL film upon absorption of organic vapors at room temperature was the main cause for the change in frequency rather than change in the mass. The sensing mechanism of a QCM sensor using ILs is based on the fact that the viscosity of the IL membrane decreases rapidly due to solvation of the analytes in the ILs. The change in viscosity, which varies with the chemical species of the vapors and the type of ILs, results in a frequency shift of the quartz crystal.

However, it should also be noted that, since the frequency change in QCM-based devices apparently depends on both the mass load as well as the change in viscosity of the IL, it would be necessary to understand in detail to what extent the change in sensor signal is due to viscosity changes and to what degree it is due to changes in mass load when such a system is used in quantitative determinations (Wei and Ivaska 2008). A straightforward interpretation of the sensor response might be demanding, in particular when the sensor is exposed to multicomponent mixtures.

Sensitivity to air humidity can also be considered as a disadvantage of ILs. Buzzeo et al. (2003, 2004d) believe that high viscosity and low diffusion coefficients are shortcomings of ILs as well. As a result of the strong electrostatic and other interaction forces, the viscosity of ILs is typically 10–100 times higher than that of water or organic solvents (Seddon et al. 2002). The viscosity of IL is affected by the nature of both the cations and anions. Alkyl chain lengthening in the cation leads to an increase in viscosity. This is due to stronger van der Waals forces between cations leading to increase in the energy required for molecular motion. The nature of the anion also affects the viscosity of the IL, particularly through relative basicity and the ability to participate in hydrogen bonding. The relatively low conductivity manifested by the inherently high viscosity of ILs and the much smaller diffusion coefficients of gas molecules in ILs usually lead to slow responses and small limiting currents (Jin et al. 2007). The conductivity of any solution not only depends on the number of charge carriers but also on their mobility. The large constituent ions of ILs reduce the ion mobility which, in turn, leads to lower conductivities. Ion pair formation and/or ion aggregation in ILs also lead to reduced conductivity. To overcome this deficiency, great attempts are made to facilitate the diffusion of gas analytes in IL electrolytes. The most efficient strategy involves the formation of thin IL layers on a variety of planar sensor arrays, including microfabricated electrode arrays (Huang et al. 2010; Zevenbergen et al. 2011) and conventional screen-printed carbon electrodes (SPCE) (Xiong et al. 2011). This strategy can produce IL layers with thicknesses up to several micrometers at the sensing interfaces and therefore effectively improves the performance of the IL-based sensors. The possibility of using thin layers of ILs without any addition membranes (Huang et al. 2010; Xiong et al. 2011) makes it possible to achieve parameters of IL-based sensors comparable with parameters of commercially available sensors (Wang et al. 2011). However, the formation of ultrathin IL layers is often restricted by several factors, e.g., the relative low reproducibility due to small volume and high viscosity of added ILs (Xiong et al. 2011; Zevenbergen et al. 2011) and the instability of IL layers due to the uptake of atmospheric moisture altering the IL surface tension (Xiong et al. 2011). Therefore the development of suitable sensor platforms has become a great challenge for IL-based electrochemical gas sensors. One such sensor platform was designed by Hu et al. (2012). Since the creation of large three-phase electrolyte/electrode/gas interfaces is essential to the high sensitivity and rapid response of electrochemical gas sensors, Hu et al. (2012) proposed to use a nanoporous gold electrode prepared by inkjet printing integrated with porous support (cellulose membranes). This special structure allows the addition of electrolytes from the back of the working electrode to form as thin as possible electrolyte layers for creating large three-phase interfaces.

Silvester (2011) have also noted that ILs have poor detection limits. For example, electrochemical sensing in ILs is not sensitive enough to detect the low/trace concentrations required. Therefore, for electrochemical explosives sensing, it appears that ILs have to be combined with either nanomaterials or complementary techniques. The poorer detection limits may also mean that RTILs may only be useful for higher (ppm to percentages) concentrations of gases as opposed to trace (ppb) levels. Careful control of the potential vs. a stable reference electrode in IL-based electrochemical sensors is also a problem yet to be answered (Rogers et al. 2008), and more work is needed on this topic (Silvester et al. 2008b). Humidity monitors may also need to be used to account for the varying water contents in the ILs in a range of real environments, which have been shown to affect the viscosity and the electrochemical response (Silvester and Compton 2006). Another issue is the intrinsic impurities present in ILs, such as unreacted starting material (e.g., chloride), water, and dissolved gases (e.g., oxygen), all of which can interfere with the electrochemical response of the analyte.

However, Buzzeo et al. (2004d) and Silvester (2011) believe that, despite the abovementioned challenges, clearly the tunability of ILs and their ability to be easily combined with other materials make them ideal candidates as electrolytes in a range of electrochemical devices designed for operation under more extreme conditions, i.e., high temperatures and pressures, such as in the combustion industry where traditional solvents struggle to remain chemically or physically unchanged. Although this field is still in the development stage, it is envisioned that in the next decade we will see some significant advances toward commercialization of IL-based electrochemical sensing systems, particularly when using ILs as a binder and in the field of amperometric gas sensing.

References

Ahmad S (2009) Polymer electrolytes: characteristics and peculiarities. Ionics 15:309–321

AlNashef IM, Leonard ML, Matthews MA, Weidner JW (2002) Superoxide electrochemistry in an ionic liquid. Ind Eng Chem Res 41:4475–4478

Anastas PT (2007) Introduction: green chemistry. Chem Rev 107:2167–2168

Anderson JL, Armstrong DW (2003) High-stability ionic liquids. A new class of stationary phases for gas chromatography. Anal Chem 75:4851–4858

Baker GA, Baker SN, Pandey S, Bright FV (2005) An analytical view of ionic liquids. Analyst 130:800–808

Bowlas CJ, Bruce DW, Seddon KR (1996) Liquid-crystalline ionic liquids. Chem Commun 1996(14):1625–1626

Broder TL, Silvester DS, Aldous L, Hardacre C, Compton RG (2007) Electrochemical oxidation of nitrite and the oxidation and reduction of NO_2 in the room temperature ionic liquid [C2mim][NTf2]. J Phys Chem B 111: 7778–7785

Buzzeo MC, Klymenko OV, Wadhawan JD, Hardacre C, Seddon KR, Compton RG (2003) Voltammetry of oxygen in the room-temperature ionic liquids 1-ethyl-3-methylimidazolium bis((trifluoromethyl)sulfonyl)imide and hexyltriethylammonium is((trifluoromethyl)sulfonyl)imide: One-electron reduction to form superoxide. Steady-state and transient behavior in the same cyclic voltammogram resulting from widely different diffusion coefficients of oxygen and superoxide. J Phys Chem A 107:8872–8878

Buzzeo MC, Evans RG, Compton RG (2004a) Non-haloaluminate room-temperature ionic liquids in electrochemistry—a review. Chemphyschem 5:1106–1120

Buzzeo MC, Giovanelli D, Lawrence NS, Hardacre C, Seddon KR, Compton RG (2004b) Elucidation of the electrochemical oxidation pathway of ammonia in dimethylformamide and the room temperature ionic liquid, 1-ethyl-3-methylimidazolium bis(trifluoromethylsulfonyl)imide. Electroanalysis 16(11):888–896

Buzzeo MC, Klymenko OV, Wadhawan JD, Hardacre C, Seddon KR, Compton RG (2004c) Kinetic analysis of the reaction between electrogenerated superoxide and carbon dioxide in the room temperature ionic liquids 1-ethyl-3-methylimidazolium bis(trifluoromethylsulfonyl)imide and hexyltriethylammonium bis(trifluoromethylsulfonyl) imide. J Phys Chem B 108:3947–3954

Buzzeo M, Hardacre C, Compton RG (2004d) Use of room temperature ionic liquids in gas sensor design. Anal Chem 76:4583–4588

Cai Q, Xian YZ, Li H, Zhang YM, Tang J, Jin LT (2001) Studies on a sulfur dioxide electrochemical sensor with ionic liquid as electrolyte. Huadong Shifan Daxue Xuebao, Ziran Kexueban 2001(3):57–60

Demus D, Goodby J, Gray GW, Spiess H-W, Vill V (eds) (1998) Handbook of liquid crystals. Wiley, Weinheim, Vol. 1, Chap. 2, pp. 18–19

Dossi N, Toniolo R, Pizzariello A, Carrilho E, Piccin E, Battiston S, Bontempelli G (2012) An electrochemical gas sensor based on paper supported room temperature ionic liquids. Lab Chip 12:153–158

Earle MJ, Seddon KR (2000) Ionic liquids. Green solvents for the future. Pure Appl Chem 72(7):1391–1398

Fletcher KA, Pandey S, Storey IK, Hendricks AE, Pandey S (2002) Selective fluorescence quenching of polycyclic aromatic hydrocarbons by nitromethane within room temperature ionic liquid 1-butyl-3-methylimidazolium hexafluorophosphate. Anal Chim Acta 453:89–96

Forzani ES, Lu D, Leright MJ, Aguilar AD, Tsow F, Iglesias RA, Zhang Q, Lu J, Li J, Tao N (2009) A hybrid electrochemical-colorimetric sensing platform for detection of explosives. J Am Chem Soc 131:1390–1391

Giovanelli D, Buzzeo MC, Lawrence NS, Hardacre C, Seddon KR, Compton RG (2004) Determination of ammonia based on the electro-oxidation of hydroquinone in dimethylformamide or in the room temperature ionic liquid, 1-ethyl-3-methylimidazolium bis(trifluoromethylsulfonyl)imide. Talanta 62:904–911

Goubaidoulline I, Vidrich G, Johannsmann D (2005) Organic vapor sensing with ionic liquids entrapped in alumina nanopores on quartz crystal resonators. Anal Chem 77:615–619

Hu C, Bai X, Wang Y, Jin W, Zhang X, Hu S (2012) Inkjet printing of nanoporous gold electrode arrays on cellulose membranes for high-sensitive paper-like electrochemical oxygen sensors using ionic liquid electrolytes. Anal Chem 84:3745–3750

Huang X-J, Silvester DS, Streeter I, Aldous L, Hardacre C, Compton RG (2008) Electroreduction of chlorine gas at platinum electrodes in several room temperature ionic liquids: evidence of strong adsorption on the electrode surface revealed by unusual voltammetry in which currents decrease with increasing voltage scan rates. J Phys Chem C 112:19477–19483

Huang XJ, Aldous L, O'Mahony AM, del Campo FJ, Compton RG (2010) Toward membrane-free amperometric gas sensors: a microelectrode array approach. Anal Chem 82:5238–5245

Huddleston JG, Visser AE, Reichert WM, Willauer HD, Broker GA, Rogers RD (2001) Characterization and comparison of hydrophilic and hydrophobic room temperature ionic liquids incorporating the imidazolium cation. Green Chem 3(4):156–164

Jiang YY, Zhou Z, Jiao Z, Li L, Wu YT, Zhang ZB (2007) SO_2 gas separation using supported ionic liquid membranes. J Phys Chem B 111:5058–5061

Jin X, Yu L, Garcia D, Ren RX, Zeng X (2006) Ionic liquid high temperature gas sensor array. Anal Chem 78:6980–6989

Jin H, Baker GA, Arzhantsev S, Dong J, Maroncelli M (2007) Survey of solvation and rotational dynamics of Coumarin 153 in a broad range of ionic liquids and comparisons to conventional solvents. J Phys Chem B 111:7291–7302

Kou Y, Xiong W, Tao G, Liu H, Wang T (2006) Absorption and capture of methane into ionic liquid. J Nat Gas Chem 15:282–286

Kroon MC, Buijs W, Peters CJ, Witkamp GJ (2007) Quantum chemical aided prediction of the thermal decomposition mechanisms and temperatures of ionic liquids. Thermochim Acta 465(1–2):40–47

Lee YG, Chou TC (2004) Ionic liquid ethanol sensor. Biosens Bioelectron 20(1):33–40

Liang C, Yuan CY, Warmack RJ, Barnes CE, Dai S (2002) Ionic liquids: a new class of sensing materials for detection of organic vapors based on the use of a quartz crystal microbalance. Anal Chem 74:2172–2176

Mank M, Stahl B, Boehm G (2004) 2,5-Dihydroxybenzoic acid butylamine and other ionic liquid matrixes for enhanced MALDI-MS analysis of biomolecules. Anal Chem 76:2938–2950

Mantz AR, Trulove PC (2003) Physicochemical properties of ionic liquids. In: Wasserscheid P, Welton T (eds) Ionic liquids in synthesis. Wiley, Berlin, pp 75–143

O'Mahony AM, Silvester DS, Aldous L, Hardacre C, Compton RG (2008) The electrochemical reduction of hydrogen sulfide on platinum in several room temperature ionic liquids. J Phys Chem C 112:7725–7730

Oter O, Ertekin K, Topkaya D, Alp A (2006a) Room temperature ionic liquids as optical sensor matrix materials for gaseous and dissolved CO_2. Sens Actuators B Chem 117:295–301

Oter O, Ertekin K, Topkaya D, Alp S (2006b) Emission-based optical carbon dioxide sensing with HPTS in green chemistry reagents: room-temperature ionic liquids. Anal Bioanal Chem 386:1225–1234

Oter O, Ertekin K, Derinkuyu S (2008) Ratiometric sensing of CO_2 in ionic liquid modified ethyl cellulose matrix. Talanta 76:557–563

Oter O, Ertekin K, Derinkuyu S (2009) Photophysical and optical oxygen sensing properties of tris(bipyridine) ruthenium(II) in ionic liquid modified sol–gel matrix. Mater Chem Phys 113:322–328

Peng JF, Liu JF, Hu XL, Jiang GB (2007) Direct determination of chlorophenols in environmental water samples by hollow fiber supported ionic liquid membrane extraction coupled with high-performance liquid chromatography. J Chromatogr A 1139:165–170

Pinkert A, Marsh KN, Pang S, Staiger MP (2009) Ionic liquids and their interaction with cellulose. Chem Rev 109:6712–6728

Qi S, Cui S, Cheng Y, Chen X, Hu Z (2006) Rapid separation and determination of aconitine alkaloids in traditional Chinese herbs by capillary electrophoresis using 1-butyl-3-methylimidazoium-based ionic liquid as running electrolyte. Biomed Chromatogr 20:294–300

Rogers EI, Silvester DS, Poole DL, Aldous L, Hardacre C, Compton RG (2008) Voltammetric characterization of the ferrocene|ferrocenium and cobaltocenium|cobaltocene redox couples in RTILs. J Phys Chem C 112:2729–2735

Scott MP, Brazel CS, Benton MG, Mays JW, Holbrey JD, Rogers RD (2002) Application of ionic liquids as plasticizers for poly(methyl methacrylate). Chem Commun 2002:1370–1371

Seddon KR (1997) Ionic liquids for clean technology. J Chem Tech Biotechnol 68:315–316

Seddon KR, Stark A, Torres MJ (2000) Influence of chloride, water, and organic solvents on the physical properties of ionic liquids. Pure Appl Chem 72:2275–2287

Seddon KR, Stark A, Torres MJ (2002) Viscosity and density of 1-alkyl-3-methylimidazolium ionic liquids. ACS Symp Ser 819:34–49

Seyama M, Iwasaki Y, Tate A, Sugimoto I (2006) Room-temperature ionic-liquid-incorporated plasma-deposited thin films for discriminative alcohol-vapor sensing. Chem Mater 18(11):2656–2662

Silvester DS, Compton RG (2006) Electrochemistry in room temperature ionic liquids: a review and some possible applications. Z Phys Chem (N F) 220:1247–1274

Silvester DS, Ward KR, Aldous L, Hardacre C, Compton RG (2008a) The electrochemical oxidation of hydrogen at activated platinum electrodes in room temperature ionic liquids as solvents. J Electroanal Chem 618:53–60

Silvester DS, Rogers EI, Compton RG, McKenzie KJ, Ryder KS, Endres F, MacFarlane D, Abbott AP (2008b) Reference electrodes for use in room-temperature ionic liquids. In: Endres F, MacFarlane DR, Abbot A (eds) Electrodeposition from ionic liquids. Wiley, Weinheim, pp 296–309

Silvester DS (2011) Recent advances in the use of ionic liquids for electrochemical sensing. Analyst 136:4871–4882

Singh VV, Nigam AK, Batra A, Boopathi M, Singh B, Vijayaraghavan R (2012) Applications of ionic liquids in electrochemical sensors and biosensors. Int J Electrochem 2012:165683

Stetter JR, Korotcenkov G, Zeng X, Liu Y, Tang Y (2011) Electrochemical gas sensors: fundamentals, fabrication and parameters. In: Korotcenkov G (ed) Chemical sensors: comprehensive sensor technologies, vol 5, Electrochemical and optical sensors. Momentum Press, New York, pp 1–89

Sun P, Armstrong DW (2010) Ionic liquids in analytical chemistry. Anal Chim Acta 661:1–16

Wang R, Okajima T, Kitamura F, Ohsaka T (2004) A novel amperometric O_2 gas sensor based on supported room-temperature ionic liquid porous polyethylene membrane-coated electrodes. Electroanalysis 16:66–72

Wang Z, Lin P, Baker GA, Stetter J, Zeng X (2011) Ionic liquids as electrolytes for the development of a robust amperometric oxygen sensor. Anal Chem 83:7066–7073

Wasserschied P, Welton T (2003) Ionic liquids in synthesis. Wiley, Weinheim

Wei D, Ivaska A (2008) Applications of ionic liquids in electrochemical sensors. Anal Chim Acta 607:126–135

Welton T (1999) Room-temperature ionic liquids. Solvents for synthesis and catalysis. Chem Rev 99:2071–2084

Xiong SQ, Wei Y, Guo Z, Chen X, Wang J, Liu JH, Huang XJ (2011) Toward membrane-free amperometric gas sensors: an ionic liquid–nanoparticle composite approach. J Phys Chem C 115:17471–17478

Yu L, Diego G, Ren XR, Zeng X (2005) Ionic liquid high temperature gas sensors. Chem Commun 2005:2277–2279

Yu L, Jin X, Zeng X (2008) Methane interactions with polyaniline/butylmethylimidazolium camphorsulfonate ionic liquid composite. Langmuir 24:11631–11636

Zevenbergen MAG, Wouters D, Dam VAT, Brongersma SH, Crego-Calama M (2011) Electrochemical sensing of ethylene employing a thin ionic-liquid layer. Anal Chem 83:6300–6307

Chapter 8
Silicate-Based Mesoporous Materials

According to the International Union of Pure and Applied Chemistry (IUPAC), the prefix meso- refers to a region 2–50 nm, macro- is a region >50 nm, and micro- is a region <2 nm. The small mesopores limit the kinds of ions and molecules that can be admitted to the interior of the materials. In addition, control over the pore size offers the possibility of molecular sieving or molecular selectivity. Mesoporosity can also endow a material with a high surface area exceeding 1,000 m^2/g and pore volume greater than 1 cm^3/g. This greatly expands the potential of the materials for application to adsorption and as supports for immobilized catalytic or sensing moieties (Moos et al. 2006, 2009; Xu et al. 2006; Carrington and Xue 2007; Slowing et al. 2007; Basabe-Desmonts et al. 2007; Ariga et al. 2007 Walcarius 2008; Melde et al. 2008; Sahner et al. 2008; Zheng et al. 2012). Several examples of materials with different porosity are presented in Table 8.1.

8.1 Mesoporous Silicas

We need to say that there are two types of mesoporous silicas: random mesoporous structures and ordered mesoporous structures (Galarneau et al. 2001). Mesoporous silicas, especially those exhibiting ordered pore systems and uniform pore diameters, have shown great potential for sensing applications in recent years (Melde et al. 2008).

Sol–gel chemistry is frequently employed in designing random mesoporous structures of silicates (Brinker and Scherer 1990; Corma 1997). Liquid silicon alkoxide precursors ($Si(OR)_4$) are hydrolyzed and condensed to form siloxane bridges, a process that is often described as inorganic polymerization and is represented below:

$$\begin{aligned}
&\text{Hydrolysis: } Si(OR)_4 + nH_2O \rightarrow HO_n - Si(OR)_{4-n} + nROH \\
&\text{Condensation: } (RO)_3Si - OH + HO - Si(OR)_3 \rightarrow (RO)_3Si - O - Si(OR)_3 + H_2O \\
&\text{and / or } (RO)_3Si - OR + HO - Si(OR)_3 \rightarrow (RO)_3Si - O - Si(OR)_3 + ROH
\end{aligned} \quad (8.1)$$

The most commonly used precursors are tetraethoxysilane (TEOS) and tetramethoxysilane (TMOS). A colloidal sol of condensed silicate species can eventually interconnect as an immobile 3D network encompassing the space of its reaction container. Drying a gel under ambient conditions or with heat will typically cause shrinkage as solvent leaves the micropores of the silicate network. This type of material is called a xerogel. Alternatively, supercritical drying can be applied to remove solvent, yielding a product that is more similar to the size and shape of the original gel. Such aerogels may have low solid volume

G. Korotcenkov, *Handbook of Gas Sensor Materials*, Integrated Analytical Systems,
DOI 10.1007/978-1-4614-7388-6_8, © Springer Science+Business Media New York 2014

Table 8.1 Different types of porous materials

Type of material	Pore size (Å)	Examples	Pore size range (Å)
Macroporous	>500	Macroporous monolithic polymers	>500
		Porous glasses	>500
Mesoporous	20–500	Pillared layered clays	20–400
		M41S	16–100
		SBA-15	80–100
		SBA-16	50
		Diatom biosilica	20–500
		Mesoporous alumina	20
Microporous	<20	Carbon aerogels	<20
		Zeolites	<14.2
		Activated carbon	6
		ZSM-5	4.5–6
		Zeolite A	3–4.5
		Beta and Mordenite-zeolites	6–8
		Faujasite	7.4
		Cloverite	6–1.32

Source: Naik and Ghosh (2009)

fractions near 1 % and, therefore, very high pore volumes. The use of basic pH and an excess of water can result in particulate precipitation. Gels can also be deposited, permitting the generation of thin films or membranes. The isoelectric point of silica is in the pH range 1–3. This value determines the surface charge of a condensing silicate or material in solution due to protonation and deprotonatio of silanol groups (Si-OH).

Ordered mesoporous materials are made using a combination of using self-assembled surfactants as template and simultaneous sol–gel condensation around template (micelles) (Corma 1997; Galarneau et al. 2001). Surfactants are organic molecules, which comprise two parts with different polarities. One part is a hydrocarbon chain (often referred to as polymer tail), which is nonpolar and hence hydrophobic and lipophilic, whereas the other is polar and hydrophilic (often called hydrophilic head). Because of such a molecular structure, surfactants tend to enrich at the surface of a solution or interface between aqueous and hydrocarbon solvents, so that the hydrophilic head can turn toward the aqueous solution, resulting in a reduction of surface or interface energy. Such concentration segregation is spontaneous and thermodynamically favorable. Surfactant molecules can be generally classified into four families, and they are known as anionic, cationic, nonionic, and amphoteric surfactants, which are briefly discussed below:

1. Typical anionic surfactants are sulfonated compound with a general formula $R\text{-}SO_3Na$ and sulfated compounds of $R\text{-}OSO_3Na$, with R being an alkyl chain consisting of 11–21 carbon atoms.
2. Cationic surfactants commonly comprise an alkyl hydrophobic tail and a methyl-ammonium ionic compound head, such as cetyl trimethyl ammonium bromide (CTAB), $C_{16}H_{33}N(CH_3)_3Br$, and cetyl trimethyl ammonium chloride (CTAC), $C_{16}H_{33}N(CH_3)_3Cl$.
3. Nonionic surfactants do not dissociate into ions when dissolved in a solvent as both anionic and cationic surfactant. Their hydrophilic head is a polar group with structure similar to ether, R-O-R, alcohol, R-OH, carbonyl, R-CO-R, and amine, R-NH-R.
4. Amphoteric surfactants have properties similar to either nonionic surfactants or ionic surfactants. Examples are betaines and phospholipids.

When surfactants dissolve in a solvent forming a solution, the surface energy of the solution will decrease rapidly and linearly with increasing concentration. This decrease is due to the preferential enrichment and ordered arrangement of the surface of surfactant molecules on the solution surface, i.e., hydrophilic heads inside the aqueous solution and/or away from the nonpolar solution or air.

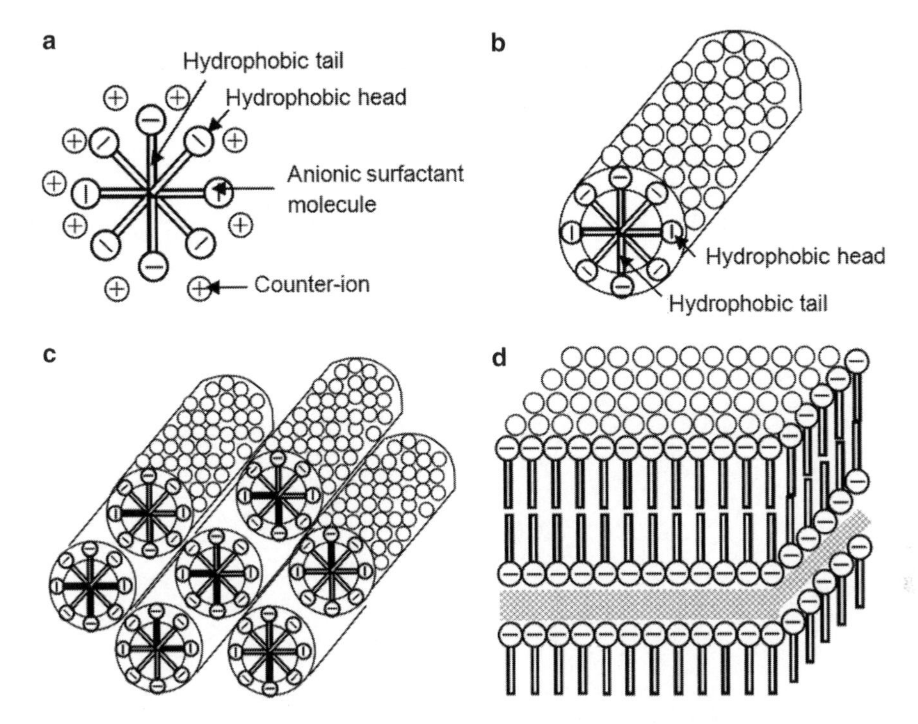

Fig. 8.1 Schematics of various micelles formed at various surfactant concentrations above the CMC: (**a**) Spherical micelle forms first as the concentration of surfactants is above the CMC; (**b**) Individual cylindrical micelle forms at the concentration of surfactants increases further: (**c**) Further increased concentration of surfactants results in the formation of hexagonally packed cylindrical micelles: (**d**) Lamellar micelles would form when the concentration of surfactants rises even further

However, such a decrease stops when a critical concentration is reached, and the surface energy remains constant with further increase in the surfactant concentration. This critical concentration is known as the *critical micellar concentration*, or CMC. Below the CMC, the surface energy decreases due to an increased coverage of surfactant molecules on the surface as the concentration increases. At the CMC, the surface has been fully covered with the surfactant molecules. Above the CMC, further addition of surfactant molecules leads to phase segregation and formation of colloidal aggregates or micelles (Mittal and Fendler 1982). The initial micelles are spherical and individually dispersed in the solution (Fig. 8.1a), and would transfer to a cylindrical rod shape (Fig. 8.1b) with further increased surfactant concentration. Continued increase of surfactant concentration results in an ordered parallel hexagonal packing of cylindrical micelles (Fig. 8.1c). At a still higher concentration, lamellar micelles would form (Fig. 8.1d).

The process of ordered mesoporous structures synthesis is conceptually straightforward and can be briefly described below (Corma 1997). Surfactants with a certain molecule length are dissolved in a polar solvent at a concentration exceeding its CMC, mostly at a concentration at which hexagonal or cubic packing of cylindrical micelles is formed. At the same time, the precursors for the formation of silica are also dissolved in the same solvent, together with other necessary chemicals such as catalyst. Inside the solution, several processes proceed simultaneously. Surfactants segregate and form micelles, whereas precursors undergo hydrolysis and condensation around the micelles simultaneously. Surfactants are often removed by calcination, or burning, to produce molecular sieves with narrow pore size distributions and highly ordered mesostructures. When extraction of templates is used instead of calcinations, organic functional groups can be incorporated into the materials during synthesis.

First-ordered mesoporous structure was realized in 1992 with the reports of the M41S materials (Kresge et al. 1992; Beck et al. 1992), followed by the introduction of FSM-16 (Inagaki et al. 1993).

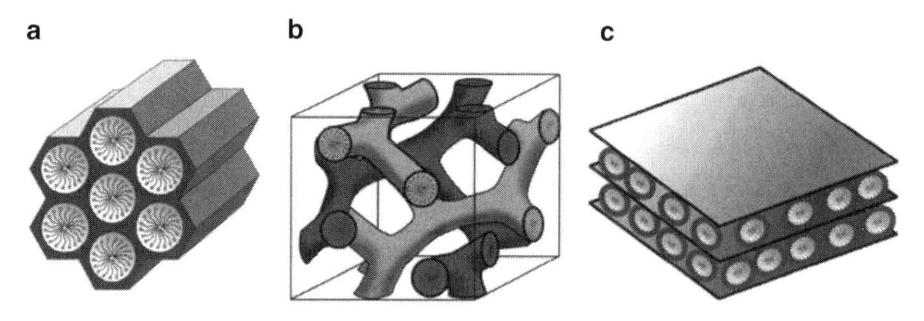

Fig. 8.2 Typical Structures of the mesoporous silica material: (**a**) MCM-42 (2D hexagonal); (**b**) MCM-48 (cubic); (**c**) MCM-50 (lamellar)

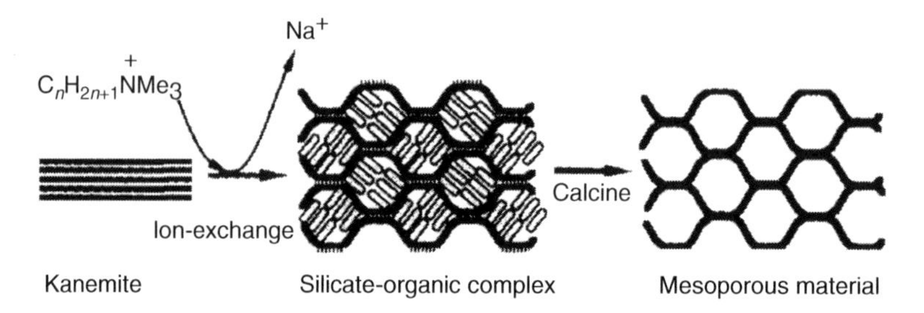

Fig. 8.3 Schematic model for the formation of the mesoporous material from kanemite (Reprinted with permission from Inagaki et al. (1993). Copyright 1993 Royal Society of Chemistry)

Syntheses of M41S materials employ cationic alkylammonium surfactants in amounts above their critical micelle concentrations. These surfactants cooperatively assemble with inorganic precursors to produce a silicate matrix. The most well-known representatives of this class include the silica solids MCM-41 (with a hexagonal arrangement of the mesopores, space group *p6mm*), MCM-48 (with a cubic arrangement of the mesopores, space group $Ia\bar{3}d$), and MCM-50 (with a laminar structure, space group *p2*) (see Fig. 8.2). Since the first reports of M41S, many different surfactants, precursors, and combinations of the two have been studied. A schematic model for the formation of the mesoporous materials from kanemite is shown in Fig. 8.3. The silicate layers of kanemite can wind around the exchanged alkyltrimethylammonium ions. This causes the condensation of silanol groups on the adjacent silicate layers of kanemite, since the silicate layers have the flexibility to wind due to its single-layered structure.

One can find in review papers of Huo et al. (1996), Corma (1997), Soler-Illia et al. (2002), Hoffmann et al. (2006), and Naik and Ghosh (2009) additional information regarding synthesis and properties of mesoporous silica, including as-synthesized and functionalized materials.

It should be noted that the described approach to synthesis of mesoporous silica can be used for the synthesis of other metal oxides, including conductive metal oxides designed for gas sensor applications.

8.1.1 Gas Sensor Applications of Mesoporous Silicas

Silica is an attractive material for many sensing applications because of its high surface area, stability over a fairly wide range of pH (excluding alkaline), relative inertness in many environments, transparency in the UV–visible spectrum, and its morphological control (Melde et al. 2008).

Table 8.2 Silicate-based optical oxygen sensors

Dye incorporated in silicas	Reference
Pt (II) 2,3,7,8,12,13,17,18-octaethyl porphine	Han et al. (2005)
Pd (II) 2,3,7,8,12,13,17,18-octaethyl porphine	
Pt(II) *meso*-tetraphenylporphine	
Pt (II) *meso*-tetra (pentafluorophenyl)porphine	
$[Ru(dpp)_3]Cl_2$	
$[Ru(phen_2phenCH_3)]Cl_2$	
$[Ru(phen)_3]Cl_2$	
$[Ru(bpy)_3]Cl_2$	
Pt(II) *meso*-tetra(4-*N*-pyridyl)porphyrin	Zhang et al. (2005)
Pt(II) *meso*-tetra(3,5-dihydroxyphenyl)porphyrin	Huo et al. (2006)
Pt(II) *meso*-tetra(3,5-di[(*N*-carbazyl)-*n*-octyloxyphenyl]) porphyrin	
Pt(II) *meso*-tetra(3,5-di[(*N*-carbazyl)-*n*-hexyloxyphenyl]) porphyrin	
Pt(II) *meso*-tetra(3,5-di[(*N*-carbazyl)-*n*-butyloxyphenyl]) porphyrin	
N-(3-Trimethoxysilylpropyl)-2,7-diazapyrenium bromide	Leventis et al. (1999)
Ru(II) complex dyads ($[Ru(phen)_3]^{2+}$)	Leventis et al. (2004)
4-Benzoyl-*N*-methylpyridinium-based dyads	
N-Benzyl-*N*-methyl viologen	
Tris(bipyridine)ruthenium(II) ($[Ru(bpy)_3]^{2+}$)	Zhang et al. (2002)
$[Ru(bpy)_2phen]^{2+}$	Lei et al. (2006)

Moreover, the silane chemistry makes it possible to attach covalently molecular probes to the pore walls. Therefore, spectrophotometrically active molecular probes can be entrapped in sol–gel glass and used for heterogeneous detection of analytes in solution or gas. Bulk materials may be applied as synthesized (e.g., batch adsorption of an analyte from solution) or as part of a surface coating. Gels can be used to form monolithic materials or thin films on a wide variety of substrates by spin- and dip-coating techniques. They can be deposited as or embedded in a specialized coating on an electrode and active area of various gas-sensing devices. Pore size can be controlled, and surface properties can be altered (e.g., grafting hydrophobic groups) to encourage the entrance of a specific analyte over that of similar species.

At present there are several approaches to using mesoporous silicates in gas sensors. In particular, humidity sensors can be designed based on silicates (Innocenzi et al. 2001, 2005; Domansky et al. 2001; Falcaro et al. 2004; Bertolo et al. 2005). Water molecules interact with hydroxyl sites providing a base for physisorption of water layers as relative humidity increases. For a dry surface at relatively low humidity, conductance occurs through proton "hopping" between the adsorption sites. At higher humidity, water concentrates to form multilayers or condenses to fill a pore. Proton mobility, therefore, becomes more facile, and conductivity increases with protons moving from molecule to molecule (Grotthuss chain reaction model). A mesopore structure increases the surface area and number of hydroxyl groups available for water adsorption. Factors that affect sensor response include the size and accessibility of mesopores, film thickness, number of hydroxyl sites, and organic matter within the pores. Organic matter refers to residual surfactant from a templating process or polymer introduced either during or post-synthesis.

Optical sensors based on UV–visible and fluorescence spectrophotometry and a visual color change in a material are other directions for mesoporous silicates application (Melde et al. 2008). Several examples of such sensors are presented in Table 8.2. Usually optical detection of gases in mesoporous silica-based sensors takes place through the use of an incorporated dye. In particular, oxygen sensing

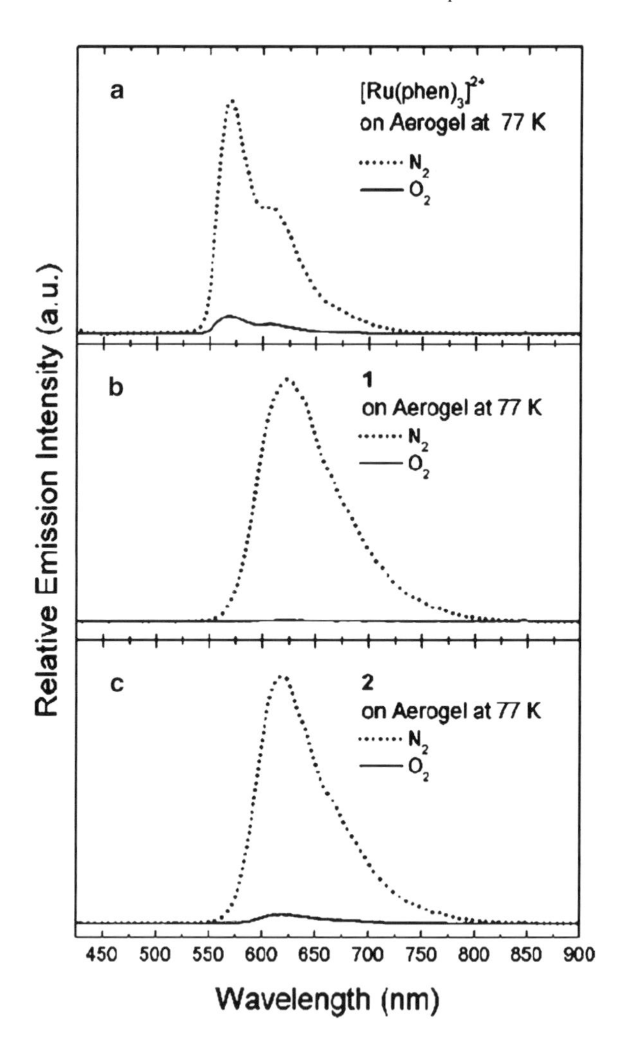

Fig. 8.4 Differences in the emission spectra at 77 K of silica aerogels doped with $[Ru(phen)_3]^{2+}$ (**a**), 4-benzoyl-*N*-methylpyridinium-based dyads (1) (**b**), and *N*-benzyl-*N*¢-methyl viologen (2) (**c**) under nitrogen and under oxygen. A pure oxygen environment quenches the photoemission of aerogels doped 4-benzoyl-*N*-methylpyridinium-based dyads more efficiently than it quenches the photoemission of aerogels doped with $[Ru(phen)3]^{2+}$ (Reprinted with permission from Leventis et al. (2004). Copyright 2004 American Chemical Society)

has been accomplished through quenching of the fluorescence of *N*-(3-trimethoxysilylpropyl)-2,7-diazapyrenium bromide incorporated into a highly porous aerogel through direct co-condensation and post-synthesis grafting (Leventis et al. 1999, 2004) (see Fig. 8.4). It was found that co-condensation provided a more uniform distribution of the dye and more effective interrogation of the resulting photoluminescence. The detection of oxygen through interrogation of fluorescence intensities as well as through phosphorescent lifetime measurements can be achieved using platinum and palladium metalloporphyrins incorporated in silica matrix as well (Han et al. 2005; Zhang et al. 2005; Huo et al. 2006). The incorporation of ruthenium complexes into a variety of materials can also be applied to detect and quantify oxygen (Zhang et al. 2002; Leventis et al. 2004; Han et al. 2005; Lei et al. 2006; Wang et al. 2008). In the case of ruthenium complexes, covalent modification techniques were found to yield notably more linear Stern–Volmer plots (I_0/I vs. oxygen concentration) than techniques that employed entrapment of the dyes (Lei et al. 2006).

Experiments have shown that other types of gas sensors can also use mesoporous silicas as the sensing layer. For example, selective sensing of nitric oxide has been accomplished in the presence of carbon monoxide through the application of quartz crystal microbalance technology employing a cobalt phthalocyanine-modified sol–gel thin film (Palaniappan et al. 2006, 2008). Electrochemical detection of nitrite has been accomplished using a metalloporphyrin-modified silicate material

(Xie et al. 2004; Cardoso and Gushikem 2005; Cardoso et al. 2005). Gaseous hydrogen peroxide and hydrazine can also be detected electrochemically using nanostructured silica as the electrode (Holmstrom and Cox 1998; Holmstrom et al. 2000). An ordered mesoporous film incorporated in a metal–insulator–semiconductor (MIS) made it possible to detect NO and NO_2 (Yamada et al. 2002, 2004; Yuliarto et al. 2004, 2006). The MIS structure consisted of a silicon wafer with a silicon dioxide layer and a Si_3N_4 layer over that. An ordered mesoporous film was spin coated as an insulating layer and calcined over the Si_3N_4. Al was evaporated on the bottom of the Si, and Au was sputtered on top of the mesoporous silica to work as electrodes. An LED light irradiated the Si and induced an AC photocurrent while a DC bias voltage was applied; adsorption of gas changed the dielectric constant of the insulating layer and the measured photocurrent response. It was found that such sensors can have a detection limit of 100 ppb NO_2.

Mesoporous silica can also be employed as a hard template for the synthesis of a mesoporous material of a composition valuable for sensing (Melde et al. 2008). Ordered, well-defined mesostructures are particularly suited to these applications. In particular, mesoporous silicates have been used to template carbons, metals, and metal oxides. The silicate framework is usually removed following templating by dissolving with hydrofluoric acid or a strong base. For example, Wang et al. (2005) used 3D cubic mesoporous silica thin films to create Pt nanowire networks by electrodeposition. The Pt networks had an electrochemically active surface area ca. 27 m^2/g and, when applied as an electrode, exhibited higher current densities for the oxidation of methanol than a nonporous polycrystalline Pt electrode. Mesostructured tungsten oxide has been templated by impregnation of 2D hexagonal and 3D cubic mesoporous materials with phosphotungstic acid (Rossinyol et al. 2007a, b). A similar technique was applied to the generation of In_2O_3- and CaO-loaded In_2O_3 materials with SBA-15 (Prim et al. 2007). These materials were applied to the detection of CO_2. Wagner et al. synthesized a mesoporous ZnO for sensing CO and NO_2 by a double hard templating route (Wagner et al. 2007). First a mesostructured carbon was synthesized by impregnating a mesoporous material with sucrose, pyrolyzing, and removing the silica. The carbon mesostructure was then impregnated with zinc nitrate and heated to convert to ZnO and combust the carbon.

In addition, mesoporous materials can be used for adsorption and preconcentration of analytes in order to attain detectable concentrations for a particular sensor system. Silicate-based layers with incorporated catalysts can act as a protective catalytic filter to eliminate an interferent as well. One can find in some reviews focused on the application of hybrid sol–gel films and monoliths for optical and electrochemical sensing of inorganic species other information related to silicas (Carrington and Xue 2007; Melde et al. 2008), mesoporous silica nanoparticles for biosensing (Slowing et al. 2007; Walcarius 2008), mesoporous silicates for electrochemical detection (Walcarius 2008), and sol–gels and templated mesoporous materials for fluorescence-based sensing (Basabe-Desmonts et al. 2007; Ariga et al. 2007), etc.

8.2 Aluminosilicates (Zeolites)

Zeolites are another group of compounds that have generated some interest (Rolison 1990; Walcarius 1999; Kulprathipanja 2010). Chemically, they are represented by the empirical formula

$$\mathrm{M}_{x/m}\left[(AlO_2)_x(SiO_2)_y\right]z\mathrm{H_2O} \tag{8.2}$$

where M is the cation with valence *m*, *z* is the number of water molecules in each unit cell, and *x* and *y* are integers such that *y*/*x* is greater than or equal to 1 (Jacobs 1977). Great interest in zeolites is not surprising considering that they comprise a microporous open framework structure which is accessible

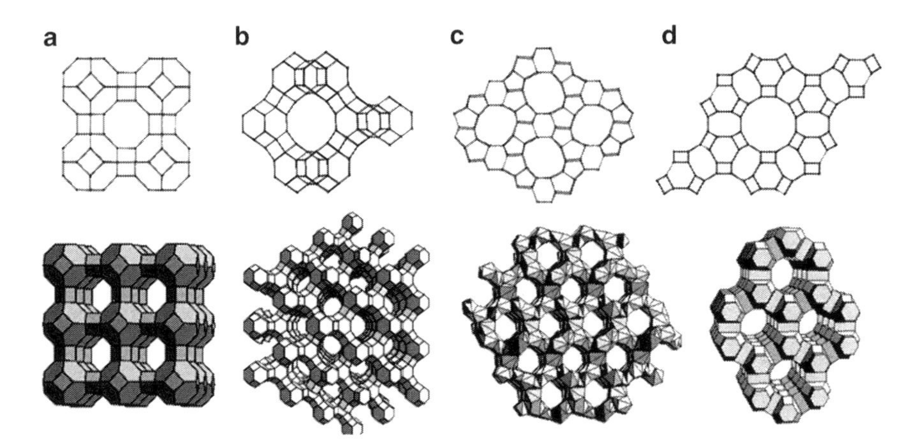

Fig. 8.5 Representative zeolite frameworks (with pore openings). (**a**) Zeolite A (3D, 4.2 Å). (**b**) Zeolite Y (3D, 7.4 Å). (**c**) Zeolite L (1D, 7.1 Å). (**d**) ZSM-5 (silicalite) (2D, 5.3×5.6 Å, 5.1×5.5 Å): D—dimensions of channel system (Reprinted from Zheng et al. 2012. Published by MDPI)

to certain guest molecules. The initial building blocks of the zeolite crystal lattice are AlO_4 or SiO_4 tetrahedra that are interconnected via oxygen bridges. Thus, a well-defined, 3D framework is created in which micro- and mesopores are linked by channels (see Fig. 8.5). At the same time, this high degree of open porosity gives rise to an exceptionally high surface area. Each AlO_4 tetrahedron in the framework bears a net negative charge which is balanced by an extra-framework cation. Intracrystalline channels or interconnected voids are occupied by the cations and water molecules. The cations are mobile and ordinarily undergo ion exchange. The water may be removed reversibly, generally by the application of heat, which leaves intact a crystalline host structure permeated by the micropores and voids which may amount to 50 % of the crystals by volume. The intracrystalline channels or voids can be 1, 2, or 3D (Flanigen et al. 2010).

Zeolites are usually synthesized under hydrothermal conditions in the low temperature range (70–300 °C) from solutions of sodium aluminate, sodium silicate, or sodium hydroxide (Yu 2007). The precise zeolite formed is determined by the reactants and the particular synthesis conditions used, such as temperature, time, pH (usually pH > 10), and templating ion. The templating ion is usually a cation around which the aluminosilicate lattice is formed, so that the tunnel size is determined by the templating cation (Kulprathipanja 2010). The zeolite synthesis is carried out with inorganic as well as organic precursors. The inorganic precursors yielded more hydroxylated surfaces, whereas the organic precursors easily incorporated the metals into the network. Temperatures higher than 200–300 °C often give denser materials. The addition of fluoride to the reactive gel led to more perfect and larger crystals of known molecular sieve structures as well as new structures and compositions (Villaescusa and Camblor 2003). The fluoride ion is also reported to serve as a template or SDA in some cases. Fluoride addition extends the synthesis regime into the acidic pH region.

At present more than 180 distinct framework structures of zeolites are known. Several of them are listed in Table 8.3 and shown in Fig. 8.6. Some of the more important zeolite types, most of which have been used in commercial applications, include: the zeolite minerals mordenite, chabazite, erionite, and clinoptilolite; the synthetic zeolite types A, X, Y, L, "Zeolon" mordenite, ZSM-5, beta, and MCM-22; and the zeolites F and W. They exhibit pore sizes from 0.3 to 1.0 nm and pore volumes from about 0.10–0.35 cm^3/g. Typical zeolite pore sizes include: (1) small-pore zeolites with eight-ring pores, free diameters of 0.30–0.45 nm (e.g., zeolite A); (2) medium-pore zeolites with 10-ring pores, 0.45–0.60 nm in free diameter (ZSM-5); (3) large-pore zeolites with 12-ring pores of 0.6–0.8 nm (e.g., zeolites X, Y (see Fig. 8.6)); and (4) extra-large-pore zeolites with 14-ring pores (e.g., UTD-1) (Flanigen et al. 2010). It should be noted that the zeolite framework should be viewed as somewhat flexible, with the size and

Table 8.3 Typical oxide formula of some synthetic zeolites

Framework	Cationic form	Formula of typical unit cell	Window	Effective channel diameter (nm)
	Na	$Na_{12}[(AlO_2)_{12}(SiO_2)_{12}]$	8-ring	0.38
A	Ca	$Ca_2Na_2[(AlO_2)_{12}(SiO_2)_{12}]$	8-ring	0.44
	K	$K_{12}[(AlO_2)_{12}(SiO_2)_{12}]$	8-ring	0.29
	Na	$Na_{86}[(AlO_2)_{86}(SiO_2)_{106}]$	12-ring	0.84
X	Ca	$Ca_{40}Na_6[(AlO_2)_{86}(SiO_2)_{106}]$	12-ring	0.80
	Sr, Ba	$Sr_{21}Ba_{22}[(AlO_2)_{86}(SiO_2)_{106}]$	12-ring	0.80
	Na	$Na_{56}[(AlO_2)_{56}(SiO_2)_{136}]$	12-ring	0.80
Y	K	$K_{56}[(AlO_2)_{56}(SiO_2)_{136}]$	12-ring	0.80
Mordenite	Ag	$Ag_8[(AlO_2)_8(SiO_2)_{40}]$	12-ring	0.70
	H	$H_8[(AlO_2)_8(SiO_2)_{40}]$		
Silicalite	–	$(SiO_2)_{96}$	10-ring	0.60
ZSM-5	Na	$Na_3[(AlO_2)_3(SiO_2)_{93}]$	10-ring	0.60

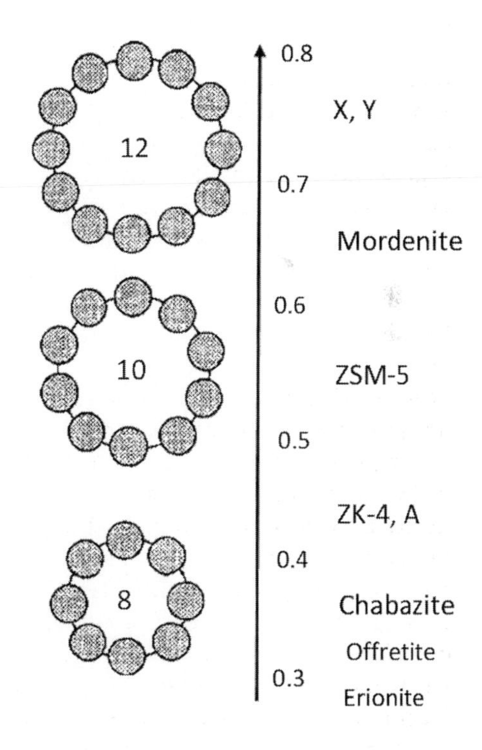

Fig. 8.6 Dimensions of the pores of different zeolites (in nanometers) (Reprinted with permission from Ribeiro (1993). Copyright 1993 Springer)

shape of the framework and pore responding to changes in temperature and guest species. For example, ZSM-5 with sorbed neopentane has a near-circular pore of 0.62 nm, but with substituted aromatics as the guest species, the pore assuming an elliptical shape, 0.45–0.70 nm in diameter.

It was established that the ratio of silicon to aluminum atoms in the lattice, x/y, is a very important parameter for zeolite characterization. The more aluminum-based units present in the zeolite lattice, i.e., the higher y, the more cations are needed for charge compensation. As a consequence, zeolites with high aluminum content are highly polar materials with excellent ion-exchange capacity and potential for high ion conductivity. This means that by changing the framework Si/Al ratio of the zeolite, the ion-exchange capacity, and conductivity, the interaction between the zeolite and the adsorbed molecules and the modification of hydrophilic or hydrophobic properties can all be changed (Xu et al. 2006). The zeolites with low silica contents are hydrophilic, and are usually used as drying agents for absorbing steam, whereas hydrophobic highly siliceous zeolites are used for absorbing organic molecules from humid air or water. Thus, varying the framework Si/Al ratio of zeolites greatly changes the adsorption selectivity toward molecules with different polarity. In addition, the aluminum ions form catalytically active sites, since they may either act as Brunstedt or Lewis acids. Such sites are of great interest in a number of catalyzed organic reactions. The x/y ratio indicates the amount of acidic centers per unit cell as well as the content of mobile cations.

Their surface and structural properties can easily be modified, which makes them ideal candidates for the selective adsorption of various volatile hydrocarbons and small organic molecules (Pejcic et al. 2007). For example, various post-synthesis steps were developed to modify the zeolite chemically by incorporating catalytically active metal clusters such as Pt, Fe, or Cu, or by anchoring organic dyes within the pore structure.

8.2.1 Zeolite-Based Gas Sensors

As shown earlier, zeolites have several physical and structural features that can be exploited for sensing (Zheng et al. 2012). The internal microporosity of zeolites that gives rise to the high surface area also provides sites for adsorption of molecules. The negative aluminosilicate framework of zeolites necessitates the presence of neutralizing ion-exchangeable cations within the framework. These cations can influence adsorption, diffusion, and catalytic properties of zeolites and, thereby, influence sensing behavior. The extra-framework cations are bound electrostatically at preferential sites and can perform energy-activated motion between these sites. The presence of guest molecule within the zeolite can interfere with this motion and has been used as the basis of sensing. Interaction of the molecule with the cation is manifested in the change in impedance/capacitance as measured by the frequency-dependent impedance spectra. Upon adsorption, mass changes as well as optical/electrical properties are altered, which can be used for sensor transduction. Selectivity toward analytes has also been observed by adsorption of species within a zeolite. Microporous spaces within the zeolite can serve as hosts for guest species.

A number of review papers on zeolite-based gas sensors are available in the literature (Moos et al. 2006, 2009; Xu et al. 2006; Walcarius 2008; Sahner et al. 2008; Zheng et al. 2012). They focus on various aspects of this materials class and mostly classify devices according to the analytes to be detected or the type of sensor transduction. When used in sensor elements, zeolites can take various roles, which fall into two major categories (see Fig. 8.7).

In a great number of cases, zeolites are used as auxiliary elements. They may act either as a framework to stabilize the sensor material, as filter layers (either catalytic or size restrictive) to enhance selectivity of a sensitive film, or as a preconcentrator of specific analytes from diluted solutions. For example, due to excellent chemical and thermal stability, zeolites can be used as a substrate to prepare compounds and devices with desirable fundamental physical and chemical properties (Xu et al. 2006). For example, inorganic or organic compounds, metal and metal–organic compounds, and their clusters can be assembled into the pores and cages in zeolites. Some nanosized metal or metal oxide particles have been successfully inserted into the caves and the pores or highly dispersed on the external surface of zeolites.

The second group encompasses devices in which the zeolite itself is the main functional material leading to a sensor effect. Such detection principles rely directly on adsorptive, catalytic, or conductive properties of one specific zeolite that are subject to well-defined changes depending on the composition of the gaseous surroundings (Alberti and Fetting 1994; Xu et al. 2006; Sahner et al. 2008).

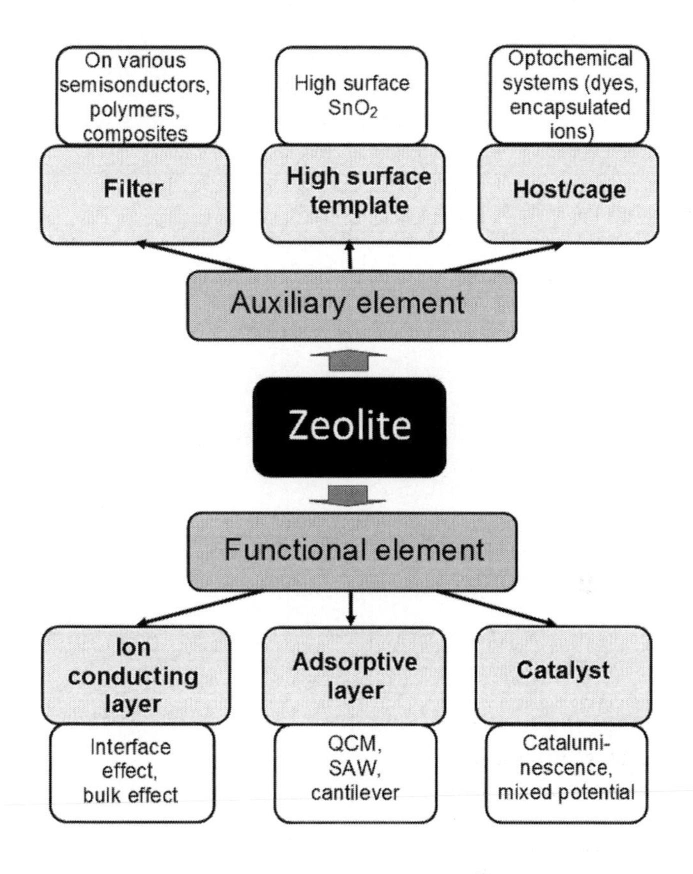

Fig. 8.7 Classification of zeolite-based gas sensors (Adapted with permission from Sahner et al. (2008). Copyright 2008 Elsevier)

For example, the encapsulating ruthenium(II) complexes inside zeolite supercages make it possible to design oxygen sensors, the insertion of methylene blue or LiCl into zeolites is effective for humidity sensing (Zou et al. 2004), and the fixing of zeolites on the surface of conductometric, quartz crystal microbalance or SAW-based sensors improves selectivity of these devices (Moos et al. 2009; Zheng et al. 2012). It was established that better selectivity can be achieved through the correct choice of both geometrical properties, i.e., pore sizes and types of porous network, and acidic properties of zeolites, which can be controlled, for example, by the Si/A1 ratio and the nature and quantity of the compensation cations (Ribeiro 1993; Moos et al. 2009). Only molecules of a certain size are able to be absorbed by a given zeolite material, or pass through its pores, while molecules of bigger sizes cannot (see Table 8.4). The size of the pores can be chosen from large-pore zeolites, like zeolite Y and ZSM-20, which are very useful in the transformation of large molecules, to intermediate-pore zeolites, like ZSM-5, which already present shape selectivity toward molecules, such as ramified paraffins, down to small-pore zeolites that have many applications in the separation of very small molecules like oxygen and nitrogen. Ammonia gas sensors using proton conductive zeolites and measuring impedance spectroscopy have been developed as well (Rodriguez-Gonzalez et al. 2005; Moos et al. 2009). Various composites with specific properties can also be designed based on zeolites. For example, Chuapradit et al. (2005) reported on polyaniline/zeolite composites responsive to CO. Several examples of zeolites applications in gas sensors are presented in Table 8.5.

However, zeolites do have a strong tendency to adsorb water preferentially relative to analyte, and this appears to be hindering their application in environments with elevated moisture levels. Moreover, due to the above-mentioned effect, thermal treatment for zeolites activation is required. We also need to take into account that, due to small-pore size diffusion, a limitation

Table 8.4 Molecular sieving effect in zeolites

Typical molecules rapidly occluded at RT or below	Typical molecules moderately, rapidly or slow occluded at RT or above	Typical molecules not appreciably occluded at RT or above
1	2	3
I. Zeolites with pore size from 0.489 to 0.558 nm		
He; Ne; Ar; H_2; N_2; O_2; CO; CO_2; COS; CS_2; H_2O; HCl; HBr; NO; NH_3; CH_3NH_2; H_2S; CH_3OH; CH_3CN; NCN; Cl_2; CH_3Cl; CH_3Br; CH_3F; CH_2Cl_2; CH_2F_2; CH_4; C_2H_6; C_2H_2; CH_2O; CH_3SH	C_3H_8; CH_3CH_2OH; $CH_3CH_2NH_2$; CH_3CH_2F; CH_3CH_2Cl; CH_3CH_2Br; I_2; HI; CH_2Br_2; CH_3I; CH_3CH_2CN; CH_3CH_2SH; HCO_2CH_3; $HCO_2CH_2CH_3$; CH_3COCH_3; $(CH_3)NH$; $(CH_3CH_2)_2NH$	Aromatic hydrocarbons, cyclo- and isoparaffins; heterocyclic compounds (e.g., thiophene, pyrrole, pyridine); $CHCl_3$; CCl_4; $CHCl{:}CCl_2$; CH_3CHCl_2; $CHCl_2CCl_3$; CCl_3CCl_3; secondary straight-chain alcohols, thiols, nitriles, and halides; primary amines with NH_2 group attached to secondary carbon atom; tertiary amines; branched-chain ethers, thioethers, and secondary amines
II. Zeolites with pore size from 0.4 to 0.489 nm		
He; Ne; Ar; H_2; O_2; N_2; CO; NH_3; H_2O	CH_4; C_2H_6; CH_3OH; CH_3NH_2; CH_3CN; CH_3Cl; CH_3F; NCN; Cl_2	All classes of molecules in cols. 2 and 3 for section I
III. Zeolites with pore size from 0.384 to 0.4 nm		
He; Ne; H_2; O_2; N_2; H_2O	Ar; HCl; NH_3	All classes of molecules in cols. 2 and 3 for section I

Source: Data from Barrer (1949)

Table 8.5 Types of zeolite-based materials for gas sensors

Material	Sensor type/function of zeolite	Target gas
Ru(II) complex/zeolite	Optical/support (guest/host)	O_2
methylene blue/zeolite; LiCl/zeolites (FAU)		H_2O
Zeolite films (zeolite A, FAU, ZSM-5, etc.)	QCM; SAW; microcantilever/change of mass(absorption)	NO; SO_2; H_2O; NH_3; acetone; pentane; hexane; Freon; dimethyl methylphosphonate
Ion conductive zeolite (ZSM-5, etc.)	Conductometric; potentiometric/conductivity change (adsorption)	NH_3; hydrocarbons; methanol; 2-propanol and 3-pentanol
Zeolite films (LTA, etc.)	Chemiresistors; optical/filter (absorption)	O_2; CO_2; CO; organics; butylamine
Composites: polyaniline/zeolite; ZrO_2:Y/zeolite; Pt/Y-zeolite; $SrTi_{1-x}Fe_xO_3$/zeolite; TiO_2/zeolite; CrTiO/zeolite; SnO_2/zeolite	Conductometric; optical/support	CO; NO; H_2; ethanol; hydrocarbons (methane; butane; propane)

Source: Reprinted from Xu et al. (2006). Published by MDPI

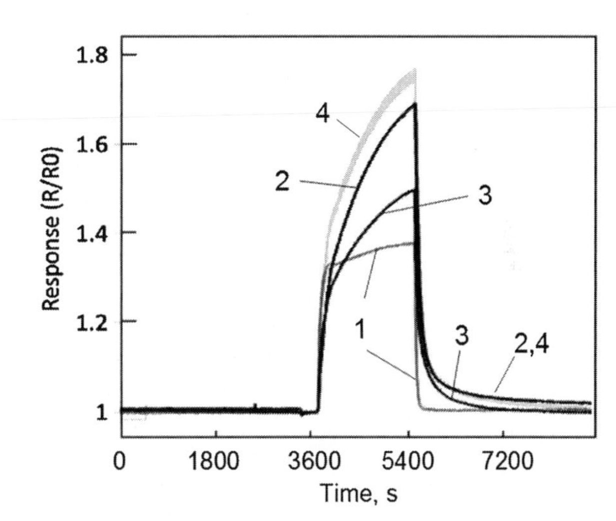

Fig. 8.8 Resistive response of WO_3 sensors to 400 ppb NO_2 in dry air at an operating temperature of 350 °C: 1—WO_3; 2—WO_3+H-ZSM-5; 3—WO_3+Cr-ZSM-5; 4—WO_3+Cr-LTA (Adapted with permission from Varsani et al. (2011). Copyright 2011 Elsevier)

will be present in kinetics of zeolite-based gas sensors. This means that response and recovery will be slow. Figure 8.8 shows this effect for zeolite-modified WO_3-based gas sensors. It is believed that the use of very thin zeolitic films (nanometer dimensions) can only improve response and recovery times. In addition, we need to take into account that the main problems with the use of zeolites with small sizes (nanosizes) are their low synthesis yields and inconsistent reproducibility. Chiral zeolites have better reproducibility, but the difficulties in resolving enantiopure zeolites still persist.

References

Alberti K, Fetting F (1994) Zeolites as sensitive materials for dielectric gas sensors. Sens Actuators B Chem 21:39–50

Ariga K, Vinu A, Hill JP, Mori T (2007) Coordination chemistry and supramolecular chemistry in mesoporous nanospace. Coord Chem Rev 251:2562–2591

Barrer RM (1949) Separation using zeolitic materials. Discuss Faraday Soc 7:135–141

Basabe-Desmonts L, Reinhoudt DN, Crego-Calama M (2007) Design of fluorescent materials for chemical sensing. Chem Soc Rev 36:993–1017

Beck JS, Vartuli JC, Roth WJ, Leonowicz ME, Kresge CT, Schmitt KD, Chu CT-W, Olson DH, Sheppard EW, McCullen SB, Higgins JB, Schlenker JL (1992) A new family of mesoporous molecular sieves prepared with liquid crystal templates. J Am Chem Soc 114:10834–10843

Bertolo JM, Bearzotti A, Generosi A, Palummo L, Albertini VR (2005) X-Rays and electrical characterizations of ordered mesostructured silica thin films used as sensing membranes. Sens Actuators B Chem 111–112:145–149

Brinker CJ, Scherer GW (1990) Sol–gel science: the physics and chemistry of sol–gel processing. Academic, San Diego

Cardoso WS, Gushikem Y (2005) Electrocatalytic oxidation of nitrite on a carbon paste electrode modified with Co(II) porphyrin adsorbed on SiO_2/SnO_2/phosphate prepared by the sol–gel method. J Electroanal Chem 583:300–306

Cardoso WS, Francisco MSP, Landers R, Gushikem Y (2005) Co(II) porphyrin adsorbed on SiO_2/SnO_2/phosphate prepared by the sol–gel method. Application in electroreduction of dissolved dioxygen. Electrochim Acta 50:4378–4384

Carrington NA, Xue ZL (2007) Inorganic sensing using organofunctional sol–gel materials. Acc Chem Res 40:343–350

Chuapradit C, Wannatong LR, Chotpattananont D, Hiamtup P, Sirivat A, Schwank J (2005) Polyaniline/zeolite LTA composites and electrical conductivity response towards CO. Polymer 46:947–953

Corma A (1997) From microporous to mesoporous molecular sieve materials and their use in catalysis. Chem Rev 97:2373–2419

Domansky K, Liu J, Wang L-Q, Engelhard MH, Baskaran S (2001) Chemical sensors based on dielectric response of functionalized mesoporous silica films. J Mater Res 16:2810–2816

Falcaro P, Bertolo JM, Innocenzi P, Amenitsch H, Bearzotti A (2004) Ordered mesostructured silica films: effect of pore surface on its sensing properties. J Sol-Gel Sci Technol 32:107–110

Flanigen EM, Broach RW, Wilson ST (2010) Introduction. In: Kulprathipanja S (ed) Zeolites in industrial separation and catalysis. Wiley-VCH, Weinheim

Galarneau A, Di Renzo F, Fajula F, Vedrine J (eds) (2001) Zeolites and mesoporous materials at the dawn of the 21st century. Elsevier, Amsterdam

Han B-H, Manners I, Winnik MA (2005) Oxygen sensors based on mesoporous silica particles on layer-by-layer self-assembled films. Chem Mater 17:3160–3171

Hoffmann F, Cornelius M, Morell J, Froba M (2006) Silica-based mesoporous organic–inorganic hybrid materials. Angew Chem Int Ed 45:3216–3251

Holmstrom SD, Cox JA (1998) Solid-state voltammetric determination of gaseous hydrogen peroxide using nanostructured silica as the electrode. Electroanalysis 10:597–601

Holmstrom SD, Sandlin ZD, Steinecker WH, Cox JA (2000) Mediated oxidation and determination of gaseous monomethyl hydrazine in a solid-state voltammetric cell employing a sol–gel electrolyte. Electroanalysis 12:262–266

Huo Q, Margolese DI, Stucky G (1996) Surfactant control of phases in the synthesis of mesoporous silica-based materials. Chem Mater 8:1147–1160

Huo C, Zhang H, Zhang H, Zhang H, Yang B, Zhang P, Wang Y (2006) Synthesis and assembly with mesoporous silica MCM-48 of platinum(II) porphyrin complexes bearing carbazyl groups: spectroscopic and oxygen sensing properties. Inorg Chem 45:4735–4742

Inagaki S, Fukushima Y, Kuroda K (1993) Synthesis of highly ordered mesoporous materials from a layered polysilicate. J Chem Soc Chem Commun 8:680–682

Innocenzi P, Martucci A, Guglielmi M, Bearzotti A, Traversa E (2001) Electrical and structural characterisation of mesoporous silica thin films as humidity sensors. Sens Actuators B Chem 76:299–303

Innocenzi P, Falcaro P, Bertolo JM, Bearzotti A, Amenitsch H (2005) Electrical responses of silica mesostructured films to changes in environmental humidity and processing conditions. J Non-Cryst Solids 351:1980–1986

Jacobs PA (1977) Carboniogenic activity of zeolites. Elsevier Scientific, New York

Kresge CT, Leonowicz ME, Roth WJ, Vartuli JC, Beck JS (1992) Ordered mesoporous molecular sieves synthesized by a liquid-crystal template mechanism. Nature 359:710–712

Kulprathipanja S (ed) (2010) Zeolites in industrial separation and catalysis. Wiley-VCH, Weinheim

Lei B, Li B, Zhang H, Lu S, Wenlian Z, Wang Y (2006) Mesostructured silica chemically doped with Ru(II) as a superior optical oxygen sensor. Adv Funct Mater 16:1883–1891

Leventis N, Elder IA, Rolison DR, Anderson ML, Merzbacher CI (1999) Durable modification of silica aerogel monoliths with fluorescent 2.7-diazapyrenium moieties. Sensing oxygen near the speed of open-air diffusion. Chem Mater 11:2837–2845

Leventis N, Rawashdeh A-MM, Elder IA, Yang J, Dass A, Sotiriou-Leventis C (2004) Synthesis and characterization of Ru(II) incorporating the 4-benzoyl-*N*-methylpyridinium cation or *N*-benzyl-*N* -methyl viologen. Improving the dynamic range, sensitivity, and response time of sol–gel based optical oxygen sensors. Chem Mater 16:1493–1506

Melde BJ, Johnson BJ, Charles PT (2008) Mesoporous silicate materials in sensing. Sensors 8:5202–5228

Mittal KL, Fendler EJ (eds) (1982) Solution behavior of surfactants. Plenum, New York

Moos R, Sahner K, Hagen G, Dubbe A (2006) Zeolites for sensors for reducing gases. Rare Metal Mater Eng Suppl B 35:447–451

Moos R, Sahner K, Fleischer M, Guth U, Barsan N, Weimar U (2009) Solid state gas sensor research in Germany—a status report. Sensors 9:4323–4365

Naik B, Ghosh NN (2009) A review on chemical methodologies for preparation of mesoporous silica and alumina based materials. Recent Pat Nanotechnol 3:213–224

Palaniappan A, Li X, Tay FEH, Li J, Su X (2006) Cyclodextrin functionalized mesoporous silica films on quartz crystal microbalance for enhanced gas sensing. Sens Actuators B Chem 119:220–226

Palaniappan A, Moochhala S, Tay FEH, Su X, Phua NCL (2008) Phthalocyanine/silica hybrid films on QCM for enhanced nitric oxide sensing. Sens Actuators B Chem 129:184–187

Pejcic B, Eadington P, Ross A (2007) Environmental monitoring of hydrocarbons: a chemical sensor perspective. Environ Sci Technol 41(18):6333–6342

Prim A, Pellicer E, Rossinyol E, Peiró F, Cornet A, Morante JR (2007) A novel mesoporous CaO-loaded In_2O_3 material CO_2 sensing. Adv Funct Mater 17:2957–2963

Ribeiro FR (1993) Adaptation of the porosity of zeolites for shape selective reactions. Catal Lett 22:107–121

Rodriguez-Gonzalez L, Franke ME, Simon U (2005) Electrical detection of different amines with proton-conductive H-ZSM-5. Stud Surf Sci Catal 158:2049–2056

Rolison DR (1990) Zeolite-modified electrodes and electrode-modified zeolites. Chem Rev 90(5):867–878

Rossinyol E, Prim A, Pellicer E, Arbiol J, Hernández-Ramírez F, Peiró F, Cornet A, Morante JR, Solovyov LA, Tian B, Bo T, Zhao D (2007a) Synthesis and characterization of chromium-doped mesoporous tungsten oxide for gas-sensing applications. Adv Funct Mater 17:1801–1806

Rossinyol E, Prim A, Pellicer E, Rodríguez J, Peiró F, Cornet A, Morante JR, Tian B, Bo T, Zhao D (2007b) Mesostructured pure and copper-catalyzed tungsten oxide for NO_2 detection. Sens Actuators B Chem 126:18–23

Sahner K, Hagen G, Schönauer D, Reiß S, Moos R (2008) Zeolites—versatile materials for gas sensors. Solid State Ionics 179(40):2416–2423

Slowing II, Trewyn BG, Giri S, Lin VS-Y (2007) Mesoporous silica nanoparticles for drug delivery and biosensing applications. Adv Funct Mater 17:1225–1236

Soler-Illia GJ, Sanchez C, Lebeau B, Patarin J (2002) Chemical strategies to design textured materials: from microporous and mesoporous oxides, to nanonetworks and hierarchical structures. Chem Rev 102:4093–4138

Varsani P, Afonja A, Williams DE, Parkin IP, Binions R (2011) Zeolite-modified WO_3 gas sensors-enhanced detection of NO_2. Sens Actuators B Chem 160:475–482

Villaescusa L, Camblor M (2003) The fluoride route to new zeolites. Recent Res Dev Chem 1:93–141

Wagner T, Waitz T, Roggenbuck J, Fröba M, Kohl C-D, Tiemann M (2007) Ordered mesoporous ZnO for gas sensing. Thin Solid Films 515:8360–8363

Walcarius A (1999) Zeolite-modified electrodes in electrochemical chemistry. Anal Chim Acta 384:1–16

Walcarius A (2008) Electroanalytical applications of microporous zeolites and mesoporous (organo)silicas: recent trends. Electroanalysis 20(7):711–738

Wang D, Kou R, Gil MP, Jakobson HP, Tang J, Yu D, Lu Y (2005) Templated synthesis, characterization, and sensing application of macroscopic platinum nanowire network electrodes. J Nanosci Nanotechnol 5:1904–1909

Wang B, Liu Y, Li B, Yue S, Li W (2008) Optical oxygen sensing materials based on trinuclear starburst ruthenium(II) complexes assembled in mesoporous silica. J Lumines 128:341–347

Xie F, Li W, He J, Yu S, Yang H (2004) Directly immobilize polycation bearing Os complexes on mesoporous material MAS-5 and its electrocatalytic activity for nitrite. Mater Chem Phys 86:425–429

Xu X, Wang J, Long Y (2006) Zeolite-based materials for gas sensors. Sensors 6:1751–1764

Yamada T, Zhou HS, Uchida H, Tomita M, Ueno Y, Honma I, Asai K, Katsube T (2002) Application of a cubic-like mesoporous silica film to a surface photovoltage gas sensing system. Micropor Mesopor Mater 54:269–276

Yamada T, Zhou HS, Uchida H, Honma I, Katsube T (2004) Experimental and theoretical NO_x physisorption analyses of mesoporous film (SBA-15 and SBA-16) constructed surface photo voltage (SPV) sensor. J Phys Chem B 108:13341–13346

Yu J (2007) Synthesis of zeolites. In: Cejka J, Van Bekkum H, Corma A, Schuth F (eds) Introduction to zeolite science and practice, vol 168, 3rd edn, Studies in surface science and catalysis. Elsevier, Amsterdam, pp 39–103

Yuliarto B, Zhou HS, Yamada T, Honma I, Katsumura Y, Ichihara M (2004) Effect of tin addition on mesoporous silica thin film and its application for surface photovoltage NO_2 gas sensor. Anal Chem 76:6719–6726

Yuliarto B, Honma I, Katsumura Y, Zhou H (2006) Preparation of room temperature NO_2 gas sensors based on W- and V-modified mesoporous MCM-41 thin films employing surface photovoltage technique. Sens Actuators B Chem 114:109–119

Zhang P, Guo J, Wang Y, Pang W (2002) Incorporation of luminescent tris(bipyridine)ruthenium(II) complex in mesoporous silica spheres and their spectroscopic and oxygen-sensing properties. Mater Lett 53:400–405

Zhang H, Sun Y, Zhang P, Wang Y (2005) Oxygen sensing materials based on mesoporous silica MCM-41 and Pt(II)-porphyrin complexes. J Mater Chem 15:3181–3186

Zheng Y, Li X, Dutta PK (2012) Exploitation of unique properties of zeolites in the development of gas sensors. Sensors 12:5170–5194

Zou J, He HY, Dong JP, Long YC (2004) A guest/host material of LiCl/H-STI (stilbite) zeolite assembly: preparation, characterization and humidity-sensitive properties. J Mater Chem 14:2405–2411

Chapter 9
Cavitands

9.1 Cavitands: Characterization

In the last 2 decades, researchers active in the field of supramolecular chemistry have designed and prepared an amazing number of different synthetic receptors for binding and recognition of neutral molecules (Vögtle 1996). Cavitands, synthetic organic compounds with a container shape, are extremely interesting and versatile molecular receptors (Cram 1983). Possible structures of cavitands are shown in Fig. 9.1. The specific interactions between the cavitand and guest molecules are mainly based on their bucket-like conformation. The cavity of the cavitand allows it to engage in host–guest chemistry with guest molecules of a complementary shape and size. As seen in Fig. 9.1, cavitands have different shapes and sizes of their cavities, which are easy to modify, and therefore they have different complexing abilities with target molecules.

Phosphonate cavitands represent one class of molecular receptors that have been studied in detail to reveal the factors that lead to selective binding using alcohols as model analyte vapors.

It was established that these factors include: (1) simultaneous hydrogen bonding with a P=O group and CH–π interactions with the π-basic cavity; (2) a rigid cavity that provides a permanent free volume for the analyte around the inward facing P=O groups, which is essential for effective hydrogen bonding; and (3) a network of energetically equivalent hydrogen-bonding sites available to the analyte (Melegari et al. 2008). The nonspecific dispersion interactions can be much stronger than the specific interactions and can depend on the chain length of sensed alcohols and their concentration (Pinalli et al. 2004). Thus, the main specific interactions responsible for recognition are H-bonding, CH–π, and dipole–dipole interactions (Hartmann et al. 1994; Dickert and Schuster 1995). This strategy for selectivity enhancement is fundamentally different from that used in other chemically selective coatings (e.g., polymers) that rely on the solubility of the targets with the coating layer.

Examples of cavitands include cyclodextrins (CD), calixarenes, pillarenes, and cucurbiturils. However, in gas sensors, cyclodextrins and calixarenes are usually used. These compounds are a relatively new family of ion receptors that are receiving increasing attention due to their ease of synthesis and multiple sites for structural modification. For example, the calix[*n*]arenes are a class of cyclooligomers, cyclic supramolecules, synthesized via a phenol–formaldehyde condensation. The *n* in calix[*n*]arenes represents the number of aryl units in the macrocyclic ring which are linked to each other through methylene bridges. A great number of calix[*n*]arenes varying in shape and diameter of the nano-cavity (cylinder, truncated cone) as well as in the type of peripheral functional groups have been developed (Parker 1996). One of the examples of calixarenes is shown in Fig. 9.2. It was found

G. Korotcenkov, *Handbook of Gas Sensor Materials*, Integrated Analytical Systems,
DOI 10.1007/978-1-4614-7388-6_9, © Springer Science+Business Media New York 2014

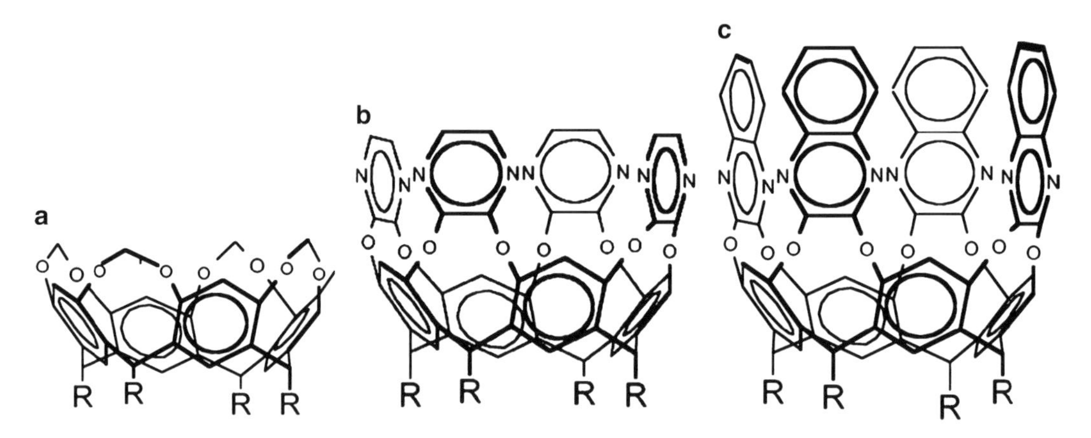

Fig. 9.1 Three cavitands: (**a**) methylene bridged (Me-Cav), (**b**) pyrazine bridged (Py-Cav), and (**c**) quinoxaline bridged (Qx-Cav) (Reprinted with permission from Feresenbet et al. 2004, Copyright 2004 Elsevier)

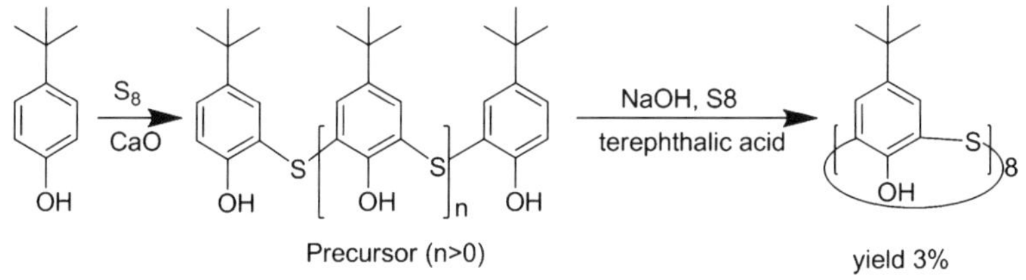

Fig. 9.2 Synthesis of *p-tert*-butyl-octathiacalix[8]arene

that calixarenes form cavities of various diameters and are able to capture metal ions and organic molecules into these cavities ("host–guest" complexation) (Gutsche 1989). Forster (1998) have shown that the nature of the ionophoric activity displayed by derivatized calixarenes is strongly dependent on the cavity size, which can be conveniently altered by varying the reaction conditions and the number of phenyl units (n=4–20) in the macrocycle. Taking into account the synthetic flexibility of the calix[n]arenes allowing direct control, the number of phenolic units present within a single supramolecule, and the nature and length of the spacer units, we make it possible to manage ionophoric activity of calixarenes over this cavity size. It was established that prominent calix[n]arenes (called major calixarenes when n=4, 6, or 8) can be prepared in excellent yields with high purity and at the multigram scale. Calix[n]arenes containing five and seven aryl units (called minor calixarenes) have been obtained in low yields. Calix[n]arenes containing more than eight aryl units (referred to as higher calixarenes) have been discovered only recently and have not yet become readily available. One can find in Chawla et al. 2011 and Sharma and Cragg 2011 more detailed information about the synthesis, properties, and applications of calix[n]arenes in chemical sensing.

Cyclodextrins (CD) possess the same features. They are natural-occurring cyclic oligosaccharides, which have a rigid torus shape, with an inner hydrophobic cavity and an outer hydrophilic one (Szetjli 1998). Similarly to calix[n]arenes, they possess the remarkable ability of forming inclusion complexes with host molecules. The inner diameter of the cavity, hydrophobic properties, and the weak van der Waals forces are the factors that decide the bonding between CD molecules and guest molecules. Moreover, sulfated β-cyclodextrin, in particular, shows a high solubility in water and an anionic behavior in aqueous solution, so it can be electrochemically incorporated in a polymer matrix during an oxidative process. The structure of β-cyclodextrin is shown in Fig. 9.3.

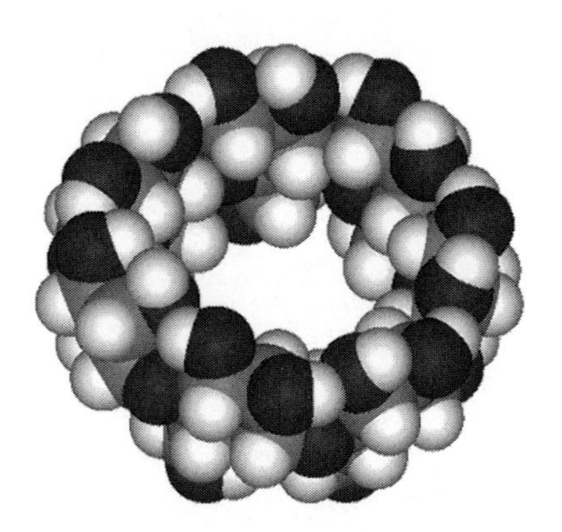

Fig. 9.3 Model of β-cyclodextrin (http://en.wikipedia.org)

9.2 Cavitands as a Material for Gas Sensors

The above-mentioned features of calixarenes and cyclodextrins reveal multiple possibilities for using cavitand films as sensitive layers for various kinds of sensors. By functionally modifying either the upper or lower rims it is possible to prepare various derivatives with differing selectivities for various guest ions and small molecules (Diamond and Nolan 2001; Rudkevich 2007). For example, calixarene derivatives can be incorporated into plasticized poly(vinylchloride) membranes to produce functioning ion-selective electrodes (ISEs) and fluorescence-based matrices (Forster 1998; Ludwig and Dzung 2002) in electrochemical and optical gas sensors. It was found that calixarenes are also useful building platforms in the design of multichromophoric systems in which photoinduced phenomena (electron, charge, and proton transfers, excimer formation, and resonance energy transfer) are controlled by ions. The outstanding selectivities offered by calixarene-based ligands are of major interest for sensing ions (Valeur and Leray 2007). This possibility was realized in Cl_2, NH_3, HCl, and NO_2 calixarene-based gas sensors (Lavrik et al. 1996; Rudkevich 2007; Ohira et al. 2009). In particular, Grady et al. (1997) designed calixarenes-based optical sensors for gaseous ammonia detection in fish samples. The optical detector was based on a calix[4]arene to which a nitrophenylazophenol chromophore was attached. Similarly, it has been observed that the calixarene derivatives respond very strongly to chloroform vapors (Wang et al. 2002). The specific and selective formation of a colored complex of alkylated calixarenes has been utilized to develop a fiber-optic-based colorimetric NO_2 sensor (Ohira et al. 2009). Maffei et al. (2011) established that fluorescent phosphonate cavitands are also good basis for designing selective optical sensing of alcohol vapors. Moreover, they demonstrated that it is possible to achieve high selectivity in chemical vapor sensing by harnessing the binding specificity of a cavitand receptor.

Shenoy (2005) has shown that, by using a real-time, label-free, optical technique called surface plasmon resonance (SPR), refractive index changes induced by analyte–cavitand interactions provide selective signals for sensitive chemical vapor detection as well. For sensor fabrication, cavitand solutions (0.38 mM) in chloroform were spin coated onto SPR substrates (50-nm-thick gold-coated cover glass). Feresenbet et al. (2004) established that the methylene-bridged cavitands (Me-Cav) (see Fig. 9.1) with shallow cavities do not complex aromatic vapors, the pyrazine-bridged cavitands (Py-Cav) show intermediate selectivity, whereas the quinoxaline-bridged cavitands (Qx-Cav) with the deepest cavities show the best selectivity for aromatic vapors. A comparison with polymer coatings, polyepichlorohydrin (PECH) and polyisobutylene (PIB), shows that cavitands have higher selectivity despite the fact that the polymer coatings are more than twice the thickness of the spin-coated cavitand films (Shenoy 2005).

However, the application of calixarenes and other cavitands such as the phosphorus-bridged ones or β-cyclodextrin in low-temperature gas sensors, particularly quartz crystal microbalance (QCM)- and

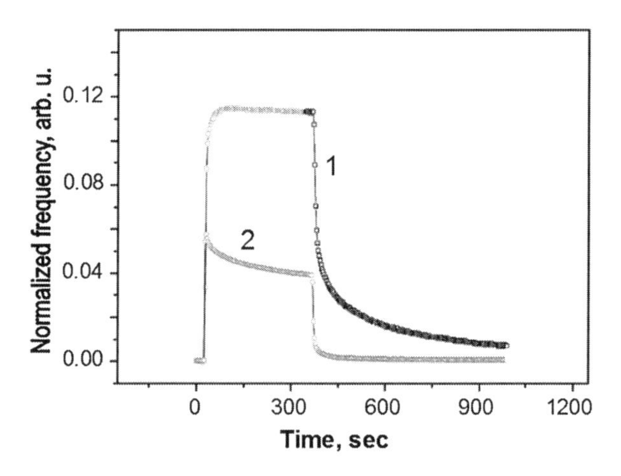

Fig. 9.4 Experimental response of calixarene-based QCM sensors to injection of acetone vapor: sensitive layers are 1 (calix[4]arene), 2 (calix[4]arene) containing single O=P(OPr*i*)Ph functional group (Reprinted with permission from Kalchenko et al. 2002, Copyright 2002 Springer)

surface acoustic wave (SAW)-based sensors, is also possible (Grady et al. 1997; Dickert et al. 1997; Rosler et al. 1998; Li and Ma 2000). For example, Hartmann et al. (1996) and Kalchenko et al. (2002) reported about cavitand-based QCM sensor arrays aimed at detection of volatile organic vapors including pentane, heptane, benzene, chloroform, methanol, acetonitrile, tetrachloroethylene, diethylamine, ethanol, and nitrobenzene (see Fig. 9.4). Dickert et al. (1997) have shown that, in combination with aliphatic spacers, this material fulfills the required demands of high sensitivity and short response times, and the detection of solvents in the gas phase to 2.5 ppm can be realized. Moreover, according to Dickert et al. (1997), the molecular structure of calixarenes could be modified to tune the density and porosity of the coating to the special requirements of SAW and QCM devices. For example, this could be realized on the one hand by using various aldehydes to create different basic calix[4]resorcinarene cavities with variable spacers, such as alkyl chains or alkyl thiolates for self-assembling on gold surfaces, and on the other via bridging of two resorcin molecules of the cavitand by forming cyclic ethers (Davis and Stirling 1995). Sensitivity and selectivity of the host molecules could be varied within a large range in this way. Pinalli et al. (2004) have shown that phosphate and phosphonate cavitands are also sensitive to ethanol, methanol, and benzene vapors. The same result was observed for β-cyclodextrin-based QCM sensors (Wang et al. 2001; Palaniappan et al. 2006). In this study, alkenyl-β-CD was used as the sensing material because guest molecules such as benzene bind tightly to β-CD when compared to α-CD or γ-CD. In addition, it was found that the presence of alkenyl groups in β-CD ensures a covalent attachment of β-CD to the silica matrix used (Palaniappan et al. 2006). Clathrate materials that crystallize in phases with channels or cavities containing solvent molecules can also be used as sensing materials (Ehlen et al. 1993; Finklea et al. 1998; Yakimova et al. 2008; Cha et al. 2009). It was shown that these materials were ~100 times more sensitive to VOCs than polymer-coated thickness shear mode (TSM) devices at low concentrations (Finklea et al. 1998).

Cavitand-based devices have advantages and disadvantages similar to those of polymer-based devices. In particular, cavitand-based sensors are low-temperature devices. For example, the melting temperatures of calix[*n*]arenes usually vary from 200 to 450 °C. Experiments carried out with cavitand-based QCM sensors have shown that the increase of operating temperature is accompanied by strong decrease of sensor signal (see Fig. 9.5). Ferrari et al. (2004) believe that the observed temperature dependence of response is due to the partition coefficient behavior. Another characteristic feature of calixarenes and tetrathiacalixarenes, which limits technological possibilities of sensor fabrication, is their insolubility in water as well as in aqueous bases and their very low solubility in organic solvents. The solubility of calixarenes can be substantially modified only via derivatization (Asfari and Vicens 1988). We should also note that the response of cavitand-based gas sensors is usually slow. Response and recovery times for various sensors vary within the range 10 s to 10 min (Ferrari et al. 2004).

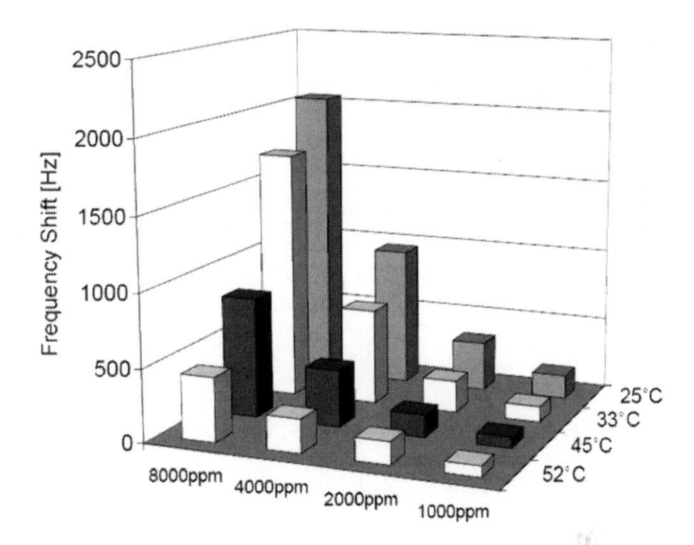

Fig. 9.5 Steady-state frequency shift (absolute value) vs. toluene concentrations at different substrate temperatures for a (Me-Cav)-based QCM sensor (Reprinted with permission from Ferrari et al. 2004, Copyright 2004 Elsevier)

Liu et al. (2005) established that the recovery during the N_2 purging cycle can even be irreversible at room temperature.

In addition we have to recognize that, despite all attempts, a fully specific supramolecular sensor, in which nonspecific interactions and competitive binding by undesired analytes have been eliminated, has not yet been obtained. It was found that selectivity response patterns of cavitands to small molecules can be similar to those of common amorphous polymers (Grate et al. 1996), indicating that cavitand sensors can respond not only to molecules with an ideal fit in the cavity but also to sorbed molecules occupying both intracavity and intercavity sites (Dickert and Schuster 1995; Grate 2000; Pirondini and Dalcanale 2007; Schneider 2009). Grate et al. (1996) have shown that the binding and selectivity in the examples cited are governed primarily by general dispersion interactions and not by specific-oriented interactions that could lead to molecular recognition. Nevertheless, the presence of a preorganized cavity in cavitands promises an advantage in sensitivity compared to amorphous polymers, especially if applied to the sensor in multilayers (Grate 2000). Moreover, Pirondini and Dalcanale (2007) believe that a truly specific receptor for a given molecule can be designed and prepared. According to Pirondini and Dalcanale (2007), two different strategies can be envisioned to avoid nonspecific interactions. From the receptor side, the challenge is to design a host incorporating a suitable transduction group (i.e., a chromophore), which can be activated exclusively by the molecular recognition event. Alternatively, the collective behavior of self-organizing materials can be tapped to amplify the molecular recognition phenomena at the macroscopic level.

References

Asfari Z, Vicens J (1988) Preparation of series of calix[6]arenes and calix[8]arenes derived from *p-n-* alkylphenols. Tetrahedron Lett 29(22):2659–2660

Cha J-H, Lee W, Lee H (2009) Hydrogen gas sensor based on proton-conducting clathrate hydrate. Angew Chem Int Ed 48:8687–8690

Chawla HM, Pant N, Kumar S, Black DSC, Kumar N (2011) Calixarene-based materials for chemical sensors. In: Korotcenkov G (ed) Chemical sensors: fundamentals of sensing materials. Vol. 3. Polymers and other materials. Momentum, New York, pp 117–200

Cram DJ (1983) Cavitands: organic hosts with enforced cavities. Science 219(4589):1177–1183

Davis F, Stirling CJM (1995) Spontaneous multilayering of calix[4]resorcinarenes. J Am Chem Soc 117: 10385–10386

Diamond D, Nolan K (2001) Calixarenes: designer ligands for chemical sensors. Anal Chem 73(1):22A–29A
Dickert FL, Schuster O (1995) Supramolecular detection of solvent vapours with calixarenes: mass-sensitive sensors, molecular mechanics and BET studies. Mikrochim Acta 119(1–2):55–62
Dickert FL, Balumler UPA, Stathopulos H (1997) Mass-sensitive solvent vapor detection with calix[4]resorcinarenes: tuning sensitivity and predicting sensor effects. Anal Chem 69:1000–1005
Ehlen A, Wimmer C, Weber E, Bargon J (1993) Organic clathrate-forming compounds as highly selective sensor coatings for the gravimetric detection of solvent vapors. Angew Chem Int Ed 32:110–112
Feresenbet E, Dalcanale E, Dulcey C, Shenoy DK (2004) Optical sensing of the selective interaction of aromatic vapors with cavitands. Sens Actuators B 97:211–220
Ferrari M, Ferrari V, Marioli D, Taroni A, Suman M, Dalcanale E (2004) Cavitand-coated PZT resonant piezo-layer sensors: properties, structure, and comparison with QCM sensors at different temperatures under exposure to organic vapors. Sens Actuators B 103:240–246
Finklea HO, Phillippi MA, Lompert E, Grate JW (1998) Highly sorbent films derived from $Ni(SCN)_2(4\text{-picoline})_4$ for the detection of chlorinated and aromatic hydrocarbons with quartz crystal microbalance sensors. Anal Chem 70:1268–1276
Forster RJ (1998) Miniaturized chemical sensors. In: Diamond D (ed) Principles of chemical and biological sensors. Wiley, New York, p 243
Grady T, Butler T, MacCraith BD, Diamond D, McKervey MA (1997) Optical sensor for gaseous ammonia with tunable sensitivity. Analyst 122(8):803–806
Grate JW (2000) Acoustic wave microsensor arrays for vapor sensing. Chem Rev 100:2627–2647
Grate JW, Patrash SJ, Abraham MH, Du CM (1996) Selective vapor sorption by polymers and cavitands on acoustic wave sensors: is this molecular recognition? Anal Chem 68:913–917
Gutsche CD (1989) Calixarene. The Royal Society of Chemistry, Cambridge
Hartmann J, Auge J, Hauptmann P (1994) Using the quartz crystal microbalance principles for gas detection with reversible and irreversible sensors. Sens Actuators B 18–19:429–433
Hartmann J, Hauptmann P, Levi S, Dalcanale E (1996) Chemical sensing with cavitands: influence of cavity shape and dimensions on the detection of solvent vapors. Sens Actuators B 35–36:154–157
Kalchenko VI, Koshets IA, Matsas EP, Kopulov ON, Solovyov A, Kazantseva ZI, Shitshov YM (2002) Calixarene-based QCM sensors array and its response to volatile organic vapours. Mater Sci 20(3):73–88
Lavrik NV, DeRossi D, Kazantseva ZI, Nabok AV, Nesterenko BA, Piletsky SA, Kalchenko VI, Shivaniuk AN, Markovskiy LN (1996) Composite polyaniline/calixarene Langmuir–Blodgett films for gas sensing. Nanotechnology 7(4):315–319
Li DQ, Ma M (2000) Surface acoustic wave microsensors based on cyclodextrin coatings. Sens Actuators B 69:75–84
Liu CJ, Lin JT, Wang SH, Jiang JC, Lin LG (2005) Chromogenic calixarene sensors for amine detection. Sens Actuators B 108:521–527
Ludwig R, Dzung NTK (2002) Calixarene-based molecules for cation recognition. Sensors 2:397–416
Maffei F, Betti P, Genovese D, Montalti M, Prodi L, De Zorzi R, Geremia S, Dalcanale E (2011) Highly selective chemical vapor sensing by molecular recognition: specific detection of C1–C4 alcohols with a fluorescent phosphonate cavitand. Angew Chem Int Ed 50:4654–4657
Melegari M, Suman M, Pirondini L, Moiani D, Massera C, Ugozzoli F, Kalenius E, Vainiotalo P, Mulatier J-C, Dutasta J-P, Dalcanale E (2008) Supramolecular sensing with phosphonate cavitands. Chem Eur J 14(19):5772–5779
Ohira SI, Wanigasekar E, Rudkevich DM, Dasgupta PK (2009) Sensing parts per million levels of gaseous NO_2 by a optical fiber transducer based on calix[4]arenes. Talanta 77(5):1814–1820
Palaniappan A, Li X, Tay FEH, Li J, Su X (2006) Cyclodextrin functionalized mesoporous silica films on quartz crystal microbalance for enhanced gas sensing. Sens Actuators B 119:220–226
Parker D (ed) (1996) Macrocycle synthesis. A practical approach. Oxford University Press, Oxford
Pinalli R, Suman M, Dalcanale E (2004) Cavitands at work: from molecular recognition to supramolecular sensors. Eur J Org Chem 2004:451–462
Pirondini L, Dalcanale E (2007) Molecular recognition at the gas–solid interface: a powerful tool for chemical sensing. Chem Soc Rev 36:695–706
Rosler S, Lucklum R, Borngraber R, Hartmann J, Hauptmann P (1998) Sensor system for the detection of organic pollutants in water by thickness shear mode resonators. Sens Actuators B 48:415–424
Rudkevich DM (2007) Progress in supramolecular chemistry of gases. Eur J Org Chem 20:3255–3270
Schneider H-J (2009) Binding mechanisms in supramolecular complexes. Angew Chem Int Ed 48:3924–3977
Sharma K, Cragg PJ (2011) Calixarene based chemical sensors. Chem Sensors 1(9):1–18
Shenoy DK (2005) Cavitands: container molecules for surface plasmon resonance (SPR)-based chemical vapor detection. Mater Sci Technol 2005 NRL review:171–173
Szetjli J (1998) Introduction and general overview of cyclodextrin chemistry. Chem Rev 98:1743–1753
Valeur B, Leray I (2007) Ion-responsive supramolecular fluorescent systems based on multichromophoric calixarenes: a review. Inorg Chim Acta 360:765–774

Vögtle F (ed) (1996) Comprehensive supramolecular chemistry. Pergamon, Oxford

Wang C, Chen F, He XW, Kang SZ, You CC, Liu Y (2001) Cyclodextrin derivative-coated quartz crystal microbalances for alcohol sensing and application as methanol sensors. Analyst 126:1716–1720

Wang C, Chen F, He X-W (2002) Kinetic detection of benzene/chloroform and toluene/chloroform vapors using a single quartz piezoelectric crystal coated with calix[6]arene. Anal Chim Acta 464:57–64

Yakimova LS, Ziganshin MA, Sidorov VA, Kovalev VV, Shokova EA, Tafeenko VA, Gorbatchuk VV (2008) Molecular recognition of organic vapors by adamantylcalix[4]arene in QCM sensor using partial binding reversibility. J Phys Chem B 112:15569–15575

Chapter 10
Metallo-Complexes

Because transition metals are capable of establishing reversible interactions with other atoms, they can be exploited to form metallo-complexes (MCs), which may act as receptors for different types of analytes. It was shown that macrocyclic compounds such as crown ethers, cyclodextrins, calixarenes, cyclophanes, cavitands, cryptands, spherands, carcerands, cyclopeptides, and other structurally related species can be incorporated in these metallosupramolecules (Atwood et al. 1996). Metal phthalocyanines (MPcs), aromatic macrocyclic compound which have semiconductor properties, can also be referred to as this class of materials (Schollhorn et al. 1998; Fietzek et al. 1999; Ceyhan et al. 2006). Phthalocyanine (Pc) ligands can coordinate with various metal ions, and the central metals can interact with small molecules through a coordination bond (see Fig. 10.1). The phthalocyanines are stable up to 450 °C; at this temperature, the materials decompose but do not melt. Because of their high decomposition temperature, they can be vacuum evaporated to produce thin films. Porphyrin molecules can also be assembled into nanostructures using several methods (Kosal et al. 2002; Medforth et al. 2009). Several reviews are available analyzing the performance of porphyrins and cyanines in gas sensing (Di Natale et al. 1998, 2007; Ozturk et al. 2009; Nardis et al. 2011; Trogler 2012). In metalloporphyrins, metallophthalocyanines, and related macrocycles, gas sensing is accomplished either by π-stacking of the gas into organized layers of the flat macrocycles or by gas coordination to the metal center without the cavity inclusion. In particular, metalloporphyrins provide several mechanisms of gas response including hydrogen bonding, polarization, polarity interactions, metal center coordination interactions, and molecular arrangements (Di Natale et al. 2007; Nardis et al. 2011).

Harbeck et al. (2011) and Sen et al. (2011) have shown that metal complexes of *vic*-dioximes can be characterized as candidate materials for volatile organic compound sensing with sorption-based chemical gas sensors as well. The *vic*-dioximes are known to form stable complexes with a variety of metals such as Ni^{2+}, Pd^{2+}, Cu^{2+}, Co^{2+}, Zn^{2+}, or Cd^{2+}. Furthermore, they can be modified easily in the substituent structure. Metallodendrimers are representative of metallo-complexes as well (Albrecht and van Koten 1999; Hwang et al. 2007). Metallodendrimers can be categorized (see Fig. 10.2), such as when metal centers are positioned at the infrastructure's core and connectors positioned between branching centers, or act as terminal groups. Metal centers can also be integrated as structural auxiliary points within the dendritic framework by their incorporation after dendrimer construction. This means that there are numerous combinations in metallo-complexes. It is important that all these supramolecules possess specific host–guest behavior with different luminescent or electronic properties, which can be exploited for sensing purposes. The necessary components can be incorporated in metallo-complex according to the needs of the analyte to generate analytically useful and observable signals.

G. Korotcenkov, *Handbook of Gas Sensor Materials*, Integrated Analytical Systems,
DOI 10.1007/978-1-4614-7388-6_10, © Springer Science+Business Media New York 2014

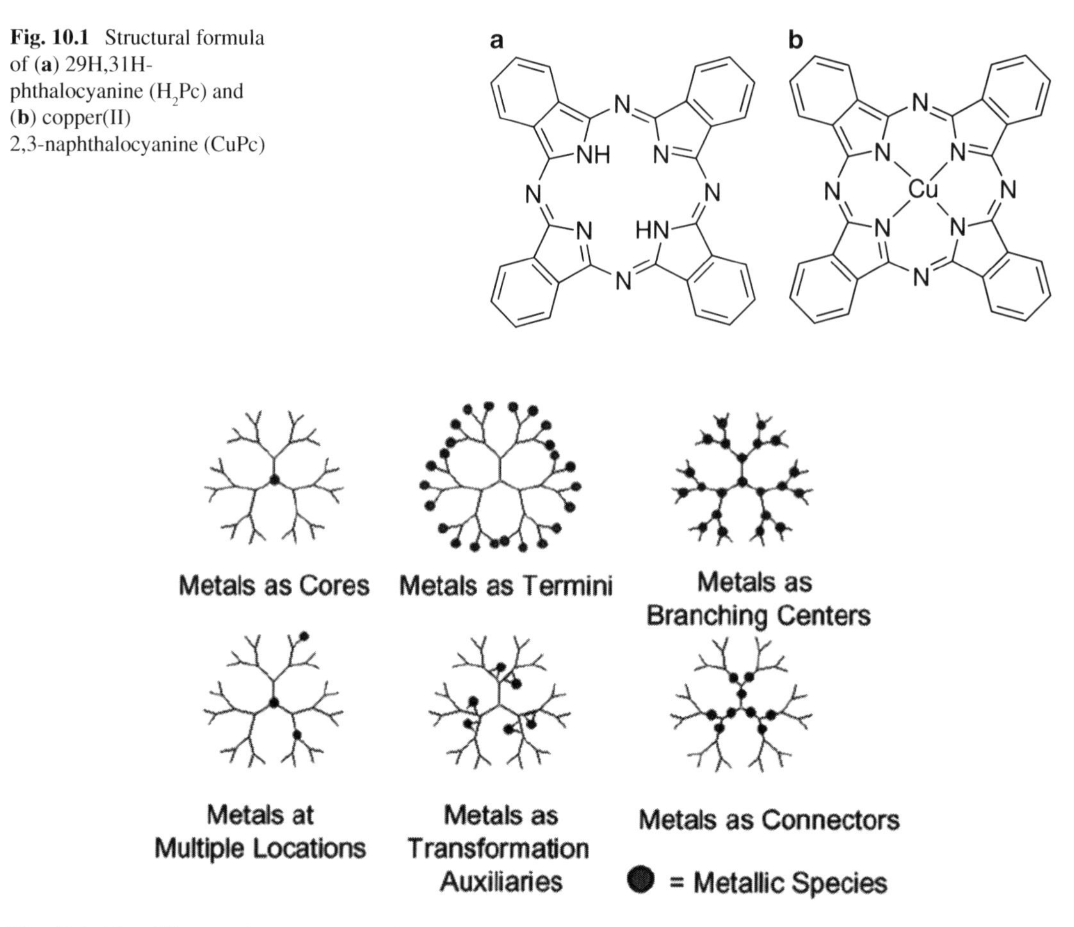

Fig. 10.1 Structural formula of (**a**) 29H,31H-phthalocyanine (H_2Pc) and (**b**) copper(II) 2,3-naphthalocyanine (CuPc)

Fig. 10.2 The different roles metals can play in metallodendrimers (Reprinted with permission from Hwang et al. 2007, Copyright 2007 Royal Society of Chemistry)

10.1 Gas Sensor Applications of Metallo-Complexes

Of course, metallo-complexes are preferable for cation recognition (De Silva et al. 1997; Bergonzi et al. 1998; Bren 2001). However, the application of MCs for detection of gas molecules sensing using different transduction principles such as impedometric, optical, and mechanical is also promising. One of the possible examples of metallo-complexes applied in luminescence- and QCM-based VOCs sensors is shown in Fig. 10.3. It was established that, first, the metallo-complex is generally more sensitive and responsive on electro- and photochemical stimuli compared to metal-free organic macrocyclic molecules (Kumar et al. 2008). According to Albrecht and van Koten (1999), a proper choice of the (transition) metal center and the corresponding ligand array is crucial for gas sensor design since the metal center generally exhibits a high selectivity for particular substances. This enables the preparation of detector materials of high selectivity. Second, various functionalities can easily be introduced into metallo-complex structure by employing functional ligands. For example, experiment has shown that a wide range of chromophores, fluorophores, and redox-active functionalities have been successfully incorporated into supramolecular frameworks (Holliday and Mirkin 2001). Furthermore, ligand tuning in organometallic complexes can be used as a method to modify and optimize selectivity and sensitivity of the detector units through electronic and steric effects. Ligand fragments that have minor

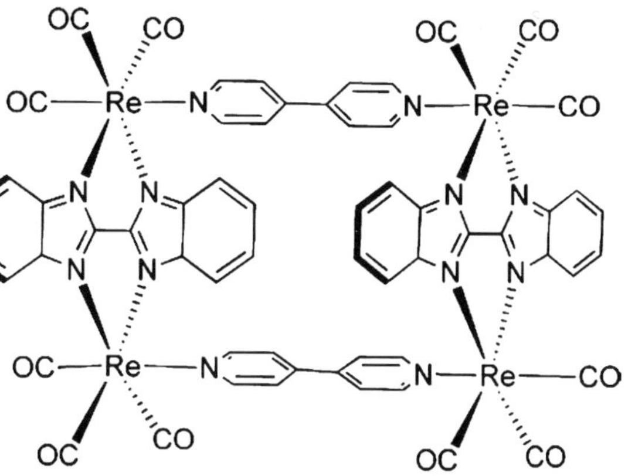

Fig. 10.3 Metallo-complex sensitive to volatile organic compounds (VOCs) (Reprinted with permission from Kumar et al. 2008, Copyright 2008 Elsevier)

or no influence on the sensor activity may also serve as potential anchoring points to immobilize the sensor sites on an appropriate support, e.g., on dendrimers or polymers (Albrecht and van Koten 1999). In addition, they provide sites for the introduction of signal transduction and amplification devices. Through a suitable fixation of the sensing unit, recovery of the sensors by common separation techniques is facilitated.

In addition, metallo-complexes may play the role of catalysts (Hwang et al. 2007). In particular, MPcs and derivatives are shown to catalyze reactions like the reduction of oxygen and CO_2 (Ceyhan et al. 2006). The catalytic behavior of MPc complexes is associated with the redox activity of the central metal which undergoes oxidation and reduction. The values of the oxidation or reduction potentials of the central metal in an MPc thus strongly influence the catalytic behavior of the complexes which is essential for the design of more efficient gas sensors.

The principal advantage of using metal complexes as the sensing layer is the reversible specific reactions between the analytes and the devices. The coordination bonds formed during the detection process can be broken by increasing the temperature or changing the chemical environment of the sensor. One can assume that, due to specific interactions in metallo-complexes, MCs can provide new opportunities to develop novel devices with improved sensitivity and selectivity. The most significant features, which arise from the architecture of metallo-complexes, include encapsulation of guest molecules, luminescence, and redox activity. Jimenez-Cadena et al. (2007) believe that, due to the indicated properties, metallo-complexes can also find application as specific receptors in functionalizing processes that use other nanostructured materials as transducers. One can find in review papers by De Silva et al. (1997), Leininger et al. (2000), Donilfo and Hupp (2001), Holliday and Mirkin (2001), and Kumar et al. (2008) more detailed descriptions of metallo-complexes. Several examples of metallo-complexes, which are usually used in various types of gas sensors, are listed in Table 10.1. Phthalocyanines tested as gas-sensing material are listed in Table 10.2.

As research has shown, the approach, based on the incorporation of metallo-complexes in gas-sensing devices, found application mainly in optical- (Del Bianco et al. 1993; De Silva et al. 1997; Albrecht and van Koten 1999; Elosua et al. 2006), SAW-, and QCM-based sensors (Nieuwenhuizen and Harteveld 1994; Benkstein et al. 2000; Kimura et al. 2010). In particular, Elosua et al. (2006) reported an optical fiber sensor coated with a complex of Au and Ag $\{[Au_2Ag_2(C_6F_5)_4(C_6H_5N)_2]\}$ for detecting volatile alcoholic compounds. The recognition layer of the sensor was a nanometer-scale Fizeau interferometer, doped with the complex which, in the presence of methanol, ethanol, and isopropanol vapors, exhibits vapochromic behavior. The complex was incorporated in the last layer of the cationic polymer. Albrecht and van Koten (1999) have shown that organoplatinum complexes containing a terdentate coordinating monoanionic "pincer" ligand $[C_6H_3(CH_2NMe_2)_2\text{-}2,6]^-$, abbreviated as NCN, are prime candidates for SO_2 detection as these complexes reversibly bind SO_2 in the

Table 10.1 Metallo-complexes and gas sensors, which can be based on this material

Me-complex	Sensor type	Target gas	Principle	References
Au–Ag complex $[Au_2Ag_2(C_6F_5)_4(C_6H_5N)_2]$	Fiber optic	Methanol; ethanol; isopropanol; acetic acid	Transmittance (vapochromic)	Elosua et al. (2006); Casado-Terrones et al. (2006)
Bis(histidinato)cobalt(II) $[Co(His)_2]$	Fiber optic	O_2	Transmittance (vapochromic)	Del Bianco et al. (1993)
Arylplatinum(II) complex $[PtX(4\text{-}E\text{-}2,6\text{-}\{CH_2NRR'\}_2\text{-}C_6H_2)]$	Optical	SO_2	Transmittance (vapochromic)	Albrecht et al. (2000)
Ir(III) complex [Ir(2-phenylpyridine)2(4,4′-bis(2-(4-*N,N*-methylhexylaminophenyl) ethyl)-2-2′-bipyridine)Cl]	Optical	O_2	Phosphorescent quenching	Medina-Castillo et al. (2007)
$Ru(byp)_3^{2+}$	Optical	O_2 in hexane	Luminescence quenching	García et al. (2005)
4,4′-Bipyridine-bridged [Mn(I) and Re(I)] molecular rectangles	QCM	Toluene; 4-fluorotoluene; benzene; fluorobenzene	Adsorption	Benkstein et al. (2000)
Copper octanediylbis(phosphonate) and *p*-xylylenediamine (pXDA)	SAW	CO_2	Adsorption	Brousseau et al. (1997)
La(III) 2-bis(carboxymethyl)amino hexadecanoic acid (LaBHA)	SAW	DMMP; NO_2; ethanol; NH_3; n-hexane; toluene	Adsorption	Nieuwenhuizen and Harteveld (1994)
$C_{280}H_{362}N_{34}S_{24}O_4Zn_4$	Resistive	VOCs	Redox	Ceyhan et al. (2006)

Table 10.2 Metal phthalocyanine materials used for gas sensor design

Me-complex	Sensor type	Target gas	Principle	References
Sulfonated CoPc	QCM	VOCs	Adsorption	Kimura et al. (2010)
Copper phthalocyanine (CuPc)	SAW	NO_2	Adsorption	Hechner and Soluch (2002)
	FET	NO_2	Work function	Oprea et al. (2006)
	Resistive	NO_2	Redox	Moriya et al. (1993)
Titanyl phthalocyanine (TiOPc)	Resistive	NO_2	Redox	Liu et al. (1998)
Lead phthalocyanine (PbPc)	Resistive	NO_2	Redox	Bott and Jones (1984); Sadaoka et al. (1990a, b); Liu et al. (1996)
CoPc, CuPc, NiPc, and PbPc	Resistive	NO_2, NH_3, water	Redox	Belghachi and Collins (1990)
(AlPcF)*n*	Resistive	O_2, NO_2,	Redox	Passard et al. (1994)
1,4-Bis(4-hydroxybutyl)-8,11,15,18,22,25-hexahexylphthalocyanine	Resistive	NO_2, Cl_2, CO,	Redox	Crouch et al. (1994)
Zinc hexadecafluorophthalocyanine ($ZnF_{16}Pc$)	Resistive	NH_3, H_2	Redox	Schollhorn et al. (1998)
Bis-metallo [Co(II)] phthalocyanines	Resistive	CO_2	Redox	Altun et al. (2008)

solid state and in solution (see Fig. 10.4). Furthermore, the color change, upon coordination of SO_2, can serve as a signal transduction device for the molecular recognition of SO_2.

It was found, however, that metallo-complexes that contain transition metal centers are especially convenient for fluorescence sensors as they undergo fast and kinetically uncomplicated one-electron redox changes (Bergonzi et al. 1998). Moreover, transition metals tend to participate in luminescence quenching of both the electron transfer, eT, and energy transfer, ET, varieties (see Fig. 10.5). Finally, the potential of the metal-centered redox couple can be modulated by varying the nature of the hosting

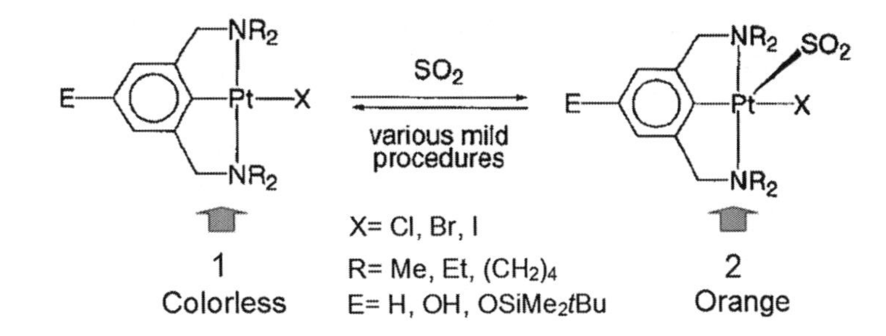

Fig. 10.4 Reversible gas adsorption and desorption of PtII complexes 1 containing the "pincer" ligand via formation of a colored pentacoordinated adduct 2 (Reprinted with permission from Albrecht and van Koten 1999, Copyright 1999 WILEY-VCH Verlag)

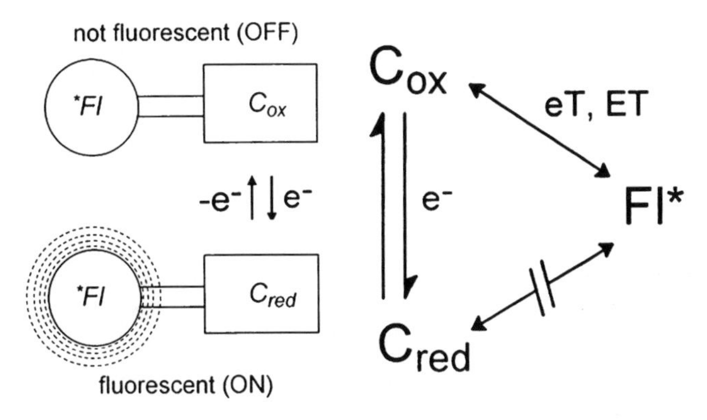

Fig. 10.5 The basis of an OFF/ON redox switch of fluorescence. Either an energy transfer (ET) or an electron transfer (eT) mechanism can be responsible for the quenching of the light emitting fragment fluorophore (Fl*) in a multicomponent redox unit-spacer-fluorophore system. Switch efficiency requires that the control unit C in its oxidized form, C_{ox}, quenches the proximate photo-excited fluorophore Fl* and the reduced form C_{red} does not (OFF/ON switch). The other favorable on/off situation can be obtained when C_{red} quenches Fl* and C_{ox} does not (Reprinted with permission from Bergonzi et al. 1998, Copyright 1998 Elsevier)

coordinative environment. We should note that fluorescence dyes, which are used in optical gas sensors and will be discussed in Chap. 15 (Vol. 2), are metallo-complexes as well.

One can find in Kimura et al. (2010) an example of QCM sensors based on metallo-complexes. One can find in a review by Wright (1989) a detail discussion of gas adsorption on metal phthalocyanines. Analyzing the response of the QCM sensor with an MPc-sensing layer, Kimura et al. (2010) has shown that the incorporation of sterically protected MPc within the polymer brushes on QCMs allows them to work as molecular receptors to recognize the chemical properties of VOC vapors based on their size and polarity. The selectivity and sensitivity of the sensing layer on QCMs can be tuned by modifying both molecular recognition receptors, which depend on the structure of peripheral substituents and the central metals, and polymer brushes. In particular, they found that sulfonated CoPc shows a stronger affinity for acetone, ethanol, and pyridine vapors relative to other MPcs (NiPc, CuPc, ZnPc). These coordinative VOC vapors can form a weak coordination interaction with the central metal, and the formation of coordination bonds enhances the selectivity of sensors.

However, the application of metallo-complexes, mainly MPcs, in conductometric gas sensors is also possible. One can find in papers published by Wright (1989), Schollhorn et al. (1998), and Germain et al. (1998) detailed analyses of MPc-based conductometric gas sensors. We must say that metal phthalocyanines, which are organic semiconductors (Wright 1989), can also be used for designing gas-sensitive thin-film transistors. This application was discussed in Chap. 20 (Vol. 1). Regarding MPc-based conductometric gas sensors, that many reports have been published on the response of MPc films to electron donor and electron acceptor gases such as NH_3, H_2, NO_x, VOCs, and halogens

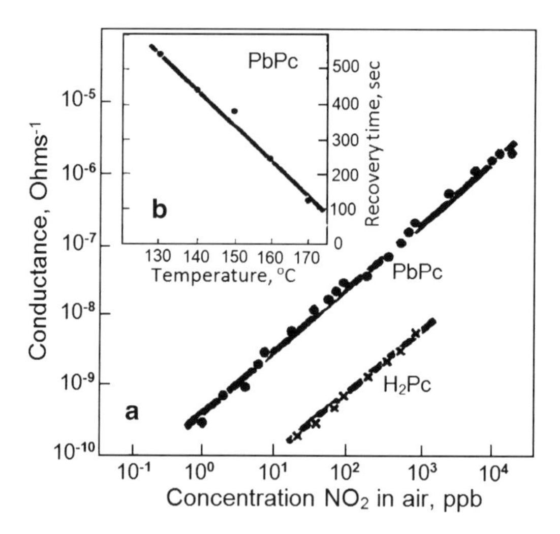

Fig. 10.6 (**a**) Variation of conductance of PbPc and H_2PC films with concentration of NO_2 in air at 150 °C. (**b**) Influence of operation temperature on recovery time of PbPc-based sensors measured after interaction with 50 ppb NO_2 (Data from Bott and Jones 1984)

(Schollhorn et al. 1998; Ceyhan et al. 2006) (see Table 10.2). For explanation of MPc gas sensitivity the approach based on gas molecule interaction with the π-electron network of the phthalocyanines is normally used (Roisin et al. 1992). For example, in the case of VOCs sensors the VOCs would play an electron acceptor role. When the VOC molecules interact with the π-electron network of the phthalocyanines, it causes the transfer of an electron from the phthalocyanine ring to the VOC molecule (Schollhorn et al. 1998). Thus, the induced positive holes on the film surface give rise to an increase in the *p*-type conductivity of the film.

According to Germain et al. (1998), gas sensitivities of metallophthalocyanines can be related to the values of their redox potentials. Analyzing gas-sensing properties of $LuPc_2$, CuPc, ZnPc, (AlPcF)*n*, PNnNN82nLu, and $ZnF_{16}Pc$, Germain et al. (1998) found that $LuPc_2$ and PNnNN82nLu are highly sensitive to oxidizing gases. The energy E_{CT} required for the charge transfer process between a phthalocyanine and the oxidizing gas NO_2 is lower in the case of PNnNN82n and $LuPc_2$ than in the case of classic phthalocyanines. Due to their low reduction energy, they should also exhibit sensitivity to reducing gases, but the effect is hidden by their high intrinsic conductivity due to a low energy gap and by the significant action of oxygen. ZnF16Pc is more interesting than $LuPc_2$ and PNnNN82nLu for gas-sensing application because, if it is sensitive to reducing gases, its sensitivity to oxidants is low and its energy gap is wide enough so that intrinsic conductivity does not noticeably interfere with gas effects. Sensitivity of CuPc, ZnPc, and (AlPcF)*n* to reducing gases is weak because of their high reduction energy levels, but they are more sensitive than ZnF16Pc to oxidizing gases, and, like it, they exhibit a negligible intrinsic conductivity. Bott and Jones (1984) also tested several metal phthalocyanines and found that (Mg, Co, NI, Cu, and Zn) PCs were less sensitive than PbPc to oxidizing gas NO_2. Conductivity changes of PbPc and H_2Pc films in NO_2 atmosphere are shown in Fig. 10.6a. Other common atmospheric pollutants, namely H_2S (50 ppm), NH_3 (40 ppm), SO_2 (10 ppm), H_2 (100 ppm), CH_4 (1 %), CO (100 ppm), and H_2O (50 % RH), and variation in oxygen pressures, had comparatively little effect (<20 % change) on the conductance of PbPc sensors. Chlorine, however, had an effect comparable to NO_2.

However, we need to recognize that the indicated sensors in comparison with conventional metal oxide gas sensors have low sensitivity to many specific gases and slow responses and recovery processes even at T_{oper} = 180 °C (see Figs. 10.6b and 10.7). Sadaoka et al. (1990a, b) found that post-deposition annealing at 330 °C could shorten the response time. However, this improvement is not cardinal. The failure to return to baseline after the VOC vapors or other gas molecules are removed originates from the strong bonding of the gas molecules on the surface. This appreciably limits the

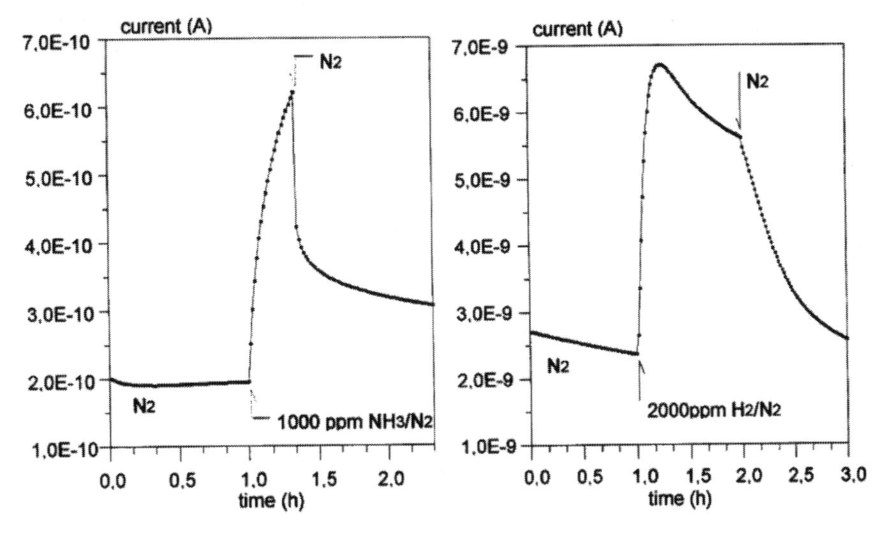

Fig. 10.7 Current variation of a $ZnF_{16}Pc$ layer exposed to the gaseous run: $N_2 \rightarrow NH_3$ in $N_2 \rightarrow N_2$ and $N_2 \rightarrow H_2$ in $N_2 \rightarrow N_2$. $T_{oper} = 180$ °C. $V = 1$ V (Reprinted with permission from Schollhorn et al. 1998, Copyright 1998 Elsevier)

application of these materials to devices for the sensor market. Operation characteristics of MePc-based devices are also sensitive to water vapor, especially at low temperatures (Belghachi and Collins 1990). NiPc is found to display much greater humidity sensitivity than the other metal Pcs studied.

10.2 Approaches to Improvement of Gas Sensor Parameters and Limitations

According to Schollhorn et al. (1998), the selectivity and sensitivity of metal complexes to gases could be improved by molecular engineering, which can be done in three ways:

1. *Change of the macrocycle structure: porphyrins, phthalocyanines*, etc. No structure could be found yet that immediately fulfills the requirements of high sensitivity to a gas, selectivity to a gas toward interferents, and good reversibility. However, owing to their properties (good chemical stability, for instance), phthalocyanines appear as good materials for gas-sensing purpose.
2. *Change of central metal ion M*. The exchange of the metal in metallophthalocyanine complexes (MPc) results in variations of sensitivity to gases that can be assigned both to changes in oxidation potentials and to morphological modifications. Unfortunately, the selectivity toward gases cannot be significantly improved this way.
3. *Change of peripheral substituents of the macrocycle*. Addition of electron-withdrawing/donating substituents on a metallophthalocyanine macrocycle decreases/increases the electron density of the conjugated cycle and increases/decreases the oxidation potential of the macrocycle. The sensitivity to reducing/oxidizing gases should be increased by substitution of the cycle by electron-withdrawing/donating groups.

For example, Schollhorn et al. (1998) found that the substitution of ZnPc by electron-withdrawing fluorine atoms makes it sensitive to the reducing gases NH_3 and H_2, when unsubstituted ZnPc is not or only weakly sensitive. In addition, $ZnF_{16}Pc$ is less sensitive to oxidizing gases than ZnPc. Using the same approach, Ceyhan et al. (2006) designed novel multinuclear metallophthalocyanines (Zn, Co) with alkylthio substituents, which provided acceptable sensitivity to VOCs. The use of convention MPcs

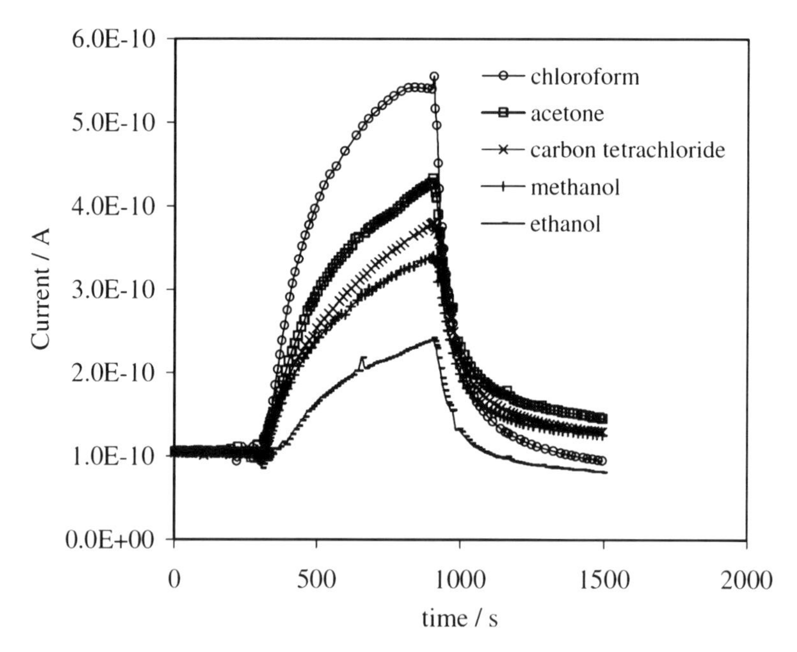

Fig. 10.8 Current vs time of a spin-coated film of $C_{280}H_{362}N_{34}S_{24}O_4Zn_4$ exposed to 200 ppm of five different VOC vapors (Reprinted with permission from Ceyhan et al. 2006, Copyright 2006 Elsevier)

for the detection of VOCs vapors is limited due to the low reactivities of organic gases compared with inorganic gases. 2-Nitro-2-methyl-1,3-bis(3,4-dicyanophenoxymethyl)propane was used as a key compound for the preparation of the above-mentioned MPcs. Operating characteristics of the designed sensors are shown in Fig. 10.8. It was found that the substitution of MPcs can improve their structural properties as well. It is known that nonsubstituted Pcs do not dissolve in any solvents owing to their strong aggregation tendency. The aggregation between MPcs hindered the access of small molecules into the central metal. To diminish the intermolecular aggregation of MPcs, bulky substituents have been introduced at the peripheral positions of MPcs (Walsh and Mandal 2000). When bulky oligophenylene units were introduced at the peripheral positions of the Pc ring, they served to prevent direct interaction between the MPcs.

However, we need to note that the synthesis of these metallo-complexes can be expensive. It is known that as the scale and complexity of target molecules increase, the stepwise synthesis of large discrete supermolecules from molecular building blocks becomes increasingly difficult, often low yielding, and specific to only a few approaches. The synthesis of large symmetric structures is long and requires careful consideration of the entropic and enthalpic costs involved. This process includes the time required for the linear, stepwise synthesis of complex macromolecules and molecular assemblies composed of hundreds or even thousands of subunits as well as a decrease of the overall yield in multistep reactions. The purification of the larger molecules also still poses some problems (Leininger et al. 2000). This means that the reproducibility of sensor parameters can be low. Advantages and disadvantages of various approaches, such as directional-bonding approach, symmetry-interaction approach, and weak-link synthetic strategy, usually used for synthesis of metallo-complexes, were reviewed in detail by Holliday and Mirkin (2001). We also note that the problem related to sensor stability is also present for the discussed sensors. For example, Elosua et al. (2006) reported only about 3 months stability of optical gas sensors based on (Au–Ag) $[Au_2Ag_2(C_6F_5)_4(C_6H_5N)_2]$ complex. As we noted before, in principle, metallo-complexes are promising materials for designing selective gas sensors because each receptor can be designed to interact with a specific analyte. Unfortunately, till now, such gas sensors have not been developed.

References

Albrecht M, van Koten G (1999) Gas sensor materials based on metallodendrimers. Adv Mater 11(2):171–174

Albrecht M, Gossage RA, Lutz M, Spek AL, van Koten G (2000) Diagnostic organometallic and metallodendritic materials for SO_2 gas detection: reversible binding of sulfur dioxide to Arylplatinum(II) complexes. Chem Eur J 6:1431–1445

Altun S, Altındal A, Rıza OA, Bulut M, Bekaroglu O (2008) Synthesis, characterization, electrochemical and CO_2 sensing properties of novel mono and ball-type phthalocyanines with four phenolphthalein units. Tetrahedron Lett 49:4483–4486

Atwood JL, Davis JED, McNicol DD, Vogtle F, Lehn J-M (eds) (1996) Comprehensive supramolecular chemistry, vols 1–11. Pergamon, Oxford

Belghachi A, Collins RA (1990) The effects of humidity on phthalocyanine NO_2, and NH_3, sensors. J Phys D: Appl Phys 23:223–227

Benkstein KD, Hupp JT, Stern CL (2000) Luminescent mesoporous molecular materials based on neutral tetrametallic rectangles. Angew Chem Int Ed 39:2891–2893

Bergonzi R, Fabbrizzi L, Licchelli M, Mangano C (1998) Molecular switches of fluorescence operating through metal centred redox couples. Coord Chem Rev 170:31–46

Bott B, Jones TA (1984) A highly sensitive NO_2 sensor based on electrical conductivity changes in phthalocyanine films. Sensor Actuator 5:43–53

Bren VA (2001) Fluorescent and photochromic chemosensors. Rus Chem Rev 70(12):1017–1036

Brousseau L, Aurentz D, Benesi A, Mallouk T (1997) Molecular design of intercalation-based sensors. 2. Sensing of carbon dioxide in functionalized thin films of copper octanediylbis(phosphonate). Anal Chem 69:688–694

Casado-Terrones S, Elosua-Aguado C, Bariain C, Segura-Carretero A, Matias-Maestro IR, Fernández-Gutiérrez A, Luquin A, Garrido J, Laguna M (2006) Volatile-organic-compound optic fiber sensor using a gold-silver vapochromic complex. Opt Eng 45:044401/1–044401/7

Ceyhan T, Altındal A, Erbil MK, Bekaroglu O (2006) Synthesis, characterization, conduction and gas sensing properties of novel multinuclear metallo phthalocyanines (Zn, Co) with alkylthio substituents. Polyhedron 25:737–746

Crouch D, Thorpe SC, Cook MJ, Chambers I, Ray AK (1994) Langmuir Blodgett film of asymmetrically substituted phthalocyanine improved gas sensing properties. Sens Actuators B 18–19:411–414

De Silva AP, Gunaratne HQN, Gunnlaugsson T, Huxley AJM, McCoy CP, Rademacher JT, Rice TE (1997) Signaling recognition events with fluorescent sensors and switches. Chem Rev 97:1515–1566

Del Bianco A, Baldini F, Bacci M, Klimant I, Wolfbeis OS (1993) A new kind of oxygen sensitive transducer based on an immobilised metallo-organic compound. Sens Actuators B 11:347–350

Di Natale C, Macagnano A, Repole G, Saggio G, D'Amico A, Paolesse R, Boschi T (1998) The exploitation of metalloporphyrins as chemically interactive material in chemical sensors. Mater Sci Eng C 5:209–215

Di Natale C, Paolesse R, D'Amico A (2007) Metalloporphyrins based artificial olfactory receptors. Sens Actuators B 121:238–246

Donilfo PH, Hupp JT (2001) Supramolecular coordination chemistry and functional microporous molecular materials. Chem Mater 13:3113–3125

Elosua C, Bariain C, Matias I, Arregui F, Luquin A, Laguna M (2006) Volatile alcoholic compounds fibre optic nano sensor. Sens Actuators B 115:444–449

Fietzek C, Bodenhofer K, Haisch P, Hees M, Hanack M, Steinbrecher S, Zhou F, Plies E, Gopel W (1999) Soluble phthalocyanines as coatings for quartz-microbalances: specific and unspecific sorption of volatile organic compounds. Sens Actuators B 57:88–98

García EA, Fernández RG, Díaz-García ME (2005) Tris(bipyridine)ruthenium(II) doped sol–gel materials for oxygen recognition in organic solvents. Micropor Mesopor Mat 77:235–239

Germain JP, Pauly A, Maleysson C, Blanc JP, Schollhorn B (1998) Influence of peripheral electron-withdrawing substituents on the conductivity of zinc phthalocyanine in the presence of gases. Part 2: oxidizing gases. Thin Solid Films 333:235–239

Harbeck N, Sen Z, Gurol I, Gumus G, Musluoglu E, Ahsen V, Ozturk ZZ (2011) Vic-dioximes: a new class of sensitive materials for chemical gas sensors. Sens Actuators B 56:673–679

Hechner J, Soluch W (2002) Effect of copper phthalocyanine layer thickness on properties of SAW NO_2 sensor. Electron Lett 38(15):841–842

Holliday BJ, Mirkin CA (2001) Strategies for the construction of supramolecular compounds through coordination chemistry. Angew Chem Int Ed 40:2022–2043

Hwang S-H, Shreiner CD, Moorefield CN, Newkome GR (2007) Recent progress and applications for metallodendrimers. New J Chem 31:1192–1217

Jimenez-Cadena G, Riu J, Rius FX (2007) Gas sensors based on nanostructured materials. Analyst 132:1083–1099

Kimura M, Sugawara M, Sato S, Fukawa T, Mihara T (2010) Volatile organic compound sensing by quartz crystal microbalances coated with nanostructured macromolecular metal complexes. Chem Asian J 5:869–876

Kosal ME, Chou J-H, Wilson SR, Suslick KS (2002) A functional zeolite analogue assembled from metalloporphyrins. Nat Mater 1:118–121

Kumar A, Sun S-S, Lees AJ (2008) Directed assembly metallocyclic supramolecular systems for molecular recognition and chemical sensing. Coord Chem Rev 252:922–939

Leininger S, Olenyuk B, Stang PJ (2000) Self-assembly of discrete cyclic nanostructures mediated by transition metals. Chem Rev 100:853–908

Liu CJ, Hsieh JC, Ju YH (1996) Response characteristics of lead phthalocyanine gas sensor: effect of operating temperature and deposition annealing. J Vac Sci Technol A 14:753–756

Liu CJ, Peng CH, Ju YH, Hsieh JC (1998) Titanyl phthalocyanine gas sensor for NO_2 detection. Sens Actuators B 52:264–269

Medforth CJ, Wang Z, Martin KE, Song Y, Jacobsen JL, Shelnutt JA (2009) Self-assembled porphyrin nanostructures. Chem Commun 2009:7261–7277

Medina-Castillo AL, Fernández-Sánchez JF, Klein C, Nazeeruddin MK, Segura-Carretero A, Fernández-Gutiérrez A, Graetzel M, Spichiger-Keller UE (2007) Engineering of efficient phosphorescent iridium cationic complex for developing oxygen sensitive polymeric and nanostructured films. Analyst 132:929–936

Moriya K, Enomoto H, Nakamura Y (1993) Characteristics of the substituted metal phthalocyanine NO_2 sensor. Sens Actuators B 13–14:412–415

Nardis S, Pomarico G, Tortora L, Capuano R, D'Amico A, Di Natale C, Paolesse R (2011) Sensing mechanisms of supramolecular porphyrin aggregates: a teamwork task for the detection of gaseous analytes. J Mater Chem 21:18638–18644

Nieuwenhuizen MS, Harteveld JLN (1994) An automated SAW gas sensor testing system. Sens Actuators A 44:219–229

Oprea A, Weimar U, Simon E, Fleischer M, Frerichs H-P, Wilbertz C, Lehmann M (2006) Copper phthalocyanine suspended gate field effect transistors for NO_2 detection. Sens Actuators B 118:249–254

Ozturk ZZ, Kilinc N, Atilla D, Gurek AG, Ahsen V (2009) Recent studies of chemical sensors based on phthalocyanines. J Porphyr Phthalocya 13:1179–1187

Passard M, Pauly A, Blanc JP, Dogo S, Germain JP, Maleysson C (1994) Doping mechanisms of phthalocyanines by oxidizing gases—application to gas sensors. Thin Solid Films 237:272–276

Roisin P, Wright JD, Nolte RJM, Sielcken OE, Thorpe SC (1992) Gas-sensing properties of semiconducting films of crown-ether-substituted phthalocyanines. J Mater Chem 2:131–137

Sadaoka Y, Jones TA, Gopel W (1990a) Fast NO_2 detection at room temperature with optimized lead phthalocyanine thin film structure. Sens Actuators B 1:148–153

Sadaoka Y, Jones TA, Revell GS, Gopel W (1990b) Effects of morphology on NO2 detection in air at room temperature with phthalocyanine thin films. J Mater Sci 25:5257–5268

Schollhorn B, Germain JP, Pauly A, Maleysson C, Blanc JP (1998) Influence of peripheral electron-withdrawing substituents on the conductivity of zinc phthalocyanine in the presence of gases. Part 1: reducing gases. Thin Solid Films 326:245–250

Sen Z, Gumus G, Gurola I, Musluoglu E, Ozturk ZZ, Harbecka M (2011) Metal complexes of *vic*-dioximes for chemical gas sensing. Sens Actuators B 160:1203–1209

Trogler WC (2012) Chemical sensing with semiconducting metal phthalocyanines. Struct Bond 142:91–118

Walsh CJ, Mandal BK (2000) A novel method for the peripheral modification of phthalocyanines. Synthesis and third-order nonlinear optical absorption of β-tetrakis(2,3,4,5,6-pentaphenylbenzene)phthalocyanine. Chem Mater 12:287–289

Wright JD (1989) Gas adsorption on phthalocyanines and its effects on electrical properties. Prog Surf Sci 31:1–60

Chapter 11
Metal–Organic Frameworks

11.1 General Consideration

Over the past few decades, a myriad of solids have been described that contain metal ions linked by molecular species. In particular, metallo-complexes (MCs) discussed in the previous chapter are related to such materials. Metal–organic frameworks (MOFs) can be ascribed to this class of materials as well. MOFs, also referred to as porous coordination polymers (PCPs) (Rowsell and Yaghi 2004; Fang et al. 2010), are relatively new highly porous hybrid organic–inorganic crystalline supramolecular materials composed of ordered networks formed from organic electron-donor linkers and metal cations via coordination bonds (Fang et al. 2010; MacGillivray 2010; Meek et al. 2011). Depending on the metal ion and its oxidation state, coordination numbers could commonly be 2–6 for transition metals, or 6–12 for lanthanides. Different coordination numbers result in various geometries, which can be linear, T- or Y-shaped, tetrahedral, square–planar, square–pyramidal, trigonal–bipyramidal, octahedral, trigonal prismatic, pentagonal–bipyramidal, or polyhedral coordination geometry, and the corresponding distorted forms (Kitagawa et al. 2004). Besides crystallinity, one great advantage of MOFs is that, given a starting framework geometry, it is possible to build frameworks that have the same topology but that differ by the presence of functional groups and by the size of the organic building blocks. This concept, called isoreticularity (Eddaoudi et al. 2002; Cavka et al. 2008; Garibay and Cohen 2010), allows one to tune the pore size of the material and adds the possibility of introducing functional groups within the framework. Moreover, if two or more isoreticular organic linkers are employed, frameworks bearing different functionalities that are randomly and homogeneously distributed within the framework are produced by exploiting the concept of multivariable or mixed MOFs (MTV-MOFs or MIXMOFs) (Burrows et al. 2008; Kleist et al. 2009; Deng et al. 2010).

The potential to construct porous structures of coordination polymers by the coordination bonds was initially proposed in 1989 by Hoskins and Robson (1989); however, it took almost 10 years to realize the first few porous MOFs with permanent porosity established by gas adsorption studies (Kondo et al. 1997), as exemplified by MOF-5 in 1999 with significantly high surface area of greater than 3,000 m^2/g (Li et al. 1999; Chui et al. 1999). The availability of various building blocks of metal ions and organic linkers makes it possible to prepare an infinite number of new MOFs with diverse structures, topologies, and porosity. Several examples of MOFs with their characterization are presented in Table 11.1. The typical structures of MOFs are shown in Fig. 11.1.

It is known that mesoporous silica, porous carbon, and other related materials can also have very large apertures (up to 100 nm), and their pore size can be varied in the scale of a few nanometers (Barbour 2006). Unlike these mesoporous materials, the formation of MOFs is governed by the precise linkage of organic struts with the metal atoms to form the secondary building unit (SBUs) (Yaghi et al. 2003). The formation of the SBUs imposes the precise disposition of the links. In this way, the

G. Korotcenkov, *Handbook of Gas Sensor Materials*, Integrated Analytical Systems,
DOI 10.1007/978-1-4614-7388-6_11,

Table 11.1 Characteristic data of several MOFs

MOFs	Formula	BET surface area (m^2/g)	Pore/channel diameter (Å)	Window diameter (Å)	Open metal sites	Thermostability (°C)	Moisture
MOF-74	M_2(2,5-DOT) (M is Zn^{2+}, Mg^{2+})		10, 14			300	
IRMOF-74-I to -XI	M_2(2,5-DOT) (M is Zn^{2+}, Mg^{2+})	1,350–2,510	10, 14–85, 98			300	
MIL-101(Cr)	$Cr_3O(H_2O)_2F(BDC)_3$	2,736–2,907	29, 34	12, 16	Yes	300–330	Yes
MIL-100(Cr)	$Cr_3O(H_2O)_2F(BTC)_2$	1,595	25, 29	5.6, 8.6	Yes	350	Yes
MOF-5, IRMOF-1	$Zn_4O(BDC)_3$	630–2,900	11, 15	7.5, 11.2	No	400–480	No
HKUST-1, MOF-199	$Cu_3(BTC)_2$	1,000–1,458	12	8, 9	Yes	280	No
IRMOF-3	$Zn_4O(NH_2–BDC)_3$	1,957		9.6	No	320	No
ZIF-8	Zn(2-methylimidazole)$_2$	1,504	11.4	3.4	No	380–550	Yes
MIL-53(Al)	Al^{III}(OH)(BDC)	940–1,038	8.5	8.5	No	330	Yes
MIL-47(V)	V^{IV}O(BDC)	800	8.5	8.5	No	350	Yes
ZIF-7	Zn(benzimidazolate)$_2$		4.3	2.9	No	480	Yes
Copper(II) isonicotinate	$Cu(4-C_5H_4N–COO)_2(H_2O)_4$	146			No		Yes

Source: Data from Gu et al. (2012), Deng et al. (2012)
BDC terephthalic acid, *BTC* 1,3,5-benzenetricarboxylate, *DOT* dioxidoterephthalate

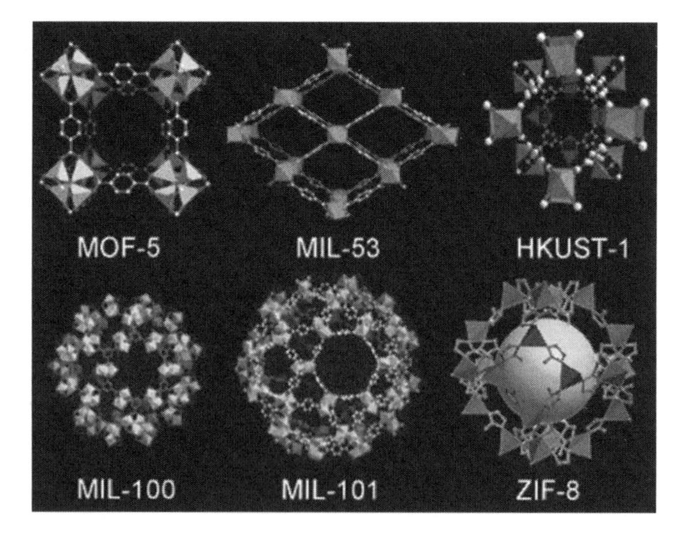

Fig. 11.1 Structures of typical MOFs used in analytical chemistry (Reprinted with permission from Gu et al. 2012, Copyright 2012 American Chemical Society)

pore aperture of MOFs can be controlled at the Ångstrom level through the gradual increase in the number of atoms in the organic links used in the MOF design. These features, coupled with the flexibility in which the MOFs' composition and structure metrics can be varied (Rowsell and Yaghi 2004; Shekhah et al. 2011; Deng et al. 2012), make them highly desirable for well-defined inclusion processes and indeed distinguish them from other mesoporous materials. In addition, the surface areas of MOFs are much higher as compared with those found in mesoporous silica, porous carbon, and zeolites, thus providing more readily available surfaces for interaction with large guest molecules.

It should be noted that, due to their great similarity to zeolites, MOFs can have the same area of applications. In particular, owing to their fascinating structures and unusual properties, such as permanent nanoscale porosity (up to 90 % free volume), high surface area, tunable pore size, adjustable internal surface properties, good thermostability, uniform structured cavities, and unique sieving properties, MOFs have great potential for diverse applications in clean energy, most significantly as storage media for gases such as hydrogen and methane and as high-capacity adsorbents to meet various separation needs. Additional applications in membranes, thin-film devices, catalysis, analytical chemistry, chemical sensors, and biomedical imaging are increasingly gaining in importance (Gu et al. 2012; He et al. 2012). However, compared to crystalline and microporous fully inorganic zeolites, MOFs have much broader synthetic flexibility facilitated by the coordination environment provided by the metal ion and the geometry of the organic "linker" groups (Meek et al. 2011). The combination of the two components of an MOF, the metal ion or cluster and the organic linker, provides endless possibilities. The sum of the physical properties of the inorganic and organic components and the possible synergistic play between the two provide intriguing properties for an MOF. In addition, the pore size in MOFs can be changed over a bigger range. In contrast to nanoporous materials such as zeolites and carbon nanotubes, the MOFs have the ability to tune the structure and functionality of MOFs directly during synthesis. This tunability is significantly different from that of traditional zeolites whose pores are confined by rigid tetrahedral oxide skeletons that are difficult to alter. However, at the same time, we need to recognize that MOFs have worse stability in comparison with zeolites. This limits application of these materials in high-temperature devices.

11.2 MOFs Synthesis

It was established that various synthetic methods, including microwave, electrochemical, mechanochemical, ultrasonic, and high-throughput syntheses, can be applied to MOFs synthesis (Meek et al. 2011). However, it was found that MOFs and zeolites alike are produced almost exclusively by hydrothermal or solvothermal techniques, where crystals are slowly grown from a hot solution of metal precursor, such as metal nitrates, and bridging ligands (Li et al. 1999; Yaghi and Li 1995; Yaghi et al. 2003; Pichon et al. 2006). We should note that MOFs and zeolites have very similar synthetic techniques. Ligands (see Table 11.2), the organic units used for MOFs synthesis, are typically mono-, di-, tri-, or tetravalent. This means that the pores can be tuned by the organic linkers of different length and/or space. Thus, similar to the synthesis of organic copolymers, the building blocks of an MOF should be chosen carefully (Rowsell and Yaghi 2004). Whereas the nature and concentration of the monomers in an organic polymer determine its processability and physical and optical characteristics, it is the network connectivity of the building units that largely determines the properties of an MOF, for example, the definition of large channels available for the passage of molecules. The choice of metal has significant effects on the structure and properties of the MOF as well. For example, the metal's coordination preference influences the size and shape of pores by dictating how many ligands can bind to the metal and in which orientation. Thus, interchangeable linkers and coordinating metal ions offer great flexibility in framework design (see Fig. 11.2). This allows judicious manipulation of the pore or channel sizes, surface area, and type of metal sites in the MOFs (Li et al. 1999; Eddaoudi et al. 2002; Rowsell and Yaghi 2004; Rosseinsky 2004; Sudik et al. 2005; Kitagawa et al. 2006). Rowsell and Yaghi (2004) believe that several factors must be borne in mind when approaching the synthesis of a new MOF, aside from the geometric principles that are considered during its design. By far the most important is the maintenance of the integrity of the building blocks. Quite often a great deal of effort has been expended on the synthesis of a novel organic link, and conditions must be found that are mild enough to maintain the functionality and conformation of this moiety, yet keep it reactive enough to establish the metal–organic bonds. The inclusions of chiral centers or reactive sites within an open framework are also active goals for generating functional materials (Rowsell and Yaghi 2004). It was found that the pore surfaces can be functionalized by the immobilization of functional sites, such as -NH_2 and -OH, into their isostructural MOFs (Eddaoudi et al. 2002; Chen et al. 2010a, b; Vaidhyanathan et al. 2010).

Table 11.2 Common ligands in MOFs

Common name	IUPAC name	Chemical formula	Structural formula
Bidentate carboxylics			
Oxalic acid	Ethanedioic acid	HOOC–COOH	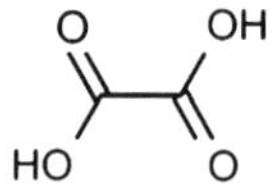
Malonic acid	Propanedioic acid	HOOC–(CH_2)–COOH	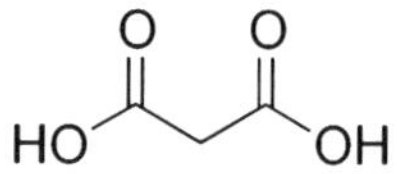
Succinic acid	Butanedioic acid	HOOC–$(CH_2)_2$–COOH	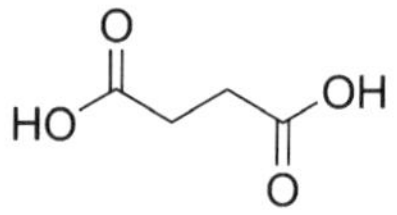
Glutaric acid	Pentanedioic acid	HOOC–$(CH_2)_3$–COOH	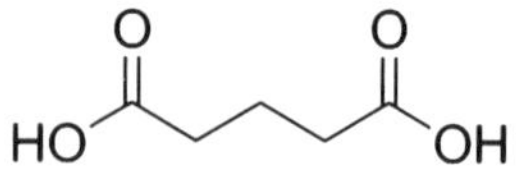
Phthalic acid	Benzene-1,2-dicarboxylic acid *o*-phthalic acid	$C_6H_4(COOH)_2$	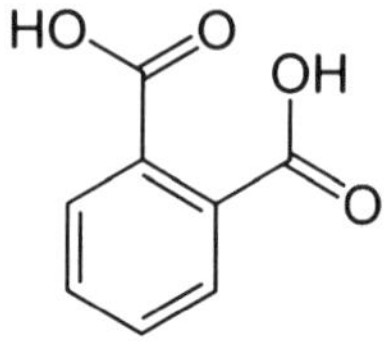
Isophthalic acid	Benzene-1,3-dicarboxylic acid *m*-phthalic acid	$C_6H_4(COOH)_2$	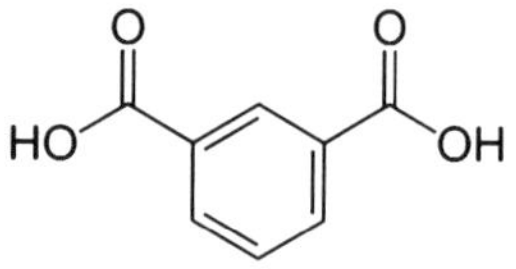
Terephthalic acid	Benzene-1,4-dicarboxylic acid *p*-phthalic acid	$C_6H_4(COOH)_2$	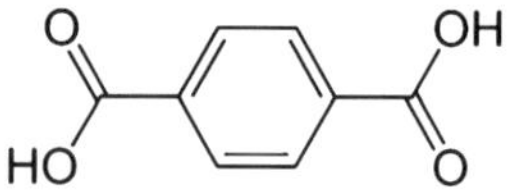
Tridentate carboxylates			
Citric acid	2-Hydroxy-1,2,3-propanetricarboxylic acid	(HOOC)CH_2C(OH)(COOH)CH_2(COOH)	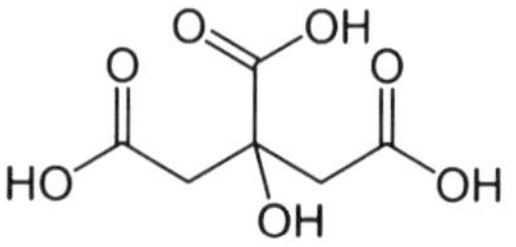
Trimesic acid	Benzene-1,3,5-tricarboxylic acid	$C_9H_6O_6$	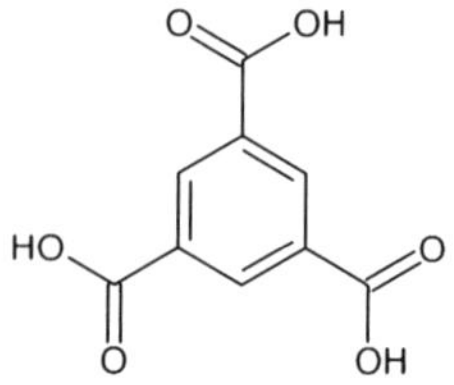
Azoles			
1,2,3-Triazole	1*H*-1,2,3-triazole	$C_2H_3N_3$	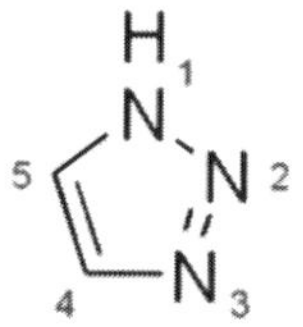
Pyrrodiazole	1*H*-1,2,4-triazole	$C_2H_3N_3$	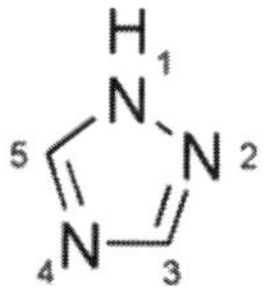

(continued)

Table 11.2 (continued)

Common name	IUPAC name	Chemical formula	Structural formula
Other			
Squaric acid	3,4-Dihydroxy-3-cyclobutene-1,2-dione	$C_4H_2O_4$	HO, O, HO, O

Source: http://www.absoluteastronomy.com/topics/Metal-organic_framework; http://en.wikipedia.org/wiki/Metal-organic_framework

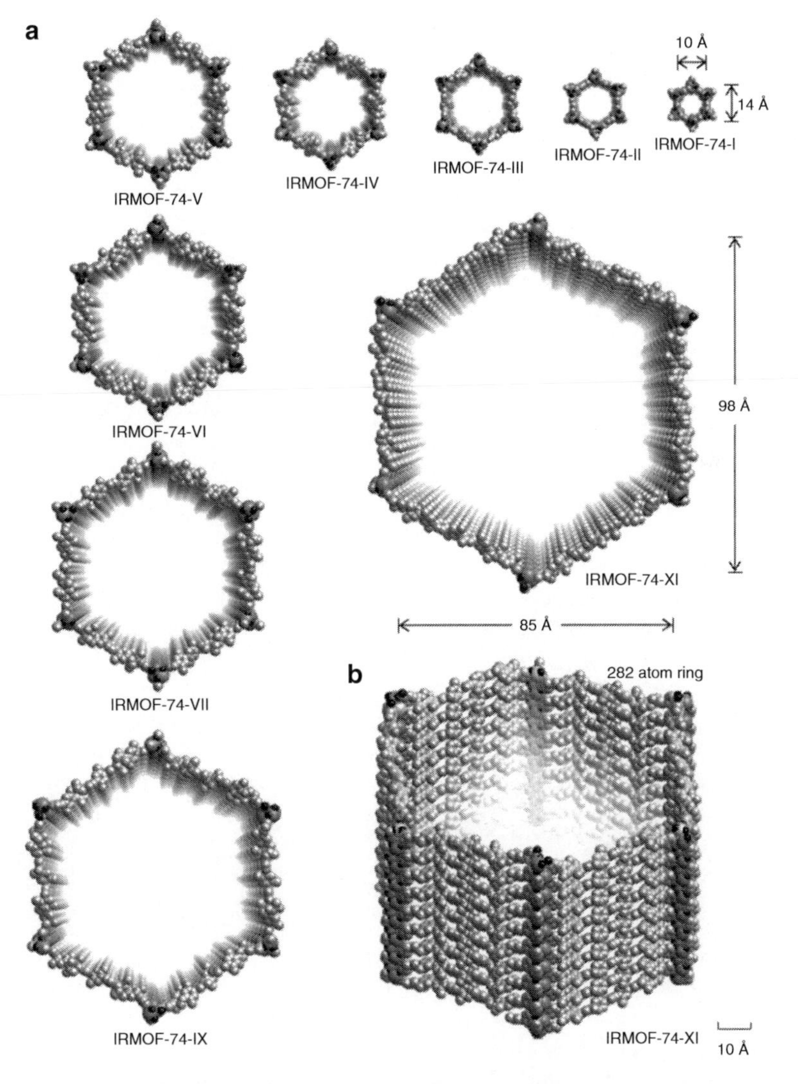

Fig. 11.2 Crystal structures of IRMOF-74 series. (**a**) Perspective views of a single one-dimensional channel shown for each member of IRMOF series, starting from the smallest (*top right*). Pore aperture is described by the length of the diagonal and the distance between the two opposite edges in the regular hexagonal cross section. Hexyl chains as well as hydrogen atoms are omitted for clarity. C atoms are shown in *light gray*, O atoms in *black*, Mg atoms in *dark gray*. IRMOF-74-VI instead of Mg includes Zn in *dark gray*. (**b**) Perspective side view of the hexagonal channel, showing the ring of 282 atoms (*highlighted in gold*) that define the pore aperture of the largest member of the series, IRMOF-74-XI (Reprinted with permission from Deng et al. 2012, Copyright 2012 AAAS)

Zeolite synthesis often makes use of a variety of templates, or structure-directing compounds, and a few examples of templating, particularly by organic anions, are seen in the MOF literature as well. A particular templating approach that is useful for MOFs intended for gas storage is the use of metal-binding solvents such as *N,N*-diethylformamide and water. In these cases, metal sites are exposed when the solvent is fully evacuated, allowing hydrogen to bind at these sites. A solvent-free synthesis of a range of crystalline MOFs can also be used (Pichon et al. 2006).

It should be noted that one important issue during MOFs preparing is the activation of MOFs after synthesis. Solvents used during synthesis usually remain in the pores of the materials, and activation by heating is usually required to remove these solvent molecules. Studies have shown that activation at elevated temperatures can cause sample decomposition, whereas activation at lower temperatures greatly minimizes the danger of reducing metal ions (Liu et al. 2007).

Post-synthetic functionalization of MOFs opens up another dimension of structural possibilities that might not be achieved by conventional synthesis (Ranocchiari et al. 2012). This post-synthetic modification of MOFs gives a large additional variety of materials with different chemical and physical properties (Wang et al. 2009; Cohen 2011; Tanabe and Cohen 2009). A great deal of recent work explores covalent modification of the bridging ligands. Of particular interest to MOFs for gas sensors are modifications which expose metal sites. This has been demonstrated with post-synthetic coordination of additional metal ions to sites on the bridging ligands and addition and removal of metal atoms to the metal site.

11.3 Gas Sensor Applications

We have to recognize that MOFs were designed first of all for application as hydrogen storage, sorbents, and membranes for gas separation (Eddaoudi et al. 2002; Sudik et al. 2005; Kuppler et al. 2009; Thomas 2009; Shekhah et al. 2011; He et al. 2012; Gu et al. 2012; Shah et al. 2012). Experiment has shown that, in these materials, MOFs, the specific recognition of gases is accomplished through several types of interactions that include van der Waals interactions of the framework surface with gases, coordination of the gas molecules to the central metal ion, and hydrogen bonding of the framework surface with gases (Kuppler et al. 2009; MacGillivray 2010; Chen et al. 2010a, b; Zacher et al. 2011). However, unique properties of MOFs such as the unusually high surface area (>3,000–6,000 m^2/g) (Ferey et al. 2005; Furukawa et al. 2010), the ability for tuning pore size, controlling the interaction with guest molecules by varying the nature of the organic linker molecules as well as the coordination environment of the constituting metal ions, chemical functionality, and post-synthetic modifications make these materials attractive for gas sensing using different transduction principles as well. Generally speaking, MOFs with pores functionalized by both tuning the pore/window sizes to make use of size-exclusive effects and immobilizing functional sites such as Lewis basic or acidic and open metal sites within porous MOFs to introduce their specific interaction with gas molecules can demonstrate great abilities for specific and unique gas molecular recognition (Allendorf et al. 2008). The pore apertures dictate the size of the molecules that may enter the pores, which provide the surface and space to carry out these functions (Deng et al. 2012).

Taking into account high surface area and high porosity, one can assume that selective filters, pre-concentrators (Gu et al. 2010), and sorptive layers in mass-sensitive gas sensors (Biemmi et al. 2008; Zybaylo et al. 2010; Khoshaman and Bahreyni 2012) are the best area for MOF applications. Highly porous materials such as MOFs should be inherently sensitive for gas or vapor detection using mass-sensitive devices because they effectively concentrate analyte molecules at higher levels than are present in the external atmosphere. For example, Gu et al. (2010) reported that MOF-5 has good performance for in-field sampling and preconcentration of formaldehyde from air samples with a relative humidity less than 45 %. The collected formaldehyde on MOF-5 sorbent was stable for at least 72 h

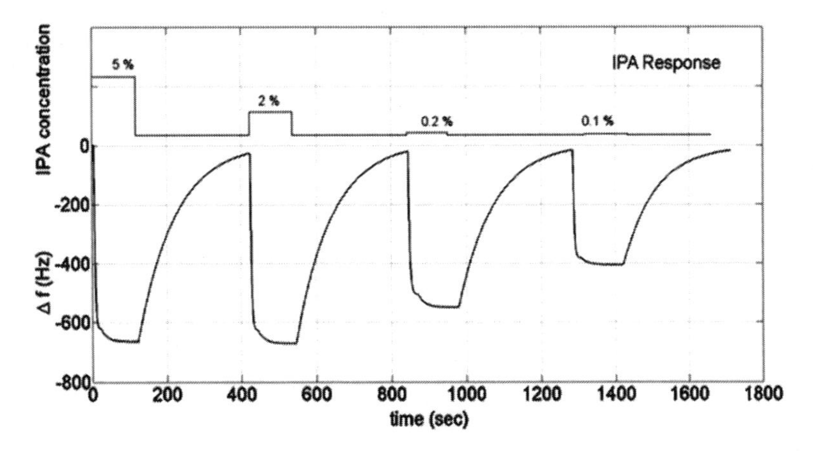

Fig. 11.3 Response to isopropyl alcohol (IPA) of QCM-based sensors with electrosprayed MOF film (Reprinted with permission from Khoshaman and Bahreyni 2012, Copyright 2012 Elsevier)

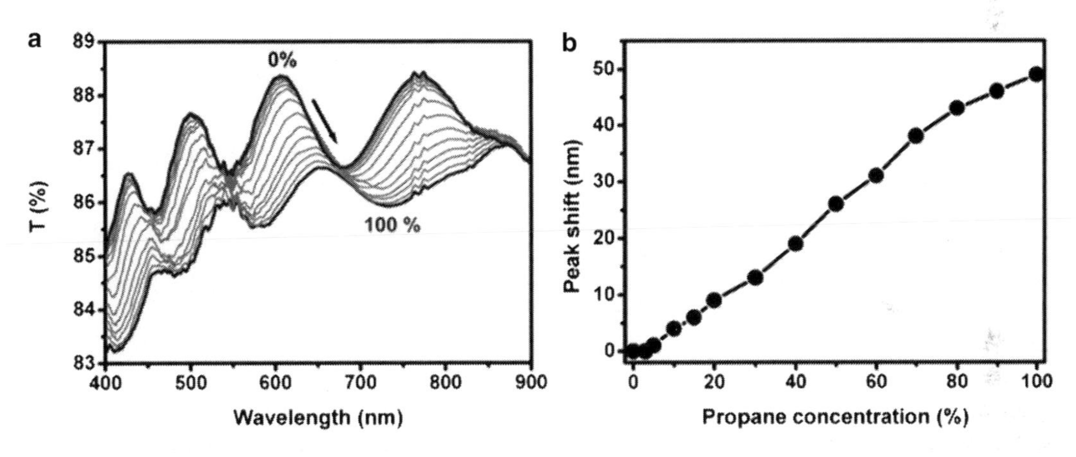

Fig. 11.4 (**a**) UV–vis transmission spectra of 10-cycle ZIF-8 film grown on glass substrate after exposure to propane of various concentrations from 0 to 100 % and (**b**) corresponding interference peak (originally at 612 nm) shift vs. propane concentration. The propane concentration is expressed as a percentage of the total gas flow where nitrogen is used as diluents (Reprinted with permission from Lu and Hupp 2010, Copyright 2010 American Chemical Society)

at room temperature before TD-GC/MS analysis. Typical response of MOF-based QCM sensors is shown in Fig. 11.3. Khoshaman and Bahreyni (2012) have found that these sensors can detect organic vapors of acetone, tetrahydrofuran, and isopropyl alcohol with concentrations of 50, 10, and 2 ppm, respectively.

However, it was established that the application of MOFs for detection of gas molecule sensing using other transduction principles such as impedometric (Achmann et al. 2009a), optical (Lu and Hupp 2010; Kreno et al. 2010; Hinterholzinger et al. 2012), and mechanical (Allendorf et al. 2008; Venkatasubramanian et al. 2010) is also possible. In particular, Fig. 11.4 shows the changes in UV–vis transmission spectra of ZIF-8 film under the influence of vapors of propane. ZIF-8 coatings of various thicknesses were also used to fabricate a Fabry–Perot interferometer. The resulting colorimetric signal was successfully used to detect various hydrocarbons and EtOH (Lu and Hupp 2010). The sensor is unresponsive to water due to the hydrophobic nature of the ZIF-8 pores. The changes of optical properties of MOFs under the influence of vapors of solvents were also discussed by Beauvais

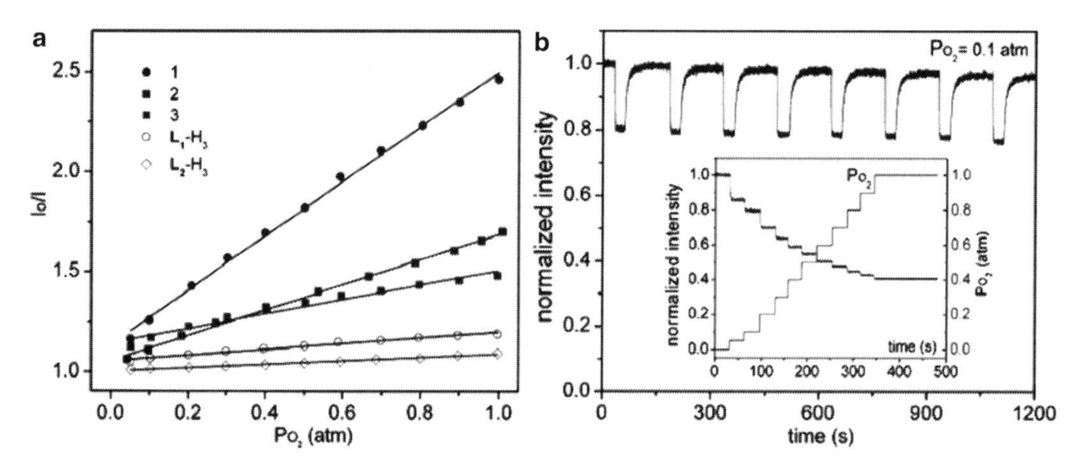

Fig. 11.5 (**a**) Stern–Volmer plot showing I_0/I vs. O_2 partial pressure for Ir complexes (**L1**-H_3 and **L2**-H_3) and $[Zn_4O(carboxylate)_6]$ coordination polymers (**1**–**3**). (**b**) Reversible quenching of phosphorescence of $Ir(ppy)_3$ derivatives- $[Zn_4O(carboxylate)_6]$ complex upon alternating exposure to 0.1 atm O_2 and application of vacuum. The *inset* shows rapid equilibration of phosphorescence after each dose of O_2 (Reprinted with permission from Xie et al. 2009, Copyright 2009 American Chemical Society)

et al. (2000), Lu et al. (2011), and Lee et al. (2011). For example, Beauvais et al. (2000) showed that exposing Co^{2+} MOFs to various vapors could shift the optical absorption across the visible region. The explanation was a change in coordination environment from the as-synthesized octahedral to a tetrahedral geometry.

Although luminescent MOFs have yet to be incorporated into actual gas-sensing devices, Allendorf et al. (2009) believe that the wide variety of fluorescent MOFs and the synthetic versatility inherent in these materials would seem to make them ideal for molecular recognition. This prediction is sound, because luminescent frameworks are by far the most widely explored type of MOF sensor to date (Qiu et al. 2008; Zou et al. 2009; Chen et al. 2010a, b; An et al. 2011; Kent et al. 2011). The popularity of luminescence over other transduction mechanisms is a consequence of several key elements, such as the production of a signal that is visible by eye. For example, recently, Li and co-workers (Lan et al. 2009; Pramanik et al. 2011) reported two fluorescent Zn-based MOFs capable of sensing nitro-containing molecules relevant to detection of explosives. Xie et al. (2009) have shown that luminescent MOFs-based sensors can be used for oxygen sensing (see Fig. 11.5). Xie et al. (2009) incorporated phosphorescent complexes of iridium(III) into an MOF as struts and showed that emission is quenched by energy transfer to O_2.

Temporal response of the cantilever-based sensor to water vapor diluted in N_2 is shown in Fig. 11.6. Allendorf et al. (2008) established that the energy of molecular adsorption within a porous MOF structure could be transformed into mechanical energy which could be utilized to create a responsive, reversible, and selective sensor. Allendorf et al. (2011) believe that this application is limited only by the development of MOFs with high chemical selectivity and the ability to grow these onto the desired substrate.

However, it seems that the application of MOFs in localized surface plasmon resonance (LSPR)-based optical gas sensors is especially promising. Despite its high refractive index sensitivity, LSPR spectroscopy has been generally restricted to large biological analytes. Kreno et al. (2010) have shown that in the case of MOF usage, LSPR-based sensors can be utilized to detect hazardous or toxic gases. MOF, $Cu_3(BTC)_2(H_2O)_3$, used for coating the plasmonic substrate, was grown on Ag nanoparticles using a layer-by-layer method in order to control the MOF thickness. Kreno et al. (2010) believe that, in principle, because the sensing signal originates in the nanoparticle extinction spectrum and not in

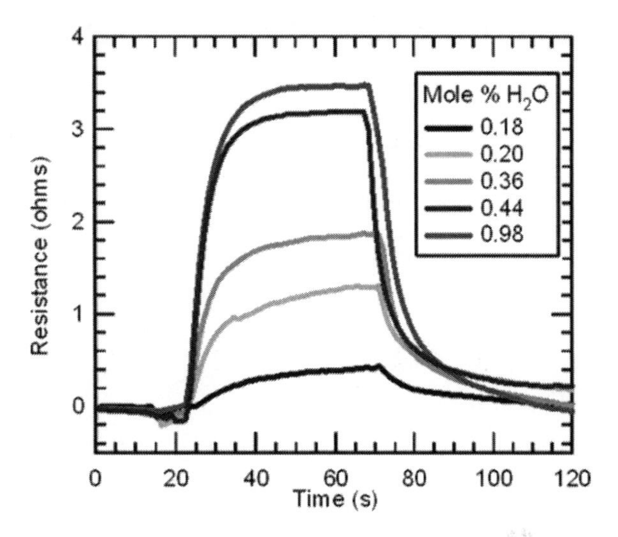

Fig. 11.6 Temporal response of the sensor to water vapor diluted in N_2. The device coated with HKUST-1 responds rapidly to water vapor but has no response to N_2 or O_2 (Reprinted with permission from Allendorf et al. 2008, Copyright 2008 American Chemical Society)

the MOF itself, MOFs can be tailored for sorbing different analytes, making them ideal materials for this amplification strategy. In particular, they established that preferential concentration of CO_2 within the MOF pores produces a 14-fold signal enhancement for CO_2 sensing.

Although electrical and electrochemical methods have been widely used in sensors based on solid electrolytes, chemiresistive metal oxides, and metal oxide semiconductor field-effect transistors, they have been minimally explored for MOFs. This is most likely because the majority of MOFs are insulating (Kreno et al. 2012). To our knowledge, only one group has reported measurement of MOF electrical properties as a means of gas sensing. For example, Achmann et al. (2009a, b) found that M-BTC (M-Al, Fe, Cu) MOF did not show any signal to O_2 (10 vol.%), CO_2 (10 vol.%), C_3H_8 (1,000 ppm), NO (1,000 ppm), and H_2 (1,000 ppm). Noticeable effects that were observed only for hydrophilic gases like ethanol (0–18 vol.%), methanol (0–35 vol.%), and water (0–2.5 vol.%) were applied. However, even in this case, impedometric gas sensors had low sensitivity.

Several recent reviews summarize the gas adsorption isotherms and gas-sensing applications of MOFs (Thomas 2009; Chen et al. 2010a; Fang et al. 2010; Allendorf et al. 2011; Keskin and Kizilel 2011; Meek et al. 2011; Shekhah et al. 2011; Khoshaman and Bahreyni 2012; Kreno et al. 2012). Table 11.3 summarizes results of research related to application of MOFs in gas sensors.

It should be noted that, in spite of extraordinary properties of MOFs, we have to recognize that the exploration of porous MOFs for sensing functions is still at a very early stage. Therefore, to date, we do not have any MOFs-based gas sensors acceptable for the sensor market. New approaches to film deposition and sensor fabrication are required and new effective methods should be designed to reduce the amount of drift and improve the reproducibility in order to make them suitable for long-term applications in various environmental conditions. It was found that many MOFs tend to decompose once exposed to humid air (Huang et al. 2003; Greathouse and Allendorf 2006; Kaye et al. 2007; Kusgens et al. 2009). In order to improve the sensitivity and stability of sensor response, suitable methods should be employed to provide good control over the uniformity and surface properties of the sensing materials (Khoshaman and Bahreyni 2012). Other limitations of MOFs-based gas sensors are the following (Allendorf et al. 2011; Kreno et al. 2012): (1) expensive technology, i.e., high cost; (2) extremely slow growth rates, typically requiring up to a day to produce films on the order of 100 nm thick; (3) low thermal stability; (4) limited long-term stability; (5) the absence of large-scale manufacturing—present methods of growing bulk samples of MOFs produce no more than ~1 g of material; (6) limited number of gases which can be tested by MOF-based sensors—for example, MOF devices are not sensitive to toxic gases; (7) strong sensitivity to air humidity—like other nanoporous materials, MOFs have the ability to adsorb large quantities of water; for example, it is

Table 11.3 Examples of MOFs application in gas sensors

MOF	Sensor type/T_{oper}	Gas tested	Characterization	References
$Zn_4O(BDC)_3$ (MOF-5)	Preconcentrator	Formaldehyde	RH<45 %. One tube packed with 300 mg of MOF-5 lasted 200 cycles of adsorption/TD without significant loss of collection efficiency	Gu et al. (2010)
$Zn(mim)_2$ (ZIF-8)	Optical; RT (Fabry–Perot interference)	H_2O, EtOH, *n*-hexane, propane	Not sensitive to H_2O	Lu and Hupp (2010)
ZIF-8–TiO_2 (three-bilayer Bragg stack)	Optical (reflection)	EtOH, MeOH, isobutanol, pentanol (0–100 % +N_2)	Maximum sensitivity to EtOH with maximal hydrophobicity	Hinterholzinger et al. (2012)
Ag–$Cu_3(BTC)_2(H_2O)_3$	Optical (LSPR)	CO_2	High sensitivity	Kreno et al. (2010)
$Ir(ppy)_3$ derivatives-[$Zn_4O(carboxylate)_6$] complex	Optical (luminescent)	O_2 (0–1 atm)	Linear dependence	Xie et al. (2009)
Cu(II)–$BTC_2(H_2O)_3$ (HKUST-1)	Mechanical; RT (microcantilever)	N_2, O_2, CO_2, H_2O	Sensitive only to water	Allendorf et al. (2008)
	Mechanical; 30–70 °C (microcantilever)	CO_2 (0–10 % in dry N_2)	Microfabrication technology	Venkatasubramanian et al. (2010)
$Cu_3(BTC)_2(H_2O)_3 \cdot xH_2O$ (HKUST-1)	QCM; RT	Acetone, THF, IPA	Electrospraying gave better results than drop-casting method	Khoshaman and Bahreyni (2012)
	QCM; 294–343 K	H_2O	Good sensitivity. Hysteresis is absent at T_{oper}=343 K	Biemmi et al. (2008)
Fe–BTC; Cu–BTC; Al–BDC	Impedometric; 120 °C	O_2, C_3H_8, NO, CO_2, H_2, H_2O, MeOH, EtOH	Sensitivity only for hydrophilic gases (H_2O, MeOH, EtOH)	Achmann et al. (2009a, b)

BTC 3,5-benzenetricarboxylate, *BDC* 1,4-benzenedicarboxylate, *THF* tetrahydrofuran, *mim* 2-methylimidazolate, *ndc* 2,6-naphtalenedicarboxylic acid, *ur* urotropin, *DMF* dimethylformamide; *LSPR* localized surface plasmon resonance, *IPA* isopropyl alcohol

reported that HKUST-1 can adsorb as much as 40 wt% water (Chui et al. 1999; Wang et al. 2002)—as was shown by (Mintova et al. 2001; Allendorf et al. 2008), this makes them attractive for humidity sensing using different platforms; however, water vapor is also a common interfering gas and must be addressed in the design of MOF-based sensing systems—there are several approaches to resolving this problem, for example, hydrophobic MOFs, such as the ZIF materials (Lu and Hupp 2010), could be used; unfortunately, ZIF materials cannot be used in all devices; (8) low selectivity—though the potential selectivity of MOF materials for specific analytes, or classes of analytes, is substantial, as yet, it is not highly developed.

References

Achmann S, Hagen G, Kita J, Malkowsky IM, Kiener C, Moos R (2009a) Metal-organic frameworks for sensing applications in the gas phase. Sensors 9:1574–1589

Achmann S, Hagen G, Moos R, Malkowsky I, Kiener C (2009b) Metal-organic framework for sensing applications in the gas phase. In: Proceedings of Sensor+Test conference-sensor 2009, vol 2, pp 417–420

Allendorf MD, Houk RJT, Andruszkiewicz L, Talin AA, Pikarsky J, Choudhury A, Gall KA, Hesketh PJ (2008) Stress-induced chemical detection using flexible metal–organic frameworks. J Am Chem Soc 130:14404–14405
Allendorf MD, Bauer CA, Bhakta RK, Houk RJT (2009) Luminescent metal–organic frameworks. Chem Soc Rev 38:1330–1352
Allendorf MD, Schwartzberg A, Stavila V, Talin AA (2011) A roadmap to implementing metal–organic frameworks in electronic devices: challenges and critical directions. Chem Eur J 17:11372–11388
An JY, Shade CM, Chengelis-Czegan DA, Petoud S, Rosi NL (2011) Zinc-adeninate metal–organic framework for aqueous encapsulation and sensitization of near-infrared and visible emitting lanthanide cations. J Am Chem Soc 133:1220–1223
Barbour LJ (2006) Crystal porosity and the burden of proof. Chem Commun 2006:1163–1168
Beauvais LG, Shores MP, Long JR (2000) Cyano-bridged Re6Q8 (Q = S, Se) cluster-cobalt(II) framework materials: versatile solid chemical sensors. J Am Chem Soc 122:2763–2772
Biemmi E, Darga A, Stock N, Bein T (2008) Direct growth of $Cu_3(BTC)_2(H_2O)_3 \cdot xH_2O$ thin films on modified QCM-gold electrodes—water sorption isotherms. Microporous Mesoporous Mater 114:380–386
Burrows AD, Frost CG, Mahon MF, Richardson C (2008) Post-synthetic modification of tagged metal-organic frameworks. Angew Chem 47:8610–8614
Cavka JH, Jakobsen S, Olsbye U, Guillou N, Lamberti C, Bordiga S, Lillerud KP (2008) A new zirconium inorganic building brick forming metal organic frameworks with exceptional stability. J Am Chem Soc 130:13850–13851
Chen B, Xiang S, Qian G (2010a) Metal–organic frameworks with functional pores for recognition of small molecules. Acc Chem Res 43:1115–1124
Chen Z, Xiang S, Arman HD, Li P, Zhao D, Chen B (2010b) A microporous metal–organic framework with immobilized–OH functional groups within the pore surfaces for selective gas sorption. Eur J Inorg Chem 24:3745–3749
Chui SS-Y, Lo SM-F, Charmant JPH, Orpen AG, Williams ID (1999) A chemically functionalizable nanoporous material $[Cu_3(TMA)_2(H_2O)_3]n$. Science 283:1148–1150
Cohen SM (2011) Postsynthetic methods for the functionalization of metal-organic frameworks. Chem Rev 112:970–1000
Deng H, Doonan CJ, Furukawa H, Ferreira RB, Towne J, Knobler CB, Wang B, Yaghi OM (2010) Multiple functional groups of varying ratios in metal-organic frameworks. Science 327:846–850
Deng H, Grunder S, Cordova KE, Valente C, Furukawa H, Hmadeh M, Gándara F, Whalley AC, Liu Z, Asahina S, Kazumori H, O'Keeffe M, Terasaki O, Stoddart JF, Yaghi OM (2012) Large-pore apertures in a series of metal-organic frameworks. Science 336:1018–1023
Eddaoudi M, Kim J, Rosi N, Vodak D, Wachter J, Keeffe MO, Yaghi OM (2002) Systematic design of pore size and functionality in isoreticular MOFs and their application in methane storage. Science 295:469–472
Fang Q-R, Makal TA, Young MD, Zhou H-C (2010) Recent advances in the study of mesoporous metal-organic frameworks. Commun Inorg Chem 31:165–195
Ferey C, Mellot-Draznieks C, Serre C, Millange F, Dutour J, Surble S, Margiolaki IA (2005) Chromium terephthalate-based solid with unusually large pore volumes and surface area. Science 309:2040–2042
Furukawa H, Ko N, Go YB, Aratani N, Choi SB, Choi E, Yazaydin AO, Snurr RQ, O'Keeffe M, Kim J, Yaghi OM (2010) Ultrahigh porosity in metal-organic frameworks. Science 329:424–428
Garibay SJ, Cohen SM (2010) Isoreticular synthesis and modification of frameworks with the UiO-66 topology. Chem Commun 46:7700–7702
Greathouse JA, Allendorf MD (2006) The interaction of water with MOF-5 simulated by molecular dynamics. J Am Chem Soc 128(40):13312
Gu Z-Y, Wang G, Yan X-P (2010) MOF-5 metal–organic framework as sorbent for in-field sampling and preconcentration in combination with thermal desorption GC/MS for determination of atmospheric formaldehyde. Anal Chem 82:1365–1370
Gu Z-Y, Yang C-X, Chang N, Yan X-P (2012) Metal-organic frameworks for analytical chemistry: from sample collection to chromatographic separation. Acc Chem Res 45(5):734–745
He Y, Zhang Z, Xiang S, Fronczek FR, Krishna R, Chen B (2012) A microporous metal–organic framework for highly selective separation of acetylene, ethylene, and ethane from methane at room temperature. Chem Eur J 18:613–619
Hinterholzinger FM, Ranft A, Feckl H, Bein T, Lotsch BT (2012) One-dimensional metal-organic framework photonic crystals used as platforms for vapor sensing. J Mater Chem 22:10356–10362
Hoskins BF, Robson R (1989) Infinite polymeric frameworks consisting of three dimensionally linked rod-like segments. J Am Chem Soc 111:5962–5964
Huang LM, Wang HT, Chen JX, Wang ZB, Sun JY, Zhao DY, Yan YS (2003) Synthesis, morphology control, and properties of porous metal-organic coordination polymers. Microporous Mesoporous Mater 58:105–114
Kaye SS, Dailly A, Yaghi OM, Long JR (2007) Impact of preparation and handling on the hydrogen storage properties of $Zn_4O(1,4\text{-benzenedicarboxylate})_3$ (MOF-5). J Am Chem Soc 129:14176–14177
Kent CA, Liu D, Ma L, Papanikolas JM, Meyer TJ, Lin W (2011) Light harvesting in microscale metal–organic frameworks by energy migration and interfacial electron transfer quenching. J Am Chem Soc 133(33):12940–12943
Keskin S, Kizilel S (2011) Biomedical applications of metal organic frameworks. Ind Eng Chem Res 50:1799–1812

Khoshaman AH, Bahreyni B (2012) Application of metal organic framework crystals for sensing of volatile organic gases. Sens Actuators B 162:114–119

Kitagawa S, Kitaura R, Noro S (2004) Functional porous coordination polymers. Angew Chem Int Ed 43:2334–2375

Kitagawa S, Noro S-I, Nakamura T (2006) Pore surface engineering of microporous coordination polymers. Chem Commun 2006:701–707

Kleist W, Jutz F, Maciejewski M, Baiker A (2009) Mixed-linker metal-organic frameworks as catalysts for the synthesis of propylene carbonate from propylene oxide and CO_2. Eur J Inorg Chem 2009:3552–3561

Kondo M, Yoshitomi T, Seki K, Matsuzaka H, Kitagawa S (1997) Three-dimensional framework with channeling cavities for small molecules: $\{[M_2(4,4¢\text{-bpy})_3(NO_3)_4] \cdot xH_2O\}n$ (M) Co, Ni, Zn. Angew Chem Int Ed Engl 36:1725–1727

Kreno LE, Hupp JT, Van Duyne RP (2010) Metal–organic framework thin film for enhanced localized surface plasmon resonance gas sensing. Anal Chem 82:8042–8046

Kreno LE, Leong K, Farha OK, Allendorf M, Van Duyne RP, Hupp JT (2012) Metal-organic framework materials as chemical sensors. Chem Rev 112:1105–1125

Kuppler RJ, Timmons DJ, Fang Q-R, Li J-R, Makal TA, Young MD, Yuan D, Zhao D, Zhuang W, Zhou H-C (2009) Potential applications of metal-organic frameworks. Coord Chem Rev 253:3042–3066

Kusgens P, Rose M, Senkovska I, Frode H, Henschel A, Siegle S, Kaskel S (2009) Characterization of metal-organic frameworks by water adsorption. Microporous Mesoporous Mater 120:325–330

Lan A, Li K, Wu H, Olson DH, Emge TJ, Ki W, Hong M, Li J (2009) A luminescent microporous metal-organic framework for the fast and reversible detection of high explosives. Angew Chem Int Ed 48:2334–2338

Lee H, Jung SH, Han WS, Moon JH, Kang S, Lee JY, Jung JH, Shinkai SA (2011) Chromo-fluorogenic tetrazole-based $CoBr_2$ coordination polymer gel as a highly sensitive and selective chemosensor for volatile gases containing chloride. Chem-Eur J 17:2823–2827

Li H, Eddaoudi M, O'Keeffe M, Yaghi OM (1999) Design and synthesis of an exceptionally stable and highly porous metal-organic framework. Nature 402:276–279

Liu J, Culp JT, Natesakhawat S, Bockrath BC, Zande B, Sankar SG, Garberoglio G, Johnson JK (2007) Experimental and theoretical studies of gas adsorption in $Cu_3(BTC)_2$: an effective activation procedure. J Phys Chem C 111:9305–9313

Lu G, Hupp JT (2010) Metal–organic frameworks as sensors: a ZIF-8 based Fabry–Pérot device as a selective sensor for chemical vapors and gases. J Am Chem Soc 132:7832–7833

Lu Z-Z, Zhang R, Li Y-Z, Guo Z-J, Zheng H-G (2011) Solvatochromic behavior of a nanotubular metal–organic framework for sensing small molecules. J Am Chem Soc 133:4172–4174

MacGillivray LR (ed) (2010) Metal-organic frameworks: design and application. Wiley, Hoboken, NJ

Meek ST, Greathouse JA, Allendorf MD (2011) Metal-organic frameworks: a rapidly growing class of versatile nanoporous materials. Adv Mater 23:249–267

Mintova S, Mo SY, Bein T (2001) Humidity sensing with ultrathin LTA-type molecular sieve films grown on piezoelectric devices. Chem Mater 13:901–905

Pichon A, Lazuen-Garay A, James SL (2006) Solvent-free synthesis of a microporous metal organic framework. CrystEngComm 8:211–214

Pramanik S, Zheng C, Zhang X, Emge TJ, Li J (2011) New microporous metal–organic framework demonstrating unique selectivity for detection of high explosives and aromatic compounds. J Am Chem Soc 133:4153–4155

Qiu L-G, Li Z-Q, Wu Y, Wang W, Xu T, Jiang X (2008) Facile synthesis of nanocrystals of a microporous metal-organic framework by an ultrasonic method and selective sensing of organoamines. Chem Commun 2008:3642–3644

Ranocchiari M, Lothschütz C, Grolimund D, van Bokhoven JA (2012) Single-atom active sites on metal-organic frameworks. Proc R Soc A 2012(0078):1–15

Rosseinsky MJ (2004) Recent developments in metal-organic framework chemistry: design, discovery, permanent porosity and flexibility. Microporous Mesoporous Mater 73:15–30

Rowsell JLC, Yaghi O (2004) Metal–organic frameworks: a new class of porous materials. Microporous Mesoporous Mater 73:3–14

Shah M, McCarthy MC, Sachdeva S, Lee AK, Jeong H-K (2012) Current status of metal-organic framework membranes for gas separations: promises and challenges. Ind Eng Chem Res 51:2179–2199

Shekhah O, Liu J, Fischer RA, Woll C (2011) MOF thin films: existing and future applications. Chem Soc Rev 40:1081–1106

Sudik AC, Millward AR, Ockwig NW, Côté AP, Kim J, Yaghi OM (2005) Design, synthesis, structure, and gas (N_2, Ar, CO_2, CH_4, and H_2) sorption properties of porous metal-organic tetrahedral and heterocuboidal polyhedra. J Am Chem Soc 127:7110–7118

Tanabe KK, Cohen SM (2009) Engineering a metal-organic framework catalyst by using postsynthetic modification. Angew Chem Int Ed 48:7424–7427

Thomas KM (2009) Adsorption and desorption of hydrogen on metal-organic framework materials for storage applications: comparison with other nanoporous materials. Dalton Trans (9):1487–1505

Vaidhyanathan R, Iremonger SS, Shimizu GKH, Boyd PG, Alavi S, Woo TK (2010) Direct observation and quantification of CO_2 binding within an amine-functionalized nanoporous solid. Science 330:650–653

Venkatasubramanian A, Lee J-H, Houk RJ, Allendorf MD, Nair S (2010) Characterization of HKUST-1 crystals and their application to MEMS microcantilever array sensors. ECS Trans 33(8):229–238

Wang QM, Shen DM, Bulow M, Lau ML, Deng SG, Fitch FR, Lemcoff NO, Semanscin J (2002) Metallo-organic molecular sieve for gas separation and purification. Microporous Mesoporous Mater 55:217–230

Wang L, Reis A, Seifert A, Philippi T, Ernst S, Jia M, Thiel WR (2009) A simple procedure for the covalent grafting of triphenylphosphine ligands on silica: application in the palladium catalyzed Suzuki reaction. Dalton Trans 17:3315–3320

Xie Z, Ma L, de Krafft KE, Jin A, Lin W (2009) Porous phosphorescent coordination polymers for oxygen sensing. J Am Chem Soc 132:922–923

Yaghi OM, Li HL (1995) Hydrothermal synthesis of a metal organic framework containing large rectangular channels. J Am Chem Soc 117:10401–10402

Yaghi OM, O'Keeffe MO, Ockwig NW, Chae HK, Eddaoudi M, Kim J (2003) Reticular synthesis and the design of new materials. Nature 423:705–714

Zacher D, Schmid R, Woll C, Fischer RA (2011) Surface chemistry of metal–organic frameworks at the liquid–solid interface. Angew Chem Int Ed 50:176–199

Zou X, Zhu G, Hewitt IJ, Sun F, Qiu S (2009) Synthesis of a metal-organic framework film by direct conversion technique for VOCs. Dalton Trans 2009:3009–3013

Zybaylo O, Shekhah O, Wang H, Tafipolsky M, Schmid R, Johannsmann D, Woll C (2010) A novel method to measure diffusion coefficients in porous metal–organic frameworks. Phys Chem Chem Phys 12:8092–8097

Part III
Nanocomposites

Chapter 12
Nanocomposites in Gas Sensors: Promising Approach to Gas Sensor Optimization

Nanocomposites provide another promising direction in the development of materials for gas sensors (Ferroni et al. 1999; Gas'kov and Rumyantseva 2001; Comini et al. 2002; Galatsis et al. 2002; Sadek et al. 2006; Rumyantseva et al. 2006; Yang 2011). Nanocomposite materials have recently attracted increasing interest because of the possibility of synthesizing materials with unique physical–chemical properties (Gas'kov and Rumyantseva 2001; Zhang et al. 2003a, b). It was established that highly sophisticated surface-related factors important for gas sensor applications such as optical, electronic, catalytic, mechanical, and chemical properties can be obtained by advanced nanocomposites synthesized from various materials. Recently, various nanocomposite films consisting of either metal–metal oxide, mixed metal oxides, polymers mixed with metals or metal oxides, or carbon nanotubes mixed with polymers, metals, or metal oxides have been synthesized and investigated for their application as active materials for gas sensors. For example, it was established that metallic and metal-oxide nanoparticles incorporated in various matrixes are capable of increasing the activities for many chemical reactions due to the high ratio of surface atoms with free valences to the cluster of total atoms. As a result, we can obtain an ideal platform for gas sensor design. In particular, we should note that most of the chemiresistors are devices based on metal oxide–noble metal nanocomposites. Nanoclusters of noble metals such as Pd, Pt, Au, Rh, and Ag incorporated in a metal-oxide matrix increase catalytic activity of gas-sensing materials, change adsorption/desorption parameters, and promote the reducing operation temperature, increasing sensor response, enhancing the response rate, and improving sensor selectivity. More detailed reviews of the effects of doping on metal oxide gas sensors are available elsewhere (Kohl 1990; Korotcenkov 2005, 2007; Miller et al. 2006).

It was found that transition to nanocomposites could also improve mechanical properties and promote stabilization of the basic material's parameters (Konig 1987). For example, it was established that in CNTs–polymer composites, the presence of carbon nanotubes inside the polymeric matrix can provide a mechanical support to the polymeric chain's conformational rearrangement. CNTs are hollow nanopipes, and therefore the incorporation of CNTs in a metal oxide matrix can provide better gas permeability for sensing materials and thus enhance gas diffusion into the bulk film. Thus, combination of CNTs with metal oxide (see Fig. 12.1) can lead to development of gas sensors with improved rate of response.

Nanocomposites also provide more possibilities for control of the catalytic activity of the sensing matrix. For example, it has been shown that the introduction of TiO_2 nanoparticles into the polymer matrix of poly(*p*-phenylenevinylene) (PPV) changes the adsorption properties of the matrix. Adsorption of oxygen was found to be stronger on the PPV–TiO_2 nanocomposite than on pure PPV (Baraton et al. 1997). In some cases, a catalyst incorporated in the conducting polymer film can help in detecting some inert analytes. Conducting polymers have difficulties with detection inert analytes.

G. Korotcenkov, *Handbook of Gas Sensor Materials*, Integrated Analytical Systems,
DOI 10.1007/978-1-4614-7388-6_12, © Springer Science+Business Media New York 2014

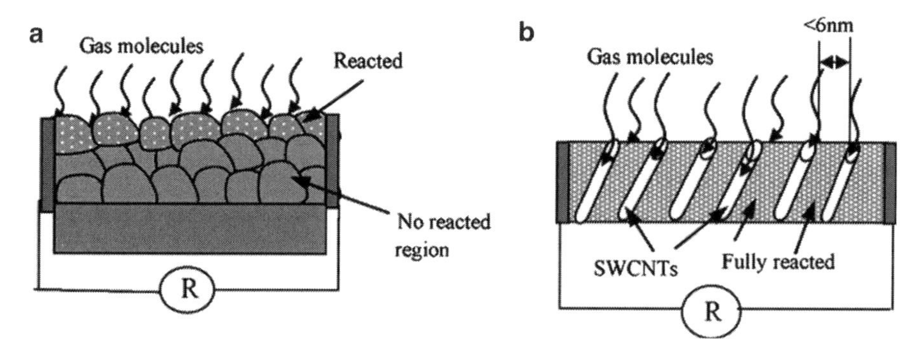

Fig. 12.1 (**a**) Sketch of pure SnO_2 film, only top surface portion reacted. (**b**) Sketch of gas-sensing mechanism between pure SnO_2 film and SWCNT/SnO_2 film (Reprinted with permission from Gong et al. 2008, Copyright 2008 Elsevier)

Table 12.1 Nanocomposites used in gas sensor design

Nanocomposite	Sensor type	Functions, effect
Insulating polymer—black carbon (PMMA, PPy, ethyl cellulose (EC))	Chemiresistors (sensing material)	Polymer—gas-sensing matrix, swelling effect Black carbon—percolation conductivity
Insulating polymer–CNTs (PMMA, PPy, ethyl cellulose (EC))	Chemiresistors (sensing material)	Polymer—gas-sensing matrix, swelling effect
	SAW, QCM	CNTs—percolation conductivity
Conducting polymer-CNTs (Pt, Au, or Cu) (PDPA, Nafion)	Electrochemical (modified electrode)	Polymer—gas penetrable matrix CNTs—conducting component Noble metal—catalyst
Polymer–black carbon (PMMA, PPy, ethyl cellulose (EC))	Chemiresistors (sensing material)	Polymer—gas-sensing matrix, swelling effect Carbo—percolation conductivity
Polymer–metal nanoparticles (Ag, Au, Cu, etc.) (phthalocyanine, Nafion, etc.)	Chemiresistors (sensing material)	Polymer—gas sensing matrix, swelling effect Metal—percolation conductivity
	Optical	Polymer—gas penetrable matrix Metal—surface plasmon resonance
	Electrochemical (gas diffusion electrode)	Polymer—gas penetrable matrix Noble metal—catalyst
Polymer (1)–polymer (2) (PVA-PPy)	Chemiresistors (humidity sensors)	Polymer (1)—gas penetrable matrix, swelling effect Polymer (2)—conductivity
Polymer dyes	Opto chemical	Polymer—gas penetrable matrix Dye—fluorescence agent
Polymer–metal oxide (PPV–TiO_2, etc.)	QCM, SAW (adsorbent)	Polymer—gas-sensing matrix Metal oxide—promoter
Metal oxide–noble metals (SnO_2–Ag; SnO_2–Pd, Pt; etc.)	Chemiresistors (sensing material)	Metal oxide—gas-sensing matrix Noble metal—catalyst, surface reactivity improvement
Metal oxide (1)–metal oxide (2) (SnO_2–Fe_2O_3; SnO_2–In_2O_3; SnO_2–ZnO; SnO_2–SiO_2; etc.)	Chemiresistors (sensing material)	Metal oxide (1)—gas-sensing matrix Metal oxide (2)—structure modifier (grain size decrease, structure optimization)
Metal oxide (1)–metal oxide (2) (SnO_2–Cu_2O; SnO_2–Ag_2O)		Metal oxide (1)—conductive matrix Metal oxide (2)—chemically reactive
CNTs (graphene)—noble metals; metal oxide; polymer		CNTs (graphene)—gas-sensing element Noble metals; metal oxide; polymer—surface functionalization

Athawale et al. (2006) prepared nanocomposite of Pd/PAni and found its electrical resistance responses rapidly and reversibly in the presence of methanol. Ram et al. (2005) studied PAni/SnO_2 composite and established that CO is able to oxide PAni with the assistance of SnO_2.

We should note that a full description of the possible methods of fabricating nanocomposites with particular properties is not possible in this brief review. There are too many nanocomposites with great differences in properties obtained as a result of nanocomposite elaborations. Interested readers may refer to the literature for details (Gas'kov and Rumyantseva 2001, 2009; Sanchez et al. 2005; Viswanathan et al. 2006; Moya et al. 2007; Rozenberg and Tenne 2008; Cury Camargo et al. 2009; Li and Zhang 2009; Yang 2011). The most popular composites used in gas sensors are listed in Table 12.1. As can be seen, research focused on the elaboration of nanocomposite materials is being carried out at present in various directions. Composites can be based on carbon nanotubes and fullerenes, metal oxides, noble metals, and polymers. However, we have to recognize that polymer-based or organic–inorganic composites are the most popular and may be the most promising for application in gas sensor.

References

Athawale AA, Bhagwat SV, Katre PP (2006) Nanocomposite of Pd-polyaniline as a selective methanol sensor. Sens Actuators B 114:263–267

Baraton MI, Merhari L, Wang J, Gonsalves KE (1997) Dispersion of metal oxide nanoparticles in conjugated polymers: investigation of the TiO_2/PPV nanocomposite. MRS Symp Proc 501:59–64

Comini E, Ferroni M, Guidi V, Faglia G, Martinelli G, Sberveglieri G (2002) Nanostructured mixed oxides compounds for gas sensing applications. Sens Actuators B 84:26–32

Cury Camargo PH, Satyanarayana KG, Wypych F (2009) Nanocomposites: synthesis, structure, properties and new application opportunities. Mater Res 12(1):1–39

Ferroni M, Boscarino D, Comini E, Gnani D, Guidi V, Martinelli G, Nelli P, Rigato V, Sberveglieri G (1999) Nanosized thin films of tungsten-titanium mixed oxides as gas sensors. Sens Actuators B 58:289–294

Galatsis K, Li YX, Wlodarski W, Comini E, Sberveglieri G, Cantalini C, Santucci S, Passacantando M (2002) Comparison of single and binary oxide MoO_3, TiO_2 and WO_3 sol–gel gas sensors. Sens Actuators B 83:276–280

Gas'kov AM, Rumyantseva MN (2001) Nature of gas sensitivity in nanocrystalline metal oxides. Russ J Appl Chem 74(3):440–444

Gas'kov A, Rumyantseva M (2009) Metal oxide nanocomposites: synthesis and characterization in relation with gas sensing phenomena. In: Baraton MI, Baraton MI (eds) Sensors for environment, health and security. Springer Science + Business Media B.V., Dordrecht, The Netherlands, pp 3–29

Gong J, Sun J, Chen Q (2008) Micromachined sol–gel carbon nanotube/SnO_2 nanocomposite hydrogen sensor. Sens Actuators B 130:829–835

Kohl D (1990) The role of noble metals in the chemistry of solid state gas sensors. Sens Actuators B 1:158–165

Konig U (1987) Deposition and properties of multicomponent hard coating. Surf Coat Technol 33:91–103

Korotcenkov G (2005) Gas response control through structural and chemical modification of metal oxides: state of the art and approaches. Sens Actuators B 107:209–232

Korotcenkov G (2007) Practical aspects in design of one-electrode semiconductor gas sensors: status report. Sens Actuators B 121:664–678

Li J, Zhang JZ (2009) Optical properties and applications of hybrid semiconductor nanomaterials. Coord Chem Rev 253:3015–3041

Miller TA, Bakrania SD, Perez C, Wooldridge MS (2006) Nanostructured tin dioxide materials for gas sensor applications. In: Geckeler KE, Rosenberg E (eds) Functional nanomaterials. American Scientific Publishers, Stevenson Ranch, CA, pp 1–24

Moya JS, Lopez-Esteban S, Pecharroman C (2007) The challenge of ceramic/metal microcomposites and nanocomposites. Prog Mater Sci 52:1017–1090

Ram MK, Yavuz O, Lahsangah V, Aldissi M (2005) CO gas sensing from ultrathin nano-composite conducting polymer film. Sens Actuators B 106:750–757

Rozenberg BA, Tenne R (2008) Polymer-assisted fabrication of nanoparticles and nanocomposites. Prog Polym Sci 33:40–112

Rumyantseva M, Kovalenko V, Gaskov A, Makshina E, Yuschenko V, Ivanova I, Ponzoni A, Faglia G, Comini E (2006) Nanocomposites SnO_2/Fe_2O_3: sensor and catalytic properties. Sens Actuators B 118:208–214
Sadek Z, Wlodarski W, Shin K, Kaner RB, Kalantar-zadeh K (2006) A layered surface acoustic wave gas sensor based on a polyaniline/In_2O_3 nanofibre composite. Nanotechnology 17:4488–4492
Sanchez C, Julian B, Belleville P, Popall M (2005) Applications of hybrid organic–inorganic nanocomposites. J Mater Chem 15:3559–3592
Viswanathan V, Laha T, Balani K, Agarwal A, Seal S (2006) Challenges and advances in nanocomposite processing techniques. Mater Sci Eng R 54:121–285
Yang D (2011) Nanocomposite films for gas sensing. In: Reddy B (ed) Advances in nanocomposites—synthesis, characterization and industrial applications. InTech, Ch. 37, pp 857–882. http://www.intechopen.com
Zhang S, Yongqing DS, Du FH (2003a) Recent advances of superhard nanocomposite coatings: a review. Surf Coat Technol 167:13–119
Zhang W, Chen D, Zhao Q, Fang Y (2003b) Effects of different kinds of clay and different vinyl acetate content on the morphology and properties of EVA/clay nanocomposites. Polymer 44(1):7953–7961

Chapter 13
Polymer-Based Composites

At present it is known that various types of polymer-based composites can be applied to design conductometric gas sensors capable of operating at room temperature (Wohltjen et al. 1985; Middlehoek and Audet 1989; Unde et al. 1996; Cespedes et al. 1996; Suri et al. 2002; Densakulprasert et al. 2005; Santhanam et al. 2005; Watcharaphalakorn et al. 2005; Wojkiewicz et al. 2011; Potts et al. 2011; Hands et al. 2012). Several examples of such composites are shown in Table 13.1. Several specific combinations of polymers applied for design NH_3 sensors are listed in Table 13.2.

The functions of incorporating another component into the conducting polymers are manifold (Bai and Shi 2007). In some cases, the second components play an important role in the sensing process. They may improve the properties of the sensing film both through the partition coefficient (Hwang et al. 1999), which helps in electron or proton transfers (Ogura et al. 1997; Athawale et al. 2006) and through direct interaction with analytes, influencing on the polymer matrix swelling (Segal et al. 2005) or electron/proton exchange (Tongpool and Yoriya 2005). In other cases, the second components are incorporated only to improve the device configuration, for example, to change the film morphology (i.e., act as a porous matrix (Silverstein et al. 2005)), to improve the mechanical properties (Brady et al. 2005), or to protect the sensing film (Kim et al. 2005).

13.1 Conductometric Gas Sensors Based on Polymer Composites

It should be noted that with the nanocomposites discussed earlier, the polymer mainly acts as the insulating matrix, while dispersed conducting particles provide the conducting path for sensing (Cho et al. 2004). Due to adsorption of interested analyte (gas or vapor) in such nanocomposites there are volumetric changes of the matrix polymer (swelling), which can lead to a distinct change in percolation-type conductivity. With that maximum effect takes place around a certain critical composition of the material, which is called the "percolation threshold" (see Fig. 13.1). When the concentration of conducting particles is very high, the particles pack very closely in the composite and form conductive pathways, which impart a low-resistance response to the sensor. As the concentration of conducting particles decreases (e.g., in a swollen polymer), the distances between particles increase and the resistance of the composite gradually increases. When the concentration of conducting particles decreases to a point at which the conductive pathways are disrupted, the resistance of the composite increases sharply (by many orders of magnitude). This point is called the percolation threshold. Therefore, if the concentration of conducting particles is only slightly higher than the percolation threshold, a small amount of swelling may cause a dramatic change in the sensor resistance. In particular, absorption of organic vapors by composites loaded at or above the percolation threshold, due

G. Korotcenkov, *Handbook of Gas Sensor Materials*, Integrated Analytical Systems,
DOI 10.1007/978-1-4614-7388-6_13, © Springer Science+Business Media New York 2014

Table 13.1 Conducting polymer composite used in gas sensors

Conducting polymer	Second component	
	Polymer	Other materials
PPy	PS (B); high-density polyethylene (HDPE) (B); PEO (B); PVA (B); PMMA (B); PMMA (C); poly(etheretherketone) (PEEK) (C); PVDF (B); PVAc (B); PVC (B); poly(acrylonitrile-cobutadiene-*co*-styrene) (ABS) (B); polyurethane (PU) (C)	C_{60} (B); SWNT (C); Nafion®/metal (C); calixarene (B); Pb-phthalocyanine (B); SnO_2 (B); Fe_2O_3 (B); MoO_3 (L); WO_3 (B); ZnO_2 (B)
PAni	PS (B); PVA (B); PMMA (B); PVDF (C); poly(butyl acrylate-*co*-vinyl acetate)(PBuA-VAc) (B); PP + carbon black + thermoplastic PU (B); PS + carbon black + thermoplastic PU (B); nylon 6 (C); ethylene vinyl acetate copolymer (EVA)/copolyamide (CoPA) (B); polyimide (PI) (B); PEDOT (C)	Carbon black (B) (C); MWNT (B); Cu(II)-exchanged hectorite (B); PtO_2 (B); TiO_2 (B); SnO_2 (B); MoO_3 (L); $CuCl_2$ (B); CeO_2 (B); In_2O_3 (B); zeolite and Cu^{2+} (B); Nafion®/metal (C); Cu (B); Pd (B)
PTh	4-*tert*-Butyl-Cu-phthalocyanine (B)	MWCN (B); SnO_2 (B); Cu (B); Pd (B)

Source: Reprinted from Bai and Shi (2007). Published by MDPI
B blend, *C* coated, *L* layered

Table 13.2 PANI-based polymer composites used for design of NH_3 gas sensors operated at room temperature

Sensing material	Detection limit (ppb)	Response time (min)	Sensitivity (% ppm^{-1})	References
PANI-based				
PANI	10,000	8 at 100 ppm		Matsuguchi et al. (2002)
PANI nanofibrous films	500	1–1.5		Sutar et al. (2007)
PANI nanofibers	1,000			Chen et al. (2010)
PANI(DBSA) film	2,000	0.55		Pan et al. (2005)
Composites				
PVDF–PANI(DBSA)	100	2.5 at 1 ppm	17	Wojkiewicz et al. (2011)
PBuA–PANI(HCl)	250	2.5 at 1 ppm	10	
PU–PANI(CSA)	20	5 at 1 ppm	0.8	
PANI-coated MWCNTs	200	0.5		He et al. (2009)

For comparison, parameters of PANI-based sensors are presented as well
PANI poly(methyl methacrylate), *PVDF* polyvinylidene fluoride, or polyvinylidene difluoride, *PBuA* poly(butyl acrylate), *PU* polyurethane, *CSA* camphorsulfonic acid, *DBSA* dodecylbenzene sulfonic acid

to swelling of the host polymer, leads to an increase in sample resistance. While in many instances the response is modest, there are examples of very large changes in resistance on exposure to saturated vapors. Increases in resistance by factors of more than 10^6 have been reported (see Fig. 13.1).

Weak Van der Waals forces between the polymer and the target gas molecules are responsible for the swelling of the sensing layer; therefore, the change is purely physical and is reversible (see Fig. 13.1). However, some hysteresis can occur when the sensor is exposed to high concentrations.

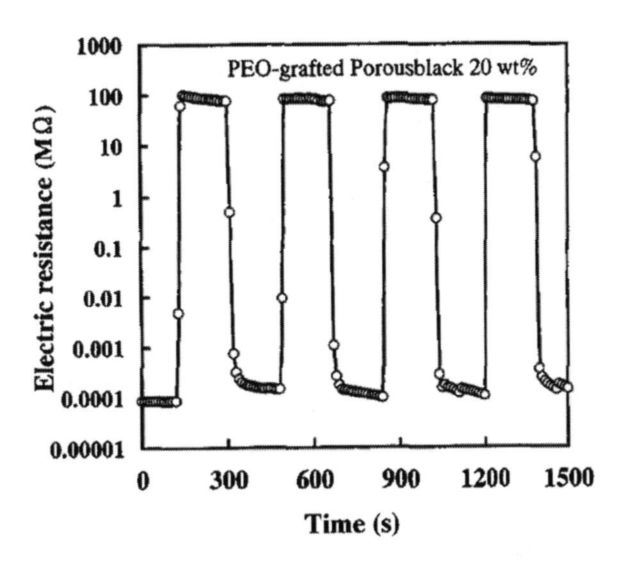

Fig. 13.1 Typical responsiveness of electric resistance of the composite from poly(ethylene oxide) (PEO) and PEO-grafted carbon black (CB) on four repeated exposures to the saturated chloroform vapor at 25 °C (Reprinted with permission from Chen and Tsubokawa 2000, Copyright 2000 The Society of Polymer Science, Japan)

It was found that as a rule the amount of swelling corresponds to the concentration of the chemical vapor in contact with the absorbent, and therefore this effect can be used for determination of gas or vapor concentration. Of course, chemiresistors with different polymer matrices will respond differently to a given VOC. The swelling of the polymer matrix is greatest when there is a match between the solubility parameter of the polymer and that of the vapor (Belmaraes et al. 2004). This means that the selectivity and the sensitivity of the conductometric sensors designed on the basis of the swelling effect will be governed by conducting particles and polymer proper components selected for sensor design. For example, poly(*N*-vinyl-pyrrolidone) is hydrophilic, so it swells substantially (and increases the sensor resistance) in water vapor, but not in toluene vapor, whereas polyisobutylene is hydrophobic, so it swells (and increases the sensor resistance) in toluene vapor, but not in water vapor.

Experiment has shown that for design of the above-mentioned composites, various fillers such as metal particles (Middlehoek and Audet 1989; Yang et al. 2010; Hands et al. 2012), expanded graphite (Li et al. 2007), carbon aerogel (Zhang et al. 2008a), carbon nanotubes (CNTs) (Zhang et al. 2008b), carbon black (Lee et al. 2010), and mixtures of carbon black and nanotubes (Zhang et al. 2006) can be used. Regarding the polymer, phthalocyanines are commonly used for preparing based composites due to their stability toward oxidation at high temperatures (Wohltjen et al. 1985). For example, Middlehoek and Audet (1989) used Cu-substituted phthalocyanine conductive polymer as a CCl_4 sensor. Other insulating polymers also are acceptable for preparing composites. Various insulating polymers have been explored actively by Lewis and co-workers (Freund and Lewis 1995; Lonergan et al. 1996; Albert et al. 2000), Ho and Hughes (2002) Ho et al. (2003) (Rivera et al. 2003) and others (Lundberg and Sundqvist 1986; Grate and Abraham 1991; Marquez et al. 1997; Tsubokawa et al. 2001; Zee and Judy 2001; Carrillo et al. 2002).

Generally, the percolation threshold is dependent on the shape of the conducting particle. As a rule, a composite consisting of particles with higher aspect ratio shows a lower threshold and higher sensitivity (Abraham et al. 2004). Experiments carried out have shown that CNTs, with almost one-dimensional threadlike structure and good conductivity, are ideal as the dispersed conducting particles in the insulating matrix for gas-sensing systems. It was established that addition of CNTs to a polymer matrix leads to a very low electrical percolation threshold (see Fig. 13.2) and allows one to obtain – with only very small amounts of CNTs – an electrical conductivity sufficient to provide an electrostatic discharge (Flahaut et al. 2000). For example, Hu et al. (2006) reported that the percolation threshold of 0.9 wt% for electrical conductivity in CNT/poly(vinylidene fluoride) (PET) composite has

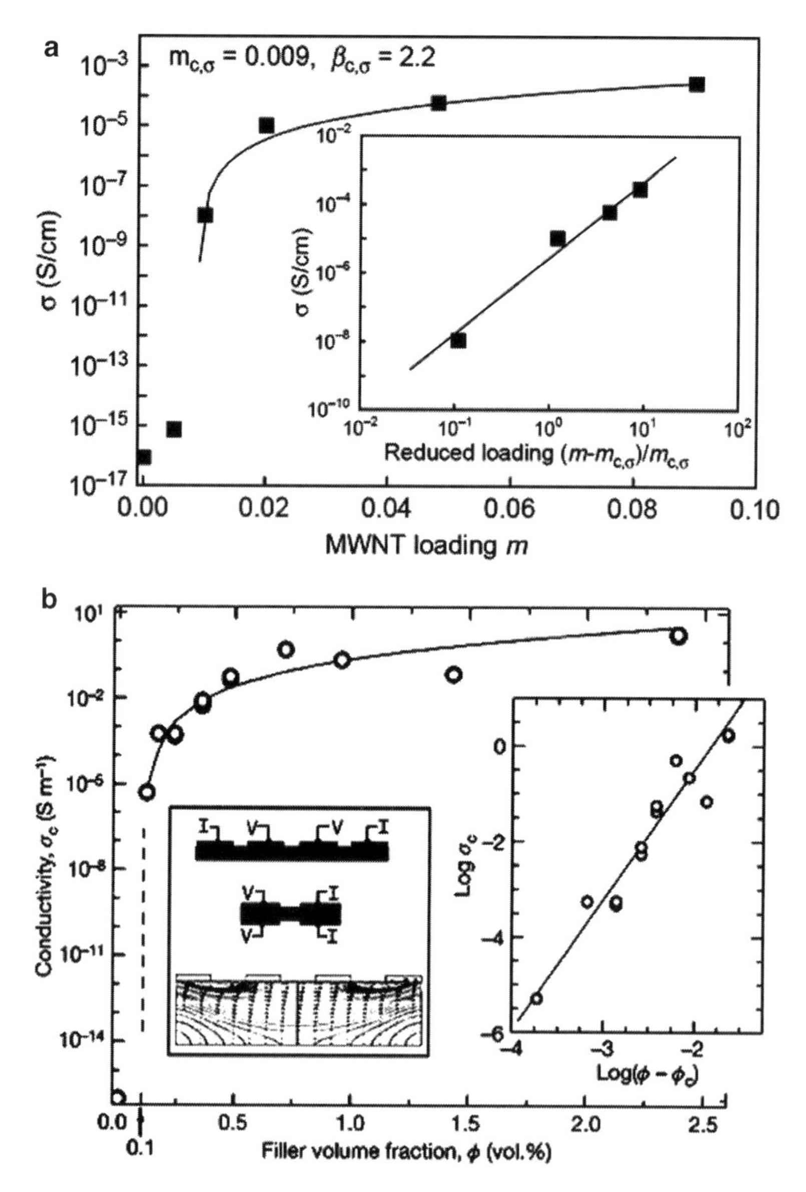

Fig. 13.2 Electrical conductivity (s) of the (**a**) PET/MWNT and (**b**) RG-O/epoxy nanocomposites as a function of MWNT and graphene loading. *Inset*: a log–log plot of electrical conductivity vs. reduced MWNT and graphene loading. The *solid lines* are fits to a power law dependence of electrical conductivity on the reduced fillers loading (13.1) ((**a**) Reprinted with permission from Hu et al. (2006), Copyright 2006 Elsevier; (**b**) reprinted with permission from Potts et al. (2011), Copyright 2011 Elsevier)

been found. In this research, pristine MWNT with diameter 10–20 nm and length 5–15 μm was used. It was found that electrical conductivity of CNT/PET composite follows the classical percolation theory and can be described by a power law relationship:

$$\sigma \propto \left(m - m_{c,\sigma}\right)^{b_{c,\sigma}} \tag{13.1}$$

where m is the volume fraction of the filler, $m_{c,\sigma}$ is the volume fraction of percolation threshold, and $\beta_{c,\sigma}$ is the critical exponent, which is related to the system dimension. It should be noted that several

Table 13.3 Values of the lowest electrical percolation thresholds which have been reported in the literature for GNP and graphene-based nanocomposites for selected polymer matrices

Matrix polymer	Filler type	Lowest percolation threshold (wt%)
Epoxy	Funct. G	1.0
Nylon-6	GO	0.5
Poly(aniline) (doped)	GNP	0.7
Polycarbonate	TEGO	0.3
Poly(ethylene)	Funct. G-O	0.2
Poly(ethylene terephthalate)	TEGO	1.0
Poly(methyl methacrylate)	GNP	0.7
Poly(propylene)	GNP	0.7
Poly(styrene)	Funct. G-O	0.2
Poly(vinyl alcohol)	RG-O	0.5
Poly(vinyl chloride)	GNP	1.4
Poly(vinylidene fluoride)	TEGO	2.0

Source: Reprinted with permission from Potts et al. (2011), Copyright 2011 Elsevier
Minimum resistance reported, ~200□
G graphene, *G-O* graphene oxide, *GO* graphite oxide, *GNP* graphite nanoplatelet fillers, *TEGO* thermally expanded graphite oxide

authors have established that the percolation threshold for electrical conductivity of epoxy composites containing CNTs can be even smaller. For example, Sandler et al. (2003) reported an electrical percolation threshold at 0.0025 wt% MWCNT. Bryning et al. (2005) observed the percolation threshold for conductivity of epoxy-based composites containing single-walled CNTs at 0.00005 vol. fraction of SWCNT (0.005 vol%). For comparison, water-dispersed carbon black particles (42 nm) added to acrylic emulsions yielded electrical conductivity percolation levels as low as 1.5 vol.% in dried films (Kim et al. 2008). Taking into account the above-mentioned behavior, CNTs/polymer composites have been intensively studied for gas sensors (Wienecke et al. 2003; Valentini et al. 2004; Wanna et al. 2006).

Recent studies showed that graphite nanoplatelets (GNP) or graphene could be used as a viable and inexpensive filler substitute for CNTs (Fukushima and Drzal 2003). Typical values of the electrical percolation thresholds, which have been reported in the literature for graphene-based nanocomposites for selected polymer matrices, are presented in Table 13.3. The influence of graphene loading on the conductivity of one of the composites presented in Table 13.3 is shown in Fig. 13.2b. One can see that the electrical percolation thresholds achieved with graphene-based nanocomposites are often compared with those reported for CNT/polymer composites.

However, we believe that, in spite of the above-mentioned results, graphene will not change CNT in nanocomposites designed for gas sensor applications. First, it is difficult to achieve complete and homogeneous dispersion of individual graphene sheets in various solvents (Kotov 2006). Second, it was established that the incorporation of GNPs and GO-derived fillers can significantly reduce gas permeation through a polymer composite relative to the neat matrix polymer. As is known, high gas penetrability is one of the main requirements for gas-sensing material. A percolating network of platelets can provide a "tortuous path" which inhibits molecular diffusion through the matrix, thus resulting in significantly reduced permeability (see Fig. 13.3) (Paul and Robeson 2008). Orientation of the platelets may further enhance barrier properties perpendicular to their alignment, while higher platelet aspect ratios correlate with increased barrier resistance (Paul and Robeson 2008).

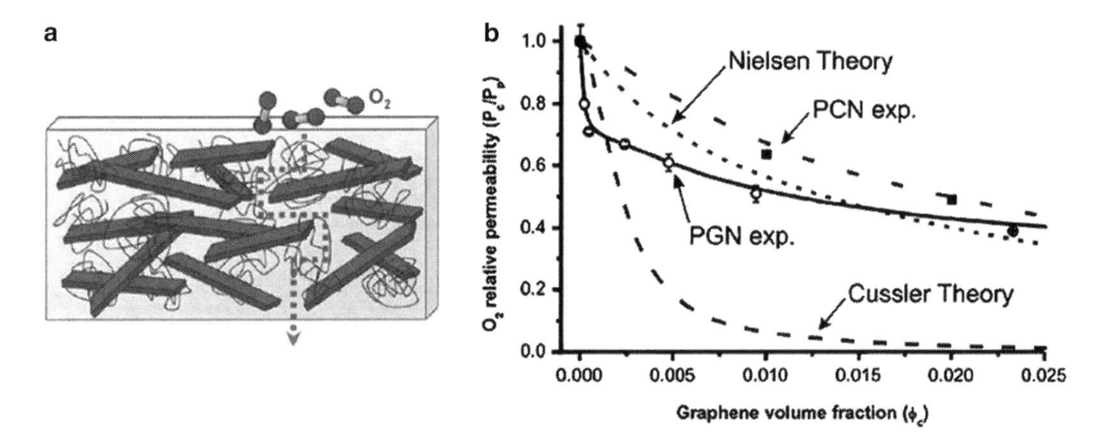

Fig. 13.3 (**a**) Illustration of formation of a "tortuous path" of platelets inhibiting diffusion of gases through a polymer composite (Nielsen model). (**b**) Measurements of oxygen permeability of CMG/PS ("PGN") and montmorillonite/PS ("PCN") composites as a function of filler loading compared with two theoretical models of composite permeability (Reprinted with permission from Potts et al. 2011, Copyright 2011 Elsevier)

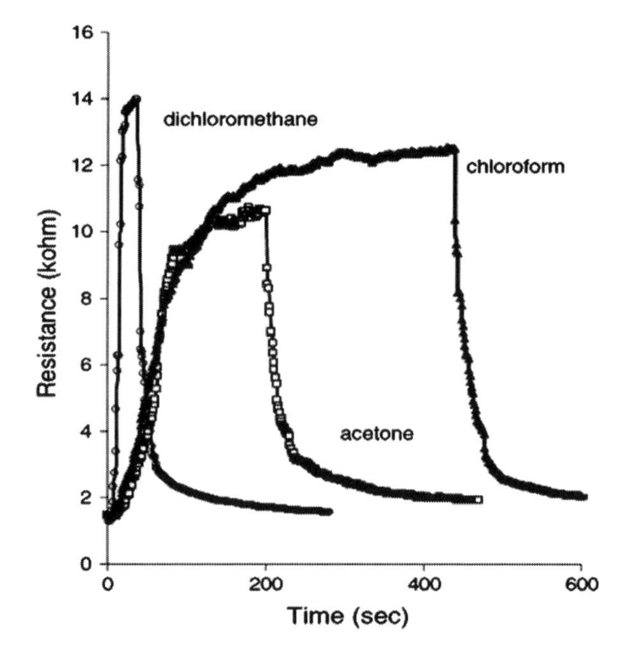

Fig. 13.4 Room temperature response of the CNT/PMMA (polymethylmethacrylate) composite to dichloromethane, chloroform, and acetone vapors (Reprinted with permission from Philip et al. 2003, Copyright 2003 Elsevier)

Experiments related to gas sensor testing have shown that while the absolute response is large, response times tend to be slow, typically of the order of minutes (see Fig. 13.4). Thin films of composite may respond more quickly. A good response can also be obtained by operating the sensor close to or in the percolation regime (Lonergan et al. 1996; Martin et al. 2003). However, since close to the percolation threshold small changes in filler loading have a large effect on sensor resistance, it is difficult to obtain precise and reproducible devices (Heaney 1995). Therefore, practical chemiresistors have generally been fabricated using commercially available polymers loaded with conducting particles at levels above the percolation transition (Sisk and Lewis 2003).

Table 13.4 Detect limits of sensors based on conducting polymers and their composites to several gas analytes

Analyte	Sensing material	Detect limit	Sensor type
NH_3	PAni/SWNT	50 ppb	Chemiresistor
NO_2	PPy/PET	<20 ppm	Chemiresistor
	PTh/CuPc	4.3 ppm	Chemiresistor
	PAni/In_2O_3	<0.5 ppm	SAW
HCl	PAni/FeAl	0.2 ppm	Chemiresistor
H_2S	PAni/heavy metal salts	<10 ppm	Chemiresistor
CO	PAni/FeAl	10 ppm	Chemiresistor
	PAni/In_2O_3	<60 ppm	SAW
Methanol	PAni/Pd	<1 ppm	Chemiresistor
Methane halide	Poly(3-methylthiophene)/ MWNT	Several ppm	Chemiresistor
	PAni/Cu	<10 ppm	Chemiresistor
Acetone	PTh copolymer	200 ~ 300 ppm	Chemiresistor
Toluene	PTh copolymer	20 ppm	Chemiresistor
Butylamine	Poly(anilineboronic acid)	10 ppb	Chemiresistor

Source: Data from Bai and Shi (2007)

Of course, polymer-based composites can be formed using conductive polymers as well. For example, Al-Mashat et al. (2010) proposed graphene/polyaniline (PANI) nanocomposite for hydrogen sensing. As discussed before, electronic structure of conducting polymers is comprised of conjugated π-bonds that can undergo changes under the influence of chemical species adsorbed onto their surfaces due to a redox- or acid–base-type interaction between the polymer and the chemical species. Al-Mashat et al. (2010) have shown that nanocomposite-based RT sensors had better sensitivity in comparison with sensors based on only graphene sheets and PANI nanofibers. However, the magnitude of this sensitivity was very small in comparison with the response of metal oxide gas sensors. Composites based on conducting polymers can be used for design gas sensors aimed for detection of other gases as well. Several examples of such sensors are listed in Table 13.4. It is seen that the detection limits are rather low for sensors based on conducting polymers. For redox-active or acid–base-active analytes, the detection limit is smaller than 1 ppm, and for inert organic analytes, that limit is about several ppm or lower. The response times of these sensors are usually hundreds of seconds, and, especially for some ultrathin-film sensors, this time can be as short as about several seconds (Bai and Shi 2007).

13.2 Problems Related to Application of Polymer-Based Composites in Gas Sensors

We should note that in spite of the presence of correlation between sensor response and percolation theory (Lonergan et al. 1996), percolation theory cannot describe the complete sensor performance. The relationships between composite resistance, resistivity, conducting particle concentration, polymer swelling, and vapor concentration are not well known. Thus, it is difficult to design sensors and to optimize polymer selection for an anticipated sample or set of samples (Lei et al. 2007). The agglomeration of conducting particles also has a strong influence on sensor parameters and their reproducibility and stability. Due to aggregation, the distance between conducting particles can become very large and so the conducting network cannot be built up. For example, since CNTs usually agglomerate due to Van der Waals force, they are extremely difficult to disperse and align in a polymer matrix. It is possible that the difference in the degrees of agglomeration of conducting particles in the polymer matrix is one of the main reasons for so great a range in the values of percolation thresholds (from 0.0025 wt% to 11 wt%), which have been obtained by different research groups for CNT/polymer composites (Du et al. 2007). Of course, the electrical properties of CNTs used by different research groups are generally significantly different depending on their purity, morphology, aspect ratio, etc. However, the electrical properties of CNTs should influence the magnitude of conductivity only. Thus, a significant challenge in developing high-performance polymer/CNTs composites is to introduce the individual CNTs in a polymer matrix in order to achieve better dispersion and alignment and strong interfacial interactions, to improve the load transfer across the CNT/polymer matrix interface (Sahoo et al. 2010). In addition, we need to take into account that CNTs are normally randomly oriented and entangled with each other. CNTs prepared by electric arc discharge particularly tend to form bundles. It is very difficult to break up these entangled and bundled CNTs to produce individual ones and then disperse them uniformly in polymer matrices. Therefore, revolutionary progress of CNT application can only be realized when we know how to break up the entangled and bundled as-prepared CNT products to produce individual CNTs without damaging their properties or to produce directly individual CNTs (Du et al. 2007). The preparation of structure-controllable CNTs with high purity, geometrical uniformity, and consistently dependable high performance are also important for achievement of highly reproducible parameters of composites.

Research has shown that the surface functionalization of carbon-based fillers, which can both maximize interfacial adhesion between carbon-based fillers and the polymer matrix and increase the dispersion of CB, CNTs, and graphene in polymer matrix, is one of the best approaches to achieve good dispersion of conducting particles in polymer matrix. At present there are several approaches for functionalization of carbon-based materials including defect functionalization, covalent functionalization, and non-covalent functionalization (Gong et al. 2000; Hirsch 2002). Some functional groups, which can improve the interaction between carbon-based fillers and polymer matrix, are covalently bonded directly to the surface of carbon. These functionalization methods will be discussed in Chap. 25 (Vol. 2).

However, we need to take into account that surface treatment of carbon fillers is not suitable for a polymer-based functional composite in many cases (Du et al. 2007). For example, surface functionalization may disrupt the bonding between graphene sheets and thereby reduce the properties of functionalized graphene and CNTs in the final composites. Another method is to have carbon-based fillers physically coated by some surfactants, which is a non-covalent approach and may be a more facile and practical processing technique. In particular, experimental results showed that CNTs were dispersed more evenly with the aid of the surfactant (Gong et al. 2000). Nevertheless, CNTs treated by different surfactants may decrease the contact between CNTs and thereby reduce the conductive properties of CNT/polymer composites (Du et al. 2007). Of course, such consequences for composites with conductive properties are negative ones. In fact, for a carbon-based fillers/polymer functional composite, as used in gas sensors, the effective conducting path is primary, and good contact between individual conducting particles is more important. It is known that an existing CNT conducting path does not guarantee good conductivity if the contact resistance between the individual CNTs is too high. Only when excellent electrical contact is established between the individual CNTs, good electrical signals can be obtained in nanocomposite-based gas sensors. Of course, the intrinsic transport properties of CNTs, the orientation of CNTs within the matrix, and the way that one CNT contacts with the others are also important for achievement of high conductivity of composites. However, in many cases the presence of an insulating layer coating around the CNT with thickness determined by the type of CNT or polymer, the surface treatment of CNTs, and the processing method of the composites, can be more influential factors. Experiments have shown that it is possible to decrease the contact resistivity and improve the transport properties between individual CNTs by some measures (Du et al. 2007). High-temperature treatment for CNTs to improve their intrinsic transport properties through improving their structural integrity and reducing the functional groups on their surface is preferable. However, to design and construct the contact means between CNTs by adjusting the content and orientation of CNTs within a matrix is also an effective measure. For example, the contact resistance can be reduced by the overlaps of CNTs in a CNT network.

In published reports the common fabricating methods of polymer-based composites are solution mixing, melt blending, and in situ polymerization. In solution mixing, conducting particles are generally dispersed in solvent and then mixed with polymer solution by mechanical mixing, magnetic agitation, or high-energy sonication. Subsequently, the polymer-based composites can be obtained by vaporizing the solvent at a certain temperature. This method is considered to be an effective measure to prepare composites with homogenous filler distribution and is often used to prepare composite films. Melt blending is a versatile and commonly used method to fabricate polymeric materials, especially for thermoplastic polymers. Fillers are dispersed within the polymer matrix by rheological shear stress generated from the blending of CNTs with melt polymer. Its well-known disadvantage is that CNTs, for example, can easily be damaged to a certain extent or broken in some cases. The melt blending method is frequently used to prepare polymer-based composite fibers. In situ polymerization is considered to be a very efficient method to improve significantly the dispersion and the interaction between conducting particles and polymer matrix. Generally, conducting particles are first mixed with monomers and then polymer-based composites can be obtained by polymerizing the monomers under certain conditions. One can find in review papers of Zeng (2003), Du et al. (2007), Paul and Robeson (2008), Sahoo et al. (2010), Potts et al. (2011), etc., descriptions of the above-mentioned methods.

References

Abraham JK, Philip B, Witchurch A, Varadan VK, Channa RC (2004) A compact wireless gas sensor using a carbon nanotube/PMMA thin film chemiresistor. Smart Mater Struct 13(5):1045–1049

Albert KJ, Lewis NS, Schauer CL, Sotzing GA, Stitzel SE, Vaid TP, Walt DR (2000) Cross-reactive chemical sensor arrays. Chem Rev 100:2595–2626

Al-Mashat L, Shin K, Kalantar-zadeh K, Plessis JD, Han SH, Kojima RW, Kaner RB, Li D, Gou X, Ippolito SJ, Wlodarski W (2010) Graphene/polyaniline nanocomposite for hydrogen sensing. J Phys Chem C 114:16168–16173

Athawale AA, Bhagwat SV, Katre PP (2006) Nanocomposite of Pd-polyaniline as a selective methanol sensor. Sens Actuators B 114:263–267

Bai H, Shi G (2007) Gas sensors based on conducting polymers. Sensors 7:267–307

Belmaraes M, Blanco M, Goddard WA II, Ross RB, Caldwell G, Chou S-H, Pham J, Olofson PM, Thomas C (2004) Hildebrand and Hansen solubility parameters from molecular dynamics with applications to electronic nose polymer sensors. J Comput Chem 25:1814–1826

Brady S, Lau KT, Megill W, Wallace GG, Diamond D (2005) The development and characterisation of conducting polymeric-based sensing devices. Synth Met 154:25–28

Bryning MB, Islam MF, Kikkawa JM, Yodh AG (2005) Very low conductivity threshold in bulk isotropic single-walled carbon nanotube-epoxy composites. Adv Mater 17(9):1186–1191

Carrillo A, Martin-Dominguez IR, Rosas A, Marquez A (2002) Numerical method to evaluate the influence of organic solvent absorption on the conductivity of polymeric composites. Polymer 43:6307–6313

Cespedes F, Martinez-Fabregas E, Alegret S (1996) New materials for electrochemical sensing I. Rigid conducting composites. Trends Anal Chem 15(7):296–304

Chen J, Tsubokawa N (2000) Electric properties of conducting composite from poly(ethylene oxide) and poly(ethylene oxide)-grafted carbon black in solvent vapor. Polym J 32(9):729–736

Chen J, Yang J, Yan X, Xue Q (2010) NH_3 and HCl sensing characteristics of polyaniline nanofibers deposited on commercial ceramic substrates using interfacial polymerization. Synth Metals 160:2452–2458

Cho SM, Kim YJ, Kim YS, Yang Y, Ha S-C (2004) The application of carbon nanotube—polymer composite as gas sensing materials. In: Proceedings of IEEE sensors conference, sensors 2004, vol. 2. pp 701–704

Densakulprasert N, Wannatong L, Chotpattananont D, Hiamtup P, Sirivat A, Schwank J (2005) Electrical conductivity of polyaniline/zeolite composites and synergetic interaction with CO. Mater Sci Eng B 117:276–282

Du J-H, Bai J, Cheng H-M (2007) The present status and key problems of carbon nanotube based polymer composites. eXPRESS Polym Lett 1(5):253–273

Flahaut E, Peigney A, Laurent C, Marliere C, Chastel F, Rousset A (2000) Carbon nanotube–metal oxide nanocomposites: microstructure, electrical conductivity and mechanical properties. Acta Mater 48:3803–3812

Freund MS, Lewis NS (1995) A chemically diverse conduction polymer-based "electronic nose". Proc Natl Acad Sci USA 92:2652–2656

Fukushima H, Drzal LT (2003) A carbon nanotube alternative: graphite nanoplatelets as reinforcements for polymers. Annu Tech Conf Soc Plast Eng 61:2230–2234

Hands PJW, Laughlin PJ, Bloor D (2012) Metal–polymer composite sensors for volatile organic compounds: part 1. Flow-through chemiresistors. Sens Actuators B 162:400–408

He L, Jia Y, Meng F, Li M, Liu J (2009) Gas sensors for ammonia detection based on polyaniline coated multiwall carbon nanotubes. Mater Sci Eng B 163:76–81

Heaney MB (1995) Measurement and interpretation of nonuniversal critical exponents in disordered conductor–insulator composites. Phys Rev B 52:12477–12480

Hirsch A (2002) Functionalization of single-walled carbon nanotubes. Angew Chem Int Ed 41:1853–1859

Ho CK, Hughes RC (2002) In situ chemiresistor sensor package for real-time detection of volatile organic compounds in soil and groundwater. Sensors 2:23–34

Ho CK, Lindgren ER, Rawlinson KS, McGrath LK, Wright JL (2003) Development of a surface acoustic wave sensor for in situ monitoring of volatile organic compounds. Sensors 3:236–247

Hwang BJ, Yang JY, Lin CW (1999) A microscopic gas-sensing model for ethanol sensors based on conductive polymer composites from polypyrrole and poly(ethylene oxide). J Electrochem Soc 146:1231–1236

Hu G, Zhao C, Zhang S, Yang M, Wang Z (2006) Low percolation thresholds of electrical conductivity and rheology in poly(ethylene terephthalate) through the networks of multi-walled carbon nanotubes. Polymer 47:480–488

Gong XY, Liu J, Baskaran S, Voise RD, Young JS (2000) Surfactant-assisted processing of carbon nanotube/polymer composites. Chem Mater 12:1049–1052

Grate JW, Abraham MH (1991) Solubility interactions and the design of chemically selective sorbent coatings for chemical sensors and arrays. Sens Actuators B 3:85–111

Kim JS, Sohn SO, Huh JS (2005) Fabrication and sensing behavior of PVF2 coated-polyaniline sensor for volatile organic compounds. Sens Actuators B 108:409–413

Kim YS, Wright JB, Grunlan JC (2008) Influence of polymer modulus on the percolation threshold of latex-based composites. Polymers 49:570–578

Kotov NA (2006) Materials science: carbon sheet solutions. Nature 442:254–255

Lee J, Choi J, Hong J, Jung D, Shim SE (2010) Conductive silicone/acetylene black composite films as chemical vapour sensors. Synth Met 160:1030–1035

Lei H, Pitt WG, McGrath LK, Ho CK (2007) Modeling carbon black/polymer composite sensors. Sens Actuators B 125:396–407

Li L, Luo Y, Li Z (2007) The preparation and vapour-induced response of a conductive nanocomposite based on poly(methyl acrylic acid)/expanded graphite by in situ polymerization. Smart Mater Struct 16:1570–1574

Lonergan MC, Severin EJ, Doleman BJ, Beaber SA, Grubbs RH, Lewis NS (1996) Array-based vapor sensing using chemically sensitive, carbon black-polymer resistors. Chem Mater 8:2298–2312

Lundberg B, Sundqvist B (1986) Resistivity of a composite conducting polymer as a function of temperature, pressure, and environment—applications as a pressure and gas concentration transducer. J Appl Phys 60:1074–1079

Marquez A, Uribe J, Cruz R (1997) Conductivity variation induced by solvent swelling of an elastomer-carbon black-graphite composite. J Appl Polym Sci 66:2221–2232

Martin JE, Anderson RA, Odinek J, Adolf D, Williamson J (2003) Controlling percolation in field structured particle composites: observations of giant thermoresistance, piezoresistance and chemiresistance. Phys Rev B 67:094207

Matsuguchi M, Gsugiyama JI, Sakai Y (2002) Effect of NH_3 gas on the electrical conductivity of polyaniline blend films. Synth Metals 128:15–19

Middlehoek S, Audet SA (1989) Silicon sensors for chemical signals. Academic, Boston

Ogura K, Saino T, Nakayama M, Shiigi H (1997) The humidity dependence of the electrical conductivity of a soluble polyaniline-poly(vinyl alcohol) composite film. J Mater Chem 7:2363–2366

Pan W, Yang SL, Li G, Jiang JM (2005) Electrical and structural analysis of conductive polyaniline/polyacrylonitrile composites. Eur Polym J 41:2127–2133

Paul DR, Robeson LM (2008) Polymer nanotechnology: nanocomposites. Polymer 49:3187–3204

Philip B, Abraham JK, Chandrasekhar A, Varadan VK (2003) Carbon nanotube/PMMA composite thin films for gas-sensing applications. Smart Mater Struct 12:935–939

Potts JR, Dreyer DR, Bielawski CW, Ruof RS (2011) Graphene-based polymer nanocomposites. Polymer 52(1):5–25

Rivera D, Alam MK, Davis CE, Ho CK (2003) Characterization of the ability of polymeric chemiresistor arrays to quantitate trichloroethylene using partial least squares (PLS): effects of experimental design, humidity, and temperature. Sens Actuators B 92:110–120

Sahoo NG, Rana S, Cho JW, Li L, Chan SH (2010) Polymer nanocomposites based on functionalized carbon nanotubes. Prog Polym Sci 35:837–867

Sandler JKW, Kirk JE, Kinloch IA, Shaffer MSP, Windle AH (2003) Ultra-low electrical percolation threshold in carbon-nanotube-epoxy composites. Polymer 44:5893–5899

Santhanam KSV, Sangoi R, Fuller L (2005) A chemical sensor for chloromethanes using a nanocomposite of multiwalled carbon nanotubes with poly(3-methylthiophene). Sens Actuators B 106:766–771

Segal E, Tchoudakov R, Narkis M, Siegmann A, Wei Y (2005) Polystyrene/polyaniline nanoblends for sensing of aliphatic alcohols. Sens Actuators B 104:140–150

Silverstein MS, Tai HW, Sergienko A, Lumelsky YL, Pavlovsky S (2005) PolyHIPE: IPNs, hybrids, nanoscale porosity, silica monoliths and ICP-based sensors. Polymer 46:6682–6694

Sisk BC, Lewis NS (2003) Estimation of chemical and physical characteristics of analyte vapours through analysis of the response data of arrays of polymer-carbon black composite vapour detectors. Sens Actuators B 96:268–282

Suri K, Annapoorni S, Sarkar AK, Tandon RP (2002) Gas and humidity sensors based on iron oxide-polypyrrole nanocomposites. Sens Actuators B 81:277–282

Sutar DS, Padma N, Aswal DK, Deshpande SK, Gupta SK, Yakhmi JV (2007) Preparation of nanofibrous polyaniline films and their application as ammonia gas sensor. Sens Actuators B 128:286–292

Tongpool R, Yoriya S (2005) Kinetics of nitrogen dioxide exposure in lead phthalocyanine sensors. Thin Solid Films 477:148–152

Tsubokawa N, Tsuchida M, Chen J, Nakazawa Y (2001) A novel contamination sensor in solution: the response of the electric resistance of a composite based on crystalline polymer-grafted carbon black. Sens Actuators B 79:92–97

Valentini L, Bavastrello V, Stura E, Armentano I, Nicolini C, Kenny JM (2004) Sensors for inorganic vapor detection based on carbon nanotubes and poly(*o*-anisidine) nanocomposite material. Chem Phys Lett 383(5–6):617–622

Unde S, Ganu J, Radhakrishnan S (1996) Conducting polymer-based chemical sensor: characteristics and evaluation of polyaniline composite films. Adv Mater Opt Electron 6:151–157

Wanna Y, Srisukhumbowornchai N, Tuantranont A, Wisitsoraat A, Thavarungkul N, Singjai P (2006) The effect of carbon nanotube dispersion on CO gas sensing characteristics of polyaniline gas sensor. J Nanosci Nanotechnol 6(12):3893–3896

Watcharaphalakorn S, Ruangchuay L, Chotpattahanont D, Sirivat A, Schwank J (2005) Polyaniline/polyimide blends as gas sensors and electrical conductivity response to CO-N_2 mixtures. Polym Int 54:1126–1133

Wienecke M, Bunescu M-C, Pietrzak M, Deistung K, Fedtke P (2003) PTFE membrane electrodes with increased sensitivity for gas sensor applications. Synthetic Met 138(1–2):165–171

Wohltjen H, Barger WR, Snow AW, Jarvis NL (1985) A vapor-sensitive chemiresistor fabricated with planar micro electrodes and a Langmuir–Blodgett organic semiconductor film. IEEE Trans Electron Dev ED-32:1170–1174

Wojkiewicz JL, Bliznyuk VN, Carquigny S, Elkamchi N, Redon N, Lasri T, Pud AA, Reynaud S (2011) Nanostructured polyaniline-based composites for ppb range ammonia sensing. Sens Actuators B 160:1394–1403

Yang X, Li L, Yan F (2010) Polypyrrole/silver composite nanotubes for gas sensors. Sens Actuators B 145:495–500

Zee F, Judy JW (2001) Micromachined polymer-based chemical array. Sens Actuators B 72:120–128

Zhang B, Fu R, Zhang M, Dong X, Zhao B, Wang L, Pitman CU Jr (2006) Studies of the vapour-induced sensitivity of hybrid components fabricated by filling polystyrene with carbon black and carbon nanofibers. Composites A 37:1884–1889

Zhang B, Dong X, Song W, Wu D, Fu R, Zhao B, Zhang M (2008a) Electrical response and adsorption performance of novel composites from polystyrene filled with carbon aerogel in organic vapors. Sens Actuators B 132:60–66

Zhang B, Dong X, Fu R, Zhao B, Zhang M (2008b) The sensibility of the composites fabricated from polystyrene filling multi-walled carbon nanotubes for mixed vapours. Comp Sci Technol 68:1357–1362

Zeng HC (2003) Carbon nanotube-based nanocomposites. In: Nalwa HS (ed) Handbook of organic hybrid materials and nanocomposites. American Scientific Publisher, Valencia, CA, pp 151–180

Chapter 14
Metal Oxide-Based Nanocomposites for Conductometric Gas Sensors

14.1 Metal–Metal Oxide Composites

In principle, almost all metal oxide gas sensors can be considered as composite-based sensors because they usually consist of noble metal nanoparticles incorporated in a metal oxide gas-sensing matrix. The metal nanoparticles play both passive and active roles in the sensing process (Kohl 1990; Korotcenkov 2005, 2010; Korotcenkov and Sysoev 2011; Yang 2011). First, metal nanoparticles such as platinum (Pt), palladium (Pd), gold (Au), and silver (Ag) show catalytic properties that can modify the analyte–metal oxide chemical interactions, increase catalytic activity, and enhance the sensing process. Metallic nanoparticles, presented on the surface of metal oxide, activate or dissociate the detected gas. As a result, these activated products are easier to react with the adsorbed oxygen species on the metal oxide surface, resulting in a change of the resistance. Second, the interfacial region between metal nanoparticles and metal oxide also has a very different electron band structure from than inside the bulk semiconducting metal oxide, which also contributes to the unique gas-sensing properties of this type of nanocomposite. Direct exchange of electrons between the semiconductor metal oxide and metallic nanoparticles causes a change in the width of the depletion layer of the semiconductor oxide, leading to a change in sensing properties. It was established that metal nanoparticles embedded in a metal oxide matrix can reduce the sensing temperatures, improve the selectivity, decrease response and recovery times, and increase the response of sensors (Korotcenkov et al. 2003; Korotcenkov 2005, 2010). In addition, metal nanoparticle can reduce the electrical resistance and increase the optical absorption of metal oxide. One can find in Chap. 10 (Vol. 1) a more detailed description of the role of metal nanoparticles in gas-sensing effects. However, despite a number of long-term efforts to quantify the effect of the foreign metal phase on the gas-sensing properties of the MOXs, there is still a need to understand better the underlying fundamental mechanisms, because the effect of each additive in the metal oxide matrix is complex and, till now, has not been well studied. Thorough characterization of even such simple systems as stable metal clusters on the oxides requires experimental data on the cluster sizes, their surface orientation, deficiency, mobility, etc.

For preparing metal–metal oxide composites, various methods can be used. For example, using ceramic and thick-film technologies, noble metal catalysts can be incorporated into a MOX by: (1) impregnating the pristine MOX powder with a noble metal chloride such as $PtCl_4$ and $PdCl_2$ solution, followed by drying and calcination (Matsushima et al. 1992); (2) mixing the pristine MOX powder with a colloid of noble metal (Nakao 1995); and (3) chemical bonding of noble metal complexes such as $PdCl_4^{2-}$ with surface hydroxyls at the pristine MOX in solution (Kaji et al. 1980). The impurity can also be introduced via sputtering of a thin intermediate layer (Gutierrer et al. 1993; Sayago et al. 1995). In this case, the profile of additive concentrations over the MOX structure is driven by

G. Korotcenkov, *Handbook of Gas Sensor Materials*, Integrated Analytical Systems,

DOI 10.1007/978-1-4614-7388-6_14,

temperature and time of annealing. Specific profiles of additive concentration can be created by using ion implantation techniques via adjusting the density of the ion current and the time of implantation (Sulz et al. 1993; Rosenfeld et al. 1993; Rastomjee et al. 1996). PVD, impregnation, sol–gel, successive ionic layer deposition, spray pyrolysis, electroless plating, etc., can also be used for preparing metal–metal oxide composites (Korotcenkov and Cho 2010).

14.2 Metal Oxide–Metal Oxide Composites

Metal oxide–metal oxide-based nanocomposites, $Me^{I}O–Me^{II}O$, are also interesting for gas sensor design (Yamazoe et al. 1983; Yamazoe 1991; Ferroni et al. 1999; Yamaura et al. 2000; Comini et al. 2002; Korotcenkov 2007; Gas'kov and Rumyantseva 2009). It was established that one of the ways for improving selectivity and stability of metal oxide conductometric gas sensors is the modification of metal oxide, $Me^{I}O$ by the introduction of catalytic or structure modifiers, $Me^{II}O$, in the nanostructured metal oxide matrix and, thereby, the development of nonhomogeneous complex materials, i.e., nanocomposites $Me^{I}O–Me^{II}O$. It was also expected that other highly sophisticated surface-related properties important for gas sensor applications such as optical, electronic, catalytic, mechanical, and chemical can also be obtained in complex metal oxides and composites.

It is known (Yamazoe 2005), despite the simple working principle of conductometric gas sensors, that the gas-sensing mechanism involved is fairly complex. Sensing performances, especially sensitivity, are controlled by three independent factors of receptor (recognition) function, transducer function, and utility, as illustrated in Fig. 14.1. Receptor function concerns the ability of the oxide surface to interact with the target gas. Chemical properties of the surface oxygen of the oxide itself are responsible for this function in a neat oxide device, but this function can be largely modified to induce a large change in sensitivity when an additive (noble metals, acidic or basic oxides) is loaded on the oxide surface. Transducer function concerns the ability to convert the signal caused by chemical interaction of the oxide surface (work function change) into an electrical signal. This function is played by each boundary between grains, to which a double-Schottky barrier model can be applied. The resistance depends on the

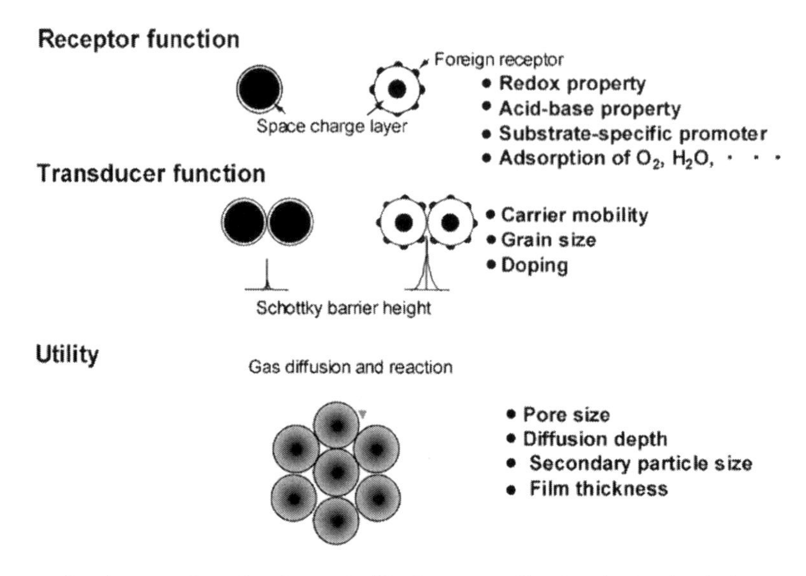

Fig. 14.1 Receptor function, transducer function, and utility factor as well as the physicochemical or material properties involved for semiconductor gas sensors (Reprinted with permission from Yamazoe 2005, Copyright 2005 Elsevier)

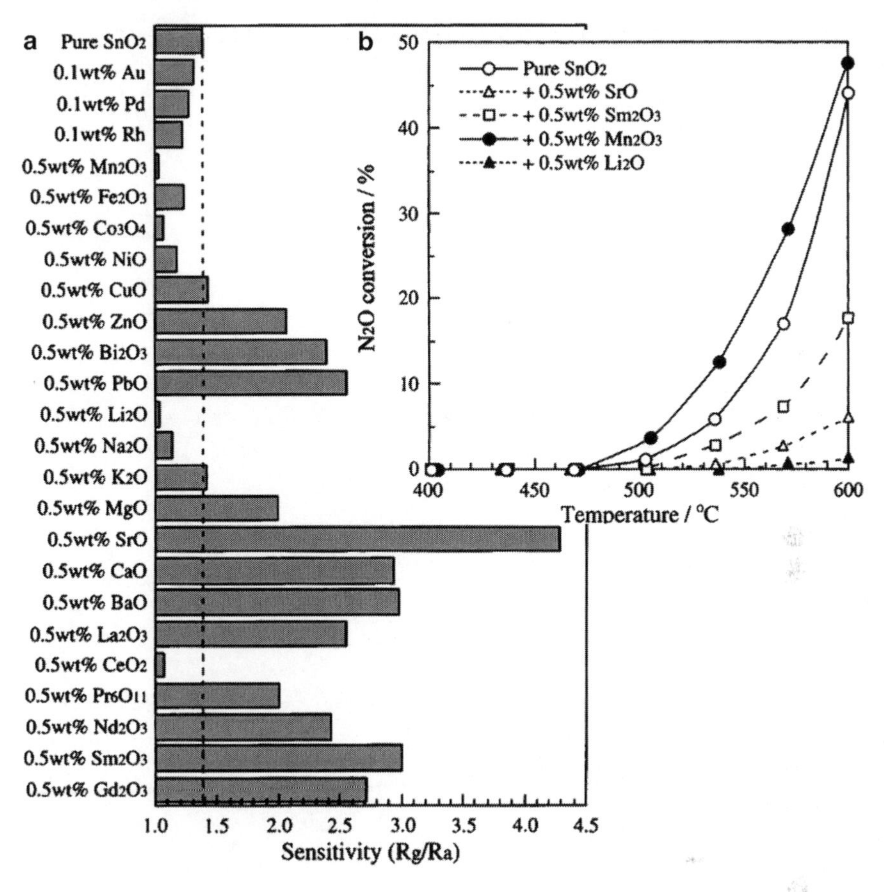

Fig. 14.2 Influence of additives to SnO_2 on (**a**) sensitivity to 300 ppm N_2O (T_{oper} = 500 °C) and (**b**) N_2O conversion. SnO_2-based composites were prepared using conventional sol–gel method. Tested sensors were Figaro type (Adapted with permission from Kanazawa et al. 2001, Copyright 2001 Elsevier)

barrier height and then on the concentration of the target gas. Thus, gas/solid or liquid/solid interaction phenomena are involved in the receptor function, while the microstructure of the oxide determines the transduction of the chemical stimulus in air or liquid into an electrical signal. Generally speaking, if a single-oxide system is adapted, these two functions cannot be optimized independently. Instead, by introducing into the system another material, which is very reactive to the target gas and can act as an "antenna" material, both functions may be optimized simultaneously and the sensor may become more sensitive even to low concentrations of the reactive gas. In these cases, the material acting as a unique receptor should be interfaced electronically to the transducer material, and its chemical change should sensitively modulate the semiconducting properties of the transducer oxide through the heterojunction (Capone et al. 2003). Composite-type sensors containing heterocontacts between two phases fulfill this novel concept, and experiments carried out in some laboratories have confirmed this assumption (Konig 1987; Baraton et al. 1997; Flahaut et al. 2000; Gas'kov and Rumyantseva 2001). As a result, metal oxide nanocomposites became very popular objects for study, and therefore at present in the literature one can find the description of a variety of nanocomposites based on semiconducting metal oxides such as SnO_2, ZnO, Ga_2O_3, In_2O_3, and WO_3 and metal oxide modifiers such as Fe_2O_3, La_2O_3, Cr_2O_3, Co_3O_4, V_2O_5, NiO, CuO, SiO_2, MoO_3, and CeO_2. For example, Fig. 14.2 presents results obtained by Kanazawa et al. (2001) during design of N_2O sensors. Table 14.1 presents noteworthy NO_x sensor data revealed recently using mixed (composite) metal oxides as sensing materials.

Table 14.1 NO_x sensors employing mixed (composite) metal oxides as active layers

Metal oxide	Optimal mixing ratio	Morphology	Method	T_{opt}[a] (°C)	Detection range (ppm)	Response[b, c] S	ppm	τ_{res}[c, d] (s)
$ZnO–SnO_2$	Zn:Sn (1:1)	Thick film	Wet chemical	RT	0.5	1,066	0.5	–
$ZnO–SnO_2$	Zn:Sn (4:6)	Thick film	Reverse microemulsion	200	200–1,000	13.5	500	–
$ZnO–SnO_2$	SnO_2 coating	Nanofibers	Electrospinning PLD	200	0.1–4	105	4	>100
$SnO_2–WO_3$	WO_3 (5 %; 7 %)	Thin film	RF diode sputtering	100	10	5.4×10^4	10	67
$SnO_2–WO_3$	WO_3 (5 %; 7 %)	Thin film	Sol–gel	300	0.1–3	36	1	<60
$SnO_2–WO_3$	WO_3 (20 %)	Thick film	Sol precipitation	200	0.01–40	186	200	–
$SnO_2–ZnO$	ZnO coating	Nanofibers	Electrospinning-ALD	200	1–5	1.1	1	>100
$In_2O_3–SnO_2$	SnO_2 (15 %)	Thin film	RF magnetron sputtering	327	5–200	27	50	–
$In_2O_3–SnO_2$	SnO_2 (17 %)	Nanofibers	Electrospinning	160	1–50	2.4	1	–

Source: Reprinted with permission from Afzal et al. (2012), Copyright 2012 Elsevier

[a]The temperature at which the best sensor performance, usually in terms of the highest response toward NO_x, is observed

[b]Herein, gas response (S) is always reported as $S=(R_g-R_a)/R_a$ per ppm of NO_x mentioned. This means that gas responses actually reported as $S=R_g/R_a$ in the respective paper are normalized as $S=R_g/R_a-1$ for the sake of consistency

[c]The approximate values of gas response and response times (τ_{res}), calculated from the data provided in the publications, are reported only if not explicitly mentioned in the corresponding paper

[d]The time required to reach 90 % of the total response toward specific concentration of NO_x

Table 14.2 M_1O/M_2O nanocomposites with promise for gas-sensitive materials

Me[I]O/Me[II]O	Ionic radii $r(M_1)$, nm		$r(M_2)$, nm Oxidized form		$r(M_2)$, nm Reduced form	
Cr_2O_3/SnO_2	Cr^{3+}	0.061	Sn^{4+}	0.069	Sn^{2+}	0.093
SnO_2/CuO	Sn^{4+}	0.069	Cu^{2+}	0.073	Cu^{1+}	0.096
SnO_2/MoO_3	Sn^{4+}	0.069	Mo^{6+}	0.042	Mo^{5+}	0.063
Ga_2O_3/Fe_2O_3	Ga^{3+}	0.062	Fe^{3+}	0.064	Fe^{2+}	0.077
In_2O_3/NiO	In^{3+}	0.079	Ni^{3+}	0.060	Ni^{2+}	0.070
In_2O_3/Fe_2O_3	In^{3+}	0.079	Fe^{3+}	0.064	Fe^{2+}	0.077
WO_3/TiO_2	W^{6+}	0.058	Ti^{4+}	0.060	Ti^{3+}	0.067

Source: Data from Gas'kov and Rumyantseva (2001)

Some aspects of the use of metal oxide nanocomposites in the elaboration of solid-state gas sensors were considered in detail by Gas'kov and Rumyantseva (2001, 2009). In particular, research conducted by Gas'kov and Rumyantseva (2001) has shown that for conductometric gas sensors, the advantage of metal oxide nanocomposites Me[I]O/Me[II]O, containing two metal cations (Me[I] and Me[II]), over simple nanocrystalline oxides is associated with a redistribution of Me[II] between the bulk and the surface of Me[I]O grains that depends on the redox properties of the gas phase. The appearance of additional Me[II] cations in the nanocrystalline system may result in a dramatic change in the state of the grain boundaries and in modification of the electronic properties of the material in the presence of even trace amounts (0.1–10 ppm) of reducing or oxidizing gas molecules in the gas phase. Table 14.2 lists examples of nanocomposites that are of interest for creating gas-sensitive materials. Materials fabricated in this way have certain advantages (Konig 1987; Tamaki et al. 1992; Flahaut et al. 2000; Ivanovskaya et al. 2003). It should be noted that mixed metal oxides based on stable conductive metal oxides with high gas sensitivity such as SnO_2, In_2O_3, WO_3, and ZnO, which were discussed in Chap. 2 (Vol. 1), are also composites.

Table 14.3 presents effects which can be achieved in metal oxide-based composites. It is seen that the use of nanocomposites in gas sensors really can produce great improvements in sensor parameters.

Table 14.3 Influence of additives (in oxide form) in metal oxide matrix on gas-sensing characteristics of SnO_2- and In_2O_3-based sensors

Additive	Effect	Nature
Al_2O_3, SiO_2	Increases sensor response; improves thermal stability	Decrease of grain size; decrease of area of intergrain contact; increase of porosity
Ag (Ag_2O), Cu (Cu_2O)	Increases response to H_2S, SO_2	Two-phase system; phase transformations during gas detection
Fe (Fe_2O_3)	Increases response to alcohols	Change of oxidation state
Ga (Ga_2O_3), Zn (ZnO)	Increases sensor response	Decrease of grain size; increase of porosity
P, B	Improves selectivity	Creation of new phase
Ca, K, Rb, and Mg	Increases sensor response; improves thermal stability	Decrease of grain size
La, Ba, Y, and Ce	Improves thermal stability; increases sensor response	Stabilization of grain size (creation of new phases); decrease of grain size
Transition MOXs (<0.5 wt%) (Co, Mn, Sr, Ni)	Increases sensor response; improves selectivity	Catalytic effect; change of electron concentration; change of *A*/*D* parameters; change of grain size

Nevertheless, the precise role of additives in composites is not yet known and the explanation for most of the observations is based on pure speculations. For example, in the study of Nitta and co-workers (Nitta and Haradome 1979; Nitta et al. 1980), ThO_2 was added to increase the sensitivity to CO of a Pd- and MgO-catalyzed SnO_2 sensor. These authors reported that the addition of ThO_2 increased selectivity of SnO_2-based gas sensors due to increasing the CO sensitivity and decreasing the hydrogen sensitivity. It was suggested that thoria removed the hydroxyl radicals from the SnO_2 surface, allowing more oxygen adsorption to accelerate the CO oxidation rate. However, such a reaction might be facilitated at fairly low temperatures only. Gas'kov and Rumyantseva (2009), who studied SnO_2-based composites with catalysts Fe_2O_3, MoO_3, and V_2O_5, have found that additives, in addition to having an influence on the structural properties of gas-sensing matrix, also affect redox properties and acidity/basicity of the surface. They believe that just these parameters determine catalytic and sensing properties of the composite materials. In particular, using this approach, Gas'kov and Rumyantseva (2009) explained the influence of composite composition on the sensitivity to NH_3 (see Fig. 14.3). In the case of NH_3 detection and for all the samples considered, the value of the sensor signal was in a good agreement with the density of suitable acid centers (Fig. 14.3b). However, it is necessary to take into account that in many cases improvement of gas-sensing characteristics of composite-based devices is not determined by catalytic activity of additives. For example, as shown in Fig. 14.3, additives with maximum catalytic activity to N_2O conversion do not give the maximum increase in sensor response.

It should be noted that the above-mentioned composites have disadvantages as well as advantages. First, the complex nature of these materials limits their use for integrated gas sensors. The large number of elements in these metal oxides makes it hard to deposit thin films with good and repeatable stoichiometric ratios. The reproducibility can also be worse. Too many additional factors, which can affect gas-sensing properties of materials, appear in complex metal oxides and nanocomposites. Second, nanocomposites highly sensitive to specific gases very often contain components which do not have high thermodynamic stability compared with SnO_2 (e.g., CuO, Fe_2O_3, AgO_x). This fact undoubtedly affects temporal and thermal stability of such sensors' parameters.

In addition, we have to take into account that metal oxide-based nanocomposites have specific structure. Research on such a two-phase system, in which the concentration of the second oxide phase is much less than the concentration of the base oxide, has shown that the second phase, as a rule, is finely dispersed on the surface of the base oxide grains (Szezuko et al. 2001; Pagnier et al. 2000; Carreno et al. 2002). Possible versions of segregation layers of foreign cations on the surface of SnO_2 grains are shown in Fig. 14.4 (Varela et al. 1999; Carreno et al. 2002). It has been established

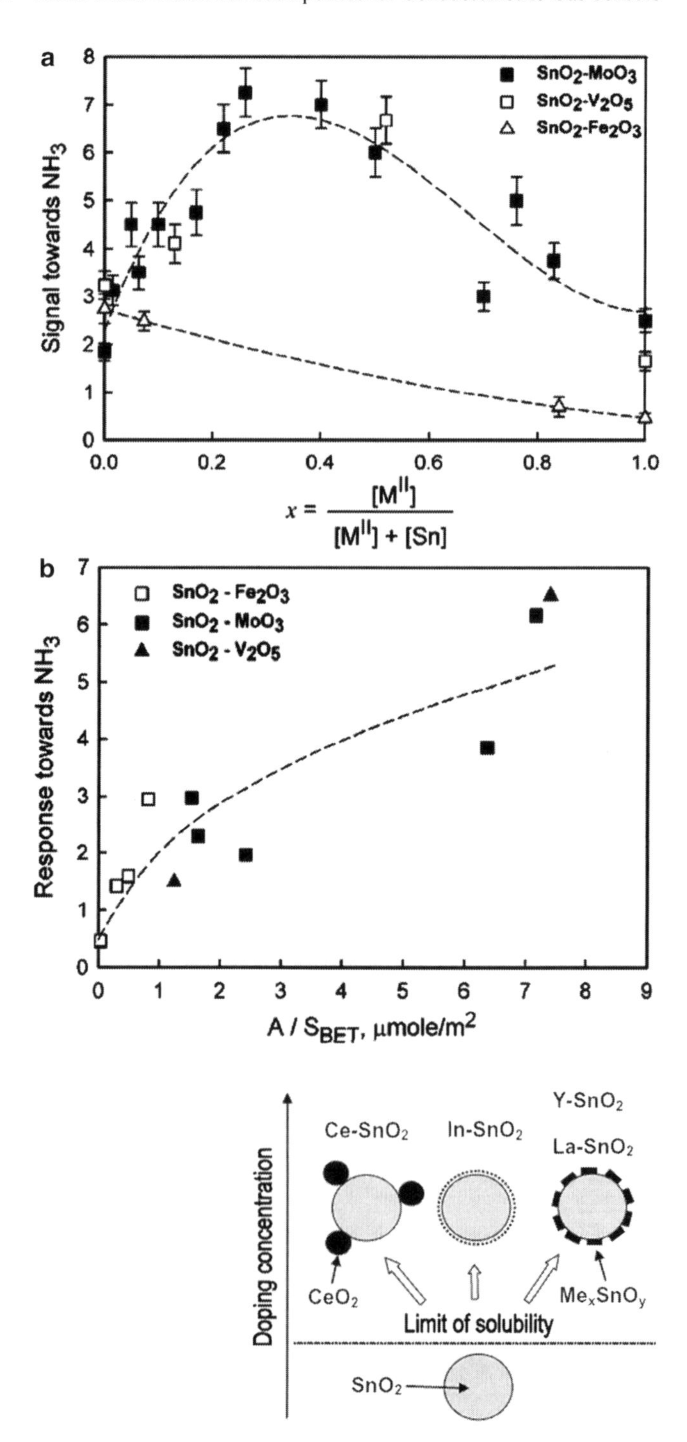

Fig. 14.3 (**a**) Sensor response of SnO_2-based composites to 500 ppm NH_3 in air at $T_{oper} = 350$ °C. (**b**) Sensor response toward NH_3 vs. density (A/S_{BET}) of acid centers for temperature range 300–500 °C (Reprinted with permission from Gas'kov and Rumyantseva 2009, Copyright 2009 Springer)

Fig. 14.4 Formation of segregation layer on the surface of SnO_2 grains as dependent on the nature of the doping elements (Reprinted with permission from Korotcenkov 2005, Copyright 2005 Elsevier)

that the degree of segregation depends strongly on the composition of the bulk phase (Barret and Dufour 1985; Dufour and Nowotny 1988), and the rate of surface segregation of additive elements is controlled by lattice diffusion, and therefore segregation equilibrium may be established at elevated temperatures.

Changes that may take place in a two-phase metal oxide matrix while a concentration of the second phase is being increased are demonstrated in Fig. 14.5. Here, for simplification, $R_{f.ph}$ and $R_{s.ph}$ are the

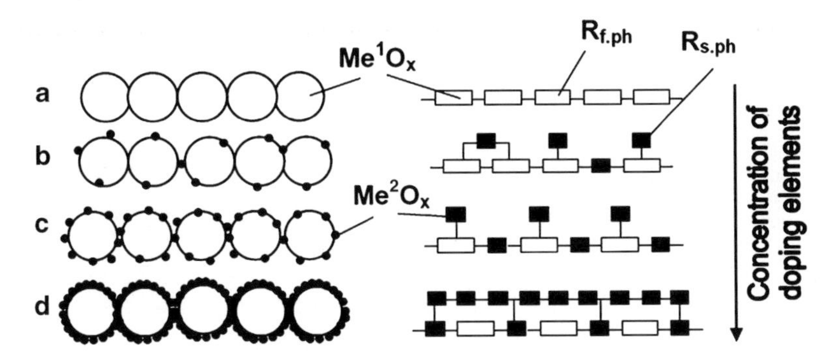

Fig. 14.5 Change of film structure and electrical scheme of film conductivity during phase modification of metal oxides (Reprinted with permission from Korotcenkov 2005, Copyright 2005 Elsevier)

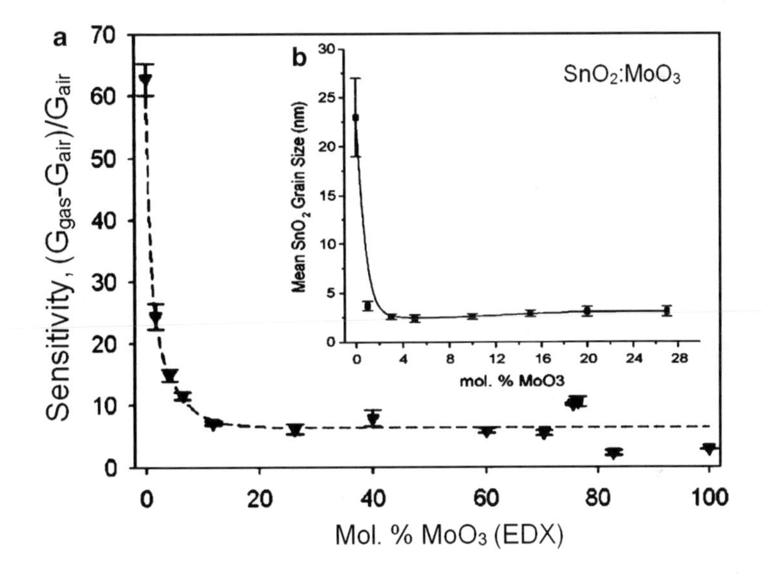

Fig. 14.6 Influence of MoO_3 contents in SnO_2:MoO_3 composite on (**a**) response to consecutive injections (1 μL) of ethanol at 275 and (**b**) SnO_2 grain size determined by TEM (Adapted with permission from Arbiol et al. 2006, Copyright 2006 Elsevier)

grains' resistance to the first and second oxide phases, including resistance of both bulk and intergrain contacts. One can see that at low concentration of additive, the second phase only modifies the surface properties of the base oxide. At somewhat higher concentration it can contribute to limiting the electroconductivity of the metal oxide matrix. At the final stage, at a certain combination of electroconductivity and gas sensitivity for two metal oxide phases in the gas-sensing matrix, the second oxide phase can produce either full blockage of interaction of the base oxide with the surrounding atmosphere or shunting of the matrix of the base oxide through a more conductive second metal oxide phase. All this will certainly lead to a significant change of both the electrophysical and gas-sensing properties of the metal oxide matrix. In this case, these properties will not be determined by the base oxide. This means that if second oxide has low gas sensitivity, the response of sensors will be decreased strongly. Thus, the precise control of composition is strongly recommended for achieving optimal gas-sensing characteristics of devices based on mixed oxides.

Experiment has shown that as a rule the optimal concentration of the second component in SnO_2- and In_2O_3-based composites to attain maximum conductivity response of gas sensors lies in the range <1–3 wt%. With higher concentration of the second component the response usually drops strongly (see Chap. 2 (Vol. 1)) even in the case of grain size decrease (see, e.g., Fig. 14.6). Only for sensors in which both "acceptor" and "transducer" functions are realized by added oxides can the concentration of the second phase attain a level of 5 wt% or more. This is realized, for example, in H_2S sensors based

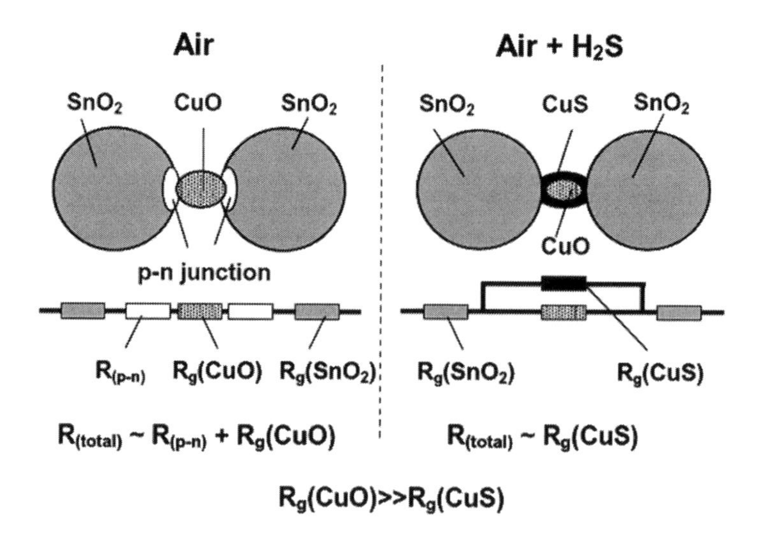

Fig. 14.7 Mechanism of SnO_2–CuO-based sensor operation

on a SnO_2–CuO system whose function is based on a change of chemical composition of the second oxide (CuO→CuS) during interaction with H_2S (Pagnier et al. 2000). The mechanism of this gas sensitivity is shown in Fig. 14.7. In these sensors, CuO is the main gas-sensing element because its interaction with the surrounding atmosphere determines the behavior of the gas sensor. The SnO_2 crystallites act as a stable conductive matrix, guaranteeing the necessary porosity of the sensing structure and the possibility to form fine, dispersed, and isolated CuO grains on their surface. However, it is necessary to remember that CuS is unstable; the crystal structure of CuS is changeable. For instance, at $T \approx 103$ °C, CuS transforms to Cu_2S, which is an ionic conductor with higher resistance, especially at temperatures >220 °C (Zhou et al. 2003). Therefore, the sensitivity of CuO–SnO_2 sensors becomes lower with increasing operating temperature. In addition CuO is not stable in an atmosphere of reducing gases such as H_2 and oxide reduction (CuO→Cu) can be observed at elevated temperatures (150–300 °C) (Rodriguez et al. 2003). High temperature or H_2 pressure and a large concentration of defects in the oxide substrate lead to a decrease in the magnitude of the induction time during this process.

The concentration of the second component can also be a higher 3–5 % in composites based on conductive metal oxides with high conductivity response to gas surrounding (Tables 14.1 and 14.2). However, even in this case sensor response has clearly shown a maximum at certain concentrations of the second phase. Data for SnO_2–In_2O_3 composite-based sensor response to CO are shown in Fig. 14.8, curve 1. Similar results were also obtained for NO_2 detection (Chen et al. 2006). As a rule, optimization of gas-sensing characteristics observed in such composites is conditioned by the improvement of structural parameters of material such as grain size and surface area (see Fig. 14.8, curve 2).

Metal oxide–metal oxide composites can be prepared using three basic approaches:

1. During the process of synthesis or deposition of initial material
2. Using layer-by-layer deposition of chosen materials with following annealing
3. Mixing already synthesized materials in certain proportions

It should be noted that the second and the third approaches assume a subsequent high-temperature treatment which will reduce the effectiveness of the methods and limit their application. It is obvious that the third approach cannot be used with thin-film technology; this approach is limited mainly to ceramics, for which high-temperature annealing is one of the usual steps in the manufacture of chemical sensors. The second and third approaches also assume that at least one of the components participating in solid-phase reactions has sufficient mobility at the annealing temperatures used. Regarding the first approach, for its realization, one could use all methods of synthesis and deposition, as described in Korotcenkov and Cho (2010). However, the most effective for this approach are sol–gel synthesis processes and aerosol-phase deposition methods.

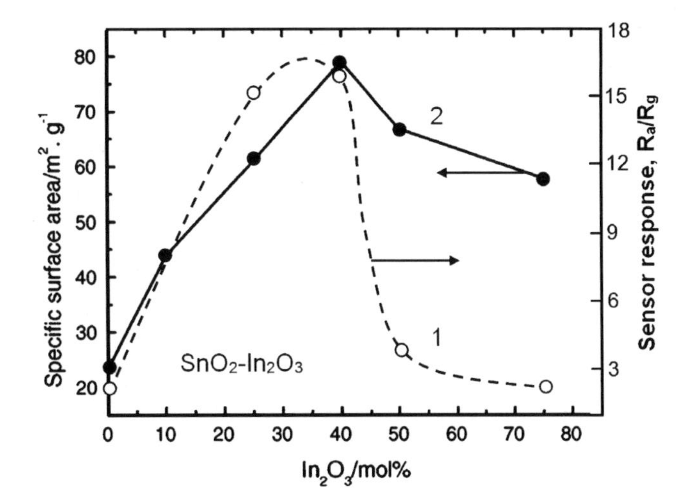

Fig. 14.8 Influence of the contents of SnO_2–In_2O_3 composites on (1) sensitivity to 1,000 ppm of CO at 250 °C operating temperature, and (2) specific surface area of various SnO_2–In_2O_3 precipitated at pH 7.8, cation concentration of 0.05 M. Nanocomposites were calcined at 600 °C in air for 4 h (data from Chen et al. 2006)

We need to remember that the incorporation of a second phase, even in small quantity, can change the conditions of base oxide growth. It is known that the sintering rate of ceramic materials is usually controlled by the grain boundary transport kinetics. This means that incorporation of a second phase, through the properties of grain boundary phases and precipitates, often has a controlling influence on the mechanism and kinetics of sintering and the final properties of the ceramic materials (Dufour and Nowotny 1988). For additional information see Chap. 2 (Vol. 1) and Chap. 23 (Vol. 2). For example, doping of SnO_2 by Nb (0.1–4 mol%) caused a decrease of crystallite size from 220 nm for pure SnO_2 to about 30 nm for Nb (0.1 mol%)-doped samples (temperature of calcination was 900 °C) (Szezuko et al. 2001). Doping of SnO_2 by Ce, Y, and La (Carreno et al. 2002) and by Ca and K (Choi and Lee 2001) led to the same effect, i.e., a reduction in grain size. The same effect was observed for In_2O_3 (Ivanovskaya and Bogdanov 2001; Ivanovskaya et al. 2001) and TiO_2 (Ferroni et al. 2001). This means that Ca, Ce, La, Nb, and Y are very effective inhibitors of metal oxide crystallite growth above 350 °C. The strong change of the bulk properties of metal oxides due to incorporation of additives in the lattice of base oxide can also take place during synthesis of nanocomposites. During the fabrication of ceramic and thick-film gas sensors, the problem of additives influencing grain size and bulk concentration of free charge carriers can be resolved by blending in additional components in a final stage of either ceramic or paste preparation. For example, this process may include mechanical mixing of components, such as SnO_2 and Al_2O_3 powders (Ihokura and Watson 1994). In this case, there is a small possibility that the properties of SnO_2 will be determined by incorporation of doping impurities (Al^{3+}) into the tin dioxide lattice. In such a system, Al_2O_3 added to tin dioxide probably plays the role of diluents, which only separate SnO_2 crystallites, preventing their sintering. As a result, both stability and sensitivity of sensors may be increased. In thin-film technology it is impossible to avoid the mutual influence of additives and base oxide during deposition of mixed metal oxides or composites (Korotcenkov et al. 2008; Korotcenkov and Han 2009). This, of course, presents problems in selecting an optimal composition for the gas-sensing matrix.

References

Afzal A, Cioffi N, Sabbatini L, Torsi L (2012) NO_x sensors based on semiconducting metal oxide nanostructures: progress and perspectives. Sens Actuators B 171–172:25–42

Arbiol J, Morante JR, Bouvier P, Pagnier T, Makeeva EA, Rumyantseva MN, Gaskov AM (2006) SnO_2/MoO_3-nanostructure and alcohol detection. Sens Actuators B 118:156–162

Baraton MI, Merhari L, Wang J, Gonsalves KE (1997) Dispersion of metal oxide nanoparticles in conjugated polymers: investigation of the TiO_2/PPV nanocomposite. MRS Symp Proc 501:59–64

Barret P, Dufour LC (eds) (1985) Reactivity of solids. Elsevier, Amsterdam

Capone S, Forleo A, Francioso L, Rella R, Siciliano P, Spadavecchia J, Presicce DS, Taurino AM (2003) Solid state gas sensors: state of the art and future activities. J Optoelectron Adv Mater 5:1335–1348

Carreno NLV, Maciel AP, Leite ER, Lisboa-Filho PN, Longo E, Valentino A, Probst LED, Paiva-Santos CO, Schreiner WH (2002) The influence of cations segregation on the methanol decomposition on nanostructured SnO_2. Sens Actuators B 86:185–192

Chen A, Huang X, Tong Z, Bai S, Luo R, Liu CC (2006) Preparation, characterization and gas-sensing properties of SnO_2–In_2O_3 nanocomposite oxides. Sens Actuators B 115:316–321

Choi SD, Lee DD (2001) CH_4 sensing characteristics of K-, Ca-, Mg-impregnated SnO_2 sensors. Sens Actuators B 77:335–338

Comini E, Ferroni M, Guidi V, Faglia G, Martinelli G, Sberveglieri G (2002) Nanostructured mixed oxides compounds for gas sensing applications. Sens Actuators B 84:26–32

Dufour LC, Nowotny J (eds) (1988) Surface and near-surface chemistry of oxide materials. Elsevier, Amsterdam

Ferroni M, Boscarino D, Comini E, Gnani D, Guidi V, Martinelli G, Nelli P, Rigato V, Sberveglieri G (1999) Nanosized thin films of tungsten-titanium mixed oxides as gas sensors. Sens Actuators B 58:289–294

Ferroni M, Carotta MC, Guidi V, Martinelli G, Ronconi F, Sacerdoti M, Traversa E (2001) Preparation and characterization of nanosized titania sensing film. Sens Actuators B 77:163–166

Flahaut E, Peigney A, Laurent C, Marliere C, Chastel F, Rousset A (2000) Carbon nanotube–metal–oxide nanocomposites: microstructure, electrical conductivity and mechanical properties. Acta Mater 48:3803–3812

Gas'kov AM, Rumyantseva MN (2001) Nature of gas sensitivity in nanocrystalline metal oxides. Russ J Appl Chem 74(3):440–444

Gas'kov A, Rumyantseva M (2009) Metal oxide nanocomposites: synthesis and characterization in relation with gas sensing phenomena. In: Baraton MI (ed) Sensors for environment, health and security. Springer Science+Business Media B.V., Dordrecht, pp 3–29

Gutierrer FJ, Ares L, Robla JI, Horillo MC, Sayago I, Getino JM, de Agapito JA (1993) NO_x tin dioxide sensors activities, as a function of doped materials and temperature. Sens Actuators B 15–16:354–356

Ihokura K, Watson J (1994) The stannic oxide gas sensor, principle and applications. CRC Press, Boca Raton, FL

Ivanovskaya M, Bogdanov P (2001) The role of catalytic additives in gas-sensitivity of SnO_2-Mo based thin film sensors. Sens Actuators B 77:268–274

Ivanovskaya M, Bogdanov P, Faglia G, Nelli P, Sberveglieri G, Taroni A (2001) On the role of catalytic additives in gas-sensitivity of SnO_2-Mo based thin film sensors. Sens Actuators B 77:268–274

Ivanovskaya M, Kotsikau D, Faglia G, Nelli P (2003) Influence of chemical composition and structural factors of Fe_2O_3/In_2O_3 sensors on their selectivity and sensitivity to ethanol. Sens Actuators B 96:498–503

Kaji T, Oono H, Nakahara T, Yamazoe N, Seiyama T (1980) Fixation of palladium(II) and copper(II) complexes on the surface of stannic oxide (SnO_2) and their catalytic activity for propylene oxidation. J Chem Soc Jpn 7:1088–1093

Kanazawa E, Sakai G, Shimanoe K, Kanmura Y, Teraoka Y, Miura N, Yamazoe N (2001) Metal oxide semiconductor N_2O sensor for medical use. Sens Actuators B 77:72–77

Kohl D (1990) The role of noble metals in the chemistry of solid state gas sensors. Sens Actuators B 1:158–165

Konig U (1987) Deposition and properties of multicomponent hard coating. Surf Coat Technol 33:91–103

Korotcenkov G (2005) Gas response control through structural and chemical modification of metal oxides: state of the art and approaches. Sens Actuators B 107:209–232

Korotcenkov G (2007) Practical aspects in design of one-electrode semiconductor gas sensors: status report. Sens Actuators B 121:664–678

Korotcenkov G (2010) Methods of sensing materials' modification: material engineering of metal oxides. In: Korotcenkov G (ed) Chemical sensors: fundamentals of sensing materials, vol 1, General approaches. Momentum Press, New York, pp 303–368

Korotcenkov G, Cho BK (2010) Methods of sensing materials synthesis and deposition. In: Korotcenkov G (ed) Chemical sensors: fundamentals of sensing materials, vol 1, General approaches. Momentum Press, New York, pp 214–303

Korotcenkov G, Han SD (2009) (Cu, Fe, Co and Ni)-doped SnO_2 films deposited by spray pyrolysis: doping influence on thermal stability of SnO_2 film structure. Mater Chem Phys 113:756–763

Korotcenkov G, Sysoev V (2011) Conductometric metal oxide gas sensors. In: Korotcenkov G (ed) Chemical sensors: comprehensive sensor technologies, vol 4, Solid state devices. Momentum Press, New York, pp 39–186

Korotcenkov G, Brinzari V, Boris Y, Ivanov M, Schwank J, Morante J (2003) Surface Pd doping influence on gas sensing characteristics of SnO_2 thin films deposited by spray pyrolysis. Thin Solid Films 436:119–126

Korotcenkov G, Brinzari V, Boris I (2008) (Cu, Fe, Co or Ni)-doped SnO_2 films deposited by spray pyrolysis: doping influence on film morphology. J Mater Sci 43(8):2761–2770

Matsushima S, Maekawa T, Tamaki J, Miura N, Yamazoe N (1992) New methods for supporting palladium on a tin oxide gas sensor. Sens Actuators B 9:71–78

Nakao Y (1995) Noble metal solid sols in poly(methyl) methacrylate. J Colloid Interface Sci 171:386–391

Nitta M, Haradome M (1979) CO gas detection by ThO_2-doped SnO_2. J Electron Mater 8:571–572

Nitta M, Otani S, Haradome M (1980) Temperature dependence of resistivities of SnO_2-based gas sensors exposed to Co, H_2, and C_3H_8 gases. J Electron Mater 9:727–743

Pagnier T, Boulova M, Galerie A, Gaskov A, Lucazeau G (2000) Reactivity of SnO_2-CuO nanocrystalline materials with H_2S: a coupled electrical and Raman spectroscopic study. Sens Actuators B 71:134–139

Rastomjee CS, Dale RS, Schaffer RJ, Jones FH, Egdell RG, Georgiadis GC, Lee MJ, Tate TJ, Cao LL (1996) An investigation of doping of SnO_2 by ion implantation and application of ion-implanted films as gas sensors. Thin Solid Films 279:98–105

Rodriguez JA, Kim JY, Hanson JC, Perez M, Frenkel AI (2003) Reduction of CuO in H_2: in situ time-resolved XRD studies. Catal Lett 85(3–4):247–254

Rosenfeld D, Sanjines R, Schreiner WH, Levy F (1993) Gas sensitive and selective SnO_2 thin polycrystalline films doped by ion implantation. Sens Actuators B 15–16:406–412

Sayago I, Gutierrer FJ, Ares L, Robla JI, Horrillo MC, Getino J, Rino J, Agapito JA (1995) The effect of additives in tin oxide on the sensitivity and selectivity to NO_x and CO. Sens Actuators B 26:19–23

Sulz G, Kuhner G, Reiter H, Uptmoor G, Schweizer W, Low H, Lacher M, Steiner K (1993) Ni, In and Sb implanted Pt and V catalysed thin-film SnO_2 gas sensors. Sens Actuators B 15–16:390–395

Szezuko D, Werner J, Oswald S, Behr G, Wetzing K (2001) XPS investigations of surface segregation of doping elements in SnO_2. Appl Surf Sci 179:301–306

Tamaki J, Maekawa T, Miura N, Yamazoe N (1992) CuO-SnO_2 element for highly sensitive and selective detection of H_2S. Sens Actuators B 9:197–203

Varela JA, Cerri JA, Leite ER, Longo E, Shamsuzzoha M, Bradt RC (1999) Microstructural evolution during sintering of CoO doped SnO_2 ceramics. Ceram Int 25:253–256

Yamaura H, Moriya K, Miura N, Yamazoe N (2000) Mechanism of sensitivity promotion in CO sensors using indium oxide and cobalt oxide. Sens Actuators B 65:39–41

Yamazoe N (1991) New approaches for improving semiconductor gas sensors. Sens Actuators B 5:7–19

Yamazoe N (2005) Toward innovations of gas sensor technology. Sens Actuators B 108:2–14

Yamazoe N, Kurokawa Y, Seiyama T (1983) Effects of additives on semiconductor gas sensors. Sens Actuators 4:283–289

Yang D (2011) Nanocomposite films for gas sensing, In: Reddy B (ed) Advances in nanocomposites—synthesis, characterization and industrial applications (Ch. 37). InTech, New York, pp 857–882. http://www.intechopen.com

Zhou XH, Cao QX, Huang H, Yang P, Hu Y (2003) Study on sensing mechanism of CuO-SnO_2 gas sensors. Mater Sci Eng B 99:44–47

Chapter 15
Composites for Optical Sensors

15.1 Dye-Based Composites

Various coatings in chemioptical sensors often are nanocomposites, because usually they consist of phosphorescent dyes or oxygen-quenchable luminescent agent embedded into polymer or sol–gel matrix of high gas permeability, which is then deposited on a solid support such as a planar waveguide, the bottom of the well of a microtiter plate, or the tip of an optical fiber (Xu et al. 1994; McDonagh et al. 1998; Ramamoorthy et al. 2003). So the sensitivity of a luminescence-based sensor, for example, is influenced by the properties of both the luminescent material (indicator dye) and the gas-permeable encapsulating medium.

15.1.1 Sol–Gel Composites

Sol–gels are very popular materials for designing optical nanosensors (Shibata et al. 1997; Jain et al. 1998; Lobnik and Wolfbeis 1998; Lobnik et al. 1998; Rossi et al. 2005), because the beads can easily be manufactured, and sol–gel matrix is porous to allow an analyte to diffuse freely inside and is robust. In addition, silica beads are easy to separate via centrifugation during particle preparation, surface modification, and other solution-treatment processes because of the higher density of silica. Sol–gels are made by hydrolysis and polycondensation of tetraalkoxysilanes of the type $Si(OR)_4$, where R is an alkyl group. The resulting polysiloxane networks are largely different depending on whether acidic or alkaline catalysis has been applied in condensation and also on the temperature and dilutions employed. Sol–gels generally suffer from aging effects, which causes the response function of the resulting sensors to change with time (Reisfeld 1996). This is a strong point of the sol–gel matrix. In addition, experiments have shown that sol–gels and organically modified sol–gels ("ormosils") have excellent optical properties, are easily prepared, and are capable of retaining indicators inside their gel network (Reisfeld 1996; Fuhr et al. 1998; Von Bueltzingsloewen et al. 2002; Aubonnet et al. 2003). They can be both hydrophobic and hydrophilic and work efficiently for gas sensors such as for oxygen (Lee and Okura 1997; Aubonnet et al. 2003; Zhang et al. 2011). A sol–gel film was doped with an appropriate pH indicator and was found to be sufficiently stable to monitor pH remotely under harsh conditions as well. Metal–ligand complexes incorporated into a sol–gel form an attractive alternative to organic indicator dyes for luminescent sensing of pH (Lam et al.

G. Korotcenkov, *Handbook of Gas Sensor Materials*, Integrated Analytical Systems,
DOI 10.1007/978-1-4614-7388-6_15, © Springer Science+Business Media New York 2014

Table 15.1 Sol–gel-based cocktails for making oxygen sensors

Dye (indicator)	Matrix	Solvent
Ru(dpp) (dodecylsulfate)$_2$ (=DS)	Silicone E4	Chloroform, ethyl acetate
Ru(dpp) trimethyl-silylpropane-sulfonate (=TMS)	Silicone E4, silicone E41	Chloroform
Ru(dpp)(DS)$_2$	TMOS + Pr-TriMOS	EtOH/HCl
Ru(dpp) TMS$_2$	Ormosil (Ph-TriMOS + TMMS)	Chloroform or acetone
Pt-(fluoro)phenyl-porphyrins	Silicone	Chloroform

Source: Components proposed by Wolfbeis (www.wolfbeis.de)
Ormosil phenyl-trimethoxysilane-*co*-tetramethoxysilane, *TMOS* tetramethoxysilane, *TriMOS* phenyl-trimethoxysilane, *TMMS* trimethylmethoxysilane

2000). For example, it was found that organically modified sol–gel precursors, such as phenyl-trimethoxysilane (Ph-SiOMe$_3$) and its copolymers with other trialkoxysilanes, display distinctly improved properties in terms of long-term stability, gas permeability, and ease of deposition on wave-guide optics. Several sol–gel-based cocktails used for making oxygen sensors are presented in Table 15.1.

15.1.2 Polymer-Based Composites

According to Pauly (1998) and Mohr (2006), polymer materials used in composites, designed for optical gas sensors, have to fulfill the following requirements. First of all, the indicator dye and all additives have to dissolve well in the polymer (and must not be washed out). The analyte also has to be soluble in the polymer and must be able to diffuse fast into the polymer and within the polymer. The polymer material has to be chemically and physically stable in order to achieve a good operational lifetime. Furthermore, no crystallization/migration/reorientation of the indicator chemistry in the polymer must occur. This can happen even after weeks or months if indicator solubility in the polymer is not as high as expected. The polymer must be stable even at elevated temperatures. It should be stable against light and chemicals, and it should be nontoxic and biocompatible. The polymer should not have any intrinsic color/luminescence, and it should be optically transparent in the spectral range where measurements are being performed. The material should be mechanically stable as well. Solubility in organic solvents is another requirement. Finally, polymers should be commercially available. Polymers with acceptable solvents, which are available for incorporation in optical gas sensors, are listed in Table 15.2. One can find in the review of Amao (2003) the analysis of polymers acceptable for use in oxygen sensors. To plasticize the above-mentioned polymers one can use such plasticizers as tributyl phosphate (TBP), tris(2-ethylhexyl)phosphate (TOP), 2-(octyloxy)benzonitrile (OBN), and 2-nitrophenyl octyl ether (NPOE).

Regarding gas permeability, we can say that the permeability of a polymer to the target molecule to be detected is a major factor influencing the response of gas sensors. For example, the efficiency of oxygen quenching of an indicator dye in a polymer strongly depends on the permeability (P) to the gas of the encapsulating medium. The permeability turn depends on the solubility (S) and the diffusion coefficient (D) of the gas in the matrix according to (Pauly 1998)

$$P = D \times S \tag{15.1}$$

The higher the permeability the higher the probability of a collisional deactivation of dye luminescence by oxygen. One can draw the same conclusion for other gases as well. Data for several polymers used in optical gas sensors are given in Table 15.3. As can be seen in Table 15.3, the permeability

Table 15.2 List of polymers and solvents used for the preparation of optical gas sensors

Polymer	Acronym	Solvent
Cellulose acetate	CAc	Chloroform
Ethyl cellulose	EC	Toluene/ethanol
Poly(tetrafluor ethylene-covinylidenfluoride-*co*-propylene)	PFE-VFP	Tetrahydrofuran
Poly(styrene-*co*-acrylonitrile)	PSAN	Chloroform
Poly(4-vinyl phenol)	PVPh	Tetrahydrofuran
Poly(vinyl methyl ketone)	PVMK	Toluene/ethanol
Polysulfone	PSu	Chloroform
Poly(vinyl chloride-*co*-isobutyl vinyl ether)	PVC-iBVE	Toluene/ethanol
Poly[(octahydro-5-(methoxycarbonyl)-5-methyl-4,7-methano-1*H*-indene-1,3-diyl)-1,2-ethanediyl]	POMMIE	Toluene/ethanol
Poly(bisphenol A carbonate)	PC	Chloroform
Poly(4-*tert*-butyl styrene)	PTBS	Toluene/ethanol
Poly(acrylonitrile)	PAN	Dimethylformamide
Poly(vinyl chloride)	PVC	Tetrahydrofuran
Polystyrene	PS	Toluene/ethanol
Poly(methyl methacrylate)	PMMA	Chloroform

Source: Data from Apostolidis (2004)

Table 15.3 Permeability coefficients of selected polymers to gases

	$P \times 10^{13}$ (cm^2 s^{-1} Pa^{-1})		
Polymer	O_2	CO_2	NH_3
LDPE	2.2[a]	9.5[a]	21.0[b]
PS[b]	2.0	7.9	No data
PVC	0.034	0.12	3.7[c]
PMMA	0.116[d]	2.33[d]	No data
PAN	0.00015[a]	0.0006[a]	No data
EC	11.0[c]	84.8	529

Source: Data from Pauly (1998)
LDPE low-density polyethylene, *PS* polystyrene, *PVC* poly(vinyl chloride), *PAN* poly(acrylonitrile), *PMMA* poly(methyl methacrylate), *EC* ethyl cellulose
[a]25 °C
[b]Biaxially oriented
[c]20 °C
[d]35 °C

of a polymer to particular gases can differ significantly. The influence of oxygen permeability on sensor response is illustrated by the results contained in Table 15.4 for $[Ru(dpp)_3^{2+}(Ph_4B^-)_2]$ in a variety of different, unplasticized encapsulation media. From the results in Table 15.4, it is apparent that the higher the oxygen permeability of the encapsulating polymer the greater the oxygen sensitivity of the film. The permeability coefficient for a gas, P (cm^3 cm cm^{-2} s^{-1} Pa^{-1}), is directly related to its diffusion coefficient, D (cm^2 s^{-1}), and its solubility, S (cm^3 cm^{-3} Pa^{-1}), or Henry's constant, K_H (mol dm^{-3} atm^{-1}); K_H (mol dm^{-3} atm^{-1}) = 4,521 × S (cm^3 cm^{-3} Pa^{-1}) at 273 K (Mills 1998). Thus, the performance of a sensor for a particular gas will at least depend on the proper choice of polymer and sensitive chemistry and has to be adjusted to the application of interest. For example, for application in sensors, the indicator dyes need to meet the following criteria: (1) a sufficient sensitivity to oxygen with respect to the range of application, (2) compatibility of the luminescence excitation of the probe with solid-state light sources as LEDs or diode lasers for excitation, (3) sufficient and commercial availability or

Table 15.4 Sensitivity of $[Ru(dpp)_3^{2+}(Ph_4B^-)_2]$-based oxygen sensors encapsulated in a variety of different, non-plasticized media with different permeability coefficients

Encapsulating polymer	Oxygen permeability (10^{13} cm^3 cm cm^{-2} s^{-1} Pa^{-1})	$pO_2(S=1{:}2)$ (Torr)
Silicone rubber (RTV118)	376	30
Cellulose acetate butyrate	3.56	102
Cellulose acetate	0.585	311
Poly(methyl methacrylate)	0.116	806
Poly(vinylchloride)	0.034	3,390

Source: Reprinted with permission from Mills (1998), Copyright 1998 Elsevier
$pO_2(S=1{:}2)$ (Torr) is the partial pressure of oxygen at which the luminescence intensity is equal to $I_o/2$

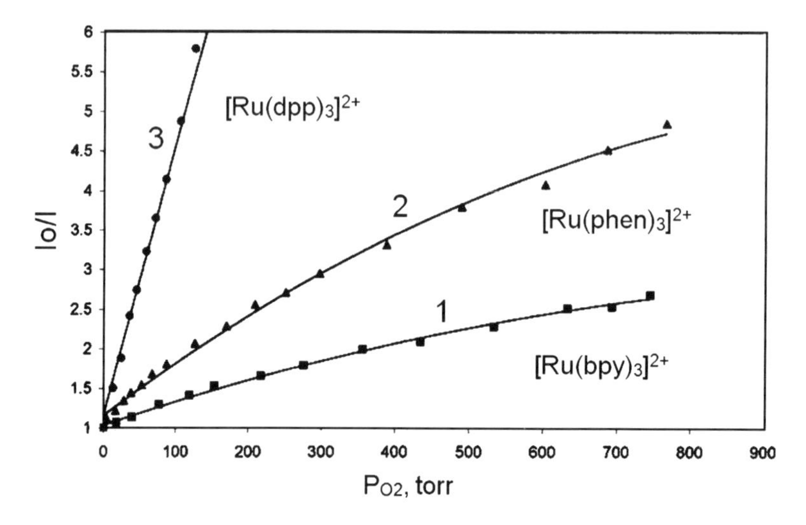

Fig. 15.1 Stern–Volmer plots of I_o/I vs. pO_2 for popular oxygen sensor lumophores: tris(bipyridyl) ruthenium (II), $[Ru(bpy)_3]^{2+}$; tris(1,10 phenanthroline) ruthenium (II), $[Ru(phen)_3]^{2+}$; and tris(4,7-diphenyl-1,10-phenanthroline) ruthenium (II), $[Ru(dpp)_3]^{2+}$ when encapsulated in the same silicone rubber medium, RTV118 (Reprinted with permission from Mills 1998, Copyright 1998 Elsevier)

accessibility, (4) high photostability, (5) solubility in organic media, and (6) ease of immobilization. This means that the tuning of the composition properties is a mandatory step for achieving the required parameters of gas sensors. This can be realized either by exchange of the indicator dye (see Fig. 15.1) and encapsulation polymer (see Table 15.4) or by tuning the gas-sensing properties of the matrix by addition of a new component with the necessary properties. For example, it was established that the quenching efficiency is strongly dependent on the type of polymer used and thus can be fine tuned by proper choice of the polymer matrix (Wolfbeis 1991, 1992; Draxler et al. 1995; Hartmann and Trettnak 1996; Mills and Thomas 1998; Mills et al. 1998; Apostolidis 2004). In particular, Mills and Thomas (1998) have shown that the quenching of the electronically excited lumophoric state of $[Ru(bpy)_3^{2+}(Ph_4B_2)_2]$ by oxygen is inversely dependent upon the viscosity of the quenching medium. For highly compatible polymer–plasticizer combinations, such as TPP–PMMA, TBP–PMMA, and DBP–PMMA, it appears that the plasticizer with the lowest viscosity (TPP in this case) produces films of the highest oxygen sensitivity. This is not too surprising given that the quenching process is likely to be near to diffusion controlled and, as a result, the quenching rate constant will be inversely proportional to the viscosity of the reaction medium. Apostolidis (2004) found that increasing the concentration of plasticizer in the polymer film resulted in an increase of sensitivity of the probes to oxygen quenching. Thus, depending on the mode of operation of the sensing mechanism and the target molecule, the composition of the sensing layer in an optical sensor can range from a simple polymer–indicator combination (Wang et al. 2002) to highly sophisticated systems. For example, optical gas sensors for the determination of oxygen, carbon dioxide, or ammonia can consist of (1) a supporting polymer, permeable to the target gas and giving the sensor physical stability, (2) a sensing chemistry, i.e., an indicator dye (indicator probe) changing its optical properties on contact with and according to the concentration of the target, (3) a softening agent (plasticizer) for adjustment of physical stability and permeability of the polymer matrix without interfering the detection signal, and (4) in the case of the carbon dioxide sensor, a lipophilic buffer system for adjusting the pH in the microenvironment of the indicator dye. This composition structure can be extended by a further additive, such as a chromoionophore or other chemical substance, according to the target and the analytical characterization method. This means that a large number of different species of each component have to be tested for their performance in a particular formulation. We need to recognize that the interactions of all the possible components in nanocomposites often are very complex and not coercively predictable. As a result, the development of sensor materials is still often a kind of empirical science.

However, experiments have shown that the approach based on using composites, in spite of the above-mentioned complications, allows the sensitivity of the sensors to be controlled over a wide range (Pang et al. 1996; Ramamoorthy et al. 2003). For example, tris(4,7-diphenyl-1,10-phenanthroline) ruthenium ($Ru(dpp)_3$) chloride embedded into a polysulfone membrane has shown good sensitivity to oxygen (Florescu and Katerkamp 2004), with a limit of detection of the order of about 2 % O_2. The luminescence character of the Ru(II) metal complexes is explained by a charge transfer from ligand to metal. The long lifetimes of the excited states is a reflection of the triplet nature due to the spin–orbit coupling with the metal center (Carraway et al. 1991). These excited states involve large changes in charge distribution and therefore the spectral properties are strongly influenced by the surrounding medium. The phosphorescence of Erythrosin B reagent immobilized in sol–gel silica deposited on one end of an optical fiber can be used for oxygen measurement as well (Chan et al. 2000). Other typical components of polymer-based composites used in optical oxygen sensors are presented in Table 15.5. Oxygen is known to act as a quencher of luminescence of many fluorophors (Sacksteder et al. 1993; Xu et al. 1994; Lakowicz 1999). Thus, an important class of O_2 sensors is based on the decrease of the luminescence signal (intensity or lifetime) of an oxygen-sensitive material, i.e., the indicator dye, as a function of oxygen partial pressure pO_2 (Klimant and Wolfbeis 1995; Amao 2003). Quenching efficiency in Table 15.5 is expressed as the ratio of signals under nitrogen and air (I_o/I_{air}). As a rule, emission is almost totally quenched at 1 % oxygen in nitrogen or argon gas. Chemical structures of several oxygen probes are shown in Fig. 15.2. It should be noted that the indicator dyes listed in Table 15.5 are known to fulfill most of the requirements mentioned above. Ruthenium diimine complexes are easily accessed via the synthesis described by Klimant and Wolfbeis (1995) and some of them are commercially available. Platinum and palladium porphyrins can be easily accessed from suppliers with a wide range of ligands (www.porphyrin-systems.de). One can find in the review of Amao (2003) a detailed analysis of indicator dyes acceptable for design oxygen sensors.

Regarding polymers likely to be promising for application as matrix in optical gas sensors, one can say that the efficiency of polymer application depends on many factors (Amao 2003; Wolfbeis 2005; Lobnik et al. 2012). For example, poly(dimethylsiloxane)s have a high gas permeability but lack the mechanical strength in thin films. On the other hand, poly(1-trimethylsilyl-1-propyne) (poly(TMSP)) has a high gas permeability of films and provides a tough and thin film. This means that an organic dye in a poly(TMSP) film may make contact with the oxygen in the gas phase and, therefore, highly sensitive gas sensors can be developed using poly(TMSP) films. In general, due to excellent optical and mechanical properties, and unique gas solubility, silicones have their main applications in sensors for oxygen and other uncharged quenchers, such as sulfur dioxide and chlorine. Silicones can also be

Table 15.5 Composition and properties of typical oxygen-sensitive fluorescent materials considered to be quite practical

Oxygen indicator (lumophores)	Encapsulating media	Excitation/emission wavelength (nm)	Quenching efficiency
Fluoranthene	Polyethylene; vycor glass		
Ru(dpp) on silica nanoparticles	Suspended in 1-component (acid releasing) silicone	455/615	2.4
$Ru(dpp)_3(TMPS)_2$	Mixture of phenyl-trimethoxysilane and tetramethoxysilane	455/620	1.6
$Ru(dpp)(TMPS)_2$ or $Ru(dpp)(LS)_2$	(a) Ethyl cellulose, (b) polystyrene	455/620	1.8–2.0
Pt–octaethylporphyrin	Poly(1-trimethylsilyl-1-propyne)	535/645	~80
Pt–tetrakis(per-fluorophenyl)-porphyrin	(a) Polystyrene (PS) (b) PS-*co*-fluoroacrylate	541/650	1.9 ~22
Pd–tetrakis(per-fluorophenyl)-porphyrin	PS-*co*-fluoroacrylate	535/655	43
Pd–octaethyl-ketoporphyrin	Teflon AF	590/660	>300
$Pb(sulfooxinate)_2$	Anion exchanger	385/625	2.8

Source: Reprinted with permission from Wolfbeis (2005), Copyright 2005 Royal Society of Chemistry
Ru(dpp) ruthenium(II)-tris(4,7-diphenyl phenanthroline), *TMPS* trimethylsilyl propanesulfonate, *LS* laurylsulfate

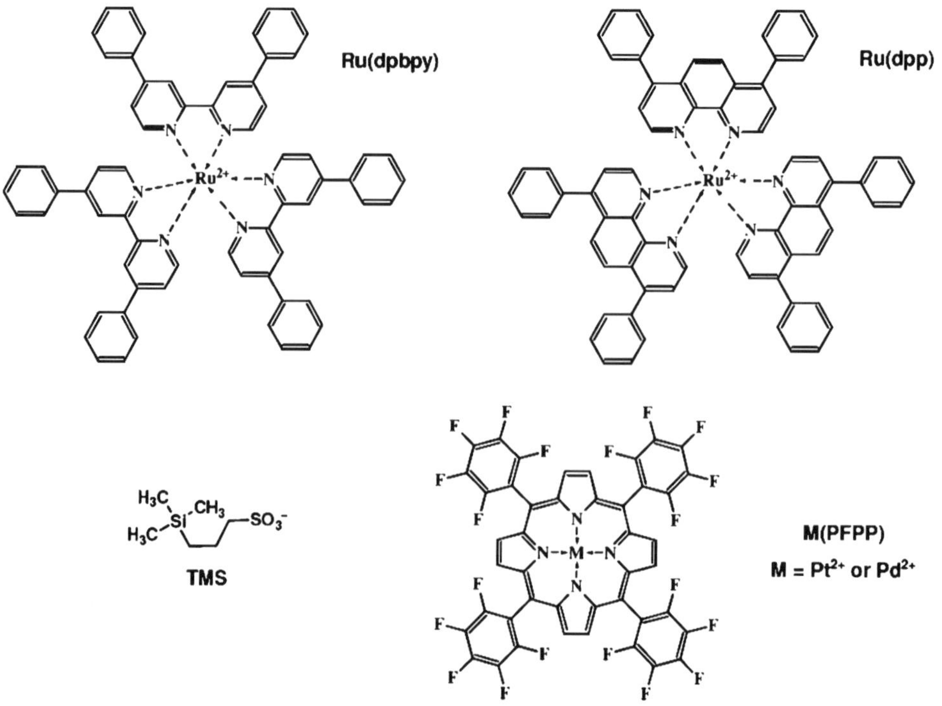

Fig. 15.2 Chemical structures of the oxygen probes used for oxygen sensor design and the trimethylsilyl (TMS) counter ion of the ruthenium complexes

used as gas-permeable covers in sensors for carbon dioxide or ammonia. Silicones cannot be easily plasticized by conventional plasticizers but form copolymers which may be used instead (Baldini et al. 2006; Wolfbeis 1991).

The fluoropolymer films also display high permeability to oxygen. Thus, fluoropolymers are suitable for gas sensor design as well (Amao 2003). Cellulose derivatives such as ethyl cellulose and cellulose acetate also provide mechanical strength in a thin film. Moreover, polymer film made from cellulose derivatives and a plasticizer (TBP) have high oxygen permeability. As a result, cellulose acetate with TBP films is widely used as a matrix for optical oxygen sensors (Mills and Lepre 1997). As organic glassy polymers such as polystyrene (PS), poly(methyl methacrylate) (PMMA), poly(isobutyl methacrylate), and poly(vinylchloride) (PVC) have lower permeability, diffusion constant, and solubility for O_2 than those of silicone polymers or fluoropolymers, these polymer films provide mechanical strength to thin films. In addition, the Stern–Volmer quenching constant (KSV) values in PS, PVC, PMMA, and poly(isobutyl methacrylate) are high. Polystyrene and poly(isobutyl methacrylate) have KSV values comparable with the other polymers. However, except for polystyrene, they are difficult to modify chemically. So, the function of these polymers is limited to the role of a "solvent" for indicators, or as a gas-permeable cover. For example, PMMA and PDMS have been selected as the optimum matrix for oxygen sensing (Lobnik et al. 2012).

The immobilization of the sensing chemistry in a polymer and sol–gel matrix as support can be achieved either by covalently binding an indicator to the matrix, e.g., via esterification, or physically, e.g., by electrostatic or van der Waals' interactions or dissolution (Wolfbeis 1991, 1992). For example, for an application in gaseous media, the sensing materials can be prepared by dissolving the indicator dye and the tuning agents in the organic polymer. In particular, the solid indicator dye can be added directly to a solution of polymer dissolved in an organic solvent or stock solutions of each are mixed in the appropriate ratio, both methods forming a homogeneous sensor cocktail. Under the stipulation that all components used are soluble in organic solvents, the automation-assisted synthesis is possible by using a liquid-dispensing robot. According to Wolfbeis (2005), one very efficient solution to the problem of the insolubility of cationic dyes in apolar solvents is to exchange the respective hard anions (such as chloride) with softer and more lipophilic anions, such as perchlorate or, even better, tetraphenylborate (TPB), trimethylsilylpropyl sulfate (TMPS), or laurylsulfate (LS) (Weigl and Wolfbeis 1995; Mohr et al. 1997). Another efficient method for permanently incorporating probes in a polymer network consists of covalent immobilization of the probes, which not only prevents aggregation and crystallization but also leaching out of the sensing material into the sample fluid. Covalent bonding may be achieved by (1) choosing the indicator that contains a functional group for covalent bonding to the polymer, which is at the same time insensitive for the target analyte, or (2) polymerizing the indicator to certain monomers to form a copolymer (Baldini et al. 2006; Lobnik et al. 1998). Experiment has shown that covalent immobilization enables the sensor to display good stability (no leaching, crystallization, and evaporation of components) and a longer operational lifetime (Lobnik et al. 2012). The disadvantage is that the covalent bonding often lowers the sensitivity for the analyte and prolongs the response time of the sensor (Lobnik et al. 1998). However, many people believe that covalent immobilization is not necessary for gas sensor design, since here no bleaching occurs like in sensors applied in liquid media where indicator can be extracted into the liquid phase. In particular, covalent immobilization is uncommon in the case of sensing oxygen (Wolfbeis 2005). Of course, this conclusion cannot be regarded as general. For example, Zhang et al. (2011) established that the covalently grafted sample showed better reversibility, higher photochemical stability, and better oxygen-sensing ability than the physically incorporated sample. Ru(II) complexes in the studied sensors were strongly covalently grafted to the Si–O network of mesoporous silica via the CH_2–Si bond. Doping can also be used for immobilization as it is not restricted to certain indicators and polymers (Lobnik et al. 2012). The sensor stability (in terms of indicator leaching) is better compared to impregnation and worse compared to covalent bonding. The response time is better than in covalent immobilization.

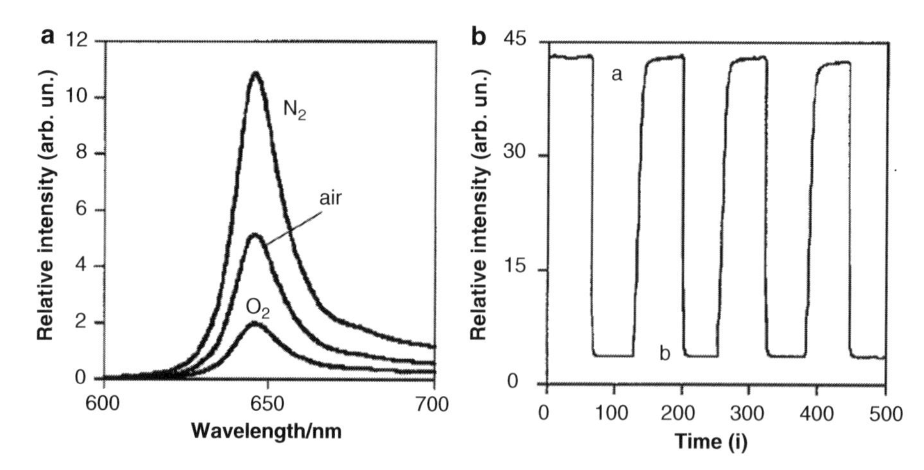

Fig. 15.3 (**a**) Room-temperature phosphorescence spectra of platinum–octaethylporphyrin (PtOEP)-doped sol–gel glass under different atmospheric conditions: N_2 (*top*), ambient conditions (*middle*), and O_2 (*bottom*). Excitation wavelength, 535 nm. (**b**) Response time, relative intensity change, and reproducibility of sensor response on switching between 100 % nitrogen (a) and 100 % oxygen (Reprinted with permission from Lee and Okura 1997, Copyright 1997 Royal Society of Chemistry)

Experiment shows that from the dyes listed in Tables 15.5 and 15.6, the most important set is the metal–porphyrin complexes, especially platinum–octaethylporphyrin (PtOEP) and palladium–octaethylporphyrin (PdOEP) (Li and Wong 1992; Douglas and Eaton 2002; Ramamoorthy et al. 2003). Due to the increased lifetime of the phosphorescent excited state in these complexes (from μs to ms), the sensitivity is normally larger than the Ru(II) complexes. In addition, the larger Stokes shift (difference in the wavelength of the exciting and emitting radiations (>100 nm)) makes the measurement easier (Papkovsky 1995). The porphyrin-based systems also exhibit faster response times as compared to the Ru complexes. A typical phosphorescence spectra and dynamic response of a PtOEP in silica glass oxygen sensor is shown in Fig. 15.3. When changing from 100 % nitrogen to 100 % oxygen, the response time is 5 s, and from O_2 to N_2 it is about 10 s (Lee and Okura 1997). The assumption that the type of polymer has an intense influence on the sensitivity of an oxygen sensor was affirmed by Apostolidis (2004). PTBS was found to be a promising alternative for PS as encapsulation polymer for Pt(PFPP) or other oxygen-sensitive dyes. The sensitivity of Pt(PFPP) in PTBS was 270 % higher than in PS showing similar physical stability of the sensor films.

The change in optical property of an immobilized colorimetric acid–base indicator, α-naphtholphthalein in ethyl cellulose, with an internal reference consisting of fluorescent porphyrin dye embedded in polystyrene may be used for CO_2 measurement (Amao and Nakamura 2004). The sensor responded rapidly and reversibly to CO_2 concentrations ranging from 0 to 100 %. Moreover, this kind of sensor is suitable for the optical sensing of CO_2 both in dry gases, with extremely short response time (in the order of few seconds), and in liquid samples. This approach for designing solid type optical $p\mathrm{CO_2}$ sensors is based on immobilizing a pH-sensitive dye directly into a hydrophobic membrane, and replacing the hydrogen carbonate buffer by a lipophilic phase transfer agent (e.g., quaternary ammonium hydroxide) was proposed by Mills et al. (1992, 1997). In other words, the determination of carbon dioxide with optical sensors utilizes pH-sensitive dyes changing color or fluorescence when the pH of the surrounding medium changes in accordance with a change of the ambient carbon dioxide level (Wolfbeis et al. 1988; Mills et al. 1992; Weigl et al. 1993). According to the acidic character of carbon dioxide, the indicator dye used has to be applied in the basic form. On exposure to CO_2 the dyes have to change absorbance reversibly due to protonation by carbonic acid produced with the traces of water in the film. Suitable indicator dyes include azo dyes, sulfonephthaleins,

Table 15.6 Data on indicator dyes applied in optical gas sensor design

Indicators		pK_D (OH) in H_2O	l_{max} (obs) in H_2O (nm)
(*A*) *Triphenyl methane dyes* (*pH indicators*)			
m-Cresol purple	MCP	8.3	580[a]
Thymol blue	TB	9.2–9.7	595[a]
Bromophenol blue	BPB	4.1	600[a]
Bromocresol green	BCG	4.9	615[a]
Phenol red	PR	8.0	560[a]
Cresol red	CR	8.5	570[a]
Alizarin red S	ARS	5.5	560[a]
Xylenol blue	XB	9.5	595[a]
Xylenol orange	XO	10.4	580[a]
Bromocresol purple	BCP	6.4	585[a]
Tetrabromophenol blue	TBPB	3.5	610[a]
Chrome azurol S	CAS	~11	600
(*B*) *Azo dyes*			
Brilliant yellow	BRY	~7	500[a]
Orange I	OI	8.3	490[a]
(*C*) *Xanthene dyes*			
Rhodamine B	RB	–	560[b]
Rhodamine 6G	R6G	–	530[b]
Eosin B	EOB	6.5	520[b]
Erythrosin B	ERB		540[b]
(*D*) *Phthalocyanine*			
Octabutoxy phthalocyanine	OBPC	–	~700[b]
(*E*) *Acridinium dyes*			
9-(4,4-Dimethylamino styryl) acridine	DMASA	4.9	618
9-(4,4-Dimethylamino cinnamyl) acridine	DMACA	4.6	640

Source: Data from Bishop (1972) and Apostolidis (2004)
[a]Dye anion
[b]Acid form of dye

nitrophenols, and phthaleins (see Table 15.6). Regarding the polymer matrix, one can say that, according to Apostolidis (2004), noncellulosic polymers can be considered as promising alternatives to the well-established EC-based carbon dioxide sensors.

The same approach can be used for design of NH_3 sensors. It was established that optical NH_3 sensors can be based on the determination of the change of pH inside the sensor matrix utilizing the pH-dependent response of a pH indicator (Wolfbeis and Posch 1986; Mills et al. 1995; Werner et al. 1995; Trinkel et al. 1997). Indicator dyes which can be used for designing optical NH_3 sensors are listed in Table 15.6. Apostolidis (2004) has shown that the indicator dyes bromophenol blue, bromocresol green, rhodamine B, and 9-(4,4-dimethylaminostyryl) acridinium perchlorate dissolved in various polymers are suitable for NH_3 detection, since they showed sensitivity to 50 ppm ammonia. Good results were obtained for materials based on ethyl cellulose and PVC, both plasticized with NPOE. Thus the combinations of these materials are promising for NH_3 gas sensor fabrication. According to Apostolidis (2004), the triphenyl methane dyes BPB and BCG performed best in various polymers. Experiments have shown that the same polymers, PSAN, PTBS, PVC, PS, and PMMA, are good polymer matrices for use in CO_2 gas sensors as well. They performed best when using phase transfer agent tetraoctylammonium hydroxide (TOA-OH) compared to the other bases investigated. Both NH_3

vapor and trace NH_3 dissolved in water can also be detected using bromocresol purple immobilized in porous silica (Tao et al. 2006). A silica optical fiber with a reagent phase consisting of $Ru(dpp)_3$ chloride encapsulated in a sol–gel structure is a basis for NO_2 fiber-optic sensor (Grant et al. 2000).

15.2 Metal Oxide-Based Nanocomposites

It should be noted that metal oxide-based colorimetric gas sensors are also designed mainly on the basis of composites because, in addition to metal oxides, they include noble metals as catalysts. Typical metal oxide-based composites used in optical sensors are listed in Table 15.7.

According to Ando et al. (2003), the enhancing effect in composite-based optical gas sensors can be of two types: (1) *absorption change of transition metal oxides* and (2) *plasmon absorption change of noble particles*. In *type 1*, the gas-sensitive optical absorption change comes simply from the transition metal oxide. The deposition of noble nanoparticles on some kinds of transition metal oxides remarkably enhances the catalytic activity for oxidation of flammable gases (Haruta 1997). Therefore, the gas-sensitive optical absorption change is enhanced by the enhancement of activity in catalytic oxidation of flammable gases. The optical gas-sensing performance of *type 1* is assumed to be closely related with the catalytic activity of the noble metal–transition metal oxide composite. In *type 2*, the gas-sensitive optical absorption change arises from the plasmon absorption change of small noble metal particles and is not related directly with the activity in catalytic oxidation of flammable gases. The state of plasmon resonance near the surface of small noble metal particles are sensitively influenced by the physical properties such as dielectric constant (ε) and refractive index (n) of the surrounding medium. Therefore, the gas-sensitive optical absorption change is enhanced if the noble metal particles are surrounded by a transition metal oxide whose ε and n largely change on exposure to flammable gases. The principle of this optical gas-sensing effect is shown in Fig. 15.4.

Features of the optical gas-sensing characteristics of the Au–transition metal oxide composites are summarized in Table 15.8. It was found that the combination of small Au particles with NiO film (Kobayashi et al. 1993; Ando et al. 1994, 1996) was effective in enhancing the optical CO sensitivity in the resulting Au–NiO composite film (type 1). In the case of the Au–Co_3O_4 film (Ando et al.

Table 15.7 Metal oxide-based composites used in optical gas sensors

Composite	Tested gas	References
Au–CuO	CO, H_2	Ando et al. (1995, 1997a, 2003)
Au–NiO	CO	Kobayashi et al. (1993), Ando et al. (1994, 1996)
Au–Co_3O_4	CO, H_2	Ando et al. (1997b)
Au–WO_3	H_2	Ando et al. (2001)
Au–$In_xO_yN_z$	CO, NO_2, H_2	Schleunitz et al. (2007)
Pd–WO_3	H_2	Smith et al. (2001, 2004)
Au–WO_{3-x}	NO	Deng et al. (2008)
Ag–WO_{3-x}		
Au–TiO_2–NiO	H_2S	Della Gaspera et al. (2010)
Au–TiO_2	Alcohol vapors	Manera et al. (2008)
	Vapor organic compounds	Fernandez et al. (2005)

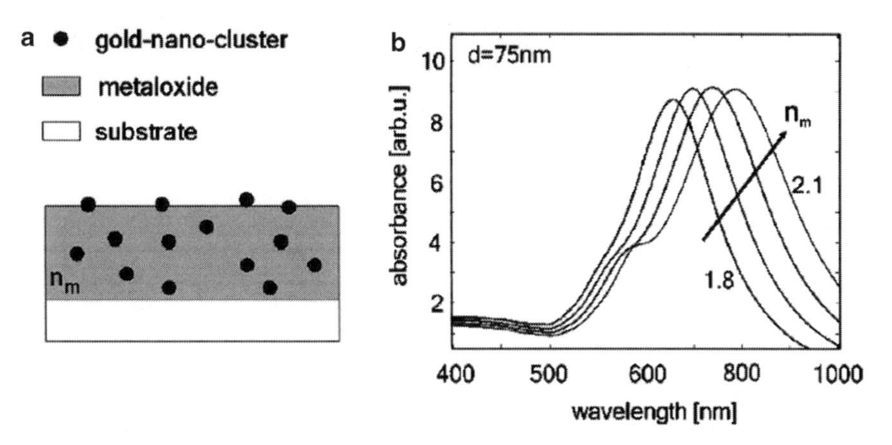

Fig. 15.4 Principle of the optical gas-sensing effect. (**a**) Schematic illustration of gold nanoparticles embedded in the volume and on the surface of a semiconducting metal oxide layer with refractive index n_m. (**b**) Shift of the absorption peak of a single gold nanocluster (75 nm in diameter) by a variation of the refractive index of the surrounding medium at exposure to a reducing or oxidizing gas (Reprinted with permission from Schleunitz et al. 2007, Copyright 2007 Elsevier)

Table 15.8 Optical sensitivities for CO and H_2 in the gold–transition metal oxide composite films

Sample	T_{oper} (°C)	λ_{max} of plasmon absorption (nm)	λ_{max} of CO sensitivity (ΔA) (nm)	λ_{max} of H_2 sensitivity (ΔA) (nm)	Type of optical gas effect[a]
Au–NiO	150–200	600	400, 700 (negative ΔA)	400, 700 (negative ΔA)	CO: 1, H_2: 1
Au–Co_3O_4	175–225	600	860 (negative ΔA)	870 (negative ΔA)	CO: 1, H_2: 1, 2
		610 (positive ΔA)		610 (positive ΔA)	
Au–WO_3	200–250	550	Insensitive	700 (negative ΔA)	H_2: 1
				550 (positive ΔA)	
Au–CuO	175–300	600	710 (positive ΔA)	710 (positive ΔA)	CO: 2, H_2: 2

Source: Reprinted with permission from Ando et al. (2003), Copyright 2003 Elsevier

[a]Type 1: enhancement of optical gas sensitivity of transition metal oxides by the enhancement of activity in catalytic oxidation of flammable gases; type 2: appearance of optical gas sensitivity arising from plasmon absorption change of small Au particles by flammable gases

1997b), the enhancement (type 1) in the optical sensitivity for CO and H_2 was found and, furthermore, the changes of plasmon absorption of small Au particles (type 2) appeared for H_2 but not for CO. This selectivity created a function to recognize CO and H_2 in the Au–Co_3O_4 film. The Au–WO_3 composite film (Ando et al. 2001) showed H_2-sensitive plasmon absorption change (type 2). CO- and H_2-sensitive plasmon absorption change was also observed for Au–CuO composite film (Ando et al. 1997a, 2003) (type 2).

Comparison of the CO-sensing performance of Au–transition metal oxide composite films carried out by Ando et al. (2003) has shown that the Au–CuO composite film shows better sensitivity and resolution than the Au–NiO (Kobayashi et al. 1993; Ando et al. 1994, 1996) and the Au–Co_3O_4 (Ando

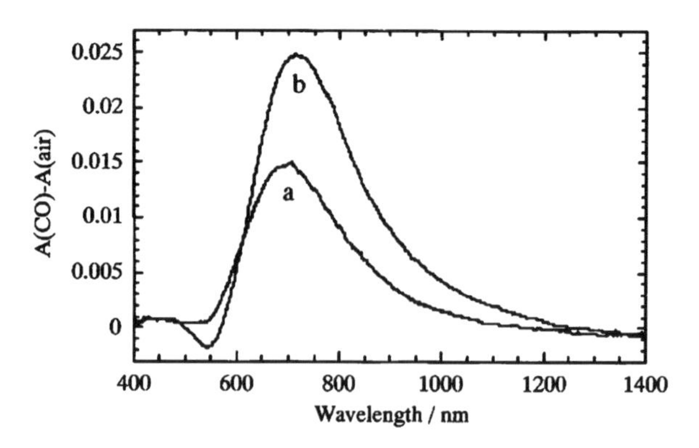

Fig. 15.5 Absorbance change (ΔA) of the Au–CuO composite film caused by CO as a function of wavelength: (**a**) by 1,000 ppm CO; (**b**) by 1 vol.% CO. Temperature: 300 °C (Reprinted with permission from Ando et al. 2003, Copyright 2003 Elsevier)

et al. 1997b) composite films. Absorbance change (ΔA) of the Au–CuO composite film caused by CO is shown in Fig. 15.5. However, it was found that the Au–CuO needs higher-operating temperature and shows slower recovery after removal of flammable gas from the atmosphere. Ando et al. (2003) speculated that the large CO sensitivity of Au–CuO film reflects the large change in ε and n of copper oxide by CO and that the stability of the reduced state of copper oxide may result in slow recovery after removal of CO from the atmosphere.

References

Amao Y (2003) Probes and polymers for optical sensing of oxygen. Mikrochim Acta 143:1–12

Amao Y, Nakamura N (2004) Optical CO_2 sensor with the combination of colorimetric change of α-naphtholphthalein and internal reference fluorescent porphyrin dye. Sens Actuators B 100:347–351

Ando M, Kobayashi T, Haruta M (1994) Enhancement in the optical CO sensitivity of NiO film by the deposition of ultrafine gold particles. J Chem Soc Faraday Trans 90:1011–1013

Ando M, Kobayashi T, Haruta M (1995) Optical CO detection by use of CuO/Au composite films. Sens Actuators B 24–25:851–853

Ando M, Zehetner J, Kobayashi T, Haruta M (1996) Large optical CO sensitivity of NO_2-pretreated Au-NiO composite films. Sens Actuators B 35–36:513–516

Ando M, Kobayashi T, Haruta M (1997a) Combined effects of small gold particles on the optical gas sensing by transition metal oxide films. Catal Today 36:135–141

Ando M, Kobayashi T, Iijima S, Haruta M (1997b) Optical recognition of CO and H_2 by use of gas-sensitive Au-Co_3O_4 composite film. J Mater Chem 7:1779–1783

Ando M, Chabicovsky R, Haruta M (2001) Optical hydrogen sensitivity of noble metal-tungsten oxide composite films prepared by sputtering deposition. Sens Actuators B 76:13–17

Ando M, Kobayashi T, Iijima S, Haruta M (2003) Optical CO sensitivity of Au–CuO composite film by use of the plasmon absorption change. Sens Actuators B 96:589–595

Apostolidis A (2004) Combinatorial approach for development of optical gas sensors: concept and application of high-throughput experimentation. PhD Thesis, University of Regensburg, Regensburg, Germany

Aubonnet S, Barry HF, von Bueltzingsloewen C, Sebattie JM, MacCraith BD (2003) Photo-patternable optical chemical sensors based on hybrid sol–gel materials. Electron Lett 39:913–914

Baldini F, Chester AN, Homola J, Martellucci S (2006) Optical chemical sensors. Springer, Dordrecht

Bishop E (1972) Indicators. Pergamon Press, New York

Carraway ER, Demas JN, Degraff BA, Bacon JR (1991) Photophysics and photochemistry of oxygen sensors based on luminescent transition-metal complexes. Anal Chem 63:337–342

Chan MA, Lawless JL, Lam SL, Lo D (2000) Fiber optic oxygen sensor based on phosphorescence quenching of erythrosin B trapped in silica-gel glasses. Anal Chim Acta 408:33–37

Della Gaspera E, Guglielmi M, Agnoli S, Granozzi G, Post ML, Bello V, Mattei G, Martucci A (2010) Au nanoparticles in nanocrystalline TiO_2-NiO films for SPR-based, selective H_2S gas sensing. Chem Mater 22(11):3407–3417

Deng H, Yang D, Chen B, Lin C (2008) Simulation of surface plasmon resonance of Au-WO_{3-x} and Ag-WO_{3-x} nanocomposite films. Sens Actuators B 134:502–509

Douglas P, Eaton K (2002) Response characteristics of thin film oxygen sensors, Pt and Pd octaethylporphyrins in polymer films. Sens Actuators B 82:200–208

Draxler S, Lippitsch ME, Klimant I, Kraus H, Wolfbeis OS (1995) Effects of polymer matrixes on the time-resolved luminescence of a ruthenium complex quenched by oxygen. J Phys Chem 99:3162–3167

Fernandez CJ, Manera MG, Spadavecchia J, Buso D, Pellegrini G, Mattei G, Martucci A, Rella R, Vasanelli L, Guglielmi M, Mazzoldi P (2005) Gold/titania nanocomposites thin films for optical gas sensing devices. Proc SPIE 5116:703–710

Florescu M, Katerkamp A (2004) Optimisation of a polymer membrane used in optical oxygen sensing. Sens Actuators B 97:39–44

Fuhr PL, Huston DR, MacCraith B (1998) Embedded fiber optic sensors for bridge deck chloride penetration measurement. J Opt Eng 37:1221–1228

Grant SA, Stacher JH, Bettencourt K Jr (2000) Development of sol–gel-based fiber optic nitrogen dioxide gas sensors. Sens Actuators B 69:132–137

Hartmann P, Trettnak W (1996) Effects of polymer matrixes on calibration functions of luminescent oxygen sensors based on porphyrin ketone complexes. Anal Chem 68:2615–2620

Haruta M (1997) Novel catalysis of gold deposited on metal oxides. Catal Surv Jpn 1:61–73

Jain TK, Roy I, De TK, Maitra A (1998) Nanometer silica particles encapsulating active compounds: a novel ceramic drug carrier. J Am Chem Soc 120:11092–11095

Klimant I, Wolfbeis OS (1995) Oxygen-sensitive luminescent materials based on silicone-soluble ruthenium diimine complexes. Anal Chem 67:3160–3166

Kobayashi T, Haruta M, Ando M (1993) Enhancing effect of gold deposition in the optical detection of reducing gases in air by metal oxide thin films. Sens Actuators B13–14:545–546

Lakowicz JR (1999) Principles of fluorescence spectroscopy, 2nd edn. Kluwer Academic/Plenum Press, New York

Lam MHW, Lee DYK, Man KW, Lau CSW (2000) A luminescent pH sensor based on a sol–gel film functionalized with a luminescent organometallic complex. J Mater Chem 10:1825–1828

Lee S-K, Okura I (1997) Optical sensor for oxygen using a porphyrin-doped sol–gel glass. Analyst 122:81–84

Li XM, Wong KY (1992) Luminescent platinum complex in solid films for optical sensing of oxygen. Anal Chim Acta 262:27–32

Lobnik A, Wolfbeis OS (1998) Sol–gel based optical sensor for dissolved ammonia. Sens Actuators B 51:203–207

Lobnik A, Oehme I, Murkovic I, Wolfbeis OS (1998) pH optical sensors based on solgels: chemical doping versus covalent immobilization. Anal Chim Acta 367:159–165

Lobnik A, Turel M, Urek SK (2012) Optical chemical sensors: design and applications. In: Wang W (ed) Advances in chemical sensors. InTech, New York, pp 3–28

McDonagh C, MacCraith BD, McEvoy AK (1998) Tailoring of sol–gel films for optical sensing of oxygen in gas and aqueous phase. Anal Chem 70:45–50

Mills A (1998) Controlling the sensitivity of optical oxygen sensors. Sens Actuators B 51:60–68

Mills A, Lepre A (1997) Controlling the response characteristics of luminescent porphyrin plastic film sensors for oxygen. Anal Chem 1997(69):4653–4659

Mills A, Thomas MD (1998) Effect of plasticizer viscosity on the sensitivity of an $[Ru(bpy)_3^{2+}(Ph_4B^-)_2]$-based optical oxygen sensor. Analyst 123:1135–1140

Mills A, Chang Q, McMurray N (1992) Equilibrium study on colorimetric plastic film sensors for carbon dioxide. Anal Chem 64:1383–1389

Mills A, Wild L, Chang Q (1995) Plastic colorimetric film sensors for gaseous ammonia. Mikrochim Acta 121:225–236

Mills A, Lepre A, Wild L (1997) Breath-by-breath measurement of carbon dioxide using a plastic film optical sensor. Sens Actuators B 38–39:419–425

Mills A, Lepre A, Wild L (1998) Effect of plasticizer-polymer compatibility on the response characteristics of optical thin film CO_2 and O_2 sensing films. Anal Chim Acta 362:193–202

Mohr GJ (2006) Polymers in optical sensors. In: Baldini F, Chester AN, Homola J, Martellucci S (eds) Optical chemical sensors, NATO series. Springer, Amsterdam, pp 297–322

Mohr GJ, Werner T, Oehme I, Preininger C, Klimant I, Kovacs B, Wolfbeis OS (1997) Novel optical sensor materials based on solubilisation of polar dyes in apolar polymers. Adv Mater 9:1108–1113

Pang Z, Gu X, Yekta A, Masoumi Z, Coll JB, Winnik MA, Manners I (1996) Phosphorescent oxygen sensors utilizing sulfur–nitrogen–phosphorus polymer matrixes. Adv Mater 8:768–771

Papkovsky DB (1995) New oxygen sensors and their application to biosensing. Sens Actuators B 29:213–218

Pauly S (1998) Permeability and diffusion data. In: Brandrup J, Immergut EH (eds) Polymer handbook, 4th edn. Wiley-VCH, New York

Ramamoorthy R, Dutta PK, Akbar SA (2003) Oxygen sensors: materials, methods, designs and applications. J Mater Sci 38:4271–4282
Reisfeld R (ed) (1996) Optical and electronic phenomena in sol–gel glasses. Springer, Berlin
Rossi LM, Shi L, Quina FH, Rosenzweig Z (2005) Stöber synthesis of monodispersed luminescent silica nanoparticles for bioanalytical assays. Langmuir 21:4277–4280
Sacksteder L, Demas JN, DeGraff BA (1993) Design of oxygen sensors based on quenching of luminescent metal complexes: effect of ligand size on heterogeneity. Anal Chem 65:3480–3483
Schleunitz A, Steffes H, Chabicovsky R, Obermeier E (2007) Optical gas sensitivity of a metaloxide multilayer system with gold-nano-clusters. Sens Actuators B 127:210–216
Shibata S, Taniguchi T, Yano T, Yamane M (1997) Formation of water-soluble dye-doped silica particles. J Sol-Gel Sci Technol 10:263–268
Smith RD II, Benson DK, Pitts RJ, Oison DL, Wildeman TR (2001) Diffusible weld hydrogen measurement by fiber optic sensors. Materialpruefung/Mater Test 43(1–2):26–29
Smith RD II, Pitts JR, Lee S-H, Tracy E (2004) Protective coatings for Pd-based hydrogen sensors. Prep Pap Am Chem Soc Div Fuel Chem 49(2):968–969
Tao S, Xu L, Fanguy JC (2006) Optical fiber ammonia sensing probes using reagent immobilized porous silica coating as transducers. Sens Actuators B 115:158–163
Trinkel M, Trettnak W, Reininger F, Benes R, O'Leary P, Wolfbeis OS (1997) Optochemical sensor for ammonia based on a lipophilised pH indicator in a hydrophobic matrix. Int J Environ Anal Chem 67:237–251
Von Bueltzingsloewen C, McEvoy AK, McDonagh C, MacCraith BG, Klimant I, Krause C, Wolfbeis OS (2002) Sol–gel based optical carbon dioxide sensor employing dual luminophore referencing for application in food packaging technology. Analyst 127:1478–1483
Wang Z, McWilliams AR, Evans CEB, Lu X, Chung S, Winnik MA, Manners I (2002) Covalent attachment of RuII phenanthroline complexes to polythionylphosphazenes: the development and evaluation of single-component polymeric oxygen sensors. Adv Funct Mater 12:415–419
Weigl BH, Wolfbeis OS (1995) New hydrophobic materials for optical carbon dioxide sensors based on ion pairing. Anal Chim Acta 302:249–254
Weigl BH, Holobar A, Rodriguez NV, Wolfbeis OS (1993) Chemically and mechanically resistant carbon dioxide optrode based on a covalently immobilized pH indicator. Anal Chim Acta 282:335–343
Werner T, Klimant I, Wolfbeis OS (1995) Ammonia-sensitive polymer matrix employing immobilised indicator ion pairs. Analyst 120:1627–1631
Wolfbeis OS (1991) Fiber optic chemical sensors and biosensors, vol 1. CRC Press, Boca Raton, FL
Wolfbeis OS (ed) (1992) Fiber optic chemical sensors and biosensors, vol 2. CRC Press, Boca Raton, FL
Wolfbeis OS (2005) Materials for fluorescence-based optical chemical sensors. J Mater Chem 15:2657–2669
Wolfbeis OS, Posch HE (1986) Fiber-optic fluorescing sensor for ammonia. Anal Chim Acta 185:321–327
Wolfbeis OS, Weis LJ, Leiner MJP, Ziegler WE (1988) Fiber-optic fluorosensor for oxygen and carbon dioxide. Anal Chem 60:2028–2030
Xu X, McDonough RC, Langsdorf B, Demas JN, De Graff BA (1994) Oxygen sensors based on luminescence quenching: interactions of metal complexes with the polymer supports. Anal Chem 66:4133–4141
Zhang H, Lei B, Mai W, Liu Y (2011) Oxygen-sensing materials based on ruthenium(II) complex covalently assembled mesoporous MSU-3 silica. Sens Actuators B 160:677–683

Chapter 16
Nanocomposites in Electrochemical Sensors

16.1 Solid Electrolyte-Based Electrochemical Sensors

Electrochemical gas sensors represent another field for nanocomposites application. Moreover, it should be stated that almost all solid electrolyte sensors have been designed based on composites. For example, YSZ (ZrO_2–Y_2O_5), NASICOM ($Na_3Zr_2Si_2PO_{12}$), and Na-β-alumina (Na_2O–Al_2O_3 system), the main sensing materials in solid electrolyte gas sensors, can be considered as composites, which have high ionic conductivity due to their specific composition (Nakayama and Sadaoka 1994). $Gd_{0.7}Ca_{0.3}CoO_{3-\delta}$ (GCC) and $Ce_{0.8}Gd_{0.2}O_{1.9}$ (CGO) (Nigge et al. 2002) tested as oxygen-permeable membranes in an amperometric sensor for NO_x detection in exhaust gases, $A_{0.7}E_{0.3}MnO_3$ (A=Gd, Y, and Pr and E=Ca and Sr) (Wiemhofer et al. 2002) used for oxygen sensors, and $BiCuVO_x$ ($Bi_2Cu_{0.1}V_{0.9}O_{5.35}$; oxygen-ion conductor) fitted with perovskite-type oxide ($La_{0.6}Sr_{0.4}Co_{0.8}Fe_{0.2}O_3$; mixed electro and ionic conductor) for fabrication mixed-potential-type gas sensors of volatile organic compounds (Kida et al. 2009) are composites as well. Kida et al. (2009) showed that $La_{0.6}Sr_{0.4}Co_{0.8}Fe_{0.2}O_3$ and $BiCuVO_x$ composite materials have good stability against humidity and CO_2. The high ionic conductivity, enhanced mechanical strength, and extensive possibilities of object-oriented control of the electrolyte properties through varying the conductance type and dopant concentration are advantages which make solid electrolyte composites very good candidates for application in electrochemical gas sensors.

It is important that more complicated ZrO_2-based composites have better performance than mono- or two-phase ones, including electrical, electrochemical, and mechanical properties and thermal stability, and therefore they can be used for fabricating oxygen sensors designed for operation at temperatures up to 1,600 °C, needed, for example, during steel production for controlling the level of oxygen dissolved in the melt (Fray 1996; Liu 1996). At present commercial oxygen sensors intended for application in this temperature range are usually based on magnesia-partially stabilized zirconia (Mg-PSZ) electrolytes. The main reason for using Mg-PSZ solid electrolytes in commercial oxygen sensors, besides their high emf signal, is their high thermal shock resistance. However, in the range of low oxygen potentials (<200 ppm) at high temperatures, the electronic conduction of Mg-PSZ electrolyte introduces errors in the emf values (Liu 1996). Experiments have shown that the $x(8MgSZ)+(1-x)(3YSZ)$ composites show thermal shock resistance and electrical conductivity values suitable for high-temperature oxygen gas detection (Caproni et al. 2008). It was established that the addition of 3YSZ (zirconia: 3 mol% yttria) improves the total conductivity of the composite. This is important because the better the electrical conductivity the larger the signal response to oxygen. Consequently, lower values of oxygen levels could be detected. In addition, 3YSZ added to 8MgSZ (80–20 wt%) suppresses the electronic contribution to the electrical conductivity at 620 °C.

G. Korotcenkov, *Handbook of Gas Sensor Materials*, Integrated Analytical Systems,
DOI 10.1007/978-1-4614-7388-6_16, © Springer Science+Business Media New York 2014

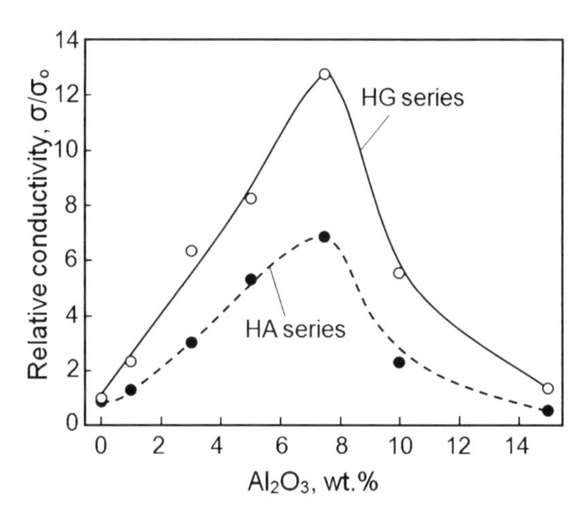

Fig. 16.1 Changes of the relative total conductivity (σ/σ_o, σ_o—conductivity of pure YSZ matrix) for HG and HA series as a function of alumina content at 400 °C. Materials originating from gamma-alumina powder were marked as HG series, while those prepared from alpha-alumina powder as HA series (Data from Bu ko 2000)

Experiment has shown that oxide ionic conductors in the system cubic calcia stabilized zirconia solid electrolytes, calcium zirconate, also seem to be promising solid electrolytes for application in electrochemical probes for controlling oxygen dissolved in molten steel. The ionic conduction limit for electrolytes based on $CaZrO_3$ is lower than that for calcia-stabilized zirconia (13CSZ). Hence, $CaZrO_3$-based materials perform better at low oxygen concentration in molten alloys (Dudek 2008a).

It was established that the introduction of Al_2O_3 additives into cubic yttria–zirconia solid solution (8YSZ) matrix also led to the improvement of electrical and mechanical properties compared to pure 8YSZ (Oe et al. 1996; Bu ko 2000). Results of Al_2O_3 content influence on the conductivity of Al_2O_3–8YSZ composite are shown in Fig. 16.1. It is seen that the maximum of conductivity has the samples containing up to 7–9 wt% of Al_2O_3. The maximum values of fracture toughness K_{Ic} was reached in the same concentration range (5–10 %). Guo (2003) summarized and analyzed the positive and negative effects of Al_2O_3 additions on the electrical properties of ZrO_2 and showed that the increase in bulk resistivity takes place mainly when Al_2O_3 content is over the solubility limit because of the formation of defect associates and insulating of ZrO_2 grains by Al_2O_3 second-phase particles. Within the solubility limit of Al_2O_3 in ZrO_2, alumina scavenges the silica-reach phase from grain boundaries, thereby decreasing the grain boundary resistivity (Bu ko 2000). The $Nd_2Ti_2O_7$ secondary phase was also able to coexist with 8YSZ matrix, and the fracture toughness K_{Ic} of 8YSZ ceramics was also significantly improved by $Nd_2Ti_2O_7$ addition (15 mol%) (Liu and Chen 2005).

It is known that ceria is a promising solid electrolyte for application in solid electrolyte sensors for monitoring hydrocarbons and other exhaust gases at intermediate-temperature range 600–800 °C (Mukundan et al. 1999). The primary problem encountered in using ceria-based electrolytes for gas sensors is the partial reduction of ceria in reducing atmospheres (Inaba and Tagawa 1996; Abrantes et al. 2003; Sameshima et al. 2006). It has been reported that the reduction of ceria can be neglected at a lower temperature—around 600–700 °C. However, such low temperatures are not suitable for using singly doped ceria, $Ce_{1-}xMxO_2$ (M=Sm, Gd, Y where x=0.15–0.20), as solid electrolyte due to the high resistivity of this material (Doshi et al. 1999; Matsui et al. 2005). Research has shown that a structural modification of ceria solid solutions is one of the possible ways to improve their electrical conductivity (Herle et al. 1999; Dudek 2008b) and stability. Co-doped ceria of $Ce_{0.85}Gd_{0.15-}xSmxO_2$, where $0.05 \leq x \leq 0.1$, showed much higher ionic conductivity at 500–700 °C. Thus, these materials seem to be more suitable electrolyte materials for application in gas sensors (Wang et al. 2004b). Maricle et al. (1992) also reported that, due to co-doping small quantities of praseodymium in $Ce_{1-}xGdxO_2$ solid solution, the application region is shifted by two orders of magnitude to lower oxygen partial pressure. The co-doping ceria with calcia and samaria also leads to an improvement in their electrolytic

Table 16.1 Typical examples of mixed-potential-type gas sensors utilizing NASICON and different oxide electrodes

Gas	Sensor structure Air, RE | electrolyte | SE, target gas	Sensitivity, mV/decade	Gas conc. (ppm)	T_{oper} (°C)
Cl_2	Air, Au | NASICON | Au–$Cd_3O_2SO_4$, Cl_2 (+air)	−392	1–10	200
Cl_2	Air | NASICON | Au | NASICON | Au–Cr_2O_3, Cl_2 (+air)	−270	1–50	300
Cl_2	Air, Au | NASICON | $CaMg_3(SiO_3)_4$–CdS, Cl_2 (+air)	−392	1–10	200
Cl_2	Air, RuO_2 | NASICON: (40 %) (Na_2O–Al_2O_3–$4SiO_2$) | RuO_2–NaCl, Cl_2 (+air)	100	1–10	400
H_2S	Air, Au | NASICON | Au–Pr_6O_{11}–SnO_2, H_2S (+air)	74	5–50	300
SO_2	Air, Au | NASICON | Au–V_2O_5–TiO_2, SO_2 (+air)	−78	1–50	300
NH_3	Air, Au | NASICON | Au–Cr_2O_3, NH_3 (+air)	−89	50–500	350
NO_2	Air, Pt | NASICON | Au–NiO, NO_2 (+air)	78	5–200	350
NO	Air, Au | NASICON | Au–NiO–WO_3, CO (+air)	70	5–500	350
CO	Air, Pt | NASICON | A–NiO–Fe_2O_3, CO (+air)	−45	10^2–10^3	350
CO	Air, Au | NASICON | Au–Y_2O_3, CO (+air)	−45	5–50	400
C_7H_8	Air, Au | NASICON | Au–Sm_2O_3, C_7H_8 (+air)	−75	5–50	350
NH_3/C_7H_8	NH_3(+air), Cr_2O_3– Au | NASICON | Au, Air, Au | NASICON | Au–ZnO–TiO_2, C_7H_8 (+air)	−91/−60	50–500/−50	350

Source: data from Liang et al. (2012), etc.
RE reference electrode, *SE* sensing electrode

properties when compared to only samaria-doped ceria (Liu et al. 2005; Dudek and Ziewiec 2006). It was also established that composite layered ceramics involving $Ce_{0.8}Sm_{0.2}O_2$–$Bi_{0.8}Eb_{0.2}O_2$ or $Ce_{0.9}Gd_{0.1}O_2$–$BaCe_{0.8}Y_{0.2}O_3$–$Ce_{0.9}Gd_{0.1}O_{1.95}$ system exhibited better electrolytic stability in gas atmospheres with low oxygen partial pressure at temperatures 600–800 °C (Wachsman et al. 1997).

Sensing electrodes and auxiliary electrodes are also composites in many cases (see Chap. 15 (Vol. 1)). The use of composites makes a real improvement to the operating characteristics of solid electrolyte gas sensors. As indicated in Chap. 15 (Vol. 1), gas diffusion electrodes (GDEs) from composites, such as noble metal/metal oxide (Guth and Zosel 2004), noble metal/carbon (Sakata et al. 2007), or metal oxide/metal oxide (Zhong et al. 2011) provide the increase of conductivity, gas penetrability, stability, catalytic activity of electrodes, and the number of the triple-phase boundaries (TPB) necessary for effective operation of electrochemical sensors (see Chap. 15 (Vol. 1)). As a result, a dramatic increase of the limiting current and shortening of the response time can be obtained. In addition, the variation in structure of composites used in sensing and auxiliary electrodes makes possible a high selectivity to the required analyte. The application of composites also greatly improves the stability of electrodes, especially those based on noble metals. For example, Westphal et al. (2001) have shown that the electrode based on Au–(20 %) Ga_2O_3 composite can be treated in reducing gases at temperatures of 850 °C without changing its characteristics.

Of course, due to the complexity of the task of electrochemical sensor design and the variety of materials and technologies which can be used, the results obtained in various labs have been different. For example, according to Guth and Zosel (2004), maximum sensitivity to NO_x was observed for gas symmetrical mixed-potential YSZ-based sensors with electrodes made from Au/Nb_2O_5 composites (T_{oper}=700 °C). At the same time, Zhong et al. (2011) found that, among the various single-oxide or composite oxide SEs tested, NiO–WO_3 composite provided the highest sensitivity of NASICOM-based sensors to NO at 350 °C. The ΔEMF value to 500 ppm NO was as high as about 140 mV. The sensor device using NiO–WO_3–SE showed a fast response–recovery rate to NO and excellent selectivity over the other interference gases. Shimizu and Yamashita (2000) believe, however, that a NASICOM-based mixed-potential-type NO_x sensor should have $Pb_2Ru_{1.9}V_{0.1}O_{7-}z$ composite as sensing electrode. For CO_2 sensors, Shimizu and Yamashita (2000) proposed using $La_{0.8}Ba_{0.2}CoO_3$ composites as sensing electrode. NASICOM-based sensors with such an electrode had improved stability and moisture resistance, but the sensitivity needs to be improved. Examples of other metal oxide composites used for fabrication of NASICOM-based gas sensors are listed in Table 16.1. We also need to take into account that the differences in results obtained by various teams can be due to the difference in operating temperatures of the sensors designed. In general, the YSZ-based sensors operate at high temperature (600–800 °C), while the mixed-potential-type sensors based on NASICON are generally used at intermediate temperatures (300–500 °C).

The use of composites leads to high temporal stability, high thermal stability, and high conductivity of ionic salts. The application of Li_2CO_3–$LiNbO_3$ (Singh et al. 2002) and Na_2CO_3–$BaCO_3$ or Li_2CO_3–$BaCO_3$ (Miura et al. 1992; Kida et al. 2001) composites in potentiometric electrochemical CO_2 sensors as electrolyte or auxiliary electrode improves overall performance relative to the sensor based on pure carbonate, including a decrease in the sensitivity to moisture and shortening of response time. Bhoga and Singh (1999) have shown that a galvanic CO_2 gas sensor using an optimized composite electrolyte (50% glass-dispersed Na_2CO_3) also has improved parameters. The dispersion of [$40Na_2O$:$50SiO_2$:$10B_2O_3$] glass in a polycrystalline Na_2CO_3 matrix not only enhances the ionic conductivity of pure Na_2CO_3 but also improves the mechanical integrity and sinterability, leading to better performance and imperviousness to thermal shocks. Na_2SO_4–$BaSO_4$ composites as auxiliary electrodes showed excellent stability in SO_2 sensors (Min and Choi 2003). The equilibrium-potential-type NO_2 sensor with indium tin oxide (ITO) and $NaNO_2$–Li_2CO_3 as the auxiliary electrode had a low detecting limit (about 2 ppm) and an excellent moisture resistance (Obata and Matsushima 2008). Other information about solid electrolyte sensors based on composites of ionic salts may be found in Chap. 6 (Vol. 1).

The review papers of Knauth (2000), Dudek (2008a), and Uvarov et al. (2010), who analyzed in details the ion conductor composites, including solid electrolytes of the type "ionic salt oxide" are recommended reading. It was shown that to improve electrical, thermodynamic, and mechanical properties as well, a dispersed heterophase can be introduced into a solid electrolyte matrix. Since the first experiments performed by Liang (1973) on LiI–Al_2O_3 composites, numerous studies have been performed on various solid electrolyte composites. The effect of conductivity increase was found in many systems based either on halides (e.g., AgCl, BaF_2, CaF_2) or oxides (ZrO_2, Ca-β-Al_2O_3) with alumina, titania, or silica inclusions (Shai and Wagner 1982; Fuijtsu et al. 1985; Vaidehi et al. 1986; Jacob and Shukla 1987; Bu ko and Róg 1995). Generally, in the composite material, the ionic conductivity increases strongly with the concentration of the dispersed phase and then decreases when after reaching maximum. For example, the addition of 2.5 mol% dispersed Al_2O_3 to CaF_2 caused an ionic conductivity enhancement up to two orders of magnitude, depending on the preparation technique, but a conductivity decrease was observed for samples with more than 5 % alumina (Rog et al. 1998). The same regularity for LiI–Al_2O_3 system is shown in Fig. 16.2. An ionic conductivity enhancement was also reported for mixtures with other Li-ion conductors. In the Li_2SO_4–Li_2CO_3 system (Singh and Bhoga 1990), the ionic conductivity maximum lay close to the eutectic concentration, where the grain size had a minimum. Brosda et al. (1996) described composites of M_2SO_4 (M=Na, K) with Al_2O_3 and also mixtures M_2CO_3–$BaCO_3$ and M_2SO_4–$BaSO_4$. Besides the enhancement in electrical conductivity, the mixed and heterogeneous doping led to both better sinterability and therefore mechanical strength and improved thermal shock resistance. In the M_2SO_4–$BaSO_4$ system, which presents a domain of solid solubility (dissolution of $BaSO_4$ up to 5 mol%), the conductivity enhancement could be interpreted by mixed homogeneous and heterogeneous doping. The maximum increase lay in the two-phase region.

To interpret the results of conductivity measurements in those composites, several theoretical models were proposed. All of them assumed interactions between mobile charged defects and the surface of the dispersed inert phase, leading to a space charge region with defect concentrations differing from the intrinsic bulk concentration (Wagner 1980; Dudney 1985; Meier 1987; Jamnik and Meier 1999; Bunde and Dieterich 2000; Knauth 2000). Uvarov et al. (1992) presented a model that considered the surface conductivity and the microstructure of composite solid electrolytes, and he used it to explain their electrical conduction behavior. He showed that enhancement in electrical conductivity for such composites is strongly dependent on the sample preparation conditions, and higher conductivities can be expected if a better contact between the solid electrolytes and the dispersoid can be achieved. Without doubt the approach to the design of ion conductor composites discussed above is promising for application in electrochemical gas sensors.

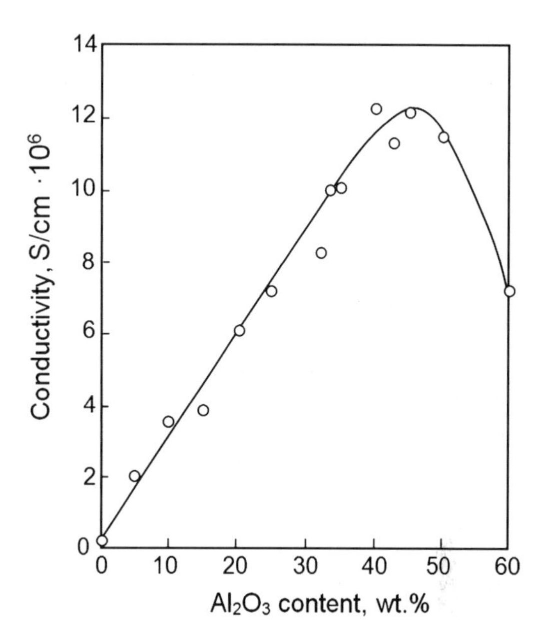

Fig. 16.2 Conductivity of $LiI–Al_2O_3$ composites as function of the alumina concentration (Reprinted with permission from Liang 1973, Copyright 1973 electrochemical society)

16.2 Electrochemical Sensors with Liquid Electrolyte

It is clear that composites in electrochemical gas sensors with liquid electrolytes can be applied only in electrodes and membranes. It was established that the preparation of hybrid composite materials has led to the achievement of modified electrode surfaces that exhibit special properties due to the synergic effect from the individual components.

16.2.1 Polymer-Modified Electrodes

We need to recognize that conducting polymer-modified electrodes have for many years been one of the most preferred approaches for the preparation of electrochemical sensors (Wang 2006). For example, the GDEs (see Chap. 15 (Vol. 1)) in electrochemical gas sensors, which are the composite materials, combine the functions of catalyst, ion and electron conductor, and membrane penetrable for gas and products of reaction (Sundmacher et al. 2005). Optimal electrode design requires a perfectly executed balance of the various above-mentioned functions. This is often achieved by preparing mixtures of ion-conducting particles (made of the membrane material, usually grained Nafion), particles of an electron conductor, and catalytic particles that are the same as the metal or different. In the indicated electrodes, Pt is usually used as catalyst and carbon as conductor. By using well-defined particle size distributions and polymer/Pt/C composition, one can adjust the electrode pore structure. The combination of the well-known characteristics of conducting polymers (good stability, reproducibility, high number of active sites, strong adherence, and homogeneity in electrochemical deposition) with characteristics of the above-mentioned materials can lead to an improved performance of the resulting sensing devices (Hong and Oh 1996; Santhosh et al. 2007). In particular, this in turn makes it possible to optimize the transport properties of the GDE with respect to the noncharged reactants. In the sensors with GDE, the three-phase boundary sites are well developed, even in the deeper region of the membrane. In addition, because carbon is conductive, the electrode can achieve an optimum combination of such properties as conductivity–porosity; carbon provides good electrical contact between the grains of the noble metal in the porous matrix (Cao et al. 1992). The incorporation of metal nanoparticles into electrode matrices

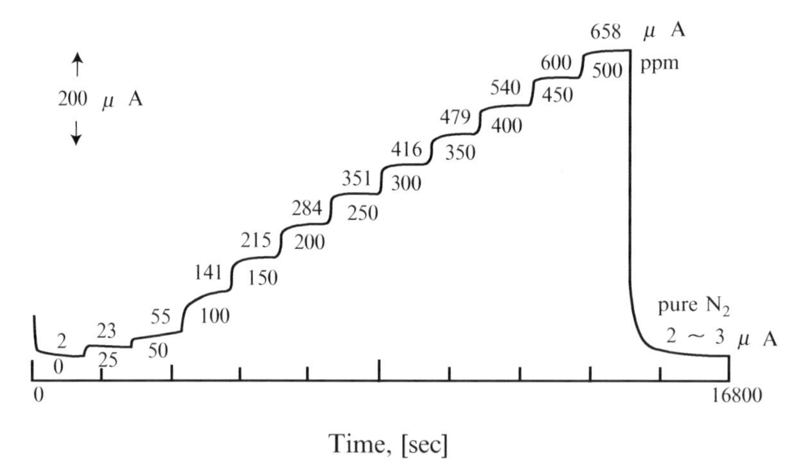

Fig. 16.3 Amperometric response of the gold/PTFE/carbon gas diffusion electrode to the addition of SO_2. After the last steady-state current has been reached, N_2 is blown onto the reverse side of the electrode. Constant electrode potential: +600 mV (vs. Ag/AgCl) (Reprinted with permission from Chiou and Chou 1996, Copyright 1996 Wiley)

enhances the electrocatalytic efficiency of many electrochemical processes taking place in gas sensors (Hrapovic et al. 2006). As a result, the catalytically active surface area of the electrode can be several orders of magnitude higher in comparison with standard electrodes, allowing species with relatively poor electroactivity to produce measurable currents. It was established that the current sensitivities of the Pt/Nafion/Pt composite electrode sensors were well correlated with the effective surface area of the Pt deposits (Hong and Oh 1996). For preparing composite-based electrodes, various technologies, including electrodeposition, pasting, magnetron sputtering, or screen printing, can be used. Recently, magnetron sputtering is the most popular method, which makes it possible to deposit carbon/noble-metal nanocomposite films on porous Teflon substrates at room temperature by sputtering of a graphite/noble-metal target in Ar (Okamoto et al. (1997); Baranov et al. (2007)). Of course, the method of electrode fabrication influences sensor parameters (Hong and Oh 1996; Baranov et al. 2007).

Note that besides Pt, carbon black, and Nafion-based composites, other composites with CNTs, metal oxides, and metals have been also reported (Wang 2006; Hrapovic et al. 2006). For example, experiment has shown that instead of Pt nanoparticles, other metal nanoparticles can also be used in nanocomposites designed for application in electrochemical gas sensors (Wang and Hu 2009). It was found that the application of different kinds of metallic nanoparticles integrated into different organic matrices allows the tailoring of controllable interactions with variable vapor materials and the development of sensors with specific functions. In fact, the incorporation of semiconductor and noble-metal nanoparticles into membrane electrodes can rapidly increase their catalysis effect and the sensitivity of relative sensors. For example, Chiou and Chou (1996) developed a composite-based GDE for SO_2 sensing. The electrode was a composite material polytetrafluoroethylene (PTFE)/carbon with gold as dispersed catalyst. Gold particles catalyzed the electrochemical oxidation of SO_2 when this gas diffused through the porous working electrode. The electrolyte (1 M H_3PO_4) filled the space between the working and counter electrodes, while the reference electrode was fixed near the working electrode in the electrolyte. This SO_2 sensor resulted in a stable device with a very fast response time (see Fig. 16.3). Okamoto et al. (1997) have also shown that gold–carbon composite, prepared by plasma sputtering and supported on gas-permeable PTFE membranes, is an excellent candidate for a thin-film working electrode of PH_3 electrochemical gas sensor. The sensors could detect PH_3 in the concentration range from 0.1 to 1.0 ppm.

Zhu et al. (2002) reported about a sensitive, selective, and stable NO microsensor with an electrode which was modified by nano-Au colloid supported on Nafion. A low detection limit, high selectivity,

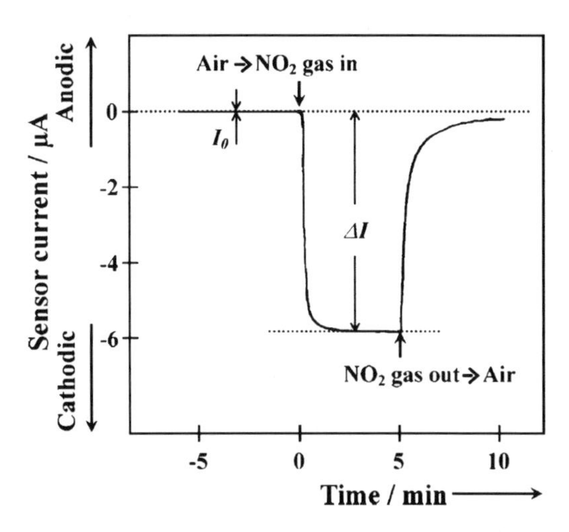

Fig. 16.4 Typical response curve obtained by the Au/C–F sensor for the air and NO_2 (3 ppm). Electrode potential, −100 mV vs. RE in the sensor. The amperometric sensor is constructed as a three-electrode cell system with 3 cm^3 of a 5 M H_2SO_4 aqueous solution (Reprinted with permission from Mizutania et al. 2005, Copyright 2005 Elsevier)

and sensitivity for NO determination could be obtained using this microsensor. Do and Chang (2001) also studied Au/Nafion-based working electrodes for NO_2 sensors and found that the use of PAn/Au/Nafion electrodes instead of Au/Nafion increases sensitivity to NO_2. Polyaniline (Pan) was prepared on Au/Nafion in 0.1 M aniline and 1.0 M $HClO_4$ aqueous solution by the constant potential electrochemical polymerization method. However, the response time of an NO_2 gas sensor for using PAn/Au/Nafion as a working electrode was longer than that when using Au/Nafion as a working electrode. The response time for sensors with an Au/Nafion electrode was about 5 min. Yu et al. (2003) prepared a novel sensor by infiltrating 4-(dimethylamino) pyridine-stabilized gold nanoparticles (DMAP-AuNP) into polyelectrolyte (PE) multilayers preassembled on ITO electrodes. The results testified that gold nanoparticles in the PE multilayers showed high electrocatalytic activity to the oxidation of NO. The sensitivity of the composite films for measuring NO could be further tailored by controlling the gold nanoparticle loading in the film. Li et al. (2002) have found that nano-Au-assembling on Pt electrode can improve the parameters of SO_2 sensors as well. Au colloid particles were 16 nm in size. It was established that in optimal conditions for determining SO_2, 1 M NaOH as internal electrolyte and +0.6 V as applied potential, other gases, such as CO, NO, NH_3, and CO_2, did not cause interference.

Subsequently, Pt–Fe(III) nanoparticles, carbon nanotube-gold nanoparticles, and copper nanoparticles were also used to design NO electrochemical sensors (Wang et al. 2004a; Wang and Lin 2005; Zhang and Oyama 2005; Polsky et al. 2006).

Polymer membranes in electrochemical sensors can also be composites. For example, Mizutania et al. (2005) have shown that the sensor with GDE using composite carbon black–fluorocarbon (C–F) membrane had a higher sensitivity to NO_2 and selectivity of NO_2 against ozone than that using conventional expanded PTFE (Nafion) membrane. The gold black electrodes (ca. 100 μm thick) were used for a reference (RE) and a counter (CE) electrode in the sensor. The superior characteristic of the C–F sensor could be due to the high gas permeability and the catalytic nature of the C–F membrane. A typical response curve for this sensor is shown in Fig. 16.4. Stetter and co-workers (Stetter and Maclay 2008) have shown that the CNT/PTFE-based composite also has high catalytic activity and can be used as working electrode for the detection of NO_2 and H_2S (see Fig. 16.5). Carbon nanotubes can greatly increase the surface area of the TPB because the carbon nanotubes have very high surface area, e.g., single-walled carbon nanotubes can achieve 1,600 m^2/g (Cinke et al. 2002). The CNT/PTFE working electrode was coupled with two Pt/TFE composite electrodes as a counter (CE) and reference electrode (RE), respectively, in a commercial amperometric gas sensor configuration (Roh and Stetter 2003). The CNT loading was 0.3 mg/cm^2 and the electrolyte was 4 M H_2SO_4. The catalyst is deposited

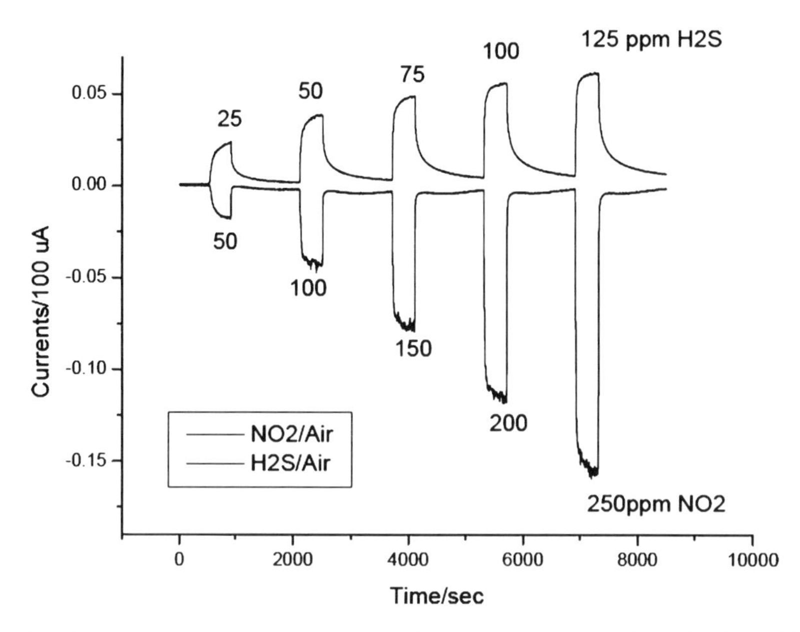

Fig. 16.5 CNT/PTFE sensor responses with 25, 50, 75, 100, and 125 ppm of H_2S/air and 50, 100, 150, 200, and 250 ppm NO_2 air at 0 mV vs. Pt/air-QRE (Data from Stetter and Maclay 2008)

electrochemically on the membrane in order to get a nanostructured surface and to increase the TPB area. This approach made it possible to improve sensitivity from ppm to ppb gas concentration levels.

16.2.2 Carbon–Ceramic Electrodes

Carbon–ceramic electrodes (CCEs) are other group of composite-based materials important for gas sensor applications. These materials have seen great development during the last decade (Rabinovich et al. 1999; Sun et al. 2007; Arguello et al. 2009; Skeika et al. 2011). The CCEs were firstly described by Lev and co-workers in 1994 (Tsionsky et al. 1994). These electrodes are basically constructed by doping a silica matrix obtained by the sol–gel method (Sakka 2003) with powdered carbon such as graphite, or other carbon materials (carbon nanotubes and glassy carbon) (Lev et al. 1997; Rabinovich and Lev 2001). In particular, Tsionsky and Lev 1995a, b prepared CCE by mixing carbon powders with sol–gel precursors, such as acidic solutions of water and tetraalkoxysilane, organotrialkoxysilane, or another organofunctional alkoxysilane. Schematically, CCEs are shown in Fig. 16.6. Similar to GDEs discussed in the previous section, the porous structure of the electrode material permits high gas permeability. Their hydrophobic surface rejects water, leaving only a very thin layer at the outermost surface in contact with the electrolyte, thus minimizing the effects of liquid side mass transfer. The carbon powder provides electric conductivity, and the catalyst guarantees selectivity and sensitivity.

The advantage of using these materials compared to other carbon-based electrodes is the combination of the sol–gel process properties (such as high surface area) and conductivity of the carbon materials, thus enabling one to obtain a renewable surface electrode similar to a carbon paste electrode, but more robust and with higher stability (Rabinovich and Lev 2001; Zou et al. 2008). An alternative to increasing the application of CCE as electrochemical sensors is the modification of these materials using electron mediator species (Tsionsky and Lev 1995a, b; Lei et al. 2004; Zheng et al. 2007; Arguello et al. 2009). Thus, the electrode can be modified by appropriate selection of the catalyst, infinite possible organic modifications on the silica backbone, choice of the type of carbon or metal powder (or fibers), and any combination of these building blocks (Tsionsky and Lev 1995a, b).

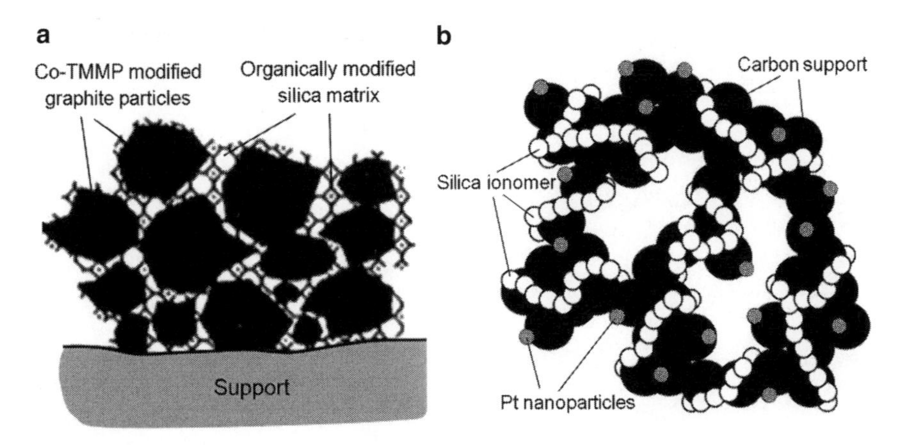

Fig. 16.6 Carbon-ceramic electrodes (CCE) presented schematically (**a**) by Tsionsky and Lev (1995a, b) and (**b**) by Anderson et al. (2002)

These mediators can provide new interesting features to these electrodes, such as the shift of peak potential of the analyte species to less positive potentials, thus increasing the sensitivity of the electrode (Lei et al. 2004; Arguello et al. 2009). CCEs can be easily prepared (by virtually one-step molding) and offer considerable versatility in tailoring electrode characteristics.

According to Lev and co-workers (Rabinovich and Lev 2001; Lev et al. 2005) there are three types of gas electrodes that can be attained by appropriate manipulation of sol–gel composition. The first type (unwetted) presents a hydrophobic blank CCE that is prepared by mixing methyltrimethoxysilane and graphite powders, casting the electrode in a short insulating tube. The methyl group entails high hydrophobicity and prevents the penetration of the electrolyte into the bulk electrode; the hydrophobic graphite grains provide electron percolation conductivity; and the siloxane network acts as a porous binder. The highly hydrophobic CCE prevents water penetration and assures the only outer section of the electrode will be wetted by the electrolyte. However, these electrodes exhibit poor electrocatalytic activity for oxygen reduction. Incorporation of a small amount of palladium chloride into the sol–gel precursor and its subsequent reduction after gel formation generated electrodes of the second type of CCE (flooded) (Rabinovich et al. 1999). Here, the catalytic activity for oxygen reduction is improved due to the presence of metallic palladium. Additionally, the presence of hydrophilic palladium oxide reduces the hydrophobicity of the electrode and increases its wetted section somewhat. Thus, a controlled thickness section of up to 100 μm of the electrode is wetted by the electrolyte. The incorporation of hydrophilic additives such as poly(ethylene glycol) or small amount of TMOS into the precursors of hydrophobic CCEs also endows a degree of hydrophilicity and a controlled section of the electrode can be wetted by the electrolyte. The wetted section is very stable and remains constant even after several weeks in an electrolyte (Lev et al. 1997). The third type of CCE (channeled) has a configuration where thin channels of electrolyte penetrate deep into the electrode and wet the carbon particles, thereby increasing the gas/liquid interface. Channeled CCEs can be prepared by impregnation of the graphite with palladium chloride, reducing it by a hydrogen stream and only then mixing the modified grains with the sol–gel precursors. In this case, the hydrophilic palladium (and palladium oxide) forms electrolyte channels that go deep (up to approximately 0.5 mm) into the CCE (Rabinovich et al. 1999).

Flooded type CCE are most suited for applications involving mass transport from/to the electrolyte and from/to the gas side, while channeled electrodes are best for situations where the gas transport is the sole rate determining step (Lev et al. 2005). The electrodes can be used as supported films spread on a porous substrate (where the gas flows through this substrate into the bulk electrode and the other side of the electrode is immersed in an electrolyte) or as an unsupported film, with one side immersed in the electrolyte and its other side connected to an external lead and exposed to the gas.

Of course, the CCEs are used mainly in fuel cells (Eastcott and Easton 2009) and electrochemical sensors designed for various ion detection applications (Wang et al. 1997; Zheng et al. 2007). Another interesting application for the carbon–ceramic composite is related to the construction of biosensors (Tsionsky et al. 1994; Skeika et al. 2011). Experiment, however, has shown that CCEs can also be used in gas sensors. For example, Lev and co-workers (Tsionsky and Lev 1995a, b; Rabinovich et al. 1997) using Co tetramethoxymesoporphyrin (Co-TMMP)-modified CCE designed SO_2 (anodic reaction) and O_2 (cathodic reaction) sensors. The choice of Co-TMMP for CCE modification was in part motivated by the high sensitivity and high stability of oxygen reduction at heat-treated Co-TMMP-supported CCE, which are imperative conditions for effective chemical sensing. All experiments were conducted in 0.5 M sulfuric acid solution.

However, we should note that modified polymer-based electrodes have much wider application in electrochemical gas sensors in comparison with CCE. A very limited number of teams use CCEs for gas sensors design. Perhaps the technology of CCE fabrication (high-temperature annealing is required) and too high gas penetrability limit their application in gas sensors.

References

Abrantes J, Perez-Coll D, Nuntez P, Frade J (2003) Electronic transport in $Ce_{0.8}Sm_{0.2}O_{1.9}$ samples. Electrochim Acta 48:2761–2766

Anderson ML, Stroud RM, Rolison DR (2002) Enhancing the activity of fuel-cell reactions by designing three-dimensional nanostructured architectures: catalyst-modified carbon-silica composite aerogels. Nano Lett 2:235–240

Arguello J, Magosso HA, Ramos RR, Canevari TC, Landers R, Pimentel VL, Gushikem Y (2009) Structural and electrochemical characterization of a cobalt phthalocyanine bulk-modified SiO_2/SnO_2 carbon ceramic electrode. Electrochim Acta 54:1948–1953

Baranov A, Fanchenko S, Calliari L, Speranza G, Minati L, Shorokhov AV, Fedoseenkov D, Nefedov A (2007) a-C/Me (Me=Pt, Au) nanocomposite films for electrochemical gas sensors. Diamond Rel Mater 16:1365–1369

Bhoga SS, Singh K (1999) A new Na^+ glass-dispersed Na_2CO_3 composite for a solid state electrochemical CO_2 gas sensor. J Solid State Electrochem 3:258–263

Brosda S, Bouwmeester HJM, Guth U (1996) Composite effect of solid electrolytes based on alkali carbonates and sulfates ionics. Ionics 2:323–328

Bu ko M (2000) Ionic conductivity of alumina–zirconia composites. Pol Ceram Bull 61:95–102

Bu ko M, Róg G (1995) Properties of ZrO_2–Ca-β-Al_2O_3 composites. In: Gusmano G, Traversa E (eds) Proceedings of the 4th European ceramic society conference, Riccione'95, vol 5. Electroceramics, pp 421–426

Bunde A, Dieterich W (2000) Percolation in composites. J Electroceram 5:81–92

Cao Z, Buttner WJ, Stetter JR (1992) The properties and applications of amperometric gas sensors. Electroanalysis 4:253–266

Caproni E, Carvalho FMS, Muccillo R (2008) Development of zirconia–magnesia/zirconia–yttria composite solid electrolytes. Solid State Ionics 179:1652–1654

Chiou CY, Chou TC (1996) Amperometric sulfur dioxide sensors using the gold-deposited gas-diffusion electrode. Electroanalysis 8:1179–1182

Cinke M, Li J, Chen B, Cassell A, Delzeit L, Han J, Meyyappan M (2002) Pore structure of raw and purified HiPco single-walled carbon nanotubes. Chem Phys Lett 365:69–74

Do J-S, Chang W-B (2001) Amperometric nitrogen dioxide gas sensor: preparation of PAn/Au/SPE and sensing behavior. Sens Actuators B 72:101–107

Doshi R, Richards V, Carter J, Wang X, Krumpelt M (1999) Development of solid-oxide fuel cells that operate at 500°C. J Electrochem Soc 146:1273–1278

Dudek M (2008a) Composite oxide electrolytes for electrochemical devices. Adv Mater Sci 8(1):15–30

Dudek M (2008b) Ceramic oxide electrolytes based on CeO_2-preparation, properties and possibility of application to electrochemical devices. J Eur Ceram Soc 28(5):965–971

Dudek M, Ziewiec K (2006) Preparation and the electrolytic properties of CaO-Sm_2O_3-CeO_2 system. Adv Mater Sci 6:53–58

Dudney N (1985) Effect of interfacial space–charge polarization on the ionic conductivity of composite electrolytes. J Am Ceram Soc 68:538–545

Eastcott JI, Easton EB (2009) Electrochemical studies of ceramic carbon electrodes for fuel cell systems: a catalyst layer without sulfonic acid groups. Electrochim Acta 54:3460–3466

Fray DJ (1996) The use of solid electrolytes as sensors for applications in molten metals. Solid State Ionics 86–88:1045–1054

Fuijtsu S, Koumoto M, Yanagida H, Kanazawa T (1985) Enhancement of ionic conduction in CaF_2 and BaF_2 by dispersion of Al_2O_3. J Mater Sci 22:2103–2109

Guo X (2003) Roles of alumina in functional ceramics. J Am Ceram Soc 86:1867–1873

Guth U, Zosel J (2004) Electrochemical solid electrolyte gas sensors: hydrocarbon and NO_x analysis in exhaust gases. Ionics 10(5–6):366–377

Herle J, Senevirate D, McEvoy A (1999) Lanthanide co-doping of solid electrolytes: AC conductivity behavior. J Eur Ceram Soc 19:837–841

Hong YJ, Oh SM (1996) Fabrication of polymer electrolyte fuel cell (PEFC) H_2 sensors. Sens Actuators B 32:7–13

Hrapovic S, Liu EY, Male K, Luong JHT (2006) Metallic nanoparticle–carbon nanotube composites for electrochemical determination of explosive nitroaromatic compounds. Anal Chem 78:5504–5512

Inaba H, Tagawa H (1996) Ceria –based solid electrolytes. Solid State Ionics 83:1–16

Jacob KT, Shukla A (1987) Kinetic decomposition of Ni_2SiO_4 in oxygen potential gradients. J Mater Res 2:338–342

Jamnik J, Meier J (1999) Defect chemistry and chemical transport involving interfaces. Solid State Ionics 119:191–198

Kida T, Shimanoe K, Miura N, Yamazoe N (2001) Stability of NASICON-based CO_2 sensor under humid conditions at low temperature. Sens Actuators B 75:179–187

Kida T, Minami T, Kishi S, Yuasa M, Shimanoe K, Yamazoe N (2009) Planar-type BiCuVO*x* solid electrolyte sensor for the detection of volatile organic compounds. Sens Actuators B 137:147–153

Knauth P (2000) Ionic conductor composites: theory and materials. J Electroceram 5:111–125

Lei CX, Hu SQ, Gao N, Shen GL, Yu RQ (2004) An amperometric hydrogen peroxide biosensor based on immobilizing horseradish peroxidase to a nano-Au monolayer supported by sol–gel derived carbon ceramic electrode. Bioelectrochemistry 65:33–39

Lev O, Wu Z, Bharathi S, Glezer V, Modestov A, Gun J, Rabinovich L, Sampath S (1997) Sol–gel materials in electrochemistry. Chem Mater 9:2354–2375

Lev O, Rizkov D, Mizrahi I, Ekeltchik I, Kipervaser ZG, Gitis V, Goifman A, Kamyshny A Jr, Modestov AD, Gun J (2005) Sol–gel derived silicate based composite electrode. In: Sakka S (ed) Handbook of sol–gel science and technology, vol. 3: processing characterization and applications. Springer, New York, pp 329–354

Li H, Wang Q, Xu J, Zhang W, Jin L (2002) A novel nano-Au-assembled amperometric SO_2 gas sensor: preparation, characterization and sensing behavior. Sens Actuators B 87:18–24

Liang C (1973) Conduction characteristic of the lithium iodide-aluminium oxide solid electrolytes. J Electrochem Soc 120:1289–1292

Liang X, Wang B, Zhang H, Diao Q, Quan B, Lu G (2012) Progress in solid electrochemical gas sensors based on NASICON and oxide electrodes. In: Proceedings of the 14th international meeting on chemical sensors, IMCS 2012, May 20–23, Nuremberg, Germany, pp 500–503. doi:10.5162/IMCS2012/5.5.4

Liu Q (1996) The development of high temperature electrochemical sensors for metallurgical processes. Solid State Ionics 86–88:1037–1043

Liu X, Chen X (2005) Toughening of 8Y-FSZ ceramics by neodymium titanate secondary phase. J Am Ceram Soc 88:456–558

Liu Y, He T, Wang J, Shu W (2005) The effect of Pr co-dopant on the performance of solid oxide fuel cells with Sm-doped ceria electrolyte. J Alloy Compd 389:317–322

Maricle DL, Swarm TE, Karavolis S (1992) Enhanced ceria–a low-temperature SOFC electrolyte. Solid State Ionics 52:173–178

Matsui T, Inaba M, Mineshige A, Ogumi Z (2005) Electrochemical properties of ceria based oxides for use in intermediate-temperature SOFCs. Solid State Ionics 176:647–654

Meier J (1987) Defect chemistry and conductivity effects in heterogeneous solid electrolytes. J Electrochem Soc 134:1524–1535

Min B-K, Choi S-D (2003) SO_2-sensing characteristics of Nasicon sensors with Na_2SO_4–$BaSO_4$ auxiliary electrolytes. Sens Actuators B 93:209–213

Miura N, Yao S, Shimizu Y, Yamazoe N (1992) High performance solid-electrolyte carbon dioxide sensor with a binary carbonate electrode. Sens Actuators B 9:165–170

Mizutania Y, Matsudaa H, Ishijia T, Furuya N, Takahashi K (2005) Improvement of electrochemical NO_2 sensor by use of carbon–fluorocarbon gas permeable electrode. Sens Actuators B 108:815–819

Mukundan E, Brosha E, Brown D, Garzon F (1999) Ceria-electrolyte-based mixed potential sensors for the detection of hydrocarbons and carbon monoxide. Electrochem Solid State Lett 2:412–414

Nakayama S, Sadaoka Y (1994) Preparation of $Na_3Zr_2Si_2PO_{12}$-sodium aluminosilicate composite and its application as a solid-state electrochemical CO_2 gas sensor. J Mater Chem 4(5):663–668

Nigge U, Wiemhofer H-D, Romer EWJ, Bouwmeester HJM, Schulte TR (2002) Composites of $Ce_{0.8}Gd_{0.2}O_{1.9}$ and $Gd_{0.7}Ca_{0.3}CoO_{3-}\delta$ as oxygen permeable membranes for exhaust gas sensors. Solid State Ionics 146:163–174
Obata K, Matsushima S (2008) NASICON-based NO_2 device attached with metal oxide and nitrite compound for the low temperature operation. Sens Actuators B 130:269–276
Oe K, Kikkawa K, Kishimoto A, Nakamura Y, Yanagida H (1996) Toughening of ionic conductive zirconia ceramics utilizing a nonlinear effect. Solid State Ionics 91:131–136
Okamoto A, Suzuki Y, Yoshitake M, Ogawa S, Nakano N (1997) Gold-carbon composite thin films for electrochemical gas sensor prepared by reactive plasma sputtering. Nucl Instrum Methods Phys Res B 121:179–183
Polsky R, Gill R, Kaganovsky L, Willner I (2006) Nucleic acid-functionalized Pt nanoparticles: catalytic labels for the amplified electrochemical detection of biomolecules. Anal Chem 78:2268–2271
Rabinovich L, Lev O (2001) Sol–gel derived composite ceramic carbon electrodes. Electroanalysis 13(4):265–275
Rabinovich L, Gun J, Tsionsky M, Lev O (1997) Fuel-cell type ceramic-carbon oxygen sensors. J Sol–Gel Sci Technol 8:1077–1081
Rabinovich L, Lev O, Tsirlina GA (1999) Electrochemical characterization of Pd modified ceramic vertical bar carbon electrodes: partially flooded versus wetted channel hydrophobic gas electrodes. J Electroanal Chem 466(1):45–59
Rog G, Kielski A, Kozlowska-Rog A, Bucko M (1998) Composite (CaF_2-Al_2O_3) solid electrolytes–preparation, properties and application to the solid oxide galvanic cells. Ceram Int 24:91–98
Roh S-W, Stetter JR (2003) Amperometric sensing of NOx with cyclic voltammetry. J Electrochem Soc 150(11):H266–H272
Sakata M, Kimura T, Goto T (2007) Ru-C nano-composite films prepared by PECVD as an electrode for an oxygen sensor. Key Eng Mater 352:319–322
Sakka S (ed) (2003) Sol–gel science and technology. Vol. 3: application of sol–gel technology. Kluwer Academic, Massachusetts
Sameshima S, Hirata Y, Ehira Y (2006) Structural change in Sm- and Nd-doped ceria under a low oxygen partial pressure. J Alloys Compd 408–412:628–631
Santhosh P, Manesh KM, Gopalan A, Lee K-P (2007) Novel amperometric carbon monoxide sensor based on multi-wall carbon nanotubes grafted with polydiphenylamine–fabrication and performance. Sens Actuators B 125:92–99
Shai K, Wagner J (1982) Enhanced ionic conduction in dispersed solid electrolyte systems (DSES) and/or multiphase systems: AgI-Al_2O_3, AgI-SiO_2, AgI-Fly ash, and AgI-AgBr. J Solid State Chem 42:107–119
Shimizu Y, Yamashita N (2000) Solid electrolyte CO_2 sensor using NASICON and perovskite-type oxide electrode. Sens Actuators B 64:102–106
Singh K, Bhoga SS (1990) On the dispersion of ionically conducting glass into the Li_2SO_4-Li_2CO_3 eutectic composite system. Solid State Ionics 40(41):1025–1027
Singh K, Ambekar P, Bhoga SS (2002) Ferroelectric dispersed composite solid electrolyte for CO_2 gas sensor. In: Chowdari BVR, Prabaharan SRS, Yahaya M, Talib IA (eds) Solid state ionics: trends in the new millennium. World Scientific Publishing, Singapore, pp 469–476
Skeika T, Zuconelli CR, Fujiwara ST, Pessoa CA (2011) Preparation and electrochemical characterization of a carbon ceramic electrode modified with ferrocenecarboxylic acid. Sensors 11:1361–1374
Stetter JR, Maclay GJ (2008) Carbon nanotubes and sensors: a review. In: Baltes H, Brand O, Fedder GK, Hierold C, Korvink JG, Tabata O (eds) Enabling technology for MEMS and nanodevices (chapter 10). Wiley-VCH Verlag GmbH, Weinheim, Germany. doi:10.1002/9783527616701.ch10
Sun D, Zhu L, Zhu G (2007) Glassy carbon ceramic composite electrodes. Anal Chim Acta 564:243–247
Sundmacher K, Rihko-Struckmann LK, Galvita V (2005) Solid electrolyte membrane reactors: status and trends. Catal Today 104:185–199
Tsionsky M, Lev O (1995a) Electrochemical composite carbon–ceramic gas sensors: introduction and oxygen sensing. Anal Chem 67:2409–2414
Tsionsky M, Lev O (1995b) Investigation of the kinetics and mechanism of Co-porphyrin catalyzed oxygen reduction by hydrophobic carbon-ceramic electrodes. J Electrochem Soc 142:2132–2138
Tsionsky M, Gun G, Giezer V, Lev O (1994) Sol–gel-derived ceramic-carbon composite electrodes–introduction and scope of applications. Anal Chem 66:1747–1753
Uvarov N, Iusupov V, Sharama V, Shukla K (1992) Effect of morphology and particle size on the ionic conductivities of composite solid electrolytes. Solid State Ionics 51:41–52
Uvarov NF, Ponomareva VG, Lavrova GV (2010) Composite solid electrolytes. Russ J Electrochem 46(7):722–733
Vaidehi N, Akila R, Shukla A, Jacob KT (1986) Enhanced ionic conduction in dispersed solid electrolyte systems CaF_2-Al_2O_3 and CaF_2-CeO_2. Mater Res Bull 21:909–916
Wachsman E, Jayaweera P, Jiang N, Lowe D, Pound B (1997) Stable high conductivity ceria/bismuth bilayered electrolytes. J Electrochem Soc 144:233–236
Wagner J (1980) Transport in compounds containing a dispersed second phase. Mater Res Bull 15:1690–1701
Wang J (2006) Analytical electrochemistry, 3rd edn. Wiley-VCH, Hoboken, NJ, p 146

Wang F, Hu S (2009) Electrochemical sensors based on metal and semiconductor nanoparticles. Microchim Acta 165:1–22

Wang SQ, Lin XQ (2005) Electrodeposition of Pt-Fe(III) nanoparticle on glassy carbon electrode for electrochemical nitric oxide sensor. Electrochim Acta 50:2887–2891

Wang J, Pamidi PVA, Nascimento VB, Angnes L (1997) Dimethylglyoxime doped sol–gel carbon composite voltammetric sensor for trace nickel. Electroanalysis 9:689–692

Wang HY, Huang YG, Tan ZA, Hu XY (2004a) Fabrication and characterization of copper nanoparticle thin-films and the electrocatalytic behavior. Anal Chim Acta 526:13–17

Wang F, Chen S, Cheng S (2004b) Gd^{3+} and Sm^{3+} co-doped ceria based electrolytes for intermediate temperature solid oxide fuel cells. Electrochem Commun 6:743–746

Westphal D, Jakobs S, Guth U (2001) Gold-composite electrodes for hydrocarbon sensors based on YSZ solid electrolyte. Ionics 7:182–186

Wiemhofer HD, Bremes HG, Nigge U, Zipprich W (2002) Studies of ionic transport and oxygen exchange on oxide materials for electrochemical gas sensors. Solid State Ionics 150:63–77

Yu AM, Liang ZJ, Cho JH, Caruso F (2003) Nanostructured electrochemical sensor based on dense gold nanoparticle films. Nano Lett 3:1203–1207

Zhang JD, Oyama M (2005) Gold nanoparticle arrays directly grown on nanostructured indium tin oxide electrodes: characterization and electroanalytical application. Anal Chim Acta 540:299–306

Zheng J, Sheng Q, Li L, Shen Y (2007) Bismuth hexacyanoferrate-modified carbon ceramic electrodes prepared by electrochemical deposition and its electrocatalytic activity towards oxidation of hydrazine. J Electroanal Chem 611:155–166

Zhong T, Liang X, Zhang H, Yang S, Li J, Quan B, Lu G (2011) Sensing characteristics of potentiometric NO sensor using NASICON and $NiWO_4$ sensing electrode. Sensor Lett 9(1):307–311

Zhu M, Liu M, Shi G, Xu F, Ye X, Chen J, Jin L, Jin J (2002) Novel nitric oxide microsensor and its application to the study of smooth muscle cells. Anal Chim Acta 455:199–206

Zou H, Wu SS, Shen J (2008) Polymer/silica nanocomposites: preparation, characterization, properties, and applications. Chem Rev 108:3893–3957

Chapter 17
Disadvantages of Nanocomposites for Application in Gas Sensors

Our short review shows that the use of nanocomposites can lead to an improvement in gas sensors parameters. However, one should remember that the complicated nature of the composition of gas-sensing matrix is always accompanied by the deterioration of sensor parameters' reproducibility. Many additional factors, which can have an effect on gas-sensing properties of materials, appear in nanocomposites. It is especially typical for nanocomposites used as sensing materials in metal oxide chemiresistors. It has been established that the change of additional component concentration has an equally powerful influence on the gas-sensing materials' parameters as does the change of deposition temperature (synthesis) or technological route used. All the main parameters of gas-sensing materials such as grain size, porosity, interphase interaction, surface and interfacial energy, catalytic activity, chemical reactivity, texture, stress, and strain, which control sensor response, significantly depend on the type and the concentration of additives (Gas'kov and Rumyantseva 2001, 2009; Korotcenkov 2005, 2007; Zhang et al. 2003a, b). Therefore, even small additives of the second phase could change parameters of metal oxide gas-sensing matrix fundamentally. However, it was established that for achievement of optimizing effect we have to find specific composition of nanocomposites, because as a rule the optimizing effect is being observed at certain concentration of one of the components only. Moreover, as one can see in Fig. 17.1, the range of doping concentration, which could be accompanied by sensor response improvement, is narrow, and deviation from this optimal concentration could lead to not an improvement but a sharp drop of sensitivity or lead to fast sensor degradation (see Fig. 17.1). For example, it was established for metal oxide-based nanocomposites that if additives do not form a conductive phase with good sensing properties, the optimizing concentrations of additives in the metal oxide matrix would not exceed 1–3 % (Tricoli et al. 2008; Liu et al. 2010). Only in the case of composites which are formed by two gas-sensing conductive materials could optimal content of the second phase reach 10–20 % (see Chap. 14 (Vol. 2)).

The same effect was also observed in gas sensors based on polymer–black carbon (metal, NCTs, or metal oxide) nanocomposites, where, due to the swelling effect, the change of conductivity near the "percolation threshold" was being used. In mathematics, percolation theory describes the behavior of clusters connected in a random graph. In our case the percolation threshold corresponds to the concentration of conductive phase, at which in composite consisted from conductive and insulating phases, the transition from insulating state to conductive one takes place, i.e., the transition from a high-resistance to a low-resistance state. Maximal sensor response, conditioned by the swelling effect, is possible only if the nanocomposite composition corresponds to the percolation threshold, fc. Only in such a composite is the switching from the conductive state to the insulated one and back possible at the expense of the swelling effect. For f below fc there are only isolated conducting clusters, and the conductivity is zero independent from changes of the surrounding atmosphere. For f above fc the composite can remain in a conductive state without switching to an insulating state even at maximal swelling effect. Experiment shows that the control of composition in a complex matrix, already at the level of 1 %, is a very hard task. However, in

G. Korotcenkov, *Handbook of Gas Sensor Materials*, Integrated Analytical Systems,
DOI 10.1007/978-1-4614-7388-6_17, © Springer Science+Business Media New York 2014

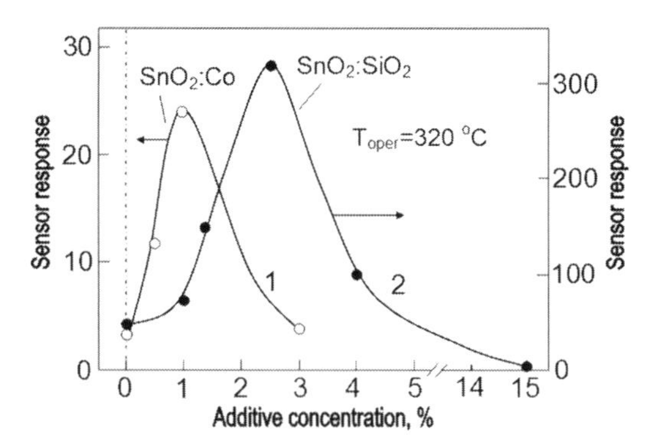

Fig. 17.1 Additives influence on response of doped SnO_2-based gas sensors. (1) Responses of pure and co-doped SnO_2 nanofibers to100 ppm H_2. (Data from Liu et al. 2010). (2) Sensor response to 50 ppm EtOH as a function of the SiO_2 content (Data from Tricoli et al. 2008)

real gas-sensing nanocomposites, we have to control the concentration of the second phase in the gas-sensing matrix at levels <0.1 %. For example, for CNT/polymer nanocomposites, as we indicated earlier, in many cases the threshold of conductivity was observed at 0.0025 wt% CNT (Sandler et al. 2003). Local agglomeration of nanoparticles incorporated in the polymer matrix as well as variation in nanoparticle diameter and length are other reasons of significant differences in the values of percolation thresholds in the same nanocomposites reported in various works. This means that the dispersion of gas sensor parameters, induced by variability of the nanocomposite composition, could be huge. Of course, the reproducibility of sensor parameters in this case will be bad as well. All the factors listed above require more careful control in a technological process. As a result, the cost of sensors could grow.

Incompatibility of materials used in nanocomposite can also be a reason for sensor parameters worsening. Wolfbeis (2005) noted that just material incompatibility has caused substantial difficulties in optical sensing. Because of this problem, the incorporation of dyes into polymers sometimes turned out to be difficult. For example, the polycyclic aromatic hydrocarbon decacyclene is compatible with longwave excitation, displays strong green fluorescence, but, unless made much more lipophilic by the introduction of *tert*-butyl groups, is completely insoluble in the material preferably used for oxygen sensing which is a silicone. The same is true for other probes for oxygen including certain ruthenium–phenanthroline complexes and platinum (or palladium)–porphyrins. It was established that, in the worst case, the fluorescent indicator probe starts to crystallize in the polymer matrix because of poor solubility, sometimes months after the sensor is manufactured. Mutual compatibility of probes and polymers has been discussed in a review (Amao 2003). According to Wolfbeis (2005), one very efficient solution to the problem of the insolubility of cationic dyes in apolar solvents is to exchange the respective hard anions (such as chloride) with softer and more lipophilic anions, such as perchlorate or, even better, tetraphenylborate (TPB), trimethylsilylpropyl sulfate (TMPS), or laurylsulfate (LS). At first glance the choice of a more polar polymer (having better solubilizing properties for the indicator) would also be the solution to the problem. However, silicones (and other apolar polymeric solvents such as polystyrene and ethyl cellulose) are preferred for their unique solubility and permeability for oxygen, but also because they act as a permeation-selective material that prevents interfering quenchers, such as metal ions, to enter a sensing membrane. A second reason for selecting these materials is based on the ease of deposition on solid supports and on good adhesion to most supports. Hydrogels, in contrast to silicones, do not adhere well on glass supports or on plastic fibers and are also permeable to ions, which may act as quenchers. Another efficient method for permanently incorporating probes in a polymer network consists of covalent immobilization of the probes, which prevents not only aggregation and crystallization but also leaching out of the sensing material into the sample fluid (Wolfbeis 2005). While covalent immobilization is not common when sensing oxygen, this turned out to be almost mandatory in the case of sensors for hydrophilic species and ions.

One should take into account that sometimes the increase of the sensitivity of devices based on nanocomposites is being attained at the expense of worsening other parameters of the sensors. For example, Tamaki et al. (1992) have shown that sensors which used SnO_2–CuO composite had a higher sensitivity to H_2S in comparison with the SnO_2 sensors. However, SnO_2–CuO gas-sensing matrix contains CuO phase (Pagnier et al. (2000)), which does not have high thermodynamic stability as compared with SnO_2. This fact undoubtedly adversely affects the temporal and thermal stability of such sensors' parameters (Korotcenkov 2007). The same could be said about SnO_2–Fe_2O_3, In_2O_3–Fe_2O_3, or SnO_2–AgO_x

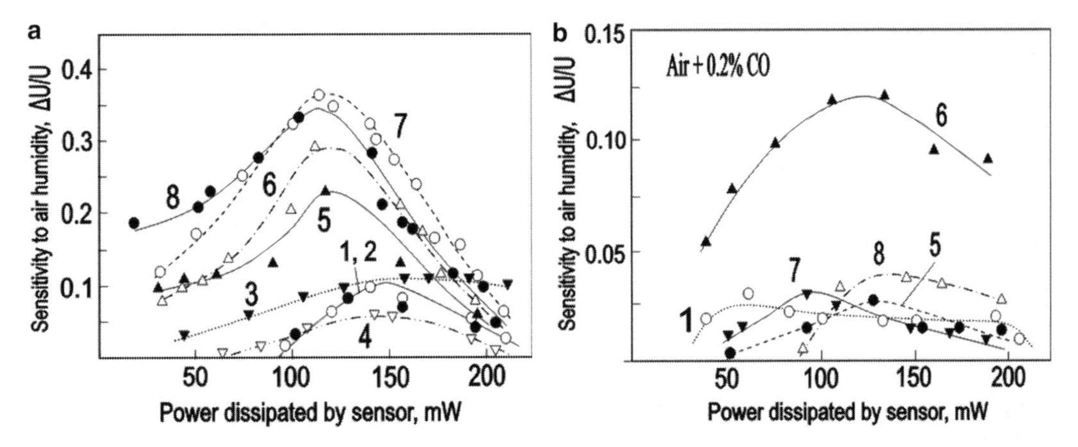

Fig. 17.2 Sensitivity of In_2O_3 one electrode sensor to air humidity in cycles (**a**) dry air ↔ wet air and (**b**) CO + dry air ↔ CO + wet air. 1, undoped In_2O_3; 2, Ga_2O_3 (2 %); 3, P_2O_5 (10 %); 4, Al_2O_3 (4 %); 5, ZnO (2 %); 6, Ga_2O_3 (4 %); 7, B_2O_3 (1 %); 8, CuO (2 %) (Reprinted with permission from Korotcenkov et al. 2004d, Copyright 2004 Elsevier)

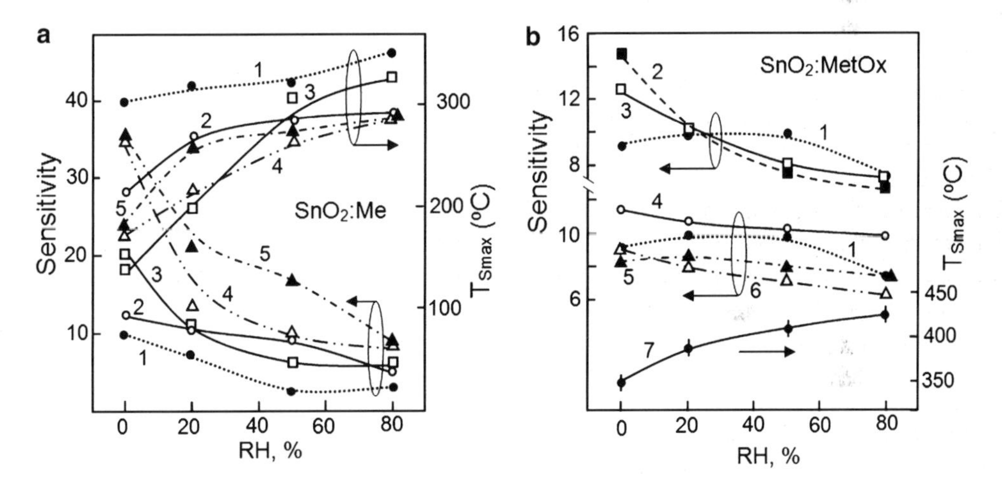

Fig. 17.3 Variation of magnitude (S_{max}) and temperature position of maximum sensitivity (T_{Smax}) to H_2 (20 ppm) of gas sensors based on (**a**) SnO_2:Me and (**b**) SnO_2:MetOx composites: (**a**) 1, SnO_2:Pt; 2, SnO_2:(Pt/Rh); 3, SnO_2:Pd; 4, SnO_2:(Pd/Pt); 5, SnO_2:Rh; (**b**) 1, SnO_2; 2, SnO_2:La_2O_3; 3, SnO_2:Y_2O_3; 4, SnO_2:TiO_2; 5, SnO_2:HfO_2; 6, SnO_2:ZrO_2; 7, T_{Smax} for all SnO_2:MetOx composites. Concentration of additives in the SnO_2-based nanocomposites was varied in the range 0.3–2.1 at.% (Data from Pavelko et al. 2012)

composites, showing high sensitivity to several specific gases and vapors (Gas'kov and Rumyantseva 2001, 2009; Ivanovskaya et al. 2003). It is known that these metal oxides, which have increased chemical reactivity, can be reduced/oxidized at low temperatures. This means that the nanocomposites mentioned above cannot be used in devices working at high temperatures and in aggressive environments.

Korotcenkov et al. (2004), using the example of In_2O_3-based nanocomposites, have also shown that the appearance of the second phase could stimulate growth of sensitivity to humidity change in the surrounding atmosphere (see Fig. 17.2). However, it is known that the increased sensitivity to air humidity is a significant disadvantage of sensors designed for environment monitoring (Korotcenkov and Cho 2011). Recently the same result was observed by Pavelko et al. (2012) for SnO_2-based hydrogen gas sensors. They found that sensors based on undoped SnO_2 had considerably less sensitivity to air humidity in comparison with composite-based sensors such as SnO_2:(Pd; Pt; Rh) and SnO_2(Y_2O_3; La_2O_3) even at small concentrations of additives. In particular, increase of RH caused a remarkable decrease of the signal and shifted the signal maximum toward higher temperatures (see Fig. 17.3). Only composite-based materials such as SnO_2:(TiO_2; HfO_2; ZrO_2), which include IVB elements, had a sensitivity to air humidity similar to undoped material.

References

Amao Y (2003) Probes and polymers for optical sensing of oxygen. Microchim Acta 143:1–12

Gas'kov AM, Rumyantseva MN (2001) Nature of gas sensitivity in nanocrystalline metal oxides. Russ J Appl Chem 74(3):440–444

Gas'kov A, Rumyantseva M (2009) Metal oxide nanocomposites: synthesis and characterization in relation with gas sensing phenomena. In: Baraton MI (ed) Sensors for environment, health and security. Springer Science + Business Media B.V, Dordrecht, The Netherlands, pp 3–29

Ivanovskaya M, Kotsikau D, Faglia G, Nelli P (2003) Influence of chemical composition and structural factors of Fe_2O_3/In_2O_3 sensors on their selectivity and sensitivity to ethanol. Sens Actuators B 96:498–503

Korotcenkov G (2005) Gas response control through structural and chemical modification of metal oxides: state of the art and approaches. Sens Actuators B 107:209–232

Korotcenkov G (2007) Practical aspects in design of one-electrode semiconductor gas sensors: status report. Sens Actuators B 121:664–678

Korotcenkov G, Cho BK (2011) Instability of metal oxide-based conductometric gas sensors and approaches to stability improvement. Sens Actuators B 156:527–538

Korotcenkov G, Boris I, Brinzari V, Luchkovsky Y, Karkotsky G, Golovanov V, Cornet A, Rossinyol E, Rodriguez J, Cirera A (2004) Gas sensing characteristics of one-electrode gas sensors on the base of doped In_2O_3 ceramics. Sens Actuators B 103:13–22

Liu L, Guo C, Li L, Wang L, Dong Q, Li W (2010) Improved H2 sensing properties of Co-doped SnO2 nanofibers. Sens Actuators B 150:806–810

Pagnier T, Boulova M, Galerie A, Gaskov A, Lucazeau G (2000) Reactivity of SnO_2-CuO nanocrystalline materials with H_2S: a coupled electrical and Raman spectroscopic study. Sens Actuators B 71:134–139

Pavelko RG, Vasiliev AA, Llobet E, Sevastyanov VG, Kuznetsov NT (2012) Selectivity problem of SnO_2 based materials in the presence of water vapors. Sens Actuators B 170:51–59

Sandler JKW, Kirk JE, Kinloch IA, Shaffer MSP, Windle AH (2003) Ultra-low electrical percolation threshold in carbon-nanotube-epoxy composites. Polymer 44:5893–5899

Tamaki J, Maekawa T, Miura N, Yamazoe N (1992) CuO-SnO_2 element for highly sensitive and selective detection of H_2S. Sens Actuators B 9:197–203

Tricoli A, Graf M, Pratsinis SE (2008) Optimal doping for enhanced SnO_2 sensitivity and thermal stability. Adv Funct Mater 18:1969–1976

Wolfbeis OS (2005) Materials for fluorescence-based optical chemical sensors. J Mater Chem 15:2657–2669

Zhang S, Yongqing DS, Du FH (2003a) Recent advances of superhard nanocomposite coatings: a review. Surf Coat Technol 167:13–119

Zhang W, Chen D, Zhao Q, Fang Y (2003b) Effects of different kinds of clay and different vinyl acetate content on the morphology and properties of EVA/clay nanocomposites. Polymer 44(1):7953–7961

Part IV
Stability of Gas Sensing Materials and Related Processes

Chapter 18
The Role of Temporal and Thermal Stability in Sensing Material Selection

The problems of stability and reliability of gas sensors operation remain dominant while designing devices for the sensor market. Devices designed for this market should provide a stable and reproducible signal for a period of at least 2–3 years (typically 17,000–26,000 h of operation). Therefore, sensing materials and conditions of their operation should be selected in consideration of the above-mentioned requirements (Korotcenkov 2007; Korotcenkov and Cho 2011). For example, the organic polymer Nafion may retain working capacity in electrochemical gas sensors for a period of up to about 1 year. To achieve this result, however, the Nafion should be wetted by a wick system connected to a reservoir (Pasierb et al. 2004). This means that dry atmosphere does not facilitate a long lifetime for polymer-based electrochemical gas sensors.

The same situation takes place with exploitation of polymer-based sensors in ozone-containing atmosphere. It has been reported that ozone and other oxidizing components in the polluted atmospheres of industrial centers can either initiate or accelerate the photochemical destruction of polymers (Razumovskii and Zaikov 1982). Thus their lifetime in sensors may be limited by the presence of atmospheric ozone. Polymer sensors used for environmental control also have a significant disadvantage in terms of their sensitivity to UV radiation. Moreover, it was found that polymer degradation is almost always faster in the presence of oxygen (air) and moisture. It was established that, as a rule, a polymer-based conductometric gas sensor can present a high sensitivity in its first post-produced day, but it can also exhibit important signal response reduction on the next few days. Experiment has shown that this effect is especially strong for chemiresistors based on doped conducting polymers. Results obtained by Lima and de Andrade (2009) for polymer-based chemiresistors and presented in Table 18.1 are a good illustration of this effect. Table 18.1 shows the normalized sensitivity loss of each sensor for different analytes. The sensors were produced by spin-coating and layer-by-layer techniques using the different conducting polymers indicated in Table 18.1. All sensors were submitted to the same different analytes during 5 consecutive days. It is seen that the sensitivity loss is different for each sensor, but these changes are considerable.

As a result of such temporal instability, in spite of the wide range of chemical sensor prototypes based on polymer films, very few have found their way onto the market. Though they may have excellent analytical qualities, the devices are often unsuitable for industrial fabrication, either because of low technological effectiveness of the fabrication process or insufficient reliability and stability. This means that to utilize the advantages of some polymers, such as a rare combination of electrical, electrochemical, and physical properties, it is necessary to increase their processability as well as their temporal, environmental, and thermal stability (Kumar and Sharma 1998).

G. Korotcenkov, *Handbook of Gas Sensor Materials*, Integrated Analytical Systems,
DOI 10.1007/978-1-4614-7388-6_18, © Springer Science+Business Media New York 2014

Table 18.1 Normalized per day sensitivity loss (%/day)

			Fragrances		
Sensor	Polymer	Ethanol	1	2	3
1	P3HT 1000 rpm_1	6	3	8	11
2	PEDOT 1000 rpm_3	20	20	16	17
3	PEDOT 3000 rpm_2	19	20	20	19
4	POMA 3000 rpm_2	20	17	12	15
5	POMA 1000 rpm_2	19	17	14	15
6	POMA/PPY 20B_3	18	13	10	15
7	POMA/FTNi 20B_3	11	55	14	3
8	PEDOT 2000 rpm_1	19	17	5	14
9	PANI/PEDOT 10B_2	7	14	8	16

Source: Data from Lima and de Andrade (2009)
P3HT poly(3-hexylthiophene), *PEDOT* poly(3,4-ethylenedioxythiophene), *POMA* poly(*o*-methoxyaniline), *PPY* polypyrrole, *PANI* polyaniline

It is clear that environment stability depends on the type of polymer material. For example, according to Tourillon and Garnier (1983), polythiophenes (PTh) are more stable in an oxygen environment than polypyrrole (PPy). However, Wang et al. (1991) established that the conductivity stability of the polypyrroles was significantly better than that of the poly(3-alkylthiophenes). Copolymers of 3-octylthiophene and 3-methylthiophene (POTMT) doped by $FeCl_3$ are more stable than doped poly (3-octylthiophene) (P3OT) (Pei and Inganas 1992). The bithiophenes are more stable than the thiophenes because the long alkyl side chains are further apart in the bithiophenes and therefore have less chance to interact (Pei et al. 1993). The oxidation–reduction potentials also have a strong effect on the stability of polymers (Kumar and Sharma 1998). Polymethylthiophene (PMeT) is stable in an environment containing moisture and oxygen because an oxidation potential is much lower than of water and a reduction potential is much greater than that of oxygen (Tourillon and Garnier 1983). Polyacetylene (PA) also has oxidation–reduction potentials close to those of PMeT, but the chemical reactivity of oxygen with double bonds leads to lower stability. The alkyl chain length plays an important role in polymer stability as well. The stability becomes worse with increasing length of the side chains (Wang and Rubner 1990). The increase of the alkyl chain length causes the oxidation–reduction potentials of polyalkylthiophenes to move further away from those of oxygen and water. Research has also shown that undoped conducting polymers have much better stability than doped ones. If undoped polymers can keep their properties without changes for a year or more, even in the presence of oxygen (air) and moisture, the conductivity of doped polymers at ambient conditions starts to change during the first few days and even hours (see Chap. 19 (Vol. 2)). For example, Rahman et al. (1991) established that storage of the powdered undoped conducting polymer poly-(*p*-phenylphosphoethynediyl) for about 10 months did not affect the conductivity value. At the same time, Br_2-doped polymer lost its doped conductivity value almost completely within 7 days. Exactly such behavior of doped conjugated polymers created a situation where most polymer-based sensors designed for real applications were based on undoped polymers forming composite with conductive inorganic materials (see Chap. 19 (Vol. 2)).

Standard covalent semiconductors such as Si, InP, GaAs, GaP, including porous semiconductors, also do not guarantee high stability of gas sensor parameters. In an oxygen atmosphere, surface oxidation takes place (see Fig. 18.1 related to Si oxidation in air and water), which inevitably leads to the change of electronic, adsorption and catalytic properties of the semiconductor surface. Therefore, gas sensors based on standard covalent semiconductors, including Si, generally need to have an aging treatment to have reliable and repeatable sensitivity. Even then, lifetimes of gas sensors based on covalent semiconductors (InP, GaAs, GaP) and especially on porous Si can be short (Han et al. 2001). Gas sensors based on ionic compounds, such as CuBr, have unstable parameters as well (Bendahan et al. 2002).

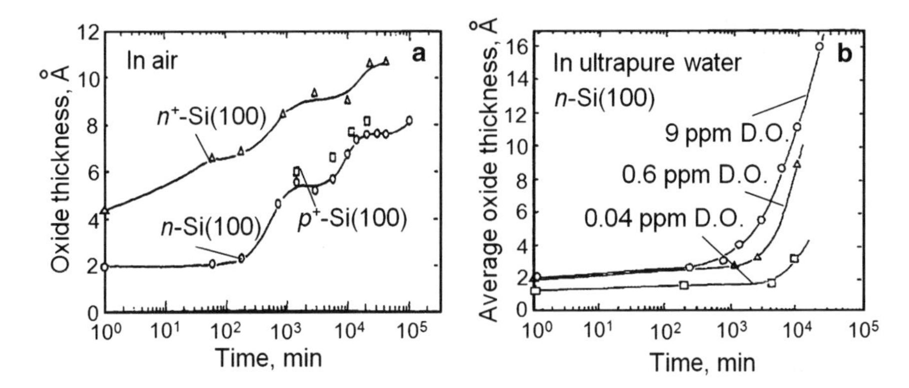

Fig. 18.1 Oxide thickness as a function of exposure time of wafers to (**a**) air at room temperature and (**b**) in ultrapure water at room temperature for different dissolved oxygen (D.O.) concentrations (Data from Morita et al. 1990)

Fig. 18.2 Schematic diagram illustrating intercorrelation between types of sensing materials and their stability

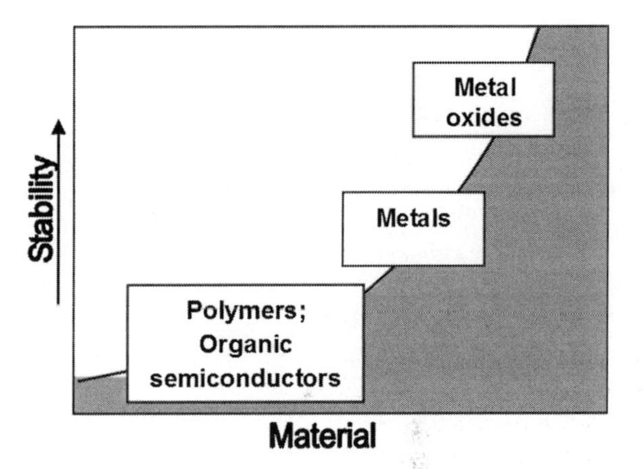

A similar situation occurs with sensors based on several types of solid electrolytes (Fergus 2008). For example, carbonate and sulfate electrolytes could be used with CO_2 and SO_2 sensors. However, those electrolytes generally do not provide adequate stability (see Chap. 6 (Vol. 1)), and therefore the most promising sensors use common electrolytes, such as Nasicon, β-alumina, and yttria-stabilized zirconia (YSZ). These electrolytes require auxiliary electrodes to provide the desired response, but they provide good stability and long operating lives. Therefore, while optimizing the reactions responsible for a gas sensor's sensitivity, one should also aim to maximize the chemical, structural and time stability of the device.

Metal oxides and wide band semiconductors, such as SiC and GaN, with dielectric covering have a much better stability of surface and bulk properties in both oxygen and water environments in comparison with polymers and standard semiconductors (see Fig. 18.2), which prepares them for wide practical use in real devices of long-term use, available in the sensor market (Kerlau et al. 2006; Connolly et al. 2005). The results presented by Badwal (Badwal 1992; Badwal et al. 2000) show how stable metal oxides could be. Zirconia-based ceramics, which belong to the group of the most stable metal oxides, kept their electro-conductivity without considerable changes even at $T > 1{,}000$ °C. For comparison, the working temperature of polymer-based sensors is limited by 100–150 °C.

Thus, comparative analysis shows that though the fact that polymers can satisfy the largest number of requirements as materials for low-temperature selective gas sensors, polymer-based devices are not expected to operate in tough conditions. At the same time, metal oxides, though having low selectivity in gas-sensing effects, have considerable advantages as materials for high-temperature gas sensors designed for long-term use in tough conditions.

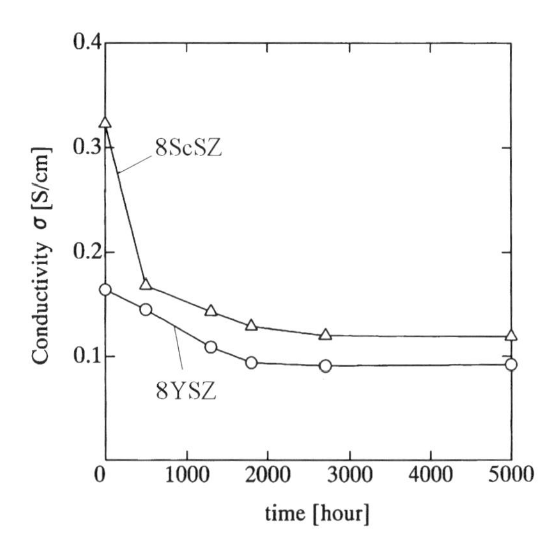

Fig. 18.3 Electrical conductivity changes of 8 % Y_2O_3–92 % ZrO_2 (8YSZ) and 8 % Sc_2O_3–92 % ZrO_2 (8ScSZ) during annealing at 1,000 °C (Reprinted with permission from Nomura et al. 2000, Copyright 2000 Elsevier)

However, it should be noted that the problem of instability and aging exists even for metal oxides with the reputation as the most stable materials. As indicated before, yttria (Y_2O_3)-stabilized zirconia (ZrO_2) (8 mol% Y_2O_3–92 mol% ZrO_2: 8YSZ) is the most commonly used electrolyte in high-temperature solid electrolyte gas sensors, because it fulfills several desired criteria, such as high thermal stability and high ionic conductivity at operating temperatures 600–1,000 °C (see Chap. 2 (Vol. 1)). However, it is also well known that the conductivity of 8YSZ degrades (see Fig. 18.3) after long-term operation (Yamamoto et al. 1995; Nomura et al. 2000). Moghadam and Stevenson (1982) have reported that the conductivity of 4.5YSZ also changes during annealing at 1,000 °C. Results of research of Carter and Roth presented in Alcock (1968) have shown that the defect-ordering process, taking place during the exploitation of YSZ-based devices, is responsible for the temporal changes observed in YSZ conductivity. Moghadam and Stevenson (1982) established that the initial decrease of conductivity in 4.5YSZ was attributable to precipitation of tetragonal zirconia from the cubic matrix and further decreases to the ordering in the cubic phase. In other words, the YSZ cubic phase is decomposed at high temperature into cubic and tetragonal phases. The increase of tetragonal phase by decomposition of the cubic phase leads to the reduction of ionic conductivity of YSZ. Kondoh et al. (1998a, b) concluded that the short-range ordering of oxide ion vacancies around Zr ions, caused by relaxation of the lattice distortion, controls the decrease in conductivity with aging. On the other hand, such an aging effect decreases with increase of Y_2O_3 content and looks to fade into the background in the case of 10YSZ (Kondoh et al. 1998b). So, it could seem that YSZ with yttria concentration higher than 10 % is preferable for high-temperature gas sensor application due to higher stability. However, research has shown that these compositions have reduced ionic conductivity. For high efficient operation of gas sensors the ionic conductivity of solid electrolyte should be high. Hattori et al. (2004) studied the electrical conductivity change with annealing at 1,000 °C in the Y_2O_3–ZrO_2 (YSZ) system, including 8YSZ, 8.5YSZ, 9YSZ, 9.5YSZ, and 10YSZ, and established that 9.5YSZ and 10YSZ showed no conductivity decrease even for the annealing period of 1,000 h, while 8YSZ and 8.5YSZ showed significant decrease with time although the initial conductivities were higher than those of 9.5YSZ and 10.0YSZ. The 9.0YSZ showed only slight conductivity decrease. The measurement of Raman spectra demonstrated that the deterioration in conductivity was related to the gradual formation of fine tetragonal phase in the cubic phase. Consequently, 9YSZ and 9.5YSZ compositions seemed to be optimum as the electrolyte material of solid oxide gas sensors from the point of stability in high conductivity.

Other research has shown that the stability of yttria-stabilized zirconia (8YSZ) with high ionic conductivity can also be improved by addition of small amounts of other oxides into YSZ during sintering. For example, Kondo et al. (2003) established that a solid solution of NiO into YSZ increases the stability of YSZ cubic phase and moderates time-dependent conductivity during high-temperature annealing. Bang (2008) has found that a small amount of Ta_2O_5 doping also is a potential promising way for improving thermal stability of YSZ without a significant sacrifice of ionic conductivity. For comparison, the excess amounts of Li_2O in 4YSZ destabilized the crystal structure and induced a tetragonal to monoclinic phase transformation that negatively impacts ionic conductivity. However, it is necessary to take into account that the selection of additional dopants and their concentration should be careful because this doping can be accompanied by other effects. For example, the electron conductivity can appear. In particular, multivalent-cation dopants, such as Ce, Pr, Tb, and Ti, introduce electronic conductivity into YSZ. It is also important to note that the introduction of electronic conductivity can degrade the oxygen-ion conductivity. For example, if the electronic conductivity is established by a cation whose predominant valency is 3^+, the oxygen-ion conductivity of YSZ can decrease. This is due to the increased oxygen-vacancy concentration necessary to compensate for the placement of the 3^+ cations on a Zr^{4+} cation sites (Worrell and Wang 2001).

So, this short preview testifies that the aging and temporal change in properties are peculiar to all materials used in gas sensors independent of their nature and these effects should be taken into account during the design of materials for gas sensors of all types. We have to recognize that the selection of optimal approach is a complicated task because of the high stability clashes very often with obtaining other important operational parameters such as sensitivity and response time.

References

Alcock CB (ed) (1968) Electromotive force measurements in high-temperature systems. Institution of Mining and Metallurgy, London, pp 125–144

Badwal SPS (1992) Zirconia-based solid electrolytes: microstructure, stability and ionic conductivity. Solid State Ion 52:23–32

Badwal SPS, Ciacchi FT, Milosevic D (2000) Scandia–zirconia electrolytes for intermediate temperature solid oxide fuel cell operation. Solid State Ion 136–137:91–99

Bang S (2008) Ionic conductivity and phase stability of yttria stabilized zirconia doped with monovalent and pentavalent cations for solid oxide fuel cell electrolyte applications. PhD Thesis. University of California, Irvine, USA

Bendahan M, Lauque P, Lambert-Mauriat C, Carchano H, Seguin JL (2002) Sputtered thin films of CuBr for ammonia microsensors: morphology, composition and ageing. Sens Actuators B 84:6–11

Connolly EJ, Timmer B, Pham HTM, Groeneweg J, Sarro PM, Olthuis W, French PJ (2005) A porous SiC ammonia sensor. Sens Actuators B 109:44–46

Fergus JW (2008) A review of electrolyte and electrode materials for high temperature electrochemical CO_2 and SO_2 gas sensors. Sens Actuators B 134:1034–1041

Han PG, Wong H, Poon MC (2001) Sensitivity and stability of porous polycrystalline silicon gas sensor. Coll Surf A 179:171–175

Hattori M, Takeda Y, Sakaki Y, Nakanishi A, Ohara S, Mukai K, Lee J-H, Fukui T (2004) Effect of aging on conductivity of yttria stabilized zirconia. J Power Sources 126:23–27

Kerlau M, Merdrignac-Conanec O, Reichel P, Barsan N, Weimar U (2006) Preparation and characterization of gallium (oxy)nitride powders: preliminary investigation as new gas sensor materials. Sens Actuators B 115:4–11

Kondo H, Sekino T, Kusunose T, Nakayama T, Yamamoto Y, Niihara K (2003) Phase stability and electrical property of NiO-doped yttria-stabilized zirconia. Mater Lett 57:1624–1628

Kondoh J, Kawashima T, Kikuchi S, Tomii Y, Ito Y (1998a) Effect of aging on yttria-stabilized zirconia. J Electrochem Soc 145:1527–1536

Kondoh J, Kikuchi S, Tomii Y, Ito Y (1998b) Effect of aging on yttria-stabilized zirconia. II. A study of the effect of the microstructure on conductivity. J Electrochem Soc 145:1536–1550

Korotcenkov G (2007) Metal oxides for solid state gas sensors. What determines our choice? Mater Sci Eng B 139:1–23

Korotcenkov G, Cho BK (2011) Instability of metal oxide-based conductometric gas sensors and approaches to stability improvement. Sens Actuators B 156:527–538

Kumar D, Sharma RC (1998) Advances in conductive polymers. Eur Polym J 34:1053–1060

Lima JPH, de Andrade AM (2009) Stability study of conducting polymers as gas sensors. In: Proceedings of 11th international conference on advanced materials, ICAM 2009, Sept. 20–25, Rio de Janeiro, Brazil, p I566

Moghadam FK, Stevenson DA (1982) Influence of annealing on the electrical conductivity of polycrystalline ZrO_2+8 wt% Y_2O_3. J Am Ceram Soc 65:213–216

Morita M, Ohmi T, Hasegawa E, Kawakami M, Ohwada M (1990) Growth of native oxide on a silicon surface. J Appl Phys 68(3):1272–1281

Nomura K, Mizutani Y, Kawai M, Nakamura Y, Yamamoto O (2000) Aging and Raman scattering study of scandia and yttria doped zirconia. Solid State Ion 132:235–239

Pasierb P, Komornicki S, Kozinski S, Gajerski R, Rekas M (2004) Long-term stability of potentiometric CO_2 sensors based on Nasicon as a solid electrolyte. Sens Actuators B 101:47–56

Pei Q, Inganas O (1992) Poly(3-octylthiophene-co-3-methylthiophene), a processible and stable conducting copolymer. Synth Met 45:353–357

Pei Q, Inganaes O, Gustafsson G, Granstrom M, Andersson M (1993) Routes toward processible and stable conducting poly(thiophene)s. Synth Met 55:1221–1226

Rahman MS, Pal U, Choudhury AK, Maiti S (1991) New conducting polymers, 3.* Doping, stability, electrical, and optical characteristics of poly-(p-phenylphosphoethynediyl). Colloid Polym Sci 269:576–582

Razumovskii SD, Zaikov GY (1982) Effect of ozone on saturated polymers. Polym Sci USSR 24(10):2305–2325

Tourillon G, Garnier F (1983) Stability of conducting polythiophene and derivatives. J Electrochem Soc 30:2042–2044

Wang Y, Rubner MF (1990) Stability studies of the electrical conductivity of various poly (3-alkylthiophenes). Synth Met 39:153–175

Wang Y, Rubner MF, Buckley J (1991) Stability studies of electrically conducting polyheterocycles. Synth Met 41–43:1103–1108

Worrell WL, Wang C (2001) The stability, mixed-conductivity and applications of cation-doped yttria-stabilized zirconia (YSZ). In: Proceedings of 2001 Joint International Meeting—the 200th Meeting of The Electrochemical Society and the 52nd Annual Meeting of the International Society of Electrochemistry, San Francisco, California, September 2–7, Abstract 1534

Yamamoto O, Arachi Y, Takeda Y, Imanishi N, Mizutani Y, Kawai M, Nakamura Y (1995) Electrical conductivity of stabilized zirconia with ytterbia and scandia. Solid State Ion 79:137–146

Chapter 19
Factors Controlling Stability of Polymers Acceptable for Gas Sensor Application

Organic materials are by nature more susceptible to chemical degradation from, e.g., oxygen and water than inorganic materials. A number of studies have been carried out, and they show that the stability/degradation issue is rather complicated and certainly not yet fully understood though progress has been made (Scurlock et al. 1995; Abdou et al. 1997; Cumston and Jensen 1998; Dam et al. 1999; Katz et al. 2001; Yang et al. 2005; Jørgensen et al. 2008).

It should be noted that, in contrast to the measure of the magnitude of sensor response, there is no single indicator for stability, and for this reason most reports on stability of polymer gas sensors employ different measures and different experimental conditions. It is thus impossible to compare stability reports directly due to the differences in materials, conditions, data acquisition, etc. Therefore, in this chapter, the analysis of polymer-based gas sensor's stability will be only general.

19.1 Polymer Degradation

Polymer degradation is a change in the properties—tensile strength, color, shape, conductivity, etc.—of a polymer or polymer-based product under the influence of one or more environmental factors such as heat, light, and chemicals. Degradation agents and polymers most susceptible to their influence are given in Table 19.1.

While polymers that contain sites of unsaturation, such as polyisoprene and the polybutadienes, are most susceptible to oxygen and ozone oxidation, most other polymers also show some susceptibility to such degradation including polystyrene, polypropylene, nylons, polyethylenes, and most natural and naturally derived polymers (Carraher 2008). Most heterochained polymers, including condensation polymers, are susceptible to aqueous-associated acid or base degradation. Some polymer deterioration reactions occur without loss in molecular weight. These include a wide variety of reactions where free radicals (most typical) or ions are formed and cross-linking or other nonchain session reaction occurs. Cross-linking discourages chain and segmental chain movement. At times this cross-link is desired such as in permanent press fabric and in elastomeric materials. Often the cross-links bring about an increased brittleness beyond that desired. Some degradation reactions occur without an increase in cross-linking or a lessening in chain length. Thus, with minute amounts of HCl, water, ester, etc., elimination can occur with vinyl polymers giving localized sites of double bond formation. Because such sites are less flexible and more susceptible to further degradation, these reactions are generally considered as unwanted.

G. Korotcenkov, *Handbook of Gas Sensor Materials*, Integrated Analytical Systems, DOI 10.1007/978-1-4614-7388-6_19,

Table 19.1 Degradation agents and polymers most susceptible to their influence

Degradation agent	Most susceptible polymer types	Examples
Acids and bases	Heterochain polymers	Polyesters, polyurethanes
Moisture	Heterochain polymers	Polyesters, nylons, polyurethanes
High-energy radiation	Aliphatic polymers with quaternary carbons	Polypropylene, LDPE, PMMA, poly(alpha-methylstyrene)
Ozone	Unsaturated polymers	Polybutadienes, polyisoprene
Organic liquids/vapors	Amorphous polymers	
Biodegradation	Heterochain polymers	Polyesters, nylons, polyurethanes
Heat	Vinyl polymers	PVC, poly(alpha-methylstyrene)
Mechanical (applied stresses)	Polymers below T_g	

Source: Data from Carraher (2008)

Taking into account the mechanism of environment influence on polymer properties, one can select the following degradation reactions taking place in polymers:

1. Thermal degradation
2. Oxidative degradation
   - Photooxidation
   - Thermal oxidation
3. Hydrolysis degradation

19.1.1 Thermal Degradation

The way in which a polymer degrades under the influence of thermal energy in an inert atmosphere is determined, on the one hand, by the chemical structure of the polymer itself and, on the other, by the presence of traces of unstable structures (impurities or additions). Thermal degradation does not occur until the temperature is so high that primary chemical bonds are separated. Table 19.2 summarizes data related to polymer thermal degradation. Degradation of polymers begins typically at temperatures around 150–200 °C, and the rate of degradation increases as the temperature increases (Beyler and Hirschler 1995; Van Krevelen and Nijenhuis 2009). For many polymers, thermal degradation is characterized by the breaking of the weakest bond and is consequently determined by a bond dissociation energy. According to Beyler and Hirschler (1995), there are four general classes of chemical mechanisms important in the thermal decomposition of polymers: (1) random-chain scission, in which chain scissions occur at apparently random locations in the polymer chain; (2) end-chain scission, in which individual monomer units are successively removed at the chain end; (3) chain stripping, in which atoms or groups not part of the polymer chain (or backbone) are cleaved; and (4) cross-linking, in which a bond is created between polymer chains. It was established that, in all cases, the thermal degradation is usually a multistep free radical reaction with all the general features of such reaction mechanisms: initiation, propagation, branching, and termination steps.

When designing a material, there are several techniques that can be utilized to increase the temperature at which physical transformations in polymers occur (Beyler and Hirschler 1995). These strategies are generally aimed at increasing the stiffness of the polymer or increasing the interaction between polymer chains. It is clear that increasing the crystallinity of the polymer increases the interaction between polymer chains. Cross-linking also increases the melting temperature and, like stiffening, can render a material infusible. However, we need to take into account that indicated approaches strongly decrease swelling effect, which is usually used in polymer-based gas sensors.

Table 19.2 Data related to thermal degradation of polymers

Polymer	$T_{d,o}$ (K)	$T_{d,1/2}$ (K)
Poly(vinyl acetate)	–	542
Poly(vinyl chloride)	443	543
Poly(vinyl alcohol)	493	547
Poly(*a*-methyl styrene)	–	559
Poly(propylene oxide)	–	586
Poly(isoprene)	543	596
Cellulose	500	600
Poly(methyl acrylate)	–	601
Poly(methyl methacrylate)	553	610
Poly(ethylene oxide)	–	618
Poly(isobutylene)	–	621
Poly(*m*-methyl styrene)	–	631
Poly(styrene)	600	637
Poly(vinyl cyclohexane)	–	642
Poly(chloro-trifluoro ethylene)	–	653
Poly(propylene)	593	660
Poly(vinyl fluoride)	623	663
Poly(ethylene) (branched)	653	677
Poly(butadiene)	553	680
Poly(trifluoro ethylene)	673	685
Poly(methylene)	660	687
Poly(hexamethylene adipamide)	623	693
Poly(benzyl)	–	703
Poly(ε-caproamide) (Nylon 6)	623	703
Poly(*p*-xylylene) = poly(*p*-phenylene-ethylene)	–	715
Poly(ethylene terephthalate)	653	723
Poly(acrylonitrile)	563	723
Poly(dian terephthalate)	673	~750
Poly(dian carbonate)	675	~750
Poly(2,6-dimethyl *p*-phenylene oxide)	723	753
Poly(tetrafluoro ethylene)	–	782
Poly(*p*-phenylene-terephthalamide)	~720	~800
Poly(*m*-phenylene 2,5-oxadiazole)	683	~800
Poly (pyromellitide) (Kapton)	723	~840
Poly(*p*-phenylene)	>900	>925

Source: Data from Madorsky and Straus (1961), Arnold (1979), Van Krevelen and Nijenhuis (2009)

$T_{d,o}$ is the temperature of initial decomposition, i.e., this is the temperature at which the loss of weight during heating is just measurable; $T_{d,1/2}$ is the temperature of half decomposition, i.e., this is the temperature at which the loss of weight during pyrolysis reaches 50 % of its initial value

19.1.2 Oxidative Degradation

A polymer may also be degraded by chemical changes due to reaction with components in the environment (Van Krevelen and Nijenhuis 2009). The most important of these degrading reagents is oxygen. It was established that polymer degradation is almost always faster in the presence of oxygen (air), due primarily to the auto-accelerating nature of reactions between oxygen and carbon-centered radicals. Hence, most oxidation reactions are of an autocatalytic nature. If the oxidation is induced by

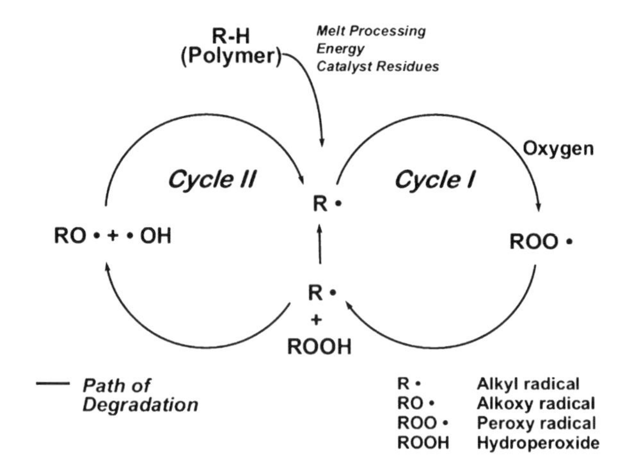

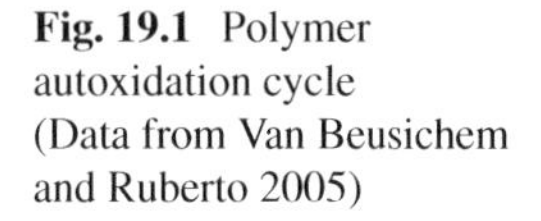
Fig. 19.1 Polymer autoxidation cycle (Data from Van Beusichem and Ruberto 2005)

Table 19.3 Products of polymer degradation

Product/function	Examples
Active products: intermediates of degradation	Alkyl radical (R•) (carbon-centered free radical)
	Peroxy radical (R–OO•) (oxygen-centered radical)
	Alkoxy radical (R–O•)
	Hydroperoxide (R–OOH → R–O• + •OH)
Inactive products: influence organoleptics	Alcohol (R–OH)
	Aldehyde (R–CHO)
	Ketone (R–C=O–R)
Modified properties	Polymer

Source: Data from Van Beusichem and Ruberto (2005)

light, the phenomenon is called photooxidation. If the oxidation is induced by purely thermal factors, the term thermal oxidation is used. In photooxidation, a radical is formed by the absorption of $h\nu$ and in thermal oxidation by $\triangle T$, shear, or even by residues of the polymerization catalysts. By the action of these factors on the polymer, free radicals are formed, which together with oxygen initiate a chain reaction. The mechanism of polymer degradation in the presence of oxygen is shown in Fig. 19.1. It is seen that interactions with oxygen lead to an increase in the concentration of polymer alkyl radicals ($R^{\bullet}$) and therefore to higher levels of scission and cross-linking products. Possible products of polymer degradation are presented in Table 19.3. We need to say that fragmentation reactions of oxygen-centered radicals ($RO^{\bullet}$) can yield new species (oxidation products) not found in polymers processed under air-free conditions.

19.1.2.1 Photochemical Oxidation

In photochemical degradation, the energy of activation is supplied by sunlight. The energies required to break single covalent bonds range, with few exceptions, from 165 to 420 kJ/mol. The indicated energies correspond to radiation of wavelengths from 720 to 280 nm. This means that the radiation in the near ultraviolet region (300–400 nm) is sufficiently energetic to break most single covalent bonds, except strong bonds such as C–H and O–H. The primary chain rupture or radical formation in the various photochemical processes is often followed by embrittlement due to cross-linking, but

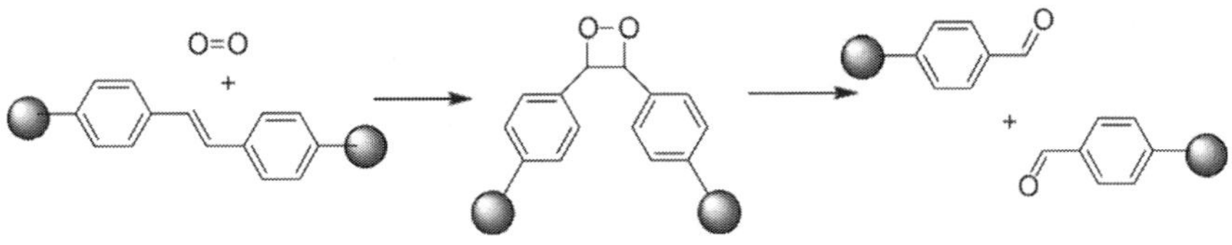

Fig. 19.2 Initial reaction of a PPV polymer with singlet oxygen. Singlet oxygen adds to the vinylene bond forming an intermediate dioxetane followed by chain scission. The aldehyde products shown can react further with oxygen (Reprinted with permission from Jørgense et al. 2008, Copyright 2008 Elsevier)

secondary reactions, especially in the presence of oxygen, cause further degradation of the polymer. The superoxide or hydrogen peroxide formed, generated due to oxygen activation by UV illumination, will then aggressively attack any organic substance present including the active polymers. As a result, properties of polymers, including optical, electrophysical, and mechanical, may deteriorate drastically. Nowadays the autocatalytic nature of the reaction is thought to be due to the decomposition of the hydroperoxides (Denisov 2000; Al-Malaika 2003):

$$ROOH \xrightarrow{hv} RO\cdot + \cdot OH \tag{19.1}$$

followed by

$$RO\cdot + RH \rightarrow ROH + R\cdot \tag{19.2}$$

$$HO\cdot + RH \rightarrow H_2O + R\cdot \tag{19.3}$$

The hydroperoxides also give rise to secondary reactions in which colored resinous products are formed (via carbonyl compounds).

Experiment has shown that some materials are more vulnerable to degradation than others, and therefore the task is to select polymers with desirable gas-sensing properties that are also resistant to chemical and photochemical degradation. For example, PPV-type polymers are especially prone to photochemical degradation. Scurlock et al. (1995) and Dam et al. (1999) investigated the singlet oxygen photodegradation of oligomers of phenylene vinylenes as model compounds for the degradation of PPVs. They believed that the singlet oxygen reacts with the vinylene groups in PPVs through a 2+2 cycloaddition reaction. The intermediate adduct may then break down, resulting in chain scission. They found that the rate of reaction with oxygen depends greatly on the nature of the substituents. Electron-donating groups enhance the rate while electron-withdrawing substituents retard the rate. A possible scheme of PPV photodegradation is shown in Fig. 19.2.

Poly-3-hexylthiophene (P3HT) is significantly more stable, but devices based on this material are also susceptible to chemical degradation (Jørgensen et al. 2008). The reaction of P3HT with oxygen has not yet been investigated in any detail. It is known however (Abdou et al. 1997) that poly (3-alkylthiophenes) form charge transfer complexes with oxygen as shown in Fig. 19.3. Most pure, organic synthetic polymers (polyethylene, polypropylene, poly(vinyl chloride), polystyrene, etc.) do not absorb at wavelengths longer than 300 nm owing to their ideal structure and hence should not be affected by sunlight. However, even these polymers, for example PPy, often do degrade when subjected to sunlight. It was found that this effect has been attributed to the presence of small amounts of impurities or structural defects, which absorb light and initiate the degradation (Rabek 1995; Fang et al. 2002). Although the exact nature of the impurities or structural defects responsible for the photosensitivity is not known with certainty, it is generally accepted that these impurities are various types of carbonyl groups (ketones, aldehydes) and peroxides. We also need to take into account that the

Fig. 19.3 Reversible formation of a charge transfer complex between poly(3-alkylthiophenes), P3AT, and oxygen. R represents an alkyl group (Reprinted with permission from Jørgense et al. 2008, Copyright 2008 Elsevier)

effect of sunlight on the rate of oxidation may be exacerbated by the presence of atmospheric pollutants capable of being activated to free radical species. This is particularly true of nitrogen and sulfur oxides, which are frequently components of industrial atmospheres. For example, the rate of physical deterioration of the polymer such as polyolefins (e.g., polypropylene) in the presence of nitrogen oxides may be accelerated by almost an order of magnitude by light. This may severely limit the use of polyolefins in the outdoor environment (Van Krevelen and Nijenhuis 2009).

It was established that the presence of heterogeneous catalysts, usually used for surface functionalizing of gas-sensing materials, accelerates the degradation of polymers as well. It was assumed that heterogeneous catalysts may also take on a role similar to that of oxygen to provide large amounts of radicals. For example, cobalt compounds were reported to accelerate the degradation of polymers and organic compounds such as polypropylene, low-density polyethylene, and cyclohexane (Perkas et al. 2001; Anipsitakis et al. 2005; Roy et al. 2005) due to their ability to produce radicals by electron transfer in the 3D subshell (Roy et al. 2005). Other transition metals such as Ni and Fe also have the potential to participate in the radical formation reactions (Osawa 1988; Perkas et al. 2001; Anipsitakis and Dionysiou 2004). However, cobalt compounds show the strongest catalytic effect in certain environments.

It should be noted that photochemical degradation strongly depends on temperature. At normal temperature, polymers generally react so slowly with oxygen that the oxidation only becomes apparent after a long time. For instance, if polystyrene is stored in air in the dark for a few years, the UV spectrum does not change perceptibly. On the other hand, if UV light under similar conditions irradiates the same polymer for 12 days, strong bands appear in the spectrum (Van Krevelen and Nijenhuis 2009). The same applies to other polymers such as polyethylene and natural rubber. Therefore, in essence, the problem is not the oxidizability as such but the synergistic action of various factors like electromagnetic radiation and thermal energy on the oxidation.

19.1.2.2 Thermal Oxidation

In the absence of light, most polymers are stable for very long periods at ambient temperatures. However, above room temperature many polymers start to degrade in an air atmosphere even without the influence of light. For example, a number of polymers show a deterioration of mechanical properties after heating for some days at about 100 °C and even at lower temperatures (e.g., polyethylene, polypropylene, poly(oxy methylene), and poly(ethylene sulfide)). Measurements have shown that the oxidation at 140 °C of low-density polyethylene increases exponentially after an induction period of 2 h. It was concluded that thermal oxidation, like photooxidation, is caused by autoxidation, the difference merely being that the radical formation from the hydroperoxide is now activated by heat. The primary reaction can be a direct reaction with oxygen (Van Krevelen and Nijenhuis 2009):

$$RH + O_2 \rightarrow R\cdot + \cdot OOH \qquad (19.4)$$

We need to note that thermal oxidation of polymers can be accompanied by combustion. Although some polymers such as PVC are not readily ignited, most organic polymers, like hydrocarbons, will burn. Some will support combustion, such as polyolefins, SBR, wood, and paper, when lit with a match or some other source of flame. Thermally, simple combustion of polymeric materials gives a complex of compounds that vary according to the particular reaction conditions. In particular, for vinyl polymers, thermal degradation in air (combustion) produces the expected products of water, carbon dioxide (or carbon monoxide if insufficient oxygen is present), and char along with numerous hydrocarbon products (Carraher 2008). Application of heat under controlled conditions can result in true depolymerization generally occurring via an unzipping. Such depolymerization may be related to the ceiling temperature of the particular polymer. Polymers such as PMMA and poly(alpha-methylstyrene) depolymerize to give large amounts of monomer when heated under the appropriate conditions. Thermal depolymerization generally results in some char and formation of smaller molecules including water, methanol, and carbon dioxide.

19.1.3 Hydrolytic Degradation

Hydrolytic degradation plays a part if hydrolysis is the potential key reaction in the breaking of bonds, as in polyesters and polycarbonates (Van Krevelen and Nijenhuis 2009). Attack by water may be rapid if the temperature is sufficiently high; attack by acids depends on acid strength and temperature. Degradation under the influence of basic substances depends very much on the penetration of the agent; ammonia and amines may cause much greater degradation than substances like caustic soda, which mainly attack the surface. The amorphous regions are attacked first and most rapidly, but crystalline regions are not free from attack. It was established that, in general, condensation polymers that contain functional groups in the polymer chain, notably polyesters, polyamides, and polyurethanes, are much more subject to hydrolytic degradation and biodegradation than polymers containing a carbon–carbon backbone.

19.1.4 Conducting Polymers Dedoping

Dedoping (undoping) of conducting polymers can also be considered as a degradation mechanism for polymer-based gas sensors. Conducting polymers are used mainly in resistive-type sensors (chemiresistors, TFT, and FET), and therefore the change in film conductivity causes both the drift of sensor baseline and the change of sensor response. Undoped conjugated polymers are high-resistance semiconductors or insulators. For example, undoped conjugated polymers, such as polythiophenes and polyacetylenes, have an electrical conductivity of around 10^{-10} to 10^{-8} S/cm. Therefore conducting polymers should be doped for application in gas sensors. The mechanism of conductivity in conducting polymers was discussed in Chap. 3 (Vol. 1).

Figure 19.4 illustrates how strong the temporal change in electroconductivity of as-prepared conducting polymers, especially during the first days, can be. Lima and de Andrade (2009) deposited POMA thin films and after annealing at 60 °C for 30 min immersed these films in an HCl solution (pH 0.8) for 60 s. After drying with nitrogen, the sensors were kept in vials. The electrical resistance was measured daily. A great decrease can be seen in the electrical resistance values in the first 5 days, followed by a gradual increase afterwards. It is necessary to note that the effect of dedoping is typical for all conducting polymers, including polymers such as PPy and PTh, which are easily dedoped when they are exposed to air. For example, Jiang et al. (2005) reported that the sensitivity of PPy/PVA composite sensor was only maintained for 2 weeks, while the sensitivity of pure PPy sensor was

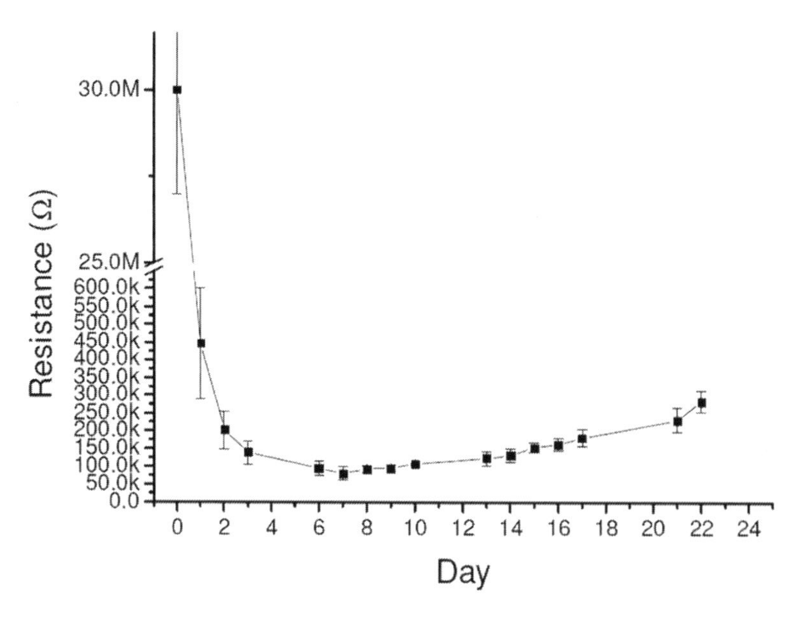

Fig. 19.4 Electrical resistance of POMA (3,000 rpm) in function of days after the chemical treatment in HCl (Data from Lima and de Andrade 2009)

maintained over 1 month. According to results obtained by Hosseini et al. (2005), the stabilities of doped polyaniline films with hydrogen halides, hydrogen cyanide, hydrogen sulfide, halogens, and halomethyl compounds were from 1 week to several months.

Of course, the stability of conductivity in air strongly depends on the dopants. For example, Wang et al. (1991) reported that polypyrroles doped with arylsulfonates were found to exhibit excellent stability in inert atmospheres but were slightly less stable in the presence of dry or humid air. Polypyrrole samples doped with the tosylate anion were found to be the most stable, while polypyrroles doped with longer sidechain-substituted benzenesulfonates exhibited poorer stability. It is believed that polymers with longer side chains are more flexible and thus more sensitive to the thermal undoping process. Wang et al. (1991) have shown that the tosylate-doped polypyrrole in humid air was approximately ten times more stable than the $FeCl_3$-doped poly(3-butylthiophene) (P3BT) in dry nitrogen. Budrowski et al. (1990) found that instability of chemically prepared PPy doped with ferric chloride at ambient conditions is caused by absorption of water molecules which leads to a weakening of hydrogen bonding between the polymer matrix and the anion. Li and Wan (1999) studied the stability of doped polyaniline films and established that over 6 months in air at room temperature the conductivity of PANI film doped with H_2SO_4, *p*-toluene sulfonic acid (*p*-TSA), and camphor sulfonic acid (CSA) changed little, while there was about a 40 % decrease in conductivity of PANI films doped with HCl and $HClO_4$. It is important to note that polypyrroles, polythiophenes, and polyaniline are considered as polymers with good stability of doped state. The stability of doped polyacetylene, polyphenylene, PPS, and PPV is considerably worse.

Research has shown that the dedoping process is accelerated strongly with increasing temperature. Rannou and Nechtschein (1999) carried out kinetic studies of the conductivity of doped PEDT thin films and found that the half-life of the conductivity in a device working in air at 100 °C is about 150 h only (see Fig. 19.5). Wang and Rubner (1990) have shown that the electrical conductivity of $FeCl_3$-doped poly(3-hexylthiophene) in a laboratory environment at $T=110$ °C is reduced to negligibly small values in approximately 1 h. Among other regularities established by Wang and Rubner (1990)

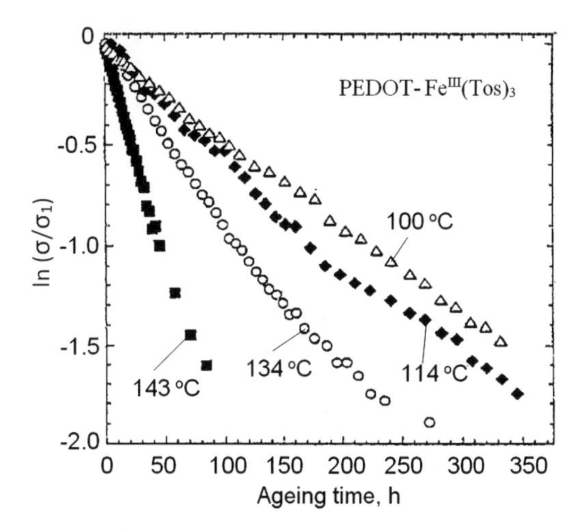

Fig. 19.5 Logarithm of the reduced conductivity vs. aging time for air-aged conductive PEDOT thin layer laid on polycarbonate substrate. Iron(III) tris-*p*-toluenesulfonate ($Fe^{III}(Tos)_3$) doped poly(3,4-ethylenedioxythiophene) (PEDOT-$Fe^{III}(Tos)_3$) thin layer had thickness of 500 nm (Reprinted with permission from Rannou and Nechtschein 1999, Copyright 1999 Elsevier)

during stability studies of P3BT, poly(3-hexylthiophene), poly(3-octylthiophene) (P3OT), and poly (3-decylthiophene), it is necessary to note the following: (1) the thermal stability of poly(alkyl thiophene)s doped with $FeCl_3$ by chemical methods is slightly better than those doped electrochemically; (2) the nature of the counterion affects the stability; and (3) the stability decreases when the length of the alkyl chain increases.

The conclusions of Truong (1992) are also important for understanding processes responsible for polymer aging. Truong (1992) studied polypyrrole films doped with *p*-toluenesulfonate and established the following. (1) At high aging temperature (120–150 °C), the oxygen diffusion into the bulk is predominant, i.e., the conductivity decay shows a $t^{1/2}$ dependence. The diffusion coefficient D derived from the experimental results is a material property independent of thickness. (2) At low aging temperatures (70–90 °C), conductivity degradation of the thick film (43 μm) apparently exhibits first-order reaction kinetics, which suggests chemical reaction-controlled degradation. This study has revealed that as the thickness increases and aging temperature decreases, thermal degradation of the PPy films exhibits a transition from a diffusion-controlled mechanism to a reaction-controlled mechanism. Therefore any extrapolation of data from high temperature to low temperature would be inappropriate.

Additional annealing of doped polymers also accelerates the process of dedoping. Li and Wan (1999) have shown that temporal change in film conductivity of doped polyaniline at room temperature is increased strongly after thermal treatment of doped polymer at $T=150$ °C. Results of this research are shown in Fig. 19.6. We need to recognize that the indicated effect limits the application of any thermal treatments during gas sensor fabrication. It was noted that the conductivity of PANI–CSA and PANI–*p*-TSA are the most stable below 200 °C, while the stability of PANI films doped with H_2SO_4 and H_3PO_4 are much better than the PANI–$HClO_4$ and PANI–HCl after thermal treatment at a high temperature (200 °C) (Li and Wan 1999).

It is important that processes of degradation and dedoping discussed above are typical for organic semiconductors as well. For example, Oester et al. (1993) studied oligothiophene thin films doped by $FeCl_3$ and found that $FeCl_3$ is not stable as a dopant in the presence of water and oxygen. After an exposure to the atmosphere for times in the order of 2 months the concentration of Cl in an organic semiconductor was beyond the detection limit of analytical instruments. For information, oligothiophenes (Fichow 1999) are a promising class of organic semiconductors, and their derivatives have been employed as the active layer in both chemiresistor and thin-film transistor gas sensors (Torsi et al. 2000; Chang et al. 2006). See also Chap. 20 (Vol. 1) and Chap. 10 (Vol. 2).

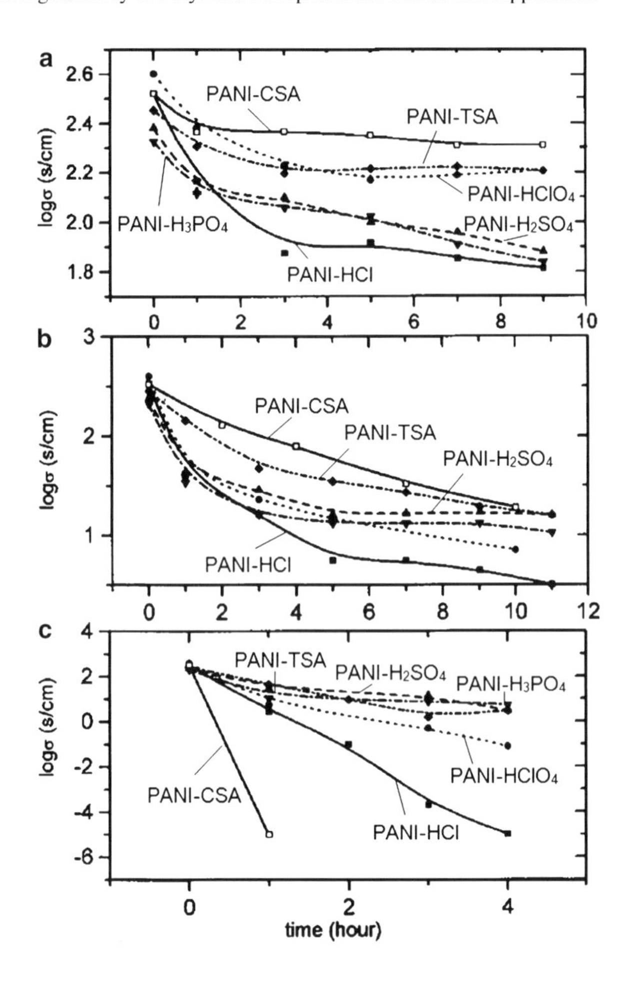

Fig. 19.6 Room-temperature conductivity of doped PANI films (PANI-HCl, PANI-$HClO_4$, PANI-H_2SO_4, PANI-H_3PO_4, PANI-TSA, and PANI-CSA) treated at different temperatures in air: (**a**) 100 °C, (**b**) 150 °C, (**c**) 200 °C. Free-standing films of doped PANI were synthesized by the doping–dedoping–redoping method (Data from Li and Wan 1999, John Wiley & Sons, Inc)

19.2 Approaches to Polymer Stabilization

It is necessary to recognize that numerous pieces of researches directed on the stabilization of polymers parameters have been carried during the last few decades and various stabilizers for polymers, which prevent various effects such as oxidation, chain scission and uncontrolled recombinations, and cross-linking reactions that are caused by photooxidation of polymers, were proposed (Scott 1981; Moss and Zweifel 1989; Al-Malaika 1994; Krohnke 2001; Van Krevelen and Nijenhuis 2009). Stabilizers are usually chemical substances which are added to polymers in small amounts (at most 1–2 w%) and are capable of trapping emerging free radicals or unstable intermediate products (such as hydroperoxides) in the course of autoxidation and transforming them into stable end products. Antioxidants, light stabilizers, antiozonants, UV absorbents, fire retardants, etc., are examples of such stabilizers. Several examples are listed in Table 19.4. Experiments have shown that activity and efficiency of the indicated stabilizers depend mainly on the following factors: (1) intrinsic stabilizer activity, which is influenced primarily by the structure of the molecule, including factors such as intramolecular interactions; (2) compatibility/mobility of the stabilizer which will again be determined by intra- and intermolecular interactions in the molecule but generally in the opposite direction to the above; and (3) volatility of the stabilizer which will be determined by molecular weight and molecular interaction in the polymer. So, for real applications, stabilizers should correspond to several

Table 19.4 Materials used as stabilizers for polymers

Material	Function	Mechanism of stabilization
Hindered phenols; aromatic amines; benzofuranones, etc.	Antioxidants	Antioxidants are used to terminate the oxidation reactions taking place due to different weathering conditions and reduce the degradation of organic materials. Antioxidants function by interfering with radical reactions that lead to polymer oxidation and, in turn, to degradation. Primary antioxidants are generally radical scavengers or H-donors. Secondary antioxidants are typically hydroperoxide decomposers
Organosulfur compounds; hindered amines; organo-phosphates, etc.	Thermal stabilizers	Thermal stabilizers are antioxidants that protect polymers during thermal treatments. For example, organosulfur compounds are efficient hydroperoxide decomposers, which thermally stabilize the polymers, while hindered amines efficiently scavenge radicals which are produced by heat
Paraffin waxes; ethylene diurea (EDU); *p*-phenylenedi-amines, etc.	Antiozonants	Antiozonants prevent or slow down the degradation of material caused by ozone gas in the air. For example, paraffin waxes form a surface barrier for ozone
Oxanilides (for polyamides); benzophenones (for PVC); benzotriazoles and hydroxyphenyltriazines (for polycarbonate); TiO_2, etc.	UV absorbers	The UV absorbers dissipate the absorbed light energy from UV rays as heat by reversible intramolecular proton transfer. This reduces the absorption of UV rays by polymer matrix and hence reduces the rate of weathering
Hindered amine	Light stabilizers	Hindered amine scavenges radicals which are produced by light. This effect may be explained by the formation of nitroxyl radicals
Nickel compounds (for polyolefins)	Quenchers	A quencher induces harmless dissipation of the energy of photoexcited states
Magnesium or aluminum hydroxide; organobromine compounds; sodium carbonate, etc.	Fire retardant	When heated, hydroxides dehydrate to form aluminum oxide, releasing water vapor in the process. This reaction absorbs a great deal of heat, cooling the material over which it is coated. Brominated flame retardants have an inhibitory effect on the ignition of combustible organic materials. Sodium carbonate, which releases carbon dioxide when heated, shields the reactants from oxygen

Source: Data from Krohnke (2001), Mark (2007), Van Krevelen and Nijenhuis (2009)

requirements such as compatibility with the polymer, nonvolatility, light fastness, heat stability, and, for some applications, resistance to water and high humidity.

There are also other approaches to polymer stabilization (Aldiss 1989). For example, it was found that the formation of composites of two conducting polymers, one of which is air stable, improves the stability of polymer materials. Experiments carried out with pyrrole/polyacetylene and polyaniline/polyacetylene composites have shown that the composites appeared to be more stable than doped polyacetylene and possessed mechanical properties similar to polyacetylene. Stabilization can also be achieved chemically by copolymerization. In particular, it was found that copolymerization of acetylene with other monomers such as styrene, isoprene, ethylene, or butadiene was accompanied by the increase of improvement of polymer stability (Aldiss 1989). Crispin et al. (2003) established that

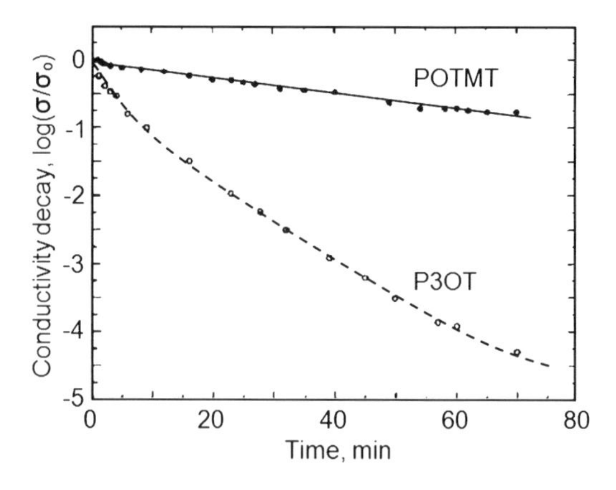

Fig. 19.7 Conductivity decay of copolymer POTMT at 110 °C in laboratory air compared with that of poly(3-octylthiophene) (P3OT). The samples were doped with $FeCl_3$ (Reprinted with permission from Pei and Inganas 1992, Copyright 1992 Elsevier)

poly(3,4-ethylenedioxythiophene) (PEDT or PEDOT) degrades under the influence of UV. UV light induces the photooxidation of the conjugated PEDT chains, which is accompanied by a reduction in electrical conductivity. Photooxidation leads to shorter conjugation lengths because of sulfone group formation and chain scission accompanied by the addition of carboxyl groups. The photooxidation mechanism of PEDT is similar to the case of polythiophene: a π–π^* transition generated by UV light is followed by energy transfer to oxygen, which consequently reaches its singlet state and reacts with the conjugated chain. The copolymerization of PEDT and poly(styrene sulfonate) (PSS) clearly helps to block this degradation pathway. However it should be noted that the degradation was not blocked completely. Due to hydroscopicity of PSS, in the presence of ambient humidity, additional effects occur on the PSS portion of the PEDT/PSS polymer blend. The degradation process as a whole leads to the appearance of nitrogen at the surface of PSS exposed to UV light in air. This occurs through the formation of ammonium sulfate salts formed in a complex chain reaction. Pei and Inganas (1992) have shown that copolymers of 3-octylthiophene and 3-methylthiophene (POTMT) doped by $FeCl_3$ are more stable than doped P3OT (see Fig. 19.7).

Surface protection gives positive results as well. It was found that when polyacetylene is doped, the charge transfer complex is stable under vacuum or inert atmosphere. Thus, if the system is protected against moisture and especially against oxygen, the system would be stable as well. A large variety of plastic polymers is available for this purpose. For instance, a 50-μm polyvinylchloride (PVC) film could keep the efficiency of a heterojunction [doped or undoped $(CH)_x$/conventional semiconductor] almost constant for several months. When a cell with the same characteristics was left in air, its properties deteriorated after 1–2 days (Aldiss 1989).

Magnoni et al. (1996) have shown that rational synthesis of polymer with a favored topology of the dopant molecule also is a very promising approach to improve stability of doped polymers. This concept based on results of simple quantum mechanical calculations (Lopez-Navarrete and Zerbi 1990) supposes that the position in space of the dopant relative to the polymer molecule is an important factor in determining the chemical stability of these systems. If the dopant can sit near the molecule in an energetically favorable position, the charge transfer (CT) bond becomes stronger, thus favoring stability. It follows that the path of approach of the donor/acceptor dopant molecule to the acceptor/donor polymer is determined by the free volume the approaching dopant molecule finds along its reaction coordinate. The steric hindrance of the side group attached to the backbone chain then plays the dominant role in affecting the strength, hence the stability, of the CT bond. In particular, the preferred site of attack is found where the steric hindrance is minimized. Following these principles, regiospecific poly(3,3″-dihexyl-2,2′:5′,2″-terthiophene) (PDHTT) was synthesized and doped with $FeCl_3$. A method for the synthesis of highly regiospecific polyalkylthiophenes with a large content of head-to-tail

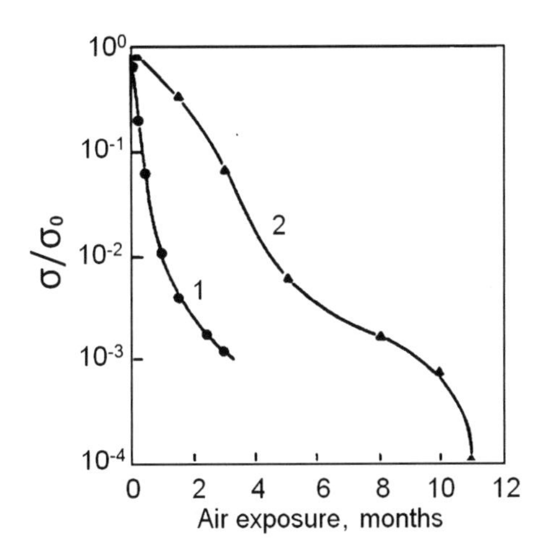

Fig. 19.8 Conductivity vs. air exposure time for I_2-doped polyacetylene (1) in the absence of and (2) in the presence of a antioxidizing agent (2-terbutyl,6-methyl phenol) (Data from Aldiss 1989)

linking was developed by McCullough et al. (1993a, b). Experiments carried out by Magnoni et al. (1996) have shown that the regular periodic chemical structure of PDHTT opens periodic pockets along the polymer backbone large and ordered enough to allow the dopant to enter and develop preferential CT interactions with the delocalized π-electrons. As a result, PDHTT in the doped state showed very great stability in air even at relatively high temperatures. In PDHTT chemically doped with $FeCl_3$, the charge carriers remained unaltered even after 1 year of standing as films in open air and at room temperature. The experiments carried out at $T=140$ °C and $T=170$ °C showed that the dedoping of samples of PDHTT doped with $FeCl_3$ is certainly much slower than that observed for other materials of the class of polyalkylthiophenes (Wang and Rubner 1990). At $T=100$ °C, the concentration of Cl in PDHTT was decreased twice during 150 h. For comparison, the concentration of I in iodine-doped PDHTT was decreased twice at $T=100$ °C during 15 min. Magnoni et al. (1996) believe that the above-mentioned approach is possibly useful for the synthesis of other stable-doped conjugated polymer materials. According to the opinion of Magnoni et al. (1996), the best way to achieve these conditions is by regiospecific synthesis, which allows the spacing between side chains to be modulated at will in order to fit the space required by the dopant molecule.

However, it should be noted that the proposed decisions only retarded the process of polymer degradation. Most polymers are susceptible to degradation under natural radiation, sunlight, and high temperatures, even in the presence of antioxidants. Results presented in Fig. 19.8 illustrate this situation. As seen, the presence of antioxidant in I_2-doped polyacetylene decreases the rate of degradation only.

Moreover, it is necessary to take into account that in many cases the incorporation of stabilizers in a polymer or additional surface protection layer is accompanied by strong deterioration of sensing characteristics, because to achieve high stability we have to decrease both polymer reactivity and gas penetrability. Therefore, stabilization of polymer material aimed for gas sensor applications continues to be an important technical field with a lot of industrial and scientific attention. In addition, we should note that there are some fundamental restrictions for the achievement of a required temporal and thermal stability of polymers with high reactivity. First, the unsaturated bonds in conducting polymers, which participate in gas-sensing effects, are often very reactive when exposed to environmental agents such as oxygen or moisture. Second, information presented in Razumovskii and Zaikov (1982) and Bhuiyan (1984) for nylon-type polymer shows that the melting point of the polymer decreases as the chain length increases. This means that a polymer's complication will inevitably be accompanied by a reduction of melting temperature (see Fig. 19.9) and, therefore, by a drop in the polymer's stability. The above-mentioned confirms again that the problem of stability and reliability of gas sensors is a determinant for practical use of any gas-sensing material.

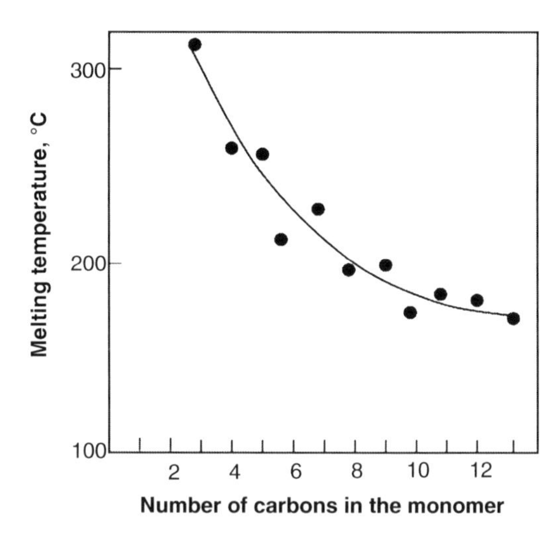

Fig. 19.9 Influence of the number of carbon atoms in the monomer on approximate melting temperature of nylon-type polymers (Experimental data were extracted from Bhuiyan 1984)

References

Abdou MSA, Orfino FP, Son S, Holdcroft S (1997) Interaction of oxygen with conjugated polymers: charge transfer complex formation with poly(3-alkylthiophenes). J Am Chem Soc 119:4518–4524

Aldiss M (1989) Inherently conducting polymers: processing, fabrication, applications, limitations. Noyes Data Corporation, Park Ridge

Al-Malaika S (1994) Some aspects of polymer stabilization. Int J Polym Mater 24(1–4):47–58

Al-Malaika S (2003) Oxidative degradation and stabilization of polymers. Intern Mater Rev 48:165–185

Anipsitakis GP, Dionysiou DD (2004) Radical generation by the interaction of transition metals with common oxidants. Environ Sci Technol 38(13):3705–3712

Anipsitakis GP, Stathatos E, Dionysiou DD (2005) Heterogeneous activation of ozone using Co_3O_4. J Phys Chem B 109(27):13052–13055

Arnold C (1979) Stability of high-temperature polymers. J Polym Sci Macromol Rev 14:265–378

Beyler CL, Hirschler MM (1995) Thermal decomposition of polymers. In: DiNenno PJ (ed) The SFPE handbook of fire protection engineering (Chaps. 1–7), 2nd edn. NFPA, Quincy, MA, pp 110–131

Bhuiyan AL (1984) Some aspects of the thermal stability action of the structure in aliphatic polyamides and polyacrylamides. Polymer 25:1699–1710

Budrowski C, Przytuski J, Kucharski Z, Suwalski J (1990) Stability of doped polypyrrole studied by Mossbauer spectroscopy. Synth Met 35:151–154

Carraher CE Jr (2008) Polymer chemistry. CRC, Boca Raton

Chang JB, Liu V, Subramanian V, Sivula K, Luscombe C, Murphy A, Liu J, Frechet MJ (2006) Printable polythiophene gas sensor array for low cost electronic noses. J Appl Phys 100:014506

Crispin X, Marciniak S, Osikowicz W, Zotti G, Van der Gon AWD, Louwet F, Fahlman M, Groenendaal L, de Schryrver F, Salaneck WR (2003) Conductivity, morphology, interfacial chemistry, and stability of poly(3,4-ethylene dioxythiophene)–poly(styrene sulfonate): a photoelectron spectroscopy study. J Polym Sci B Polym Phys 41:2561–2583

Cumston BH, Jensen KF (1998) Photooxidative stability of substituted poly(phenylene vinylene) (PPV) and poly(phenylene acetylene) (PPA). J Appl Polymer Sci 69:2451–2458

Dam N, Scurlock RD, Wang B, Ma L, Sundahl M, Ogilby PR (1999) Singlet oxygen as a reactive intermediate in the photodegradation of phenylenevinylene oligomers. Chem Mater 11:1302–1305

Denisov ET (2000) Polymer oxidation and antioxidant action. In: Hamid SH (ed) Handbook of polymer degradation (Chap. 9), 2nd edn. Marcel Dekker, New York

Fang Q, Chetwynd DG, Gardner JW (2002) Conducting polymer films by UV-photo processing. Sens Actuators A 99:74–77

Fichow D (ed) (1999) Handbook of oligo- and polythiophenes. Wiley-VCH, Weinheim

Hosseini SH, Oskooei SHA, Entezami AA (2005) Toxic gas and vapour detection by polyaniline gas sensors. Iran Polym J 14(4):333–344

Jiang L, Jun H-K, Hoh Y-S, Lim J-O, Lee D-D, Huh J-S (2005) Sensing characteristics of polypyrrole–poly(vinyl alcohol) methanol sensors prepared by in situ vapor state polymerization. Sens Actuators B 105:132–137

Jørgensen M, Norrman K, Krebs FC (2008) Stability/degradation of polymer solar cells. Sol Energy Mater Sol Cells 92:686–714

Katz EA, Faiman D, Tuladhar SM, Kroon JM, Wienk MM, Fromherz T, Padinger F, Brabec CJ, Sariciftci NS (2001) Temperature dependence for the photovoltaic device parameters of polymer-fullerene solar cells under operating conditions. J Appl Phys 90:5343–5350

Krohnke C (2001) Polymer stabilization. In: Buschow KHJ, Cahn RW, Flemings MC, Ilschner B, Kramer EJ, Mahajan S, Veyssière P (eds) Encyclopedia of materials: science and technology. Elsevier, Oxford, pp 7507–7516

Li W, Wan M (1999) Stability of polyaniline synthesized by a doping–dedoping–redoping method. J Appl Polymer Sci 71:615–621

Lima JPH, de Andrade AM (2009) Stability study of conducting polymers as gas sensors. In: Proceedings of 11th international conference on advanced materials, ICAM, Rio de Janeiro, Brazil, 2009, Sept 20–25, pp I566

Lopez-Navarrete JT, Zerbi G (1990) On the stability of doped conducting polymers: electrostatic contributions and sterical effects. Chem Phys Lett 175:125–129

Madorsky SL, Straus S (1961) High temperature resistance and thermal degradation of polymers. SCI Monograph 13, pp 60–74

Magnoni MC, Gallazzi MC, Zerbi G (1996) Search for conducting doped polymers with great chemical stability: regiospecific poly(3,3″-dihexyl-2,2′:5′,2″-terthiophene). Acta Polym 47:228–233

Mark HF (ed) (2007) Encyclopedia of polymer science and technology, vol 5, 3rd edn. Wiley-Interscience, New York

McCullough R, Lowe RD, Jayaraman M, Anderson DL (1993a) Design, synthesis, and control of conducting polymer architectures: structurally homogeneous poly(3-alkylthiophenes). J Org Chem 58:904–912

McCullough R, Tristiam-Nagle S, Williams SP, Lowe RD, Jayaraman M (1993b) Self-oriented poly(3-alkylthiophenes): new insights on structure–property relationships in conducting polymers. J Am Chem Soc 115:4910–4911

Moss S, Zweifel H (1989) Degradation and stabilization of high density polyethylene during multiple extrusions. Polym Degrad Stabil 25(2–4):217–245

Oeter D, Ziegler C, Gopel W (1993) Doping and stability of ultrapure α-oligothiophene thin films. Synth Met 61: 147–150

Osawa Z (1988) Role of metals and metal-deactivators in polymer degradation. Polym Degrad Stabil 20(3–4): 203–236

Pei Q, Inganas O (1992) Poly(3-octylthiophene-co-3-methylthiophene), a processible and stable conducting copolymer. Synth Met 45:353–357

Perkas N, Koltypin Y, Palchik O, Gedanken A, Chandrasekaran S (2001) Oxidation of cyclohexane with nanostructured amorphous catalysts under mild conditions. Appl Catal A 209(1–2):125–130

Rabek JF (1995) Polymer photodegradation: mechanism and experimental methods. Chapman & Hall, London

Rannou P, Nechtschein M (1999) Ageing of poly(3,4-ethylenedioxythiophene): kinetics of conductivity decay and lifespan. Synth Met 101:474

Razumovskii SD, Zaikov GY (1982) Effect of ozone on saturated polymers. Polymer Sci USSR 24(10):2305–2325

Roy PK, Surekha P, Rajagopal C, Chatterjee SN, Choudhary V (2005) Effect of benzil and cobalt stearate on the aging of low density polyethylene films. Polym Degrad Stabil 90(3):577–585

Scott G (1981) Mechanism of polymer stabilization. In: Scott G (ed) Developments in polymer stabilization, 4th edn. Applied Science, London, pp 276–289

Scurlock RD, Wang B, Ogilby PR, Sheats JR, Clough RL (1995) Singlet oxygen as a reactive intermediate in the photodegradation of an electroluminescent polymer. J Am Chem Soc 117:10194–10202

Torsi L, Dodabalapur A, Sabbatini L, Zambonon PG (2000) Multi-parameter gas sensors based on organic thin-film-transistors. Sens Actuators B 67:312–316

Truong V-T (1992) Thermal degradation of polypyrrole: effect of temperature and film thickness. Synth Met 52:33–44

Van Beusichem B, Ruberto MA (2005) Introduction to polymer additives and stabilization. A presentation to product quality research institute http://www.pqri.org/…/posters/Polymer_Additives_PQRI_Poster.pdf. Accessed 26 June 2011

Van Krevelen DW, Nijenhuis KT (2009) Properties of polymers. Their correlation with chemical structure; their numerical estimation and prediction from additive group contributions, 4th edn. Elsevier, Amsterdam

Wang Y, Rubner MF (1990) Stability studies of the electrical conductivity of various poly (3-alkylthiophenes). Synth Met 39:153–175

Wang Y, Rubner MF, Buckley J (1991) Stability studies of electrically conducting polyheterocycles. Synth Met 41–43:1103–1108

Yang X, Loos J, Veenstra SC, Verhees WJH, Wienk MM, Kroon JM, Michels MAJ, Janssen RAJ (2005) Nanoscale morphology of high-performance polymer solar cells. Nano Lett 5:579–583

Chapter 20
Instability of Metal Oxide Parameters and Approaches to Their Stabilization

As shown earlier, metal oxides are the most stable materials used for gas sensor design. However, one should recognize that the problem of parameters' instability also remains for metal oxides. It is known that the temporal drift of operating characteristics of conductometric gas sensors based on metal oxides (MOX), along with the low selectivity of sensor response, is considered to be a major disadvantage of these devices (Massok et al. 1995; Sayago et al. 1995; Barsan et al. 1999; Ozaki et al. 2000; Haugen et al. 2000; Romain and Nicolas 2010). According to Meixner and Lampe (1996) and Korotcenkov and Cho (2011), the main reasons for long-term instability of conductometric gas sensors are the change of the metal oxide parameters caused by structural and phase transformations taking place in these materials.

20.1 Role of Structural Transformation of Metal Oxides in Instability of Gas-Sensing Characteristics

As mentioned before, structural change in gas-sensing material, caused by the grain growth due to their coalescence, is one of the most important reasons for observed changes in the metal oxide gas sensor parameters during their fabrication and exploitation (Meixner and Lampe 1996; Pijolat et al. 1999; Korotcenkov 2007a, b; Korotcenkov and Cho 2011). For example, Nakamura (1989) established that the grain size in polycrystalline structures fabricated by SnO_2 powder sintering changed during 3 years of operation, from 5–15 to 20–40 nm. Matsuura and Takahata (1991) then observed a growth in grain size, even after 20 days of use. As shown by Korotcenkov and co-workers (Korotcenkov 2008; Korotcenkov et al. 2005a, 2009; Korotcenkov and Cho 2011), the above-mentioned changes in sensor parameters, caused by structure transformation of gas-sensing material, can be conditioned by the following processes.

Changing geometric sizes in the network of grains, forming gas-sensing matrix. This process is an unavoidable consequence of the grain growth. Due to coalescence and mass transfer from the smaller grain to the bigger one, we can observe the increase of the contact area between crystallites and the forming of necks between grains (Korotcenkov et al. 2005b; Korotcenkov 2005, 2008). One of the variants of possible transformation of intergrain contacts is shown in Fig. 20.1.

One can see in images shown in Fig. 20.2 how strong the change of morphology can be in a real situation. AFM and SEM images of the In_2O_3 films subjected to various thermal treatments are presented in Fig. 20.2. These results were presented by Korotcenkov et al. (2005a). It is seen that, depending on both temperature of thermal treatment and the crystallite size, the shape of the In_2O_3

G. Korotcenkov, *Handbook of Gas Sensor Materials*, Integrated Analytical Systems,
DOI 10.1007/978-1-4614-7388-6_20, © Springer Science+Business Media New York 2014

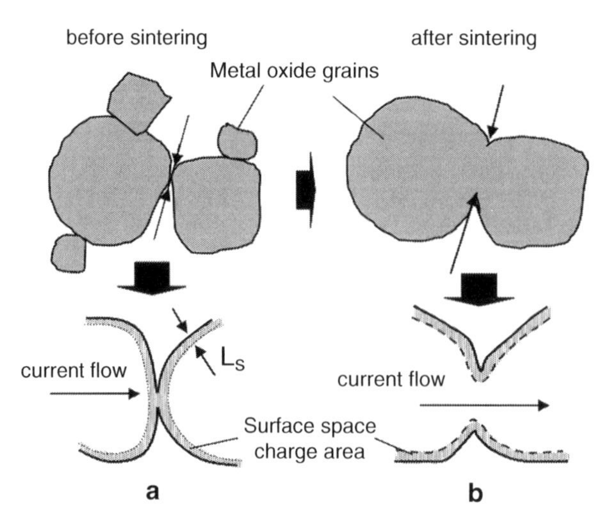

Fig. 20.1 Schematic diagram illustrating the change of the geometry of intergrain contacts in gas-sensing matrix during annealing: (**a**) the structure of as-deposited films, (**b**) after sintering

Fig. 20.2 Possible typical shapes of the In_2O_3 grains on different stages of annealing: (**a**) $d \sim 50$ nm, as deposited; (**b**) $d \sim 400$ nm, as deposited; (**c**) $d \sim 200$ nm, $T_{an} = 1{,}100$ °C (Reprinted with permission from Korotcenkov et al. 2005a, Copyright 2005 Elsevier)

crystallites can be changed completely. The same changes were observed for the SnO_2 grains (Brinzari et al. 2002; Korotcenkov et al. 2005b).

Taking into account that electroconductivity of metal oxide matrix can be represented as a network of grain conductivity and the resistance of intergrain contacts (Ciobanu et al. 1999; Barsan and Weimar 2001; Ulrich et al. 2004), it becomes clear that the structural changes in a gas-sensing matrix indicated above should strongly influence the conditions of the current transport along gas-sensing films. For example, due to the appearance of conducting channels between grains (see Fig. 20.1), the resistance of the matrix of sintered grains could be strongly decreased. The porosity of the gas-sensing matrix will change as well. As was established, parameters such as porosity and grain size are interdependent (De Souza Brito et al. 1995). Figure 20.3 shows that the increase of grains is accompanied by the increase of the pore diameter.

Changing the bulk electrophysical properties of grains. These changes also do not require special proof, because in numerous works it has been shown that grain growth could be accompanied by change of the bandgap, crystallographic structure of material, and concentration of point defects (Zhang and Liu 2000; Yang and Jiang 2006; Li and Li 2006; Rong 2005; Wang et al. 2008; Gryaznov and Trusov 1993).

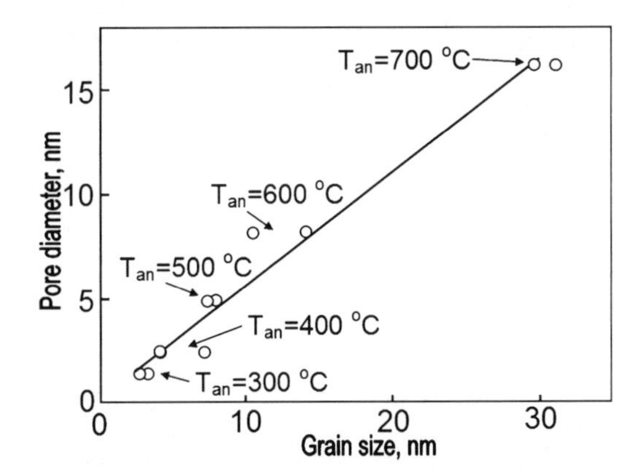

Fig. 20.3 Average pore size in the SnO_2 matrix as a function of mean crystallite size, changing during thermal annealing. SnO_2 gels were prepared from aqueous $SnC1_4{\cdot}5H_2O$, precipitated by aqueous ammonia at pH=11 (Adapted with permission from De Souza Brito et al. 1995, Copyright 1995 Elsevier)

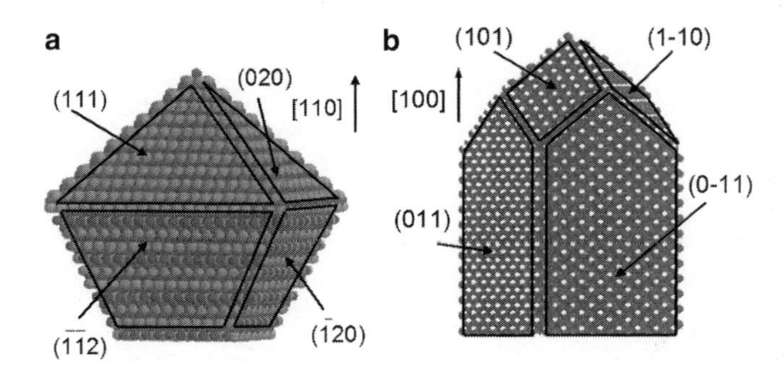

Fig. 20.4 Results of simulations on the base of XRD and HRTEM study for the SnO_2 films with the following parameters: SnO_2 films were deposited at T_{pyr}=450 °C; (**a**) thickness equaled 50–100 nm; (**b**) d~200 nm (Reprinted with permission from Korotcenkov et al. 2005b, Copyright 2005 Elsevier)

Changing electron and catalytic properties of the crystallites surface. These changes are also common enough because the reactivity of nanomaterials is mainly determined by the so-called smoothly scalable size-dependent properties, which are related to the fraction of atoms at the surface (Roduner 2006; Rao et al. 2002). As the particle size decreases, the surface-to-volume ratio increases proportionally to the inverse particle size. In addition, during the annealing process, not only crystallites' size but also their crystallographic faceting can be changed. Experimental results confirming this statement are shown in Figs. 20.4, 20.5, and 20.6. Korotcenkov et al. (2005b) have shown that the SnO_2 grains' faceting strongly depends on their size (Fig. 20.4), Gurlo (2011) has found that In_2O_3 crystallites depending on the size and technology of synthesis have (100) and (111) facets with different ratios, and Barnard and Curtiss (2005) established that the shape of metal oxide crystallites also depends on the surface chemistry (Fig. 20.5). Moreover, it was established that the change of particle size can be accompanied by changing the crystallographic structure. For example, a number of authors have found that the synthesis of nanocrystalline TiO_2 consistently resulted in anatase nanoparticles, which transformed to rutile upon reaching a particular size of ~11–17 nm (Gribb and Banfield 1997; Navrotsky 2001; Zhang et al. 2001). This means that, during the annealing process, which can be accompanied by changing of the grain size and the surface chemistry, either crystallographic planes participating in the intercrystallite contacts

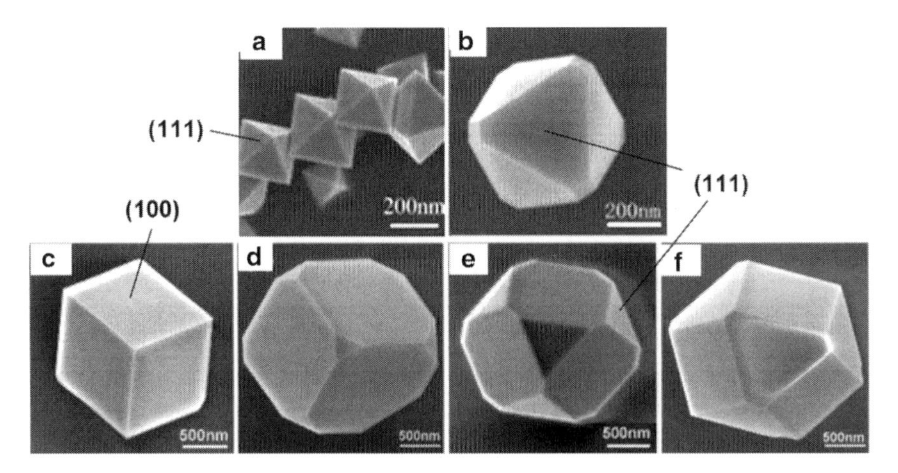

Fig. 20.5 (**a–f**) SEM images of c-In_2O_3 crystallites with corresponding schematic habits showing a transition from {111} to {100} morphology. (Reprinted with permission from Shi et al. 2007, Copyright 2007 American Chemical Society, and Gurlo 2011, Copyright 2011 Royal Society of Chemistry)

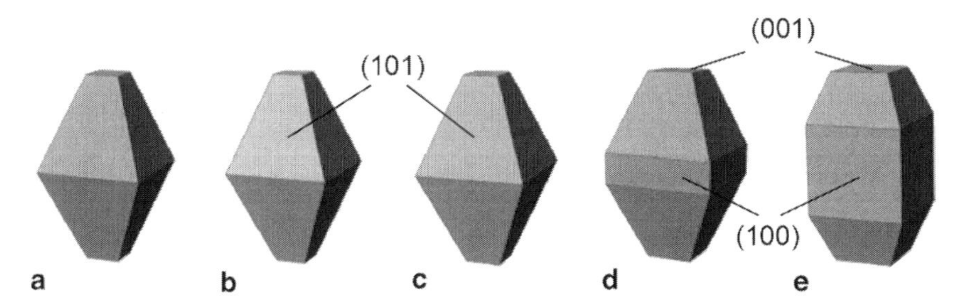

Fig. 20.6 Possible morphology predicted for anatase (TiO_2) with (**a**) hydrogenated surfaces, (**b**) with hydrogen-rich surface adsorbates, (**c**) hydrated surfaces, (**d**) hydrogen-poor adsorbates, and (**e**) oxygenated surfaces (Reprinted with permission from Barnard and Curtiss 2005, Copyright 2005 Royal Society of Chemistry)

forming or those participating in interaction with the surrounding atmosphere are being changed. It is important to note that, in some cases, nanocrystals have planes/surfaces, which are not available in bulk materials. It was established that in nanocrystals, the stability of surfaces could be reversed and high-energy surfaces, which are metastable, in bulk form become stable as the particle size decreases or surface chemistry changes (Arroyo et al. 2002).

Each crystallographic plane has its own combination of electron characteristics, determining adsorption, catalytic, and electrophysical properties (Korotcenkov 2005, 2008; Golovanov et al. 2005; Han et al. 2009a). An example of morphology influence on catalytic properties of metal oxides is shown in Fig. 20.7. It is seen that catalytic activity of ZnO nanocrystals with different morphology can change ten times.

Such strong influence of the grain faceting on electronic and catalytic properties of metal oxides means that gas-sensing properties of metal oxides are also structure sensitive. Experimentally this statement was confirmed by Han et al. (2009b). Han et al. (2009b) have shown that sensing characteristics of the SnO_2-based gas sensors are really very sensitive to crystal morphology and depend on the grain faceting (see Fig. 20.8). The same results were obtained for ZnO-based sensors as well (Han et al. 2009a). These results are presented in Fig. 20.9. One can find in a review paper (Gurlo 2011) more detailed analysis of metal oxide morphology influence on gas-sensing properties. From the above-mentioned discussions it follows that, at high dispersion in the grain size, the large set of crystallographic planes could participate in gas-sensing effects, and therefore sensor response would be

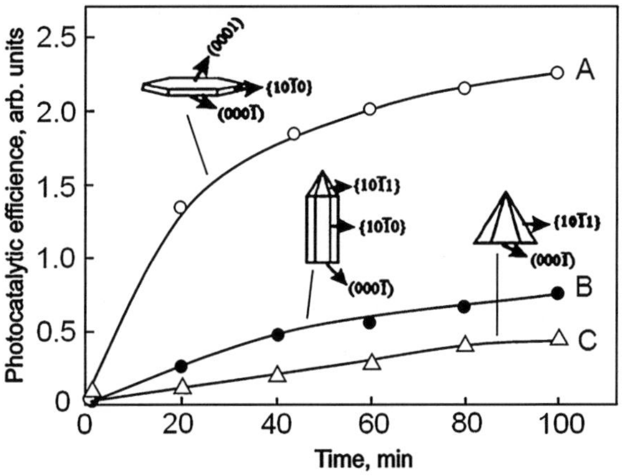

Fig. 20.7 The photocatalytic efficiencies of different ZnO samples after normalization with the surface area of the ZnO samples (Adapted with permission from Han et al. 2009a, Copyright 2009 American Chemical Society)

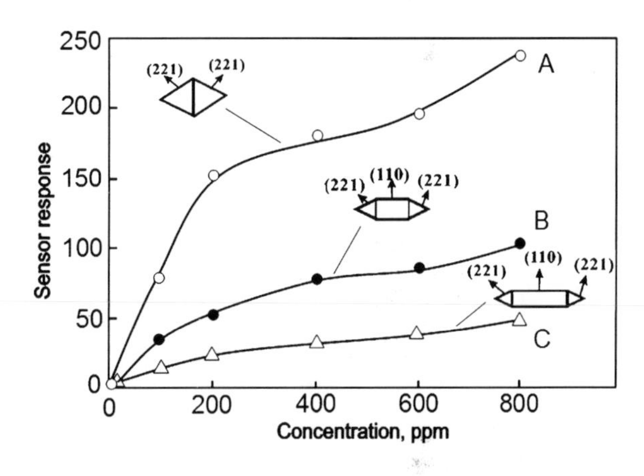

Fig. 20.8 Response to ethanol of the SnO_2 particles in dependence of crystal morphology, i.e., nature of surface exposed to ambient gas (Adapted with permission from Han et al. 2009b, Copyright 2009 John Wiley & Sons)

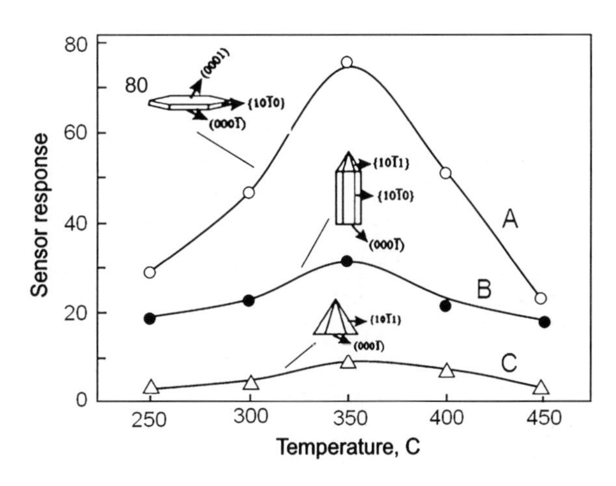

Fig. 20.9 Response to 300 ppm ethanol of the ZnO-based sensors in dependence on crystallite morphology (Adapted with permission from Han et al. 2009a, Copyright 2009 American Chemical Society)

observed in a wider temperature range. The position of the maximum of temperature dependence of sensor response depends on the activation energies of adsorption/desorption processes (Brynzari et al. 2000). It is necessary to note that the broadening of temperature dependence of sensor response lowers the selectivity of sensors.

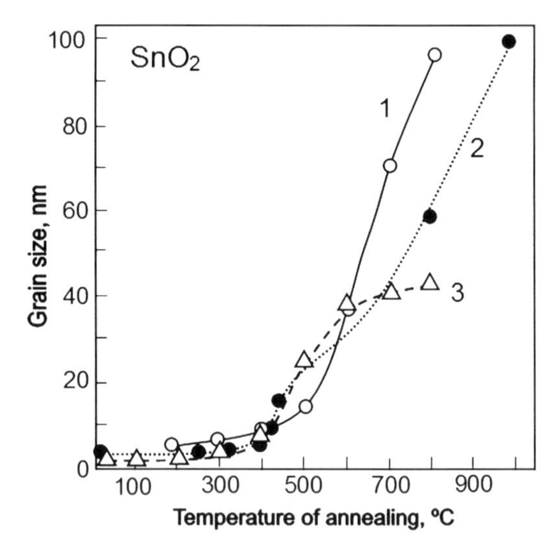

Fig. 20.10 Influence of the annealing temperature on the grain size in the SnO_2 ceramics synthesized by conventional wet-chemical methods. *1* (Data from Shek et al. 1999). *2* (Data from Dieguez et al. 1997). *3* (Data from Wu et al. 1999a)

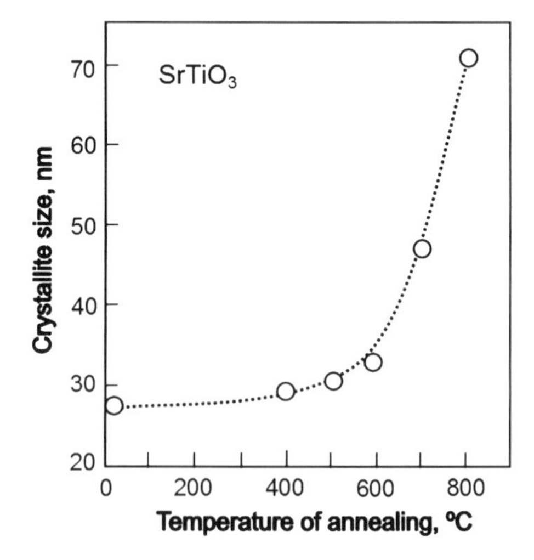

Fig. 20.11 Variation of the average grain sizes of the $SrTiO_3$ powders synthesized high-energy ball milling (120 h milling) (Adapted with permission from Hu et al. 2005, Copyright 2005 Elsevier)

While the grain size is growing, the size of noble metal clusters deposited on the surface of metal oxides for its functionalizing could also be changed (Korotcenkov 2008). The enlargement of the area of plane surfaces facilitates their migration and coalescence. Thus it was shown that for the achievement of required stability of sensor parameters, the grain size should be invariable during gas sensor fabrication and use. However, numerous studies have shown that this requirement is not applied. It was established that the grain size during thermal treatments used in sensor fabrication process could be changed several times. The longer the temperature and thermal treatment, the stronger are those changes. Figure 20.10 presents the influence of thermal treatment on the grain size in SnO_2 synthesized at different labs. For other metal oxides we have the same regularities independent of the method of preparation (Hu et al. 2005). For $SrTiO_3$, synthesized using high-energy ball milling at room temperature, such dependence is shown in Fig. 20.11. It is seen that the changes in the grain size after thermal treatment can be really huge. For example, Zhang and Liu (2000) reported that the grain size in the SnO_2 ceramics after thermal treatment at 1,000 °C has increased by more than 20 times,

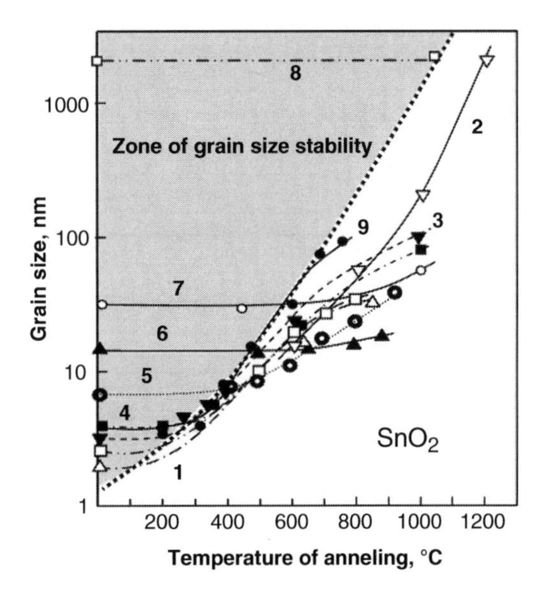

Fig. 20.12 Influence of annealing temperature on grain size in SnO_2 films and powders: *curves 1–9* were constructed on the base of experimental results obtained in various groups (Reprinted with permission from Korotcenkov et al. 2005a, Copyright 2005 Elsevier)

from 12–15 to 200–300 nm. Wu et al. (1999a) found that the annealing of SnO_2 gels at 400 °C leads to a fivefold increase in crystal size, from ~2 to 10 nm, and ~50 % loss in surface area, and hence in catalytic activity (Liu et al. 2007). It is reasonable that those changes can be accompanied by great variations in gas-sensing properties of sensors based on those materials. It was also established that the structure of metal oxide matrix is being transformed during the exploitation process.

Analyzing the influence of initial grain size on the thermal stability of metal oxide structural parameters, Korotcenkov et al. (2005a) have shown that there is strong correlation between the indicated parameters. The bigger the grain size, the more stable the initial structure, and therefore the sensors' working characteristics are more stable as well (see Fig. 20.12).In contrast, the smaller the grain size, the lower the starting temperature of the structural changes. For example, it was established that the SnO_2 grains of size ~2 nm start growing at temperatures ~200 °C, while grains of size bigger than 20 nm remain unchanged even after thermal treatment at T_{an} = 600 °C. In other words, each size of crystallite has its own critical temperature (T_{th}), above which the grain size increase becomes possible. The larger the grain size, the wider the temperature range in which crystallites retain their size and shape without changes. According to Korotcenkov et al. (2005a), this boundary temperature may be described by the expression $T_{th} = 420 \,.\, (\lg t)^{3/4}$ (°C), where t is the grain size in nanometers.

At present the decrease of the grain size in metal oxide matrix is the main approach used for the improvement of operating parameters of solid-state conductometric gas sensors (Xu et al. 1991; Franke et al. 2006; Tiemann 2007; Korotcenkov et al. 2009). Research has shown that the minimization of grain size makes extremely high sensor response possible. Data presented in these sections testify that perhaps such an approach is not optimal for sensors designed for the sensor market because commercialization requires first of all high stability and reliability but not the most attainable sensitivity (Korotcenkov and Cho 2011). Analyzing situation with gas sensor design, one can see that at present there is no problem with insufficiently high sensitivity of already-designed sensors. There are problems connected with their stability, calibrating frequency, and time of working capacity. The absence of necessary stability, but not of required sensitivity, limits commercialization of the majority of elaborated sensors. The above-mentioned results, testifying thermal instability of grains with small size, also explain why publications describing unique gas-sensing properties, as a rule, do not discuss the problems of stability and reproducibility of the results obtained. This is because ceramics with extremely small grain size cannot provide the necessary stability of structural parameters.

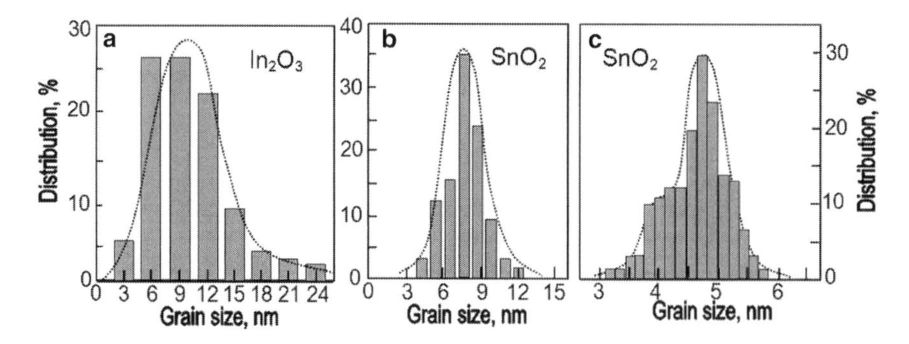

Fig. 20.13 Dispersion of the grain size in the In_2O_3 and SnO_2 films prepared using different methods. (**a**) Films deposited by spray pyrolysis at T_{pyr}=400 °C; film thickness equaled ~50 nm (Data from Korotcenkov et al. 2005a). (**b**) Hydrothermally treated SnO_2 sol suspensions (Data from Baik et al. 2000a, b, c). (**c**) Nanoparticle size distributions extracted by TEM of the SnO_2 powders obtained by the sol–gel technique (Data from Dieguez et al. 1999)

As indicated before, films of grain size 2–3 nm already start transforming at temperatures ~200 °C. This fact agrees well with the results of analysis of structural stability of films deposited at room temperatures by magnetron sputtering methods (Yan et al. 2007). As a rule, such films are amorphous—similar to those of grain size less than 1–3 nm. In those films, considerable structural changes start at temperatures exceeding 200 °C (Yan et al. 2007). Taking into account that, in resistive-type sensors, operating temperatures are in the range of 100–500 °C, one can state that a film structure with the grain size indicated above (1–4 nm) will transform during the first measuring cycle. It becomes clear that the presence of a finely-dispersed fraction with grain size smaller than 4–6 nm would lead to some structure instability of the metal oxide matrix at increased operating temperatures (at T_{oper} ~ 400–600 °C). It is important that this process will take place even in films with average grain size exceeding 100 nm.

It should be noted that the presence of a fine-dispersed phase is typical for most methods of both synthesis of metal oxide powders and deposition of metal oxide films. As research has shown (Dieguez et al. 1999; Baik et al. 2000a, b, c; Korotcenkov et al. 2005a), irrespective of the technology used, the grain size of metal oxides has a sufficiently high dispersion (see Fig. 20.13). This means that besides big crystallites films there are content small ones which are already inclined to structural transformation at low temperatures. The consequences of the indicated process are shown in Fig. 20.14. It is seen that the number of grains with minimal size strongly decreases after annealing.

This means that the thermal stability of gas-sensing material should be estimated by the presence of the fine-dispersed phase inclined to coalescence at low temperature, but not by averaged values of grain size. The greater the content of fine-dispersed phase, the stronger the changes. One should note that bad thermal stability of metal oxides with broad grain-size distribution can be explained by the theory suggested by Lifshits and Slezov (1959, 1961). According to their theory, a solid-state system with narrow particle size distribution during thermal treatment develops more slowly than that characterized by a broad size distribution. This is related to the fact that the driving force of the crystal growth is determined by the differences between the chemical potentials of the largest and the smallest particles. Thus, systems with a broad size distribution readily undergo recrystallization, while homogeneous systems are relatively inert even if they consist of very small particles. Based on the conducted analysis, one can conclude that there is no need for an extremely high sensitivity through reduction of grain size. Metal oxides containing fine-dispersed phase cannot provide the high stability and reproducibility of parameters required for the sensors. This means that a more rational approach to either the gas-sensing material synthesis or deposition is required, because the synthesis method is one of the most crucial parameters for controlling structural and sensing properties of the final metal oxide.

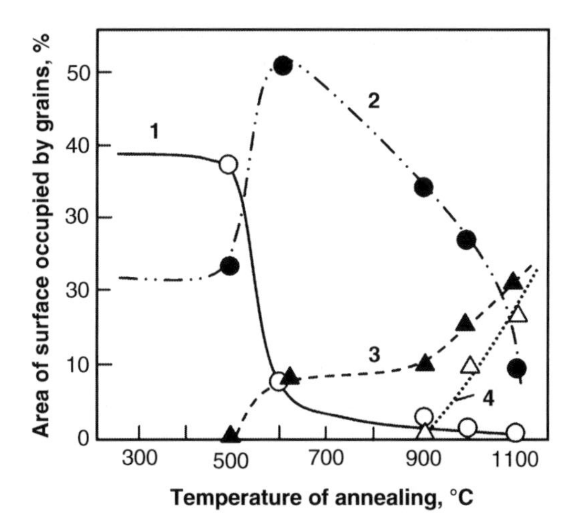

Fig. 20.14 Influence of annealing temperature on the area of the In_2O_3 film surface occupied by the grains with different sizes: (*1*) t = 10–15 nm; (*2*) t = 20–25 nm; (*3*) t = 30–35 nm; (*4*) t = 40–45 nm. In_2O_3 films were deposited at T_{pyr} = 400 °C and had the thickness d ~ 50 nm (Reprinted with permission from Korotcenkov et al. 2005a, Copyright 2005 Elsevier)

The change of film morphology may also be caused by migration and/or segregation of additives (Park and Akbar 2003; Smatko et al. 2009), although variation of the grain networks due to a mismatch of the thermal expansion coefficient can also incur changes in the sensor characteristics (Tang et al. 2001; Park and Akbar 2003). According to Sharma et al. (2001), cracking is the main reason for drift in parameters of sensor microfabricated on microhotplate platforms. When the sensing layer and substrate, which usually have different thermal expansion coefficients, are subjected to thermal cycling, the result is stress in the coating—thereby leading to cracks. Research has also shown that irreversible structural changes connected with the above factors occur, with especially strong effects in thicker films (Sharma et al. 2001; Smatko et al. 2009). During the heating and cooling cycles, the walls of the monolith cool down or warm up faster than the bulk, and this induces a thermal gradient within the ceramic, leading to fractures; bad adhesion of the sensing layer further promotes such modifications. Subsequently, as the postdeposition annealing time increases, the number of cracks in the metal oxide film increases as well, although in thin films this effect is seldom observed (Smatko et al. 2009). Structural changes in metal oxide films can also be caused by the nonuniform surface distribution of current density and the corresponding nonuniform heating of metal oxide films (Smatko et al. 2009). Interfacial reactions at the metal electrode/ceramic interface may also contribute to the instability of gas sensor parameters (Meixner and Lampe 1996).

20.2 Role of Phase Transformations in Gas Sensor Instability

It was found that under the long-term operation of gas sensors based on MOX doped with additives, these additives could segregate into separate phases (Korotcenkov 2005; Wang et al. 2007). Here, the rate of segregation is controlled by lattice diffusion, and therefore the segregation equilibrium may be rapidly established at elevated temperatures; at lower temperatures this process may be considerably slower. Research has also established that the extent of segregation strongly depends on the composition of both the bulk phase and the surrounding gas phase, which control the defect concentration both in the bulk and within the near-surface (grain boundary) region (Nowotny 1988). It was found that the limited solubility of additives in metal oxides also depends on the grain size and crystallinity; solubility of additives in metal oxides usually drastically decreases with increasing grain size (Straumal et al. 2009). This finding indicates that annealing, subsequently followed by grain growth, promotes the process of impurity segregation on the surface of those grains. Thus, the effect of segregation in metal oxides can lead to essential changes in the chemical and electrophysical properties of the near-surface (grain boundary) layer with respect to the crystalline bulk (Dufoura et al. 1997). This means that segregation processes may play a significant role in the formation of the surface layer and affect conditions of current flow through polycrystalline materials, which control the conductivity and sensor sensitivity; interfaces in such two-phase systems will be controlled by the presence of the second phase (Wang et al. 2007). A more detailed description of the processes related to grain boundary segregation in oxide ceramics can be found in Nowotny (1988), Petot-Ervas and Petot (1990), Horvath et al. (1996), Dufoura et al. (1997), and Wynblatt et al. (2003). The phase composition might also be changed due to chemical reactions between the substrate and reactive components from the surrounding atmosphere. For example, the reaction with Cl_2 is accompanied by generation of volatile chlorides. Moreover, the products of these reactions may also dope the sensing materials and/or generate either isolating or gas-insensitive layers next to the substrate. The interaction between the gas-sensing layer and substrate can also be the source of a new phase. For example, the reaction of Ga_2O_3 with Al_2O_3 at a temperature of 1,000 °C may be accompanied by the appearance of an $Al_xGa_{2-x}O_3$ ($x<2$) mixed compound (Fleischer et al. 1990). As was shown in Madou and Morrison (1989), Capone et al. (2003), and Korotcenkov and Cho (2011), poisoning of sensing material, bulk diffusion, and ionic drift, which can modify electrophysical and surface properties of metal oxide, degradation of contacts and heaters, humidity and fluctuations of temperature in the surrounding atmosphere, and interference effects, can also cause the instability of metal oxide properties. Errors in designing and in choosing intermediate materials used for manufacturing are also important factors contributing to the observed instability in sensor parameters. As is generally known, no sensor can be designed without considering the final package, and interaction of the surrounding gas with an unstable sensor casing has often been the main reason for the degradation of sensor parameters.

20.3 Approaches to the Improvement of Metal Oxide Structure Stability

It should be noted that it is impossible to determine the main reason for gas sensor instabilities in a general sense; the task is too complicated. In addition, the reasons for instability could depend on the constructive and/or technological features of the sensor fabrication or on temperature effects at the point of use. Therefore, at present there are a great number of approaches, all of which—in the author's opinion—can be effective in resolving instability problems in gas sensors. According to Korotcenkov and Cho (2011), to attain maximum gas sensor operational stability, it is necessary to stick to the following recommendations.

20.3.1 Only Chemically and Thermally Stable Materials Should Be Used in the Sensor Design

This requirement is the most important because, as said before, chemical sensors, in spite of their lack of encapsulation, should provide long-term use at high temperatures in harsh environments (Korotcenkov 2008). The better the thermodynamic stability of the sensing materials, the higher the temperature at which the chemical sensors are able to work, especially in the presence of reducing gases such as H_2 or CO (Wurzinger and Reinhardt 2004); similarly, the higher the energy of a stable oxide forming, the higher the temperature of reduction. Some parameters that have characterized thermodynamic stability of metal oxides used in gas sensor design are listed in Table 20.1.

Table 20.1 Thermodynamic stability of metal oxides

Material	Melting temperature (°C)	ΔH_f for metal oxide formation per oxygen atom, $-\Delta H_f$ @ 298 K (kJ/mol)	Temperature-programmed reduction, TPR (°C)	Thermal stability in oxygen atmosphere
MgO	2,800–2,820	601.7	N.R.	Thermally stable (T.S.)
BaO	1,923–2,015	553	330	$T>500$ °C → BaO_2
La_2O_3	2,300	699.7	468	T.S.
TiO_2	1,855	470.8	N.R.	T.S.
ZrO_2	2,690	547.4	N.R.	T.S.
CeO_2	2,727	544.6	594	T.S.
V_2O_5	690	311.9	550	$T>700$ °C, evaporates with partial dissociation
Nb_2O_5	1,512	381.1	N.R.	T.S.
Ta_2O_5	1,879	409.9	340	T.S.
MoO_3	795	251.7	575	$T>650$ °C, sublimates
WO_3	1,470	280.3	544	$T>1{,}000$ °C, sublimates
Mn_2O_3	1,347	323.9	184	$T>750$ °C, decomposes
Fe_2O_3	1,347	247.7	200	$T>1{,}400$ °C, dissociates
Co_3O_4	1,562	202.3	288	T.S.
NiO	1,957	245.2	278	T.S.
CuO	1,336	157.0	268	$T>160$ °C → Cu_2O
ZnO	1,800–1,975	348	N.R.	T.S.
Al_2O_3	2,050	558.4	N.R.	T.S.
Ga_2O_3	1,740–1,805	360	320	T.S.
In_2O_3	1,910–2,000	308.6	350	T.S.
SiO_2	1,720	429.1	N.R.	T.S.
SnO_2	1,900–1,930	290.5	500	T.S.

Source: Data from Samsonov (1973) and Korotcenkov (2007a)
N.R. no reduction detected between 150 and 700 °C

20.3.2 *Diffusion Coefficients of Both Oxygen and Any Ions in the Sensing Material Should Be Minimized*

As shown above, ionic diffusion can be one reason for the temporal drift in sensor parameters (Jamnik et al. 2002). Therefore, it is natural that minimization of the diffusion coefficients of both oxygen and any ions in the sensing material should improve the stability of sensor parameters. Previously, it was shown that the presence of structural vacancies in the lattice promotes an increase in the coefficients of metal and oxygen bulk diffusion (Neiman 1996). For example, mobility along crystallographic defects may be several orders of magnitude greater than in a structurally ordered crystal. Therefore, it can be assumed that in terms of sensor design, in which the appearance of diffusion processes diminishes the exploitation of parameters, using materials that do not contain structural defects, such as point and dimensional defects, is preferred. Reducing the concentration of uncontrolled impurities in the metal oxide due to improvement of the technological processes of synthesis or deposition can also promote a decrease in parameter instabilities, as pertaining to ionic drift. For example, research carried out by Pavelko et al. (2009) revealed that metal oxide sensors containing a lowered concentration of Na and Cl ions showed a much better parameter stability.

20.3.3 Gas-Sensing Materials with an Extremely Small Grain Size Should Not Be Used

As reported by Korotcenkov et al. (2005a, 2009), a decrease in grain size is often accompanied by a drop in the thermal stability of gas-sensing material parameters. For example, it was shown earlier that in the film with grain size smaller than 3–5 nm, structural changes are possible at temperatures below 300 °C. This means that modern approaches for gas sensor design, based on reducing the grain size to attain maximum sensor response, are not optimal. Small grain size may promote higher gas sensitivity in polycrystalline MOX layers but the structure of such layers can be unstable under real operating temperatures. Therefore, sensors intended for operation as long-term devices should not be based on the use of ultra-nanosized grains, because poor stability of their properties will be the price to pay for high sensitivity. Results presented by Min and Choi (2004) and Mandayo et al. (2003) confirmed this conclusion. This means that the search for an optimum tradeoff between high sensitivity and good stability should be a major task for sensor development when employing nanostructured materials. The growth of sensor parameter sensitivity to humidity change is also a negative factor of excessive decrease in grain size; results confirming this conclusion can be found in Korotcenkov et al. (2009) and Pavelko et al. (2009).

20.3.4 Size and Shape of the Grains That Formed a Gas-Sensing Matrix Should Be Controlled

Research has shown that, in many cases, the overall properties of polycrystalline materials employed under long-term operation at advanced temperatures are controlled by finely dispersed fractions, a subject of coalescence even at rather low temperatures, and not by the averaged value of the grain size (Korotcenkov et al. 2005a, 2009; Korotcenkov and Han 2009). The larger the fraction of nanograins, the larger the transformation of the structure of the gas-sensing layer. Taking this fact into account, it is clear that the elaboration of special technologies that ensure the gas-sensing layer comprises grains of the same size, with only minor fractions of a finely dispersed phase, seems to be a promising direction for research targeting the improvement of sensor parameter stability (Kennedy et al. 2005). In addition, technologies capable of allowing the synthesis of nanocrystals with assigned faceting have also attracted interest (Korotcenkov 2008). No doubt the development of these new technologies (and others) requires a considerable financial investment; however, the results that could be obtained while using these technologies might be worth it.

20.3.5 Elemental Composition of Gas-Sensing Material Should Be Optimized

At present, doping is the main method used for optimizing sensor parameters such as sensitivity and selectivity (Korotcenkov 2005, 2007b; Tricoli et al. 2008a; Xu et al. 2008). It was also reported that doping may be used to improve sensors' temporal stability as well. For example, Matsuura and Takahata (1991) have examined the effects of various metals and/or metal oxide additives on the stability of SnO_2 sensors. They found that considering the 31 elements that could be added, Re- or V-doped sensors showed the smallest resistance changes during a long-term stability test (20 days) and that the best stabilization was obtained using combined Re/V doping. Liu et al. (2004) then found that the addition of $BaTiO_3$ also improves the long-term stability of the SnO_2 sensor.

The same situation takes place for other metal oxides as well. For example, Anukunprasert et al. (2005) reported that Nb-doped TiO_2 at 3–5 mol% clearly hinders the anatase-to-rutile phase transition and inhibits grain growth, in comparison with pure TiO_2.

However, it does not mean that simple doping could resolve all problems connected with gas sensor instability. First, for the majority of dopants, no optimizing effect was observed. For example, in Wang et al. (2003) it was shown that SnO_2 gas sensors had better reliability and stability than SnO_2:CdO gas sensors. Moreover, in many cases, doping stimulates structural changes in metal oxides (Korotcenkov 2005; Korotcenkov et al. 2008). Second, doping has limited opportunities to influence MOX parameters. For example, it was established that CeO_2 doping did not reduce the sensitivity of SnO_2:Ce-based sensors to humidity (Lu et al. 2006). Then, in (Korotcenkov 2007; Korotcenkov et al. 2007b) it was shown that for In_2O_3-based sensors the situation is even more complicated, because doping by the majority of dopants was accompanied by an increase in sensitivity to air humidity changes. Moreover, it should be noted that dopants, especially at concentrations close to the solubility limit, could segregate at the grain surface during use, with a corresponding change in MOX properties (Yang et al. 2000; Xu et al. 2008), and that annealing, leading to a decrease in the concentration of defects (Karunagarani et al. 2002) and an increase in grain size stimulates this process. Research has also shown that an improvement of metal oxide crystallinity is accompanied by a drop in the ultimate solubility of dopants (Straumal et al. 2009). Based on these discussions, therefore, it can be concluded that the properties of metal oxides with a low doping concentration should be more stable than the properties of heavily doped ones.

The behavior of noble metals at the MOX surface, used for the activation of surface reactions, could also lead to sensor parameter instabilities (Papadopoulos et al. 1997). As a rule, interactions with an active gas, especially a reducing gas, facilitate the process of noble metal migration (Nishibori et al. 2008); Au nanoparticles have a high mobility in the presence of chlorides as well (Robertson 1960). It was shown that the inclination to migrate increases in the following order: Pt, Pd, Au, and Ag. According to this rank, the stability of parameters in as-fabricated sensors using these noble metals also changes; it was found that, as a rule, In_2O_3:Pt and SnO_2:Pt-based sensors had the most stable characteristics (Papadopoulos et al. 1997). It was also established that the smaller the initial cluster size, the bigger the probability of their migration. In Papadopoulos et al. (1997) it was shown that sensors with a thicker noble metal layer (Pd, Pt), i.e., a larger cluster of noble metals, were much more stable during interaction with all test gases. Thus, in extended use, the growth of noble metal clusters is possible. However, it is known that the reactivity of noble metals strongly depends on cluster size (Kung et al. 2003; Zanella et al. 2004). Thus, stabilizing the size of noble metal particles is important in order to achieve good stability in gas sensors designed for long-term operation (Nishibori et al. 2008). This problem can be partly resolved by introducing additional components and stabilizing the position of noble metal clusters. For example, the incorporation of additives that reinforce metal-supported interactions can affect the stability of metal clusters toward sintering. Previously, it was established that species such as C, Ca, Ba, Ce, Nb, and Ge may decrease metal atom mobility, whereas Pb, Bi, Cl, F, and S can increase it. Furthermore, in Holody et al. (2001) it was noted that Nb introduced into a Pt/SnO_2 matrix suppresses the growth of Pt clusters during annealing at <900 °C and rare earth oxides such as CeO_2 and La_2O_3 have also been suggested to “fix” noble metal atoms onto the surface of metal oxides due to a strong, localized chemical interaction (Oudet et al. 1989). However, even in this case it was not possible to exclude completely the growth of noble metal clusters. According to the results of research presented by Veltruska et al. (2001) and Skala et al. (2003), under certain conditions such as a strong surface reduction, encapsulation of noble metal clusters is even possible. This effect is an additional reason for the instability of gas-sensing characteristics in devices with a modified surface. Based on these analyses, one can suppose that bulk doping in conjunction with surface modifications is not always an optimal solution for attaining the maximal stability of sensor parameters.

20.3.6 Use Preliminary (Accelerated) Aging Prior to Sensor Tests

As mentioned above, in the presence of a finely dispersed phase, small grains can coalesce during the exploitation process with bigger grains, initiating structural changes as well as a change in the gas-sensing properties of the gas-sensing matrix. Preliminary annealing promotes the phase coalescence at this stage of fabrication, preventing similar changes during the exploitation process. Annealing also activates other processes that commonly induce a temporal drift in characteristics (de Angelis and Riva 1995; Nishibori et al. 2008; Jung et al. 2008). For example, annealing has improved the homogeneity and stabilized the size of noble metal clusters (Nishibori et al. 2008). In Karunagarani et al. (2002) it was shown that annealing decreased the density of microstrains and dislocations in the film, leading to a reduction in the concentration of lattice imperfections. In Jung et al. (2008), improvements in stability were ascribed to the reduction of trap sites in the metal oxide layer. Thus, additional annealing for a sufficiently long period promotes the stabilization of sensor properties (de Angelis and Riva 1995). In Pijolat et al. (1999) the sensors were aged for several weeks on a laboratory test bench prior to use. Then, in Jung et al. (2008), the duration of post-thermal annealing was established to be about 65 h. An important requirement of such thermal treatments is that annealing temperatures have to be higher than the operating temperature; also, it should be noted that this method is the easiest and most efficient for improving sensor stability. Of course, the artificial aging of sensors makes the technological process of their fabrication longer, because it often includes several long annealing periods in different atmospheres at various temperatures (Adamyan et al. 2009). However, the instability problem for such aged sensors is not as strong as for as-fabricated devices. Therefore, in many cases, the appearance of temporal instability of sensor parameters is considered to be a consequence of the insufficient treatment of samples prior to sensor exploitation, in terms of temperature and duration. Results presented by Papadopoulos et al. (1997) confirm this conclusion. Based on tests of In_2O_3- and SnO_2-based sensors, fabricated using sputtering technology, it was established that, if devices were not exposed to a preliminary aging, sensors only become stable after 100–150 measurement cycles, including heating up to 450 °C. No doubt it should be noted that this may be an overly simplified view of the instability problem, because annealing by itself cannot exclude all factors influencing the stability of gas sensor characteristics.

Research carried out during the last few years has shown that high-temperature annealing of as-synthesized powders can be more efficient for stabilization of metal oxide properties than annealing at low or middle temperatures. For example, Lee et al. (2010) established that the annealing of as-synthesized SnO_2 powders at $T_{an} = 1{,}000$ °C for 10 h before forming a gas-sensing layer using methods of thick-film technology considerably improved their thermal stability. For information, the conventional technology discussed above usually uses calcinations of as-synthesized metal oxide powders at temperatures lower than 400–600 °C. Jung et al. (2008) also reported that thermal annealing improved stability of gallium indium zinc oxide thin-film transistors. Itoh et al. (2008) have found that annealing in air at temperatures exceeding the operating temperature can also reduce resistance drift phenomena. Hillhouse (2005) believes that, for extremely high thermal stability of structural parameters, it is desirable to carry out the annealing at $T > 1{,}000$ °C. It is necessary to note that the above-mentioned conclusions are in good accordance with results obtained by Dieguez et al. (1997) and Kappler et al. (1998). They have shown that only annealing at temperatures higher than 500 °C results in improvement of SnO_2 crystallinity and disappearance of lattice distortion. Moreover, for full crystallization and removing tensions in crystallites, as indicated in Fig. 20.15, annealing at temperatures exceeding 700 °C is required (Cirera et al. 2000; Bagheri-Mohagheghi et al. 2008).

However, using high-temperature treatments of metal oxides, we have to take into account that this annealing can be accompanied by strong increases of both the grain size and the area of intergrain contacts. The possible influence of high-temperature annealing is shown in Figs. 20.10, 20.11, 20.12, 20.13, 20.14, and 20.15. The decrease of specific surface area is another disadvantage of the

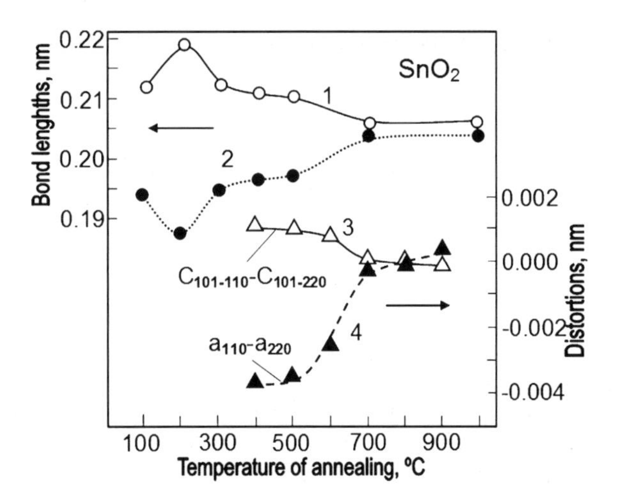

Fig. 20.15 Influence of annealing on structural parameters of SnO_2: *1* and *2* (atomic bond length (Sn–O) variations as a function of annealing temperature) (Data from Toledo-Antonio et al. 2003), *3* and *4* (distortion of the samples vs. stabilization temperature) (Data from Cirera et al. 2000)

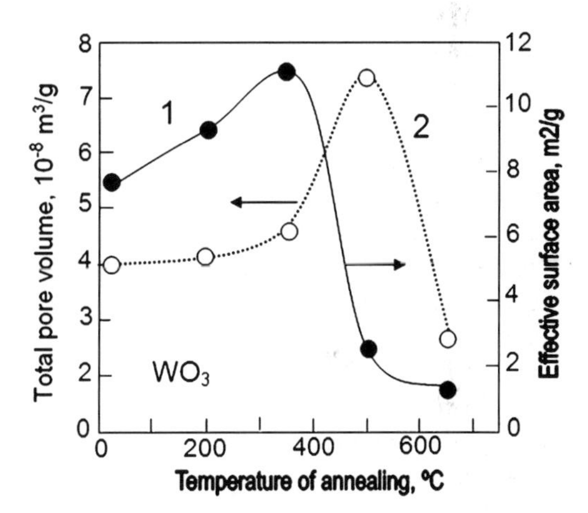

Fig. 20.16 Influence of annealing on total pore volume, effective surface area of the WO_3 films deposited on oxidized Si substrates at room temperature by reactive DC magnetron sputtering (Reprinted with permission from Liu et al. 2007, Copyright 2007 Elsevier)

high-temperature annealing of an already-created gas-sensing matrix. One of the examples of annealing influence on effective surface area and total pore volume for the SnO_2 films is shown in Fig. 20.16. The specific surface area decreased as annealing temperature was decreasing due to the strong sintering occurring in the material. In other work, Toledo-Antonio et al. (2003) reported that the specific surface area of the SnO_2 gas-sensing matrix after annealing at 1,000 °C decreased by more than 15 times. According to Antonelli and Ying (1995) and Qi et al. (1998), the destruction of the regular mesopore structures in the surfactant-templated metal oxide gels upon thermal treatment is also likely caused by extensive crystal growth. A pronounced strong change of grain-size distribution takes place as well. Such changes in the parameters of gas-sensing matrix considerably reduce the efficiency of high-temperature annealing intended for the improvement of operating characteristics of gas sensor, because improvement of thermal stability will be accompanied by a decrease of sensor response, which in some cases can be disastrous.

All the above-mentioned discussion states that resolving the thermal and temporal stability problem of sensor parameters requires new approaches to metal oxide synthesis and deposition, which could prevent both the sintering of grains and the growth of the grain size during high-temperature annealing. The problem of the grain sintering in gas-sensing material during annealing can be partially resolved at the expense of additional milling of sintered material. The milling, due to agglomerates destruction, can considerably improve the structure of gas-sensing material. However, for removing tensions in crystallites, originated during the milling, annealing at temperatures exceeding 700 °C is required (Dieguez et al. 1999; Cirera et al. 2000). In addition, the milling does not reduce the size of grains (Dieguez et al. 1999), and this method cannot be applied to already-fabricated gas sensors.

Taking into account that the growth of grains occurs due to mass transfer between the grains, one can conclude that the creation of conditions when the grains do not contact with each other during annealing is the best approach to resolving the indicated task of grain-size stabilization. The isolation of the hydrous particles upon firing may limit crystallization taking place on a very local scale, and hence, the original nanoscaled microstructure will be preserved. Unfortunately, it is very difficult to realize the indicated conditions in the frame of conventional technologies. Research has shown that metal oxide powders synthesized by conventional wet-chemical methods such as sol–gel and microwave synthesis are, as a rule, strongly agglomerated. The last fact is a considerable disadvantage of the above-mentioned methods. For example, Huang et al. (2007) found that the size of the SnO_2 crystallites synthesized by a wet-chemical method was about 9–11 nm. However, the average size of agglomerates was much bigger at ~300 nm. Dunphy Guzman et al. (2006) found that aggregates of nanocrystalline titania (TiO_2), synthesized by controlled hydrolysis, may contain from 8 to 4000 particles. No doubt such great size agglomerates facilitate mass transfer between grains during annealing and promotes a strong changes in structural parameters of the metal oxide grains and the gas-sensing matrix after high-temperature treatments. Wu et al. (1999b) established that, during annealing, the condensation of hydroxyl groups present in as-synthesized grains pulls together the constituent particles of the gel into a compact mass, and so the metal oxide crystals readily grow to a size much larger than that of the original particles.

One should note that aggregation, a common yet complex phenomenon for small, especially nanometer-sized particles, is problematic in the production and use of many chemical products (Elimelech et al. 1995; Li and Kaner 2006). In many synthetic processes for particles, especially surfactant-free chemical reactions, aggregation occurs as soon as particles are generated. In conventional studies, aggregation has simply been ascribed to the direct mutual attraction between particles via van der Waals forces or chemical bonding (Hunter 1987). Strategies for preventing aggregation mainly come from conventional colloid science in which particles are coated with foreign capping agents and/or the surface charges are tailored to separate them via electrostatic repulsions (Hunter 1987). However, in reality the aggregation during synthesis is involved in a series of complicated processes such as chemical reactions, nucleation, growth, and precipitation, which often makes it difficult to determine the actual aggregation mechanism in chemical reactions and to design the method obstructing the agglomeration. From this it follows that the design of technologies guaranteeing the formation of powders with minimum aggregation of synthesized grains is an important direction of research aimed at the achievement of improved thermal and temporal stability of semiconductor gas sensors. The literature relating to the preparation of thermally stable nanostructured SnO_2 (Srivastava et al. 2002; Wang et al. 2005; Wagner et al. 2006) indicates that this issue still poses a challenge to researchers, and further improvement is necessary.

The decrease of the grain-size dispersion, i.e., reducing the contents of the fine-dispersed phase inclined to aggregation and coalescence at low temperatures, is also one of the approaches to resolving the above-mentioned task related to stability of metal oxide properties (Kennedy et al. 2005). Unfortunately, even though the researchers understand the importance of the indicated approach, a comprehensive study in this field is absent. We were able to find just a few papers where this aspect of gas-sensing layer deposition was analyzed (Baik et al. 2000a, b, c; Panchapakesan et al. 2001). We agree with Panchapakesan et al. (2001) in that, in addition to stability improvement, narrow grain-size distribution provides more predictable and reproducible results in gas-sensing behavior from one sensor to another, and, thus, nonuniformity of the microstructure of sensing layer should be avoided.

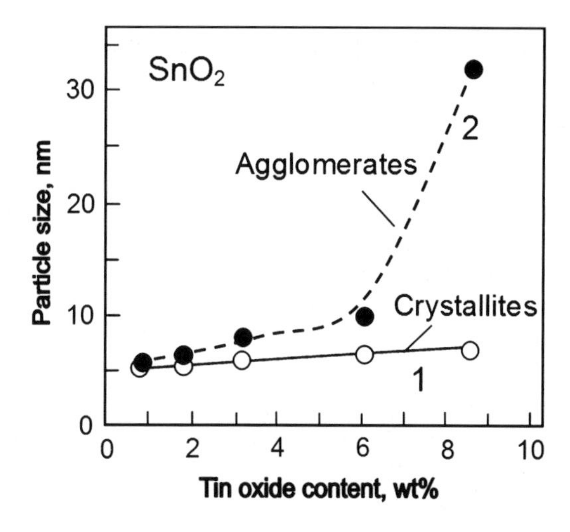

Fig. 20.17 (**1**) Crystallite size and (**2**) particle size of tin dioxide, as a function of tin oxide content in sol suspension (Adapted with permission from Baik et al. 2000a, b, c, Copyright 2000 John Wiley & Sons)

20.3.7 Use Technological Operations Contributing to the Improvement of the Temporal Stability of the Designed Structures

It was established that to achieve the required stability, technological processes should not be accompanied by doping of uncontrolled dopants and unmonitored reactions in either the gas-sensing or electrode materials. In addition, methods of deposition for gas sensing, electrodes, and isolated materials should provide devices with maximal adhesion and stability in thermal cycles. For example, in Shimizu et al. (2000) it was stated that the ion beam assistance (IBAD) used during SnO_2 and WO_3 sputtering improved the film's adhesion and in that way promoted a considerable improvement in sensor parameter stability. In Shimizu et al. (2000) it was shown that the WO_3-based thin-film sensor fabricated by IBAD kept its extremely high H_2 sensitivity during continuous heating at 300 °C for 8 months, whereas the thin-film sensor fabricated without the IBAD failed after 2 months.

As research has shown (Cirera et al. 1999; Baalousha 2009), the decrease of precursor concentration in solution used for metal oxide synthesis can significantly reduce the size of agglomerates (Fig. 20.17) and, at the expense of this effect, improve the thermal stability of the synthesized material. Examples of precursor concentration influence on the grain-size stability for two methods of the SnO_2 synthesis are shown in Fig. 20.18. It should be noted that the increase of the degree of aggregation in sols with concentration of electrolyte is a general rule of wet-chemical methods of synthesis (Brinker et al. 1992).

It has been found that the pH value of the sol–gel is a decisive factor for controlling the agglomeration as well (Sugiyama et al. 2002). Experiment shows that the maximum aggregation takes place at a pH corresponding to the isoelectric point (IEP) (Wamkam et al. 2011) or to the zero point of charge (pH_{zpc}) (Dunphy Guzman et al. 2006; Tso et al. 2010). Meanwhile, at this IEP (pH_{zpc}), particle sizes demonstrate a significant increase to maximum values (see Fig. 20.19). The reasonable explanation of these interesting phenomena is that, at this IEP (pH_{zpc}), the repulsive forces among metal oxides are zero, and nanoparticles coagulate together at this pH value. According to the Derjaguin–Landau–Verwey–Overbeek theory, when the pH is equal to or close to the IEP, nanoparticles tend to be unstable, form clusters, and precipitate.

Using specific surface-modifying agents during the SnO_2 synthesis, prohibiting aggregation of synthesized particles (Santos et al. 2004), gives the same results. It was established that organic surfactants have a long molecular chain, as large as the nanoparticles used in the preparation of the nanoscaled materials, and the volume occupied by these classical surfactants limits powder concentration

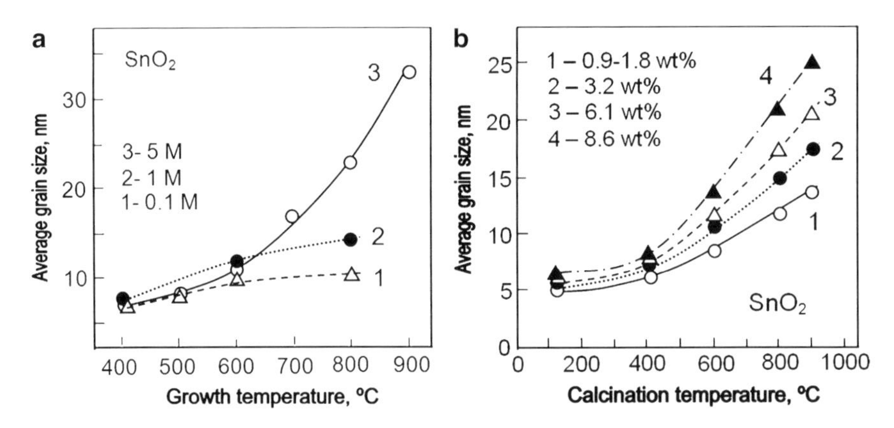

Fig. 20.18 Grain size for different concentration of precursors in solution for metal oxide synthesis vs. growth and calcinations temperature. (**a**) SnO_2 was synthesized using a method based on the pyrolytic reaction of $SnCl_4{\cdot}5H_2O$ dissolved in methanol. In contrast to the conventional spray pyrolysis technique, pyrolytic reaction does not take place during deposition on the surface of nanocrystals. The treatment in the range of 400–900 °C was carried out after drop deposition on the substrate (Adapted with permission from Cirera et al. 1999, Copyright 1999 Elsevier). (**b**) SnO_2 powders were synthesized by hydrothermal method (Adapted with permission from Baik et al. 2000a, b, c, Copyright 2000: John Wiley & Sons)

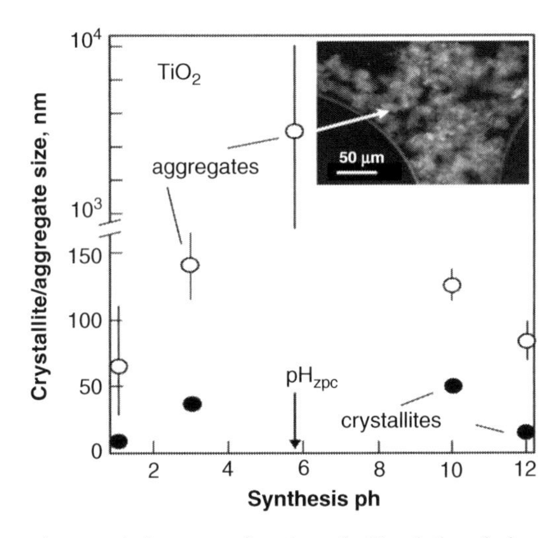

Fig. 20.19 Particle/aggregate size populations as a function of pH solution during nanocrystalline titania (TiO_2) synthesis by controlled hydrolysis. The *bars* represent the range. *Insertion* shows the images of pH 7 TiO_2 aggregate deposits in microchannel pores (Adapted from Dunphy Guzman et al. 2006, Copyright 2006 American Chemical Society)

in suspension. Steric hindrance and electrostatic repulsion also contribute to the dispersion and stability of colloidal suspensions. In particular, Santos et al. (2004) have found that the preparation of redispersible tin oxide nanoparticles by the sol–gel route can be achieved by using the Tiron molecule ($(OH)_2C_6H_2(SO_3Na)_2$) as surface-modifying agent in aqueous solutions of $SnCl_4{\cdot}5H_2O$ and ammonia. After thermal treatment at different temperature, the powder characterization shows that the presence of Tiron monolayer at the nanoparticle surface minimizes the effect of grain coalescence and increases the thermal stability of the structure of microporous material. It was assumed that the presence of the Tiron-adsorbed layer decreases the surface energy and consequently reduces the surface mobility of

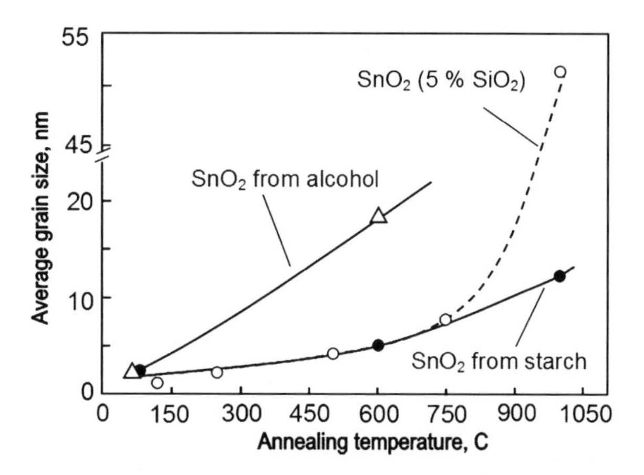

Fig. 20.20 Trends in size increase vs. thermal curing of the samples prepared by different synthetic conditions (Adapted with permission from Callone et al. 2007, Copyright 2007 Cambridge University Press)

nanoparticles. This feature can be explained by two effects of Tiron as surface modifier of metal oxide grains: (1) the condensation reaction involving surface hydroxyl groups does not occur due to their replacement by chelated molecule and (2) the nanoparticles are isolated by the Tiron-adsorbed layer restricting the grain boundaries motions. One can find in French et al. (2009), Baalousha (2009), Molina et al. (2011), and Tso et al. (2010) results related to various organic surfactants and solutions' influence on the agglomeration of other metal oxides.

Callone et al. (2006, 2007) believe that their approach for achievement of better stability of the SnO_2 grain size is also promising. They reported that starch may be successfully used in the hydrolytic route as an additive to obtain oxides and mixtures of oxides in the form of nanocrystalline powders. It was found that the polysaccharide presence offers many advantages, such as high availability at low cost, clean, easy biodegradation to soluble glucosidic units, and original particle shape maintenance during heat treatment, owing to the possible coordination of glucosidic moieties with particle surface. Moreover, the organic hydrophilic envelope of the particles facilitates the formation of colloidal suspensions with convenient binders, which are usually used to obtain bulk materials through various methods, such as micro-screen printing. In the proposed approach, starch has two important functions: (1) the stabilization of primary crystallization nuclei vs. size increase, either by mass addition of molecular species from the solution or by coalescence between particles, and (2) the preservation of nanometric dimensions during heat treatment. According to Callone et al. (2007), assembly of starch macromolecules with hydrophobic oxide particles leads to hydrophilic ensembles, which effectively hamper the particle growth. Figure 20.20 summarizes the effectiveness of all the capping agents used by Santos et al. (2004), highlighting the smaller size increase induced by the use of starch with respect to the use of other dispersants.

Results presented in Fig. 20.20 show that polysiloxanes, obtained by ordinary sol–gel processing of TEOS, are also effective as nanometric size stabilizers. The role of the SiO_2 covering in the stability of the SnO_2 grains was discussed before. However, since the current interpretation of the phenomenon requires the absorption of hydrophilic species on the solid particles, the temperature treatment reduces the Si–OH population and favors the segregation of SnO_2; this event parallels an important increase in the dimensions of metallic oxide crystallites. In contrast, pyrolysis of glucosidic moieties after H_2O_2 treatment produces carboxylic groups able to coordinate metallic ions at the particle surface, and the related size stabilization, prolonged up to 600 °C, appears exclusive of starch.

Adamyan et al. (2003) have found that their technique for SnO_2 synthesis gives an improvement in thermal stability of the SnO_2 grain size as well. Comparing parameters of the SnO_2 films based on powders obtained by two techniques, (1) tin chloride hydrolysis and (2) hydrolysis of sodium stannate with phosphoric acid, Adamyan et al. (2003) established that the second technological process gives

Table 20.2 Comparison of structural parameters of $mSnO_2$ prepared by conventional methods and various P–SnO_2 samples treated in H_3PO_4

Oxide	Treatment (P/Sn)	Annealing (°C)	Crystallite size (nm)	DET surface area (m^2/g)	Pore diameter (nm)
SnO_2 ($mSnO_2$)	No	400	4.9	97	4.9
		600	12.8	13.5	13.5
SnO_2 (P–SnO_2)	0.1	400	2.2	276	2.6
	0.5		–	406	2.3
	2.5		–	439	2.3
	0.1	600	3.5	153	3.4
	0.5		1.7	314	2.3
	2.5		–	348	2.1

Source: Reprinted with permission from Korosi et al. (2010). Copyright 2010 Elsevier

films with much better thermal stability of structural parameters. In the first technique, the 1.2 M·$SnCl_4$ solution was added dropwise in the 1 % ammonia aqueous solution at steady mixing. In accordance with the second technique, the 1 M solution of H_3PO_4 at continuous mixing was added to Na_2SnO_3 aqueous solution until neutralization of solution (pH = 7). It was established that the SnO_2 grains prepared by the second technique remained smaller than 6 nm even after calcination at 750 °C.

Korosi and Dekany (2006) and Korosi et al. (2007a, b, 2010) also carried out research in this direction. Modified tin dioxide nanoparticles of enhanced thermal stability (see data presented in Table 20.2) were prepared by the sol–gel technique. The preparation of P–SnO_2 was performed in two stages. In the first stage, tin oxide hydrate was prepared as described above. In the second stage, $SnO_2·nH_2O$ gel was treated with phosphoric acid solution. During the procedure, the molar ratio of P:Sn was varied in the range 0.01–3.4. Korosi et al. confirmed that the treatment of the SnO_2 gel in H_3PO_4, which is accompanied by chemical modification of the as-prepared hydrous metal oxides, not only resulted in the inhibition of crystal growth during calcination but also delayed the phase transformations in material during annealing. Even at a low atomic ratio P:Sn, Korosi et al. (2005) observed reasonably high thermal stability of tin dioxide particles. For example, while for P–SnO_2(P:Sn = 0.01) samples, SnO_2 particle core sizes were 3.2 nm, 7.3 nm, and 27.7 nm at 550 °C, 800 °C, and 1,000 °C, respectively, particle sizes of pure tin dioxide were 4.2 nm and 9.4 nm at 550 °C and 800 °C, respectively, and SnO_2 calcined at 1,000 °C was macrocrystalline. At this low phosphate content, the surface is only loosely covered by phosphate so that it can be supposed that sensor applications specific for tin dioxide remain possible. Earlier the same result was obtained by the group of Egashira (Hyodo et al. 2002; Hayashia et al. 2009). Applying H_3PO_4 treatment, they increased the thermal stability of $mSnO_2$ and obtained an ordered mesoporous structure with BET = 305 m^2/g after a 5-h calcination at 600 °C. The H_2 sensitivity of $mSnO_2$ sensor proved higher than that of the conventional SnO_2 sensor.

Korosi et al. (2005) concluded that through the treatment of $SnO_2·nH_2O$ with H_3PO_4, not only the porosity, crystallite size, and morphology but also the chemical properties of the surface are being modified. They found that, during H_3PO_4 treatment, phosphate species are adsorbed on the mesostructured hybrid network, and phosphorus is retained in the samples even after calcination at high temperature. As a result of the reaction between phosphorus and SnO_2, amorphous $Sn(HPO_4)_2$ may be formed on the surface, which is transformed into tin pyrophosphates at high temperature. These compounds cannot be detected by XRD up to 600 °C, assuming that they are amorphous and distributed homogenously in the sample. SnP_2O_7 with a cubic structure (JCPDS No. 29-1352) can be identified after calcination at high temperature (1,000 °C) (Korosi et al. 2005). The proposed mechanism of phosphoric acid influence on the structure of SnO_2 grains is shown in Fig. 20.21. The phosphate coverage of tin oxide hydrate particles can be controlled via the phosphoric acid concentration. The possibility of

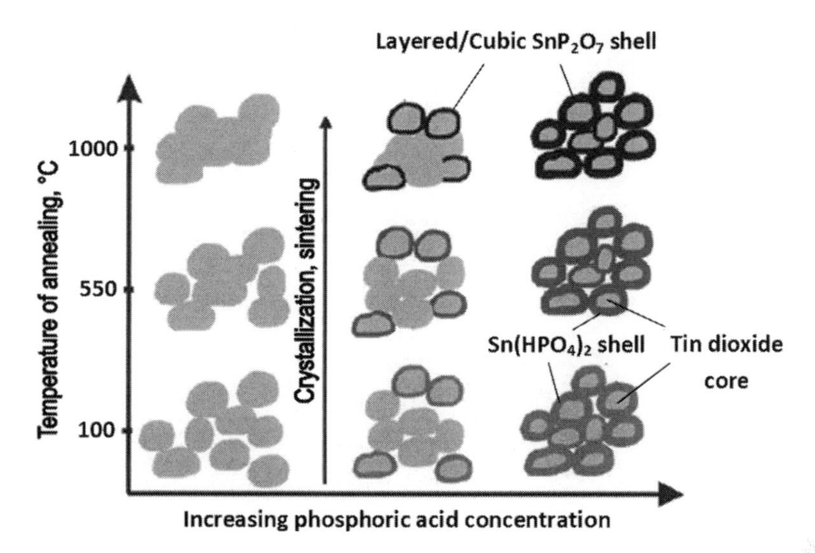

Fig. 20.21 Schematic illustration of the effect of phosphoric acid and heat treatment on $SnO_2 \cdot nH_2O$ (Adapted with permission from Korosi et al. 2005, Copyright 2005 Elsevier)

phosphates forming during metal oxide doping by phosphorus was confirmed in research carried out by Korotcenkov et al. (2004, 2007). Studying calcinated In_2O_3 powders, synthesized by the sol–gel process from solutions with addition of H_3PO_4, it was established that Raman spectra of In_2O_3:P samples contain peaks of 175, 243, 418, and 548 cm^{-1} characteristic for $InPO_4$ (see Fig. 20.22).

The improvement of stability of the SnO_2-based gas sensor parameters was also observed after SnO_2 gel treatment by thiourea (Liu et al. 2005a). The SnO_2 nanopowders were prepared by sol–gel methods as in the following procedure: a concentration of ammonia was dropped into an aqueous solution of tin chloride including citric acid as dispersant until the pH of the titrated solution reached 2 or so. After a thorough washing step, the gel was sintered at 500 °C for 2 h. The surface modification of the material was performed by dropping 0.1 mL thiourea solution in different concentration into 0.1 g SnO_2 powder, then drying for 30 min and sintering at 600 °C for 1 h. The result indicates that the stability of oxygen adsorbed on thiourea-modified surface was improved and the amount of surface hydroxyl groups adsorbed on this grain surface was decreased. The thiourea adsorbed on the SnO_2 grain surface is converted to SO_4^{2-} after being sintered at 600 °C. SO_4^{2-} species stabilize the resistance of the SnO_2 sensor. However, on the basis of present information, we cannot estimate the influence of the above-mentioned treatment on the stability of the SnO_2 grain size during annealing.

Ozaki et al. (2000) found that long term stability of SnO_2-based gas sensors can be achieved by sulfuric acid treatments as well. Sulfuric acid treatment of the sensor element was carried out in two ways: in one a sulfuric acid solution is used instead of water and is added to the mixture of SnO_2 and alumina powders used for preparing a paste, and in the other the obtained sensor element is dipped into a sulfuric acid solution for 2 s, dried for 3 min, and heated at 600 °C for 5 min. While a notable improvement in the stability and reliability of the SnO_2-based CO gas sensor has been achieved, and this is of practical importance, the mechanism of the sulfuric acid treatment is unknown at present. Ozaki et al. (2000) supposed that, due to formation of the sulfate species at the surface of the SnO_2, the formation of the surface hydroxyl groups is being blocked or reduced. The concentration change of surface hydroxyl groups is one of the main reasons of the long-term drift of the resistance of the SnO_2-based gas sensors (Korotcenkov and Cho 2011; Ihokura and Watson 1994).

However, one should note that in the field of gas sensor design, the technology of hydrothermal powder synthesis is the most usable. One can find in reviews prepared by Byrappa and co-workers

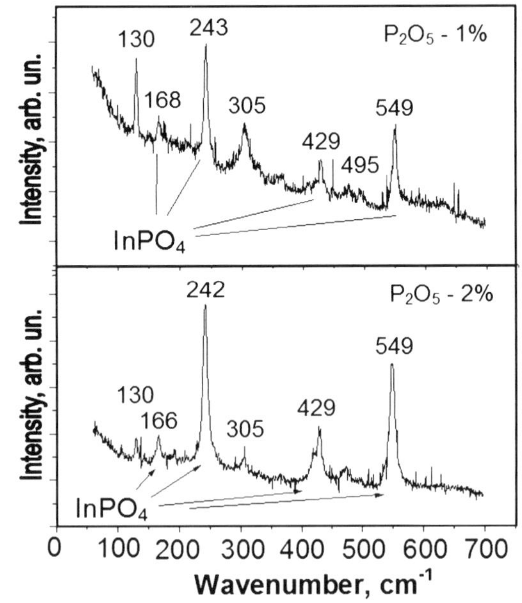

Fig. 20.22 Raman scattering spectra of In_2O_3:P ceramics (Adapted with permission from Korotcenkov et al. 2007, Copyright 2007 Elsevier)

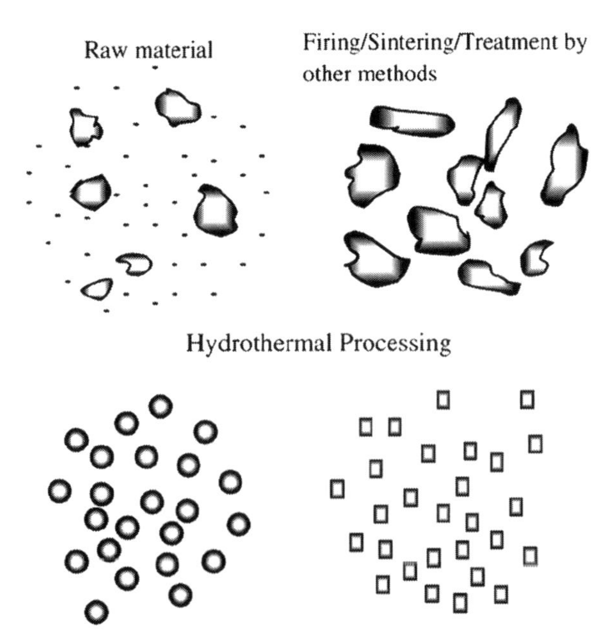

Fig. 20.23 Difference in particle processing by hydrothermal and conventional techniques (Reprinted with permission from Byrappa and Adschiri 2007, Copyright 2007 Elsevier)

(Byrappa and Yoshimura 2001; Byrappa and Adschiri 2007) a detailed description of this method. Byrappa and Yoshimura (2001) define hydrothermal as any heterogeneous chemical reaction in the presence of a solvent (whether aqueous or nonaqueous) above room temperature and at a pressure greater than 1 atm in a closed system. So, the hydrothermal processing of materials is a part of solution processing, and it can be described as superheated aqueous solution processing. This means that material processing under hydrothermal conditions requires a pressure vessel capable of containing a highly corrosive solvent at high temperature and pressure. Experiments have shown that this method has great advantages in the preparation of highly monodispersed nanoparticles with a control over size and morphology. Figure 20.23 shows the major differences in the products obtained by ball milling or sintering or firing and by the hydrothermal method.

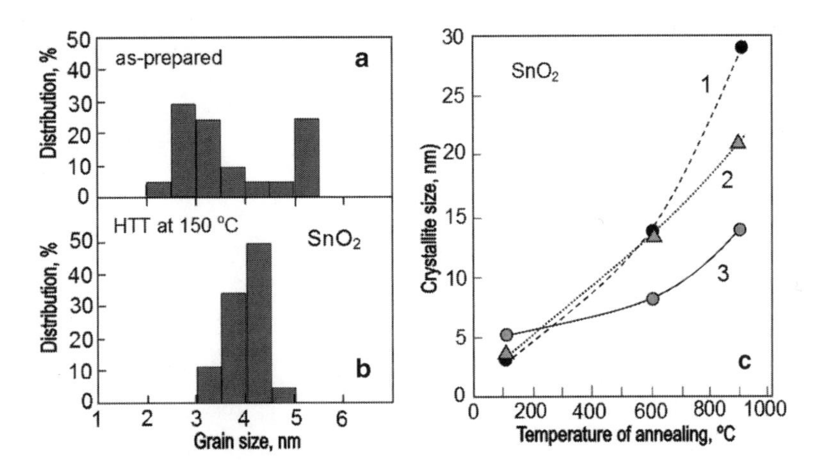

Fig. 20.24 Particle size distributions of (**a**) as-prepared SnO_2, (**b**) hydrothermally treated SnO_2 at 150 °C for 72 h (Adapted with permission from Fujihara et al. 2004, Copyright 2004 American Chemical Society). (**c**) Dependence of tin oxide crystallite size on calcination temperature: (*1*) untreated gel, (*2*) untreated sol, and (*3*) hydrothermally treated sol (Adapted with permission from Baik et al. 2000a, b, c, Copyright 2000 John Wiley & Sons)

Yamazoe and co-workers (Baik et al. 2000a, b) believe that the use of a hydrothermal method during metal oxide powder synthesis, instead of a conventional powder processing procedure, also contributes to improving the stability of gas sensor parameters. To this end, it was reported that the hydrothermal treatment of SnO_2 gel at 200 °C for 3 h with an aqueous ammonia solution was effective in stabilizing a SnO_2 crystallite against growth at elevated temperatures. The crystallite size of SnO_2 was less than 10 nm, even after calcination at 600 °C. Research carried out by Fang et al. (2008) had the same results. The composite nanoparticles prepared by the sol–gel hydrothermal route have better thermal stability against agglomeration and particle growth than those prepared by the sol–gel calcination route (see Fig. 20.24c, curve 3). It was found that, compared with the sol–gel calcination route, the sol–gel hydrothermal route led to better-dispersed spherical SnO_2 nanoparticles with narrower size distribution and larger specific surface area. For example, according to results obtained by Fujihara et al. (2004), the as-prepared sol–gel SnO_2 powders have a relatively broad size distribution with two peaks at smaller (2.5–3.0 nm) and larger (5.0–5.5 nm) sizes (Fig. 20.24a), while the HTT SnO_2 exhibits a narrower size distribution in the range of 3.0–5.0 nm (Fig. 20.24b).

The correct selection of substrates could also be used to attain positive results. For example, Fleischer et al. (1994) found that Ga_2O_3-based devices fabricated on BeO substrates had better temporal stability of operating parameters than Ga_2O_3 sensors with Al_2O_3 substrates. A search of optimal additives for doping of sensing materials is thus required in order to achieve a high stability of sensor parameters. In Tang et al. (2001) it was shown that SnO_2 films doped by Cu are subjected to less cracking; the same result was reported in Sharma et al. (2001). Then a comparison of the stability of microfabricated sensors, using SnO_2:Cu/Pt, SnO_2/Pt, and SnO_2/Pt/SnO_2/Pt gas-sensing layers, demonstrated that SnO_2:Cu/Pt-based sensors have minimal drift because SnO_2/Pt and SnO_2/Pt/SnO_2/Pt sensing layers crack more as compared to the SnO_2:Cu/Pt one. This result suggests that SnO_2 films doped by Cu are better candidates for designing devices on a hotplate platform aimed at integration with on-chip CMOS circuitry.

It has been considered that microfabrication and cointegration increase the long-term reliability and stability of the final units as well. Microfabrication and integration can improve the overall performance of the system, which needs fewer components (Laconte et al. 2005). However, prior to sensor fabrication, it is necessary to take into account that this fabrication requires a complicated technological process and high-cost equipment. In addition, the packaging of integrated sensors may also be more complicated as there is a need to reduce sensor interference with the surroundings, minimize electric crosstalk, and optimize the critical connections.

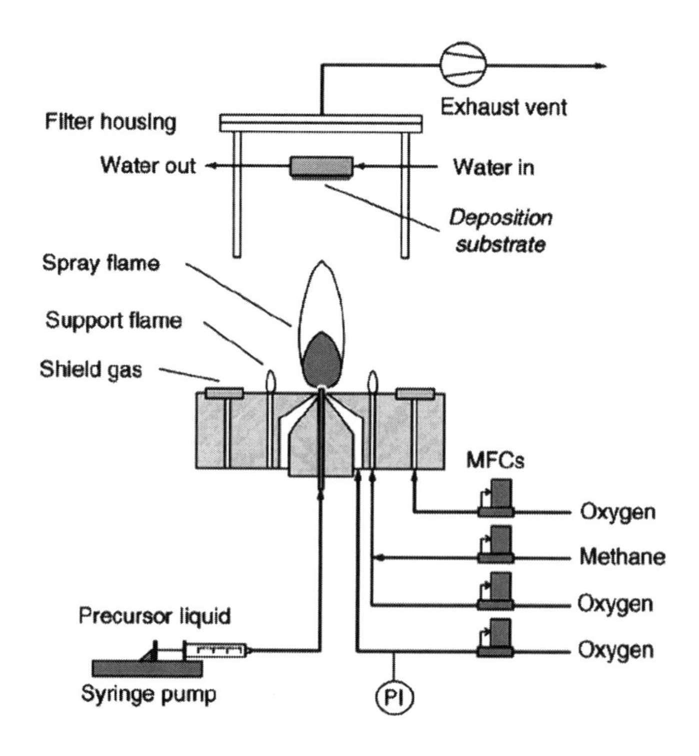

Fig. 20.25 Flame spray pyrolysis (FSP) apparatus for producing ceramic nanoparticles and directly depositing them on a cooled substrate positioned above the flame. Particles are simultaneously collected downstream on a fiber filter for comparison (Reprinted with permission from Madler et al. 2006b, Copyright 2006 Elsevier)

20.3.8 Use Novel Techniques for Metal Oxide Synthesis and Deposition Able to Produce Stable Materials

The analysis conducted above has shown that the method of metal oxide powder and film preparation strongly influences the morphology and stability of metal oxide particles. As demonstrated, correct optimization of the technological route really provides an opportunity to control the process of particle aggregation during their synthesis and affects the stability of the grain size during the thermal treatments that follow. Unfortunately the process of particle aggregation cannot be completely excluded. Thus, recently a great amount of work targeted on optimization of conventional methods and searching new progressive methods of gas-sensing layer formation has been conducted (Sahner and Tuller 2010). As a result, new approaches to gas-sensing layer formation have appeared.

The technique of flame spray pyrolysis (FSP) is one of the technologies from this list, tested for gas sensor design (Madler et al. 2006a, b; Liu et al. 2005b). It should be noted that FSP is a well-known method which is widely applied for the industrial production of powders. One can understand the nature of this method of synthesis from its name. One can find a detailed description of FSP technology in several published papers (Pratsinis 1998; Madler et al. 2002, 2006b; Tricoli et al. 2010; Strobel and Pratsinis 2007). Usually the FSP apparatus involves the scheme shown in Fig. 20.25.

The combustion CVD process discussed in several papers for the same purposes has the same nature (Liu et al. 2005b). The aggregation of growing particles is observed during both FSP and the combustion CVD process (Strobel et al. 2003; Liu et al. 2005b). Possible mechanisms of particle formation during FSP are shown in Fig. 20.26. However, due to the possibility of excluding the need for high-temperature annealing for the grain stabilization, the influence of particle aggregation on the stability of sensor parameters should be significantly reduced.

Experiments have shown that the procedure of FSP, which can be conducted in atmospheric conditions, allows fabrication of almost equal-sized nanoparticles with very good crystallinity, in which the

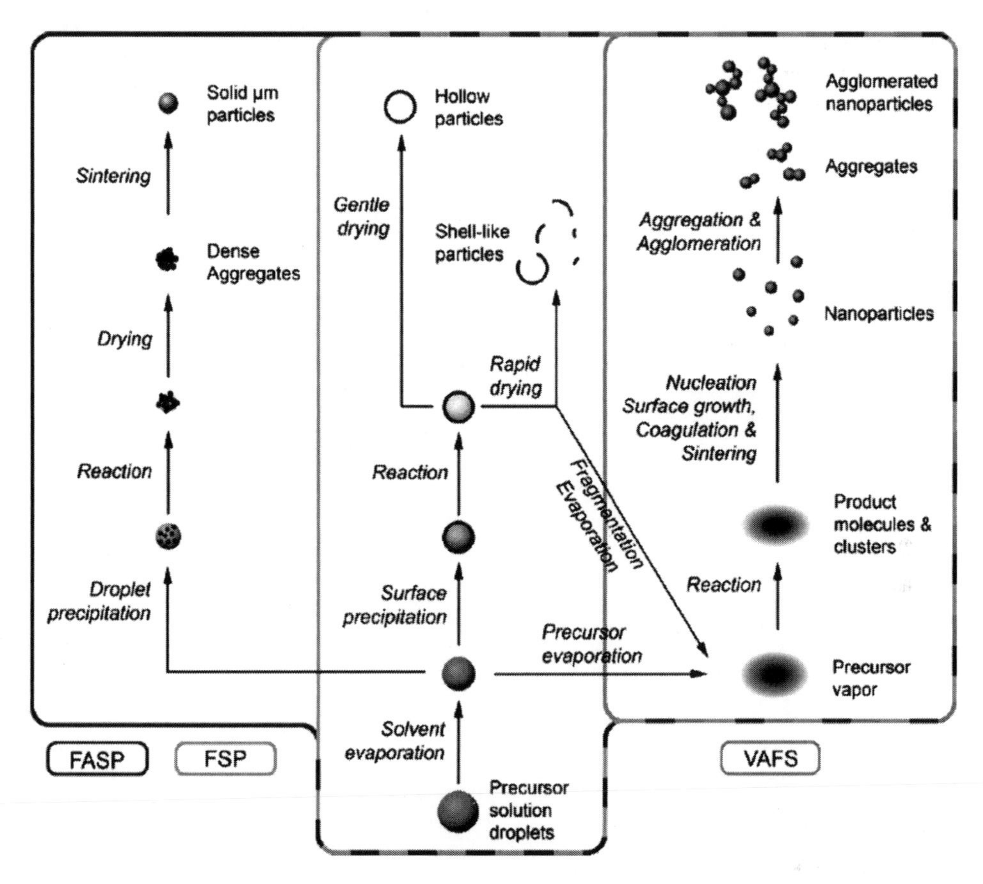

Fig. 20.26 Schematic for particle formation mechanisms during flame-assisted spray pyrolysis (FASP), FSP, and vapor-fed aerosol flame synthesis (VAFS) (Reprinted with permission from Strobel and Pratsinis 2007, Copyright 2007 Royal Society of Chemistry)

grain size could be selected independently from the substrate temperature during film formation (Kennedy et al. 2005). Furthermore, the process is clean and fast (minutes compared to days for comparable quantities) and also allows for in situ functionalization of metal oxide grains. In general, FSP has the ability to produce nanoparticles with high crystallinity of both undoped metal oxides (Madler et al. 2002) and metal oxides with surfaces functionalized by noble metals (Madler et al. 2003; Strobel et al. 2003; Tricoli et al. 2008a). Moreover, it was established that, involving pyrolysis, passing in gas phase, and correct selection of the conditions of synthesized particles, FSP deposition on the substrate is a method which, in contrast to wet-chemical methods, makes it possible to realize synthesis, annealing, and gas-sensing layer formation as a single process. For example, it was found that during FSP nanocrystalline tin oxide can be deposited directly in situ as porous films onto alumina sensor substrates by thermophoresis (Madler et al. 2006b), Brownian deposition, electrophoresis (Kim et al. 2006), and impaction (Barborini et al. 2005; Joshi et al. 2006). In the absence of restructuring, which could arise from high substrate temperature or impaction energy, the resulting film consists of the particles produced in the gas phase and thus has a homogeneous composition and morphology (Madler et al. 2006b). According to Tricoli et al. (2008b), the as-deposited FSP-made films are highly porous (e.g., 98 %) and consist of a lacelike network of nanoparticles, which have average grain and crystal sizes of approximately 10 nm. Achieved arrangement of deposited particles allows penetration of the analyte into the film and depletion of all the grain surfaces.

Based on the above discussion, one can conclude that the use of FSP considerably simplifies the process of gas-sensing matrix preparation. Moreover, high-temperature annealing during FSP is performed in-flight without contacts between grains, and therefore it is possible to keep the crystal size constant. This means that FSP guarantees the invariability of the grain size during sensor fabrication. So, due to the high-temperature processing of FSP, the use of particles synthesized by FSP does not require any post-treatment of the sensing material, which is mandatory for sensors based on powders synthesized by wet-chemistry methods. As mentioned before, this feature of FSP-based technology is extremely important for achieving either high sensitivity or high stability gas sensors.

The production of homogenous nanometer-sized SnO_2 nanoparticles by FSP for gas sensing has been demonstrated successfully by the Tubingen group (Sahm et al. 2004; Madler et al. 2006a, b). It was established that the as-obtained sensors exhibited extremely good homogeneity of the sensing film and good sensor performance (Kappler et al. 2001). Moreover, Madler et al. (2006a) have found that the application of high-temperature flame pyrolysis technology instead of the sol–gel process for synthesizing powders to be used in SnO_2 sensor design induced sufficient improvement in sensor characteristics stability. The high stability of the sensors derived from flame-made powders becomes evident when comparing with sol–gel-derived powders under the same conditions. According to the results of comparative experiments carried out by Kappler et al. (2001), in contrast to FSP-made powders, the sol–gel-made powder drifts steadily over the entire time period which was evaluated as 20 days. Madler et al. (2006a) believe that the high stability of the sensors derived from these flame-made powders may be due to high dispersion in the matrix, high homogeneity in particle size, and the well-crystalline nature of the flame-made materials. This assumption is in good accordance with previous conclusions made by us. It is important to note that FSP-based technology of gas sensor fabrication guaranteed high sensitivity even for devices based on undoped metal oxides, i.e., without using any catalytically active additives. The latter is a significant advantage of this technology because conventional wet-chemical-based methods of SnO_2 synthesis usually require the incorporation of additional catalysts (Pt, Pd, Au, etc.) in the gas-sensing matrix for achieveing acceptable sensitivity (Madler et al. 2006a; Kocemba and Rynkowski 2011). It was found that the in situ-prepared sensors of pure SnO_2 and Pt-doped SnO_2 are reproducible and have a very low detection limit for CO (down to 1 ppm) and ethanol (<0.1 ppm) with high sensor response (Madler et al. 2006b; Liu et al. 2005b). It is important to note that high gas sensitivity was observed for the SnO_2 films deposited at both low substrate temperature ($T_{sub}=500$ °C) (Madler et al. 2006b) and high temperature of substrate ($T_{sub}=850$ °C) (Liu et al. 2005b). In the first case, the SnO_2 crystallites had size ~10 nm and in the second case ~30 nm. As indicated earlier, such sizes of crystallites must guarantee high stability of sensor parameters even after high-temperature annealing. Typical SEM images of the SnO_2 films deposited by FSP at high substrate temperature are shown in Fig. 20.27. Gas-sensing characteristics are shown in Fig. 20.28. Unfortunately, the information about annealing influence on the grain-size change in the above-mentioned films is absent. Regarding the nature of the processes taking place during FSP, one can assume that the decrease of both the concentration of precursor in sprayed solution and the size of droplet will contribute to a decrease in the sizes of crystallites and agglomerates formed during FSP.

One of the main challenges of such a novel approach as FSP lies in the stabilization of mostly physically bounded porous films. It was found that mechanical stability of films deposited at low substrate temperatures is generally poor (Tricoli et al. 2008b). Mechanical properties of films deposited at high substrate temperatures are much better. There have been attempts to improve mechanical properties of films deposited at low substrate temperatures through additional thermal treatments (Madler et al. 2006b; Tricoli et al. 2008b). However, till now this problem has not been completely resolved. Low-temperature annealing ($T_{an}<500$ °C) is not effective, while high-temperature (>1,000 °C) annealing may also not be applicable, because highly porous, particulate films tend to disintegrate above a given sintering temperature (Andersen et al. 2002). It should be noted that the poor mechanical stability of FSP films is not new and was the main reason why they were discarded from industrial applications in the past (Tricoli et al. 2008a).

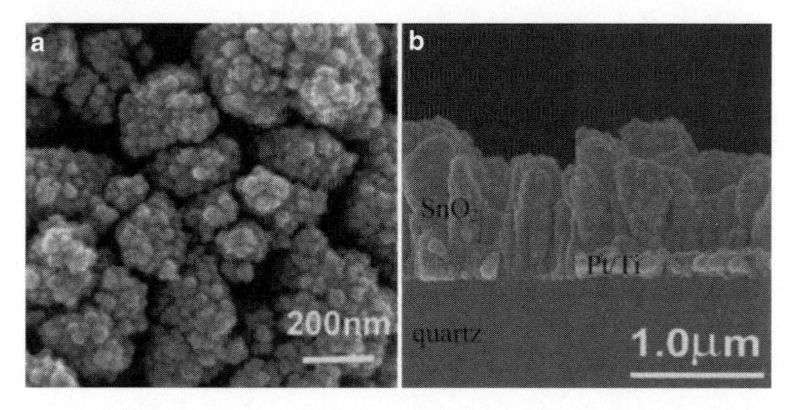

Fig. 20.27 (**a**) Top view and (**b**) cross view SEM images of nanostructured SnO_2 films deposited at 850 °C by a combustion CVD process (Adapted with permission from Liu et al. 2005b, Copyright 2005 American Chemical Society)

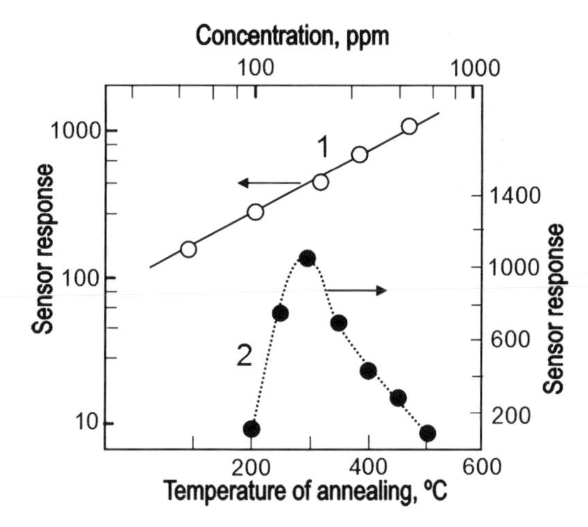

Fig. 20.28 (*1*) Sensor response to ethanol vapor concentrations for a SnO_2 sensor tested at 300 °C. (*2*) Influence operating temperature on sensor response of a SnO_2 sensor fabricated at 850 °C and tested for 500 ppm ethanol at different temperatures (Data from Liu et al. 2005b)

Electrophoretic deposition (EPD) and suspension-free spray deposition are other methods, which might have the same advantages. EPD is one of the colloidal processes in ceramic production (Heavens 1990; Zhitomirsky 2002; Besra and Liu 2007). The main differences between EPD and standard electrochemical deposition (ECD) process are presented in Table 20.3. In EPD, charged powder particles, dispersed or suspended in a liquid medium, are attracted and deposited onto a conductive substrate of the opposite charge under the influence of a DC electric field. The charge on the particles can be controlled by a variety of charging agents, such as acids, bases, and specifically adsorbed ions or polyelectrolytes, which can be added to the suspension (Zarbov et al. 2002). These additives act through different mechanisms. The main criteria for selecting a charging agent are the preferred polarity and deposition rate of the particles.

The suspension-free spray or cold spray deposition technique was proposed for the first time in 1999 as a simple method for the preparation of dense piezoelectric films at room temperature (Akedo and Lebedev 1999). The corresponding setup which can be used for suspension-free spray deposition of metal oxides consists of the parts shown diagrammatically in Fig. 20.29. In a continuously shaken

Table 20.3 Main differences between electrophoretic and electrochemical deposition techniques

Property	Deposition process	
	Electrochemical	Electrophoretic
Medium for deposition	Solution	Emulsion
Main parameter, controlling the rate of deposition	Current	Voltage
Moving species	Ions	Solid particles
Charge transfer on deposition	Ion reduction	None
The presence of surface reactions	Yes	None
Required conductance of liquid medium	High	Low
Preferred liquid	Water	Organic

Source: Adapted with permission from Besra and Liu (2007). Copyright 2007 Elsevier

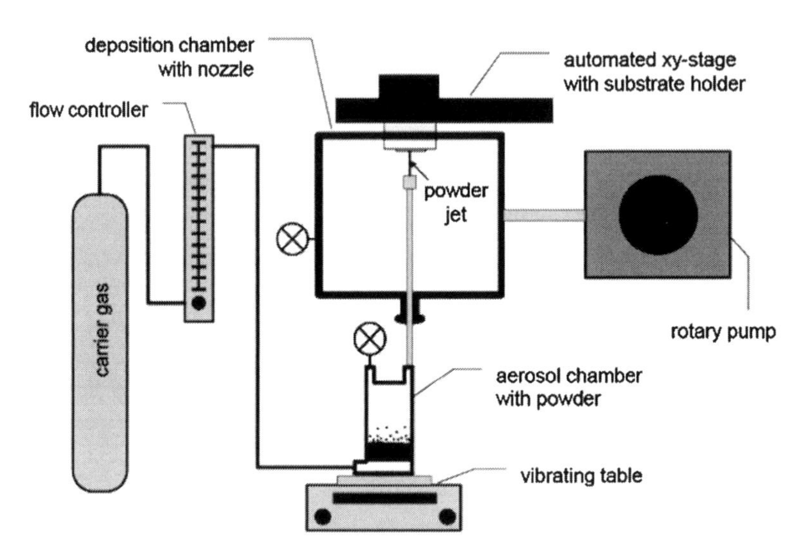

Fig. 20.29 Diagrammatical representation of the deposition setup used for suspension-free spray deposition (Reprinted with permission from Sahner et al. 2009, Copyright 2009 Elsevier)

aerosol generator, the precursor powder is fluidized. Using a vacuum pump, a pressure drop is applied between the gas inlet and outlet. When a gas—usually nitrogen or argon—is passed through the system, it is loaded with powder and accelerated toward the substrate. If the various process parameters, e.g., powder particle size, flow rate, or vacuum conditions, are carefully adjusted, the impact energy is sufficient to prepare a dense layer on the substrate without the necessity for a sintering step at high temperatures. For some materials this low-temperature deposition process might simplify the sensor preparation to a large extent. This low-temperature technique has recently been investigated for gas sensor preparation at the University of Bayreuth (Sahner et al. 2009).

Thus, the above-mentioned methods for forming sensing material can use already stabilized powders, and therefore, by way of selection of the powders used, one can influence the structural stability of the metal oxide matrix. The degree of stoichiometry and the size and the shape of grains in the deposit will be controlled by the properties of the powder used. However, the use of the indicated methods in gas sensor design was not successful. Therefore these methods will not be discussed in this chapter. Moreover, the use of lectrophoretic deposition requires a following high-temperature annealing for removing traces of solvents and improvement of mechanical properties. From this point

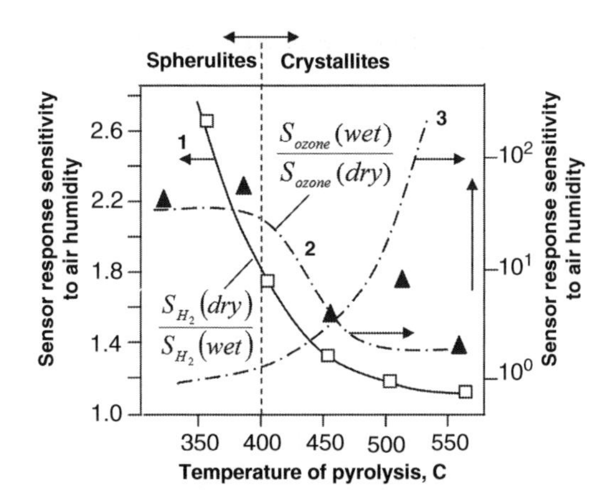

Fig. 20.30 Influence of pyrolysis temperature on the sensitivity to air humidity during (*1*) H_2 (5,000 ppm) detection by undoped In_2O_3 sensors (T_{oper} = 370 °C) and (*2*) O_3 (1 ppm) detection by undoped SnO_2 sensors (T_{oper} = 270 °C) and (*3*) on the grain size of In_2O_3- and SnO_2-based sensors. Dry air ~1 % RH, wet air ~35–45 % RH (Reprinted with permission from Korotcenkov et al. 2009, Copyright 2009 Taylor & Francis)

of view, this method does not differ from conventional methods of thick-film technology (Sahner and Tuller 2010). In addition, EPD cannot use water as the liquid medium, because application of a voltage to water causes the formation of hydrogen and oxygen gases at the electrodes, which could adversely affect the quality of the deposits formed. The difficulties of film deposition at nonconducting substrates and the deposition of ultrathin films are other important disadvantages of EPD. Films prepared by suspension-free spray deposition are too dense (Sahner et al. 2009). This fact is good for high-temperature electrochemical solid electrolyte gas sensors, but bad for adsorption-type low-temperature metal oxide sensors discussed in this chapter. In addition, the information available in the literature concerning the influence of morphology of films deposited by the above-mentioned methods on the parameters of sensors design is not enough for any analysis.

20.3.9 Materials and Processing Should Facilitate a Reduction of the Effects of Humidity

Korotcenkov et al. (2009) reported that increasing the temperature of the SnO_2 and In_2O_3 deposition substantially decreased the influence of air humidity on the sensor's response (see Fig. 20.30). It was then established that the fraction of finely dispersed grains in the layers decreased with increase in the deposition temperature (Korotcenkov et al. 2008; Korotcenkov and Han 2009). Therefore, sensors based on MOX layers deposited at high temperatures are more stable when compared to those deposited at lower temperatures and do not need additional aging prior to the tests.

In Fukui and Katsuki (2000), another approach was suggested to reduce the influence of humidity on sensor characteristics. A decrease in the dependence of SnO_2-based sensor parameters on air humidity was attained by doping; in particular, doping by oxides of Cr and, especially, cobalt (0.10 at.%) incurred a very minor humidity dependence. A low tendency to hydrate is another important requirement for materials destined for practical use. Only this property can provide a stable work capability in a wet atmosphere. Relative to parameter stability in a wet atmosphere, it is also necessary to estimate the potential of using new gas-sensing materials such as SnO_2–CuO, which has a high sensitivity to H_2S (Khanna et al. 2003). Unfortunately, high sensitivity and parameter stability have only been observed in a dry atmosphere. In a wet atmosphere, especially when the humidity is RH >75 %, CuO in the surrounding gas could be transformed into copper carbonate ($CuCO_3 \cdot Cu(OH)_2$), sulfate ($CuSO_4 \cdot 3Cu(OH)_2$), or chloride ($CuCl_2 \cdot 3Cu(OH)_2$), with a corresponding change in physical and chemical properties.

20.3.10 Increase Material Porosity and Pore Size

The relationship between material porosity and sensor sensitivity to humidity has not yet attracted the attention of most designers. Nevertheless, there are assumptions that an increase in pore size will decrease both the effect of water vapor condensation and the influence of surface contamination on gas penetration; it is known that the smaller the pore size, the more probable the occurrence of water condensation in the device (Korotcenkov 2005, 2008). Numerous capacitive-type humidity sensors were designed based on this effect. Therefore, one can suppose that during storage or in the break-in period, and when the temperature of the active zone does not exceed 20–25 °C, water condensation in small pores is possible. In this case, the long-term exposure to water could substantially modify the oxide's surface properties and thus change the sensor's performance parameters. Materials with extremely small pores also show a decreased resistance to poisoning. This is another important problem of gas-sensing materials and is directly related to the need to provide maximum temporal stability in gas sensor parameters. This problem was discussed for pellistors; simulations carried out by Gentry and Walsh (1987) for flammable gas-sensing elements showed that the poison resistance of porous sensors decreased with an increase in the parameter $h_o \sim (1/r \cdot D_p)^{1/2}$, where r is the pore radius and D_p is the pore diffusion coefficient. Accumulation of solid reaction products on the sensor surface could also block the active surface sites. With small pores, however, this process would be accompanied by a sharp decrease in the active surface area and a decrease in the sensor's response.

20.3.11 Use New Approaches to Design Sensors

For example, sensors should be designed based on individual quasi-1D structures such as nanowires, nanobelts, nanotubes, nanoribbons, nanowhiskers, nanorings, and nanorods (Sysoev et al. 2009; Dai et al. 2002). At present, the use of individual 1D nanostructures is considered one of the most promising directions of structural engineering in attempts to develop a new generation of chemical sensors. It is expected that sensors based on such structures should show better parameter stability in comparison with conventional sensing materials (Sysoev et al. 2009). However, it must be noted that quasi-1D structures also have limitations in thermal stability (see next Chapter).

References

Adamyan AZ, Adamyan ZN, Aroutiounian VM (2003) Preparation of SnO_2 films with thermally stable nanoparticles. Sensors 3:438–442

Adamyan AZ, Adamian ZN, Aroutiounian VM, Schierbaum KD, Han S-D (2009) Improvement and stabilization of thin-film hydrogen sensors parameters. Armenian J Phys 2(3):200–212

Akedo J, Lebedev M (1999) Microstructure and electrical properties of lead zirconate titanate ($Pb(Zr_{0.52}Ti_{0.48})O_3$) thick film deposited with aerosol deposition method. Jpn J Appl Phys 38:5397–5401

Andersen SK, Johannessen T, Mosleh M, Wedel S, Tranto J, Livbjerg H (2002) The formation of porous membranes by filtration of aerosol nanoparticles. J Nanopart Res 4:405–416

Antonelli DM, Ying JY (1995) Synthesis of hexagonally packed mesoporous TiO_2 by a modified sol–gel method. Angew Chem Int Ed 34:2014–2017

Anukunprasert T, Saiwan C, Traversa E (2005) The development of gas sensor for carbon monoxide monitoring using nanostructure of Nb–TiO_2. Sci Tech Adv Mater 6:359–363

Arroyo R, Cordoba G, Padilla J, Lara VH (2002) Influence of manganese ions on the anatase–rutile phase transition of TiO_2 prepared by the sol–gel process. Mater Lett 54:397–402

Baalousha M (2009) Aggregation and disaggregation of iron oxide nanoparticles: influence of particle concentration, pH and natural organic matter. Sci Total Environ 407:2093–2101

Bagheri-Mohagheghi M-M, Shahtahmasebi N, Alinejad MR, Youssefi A, Shokooh-Saremi M (2008) The effect of the post-annealing temperature on the nano-structure and energy band gap of SnO_2 semiconducting oxide nano-particles synthesized by polymerizing–complexing sol–gel method. Phys B Condens Matter 403:2431–2437

Baik NS, Sakai G, Miura N, Yamazoe N (2000a) Preparation of stabilized nanosized tin oxide particles by hydrothermal treatment. J Am Ceram Soc 83(12):2983–2987

Baik NS, Sakai G, Miura N, Yamazoe N (2000b) Hydrothermally treated sol solution of tin oxide for thin-film gas sensor. Sens Actuators B Chem 63:74–79

Baik NS, Sakai G, Shimanoe K, Miura N, Yamazoe N (2000c) Hydrothermal treatment of tin oxide sol solution for preparation of thin-film sensor with enhanced thermal stability and gas sensitivity. Sens Actuators B Chem 65:97–100

Barborini E, Bongiorno G, Forleo A, Francioso L, Milani P, Kholmanov LN, Piseri P, Siciliano P, Taurino AM, Vinati S (2005) Thermal annealing effect on nanostructured TiO_2 microsensors by supersonic cluster beam deposition. Sens Actuators B Chem 111–112:22–27

Barnard AS, Curtiss LA (2005) Prediction of TiO_2 nanoparticle phase and shape transitions controlled by surface chemistry. Nano Lett 5:1261–1266

Barsan N, Weimar U (2001) Conduction model of metal oxide gas sensors. J Electroceram 7:143–167

Barsan N, Schweizer-Berberich M, Gopel W (1999) Fundamental and practical aspects in the design of nanoscaled SnO_2 gas sensors. A status report. Fresenius J Anal Chem 365:287–304

Besra L, Liu M (2007) A review on fundamentals and applications of electrophoretic deposition (EPD). Prog Mater Sci 52:1–61

Brinker CJ, Hurd AJ, Schunk P-R, Ashley CS (1992) Review of sol–gel thin film formation. J Non-Cryst Solids 147–148:424–436

Brinzari V, Korotcenkov G, Schwank J, Lantto V, Saukko S, Golovanov V (2002) Morphological rank of nano-scale tin dioxide films deposited by spray pyrolysis from $SnCl_4{\cdot}5H_2O$ water solution. Thin Solid Films 408(1–2):51–58

Brynzari V, Korotchenkov G, Dmitriev S (2000) Theoretical study of semiconductor thin film gas sensitivity: attempt to consistent approach. J Electron Technol 33:225–235

Byrappa K, Adschiri T (2007) Hydrothermal technology for nanotechnology. Prog Cryst Growth Ch Mater 53:117–166

Byrappa K, Yoshimura M (2001) Handbook of hydrothermal technology. Noyes, Park Ridge, NJ

Callone E, Carturan G, Sicurelli A (2006) Nanopowders of metallic oxides by the hydrolytic route with starch stabilization and biological abetment. J Nanosci Nanotechnol 6(1):254–257

Callone E, Carturan G, Ischia M, Sicurelli A (2007) Size stabilization of nanoparticles by polysaccharides: effectiveness in the wet and curing steps. J Mater Res 22(12):3344–3354

Capone S, Forleo A, Francioso L, Rella R, Siciliano P, Spadavecchia J, Presicce DS, Taurino AM (2003) Solid state gas sensors: state of the art and future activities. J Optoelectron Adv Mater 5:1335–1348

Ciobanu C, Liu Y, Wang Y, Patton BR (1999) Numerical calculation of electrical conductivity of porous electroceramics. J Electroceram 3(1):17–23

Cirera A, Dieguez A, Diaz R, Cornet A, Morante JR (1999) New method to obtain stable small-sized SnO_2 powders for gas sensors. Sens Actuators B Chem 58:360–364

Cirera A, Cornet A, Morante JR, Olaizola SM, Castano E, Gracia J (2000) Comparative structural study between sputtered and liquid pyrolysis nanocrystalline SnO_2. Mater Sci Eng B 69–70:406–410

Dai ZR, Gole JL, Stout JD, Wang ZL (2002) Tin oxide nanowires, nanoribbons, and nanotubes. J Phys Chem B 106:1274–1279

De Angelis L, Riva R (1995) Selectivity and stability of a tin dioxide sensor for methane. Sens Actuators B Chem 28:25–29

De Souza Brito GE, Santilli CV, Pulcinelli SH (1995) Evolution of the fractal structure during sintering of SnO_2 compacted sol–gel powder. Colloid Surf A 97:217–225

Dieguez A, Romano-Rodriguez A, Morante JR, Barsan N, Weimar U, Gopel W (1997) Nondestructive assessment of the grain size distribution of SnO_2 nanoparticles by low-frequency Raman spectroscopy. Appl Phys Lett 71(14):1957–1959

Dieguez A, Romano-Rodriguez A, Morante JR, Kappler J, Barsan N, Gopel W (1999) Nanoparticle engineering for gas sensor optimisation: improved sol–gel fabricated nanocrystalline SnO_2 thick film gas sensor for NO_2 detection by calcination, catalytic metal introduction and grinding treatments. Sens Actuators B Chem 60:125–137

Dufoura L-C, Bertrand GL, Caboche G, Decorse P, El Anssari A, Poirson A, Vareille M (1997) Fundamental and technological aspects of the surface properties reactivity of some metal oxides. Solid State Ion 101–103:661–666
Dunphy Guzman KA, Finnegan MP, Banfield JF (2006) Influence of surface potential on aggregation and transport of titania nanoparticles. Environ Sci Technol 40:7688–7693
Elimelech M, Gregory J, Jia X, Williams RA (1995) Particle deposition and aggregation—measurement, modelling and simulation. Butterworth–Heinemann, Oxford, England
Fang LM, Zu XT, Li ZJ, Zhu S, Liu CM, Zhou WL, Wang LM (2008) Synthesis and characteristics of $Fe3^{+}$-doped SnO_2 nanoparticles via sol–gel-calcination or sol–gel-hydrothermal route. J Alloys Compd 454:261–267
Fleischer M, Hanrieder W, Meixner H (1990) Stability of semiconducting gallium oxide thin films. Thin Solid Films 190:93–102
Fleischer M, Hollbauer L, Meixner H (1994) Effect of the sensor structure on the stability of Ga_2O_3 sensors for reducing gases. Sens Actuators B Chem 18–19:119–124
Franke ME, Koplin TJ, Simon U (2006) Metal and metal oxide nanoparticles in chemiresistors: does the nanoscale matter? Small 2(1):36–50
French RA, Jacobson AR, Kim B, Isley SL, Penn RL, Baveye PC (2009) Influence of ionic strength, pH, and cation valence on aggregation kinetics of titanium dioxide nanoparticle. Environ Sci Technol 43:1354–1359
Fujihara S, Maeda T, Ohgi H, Hosono E, Imai H, Kim S-H (2004) Hydrothermal routes to prepare nanocrystalline mesoporous SnO_2 having high thermal stability. Langmuir 20:6476–6481
Fukui K, Katsuki A (2000) Improvement of humidity dependence in gas sensor based on SnO_2. Sens Actuators B Chem 65:316–318
Gentry SJ, Walsh PT (1987) The theory of poisoning of catalytic flammable gas sensing elements. In: Moseley PT, Tofield BC (eds) Solid state gas sensors. Adam Hilger, Bristol, Philadelphia, pp 32–50
Golovanov V, Pekna T, Kiv A, Litovchenko V, Korotcenkov G, Brinzari V, Cornet A, Morante J (2005) The influence of structural factors on sensitivity of SnO_2-based gas sensors to CO in humid atmosphere. Ukr Phys J 50(4):374–380
Gribb AA, Banfield JF (1997) Particle size effects on transformation kinetics and phase stability in nanocrystalline TiO_2. Am Mineral 82:717–728
Gryaznov VG, Trusov LI (1993) Size effects in micromechanics of nanocrystals. Prog Mater Sci 37:289–401
Gurlo A (2011) Nanosensors: towards morphological control of gas sensing activity. SnO_2, In_2O_3, ZnO and WO_3 case studies. Nanoscale 3:154–165
Han X-G, He H-Z, Kuang Q, Zhou X, Zhang X-H, Xu T, Xie Z-X, Zheng L-S (2009a) Controlling morphologies and tuning the related properties of nano/microstructured ZnO crystallites. J Phys Chem C 113:584–589
Han X, Jin M, Xie S, Kuang Q, Jiang Z, Jiang Y, Xie Z, Zheng L (2009b) Synthesis of tin dioxide octahedral nanoparticles with exposed high-energy 221 facets and enhanced gas-sensing properties. Angew Chem Int Ed 48:9180–9183
Haugen J-E, Tomic O, Kvaal K (2000) A calibration method for handling the temporal drift of solid state gas-sensors. Anal Chim Acta 407:23–39
Hayashia M, Hyodo T, Shimizu Y, Egashira M (2009) Effects of microstructure of mesoporous SnO_2 powders on their H_2 sensing properties. Sens Actuators B Chem 141:465–470
Heavens N (1990) Electrophoretic deposition as a *processing* route for *ceramics*. In: Binner GP (ed) Advanced ceramic processing and technology, vol 1. Noyes, Park Ridge, NJ, pp 255–283
Hillhouse HW (2005) Synthesis of ordered mesoporous tin oxide thin films displaying extremely high thermal stability: a TEM and SAXS study of structural changes during the thermal treatment. In: AIChE annual meeting conference proceedings, vol 2005, p 1
Holody PRJ, Soltis RE, Hangas J (2001) Limiting particle growth in platinum/tin oxide nanocomposites. Scripta Mater 44:1821–1824
Horvath G, Gerblinger J, Meixner H, Giber J (1996) Segregation driving forces in perovskite titanates. Sens Actuators B Chem 32:93–99
Hu Y, Tan OK, Pan JS, Huang H, Cao W (2005) The effects of annealing temperature on the sensing properties of low temperature nano-sized $SrTiO_3$ oxygen gas sensor. Sens Actuators B Chem 108:244–249
Huang JF, Xia CK, Xiong XB, Cao LY, Wu JP (2007) Preparation of SnO_2 nanocrystallites by hydrothermal liquid–solid-solution process. Mater Res Innov 11(3):118–121
Hunter RJ (1987) Foundation of colloidal science, vol 1. Oxford University Press, New York, NY
Hyodo T, Nishida N, Shimizu Y, Egashira M (2002) Preparation and gas-sensing properties of thermally stable mesoporous SnO_2. Sens Actuators B Chem 83:209–215
Ihokura K, Watson J (1994) The stannic oxide gas sensor—principles and applications. CRC Press, Boca Raton, FL
Itoh T, Matsubara I, Shin W, Izu N, Nishibori M (2008) Analytical study of resistance drift phenomena on (PANI)$x$$MoO_3$ hybrid thin films as gas sensors. Bull Chem Soc Jpn 81(10):1331–1335
Jamnik J, Kamp B, Merkle R, Maier J (2002) Space charge influenced oxygen incorporation in oxides: in how far does it contribute to the drift of Taguchi sensors? Solid State Ion 150:157–166

Joshi RK, Kruis FE, Dmitrieva O (2006) Gas sensing behavior of $SnO_{1.8}$:Ag films composed of size-selected nanoparticles. J Nanopart Res 8:797–808

Jung J-S, Son K-S, Kim T-S, Ryu M-K, Park K-B, Yoo B-W, Kwon J-Y, Lee S-Y, Kim J-M (2008) Stability improvement of gallium indium zinc oxide thin film transistors by post-thermal annealing. ECS Trans 16(9):309–313

Kappler J, Bârsan N, Weimar U, Dièguez A, Alay JL, Romano-Rodriguez A, Morante JR, Göpel W (1998) Correlation between XPS, Raman and TEM measurements and the gas sensitivity of Pt and Pd doped SnO_2 based gas sensors. Fresenius J Anal Chem 361:110–114

Kappler J, Tomescu A, Barsan N, Weimar U (2001) CO consumption of Pd doped SnO_2 based sensors. Thin Solid Films 391(2):186–191

Karunagarani B, Rajendra Kumar RT, Mangalaraj D, Narayandass SAK, Mohan RG (2002) Influence of thermal annealing on the composition and structural parameters of DC magnetron sputtered titanium dioxide thin films. Cryst Res Technol 37(12):1285–1292

Kennedy MK, Kruis FE, Fissan H, Nienhaus H, Lorke A, Metzger TH (2005) Effect of in-flight annealing and deposition method on gas-sensitive SnO_x films made from size-selected nanoparticles. Sens Actuators B Chem 108:62–69

Khanna A, Kumar R, Bhatti SS (2003) CuO-doped SnO_2 thin films as hydrogen sulfide gas sensor. Appl Phys Lett 82(24):4388–4390

Kim H, Kim J, Yang H, Suh J, Kim T, Han B, Kim S, Kim DS, Pikhitsa PV, Choi M (2006) Parallel patterning of nanoparticles via electrodynamic focusing of charged aerosols. Nat Nanotechnol 1:117–121

Kocemba I, Rynkowski J (2011) The influence of catalytic activity on the response of Pt/SnO_2 gas sensors to carbon monoxide and hydrogen. Sens Actuators B Chem 155:659–666

Korosi L, Dekany I (2006) Preparation and investigation of structural and photocatalytic properties of phosphate modified titanium dioxide. Colloid Surf A 280:146–154

Korosi L, Papp S, Meynen V, Cool P, Vansant EF, Dekany I (2005) Preparation and characterization of SnO_2 nanoparticles of enhanced thermal stability: the effect of phosphoric acid treatment on $SnO_2{\cdot}nH_2O$. Colloid Surf A 268:147–154

Korosi L, Papp S, Bert ti I, Dekany I (2007a) Surface and bulk composition, structure, and photocatalytic activity of phosphate-modified TiO_2. Chem Mater 19:4811–4819

Korosi L, Oszko A, Galbacs G, Richardt A, Zollmer V, Dekany I (2007b) Structural properties and photocatalytic behaviour of phosphate-modified nanocrystalline titania films. Appl Catal B 77:175–183

Korosi L, Papp S, Beke S, Oszko A, Dekany I (2010) Effects of phosphate modification on the structure and surface properties of ordered mesoporous SnO_2. Micropor Mesopor Mater 134:79–86

Korotcenkov G (2005) Gas response control through structural and chemical modification of metal oxides: state of the art and approaches. Sens Actuators B Chem 107:209–232

Korotcenkov G (2007a) Metal oxides for solid state gas sensors. What determines our choice? Mater Sci Eng B 139:1–23

Korotcenkov G (2007b) Practical aspects in design of one-electrode semiconductor gas sensors: status report. Sens Actuators B Chem 121:664–678

Korotcenkov G (2008) The role of morphology and crystallographic structure of metal oxides in response of conductometric-type gas sensors. Mater Sci Eng R 61:1–39

Korotcenkov G, Cho BK (2011) Instability of metal oxide-based conductometric gas sensors and approaches to stability improvement (short survey). Sens Actuators B Chem 156:527–538

Korotcenkov G, Han SD (2009) (Cu, Fe, Co and Ni)-doped SnO_2 films deposited by spray pyrolysis: doping influence on thermal stability of SnO_2 film structure. Mater Chem Phys 113:756–763

Korotcenkov G, Boris I, Brinzari V, Luchkovsky Y, Karkotsky G, Golovanov V, Cornet A, Rossinyol E, Rodriguez J, Cirera A (2004) Gas sensing characteristics of one-electrode gas sensors on the base of doped In_2O_3 ceramics. Sens Actuators B Chem 103:13–22

Korotcenkov G, Brinzari V, Ivanov M, Cerneavschi A, Rodriguez J, Cirera A, Cornet A, Morante J (2005a) Structural stability of In_2O_3 films deposited by spray pyrolysis during thermal annealing. Thin Solid Films 479:38–51

Korotcenkov G, Cornet A, Rossinyol E, Arbiol J, Brinzari V, Blinov Y (2005b) Faceting characterization of SnO_2 nanocrystals deposited by spray pyrolysis from $SnCl_4$-$5H_2O$ water solution. Thin Solid Films 471(1–2):310–319

Korotcenkov G, Boris I, Cornet A, Rodriguez J, Cirera A, Golovanov V, Lychkovsky Y, Karkotsky G (2007) Influence of additives on gas sensing and structural properties of In_2O_3-based ceramics. Sens Actuators B Chem 120:657–664

Korotcenkov G, Brinzari V, Boris I (2008) (Cu, Fe, Co or Ni)-doped SnO_2 films deposited by spray pyrolysis: doping influence on film morphology. J Mater Sci 43(8):2761–2770

Korotcenkov G, Han SD, Cho BK, Brinzari V (2009) Grain size effects in sensor response of nanostructured SnO_2- and In_2O_3-based conductometric gas sensor. Crit Rev Solid State Mater Sci 34:1–17

Kung HH, Kung MC, Costello CK (2003) Supported Au catalysts for low temperature CO oxidation. J Catal 216:425–432

Laconte J, Flandre D, Raskin JP (eds) (2005) Micromachined thin-film sensors for SOI-CMOS co-integration. Springer, Berlin

Lee G-G, Kang S-JL, Kwon J, Kim DS (2010) Effect of a sintering process on the electrical properties of SnO_2 gas sensors. J Nanosci Nanotechnol 10(1):68–73
Li D, Kaner RB (2006) Shape and aggregation control of nanoparticles: not shaken, not stirred. J Am Chem Soc 128:968–975
Li M, Li JC (2006) Size effects on the band-gap of semiconductor compounds. Mater Lett 60:2526–2529
Lifshits IM, Slezov VV (1959) Kinetics of diffusive decomposition of supersaturated solid solutions. J Exp Theor Phys 8:331–339
Lifshits IM, Slezov VV (1961) The kinetics of precipitation from supersaturated solid solutions. J Phys Chem Solids 19:35–50
Liu F, Quan B, Chen L, Yu L, Liu Z (2004) Investigation on SnO_2 nanopowders stored for different time and $BaTiO_3$ modification. Mater Chem Phys 87:297–300
Liu F, Quan B, Liu Z, Chen L (2005a) Surface characterization study on SnO_2 powder modified by thiourea. Mater Chem Phys 93:301–304
Liu Y, Koep E, Liu M (2005b) A highly sensitive and fast-responding SnO_2 sensor fabricated by combustion chemical vapor deposition. Chem Mater 17:3997–4000
Liu Z, Yamazaki T, Shen Y, Kikuta T, Nakatani N (2007) Influence of annealing on microstructure and NO_2-sensing properties of sputtered WO_3 thin films. Sens Actuators B Chem 128:173–178
Lu H, Yang X, Wu C, Li J, Qiu N (2006) Micro-type powder-sputtered thin film gas sensors with long-term stability. In: Proceedings of the 2006 IEEE international conference on mechatronics and automation, Luoyang, China, 25–28 June, pp 2111–2115
Madler L, Kammler HK, Mueller R, Pratsinis SE (2002) Controlled synthesis of nanostructured particles by flame spray pyrolysis. J Aerosol Sci 33(2):369–389
Madler L, Stark WJ, Pratsinis SE (2003) Simultaneous deposition of Au nanoparticles during flame synthesis of TiO_2 and SiO_2. J Mater Res 18(1):115–120
Madler L, Sahm T, Gurlo A, Grunwaldt J-D, Barsan N, Weimar U, Pratsinis SE (2006a) Sensing low concentrations of CO using flame-spray-made Pt/SnO_2 nanoparticles. J Nanopart Res 8:783–796
Madler L, Roessler A, Pratsinis SE, Sahm T, Gurlo A, Barsan N, Weimar U (2006b) Direct formation of highly porous gas-sensing films by in situ thermophoretic deposition of flame-made Pt/SnO_2 nanoparticles. Sens Actuators B Chem 114:283–295
Madou MJ, Morrison SR (1989) Chemical sensing with solid state devices. Academic Press, San Diego, CA, London
Mandayo GG, Castano E, Gracia FJ, Cirera A, Cornet A, Morante JR (2003) Strategies to enhance the carbon monoxide sensitivity of tin oxide thin films. Sens Actuators B Chem 95:90–96
Massok P, Loesch M, Bertrand D (1995) Comparison for the between two Figaro sensors (TGS 813 and TGS 842) for the detection of methane, in terms of selectivity and long-term stability. Sens Actuators B Chem 24–25:525–528
Matsuura Y, Takahata K (1991) Stabilization of SnO_2 sintered gas sensors. Sens Actuators B Chem 5:205–209
Meixner H, Lampe U (1996) Metal oxide sensors. Sens Actuators B Chem 33:198–202
Min B-K, Choi S-D (2004) SnO_2 thin film gas sensor fabricated by ion beam deposition. Sens Actuators B Chem 98:239–246
Molina R, Al-Salama Y, Jurkschat K, Dobson PJ, Thompson IP (2011) Potential environmental influence of amino acids on the behavior of ZnO nanoparticles. Chemosphere 83:545–551
Nakamura Y (1989) Stability of the sensitivity of SnO_2-based elements in the field. In: Seiyama T (ed) Chemical sensor technology, vol 2. Elsevier, Amsterdam, pp 71–82
Navrotsky A (2001) Thermochemistry of nanomaterials. In: Banfield JF, Navrotsky A (eds) Reviews in mineralogy and geochemistry: nanoparticles and the environment, vol 44. Mineralog Soc Am, pp 77–103
Neiman AY (1996) Cooperative transport in oxides: diffusion and migration processes involving Mo(VI), W(VI), V(V) and Nb(V). Solid State Ion 83:263–273
Nishibori M, Shin W, Tajima K, Houlet LF, Izu N, Itoh T, Matsubara I (2008) Long-term stability of Pt/alumina catalyst combustors for micro-gas sensor application. J Eur Ceram Soc 28:2183–2190
Nowotny J (1988) Surface segregation of defects in oxide ceramic materials. Solid State Ion 28–30:1235–1243
Oudet F, Vejux A, Courtine P (1989) Evolution during thermal treatment of pure and lanthanum-doped Pt/Al_2O_3 and $Pt-Rh/Al_2O_3$ automotive exhaust catalysts: transmission electron microscopy studies on model samples. Appl Catal 50:79–86
Ozaki Y, Suzuki S, Morimitsu M, Matsunaga M (2000) Enhanced long-term stability of SnO_2-based CO gas sensors modified by sulfuric acid treatment. Sens Actuators B Chem 62:220–225
Panchapakesan B, DeVoe DL, Widmaier MR, Cavicchi R, Steve SS (2001) Nanoparticle engineering and control of tin oxide microstructures for chemical microsensor applications. Nanotechnology 12:336–349
Papadopoulos CA, Vlachos DS, Avaritsiotis JN (1997) Effect of surface catalysts on the long-term performance of reactively sputtered tin and indium oxide gas sensors. Sens Actuators B Chem 42:95–101
Park CO, Akbar SA (2003) Ceramics for chemical sensing. J Mater Sci 38:4611–4637

Pavelko RG, Vasiliev AA, Llobet E, Vilanova X, Barrabes N, Medina F, Sevastyanov VG (2009) Comparative study of nanocrystalline SnO_2 materials for gas sensor application: thermal stability and catalytic activity. Sens Actuators B Chem 137:637–643

Petot-Ervas G, Petot C (1990) Surface segregation in ceramic materials during cooling or under a temperature gradient. J Eur Ceram Soc 6:323–330

Pijolat C, Pupier C, Sauvan M, Tournier G, Lalauze R (1999) Gas detection for automotive pollution control. Sens Actuators B Chem 59:195–202

Pratsinis SE (1998) Flame aerosol synthesis of ceramic powders. Prog Energy Combust Sci 24:197–219

Qi L, Ma J, Cheng H, Zhao Z (1998) Synthesis and characterization of mesostructured tin oxide with crystalline walls. Langmuir 14:2579–2581

Rao CNR, Kulkarni GU, Thomas PJ, Edwards PP (2002) Size-dependent chemistry: properties of nanocrystals. Chem Eur J 8:28–35

Robertson WM (1960) Surface diffusion of oxides. J Nucl Mater 30:30–49

Roduner E (2006) Size matters: why nanomaterials are different. Chem Soc Rev 35:583–592

Romain A-C, Nicolas J (2010) Long term stability of metal oxide-based gas sensors for E-nose environmental applications: an overview. Sens Actuators B Chem 146:502–506

Rong YH (2005) Phase transformations and phase stability in nanocrystalline materials. Curr Opin Solid State Mater Sci 9:287–295

Sahm T, Madler L, Gurlo A, Barsan N, Pratsinis SE, Weimar U (2004) Flame spray synthesis of tin dioxide nanoparticles for gas sensing. Sens Actuators B Chem 98:148–153

Sahner K, Tuller HL (2010) Novel deposition techniques for metal oxides: prospects for gas sensing. J Electroceram 24:1385–3449

Sahner K, Kaspar M, Moos R (2009) Assessment of the aerosol deposition method for preparing metal oxide gas sensors at room temperature. Sens Actuators B Chem 139:394–399

Samsonov GV (1973) The oxide handbook. IFI/Plenum, New York, NY

Santos LRB, Chartier T, Pagnoux C, Baumard JF, Santillii CV, Pulcinelli SH, Larbot A (2004) Tin oxide nanoparticle formation using a surface modifying agent. J Eur Ceram Soc 24:3713–3721

Sayago I, Gutierrez J, Ares L, Robla JI, Horrilo MC, Getino J, Rino J, Agapito JA (1995) Long-term reliability of sensors for detection of nitrogen oxides. Sens Actuators B Chem 26–27:56–58

Sharma RK, Chan PCH, Tang Z, Yan G, Hsing IM, Sin JLO (2001) Investigation of stability and reliability of tin oxide thin-film for integrated micro-machined gas sensor devices. Sens Actuators B Chem 81:9–16

Shek CH, Lai JKL, Lin GM (1999) Investigation of interface defects in nanocrystalline SnO_2 by positron annihilation. J Phys Chem Solids 60:189–193

Shi M, Xu F, Yu K, Zhu Z, Fang J (2007) Controllable synthesis of In_2O_3 nanocubes, truncated nanocubes, and symmetric multipods. J Phys Chem C 111:16267–16271

Shimizu Y, Karino S, Takao Y, Hyodo T, Baba K, Egashira M (2000) Improvement of long-term stability of thin film gas sensors by ion beam-assisted deposition. J Electrochem Soc 147(11):4379–4384

Skala T, Veltruska K, Moroseac M, Matolinova I, Korotcenkov G, Matolin V (2003) Study of Pd–In interaction during Pd deposition on pyrolytically prepared In_2O_3. Appl Surf Sci 205:196–205

Smatko V, Golovanov V, Liu CC, Kiv A, Fuks D, Donchev I, Ivanovskaya M (2009) Structural stability of In_2O_3 films as sensor materials. J Mater Sci Mater Electron 21(4):360–363

Srivastava DN, Chappel S, Palchik O, Zaban A, Gedanken A (2002) Sonochemical synthesis of mesoporous tin oxide. Langmuir 18:4160–4164

Straumal B, Baretzky B, Mazilkin A, Protasova S, Myatiev A, Straumal P (2009) Increase of Mn solubility with decreasing grain size in ZnO. J Eur Ceram Soc 29:1963–1970

Strobel R, Pratsinis SE (2007) Flame aerosol synthesis of smart nanostructured materials. J Mater Chem 17:4743–4756

Strobel R, Stark WJ, Madler L, Pratsinis SE, Baiker A (2003) Flame-made platinum/alumina: structural properties and catalytic behaviour in enantioselective hydrogenation. J Catal 213(2):296–304

Sugiyama M, Okazaki H, Koda S (2002) Size and shape transformation of TiO_2 nanoparticles by irradiation of 308-nm laser beam. Jpn J Appl Phys 41:4666–4674

Sysoev VV, Schneider T, Goschnick J, Kiselev I, Habicht W, Hahn H, Strelcov E, Kolmakov A (2009) Percolating SnO_2 nanowire network as a stable gas sensor: direct comparison of long-term performance versus SnO_2 nanoparticle films. Sens Actuators B Chem 139:699–703

Tang Z, Chan PCH, Sharma PK, Yan G, Hsing I-M, Sin JKO (2001) Investigation and control of microcracks in tin oxide gas sensor thin-film. Sens Actuators B Chem 79:39–47

Tiemann M (2007) Porous metal oxides as gas sensors. Chem Eur J 13:8376–8388

Toledo-Antonio JA, Gutierrez-Baez R, Sebastian PJ, Vazquez A (2003) Thermal stability and structural deformation of rutile SnO_2 nanoparticles. J Solid State Chem 174:241–248

Tricoli A, Graf M, Pratsinis SE (2008a) Optimal doping for enhanced SnO_2 sensitivity and thermal stability. Adv Funct Mater 18:1969–1976

Tricoli A, Graf M, Mayer F, Kuhne S, Hierlemann A, Pratsinis SE (2008b) Micropatterning layers by flame aerosol deposition-annealing. Adv Mater 20:3005–3010

Tricoli A, Righettoni M, Teleki A (2010) Semiconductor gas sensors: dry synthesis and application. Angew Chem Int Ed 49:7632–7659

Tso C-P, Zhung C-M, Shih Y-H, Tseng Y-M, Wu S-C, Doong R-A (2010) Stability of metal oxide nanoparticles in aqueous solutions. Water Sci Technol 61(1):127–133

Ulrich M, Bunde A, Kohl CD (2004) Percolation and gas sensitivity in nanocrystalline metal oxide films. Appl Phys Lett 85:242–244

Veltruska K, Tsud N, Brinzari V, Korotcenkov G, Matolin V (2001) CO adsorption on Pd clusters deposited on pyrolytically prepared SnO_2 studied by XPS. J Vacuum 61:129–134

Wagner T, Kohl C-D, Froba M, Tiemann M (2006) Gas sensing properties of ordered mesoporous SnO_2. Sensors 6:318–323

Wamkam CT, Opoku MK, Hong H, Smith P (2011) Effects of pH on heat transfer nanofluids containing ZrO_2 and TiO_2 nanoparticles. J Appl Phys 109(2):024305

Wang Y-D, Wu X-H, Zhou Z-L, Li Y-F (2003) The reliability and lifetime distribution of SnO_2- and $CdSnO_3$-gas sensors for butane. Sens Actuators B Chem 92:186–190

Wang Y, Ma C, Sun X, Li H (2005) Synthesis and characterization of ordered hexagonal and cubic mesoporous tin oxides via mixed-surfactant templates route. J Colloid Interface Sci 286:627–631

Wang Q, Varghese O, Grimes CA, Dickey EC (2007) Grain boundary blocking and segregation effects in yttrium-doped polycrystalline titanium dioxide. Solid State Ion 178:187–194

Wang N, Cai Y, Zhang RQ (2008) Growth of nanowires. Mater Sci Eng R 60:1–51

Wu N-L, Wu L-F, Rusakova IA, Hamed A, Litvinchuk AP (1999a) Evolution in structural and optical properties of stannic oxide xerogel upon heat treatment. J Am Ceram Soc 82(1):67–73

Wu N-L, Wang S-Y, Rusakova IA (1999b) Inhibition of crystallite growth in the sol–gel synthesis of nanocrystalline metal oxides. Science 285:1375–1377

Wurzinger O, Reinhardt G (2004) CO-sensing properties of doped SnO_2 sensors in H_2-rich gases. Sens Actuators B Chem 103:104–110

Wynblatt P, Rohrer GS, Papillon F (2003) Grain boundary segregation in oxide ceramics. J Eur Ceram Soc 23:2841–2848

Xu C, Tamaki J, Miura N, Yamazoe N (1991) Grain size effects on gas sensitivity of porous SnO_2-based elements. Sens Actuators B Chem 3:145–147

Xu C, Wang X, Zhu J (2008) Graphene–metal particle nanocomposites. J Phys Chem C 112:19841–19845

Yan Y, Zhou J, Wu XZ, Moutinho HR, Al-Jassim MM (2007) Structural instability of Sn-doped In_2O_3 thin films during thermal annealing at low temperature. Thin Solid Films 515:6686–6690

Yang CC, Jiang Q (2006) Size effect on the bandgap of II–VI semiconductor nanocrystals. Mater Sci Eng B 131:191–194

Yang G, Haibo Z, Biying Z (2000) Monolayer dispersion of oxide additives on SnO_2 and their promoting effects on thermal stability of SnO_2 ultrafine particles. J Mater Sci 35:917–923

Zanella R, Giorgio S, Shin CH, Henry CR, Louis C (2004) Characterization and reactivity in CO oxidation of gold nanoparticles supported on TiO_2 prepared by deposition-precipitation with NaOH and urea. J Catal 222:357–367

Zarbov M, Schuster I, Gal-Or L (2002) Methodology for selection of charging agents for electrophoretic deposition of ceramic particles. In: Proceedings of the international symposium on electrophoretic deposition: fundamentals and applications, vol 21, The Electrochemical Society Inc, USA, pp 39–46

Zhang G, Liu M (2000) Effect of particle size and dopant on properties of SnO_2-based gas sensors. Sens Actuators B Chem 69:144–152

Zhang H, Finnegan M, Banfield JF (2001) Preparing single-phase nanocrystalline anatase from amorphous titania with particle sizes tailored by temperature. Nano Lett 1:81–85

Zhitomirsky I (2002) Cathodic electrophoretic deposition of ceramic and organoceramic materials—fundamental aspects. Adv Colloid Interface Sci 97:279–317

Chapter 21
Instability of 1D Nanostructures

21.1 Stability of Metal and Semiconductor 1D Nanowires and Nanotubes

In many recent papers it has been concluded that the use of 1D metal oxide structures in conductometric GS can resolve the problem of thermal and temporal stability, i.e., the problem of temporal drift of operating characteristics (Comini 2006; Hernandez-Ramırez et al. 2007b). However, reviewed literature data testifies that the above-mentioned statement has not been proven by a detailed analysis and experimental results.

While analyzing reasons for structural instability of polycrystalline materials, one should admit that in individual nanowires some processes, indicated in Chap. 20, are impossible. That is why one can expect improvement of thermal and, therefore, temporal stability of parameters of devices based on them, in comparison with devices based on nanoparticles. For example, there are no contacts with other nanowires in devices based on individual nanowires, and therefore it is impossible for any mass transfer from one to another. However, in network of nanowires (Ponzoni et al. 2006), which are often being considered as an alternative to polycrystalline materials in metal oxide sensors, such processes are possible, and the consequences of these processes should be the same as in the case of nanoparticles.

Analysis of temporal stability of 1D-based sensors operating at low temperatures confirmed the above-mentioned statement about better temporal stability of such devices in comparison with standard nanostructured devices (Comini 2006; Hernandez-Ramırez et al. 2007b). However, all those statements regarding experimental confirmation of high stability of sensors based on 1D structures relate to sensors, elaborated on the base of nanowires with characteristic sizes exceeding 20–30 nm. For comparison, we should note that in sensors based on polycrystalline materials, where the problems of thermal and temporal stability are relevant, the grain size does not exceed to 5–10 nm. Unfortunately, at present nobody wants to work with individual nanowires sized less that 20–30 nm (Comini et al. 2002; Comini 2006; Hernandez-Ramirez et al. 2007a, b; Köck et al. 2009). Separation and manipulation of such small objects is too difficult (Hernandez-Ramırez et al. 2007b). Besides, there are technological difficulties with growing such thin nanowires ($d < 10$ nm) of greater length. Low temperatures are necessary for growing of such nanowires. However, in this case the length of nanowires will be inefficient for the creation of bridge between two bonding pads in the measurement platform (Köck et al. 2009). Creation of a reliable low-resistance electric contact with such thin nanowires is also a problem (Hernandez-Ramırez et al. 2007b). As a result, at present the response of sensors fabricated on the base of single nanowires with $d \sim 30$–50 nm is considerably smaller than the response of conventional sensors designed on the base of nanocrystallites with grain size less than 10 nm (Comini et al. 2002; Hernandez-Ramirez et al. 2007a; Köck et al. 2009). Comparable results on sensitivity were obtained only for nanowires with modulated diameters, or for network of nanowires

G. Korotcenkov, *Handbook of Gas Sensor Materials*, Integrated Analytical Systems,
DOI 10.1007/978-1-4614-7388-6_21,

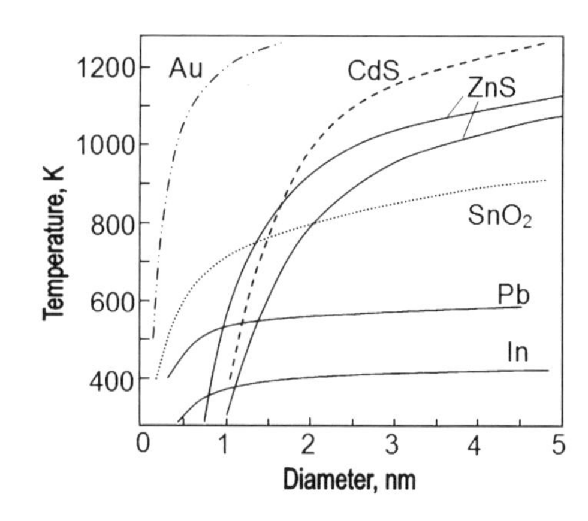

Fig. 21.1 Correlation between characteristic size and melting temperature of nanoparticles and nanowires: Au (data from Shim et al. 2002); In (data from Xie et al. 2006); CdS (data from Alivisatos 1997); ZnS (data from Li et al. 2008a); Pb (data from Jiang et al. 2004)

Table 21.1 Experimental data illustrating the influence of characteristic size on the melting temperature of nanowires

Material	Characteristic size (nm)	Temperature of phase transformation (°C)	Temperature of melting (standard) (°C)	References
Ge (nanowires)	55	650	930	Wu and Yang (2001)
Ge (nanowires)	40–100	530	930	Li et al. (2008a, b)
IrO_2 (nanowires)	10–30	700	1,070	Zhang et al. (2005)
Cu (nanowires)	160	700	1,103	Toimil Molares et al. (2004)
Cu (nanowire)	30–40	400	1,103	
Co (nanowires)	25	500	1,494	Wang et al. (2006)
ZnO (nanobelts)	2.3	882	1,976	Guisbiers and Pereira (2007)

where, like in ceramics-based sensors, the mechanism of sensitivity controlled by inter-nanowire interfaces starts functioning (Ponzoni et al. 2006; Sysoev et al. 2006, 2009).

From the graphs in Fig. 21.1 and data presented in Table 21.1 one can establish that 1D structures in the 1–10 nm size range are also unstable, as well as polycrystalline materials. With the decrease of nanowire diameter, the melting temperature drops sharply, the same as in the case of nanocrystallites. It was established that the significant decrease of melting point takes place as established for nanoparticles, i.e., it is inversely proportional to the diameter of nanowires. Moreover, it was shown that this effect takes place for all types of nanowires, including semiconductor nanowires, and especially nanotubes. As shown in Fig. 21.1 and Table 21.1, a considerable drop of melting temperature of semiconductor nanowires could be observed with noticeably larger diameters of nanowires than established for metal nanoparticles and metal oxide nanoparticles. For example, melting of Ge nanowire with a diameter of 55 nm starts at around 650 °C, while the melting point of bulk Ge is 930 °C (Wu and Yang 2001).

Due to the capability for cutting, linking, and welding nanowires at relatively low temperature, we need to recognize that thermal stability of the devices designed on the base of 1D nanowires may be limited, and this limitation should be taken into account while analyzing the prospect of nanoscale electronics evolution. For example, Guisbiers and Pereira (2007), on the basis of theoretical simulations, predicted that the use of ZnO nanotubes in electronics is restricted to sizes above 10 nm, since for smaller dimensions the energy dissipation of the circuit can raise the operating temperature close to T_m. For devices intended for operation at higher temperatures, typical for conductometric metal oxide sensors, such behavior of working material is unacceptable.

It should be noted that the reduction of thermal stability of material with a small grain size is a well-known fact for metal nanoparticles (Jiang and Shi 1998; Shim et al. 2002; Jiang et al. 2004; Xie et al. 2006; Li et al. 2008b). When the size of nanoparticles is less than 10–20 nm, melting temperature drops sharply in comparison with conventional bulk materials. This effect is a consequence of the large fraction of atoms with low coordination numbers present in solids with high surface-to-volume ratios. It is known that liquid/vapor interface energy is generally lower than the average solid/vapor interface energy (Alivisatos 1997). Thus, when the particle size decreases, its surface-to-volume ratio increases, while the melting temperature decreases as a consequence of the improved free energy at the particle surface. According to simple phenomenological modeling, the decreasing in melting temperature of a nanoparticle is inversely proportional to the particle diameter (Zhao et al. 2001; Nanda et al. 2002; Jiang et al. 2004). Usually such dependence is described by the following equation:

$$T_{\mathrm{m}}(r)/T_{\mathrm{m}}(\infty) \sim 1 - 1/r \tag{21.1}$$

One can expect that for nanowires, in comparison with nanocrystallites, the influence of effects such as surface diffusion of electrode material along nanowires and migration of a surface clusters, used for surface functionalizing, could become stronger as well (Kolmakov 2008). This diffusion can modify surface properties of metal oxides and make a contribution to the temporal instability of sensor parameters. In comparison with polycrystalline materials, nanowires do not have any steps and edges, which could restrain surface diffusion. This question has not been studied yet, because the majority of sensors designed on the base of nanowires are being tested at room temperature. However, as far as increasing working temperatures of sensors are concerned, the importance of the processes of surface diffusion indicated above could increase, especially under conditions of temperature gradients, which are constantly present in metal oxide-resistive sensors.

Thus, the above-mentioned discussion testifies that the problems of thermal stability are also peculiar to 1D structure. As is well known, the laws of physics are the same for everyone. We can only hope that these problems will not be as strong as for polycrystalline materials. Basing on the discussions presented, one can conclude that prospects of 1D structures used for elaboration of devices intended for work at higher temperature (such as conductometric gas sensors) should be estimated more realistically. Otherwise, in the future one's disappointment could be very intense, as in the case of low-temperature superconductivity.

21.2 Stability of Carbon-Based Nanotubes and Nanofibers

Experiments and theoretical simulations show that carbon nanotubes (CNTs) have higher thermal stability relative to most other organic carbons. For example, it was found that in conditions of high vacuum, multiwalled carbon nanotubes (MWCNTs) with a 60 nm diameter can withstand high temperatures, up to ~3,400 K (Wei et al. 2011). Theoretical studies predict extreme thermal stability of nanotubes up to 4,000 K (Miyamoto et al. 2002). For comparison, graphite sublimes at temperatures as low as 2,400 K (Churka and Inghram 1953). Metal oxides also usually do not possess so high a thermodynamic stability in vacuum. Of course, the decrease of CNT diameter decreases the stability of CNTs. However, it was established that CNTs even with a diameter of 0.5 nm were mechanically stable in vacuum at temperatures as high as 1,100 °C (Peng et al. 2000). According to first principle calculations, the smallest energetically stable CNT has a diameter of about 0.4 nm (Sawada and Hamada. 1992; Lucas et al. 1993; Cabria et al. 2003). This conclusion is consistent with earlier work based on empirical and semiempirical approaches, as well as with reported experimental results. For example, Peng et al. (2000) reported experimental evidence for the existence of an SWNT with a diameter of 0.33 nm. Regarding the influence of the length of CNTs, we can say that, according to Xu et al. (2010), with increase in length the thermal stabilities of CNTs increase when the diameters are 60–100 nm. However, the thermal stability of CNTs decreases appreciably when the diameters are below 60 nm. Thus, the results presented above suggest that CNT-based devices can operate at high temperatures in vacuum and that CNTs can be used as nanoscale heaters to obtain extremely high temperatures.

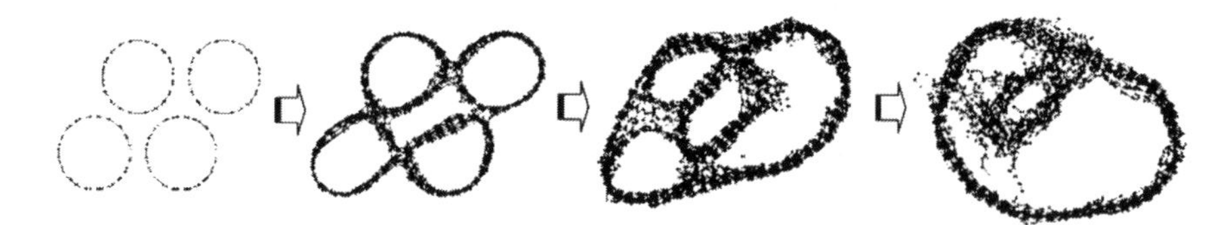

Fig. 21.2 Snapshots (top views) taken from the simulations showing the transformation of a bundle of four CNT tubes into an MWCNT consisting of two nested tubes (Reprinted with permission from Lopez et al. (2004). Copyright 2004 IEEE)

It should be noted, however, that the extremely high thermal stability indicated above was observed for individual CNTs. For a CNT array (bundle) the situation is different. It was established that, similar to metal and metal oxide nanoparticles, CNT nanotubes can coalescence and change configuration considerably. Coalescence of two SWCNTs into a larger-diameter SWCNT has been observed upon the annealing of SWCNTs at high temperatures either in the presence of H_2 (Nikolaev et al. 1997) or under electron irradiation (Terrones et al. 2000). The transformation of SWCNTs into MWCNTs under thermal treatment was also observed by Lopez et al. (2002). It was concluded that the presence of defects in the tubes, in particular vacancies, may act as a driving force for coalescence (Terrones et al. 2000). Vacancies can be produced by chemical treatment (Nikolaev et al. 1997), electron irradiation (Terrones et al. 2000), etc. Certainly, the thermal treatment of the tubes at relatively high temperatures (above 1,600 °C) will produce the required vacancies in the tubes. Due to the high temperatures, the vacancies have a high mobility and diffuse throughout the tube until they get locked in the intertube region by saturation of the associated dangling bonds, with the dangling bonds left free by a nearby vacancy in the adjacent tube. Saturation of dangling bonds between neighboring tubes leads to intertube polymerization which is the initial stage for the coalescence of tubes. According to López et al. (2004), at the experimental temperatures the creation of vacancies in the tubes requires heat treatments of the order of minutes. However, once a sufficient number of intertube links are formed, coalescence proceeds within a time scale of a few hundred picoseconds. One of the examples of possible transformations in CNTs during thermal treatment is shown in Fig. 21.2. Of course the temperature of the indicated transformation is high enough; however, the possibility of this process should be taken into account.

However, it should be noted that gas sensors do not operate in vacuum, and therefore high stability in vacuum is not a special advantage of materials used for gas sensor design. Gas sensors usually operate in an oxygen-containing atmosphere, and therefore sensor material should have high stability in such conditions. Unfortunately, in oxygen-containing atmosphere the stability of CNTs is considerably worse. It was established that the oxidation temperature of CNTs varies over the range 325–550 °C, apparently due to variations of diameter and associated metals. Oxidation for CNTs means that CNTs start to burn, i.e., the oxidation of CNTs in air results in etching away of tube caps and the thinning of tubes through layer-by-layer peeling of the outer layers, starting from the cap region (Ajayan et al. 1993). One can find in numerous papers results of research related to thermal oxidation of CNTs. They indicate that, under oxidative conditions, tubular structures are substantially less stable than graphite and most disordered carbon (Boccaleri et al. 2006). The same results were obtained by Bom et al. (2002) (see Fig. 21.3).

It was also established that smaller diameter nanotubes oxidize at lower temperatures (Joshi et al. 1990; Yao et al. 1998; Singh et al. 2010). Simulations carried out by Fathi and Forouzandeh (2009) confirmed this conclusion. The results obtained show that with increasing length of the CNT bundle and diameter of each individual CNT, relative stability increases, and the system becomes more stable. Examples of diameter influence on the oxidation temperature of MWNTs are presented in Fig. 21.4. This figure shows that T_{ox} saturates at 663 °C for MWNTs having a diameter of 30 nm and above. Different troughs in the DTA curve indicate three different rates of weight loss corresponding to three different oxidation reactions. Singh et al. (2010) believe that DTA profiles of MWNTs show two peaks at ~420–470 °C and at ~630 °C due to difference in lattice strain originating from different curvatures of inner and outer walls.

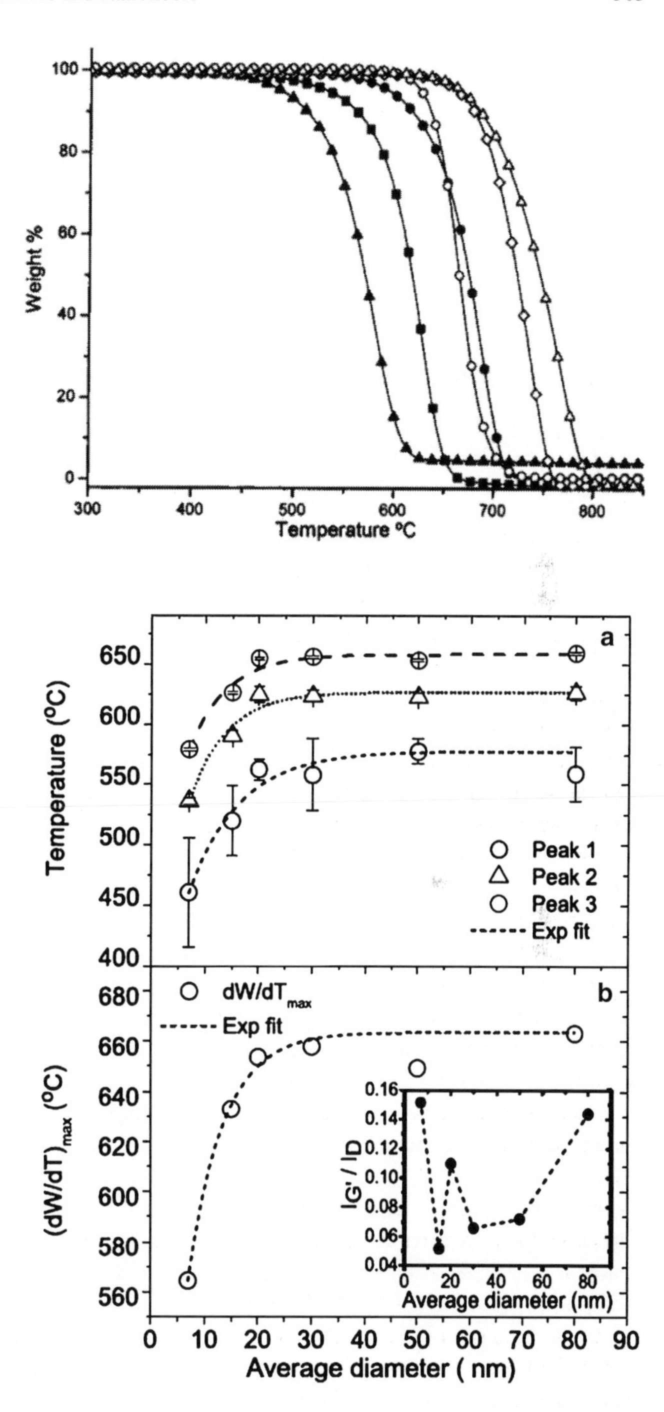

Fig. 21.3 Weight loss curves for raw MWNTs (*filled triangles*), diamond (*filled squares*), annealed diamond (*open circles*), graphite (*filled circles*), annealed MWNTs (*open diamonds*), and annealed graphite (*open triangles*); thermal annealing at 2,800 °C in vacuum (Reprinted with permission from Bom et al. 2002, Copyright 2002 American Chemical Society)

Fig. 21.4 (**a**) Variation in T_{ox} as a function of diameter for three different components in MWNTs, as obtained from experimental TGA data; (**b**) maxima of weight loss with temperature obtained from differential thermal analysis (DTA) curve for various diameters (*closed circles*) and exponential fit (*dashed line*). The *inset* shows the diameter dependence of I_G/I_D ratio, which is inversely proportional to the defect density, in MWNTs as measured from Raman studies (Reprinted with permission from Singh et al. 2010, Copyright 2010 American Institute of Physics)

On the basis of research carried out it was concluded that oxidation at lower temperature takes place due to higher curvature and associated lattice strain (Joshi et al. 1990; Yao et al. 1998; Singh et al. 2010). Bond curvature has been found to affect the oxidation of single-walled CNTs (Miyata et al. 2007). Therefore, nanotubes with smaller diameters, due to higher curvature strain, are oxidized at lower temperature. Defects and derivatization moiety in nanotube walls can also lower the thermal stability (Arepalli et al. 2004). Therefore, a higher oxidation temperature is always associated with purer, less defective samples.

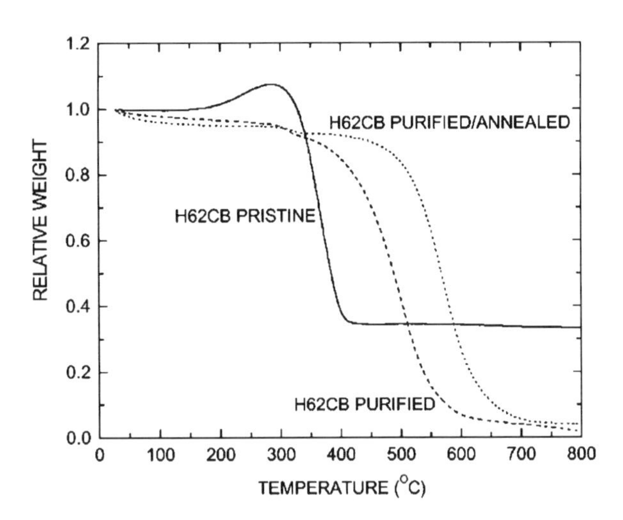

Fig. 21.5 TGA of SWNTs in flowing air. *Solid curve* is the pristine material; *dashed curve*, obtained after purification; and *dotted curve*, after a subsequent vacuum annealing at 1,150 °C (Reprinted with permission from Zhou et al. 2001, Copyright 2001 Elsevier)

Thermogravimetric analysis (TGA) studied by Bom et al. (2002) have provided evidence for defects playing a significant role in the thermal stability of MWNTs. It was established that defects such as edges, dangling bonds, vacancies, dislocations, steps, 5,7-membered ring defects, Y-junctions, and kinks are inherently present in as-grown and acid-purified CNTs and contribute to a decrease in the oxidative stability of CNTs. Most of the defect sites are particularly reactive to oxygen at elevated temperatures. In particular, it has been demonstrated that high-temperature air oxidation occurs preferably at kink sites (Lu et al. 1999). Studies carried out by Singh et al. (2010) on vacuum-annealed samples have shown that structural defects and strain influence the oxidative stability of SWNTs more strongly than that of MWNTs. Vacuum annealing studies reveal that optimum annealing temperature for strain relaxation and structural defect removal is 700 °C for SWNTs and 500 °C for MWNTs. The difference in the nature and concentration of structural defects between SWNTs and MWNTs may lead to different annealing temperatures required for defect annihilation. The same situation takes place in graphite and diamond (Bom et al. 2002). Defect sites in graphite and diamond also contribute to a decrease in the oxidative stability of these materials. Defects in graphite and diamond include edges, dangling bonds, vacancies, dislocations, and steps. These defect sites are particularly reactive to oxygen at elevated temperatures.

It was established that catalysts present in CNTs also strongly affect thermal stability of CNTs in air. Active metal particles present in the nanotube samples catalyze carbon oxidation, so the amount of metal impurity in the sample can have a considerable influence on the thermal stability. For example, Zhou et al. (2001) found that if the oxidation of as-synthesized CNTs, which contained traces of catalyst (Fe), was quite rapid and homogeneous at 350 °C due to the catalytic effect, the purified CNTs had negligible weight loss, even after annealing at 460 °C. Furthermore, the presence of Fe obscured the dependence of oxidative stability on tube diameter as discussed earlier. After removing the Fe, all tubes were more appropriate for observing diameter-dependent oxidative stability. Li et al. (2011) have found that the presence of cobalt catalysts dramatically decreases the thermal stability of CNT/peroxide-curable methyl phenyl silicone gum composites as well. This means that the presence of uncontrolled impurities in CNTs can be one of the reasons for reduced reproducibility of sensor parameters. This conclusion is confirmed by results obtained by Boccaleri et al. (2006) and Zhou et al. (2001) (see Fig. 21.5).

Research has also shown that gas-phase oxidation of CNTs, via high-temperature furnace-assisted thermal annealing in air, efficiently reduces the overall length distribution of CNTs. Marsh et al. (2007) found that heating at 500 °C for 20 or 40 min reduced the weight (length) of the MWNTs sample by 30 % or 60 %, respectively. It is reasonable to suggest that longer air annealing times result in shorter nanotubes. Moreover, TEM images demonstrated that, after heating in air at 500 °C, most nanotubes have submicron lengths as compared to tens of microns for untreated. Marsh et al. (2007) assumed that,

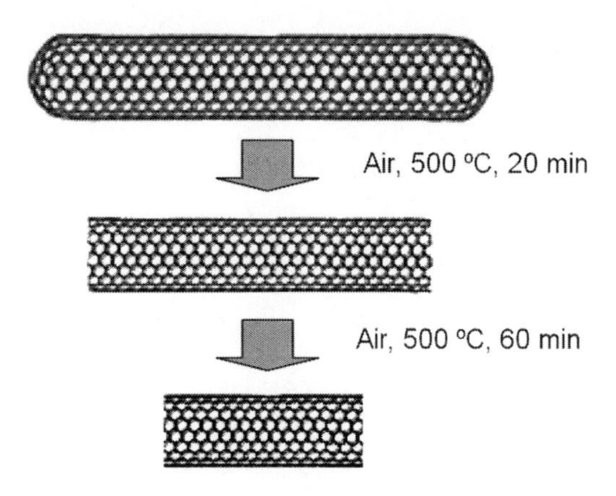

Fig. 21.6 Synthetic scheme illustrating the influence of annealing in air on the length of MWNTs (Adapted with permission from Marsh et al. 2007, Copyright 2007 Royal Society of Chemistry)

during this process, the removal of end cap on initial stage and subsequent "burning" from the ends towards the nanotube center without introducing new defects take place (see Fig. 21.6). Tsang et al. (1993) and Ajayan et al. (1993) also demonstrated that thinning and opening of arc-grown MWNTs are limited to the ends of the nanotubes when they are treated with CO_2 at 850 °C or air at 800 °C.

References

Ajayan PM, Ebbesen TW, Ichihashi T, Iijima S, Tanigaki K, Hiura H (1993) Opening carbon nanotubes with oxygen and implications for filling. Nature 362:522–525

Alivisatos AP (1997) Nanocrystals: building blocks for modern materials design. Endeavour 21(2):56–60

Arepalli S, Nikolaev P, Gorelik O, Hadjiev VG, Holmes W, Files B, Yowell L (2004) Protocol for the characterization of single-wall carbon nanotube material quality. Carbon 42:1783–1791

Boccaleri E, Arrais A, Frache A, Gianelli W, Fino P, Camino G (2006) Comprehensive spectral and instrumental approaches for the easy monitoring of features and purity of different carbon nanostructures for nanocomposite applications. Mater Sci Eng B 131:72–82

Bom D, Andrews R, Jacques D, Anthony J, Chen B, Meier MS, Selegue JP (2002) Thermogravimetric analysis of the oxidation of multiwalled carbon nanotubes: evidence for the role of defect sites in carbon nanotube chemistry. Nano Lett 2:615–619

Cabria I, Mintmire JW, White CT (2003) Stability of narrow zigzag carbon nanotubes. Int J Quantum Chem 91:51–56

Churka WA, Inghram MG (1953) Molecular species evaporating from a carbon surface. J Chem Phys 21:1313–1319

Comini E (2006) Metal oxide nano-crystals for gas sensing. Anal Chim Acta 568(1–2):28–40

Comini E, Faglia G, Sberveglieri G, Pan Z, Wang ZL (2002) Stable and highly sensitive gas sensors based on semiconducting oxide nanobelts. Appl Phys Lett 81(10):1869–1871

Fathi D, Forouzandeh B (2009) A novel approach for stability analysis in carbon nanotube interconnects. IEEE Electron Device Lett 30(5):475–477

Guisbiers G, Pereira S (2007) Theoretical investigation of size and shape effects on the melting temperature of ZnO nanostructures. Nanotechnology 18:435710

Hernandez-Ramirez F, Prades JD, Tarancon A, Barth S, Casals O, Jimenez-Diaz R, Pellicer E, Rodriguez J, Juli MA, Romano-Rodriguez A, Morante JR, Mathur S, Helwig A, Spannhake J, Mueller G (2007a) Portable microsensors based on individual SnO_2 nanowires. Nanotechnology 18:495501

Hernandez-Ramırez F, Tarancon A, Casals O, Arbiol J, Romano-Rodrıguez A, Morante JR (2007b) High response and stability in CO and humidity measures using a single SnO_2 nanowire. Sens Actuators B Chem 121:3–17

Jiang Q, Shi FG (1998) Entropy for solid–liquid transition in nanocrystals. Mater Lett 37:79–82

Jiang Q, Liang LH, Li JC (2004) Thermodynamic superheating of low-dimensional metals embedded in matrix. Vacuum 72:249–255

Joshi A, Nimmagadda R, Herrington J (1990) Oxidation kinetics of diamond, graphite, and chemical vapor deposited diamond films by thermal gravimetry. J Vac Sci Technol A 8:2137

Köck A, Tischner A, Maier T, Kast M, Edtmaier C, Gspan C, Kothleitner G (2009) Atmospheric pressure fabrication of SnO_2-nanowires for highly sensitive CO and CH_4 detection. Sens Actuators B Chem 138:160–167
Kolmakov A (2008) Some recent trends in fabrication, functionalisation and characterization of metal oxide nanowire gas sensors. Int J Nanotechnol 5:450–474
Li S, Lian JS, Jiang Q (2008a) Modeling size and surface effects on ZnS phase selection. Chem Phys Lett 455:202–206
Li S, Lian JS, Jiang Q (2008b) Thermodynamic phase stability of three nano-oxides. Mater Res Bull 43:3149–3154
Li Z, Lin W, Moon K-S, Wilkins SJ, Yao Y, Watkins K, Morato L, Wong C (2011) Metal catalyst residues in carbon nanotubes decrease the thermal stability of carbon nanotube/silicone composites. Carbon 49:4138–4148
Lopez MJ, Rubio A, Alonso JA, Lefrant S, Metenier K, Bonnamy S (2002) Patching and tearing single wall carbon-nanotube ropes into multiwall carbon nanotubes. Phys Rev Lett 89:255501-1–255501-4
López MJ, Rubio A, Alonso JA (2004) Deformations and thermal stability of carbon nanotube ropes. IEEE Trans Nanotechnol 3(2):230–236
Lu X, Ausman KD, Piner RD, Ruoff RS (1999) Scanning electron microscopy study of carbon nanotubes heated at high temperatures in air. J Appl Phys 86:186–189
Lucas AA, Lambin PH, Smalley RE (1993) On the energetic of tubular fullerenes. J Phys Chem Solids 54:587–593
Marsh DH, Rance GA, Zaka MH, Whitby RJ, Khlobystov AN (2007) Comparison of the stability of multiwalled carbon nanotube dispersions in water. Phys Chem Chem Phys 9:5490–5496
Miyamoto Y, Berber S, Yoon M, Rubio A, Tomanek D (2002) Onset of nanotube decay under extreme thermal and electronic excitations. Physica B 323:78–85
Miyata Y, Kawai T, Miyamoto Y, Yanagi K, Maniwa Y, Kataura H (2007) Bond-curvature effect on burning of single-wall carbon nanotubes. Physica Status Solidi B 244:4035–4039
Nanda KK, Sahu SN, Behera SN (2002) Liquid-drop model for the size-dependent melting of low-dimensional systems. Phys Rev A 66:013208
Nikolaev P, Thess A, Rinzler AG, Colbert DT, Smalley RE (1997) Diameter doubling of single-wall nanotubes. Chem Phys Lett 266:422–426
Peng L-M, Zhang ZL, Xue ZQ, Wu QD, Gu ZN, Pettifor DG (2000) Stability of carbon nanotubes: how small can they be? Phys Rev Lett 85(15):3249–3252
Ponzoni A, Comini E, Sberveglieri G, Zhou J, Deng S, Xu N, Ding Y, Wang Z (2006) Ultrasensitive and highly selective gas sensors using three-dimensional tungsten oxide nanowire networks. Appl Phys Lett 88:203101
Sawada S, Hamada N (1992) Energetics of carbon nanotubes. Solid State Commun 83:917–919
Shim J-H, Lee B-J, Cho YW (2002) Thermal stability of unsupported gold nanoparticle: a molecular dynamics study. Surf Sci 512:262–268
Singh DK, Iyer PK, Giri PK (2010) Diameter dependence of oxidative stability in multiwalled carbon nanotubes: role of defects and effect of vacuum annealing. J Appl Phys 108:084313-10
Sysoev V, Button B, Wepsiec K, Dmitriev S, Kolmakov A (2006) Toward the nanoscopic "electronic nose": hydrogen vs. carbon monoxide discrimination with an array of individual metal oxide nano- and mesowire sensors. Nano Lett 6(8):1584–1588
Sysoev VV, Schneider T, Goschnick J, Kiselev I, Habicht W, Hahn H, Strelcov E, Kolmakov A (2009) Percolating SnO_2 nanowire network as a stable gas sensor: direct comparison of long-term performance versus SnO_2 nanoparticle films. Sens Actuators B Chem 139:699–703
Terrones M, Terrones H, Banhart F, Charlier J-C, Ajayan PM (2000) Coalescence of single-walled carbon nanotubes. Science 288:1226–1229
Toimil Molares ME, Balogh AG, Cornelius TW, Neumann R, Trautmann C (2004) Fragmentation of nanowires driven by Rayleigh instability. Appl Phys Lett 85(22):5337–5339
Tsang SC, Harris PJF, Green MLH (1993) Thinning and opening of carbon nanotubes by oxidation using carbon dioxide. Nature 362:520–522
Wang XW, Fei GT, Wu B, Chen L, Chu ZQ (2006) Structural stability of Co nanowire arrays embedded in the PAAM. Phys Lett A 359:220–222
Wei X, Wang M-S, Bando Y, Golberg D (2011) Thermal stability of carbon nanotubes probed by anchored tungsten nanoparticles. Sci Technol Adv Mater 12:044605
Wu Y, Yang P (2001) Melting and welding semiconductor nanowires in nanotubes. Adv Mater 13(7):520–523
Xie D, Wang MP, Qi WH, Cao LF (2006) Thermal stability of indium nanocrystals: a theoretical study. Mater Chem Phys 96:418–421
Xu F, Sun LX, Zhang J, Qi YN, Yang LN, Ru HY, Wang CY, Meng X, Lan XF, Jiao QZ, Huang FL (2010) Thermal stability of carbon nanotubes. J Therm Anal Calorim 102:785–791
Yao N, Lordi V, Ma SXC, Dujardin E, Krishnan A, Treacy MMJ, Ebbesen TW (1998) Structure and oxidation patterns of carbon nanotubes. J Mater Res 13:2432–2437

Zhang F, Barrowcliff R, Hsu ST (2005) Thermal stability of IrO_2 nanowires. In: Proceedings of the international conference on MEMS, NANO and smart systems (ICMENS '05), IEEE Computer Society, Banff, Alberta, 24–27 July, pp 418–420

Zhao M, Zhou XH, Jiang Q (2001) Comparison of different models for melting point change of metallic nanocrystals. J Mater Res 16(11):3304–3308

Zhou W, Ooi YH, Russo R, Papanek P, Luzzi DE, Fischer JE, Bronikowski MJ, Willis PA, Smalley RE (2001) Structural characterization and diameter-dependent oxidative stability of single wall carbon nanotubes synthesized by the catalytic decomposition of CO. Chem Phys Lett 350:6–14

Chapter 22
Temporal Stability of Porous Silicon

22.1 Porous Silicon Aging

Research has shown that gas sensors based on porous semiconductors (PS) have extremely high sensitivity, allowing, for example, the detection of various gases at the ppm and even ppb levels in a normal atmosphere (Korotcenkov and Cho 2010a). However, the same research (Canhman 1997; Torchinskaya et al. 1997; Parkhutik 1999; Holec et al. 2002; Pancheri et al. 2004; Irajizad et al. 2004; Di Francia et al. 2005; Islam and Saha 2007) has shown that gas sensors based on porous silicon have one important disadvantage. As a chemically active material, PSi (porous silicon) can be easily oxidized. For example, it is well known that at room temperature and ambient environment about 7 nm of Si native oxide can be formed with time. Oxidation, as reported by Bjorkqvist et al. (2003), is a very slow process at room temperature and thus leads to continuous changes in the structure of PSi and its physical parameters. Therefore, the properties of highly porous silicon and PSi-based devices are not stable over time, especially at the initial stages of use after manufacturing (see Fig. 22.1), and can be significantly influenced not only by fabrication and drying conditions but also by the manner in which it is stored prior to examination or use. As stated previously, PSi slowly reacts with the ambient air, and consequently its chemical composition and its properties evolve continuously with storage time. The effects of "aging" on both the composition and structure of porous Si are now well documented (Zhu et al. 1992; Canham 1997; Lee and Tu 2007). Changes in material properties, such as electrical resistivity and strain, and optical properties, such as refractive index and photoluminescence, accompany the aging process (see Table 22.1).

As indicated before, aging depends heavily on the storage environment, operating conditions, and on how well the sensor materials are isolated from the environment (Dittrich et al. 1995; Kim et al. 2001; Islam and Saha 2007). Moreover, egging effects take place not only in air but also during storage in liquid media such as alcohols and HF. For example, Kim et al. (2001) found that the exposure of freshly formed PSi films to humid atmosphere (30 Torr) at room temperature results in a gradual increase of the PL yield during the first 2 days of storage. Simultaneously, the maximum spectral yield shifts toward red. The initial luminescence intensity may be restored by evacuating the PSi sample at a temperature of 150 °C. Treatment of PSi in a humid atmosphere for more than 2 days results in an irreversible decrease of PL intensity. It was found that the aging time may require more than 5–10 days (Setzu et al. 1998). For example, in the research of Holec et al. (2002), because of temporal instability, PSi samples were stored for about 1 month before sensing characterization. It was established that when a PSi is stored in oxidizing environments like water and H_2O_2, the formation of Si–O–Si bonds replacing Si–H_x bonds takes place (Hossain et al. 2000). The same effect was observed during PSi surface carbonization as well (Mahmoudi et al. 2007). It was observed that the rate of oxidation of the Psi layer depends on the concentration of OH– and holes in the valence band of the PSi layer.

G. Korotcenkov, *Handbook of Gas Sensor Materials*, Integrated Analytical Systems,
DOI 10.1007/978-1-4614-7388-6_22, © Springer Science+Business Media New York 2014

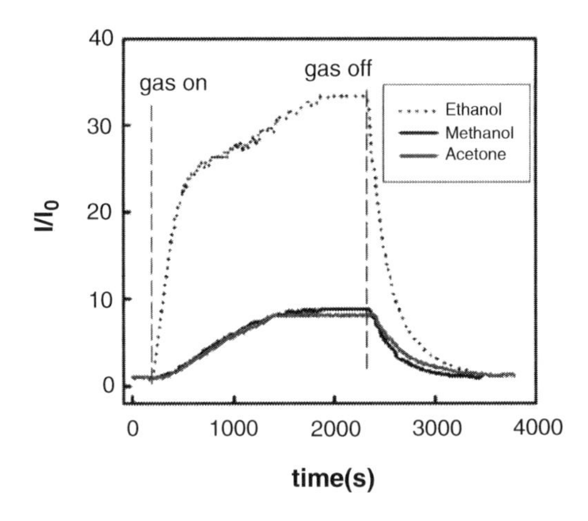

Fig. 22.1 Response of fresh, oxidized, and unpassive (boiling in CCl_4) PSi samples in the presence of 4 % ethanol (Reprinted with permission from Irajizad et al. 2004, Copyright 2004 Elsevier)

Table 22.1 Effects of varying storage conditions on properties of PSi

Storage conditions	Major effect
Ambient air (15 min–15 months)	Contaminated native oxide growth
Air, vacuum	Changes in layer strain
HF, ethanol, Freon, ether	Lowest carbon levels for HF storage
N_2, H_2, forming gas, O_2 (min–h)	Widely varying PL stability
Dry N_2, then UHV	Avoids photostimulated oxidation
Vacuum (10^{-6} Torr)	Carbon and oxygen pickup
Vacuum (10^{-3} Torr)	Heavy hydrocarbon contamination
Transport under propanol (<1 day)	Minimize oxidation by reducing air exposure
Ethanol storage and removal in UHV ($<2\times10^{-9}$ Torr)	Minimize oxidation by completely avoiding air exposure
Cooled ethylene glycol	Green PL retained
Plastic and glass containment vessels	Blue PL due to plastic box outgassing

Source: Data extracted from Canham (1997)

It was also established that the degradation of porous silicon luminescence in an oxygen atmosphere is strongly enhanced by illumination (Tischler et al. 1992) and annealing (Torchinskaya et al. 2001). Other ambient gases such as N_2, Ar_2, and H_2 did not produce an essential change of PSi properties (Xu et al. 1992). It is also interesting to note that an as-anodized PSi sample is hydrophobic in nature but the oxidized PSi (OPSi) sample is hydrophilic in nature, thereby leading to significant drift in sensor output with time (Dittrich et al. 1995; Bjorkqvist et al. 2004; Archer et al. 2005). For example, Irajizad et al. (2004) showed that annealing in air or boiling in CCl_4 strongly decreases the sensing properties of PSi-based conductometric gas sensors. XPS measurements have shown that the surface of PSi is partially oxidized after the indicated treatments. The same effect was observed by Pancheri et al. (2004) for PSi-based NO_2 sensors. High sensitivity to NO_2 was achieved only for fresh samples. Extremely poor stability was observed for PSi-based ethanol sensors as well (Han et al. 2001).

The PL parameters of porous layers turned out to be sensitive to thermo-vacuum treatments as well (Lisachenko and Aprelev 2001). It was shown that vacuum annealing at temperatures below 600 °C led to a decrease of the photoluminescence intensity (Balagurov et al. 1996) in the region of 720–880 nm. For example, Shin et al. (2003) observed that after PSi annealing in vacuum at about 550 °C, the PL centered at about 720 nm was completely quenched. The observed decrease in the photoluminescence intensity has been explained by the hydrogen desorption from monohydride and dihydride species, a decrease of the PSi band gap, and an increase of the silicon dangling bonds (Robinson et al. 1992; Kovalev et al. 1994; Ludwig 1996). IR spectroscopy data also testify to the exit of hydrogen atoms from the porous silicon surface. Dimitrov et al. (1997) established that the hydrogen desorption follows the second-order chemical kinetics.

It was also observed that the creation of nonradiative dangling bonds increases the cross-conductivity of the PSi–Si structures. The appearance in porous silicon of a large number of defects, which are the consequence of broken silicon bonds on the surface of nanostructures, is registered by the method of electron paramagnetic resonance (Meyer et al. 1993).

The slow formation of native oxide on all the pore walls stimulates strong optical effects as well (Kochergin and Foell 2006). The loss of transparency of the mesoporous silicon in the mid- and far-IR range was observed during the aging of freshly prepared PSi. The above-mentioned properties indicate that PSi-based sensors have great limitations for application in real devices. For practical applications, the decrease of the long-term drift of sensor parameters is often more important task than the increase of sensor signal and requires more attention of designers (Bjorkqvist et al. 2003). Nevertheless, the cheap mass production and the possibility of integration of the sensor element with driving and read-out electronics on the same chip push the development toward silicon-based processes.

Such behavior of PSi-based sensors requires the elaboration of special methods of stabilizing the parameters of gas sensors based on porous materials, which considerably narrows the area of their application. To minimize the variability and extent of such storage effects, there are a number of options. According to Canham (1997), to resolve the above-mentioned problem one could intentionally oxidize the material in controlled conditions (Roussel et al. 1999; Mattei et al. 2000), isolate its internal surface by capping (Loni 1997), modify its surface (Chazalviel and Ozanam 1997), or impregnate the pores (Herino 1997). Alternatively, one could simply try to optimize storage time and conditions for the given application requirement. For example, the aging effect could be avoided by storing the porous silicon under ultrahigh vacuum or in ultra-dry inert gas. However, it is clear that the conditions indicated above are not acceptable for gas sensors. As a result of such elaborations for aging effect prevention, the use of pre-aged treatment by oxidation was suggested.

22.2 Temporal Stabilization of Porous Silicon Through Oxidation

In the previous section it was concluded that, due to strong temporal instability of PSi properties, the design of the methods acceptable for their stabilization is a task of great importance for development of all types of gas sensors (Korotcenkov and Cho 2010a). After much research it was established that the oxidation is one of the most effective methods for stabilization of the properties of PSi, including optical, photoluminescence and sensing characteristics (Canhman 1997; Gelloz et al. 2005; Pirasteh et al. 2006).

The following oxidation methods have been tested and have shown good results: aging in ambient; exposure to water vapor; photochemical oxidation; anodic oxidation in a non-fluoride electrolyte; chemical oxidation using HNO_3, O_3, H_2O_2, and vapor of peridine; water solution of KNO_3 and alcohol; and thermal oxidation in wet and dry O_2 fluxes, including rapid thermal annealing were tested and shown good results (Amato et al. 1997; Canhman 1997; Maccagnani et al. 1998; Korotcenkov and Cho 2010b). Among the indicated methods, thermal oxidation is the most used. Since porous Si has a very large surface-to-volume ratio, it is oxidized at 750–800 °C (see Fig. 22.2). For example, Pap et al. (2004) have shown that after 10 h oxidation at $T=800$ °C the whole PSi structure with wall thickness ~30 nm was oxidized. At $T=950$ °C under a water vapor atmosphere, porous Si could be converted to porous SiO_2 within 60 min. However, the additional final annealing at $T=1{,}000$–$1{,}100$ °C for oxide densification is often used (Maccagnani et al. 1998). It was found that after a short-term annealing at a high oxidation temperature above 1,000 °C in wet O_2, thick OPSi films had the same properties as thermal SiO_2 of bulk silicon (Unagami 1980). Commonly used modes of thermal oxidation of porous Si are presented in Table 22.2.

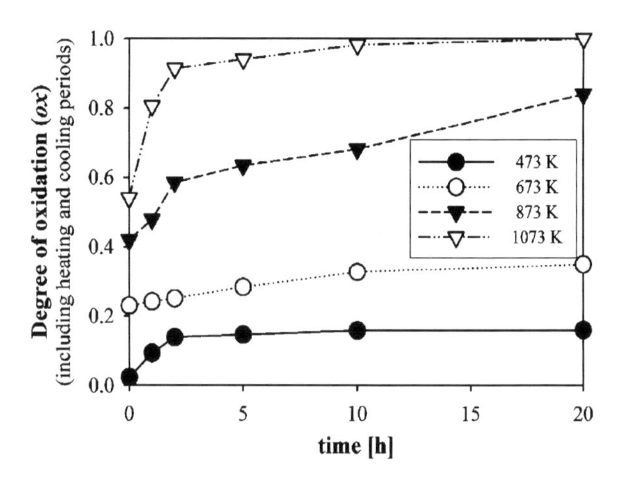

Fig. 22.2 Oxidation extent of 70 % porosity and 30 μm thick porous membranes vs. time and temperature as measured for the amount of oxide formed during the heating and cooling periods (Reprinted with permission from Pap et al. 2004, Copyright 2004 American Chemical Society)

Table 22.2 Technological parameters, usually used for thermal oxidation of porous Si

1 step	2 step	3 step
Dry air; 800 °C; 20 h		
Wet O_2; 1,050 °C; 3 h		
Dry O_2, 1,000 °C, 60 s		
Wet O_2; 300 °C; 1 h	Wet O_2; 900 °C; 1 h	
Dry O_2; 400 °C; 30 min	Dry O_2; 1,000 °C; 1 h	Wet O_2; 1,000 °C; 30 min
Dry O_2; 300 °C	Wet O_2; 850 °C	Wet O_2; 1,100 °C

Source: Reprinted with permission from Korotcenkov and Cho (2010b). Copyright 2010 Taylor and Frances

It should be noted that the use of a preliminary annealing at low temperatures sufficiently improves the parameters of OPSi. It was established that it is not desirable to anneal as-prepared porous Si at temperatures higher than 400 °C, since it could lead to a drastic restructuring of the porous layer (Yon et al. 1987). This is a consequence of the large surface area of the material: even at low temperature, surface migration of silicon atoms occurs and provokes pore coalescence. According to Yon et al. (1987), a preoxidation of porous silicon layers at low temperatures in the region of 300 °C is sufficient for creation of an oxide layer all along the pore walls, which prevents restructuring of the sample during further heating at higher temperatures up to 800–900 °C. Therefore, it was established that the procedure, preoxidation at 300 °C for structure stabilization, oxidation at 750–800 °C for complete porous silicon transformation into silicon dioxide, and final densification at 1,050–1,090 °C for obtaining a compact oxide equivalent to standard thermal silicon dioxide, is the most optimal for reproducible forming of oxide layers from porous silicon (Yon et al. 1987; Maccagnani et al. 1998).

Research has shown that for obtaining dense SiO_2 from a PSi layer, the dimensions of the pores should balance the volume expansion during the thermal oxidation by a factor of 2.27 and by a pore wall thin enough to be fully oxidized. It was found that this condition can be achieved when the PSi porosity is in the range 55–90 % (Maccagnani et al. 1998). A layer of lower porosity will never be fully oxidized, while an increased porosity would give a very fragile macro-PSi structure and consequently a very weak porous SiO_2.

Riikonen et al. (2012) also studied different approaches to PSi oxidation. They found that none of the liquid-phase oxidation methods utilized in the study were able to oxidize the surface hydrides on PSi completely. Heat treatment of PSi at a high temperature causing a rearrangement of surface atoms and a reduction in the hydride concentration was necessary for complete oxidation of the surface hydrides. In addition, it was established that the plain thermal oxidation at 700 °C (7.5 min) was easy to perform and produced a stable surface that was not observed to react chemically with the model drug cinnarizine in isothermal titration microcalorimetry (ITMC) experiments. However, it caused a relatively high decrease in the pore volume as well as in the pore diameter. The combination of thermal oxidation and annealing (thermal oxidation, air, 300 °C, 4 min; annealing, N_2, 1,000 °C, 20 min) had the smallest effect on the pore volume, although the pore size significantly increased. However, the surface was not inert and was found to react with cinnarizine and to be unstable under ambient conditions. The novel combination of thermal oxidation, annealing, and chemical oxidation (thermal oxidation in air, 300 °C, 4 min; annealing in N_2, 750 °C, 30 min; chemical oxidation in solution 1:1:6 H_2O_2 (30 %)/HCl (37 %)/H_2O, 80 °C, 15 min; rinsing in EtOH) had a very small effect on the pore size and caused a moderate decrease in the pore volume. The surface did not show chemical reactions with cinnarizine in ITMC experiments and had a large number of –OH groups on the surface available for further modification. However, this oxidation method was the most time-consuming.

For anodic oxidation of PSi, two types of electrolytes, usually based on 1 M KNO_3 and 0.1 M H_2SO_4 aqueous solutions, were used (Bsiesy et al. 1991; Cantin et al. 1996). The monitoring of the potential during the oxidation helps to assess the time at which the sample is fully oxidized by looking at the step increase in the potential. For further improvement of oxidation homogeneity, light assistance is required because anodic oxidation of silicon requires holes. PSi is depleted in holes and without illumination only the bottom layers and the crystalline silicon/PSi interface are oxidized. In this case, the oxidation stops when an insulating layer is produced at the interface, leaving most of the PSi unaffected (Bsiesy et al. 1991). With light illumination, holes are generated in PSi, and oxidation can take place in a more homogeneous way. The final result is an oxide coating of a monolayer, which covers the PSi skeleton. Larger currents yield better oxidation homogeneity as well.

Bsiesy et al. (1991) believe that electrochemical oxidation of PSi has the following advantages: (1) electrochemical oxidation of porous silicon can be achieved easily and (2) it is possible to oxidize either the lower part of the porous layer, or the whole depth, at a level which depends on the exchanged charge. This method therefore appears to be more attractive than thermal oxidation when incomplete oxidation is required. In particular, such a requirement appears during silicon (or other material) epitaxy on porous silicon. These processes generally involve temperatures above 400 °C and porous silicon must be stabilized by a preoxidation step in order to conserve its very thin microstructure. If this preoxidation is achieved by thermal oxidation, there is also oxide growth on top of the sample, which must be eliminated before subsequent epitaxy. Electrochemical oxidation, with an appropriate choice of experimental conditions, can lead to oxidation limited to the inner part of the porous layer.

The structural comparison of oxidized and initial PSi has shown that the oxidation did not modify the morphology of the porous layer. The pore density was the same and the pore size decreased in the OPSi (see Fig. 22.3). So the porosity after oxidation, of course, was lower than the porosity before oxidation. For example, after oxidation of PSi with porosity ~60 %, the porosity was estimated to be 33 % (Pirasteh et al. 2006). Furthermore, volume expansion due to oxidation was isotropic and one could observe the increase in the thickness of the layer. For a single layer, when the thickness was equal to 5 μm before oxidation, it was nearly 5.8 μm thick after oxidation (Pirasteh et al. 2006). The refractive index of porous layers decreases after oxidation. Based on the physical characterization of OPSi, it was concluded that the oxidation of PSi layers is a good way to obtain a lower optical loss of PSi waveguides by reducing both volume scattering and absorption in the near-infrared wavelength range (Pirasteh et al. 2006). Besides that, the transformation of porous silicon into OPSi also permits light transmission in the visible wavelength range. As was shown, OPSi could be successfully used instead of PSi for the elaboration of various types of sensors including gas sensors, electrochemical sensors, and biosensors promoting improvement of sensor stability (Maccagnani et al. 1998; Sakly, et al. 2006).

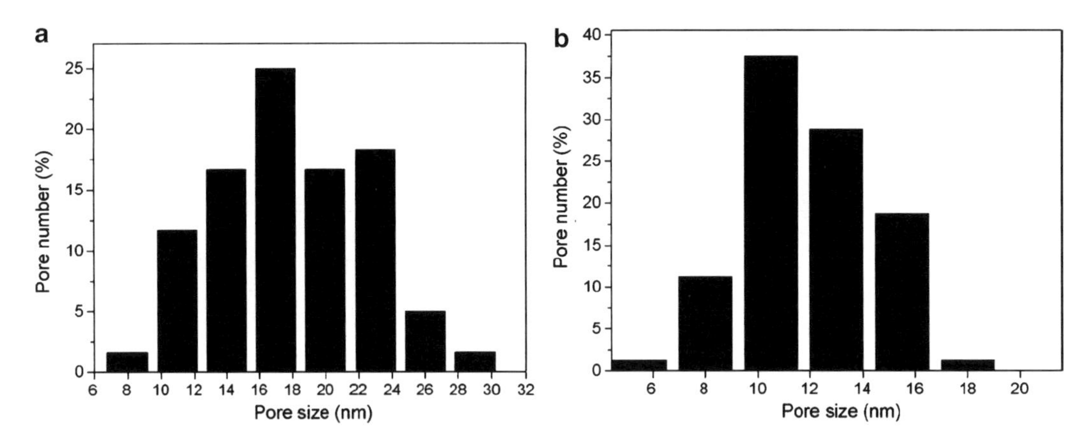

Fig. 22.3 Distribution of pore size of (**a**) porous silicon layer and (**b**) oxidized porous silicon layer (Reprinted with permission from Pirasteh et al. 2006, Copyright 2006 Elsevier)

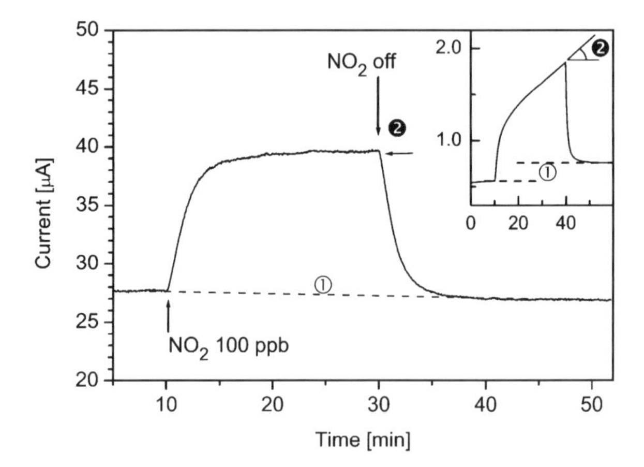

Fig. 22.4 Effect of exposure to NO_2 (concentration of NO_2: 100 ppb) on thin (3 μm), aged PSi samples (*main plot*) and on thick, fresh samples (*inset*). The relative humidity is 30 % in both cases. $T = 22$ °C (Reprinted with permission from Pancheri et al. 2004, Copyright 2004 Elsevier)

However, we need to note that oxidized porous silicon is mainly used in optical, capacitance, and mass-sensitive sensors. For conductometric- and luminescence-type sensors this method of PSi stabilization is unacceptable. Therefore, for stabilization of such sensors, only low-temperature treatments and simple pre-aging can be used. For example, Pancheri et al. (2004) showed that in aged PSi samples the changes in resistivity of the porous layer, associated with exposure to NO_2, have improved the reversibility compared to fresh samples. The stabilization of the current is also faster in aged than in fresh samples (see Fig. 22.4). It is seen that the baseline shift (1) is absent in the aged sample and significant in the fresh sample and the signal stabilization (2) is good in the aged, thin sample and poor in the fresh, thick sample. Connolly et al. (2002) proposed to use the burn-in process, which involves heating the device to ~55 °C, and repeated cycling of the RH between 5 and 95 % for sensor parameter stabilization. It was found that this treatment improved humidity sensor stability as well as reduced hysteresis effects dramatically. These results suggest that sample aging is potentially a good strategy for use of PSi in NO_2 conductometric sensors because reversibility and signal stabilization are still major unsolved limitations for such applications in freshly etched PSi sensors.

Massera et al. (2004) also used aging for stabilization of NO_2 sensor parameters. They proposed using prolonged exposure to high concentrations of NO_2. It was established that after such treatments, including a 10-h exposure to 2 ppm of NO_2 at 20 % RH in synthetic air, followed by 2 weeks of stabilization in ambient air, the devices were stable in the sub-ppm range and their electrical characteristics were greatly improved. The sensors had faster dynamics and reduced hysteresis. For the same purposes, Ben-Chorin and Kux (1994) proposed storing porous samples in a highly oxidizing medium like H_2O_2 for 48 h. It was established that initial oxidation in an H_2O_2 solution can stabilize the PSi layer to a greater extent. Holec et al. (2002) showed that methyl 10-undeacetonate derivatization of the PSi surface increases the PL time stability of PSi-based sensors as well. Boiling in HNO_3 can also be used for PSi property stabilization (Kochergin and Foell 2006).

However, it should be noted that the problem of PSi-based sensor parameter instability is still unresolved. None of the above-mentioned techniques succeeded in complete prevention of the aging effect. Even after aging, sensors have temporal drift (Korotcenkov and Cho 2011). Islam and Saha (2007) believed that the problem of temporal drift could be partially resolved through compensation using soft computing techniques. However, we need a more fundamental understanding of the nature of the drift effect, which requires additional research. According to Pancheri et al. (2004), further efforts should be aimed at the development of an oxidation treatment, which could reproduce the effects of aging on PSi conductivity. At the same time, Lewis et al. (2005) have asserted that their devices continued working for months after their initial fabrication without requiring cleaning or recalibration.

References

Amato G, Delerue C, von Bardeleben HJ (eds) (1997) Structural and optical properties of porous silicon nanostructures. Gordon & Breach, Amsterdam

Archer M, Christophersen M, Fauchet PM (2005) Electrical porous silicon chemical sensor for detection of organic solvents. Sens Actuators B 106:347–357

Balagurov LA, Yarkin DG, Petrova EA, Orlov AF, Karyagin SN (1996) Effects of vacuum annealing on the optical properties of porous silicon. Appl Phys Lett 69:2852–2854

Ben-Chorin M, Kux A (1994) Adsorbate effects on photoluminescence and electrical conductivity of porous silicon. Appl Phys Lett 67:481–483

Bjorkqvist M, Salonen J, Paski J, Laine E (2003) Comparison of stabilizing treatments on porous silicon for sensor applications. Phys Status Solidi 197:374–377

Bjorkqvist M, Salonen J, Paski J, Laine E (2004) Characterization of thermally carbonized porous silicon humidity sensor. Sens Actuators A 112:244–247

Bsiesy A, Gaspard A, Herino R, Ligeon M, Muller F, Oberlin JC (1991) Anodic oxidation of porous silicon layers formed on lightly *p*-doped substrates. J Electrochem Soc 138(11):3450–3456

Canham LT (1997) Storage of porous silicon. In: Canham L (ed) Properties of porous silicon. INSPEC, London, p 44

Canhman LT (ed) (1997) Properties of porous silicon. INSPEC, London

Cantin JL, Schoisswohl M, Grosman A, Lebib S, Ortega C, von Bardeleben HL, Vázsonyi E, Jalsovszky G, Erostyák J (1996) Anodic oxidation of p^- and p^+-type porous silicon: surface structural transformations and oxide formation. Thin Solid Films 276:76–79

Chazalviel JN, Ozanam F (1997) Surface modification of porous silicon. In: Canham L (ed) Properties of porous silicon. INSPEC, London, p 59

Connolly EJ, O'Halloran GM, Pham HTM, Sarro PM, French PJ (2002) Comparison of porous silicon, porous polysilicon and porous silicon carbide as materials for humidity sensing applications. Sens Actuators B 99:25–30

Di Francia G, Castaldo A, Massera E, Nasti I, Quercia L, Rea I (2005) A very sensitive porous silicon based humidity sensor. Sens Actuators B 111–112:135–139

Dimitrov DB, Papadimitriou D, Beshkov G (1997) Photoluminescence spectra of high temperature vacuum annealed porous silicon. In: Proceedings of 21 international conference on microelectronics, vol 1. Nis, Yugoslavia, p 91–94

Dittrich T, Flietner H, Timoshenko VY, Kashkarov PK (1995) Influence of the oxidation process on the luminescence of HF treated porous silicon. Thin Solid Films 255:149–151

Gelloz B, Kojima A, Koshida N (2005) Highly efficient and stable luminescence of nanocrystalline porous silicon treated by high pressure water vapor annealing. Appl Phys Lett 87:031107

Han PG, Wong H, Poon MC (2001) Sensitivity and stability of porous polycrystalline silicon gas sensor. Colloid Surface Physicochem Eng Aspect 179:171–175

Herino R (1997) Impregnation of porous silicon. In: Canham L (ed) Properties of porous silicon. INSPEC, London, p 66

Holec H, Chvojka T, Jelinek I, Jinndrich J, Nemec I, Pelat I, Valenta J, Dian J (2002) Determination of sensoric parameters of porous silicon in sensing of organic vapours. Mater Sci Eng C 19:251–254

Hossain S, Chakraborty S, Dutta SK, Das J, Saha H (2000) Stability in photoluminescence of porous silicon. J Lumin 91:195–202

Irajizad A, Rahimi F, Chavoshi M, Ahadian MM (2004) Characterization of porous poly-silicon as a gas sensor. Sens Actuators B 100:341–346

Islam T, Saha H (2007) Study of long-term drift of a porous silicon humidity sensor and its compensation using ANN technique. Sens Actuators A 133:472–479

Kim SJ, Lee SH, Lee CJ (2001) Organic vapour sensing by current response of porous silicon layer. J Phys Appl Phys 34:3505–3509

Kochergin VR, Foell H (2006) Novel optical elements made from porous Si. Mat Sci Eng R 52:93–140

Korotcenkov G, Cho BK (2010a) Porous semiconductors: advanced material for gas sensor applications. Crit Rev Solid State 35(1):1–37

Korotcenkov G, Cho BK (2010b) Silicon porosification: state of the art. Crit Rev Solid State 35(3):153–260

Korotcenkov G, Cho BK (2011) Instability of metal oxide-based conductometric gas sensors and approaches to stability improvement. Sens Actuators B 156:527–538

Kovalev DI, Yaroshetzkii ID, Muschik T, Petrova-Koch V, Koch F (1994) Fast and slow visible luminescence bands of oxidized porous Si. Appl Phys Lett 64:214–217

Lee M-K, Tu H-F (2007) Stabilizing light emission of porous silicon by in-situ treatment. Jpn J Appl Phys 46:2901–2903

Lewis SE, De Boer JR, Gole JL, Hesketh PJ (2005) Sensitive, selective, and analytical improvements to a porous silicon gas sensor. Sens Actuators B 110:54–65

Lisachenko AA, Aprelev AM (2001) The effect of adsorption complexes on the electron spectrum and luminescence of porous silicon. Tech Phys Lett 27(2):134–137

Loni A (1997) Capping of porous silicon. In: Canham L (ed) Properties of porous silicon. INSPEC, London, p 51

Ludwig MH (1996) Optical properties of silicon-based materials: a comparison of porous and spark-processed silicon. Crit Rev Solid State 21:265–351

Maccagnani P, Angelucci R, Pozzi P, Poggi A, Dori L, Cardinali GC, Negrini P (1998) Thick oxidized porous silicon layer as a thermo-insulating membrane for high-temperature operating thin- and thick-film gas sensors. Sens Actuators B 49:22–25

Mahmoudi B, Gabouze N, Haddadi M, Br M, Cheraga H, Beldjilali K, Dahmane D (2007) The effect of annealing on the sensing properties of porous silicon gas sensor: use of screen-printed contacts. Sens Actuators B 123:680–684

Massera E, Nasti I, Quercia L, Rea I, Di Francia G (2004) Improvement of stability and recovery time in porous-silicon-based NO_2 sensor. Sens Actuators B 102:195–197

Mattei G, Alieva EV, Petrov JE, Yakovlev VA (2000) Quick oxidation of porous silicon in presence of pyridine vapor. Phys Status Solidi 182:139–143

Meyer BK, Hofmann DM, Stadler W, Petrova-Koch V, Koch F, Omling P, Emanuelsson P (1993) Defects in porous silicon investigated by optically detected and by electron paramagnetic resonance techniques. Appl Phys Lett 63(15):2120–2122

Pancheri L, Oton CJ, Gaburro Z, Soncini G, Pavesi L (2004) Improved reversibility in aged porous silicon NO_2 sensors. Sens Actuators B 97:45–48

Pap AE, Kordas K, George TF, Leppavuori S (2004) Thermal oxidation of porous silicon: study on reaction kinetics. J Phys Chem B 108:12744–12747

Parkhutik V (1999) Porous silicon-mechanisms of growth and applications. Solid State Electron 43:1121–1141

Pirasteh P, Charrier J, Soltani A, Haesaert S, Haji L, Godon C, Errien N (2006) The effect of oxidation on physical properties of porous silicon layers for optical applications. Appl Surf Sci 253:1999–2002

Riikonen J, Salomaki M, van Wonderen J, Kemell M, Xu W, Korhonen O, Ritala M, MacMillan F, Salonen J, Lehto V-P (2012) Surface chemistry, reactivity, and pore structure of porous silicon oxidized by various methods. Langmuir 28:10573–10583

Robinson MB, Dillon AC, Haynes DR, George SM (1992) Effect of thermal annealing and surface coverage on porous silicon photoluminescence. Appl Phys Lett 61:1414–1416

Roussel P, Lysenko V, Remaki B, Delhomme G, Dittmar A, Barbier D (1999) Thick oxidised porous silicon layers for the design of a biomedical thermal conductivity microsensor. Sens Actuators A 74:100–103

Sakly H, Mlika R, Chaabane H, Beji L, Ben OH (2006) Anodically oxidized porous silicon as a substrate for EIS sensors. Mater Sci Eng C 26:232–235

Setzu S, Letant S, Solsona P, Romenstain R, Vial JC (1998) Improvement of the luminescence in *p*-type as-prepared or dye impregnated porous silicon microcavities. J Lumin 80:129–132

Shin HJ, Lee MK, Hwang CC, Kim KJ, Kang TH, Kim B, Kim GB, Hong CK, Lee KW, Kim YY (2003) Photoluminescence degradation in porous silicon upon annealing at high temperature in vacuum. J Kor Phys Soc 42(6):808–813

Tischler MA, Collins RY, Stathis JH, Tsang JC (1992) Luminescence degradation in porous silicon. Appl Phys Lett 60:639–641

Torchinskaya TV, Korsunskaya NE, Khomenkova LYu, Shenkman MK, Baran NP, Misiuk A, Surma B (1997) Complex studies of porous silicon aging phenomena. In: Proceeding of international semiconductor conference, IEEE, Sinaia, Romania, pp 173–176

Torchinskaya TV, Korsunskaya NE, Khomenkova LY, Dhumaev BR, Prokes SM (2001) The role of oxidation on porous silicon photoluminescence and its excitation. Thin Solid Films 381:88–93

Unagami T (1980) Oxidation of porous silicon and properties of its oxide film. Jpn J Appl Phys 19:231–241

Xu ZY, Gal M, Gross M (1992) Photoluminescence studies on porous silicon. Appl Phys Lett 60:1375–1377

Yon JJ, Barla K, Herino R, Bomchil G (1987) The kinetics and mechanism of oxide layer formation from porous silicon formed on p-Si substrates. J Appl Phys 62:1042–1048

Zhu WX, Gao YX, Zhang LZ, Mao JC, Zhang BR, Duan JQ, Qin GG (1992) Time evolution of the localized vibrational mode infrared absorption of porous silicon in air. Superlattice Microst 12(3):409–412

Part V
Structure and Surface Modification of Gas Sensing Materials

Chapter 23
Bulk Doping of Metal Oxides

23.1 General Approach

The doping of semiconductor MOX by various additives is one of the main methods for improving gas sensitivity and selectivity, reducing the operating temperature, enhancing the response rate, etc. (Yamazoe et al. 1983; Yamazoe 1991; Korotcenkov et al. 2005, 2007; Gas'kov and Rumyantseva 2009). Impurities also need to be added to conventional binary MOXs to stabilize their (meso)porous nanocrystal morphology (Taurino et al. 2003). For doping various impurities such as catalytically active noble metals (Pd, Pt, Au, Ag, Rh), transition metals (Fe, Co, Cu, etc.), nonmetals (Se), alkaline earth metals (Ca, Ba, Sr, Mg), metalloids (B, Si), etc., can be used. Some of these additives serve as "accelerators" or "catalysts" while others serve as "inhibitors" of various processes. However, it should be noted that the major doping additives are noble and transition metals. The additives, which are conventionally utilized to modify the properties of the gas-sensing MOXs, are summarized in Table 23.1. Here we consider mainly indium and tin oxides as the most studied gas-sensing materials. More detailed reviews of the effects of doping on MOX gas sensors are available elsewhere (Kohl 1990; Korotcenkov 2005, 2007; Miller et al. 2006).

Doping additives are brought to the bulk semiconductor as, for example, interstitials, or are placed on the surface of metal oxides. The bulk doping is done at the stage of metal oxide synthesis or deposition, while surface doping (modification) can be done following sensing layer deposition. As a rule, the surface additives form small clusters located on the surface of much larger grains of the MOX. The distribution of these small dopant particles on the surface is assumed to be more or less homogeneous. High dispersion of the bulk catalyst over the semiconductor support is also essential to obtain good performance of the conductometric gas sensors. Using ceramic and thick-film technologies, for example, noble-metal catalysts can be incorporated into an MOX by: (1) impregnating the pristine MOX powder with a noble-metal chloride such as $PtCl_4$ and $PdCl_2$ solution, followed by drying and calcination (Matsushima et al. 1992); (2) mixing the pristine MOX powder with a colloid of noble metal (Nakao 1995); and (3) chemical bonding of noble-metal complexes such as $PdCl_4^{2-}$ with surface hydroxyls at the pristine MOX in solution (Kaji et al. 1980). The impurity can also be introduced via sputtering of a thin intermediate layer (Gutierrer et al. 1993; Sayago et al. 1995). In this case, the profile of additive concentrations over the MOX structure is driven by temperature and time of annealing. Specific profiles of additive concentration can be created by using ion implantation techniques via adjusting the density of the ion current and the time of implantation (Sulz et al. 1993; Rosenfeld et al. 1993; Rastomjee et al. 1996).

It should be noted that the effect of each doping material is complex and is not well studied. For instance, a surface catalyst may increase the concentration of reactants at the MOX surface or lower

G. Korotcenkov, *Handbook of Gas Sensor Materials*, Integrated Analytical Systems,
DOI 10.1007/978-1-4614-7388-6_23, © Springer Science+Business Media New York 2014

Table 23.1 Influence of bulk doping on gas-sensing characteristics of SnO_2 and In_2O_3 sensors

Additive	Effect	Nature
Noble metals (less than 5 wt%) (Pd, Pt, Rh, Ag, Au)	Increases response to reducing gases Decreases operating temperature Decreases response time	Catalytic effect Change of *A*/*D* parameters Decrease of O_2 dissociation temperature
Al_2O_3, SiO_2	Increases sensor response Improves thermal stability	Decrease of grain size Decrease of area of intergrain contact Increase of porosity
Ag (Ag_2O), Cu (Cu_2O)	Increases response to H_2S, SO_2	Two-phase system Phase transformations during gas detection
Fe (Fe_2O_3)	Increases response to alcohols	Change of oxidation state
Ga (Ga_2O_3), Zn (ZnO)	Increases sensor response	Decrease of grain size Increase of porosity
P, B	Improves selectivity	Creation of new phase
Ca, K, Rb, Mg	Increases sensor response Improves thermal stability	Decrease of grain size
La, Ba, Y, Ce	Improves thermal stability Increases sensor response	Stabilization of grain size (creation of new phases) Decrease of grain size
Transition MOXs (<0.5 wt%) (Co, Mn, Sr, Ni)	Increases sensor response Improves selectivity	Catalytic effect Change of electron concentration Change of *A*/*D* parameters Change of grain size

the activation energy for the reaction, or it may affect both parameters. The large variety of doping effects is clearly seen in the well-studied system of Pt and TiO_2, where three cases are possible (Fisher et al. 1996): (1) the appearance of a Schottky barrier because of work-function differences ($A_{Pt} = 5.8$ eV, $A_{TiO_2(110)} = 5.3\ eV$); (2) the appearance of ohmic contact when the Pt ions diffuse into near-surface areas of the MOX and generate additional local donor states at the semiconductor gap, which makes the surface charge region (SCR) thinner and facilitates electron tunneling; and (3) the local transfer of charge from Ti^{3+} ions to Pt atoms when the latter are built-in surface point defects (e.g., surface vacancies). We also need to take into account that, while the surface modification modifies the receptor function of the semiconducting MOX layers (Chap. 2 (Vol. 1)), the bulk doping, besides influencing surface reactivity, can be accompanied by the changing of the sample microstructure (Matsuura and Takahata 1991). An example of such an effect is the influence of impurities on acceleration or limitation of the grain growth in MOXs under sintering (Fliegel et al. 1994; Behr and Fliegel 1995). The presence of the dopants can also modify the relative energy of the different crystal faces, thus yielding a way to control the crystal morphology during metal oxide synthesis, in particular crystal shape (Sinha et al. 1992; Kawamura et al. 2001a, b; Alfredsson et al. 2007), and ultimately the type and number of surface sites exposed. For example, Kawamura et al. (2001b) reported that the habitus of SnO_2 single crystals changes from short prismatic to needlelike forms or whiskers bounded by {110} by the addition of a small amount of impurity cations in various forms of oxide in an SnO_2–Cu_2O flux system. In a pure system, SnO_2 crystals have a short prismatic habit bounded by {110} and {111} faces. Theoretical simulations also show that dopants can influence the shape of crystallites (Alfredsson et al. 2007). Figure 23.1 illustrates such changes for $CaTiO_3$ crystals. It is seen that large dopants inhibit the formation of stepped surfaces, while dopants smaller than Ca^{2+} favor a more faceted morphology with an important contribution from textured {021} and {111} planes. This phenomenon is important for gas sensing effect, because exactly crystallographic places, faceting crystallites, participate in the interaction with the surrounding gas. It is known that every crystallographic plane has

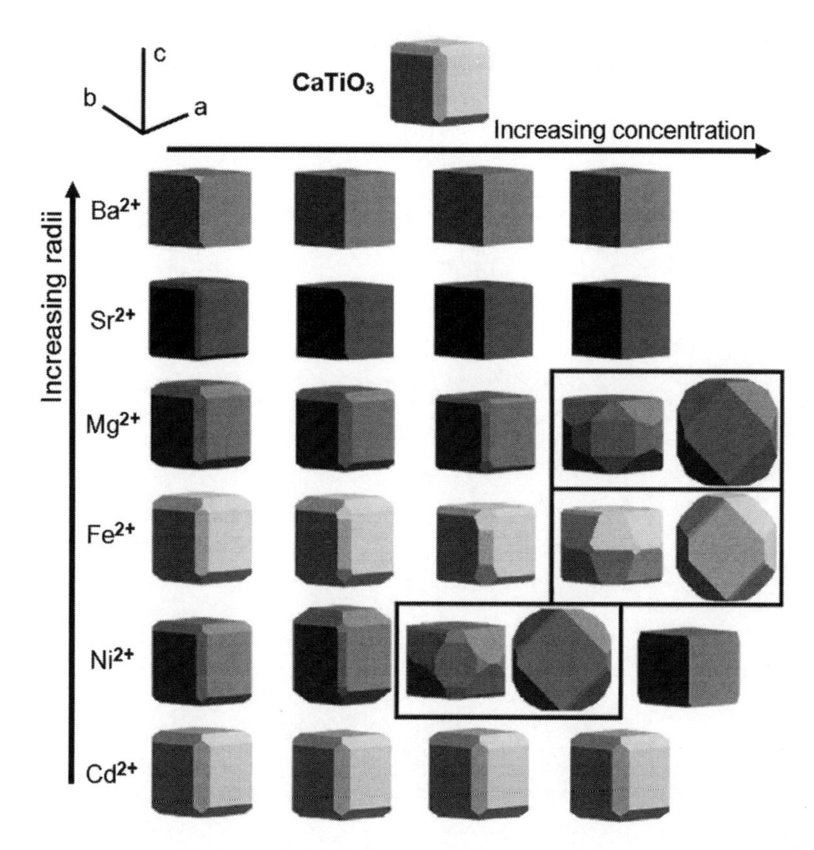

Fig. 23.1 Calculated crystal morphologies of $CaTiO_3$ as a function of dopant type and concentration; both uniform concentration of dopants on each surface and equilibrium conditions are presented. The directions of the morphologies (except those grouped by *squares*) correspond to the crystallographic coordinate system shown in the figure. Morphologies grouped with *squares* correspond to the same dopant concentration, viewed along perpendicular directions (second crystal shapes show direction along the *c*-axis according to the inserted coordinate system) (Reprinted with permission from Alfredsson et al. 2007, Copyright 2007 Elsevier)

specific combination of adsorption/desorption and catalytic properties. Experiments carried out by Korotcenkov et al. (2008, 2010) have shown that the presence of additives strongly influences the morphology of metal oxide films prepared using thin-film technologies as well. Figure 23.2 illustrates the morphology changes in SnO_2 films doped by Co, Fe, and Cu during spray pyrolysis deposition. It was established that, in addition to affecting grain size, the doping influences both the agglomeration and the twinning of grains. It was also found that agglomerates become more apparent in SEM images of SnO_2 doped by copper and cobalt. More detailed description of doping influence on the morphology of doped SnO_2 films one can find in (Korotcenkov et al. 2008).

Bulk doping due to incorporation of additives in the lattice of metal oxide can also be accompanied by a change in the concentration of charge carriers with corresponding influence on the bulk conductivity and the thickness of the space charge region. These parameters control both the resistance of intergrain contacts and conductivity response to analyte (see Chap. 2 (Vol. 1)). In addition, the changes in bulk donor/acceptor concentration result in a shift of the bulk Fermi level, which, in turn, affects the gas adsorption/desorption processes and the SCR. Figure 23.3 shows how strong this influence can be. Sb is donor and Al is acceptor for SnO_2, which has n-type conductivity. Other consequences of metal oxide doping for their parameters are shown in Fig. 23.4.

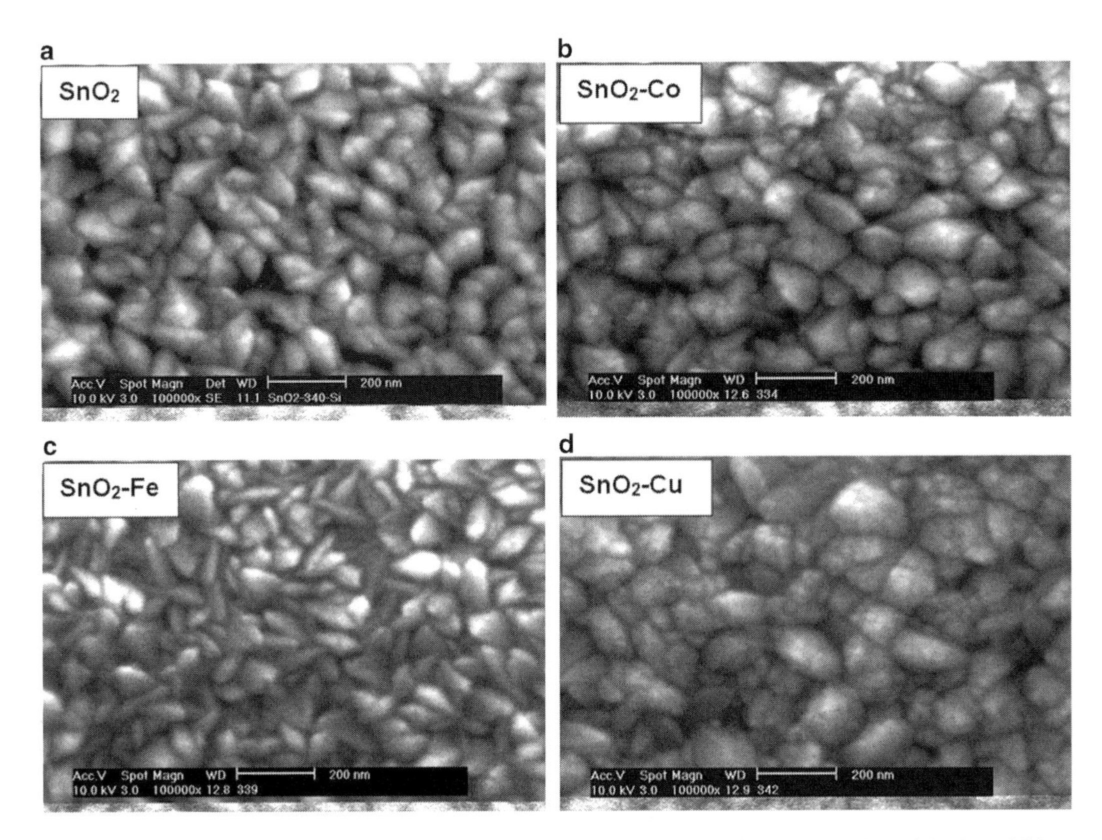

Fig. 23.2 SEM images of (**a**) undoped and doped by (**b**) Co, (**c**) Fe, and (**d**) Cu (concentration of doping additives-16 at.%) SnO_2 films ($d \sim 400$ nm) deposited by spray pyrolysis on oxidized Si substrate at $T_{pyr} = 450$ °C (Reprinted with permission from Korotcenkov et al. 2008, Copyright 2008 Springer)

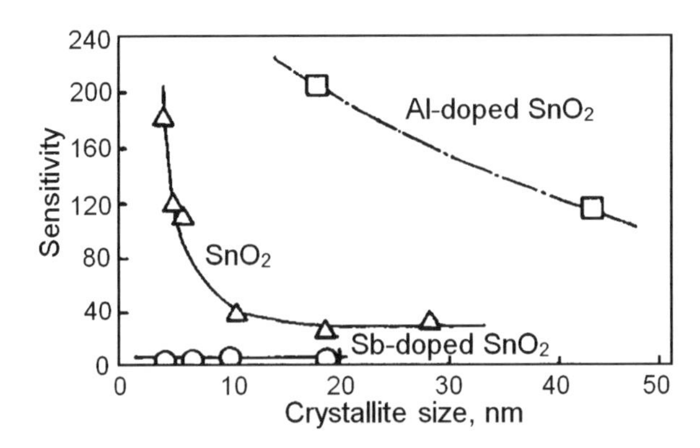

Fig. 23.3 Sensitivity of doped and undoped SnO_2 elements to 800 ppm H_2 in air as correlated with SnO_2 crystallite size (measured at 300 °C) (Reprinted with permission from Yamazoe 1991, Copyright 1991 Elsevier)

It was also found that the effect of additives on sensor performance is dependent on the synthesis method and postdeposition heat treatments employed during sensor fabrication (Miller et al. 2006). Moreover, it was established that additives do not always enhance sensor performance. For example, Min and Choi (2004) tried to control crystallite growth by introducing Ca as a growth inhibitor into the SnO_2. The additive decreased the SnO_2 crystallite size as expected but depressed the response of the gas sensor and its long-term stability. This means that, depending on the application, the doping material and its quantity have to be chosen carefully.

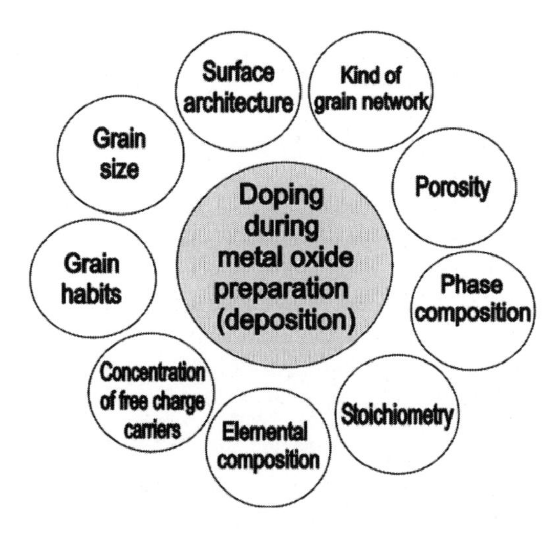

Fig. 23.4 Parameters subject to modification as a result of MOX doping (Reprinted with permission from Korotcenkov 2005, Copyright 2005 Elsevier)

Despite a number of long-term efforts to quantify the effect of the foreign phase on the gas-sensing properties of the MOXs, there is still a need to understand better the underlying fundamental mechanisms of doping influence on metal oxide properties. At present the clusters of impurities that appear on the surface of the MOXs are considered to affect the electrophysical properties by two mechanisms, usually called "chemical" and "electronic" mechanisms. Surface modification of metal oxides was discussed in detail in Chap. 10 (Vol. 1). The first mechanism corresponds to "spillover" of dissociated products from the metal clusters out to the solid surface. This effect is quite popular in the literature devoted to catalysis and is considered to occur mainly during the early stages of gas sensor development. The gas molecules adsorbed at the metal cluster are excited and dissociate (Kohl 1990) and then migrate to the MOX surface. When a reducing gas is adsorbed, these ions may interact with adsorbed or lattice surface oxygen and modulate the semiconductor conductance. In this case, the important factor is the electronegativity of the metal, which is a measure of electron attraction by the atoms (De Fresart et al. 1982). If the gas adsorption or dissociation manages the surface reaction kinetics and change of the MOX conductance, then the gas response of this sensor correlates with the electronegativity (Ionescu and Vancu 1994). The spillover effect explains the decrease in operating temperature in gas-sensing layers doped with catalytic metals. These processes also explain the appearance of conductance oscillations in, for example, SnO_2:Pt layers at low operating temperatures in the presence of reducing gases (Ionescu et al. 1994).

The "electronic" mechanism at the contact between the metal and the semiconducting oxide is driven by the SCR, whose width depends on the catalyst state (Norris 1987). If the metal is well dispersed over the oxide surface, the local SCRs underlying each metal cluster may join together, and the surface position of the Fermi level will depend entirely on the catalyst state (Tofield 1988; Sukharev et al. 1993). The change at the catalyst surface upon interaction with a reducing gas makes a difference in the electronic exchange between the surface and the bulk of the oxide support, which results in a change in its conductance (Ioannides and Verykios 1996). If a Schottky barrier appears, the gas sensitivity of the doped MOX may correlate with the work function of the metal (Vlachos et al. 1996). If so, the size of the metal clusters is of great importance. If the metal is rather thick, surface processes do not influence the bulk, and the height of the potential Schottky barrier is not changed (Madou and Morrison 1989). Therefore, the metal catalyst should be configured to be thinner to allow effective use of the potential barrier modulation due to the surface processes. Experimental evidence of the catalyst layer thickness affecting the gas sensitivity of SnO_2:Ag has been reported by Zhang and Colbow (1997); Yamazoe et al. (1983) and Kohl (1990) also believe that Ag doping of SnO_2 follows the "electronic mechanism."

However, one can assume that, in reality, we have more complicated processes of surface functionalization. For example, it was found that at low operating temperatures (around 100 °C), the influence of Pd on SnO_2 is suggested to follow the "electronic mechanism," while higher temperatures facilitate the domination of "chemical mechanisms" (Kohl 1990). In addition to the two mechanisms considered so far, Papadopoulos and Avaritsiotis (1995) suggested the appearance of surface conductance over an MOX doped with surface metal clusters due to electron tunneling between the clusters (for the example of SnO_2:Pd, SnO_2:Pt systems exposed to CO). In this case, the gas sensitivity is caused by this surface conductance, and the bulk properties of the materials play only a minor role. Moreover, in some systems, such as SnO_2:Ag, it is possible to observe mobile metal clusters which migrate under high operating temperatures over the oxide surface and change its conductivity (Sears and Love 1993; Zhang et al. 1993). Therefore, when analyzing the properties of doped MOX gas sensors, experiments should be considered to compare the contributions from different mechanisms.

23.2 Bulk Doping Influence on Response and Stability of Gas-Sensing Characteristics

As indicated before, bulk doping of metal oxides is one of approaches used for resolving a problem, connected with improvement of metal oxide gas sensor response and parameters stability (Xu et al. 1992; Zhang and Liu 2000; Korotcenkov 2005; Smatko et al. 2010). This approach is being popularized as the most promising for these purposes. It was established that introduction of special doping microadditives of various impurities into metal oxides during their synthesis could sufficiently decrease the grain size and improve both sensor response and thermal stability of the grain size in formed ceramics. It was established that for the above-mentioned purposes, different additives can be used. Examples of doping influence on the SnO_2 grain size are shown in Figs. 23.5 and 23.6.

It was found that the decrease of the grain size and its stabilization during annealing, i.e., the inhibition of particles growth, were observed for SnO_2 doped by V (Yang et al. 2003), Ce (Maciel et al. 2003), and many other metals and nonmetals (Xu et al. 1992). For example, research has shown that the impregnated foreign additives (5 at.%) could keep *D* less than 10 nm even after calcination at 900 °C, whereas pure SnO_2 underwent grain growth to have a *D* of 13 and 27 nm at 600 and 900 °C, respectively (Xu et al. 1992; Min and Choi 2004). These data for SnO_2:Ce, SnO_2:La, and SnO_2:Y are shown in Table 23.2. Reduced sintering is also reported for indium-doped (Yang et al. 1998) and antimony-doped tin dioxide (Miao et al. 2003). It has been found that the above-mentioned effects take place due

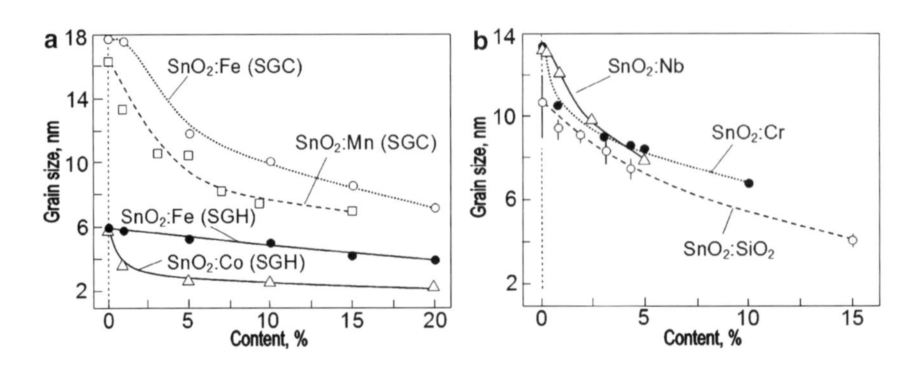

Fig. 23.5 The average crystallite size of the doped SnO_2 samples with different content of additives calculated from Scherrer formula. (**a**) Pure and Co-doped SnO_2 nanoparticles were prepared by a simple sol–gel-hydrothermal (SGH) method. $SnCl_4 \cdot 5H_2O$ and $CoCl_2 \cdot 6H_2O$ were used as tin and cobalt sources, respectively. Fe^{3+}-doped SnO_2 nanoparticles were prepared by sol–gel calcination (SGC) and SGH routes, respectively (Data from Fang et al. 2008a, b). (**b**) SnO_2 and Cr-doped SnO_2 nanoparticles with Cr content up to 10 mol% were produced by a polymer precursor method. SnO_2–SiO_2 powders were synthesized by flame spray pyrolysis technique (Data from Weber et al. 2001; Tricoli et al. 2008; Aragón et al. 2010; Azam et al. 2010)

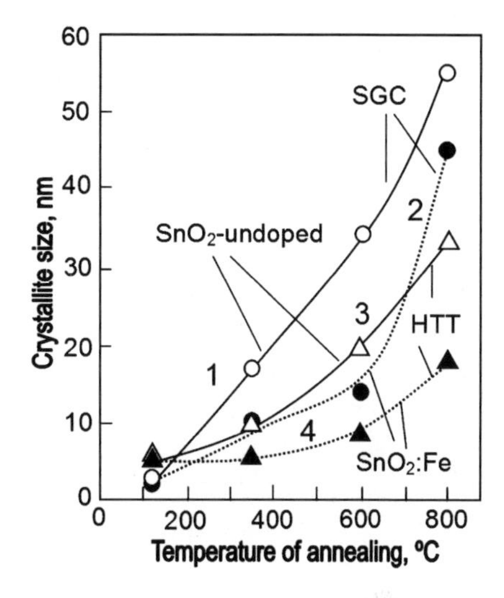

Fig. 23.6 Averaged crystallite size of the SnO_2 powders vs. temperature of annealing. Undoped SnO_2 (1, 3) and Fe-doped SnO_2 (2, 4) nanoparticles were prepared by SGC and SGH routes (Data from Fang et al. 2008b)

Table 23.2 Influence of additives (5 %) and annealing temperature on the SnO_2 grain size

	Average grain size, nm		
Samples	550 °C	900 °C	1,100 °C
SnO_2 (undoped)	12.7	44.7	158.7
SnO_2:Ce	11.7	16.9	70.3
SnO_2:La	6.2	8.7	50.4
SnO_2:Y	5.2	11.1	47.3

Source: Data extracted from Carreno et al. (2002)

to a surface segregation of the doping elements during the annealing processes (Nowotny 1988; Wynblatt et al. 2003). Foreign cations move to the particle surface and decrease particle growth rate. The surface segregation layers can strongly affect the properties of the materials obtained, first, owing to surface segregation and, second, because the created surface layers have properties different from those of the bulk material. The well-dispersed additives on the surface of the SnO_2 powders resist the mutual diffusion of SnO_2.

Xu et al. (1992) analyzed the influence of various doping elements (5 at.%) (31 metals (Li, Na, K, Rb, Cs, Mg, Ca, Ba, Sr, V, Cr, Mn, Fe, Co, Ni, Cu, Zn, Ga, Nb, Mo, In, La, Ce, Pr, Nd, Sm, Gd, W, T1, Pb, and Bi) and 3 nonmetals (S, B, and P)) on the structural properties of SnO_2 synthesized using the sol–gel method and established that the additives can be classified into two types, i.e., one type which is effective in retarding the sintering of SnO_2 particles over the whole temperature range examined (400–900 °C) (Type I) and the other effective only up to a medium temperature (Type II). Type II includes alkali metals, vanadium, sulfur, boron, copper, iron, and thallium. It was established that these additives are subject to segregation, melting, or sublimation at 700–900 °C and their disappearance from the surface of SnO_2 crystallites seems to be the main cause of the losses of stabilizing effects. In the case of Co– and Cu–SnO_2, the formation of CoO and CuO grains was detected by XRD after calcination at 900 °C (Xu et al. 1992). On the other hand, Type I consists of alkaline earth metals, some transition metals, rare-earth metals, and phosphorus, among which, phosphorus, barium, niobium, tungsten, and gallium are especially effective up to 900 °C. The nonmetal additive phosphorus is expected to be strongly bonded to the SnO_2 surface as polyoxy cluster anions, giving rise to very marked effects up to medium temperatures (600 °C). A similar explanation may be possible for high-valence metal additives such as vanadium, niobium, and tungsten which also form polyoxy anions. This means that Type I dopants appear to

Table 23.3 Effects of additives and annealing temperature on TiO_2 grain growth

Sample	Average grain size, nm		
	650 °C	850 °C	1,050 °C
TiO_2 (undoped)	16	110	214
TiO_2:Nb	24	35	109
TiO_2:Ta	18	45	63
TiO_2:V	183	397	2,012

Source: Data extracted from Ferroni et al. (2001)

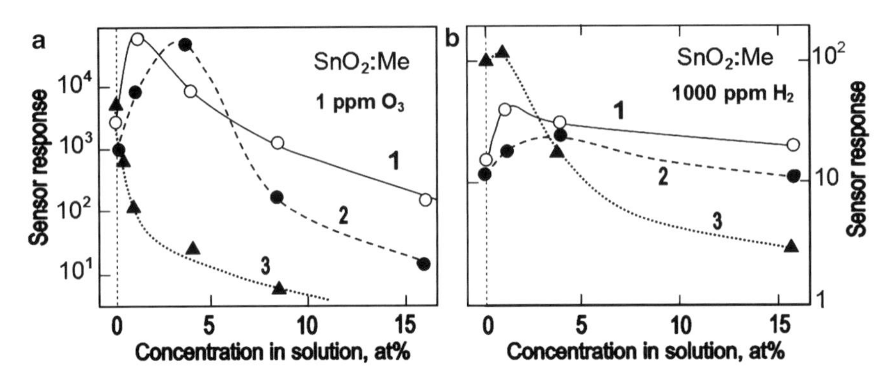

Fig. 23.7 Influence of SnO_2 doping by Co, Ni, Cu, and Fe during film deposition by spray pyrolysis on gas response to (**a**) ozone and (**b**) H_2: T_{pyr} = 410–420 °C; d ~ 45–55 nm; (1) SnO_2:Fe; (2) SnO_2:Co; (3) SnO_2:Cu (Reprinted with permission from Korotcenkov 2005, Copyright 2005 Elsevier)

stay more firmly on the surface of SnO_2 crystallites. XRD analyses showed that after calcination at 600 °C, there were no crystalline phases other than SnO_2 detected (Xu et al. 1992). The other additives having remarkable stabilizing effects at 600 °C were lead, caesium, strontium, and barium, all of which have large ionic radii. Poorly mobile species seem to be effective for the stabilization. It is suggested that additives of Type II dispersed well on the surface of SnO_2 provide a barrier for surface diffusion thus suppressing crystal growth or coagulation of SnO_2 particles during calcination. It seems that the more stable the additives on the SnO_2 surface, the more effectively could the ultrafine particles of SnO_2 be stabilized. We need to note that the above-mentioned effects are common for all metal oxides. The results of TiO_2 doping influence on thermal stability of the grain size are presented in Table 23.3.

However, it was established that for attainment of the desired effect, i.e., strong grain-size decrease and grain-size stabilization during high-temperature annealing, the concentration of those additives should usually be high enough. As a rule, those concentrations exceed 5–10 % (Weber et al. 2001; Carreno et al. 2002; Tricoli et al. 2008; Aragón et al. 2010; Azam et al. 2010). At the same time, in other work, where gas-sensing properties of such doped materials were analyzed, it was established that, as a rule, the optimum of gas-sensing properties is observed at concentrations of doping additives less that 1–2 % (Yamazoe 1991; Matko et al. 1999; Dieguez et al. 2000; Korotcenkov 2005; Ramgir et al. 2006; Yuasa et al. 2009). For example, Choi and Lee (2001) used doping at the level ~0.1 wt% for achieving better sensitivity to CH_4. Results of other authors related to doping influence on gas-sensing characteristics of the SnO_2-based sensors are shown in Figs. 23.7 and 23.8. In particular, Fig. 23.7 shows results related to influence of transition metals, and Fig. 23.8 shows results related to SnO_2 bulk doping by Pt and Pd, the main doping additives used for improving sensor response, obtained in different labs. One can see that optimal concentration of catalytic active additives providing the increase of the response of SnO_2-based sensors lies in the range <1–4 % for transition metals and 0.1–0.6 % for noble metals such as Pd and Pt. The observed spread in position of sensor response

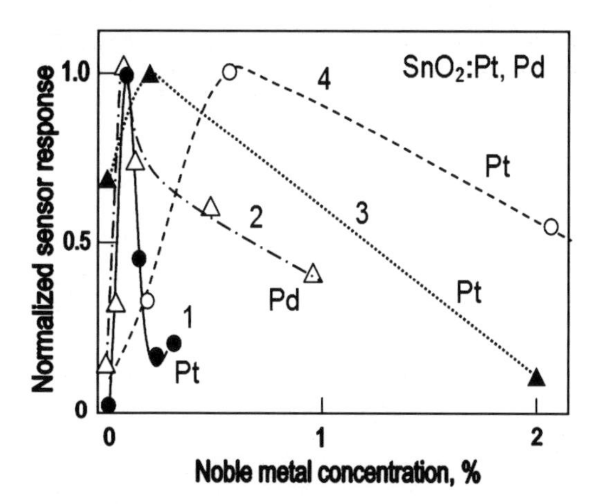

Fig. 23.8 Influence of bulk doping by Pt and Pd on normalized response of differently fabricated SnO_2-based sensors to CO. (1) SnO_2:Pt films were deposited by spray pyrolysis of hydrolyzed solution (2,500 ppm, T_{oper} = 300–350 °C) (Data extracted from Ramgir et al. 2006). (2) SnO_2:Pd powders were prepared by the reverse micelle method (200 ppm, T_{oper} = 300 °C) (Data extracted from Yuasa et al. 2009). (3) SnO_2:Pt films were synthesized in one step using the flame spray pyrolysis (50 ppm, T_{oper} = 300 °C) (Data extracted from Madler et al. 2006). (4) SnO_2:Pt films were prepared by using a submicroscopic aerosol pyrolysis method (1,300 ppm, T_{oper} ~ 250 °C) (Data extracted from Matko et al. 2002)

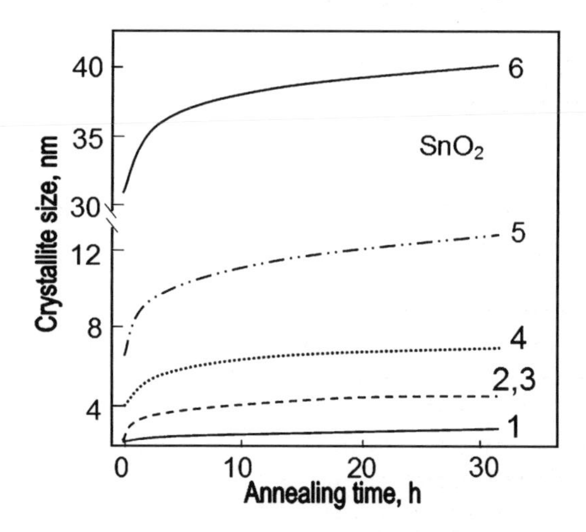

Fig. 23.9 Dependence of the SnO_2 crystallite average size on the time of isothermal heating at 700 °C: (1) SnO_2 nanopowder prepared by precipitation from tin acetate solution; (2) SnO_2 doped by S (0.036 wt%); (3) SnO_2 doped by Cl (0.038 wt%); (4) SnO_2 with surface modified by Pd (1.2 wt%); (5) SnO_2 doped by Pd (0.024 wt%); (6) commercial SnO_2 nanopowder from Sigma–Aldrich (specific surface area ~ 45 m^2/g) (Data from Pavelko et al. 2009a, b)

maxima in Fig. 23.8 is natural, since each technology has a specific influence on both the concentration and the distribution of doped additives in synthesized metal oxides. These results indicate that the introduction of high concentrations of additives, independent of their type and method of synthesis, leads to a sharp worsening of gas-sensing properties (Yamaura et al. 2000; Korotcenkov et al. 2004, 2008; Tricoli et al. 2008; Liu et al. 2010). Besides, it was established that not all impurities introduced in metal oxide matrix promote the increase of sensor response and the inhibition of the grain's growth. Pavelko et al. (2009a, b), for example, established that bulk uncontrolled impurities such as Cl^-, Na^+, SO_4^-, and Pd significantly accelerated the growth of the SnO_2 nanocrystallites (see Fig. 23.9). Pavelko

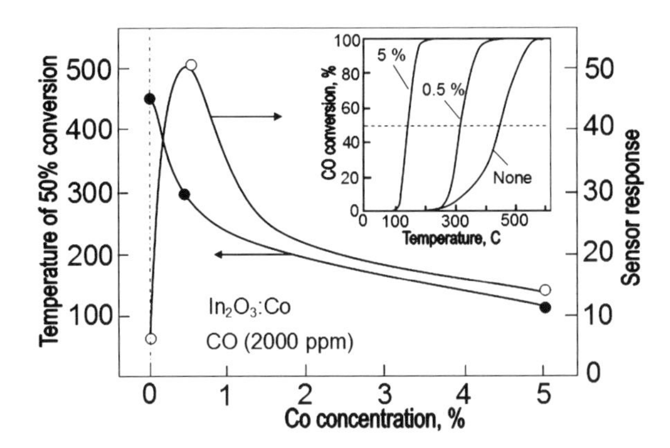

Fig. 23.10 Influence of the concentration of Co additives in the In_2O_3 on the response and temperature of 50 % CO conversion by In_2O_3:Co-based sensor and catalyst. The inset shows the temperature dependence of CO conversion by In_2O_3:Co catalyst. An aqueous solution of $InCl_3$ was hydrolyzed with ammonium hydroxide to produce In_2O_3, and the resulting precipitate was calcined at 850 °C. Metal oxides were added to the In_2O_3 powder by impregnating with an aqueous solution of each metal salt, followed by calcining at 600 °C. Catalytic oxidation of CO and H_2 was carried out in a conventional fixed bed flow reactor. CO or H_2, 2,000 ppm each with wet air, was allowed to flow through the catalyst (Data from Yamaura et al. 2000)

et al. (2009b) have also found that impurities such as Cl and Na take part in the surface oxidation processes causing remarkable signal drift of sensor response. Acceleration of the SnO_2 grain growth at high temperature also takes place for doping by Cu, Co, and V (Ferroni et al. 2001). The same effect was observed for other metal oxides including TiO_2 (see Table 23.3).

It is important that, during simultaneous study of gas-sensing and catalytic properties of doped metal oxides, it was established that, as a rule, the change of sensor response does not coincide with the change of catalytic activity of analyzed material (see Fig. 23.10) (Yamaura et al. 2000). Such behavior of tested characteristics testifies that the observed decrease of sensor response is not connected with reducing catalytic activity of sensing material, as could be assumed at first sight.

Korotcenkov et al. (2006, 2010), analyzing influence of the bulk doping on luminescence properties of SnO_2, established that sensor sensitivity drop at superfluous concentration of doping additives was a result of an increase of the concentration of structural defects. It was concluded that the high level of SnO_2 structure disordering, caused by doping, may be the reason for the increase of the concentration of surface states, pinning the surface Fermi level and limiting the Fermi level shift during interaction with the surrounding gas (Comini 2006; Hernandez-Ramirez et al. 2007a, b). It should be noted that this conclusion was confirmed in theoretical (Nowotny 1988) and experimental studies (Henshaw et al. 1996; Dieguez et al. 2000; Cabot et al. 2000). It was established that the growth of the concentration of doping additives in the SnO_2 was accompanied by the appearance of an additional spectral bands both in CL and in XRS spectra (see Fig. 23.11a). According to Cabot et al. (2000), the nature of those bands in XRS spectra is connected with structural disordering of the SnO_2 surface, which is responsible for the appearance of additional states inside a bandgap. A considerable increase in half-width of the SnO_2 Raman peaks after Pt and Pd incorporation in the SnO_2 lattice ($C_{Pt,Pd}=2$ wt%) observed by Cabot et al. (2000) is another direct confirmation that the SnO_2 lattice may be disturbed by the presence of additives at high concentrations.

According to Nowotny (1991), the depth of the interface region, involving the segregation of both intrinsic and extrinsic defects, may reach hundreds of lattice layers. With that the enrichment of extrinsic defects may reach several orders of magnitude. Both intrinsic and extrinsic defects may exhibit strong interactions within the interface layer. As a result, attractive interactions can be accompanied by an increase in the extrinsic disorder in the interface region at higher levels of nonstoichiometry. The appearance of structurally disturbed surface layer experimentally determined in the grains of heavy-doped SnO_2 (see Fig. 23.11b) corresponds to a proposed model. High-resolution

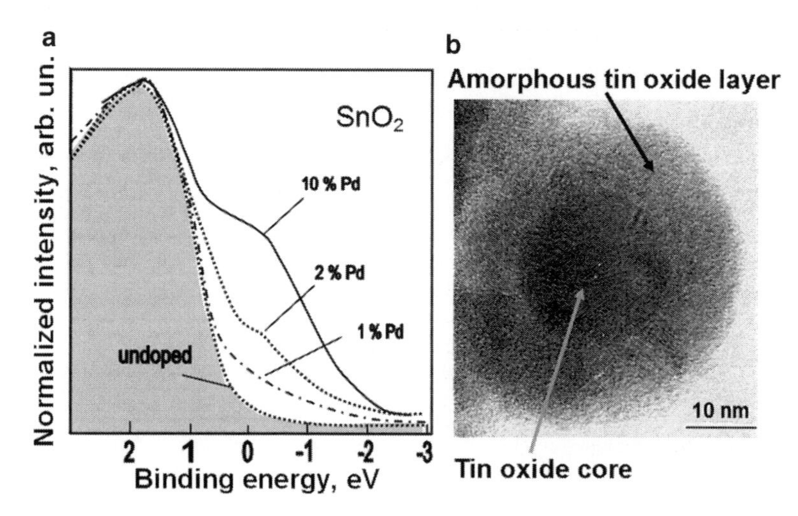

Fig. 23.11 Influence of Pd bulk doping on (**a**) XPS valence band spectra and (**b**) high-resolution transmission electron microscopy (HRTEM) images of a SnO_2 nanoparticles synthesized by wet chemical route: (**a**) (SnO_2:Pd), $T_{cal}=450$ °C (Dieguez et al. 2000; Cabot et al. 2000); (**b**) (SnO_2:Pd) (8 wt%) (Adapted with permission from Nayral et al. 2000, Copyright 2000 Elsevier)

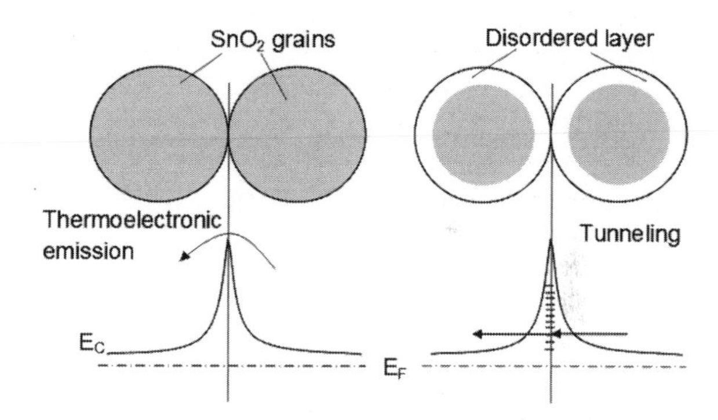

Fig. 23.12 Diagram illustrating the influence of surface states at the interface on the mechanism of current transport

transmission electron microscopy (HRTEM) images of SnO_2:Pd (8 wt%) grains presented by Nayral et al. (2000) show that highly doped SnO_2 grains consist of a well-crystallized core covered with an amorphous layer of tin oxide.

We also need to take into account that current transport across an intergrain interface, which is mainly affected by the barrier heights, can also be strongly modified in the presence of defect states in the bandgap not only by modifying the potential distribution (Klein et al. 2007) but also by adding additional transport paths involving recombination or tunneling utilizing defect states (see Fig. 23.12) (Fonash 1981; Kao and Hwang 1981). This means that the increase of structural defects at the interface can strongly decrease the role of potential barrier changing due to surface reactions in modulation of current transport in metal oxide-based gas sensors. However, we have to agree with Klein et al. (2007) that the microscopic origin of the point defects at interfaces and their individual contribution to band alignment, current transport, and electronic device properties are as yet mostly unclear and need to be resolved for different cases, including metal oxide interfaces and metal oxide-based gas sensors. In metal oxides, especially in the nanosize region and in the case of doping, the deviation from stoichiometry is essential and the picture of abrupt interfaces with a well-defined band alignment and simple transport processes neglecting defect states in the bandgap is not appropriate. This means that new approaches and new models for explanation of observed gas-sensing phenomena are required.

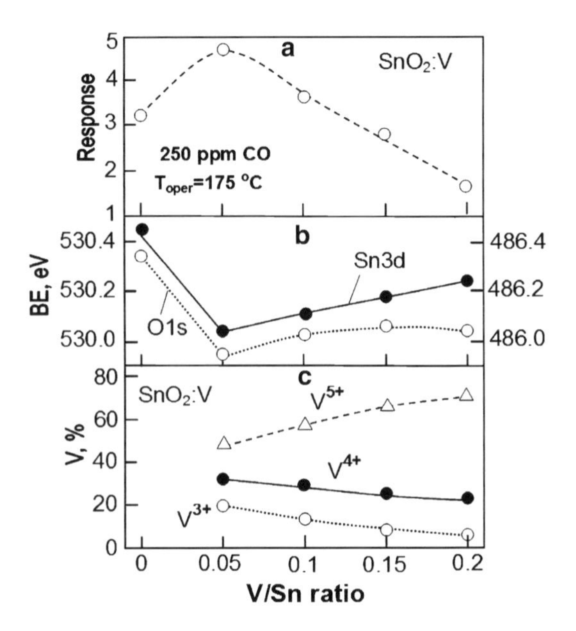

Fig. 23.13 Correlation of V/Sn ratio with (**a**) sensor response to CO and XPS data related to (**b**) Sn3d and O1s binding energy (BE) and (**c**) vanadium cation fraction (Data extracted from Wang et al. 2011a, b)

The strong correlation between impurity incorporation in the SnO_2 lattice and gas-sensing characteristics was observed by Wang et al. (2011a) for SnO_2 doped by V. These results are presented in Fig. 23.13. Wang et al. (2011a) have shown that maximum sensor response to CO is observed at minimal concentration of V (Fig. 23.13a), which corresponds to maximal concentration of V^{4+} and V^{3+}, which can be associated with V incorporated in the SnO_2 lattice, and minimal concentration of V^{5+} (Fig. 23.13c). According to Wang et al. (2011a), the appearance and the growth of the V^{5+} component indicates an increase in the degree of polymerization of the surface vanadia species, from isolated, oligomeric to polymeric forms or V_2O_5 phase, i.e., the appearance of V_2O_5 clusters (grains) in SnO_2 matrix (Wang et al. 2011b). It is important to note that the increase of the concentration of V_2O_5 phase is accompanied by remarkable broadening of UV–vis diffuse reflectance spectra of SnO_2:V (Gao and Wachs 2000), which is usually associated with structural disordering of analyzed material responsible for the appearance of additional states inside a bandgap. The binding energy (BE) values of both Sn3d5/2 and O1s core levels are minimal in the point V/Sn = 0.05 as well (Fig. 23.13b). This BE shift suggests an electronic interaction between V and Sn atoms in the V–O–Sn bond. Wang et al. (2010) believe that the V–O–Sn structure, formed during V incorporation in the SnO_2 lattice, is more chemically reducible and thus more reactive than the Sn–O–Sn structure. In addition, owing to a lower O1s BE, the O atom adjacent to the V site is considered more labile. It is important also that maximum of sensor response (V/Sn = 0.05) does not coincide with the maximum of catalytic activity. For example, the catalytic activity for methanol conversion was favored for higher vanadium loadings (V/Sn = 0.15–0.2).

Tian et al. (2008) described the same processes for SnO_2 doped by Mn. They found that Mn in SnO_2 not only lowers the crystallite size but also degrades the crystallinity of the nanoparticles. As the Mn content increases, the intensity of XRD peaks decreases and FWHM increases, which indicates the degradation of crystallinity. This means that Mn doping in SnO_2 produces crystal defects around the dopants and the charge imbalance arising from this defect changes the stoichiometry of the materials. The distortion of the SnO_2 lattice was also observed by Aragón et al. (2010) during doping by Cr. It was established that additional distortions are introduced by the substitution of Sn by Cr ions. In addition, it was found that the maximum distortion takes place for the sample doped with 3 mol% Cr content. The 3 mol% Cr is assigned as the concentration where the regime change from the solubility to surface segregation of Cr ions occurs (Aragón et al. 2010).

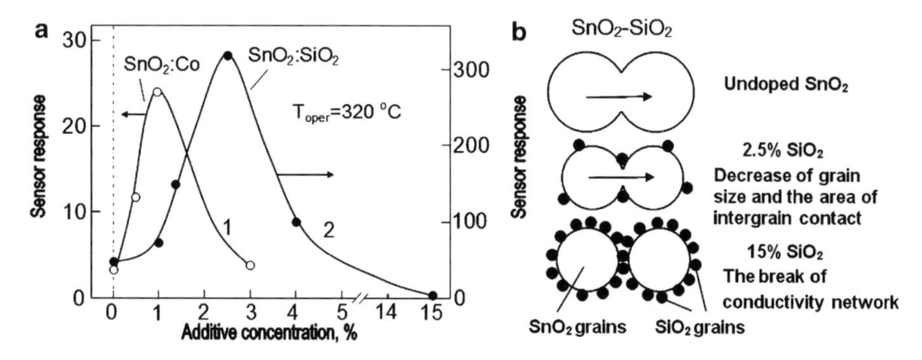

Fig. 23.14 (**a**) Additives influence on the response of doped SnO_2-based gas sensors. (1) Responses of pure and Co-doped SnO_2 nanofibers to 100 ppm H_2 (Data from Liu et al. 2010). (2) Response of SnO_2:SiO_2-based sensors to 50 ppm EtOH as a function of the SiO_2 content (Data extracted from Tricoli et al. 2008). (**b**) Diagram illustrating the mechanism of transformation of SnO_2 properties during doping by SiO_2

Similar conclusions were made earlier while studying metal oxides aimed at the design of varistors (Bueno et al. 1998). The data obtained have shown that doping the SnO_2 system with Cr_2O_3 at a level of 0.05 % builds up an optimized barrier at the grain boundary. For Cr_2O_3 concentrations higher than 0.05 % the system loses its nonlinearity. It was concluded that the annihilation of the voltage barrier at the SnO_2 intergrain contacts is conditioned by defect formation at the grain boundary. Deterioration of crystallinity of doped metal oxides was also observed by Ivanovskaya et al. (2007).

Thus, in the framework of the discussed approach, as well as in the case of sensor parameter optimization at the expense of grain-size decrease, we meet with contradiction. This means that for achievement of better thermal stability of structural properties of metal oxide ceramics we should admit the possibility of considerable worsening of their gas-sensing properties. Moreover, we need to accept that nanocomposites based on doped metal oxides with stable grain size would have properties unacceptable for gas sensor applications. For example, in case of SnO_2:Si at 15 wt% SiO_2 contents, the SnO_2:SiO_2 nanocomposite becomes insulating because insulating SiO_2 grains separate conductive SnO_2 grains (Tricoli et al. 2008). This mechanism is shown in Fig. 23.14. This means that one should choose either sensitivity or stability. However, in this case, one can question whether we need this additional doping if the same stability at better gas-sensing properties could be attained using undoped material but with bigger grain size.

Results obtained by Oswald et al. (2004) also require our attention. Their research has shown that samples of as-synthesized doped metal oxides are not in an equilibrium state. For example, according to Oswald et al., the thermodynamic equilibrium for doped SnO_2 single crystals was approached only after 24 h annealing above 850 °C and at 1,000 °C for Sb and In, respectively. This means that the presence of doping additives alone, introduced in metal oxide for the grain-size stabilization, can be the reason for sensor parameter instability. The diffusion of doping elements during exploitation with subsequent surface segregation can change significantly both the electrophysical properties of the grains and the conditions of the intergrain contacts forming, responsible for the height of intercrystallite potential barrier. Therefore, the use of maximally low concentration of doping additives and long thermal treatment during sensor fabrication, contributing to establishment of the thermodynamic equilibrium, are important conditions for achieving highly stable parameters of sensors based on doped metal oxides (Korotcenkov and Cho 2011, 2012).

It should be noted that the demixing of doped SnO_2 during high-temperature annealing, which is accompanied by an increase in the concentration of the second phase, is a common phenomenon for metal oxides (Nowotny 1988). Usually it is stated that this effect takes place at T_{an} ~ 800–900 °C. However, on the basis of results of the annealing influence on the lattice parameters of doped SnO_2

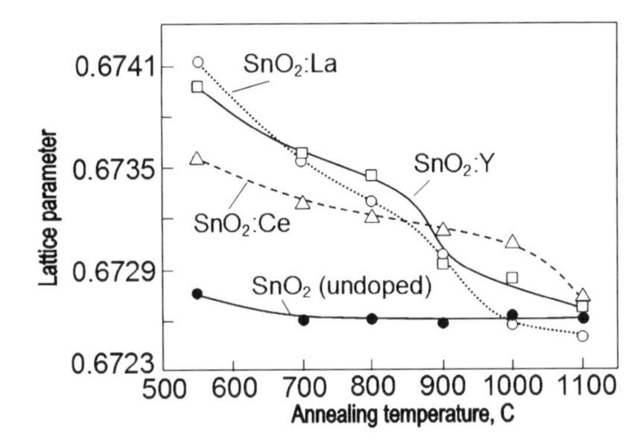

Fig. 23.15 Lattice parameter and microstrain measurements as a function of the heat-treatment temperature. Powders of stannic oxide were prepared from precipitated hydrous β-stannic acid by oxidizing tin granules with nitric acid, evaporating the liquid and firing the powders at different temperatures under oxygen atmosphere (Data from Leite et al. 2002)

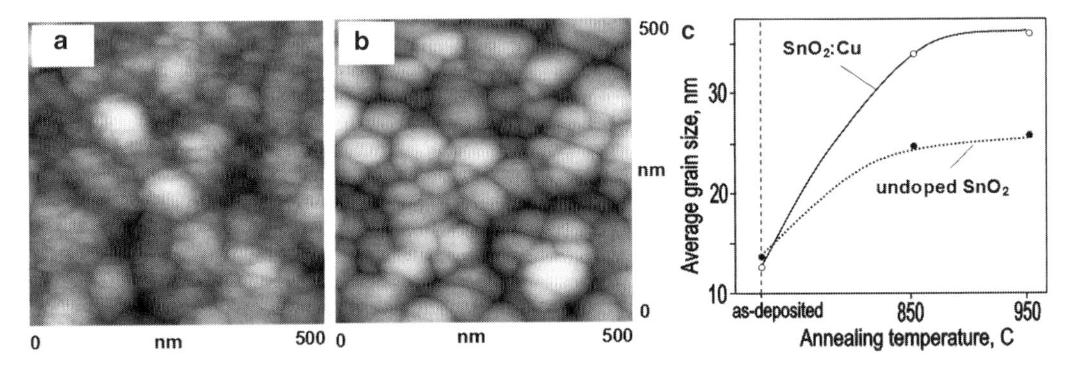

Fig. 23.16 AFM images of doped SnO_2:Cu films deposited by spray pyrolysis before (**a**) and after (**b**) annealing at $T_{an}=850$ °C. (**c**) Annealing influence on averaged grain size in undoped and Cu-doped SnO_2 films deposited by spray pyrolysis ($T_{pyr}=350$ °C; $d\sim250$ nm) (Reprinted with permission from Korotcenkov and Han 2009, Copyright 2009 Elsevier)

(see Fig. 23.15), one can conclude that demixing takes place at a lower temperature and at a low rate. It is seen that the changes of microstrains and lattice parameters started at $T_{an}\sim500$ °C.

In addition to what was said above, the fact was also established that the above-mentioned approach (grain-size stabilizing through bulk doping of metal oxide), which was designed for ceramic or thick-film technologies, has limitations in thin-film technology (Korotcenkov et al. 2008; Korotcenkov and Han 2009). It was shown that SnO_2 doping by Cu, Fe, and Co during spray pyrolysis deposition did not promote the increase of thermal stability of the films' parameters. Due to increasing contents of the fine-dispersed amorphous-like phase, greater structural changes during thermal treatments at temperatures 600–1,000 °C take place in doped films in comparison with undoped ones. We have discussed earlier in Korotcenkov et al. (2008) and Korotcenkov and Han (2009) the mechanism of the appearance of the fine-dispersed phase. According to the suggested model, the second oxide creates additional nucleation centers for the SnO_2 growth. Therefore, the growth of the SnO_2 film during deposition takes place not only due to the increasing size of crystallites incipient at the primary stage of growth but also due to the appearance of the new grains, having considerably smaller size in comparison with already present crystallites that appeared at the initial stages of the SnO_2 films' growth. This SnO_2 amorphous-like phase fills up the intercrystallite space and promotes the densification of metal oxide matrix, i.e., the decrease of gas penetrability of deposited doped SnO_2 films. It should be noted that the possibility of the simultaneous presence in tin dioxide films of both crystallites and the fine-dispersion amorphous-like phase was also experimentally proven by Jimenez et al. (1999).

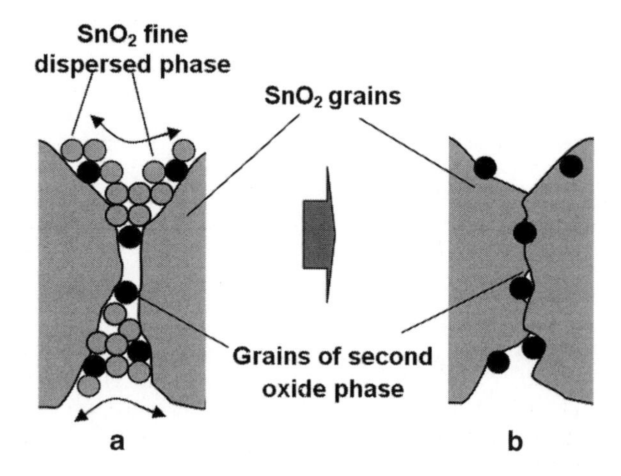

Fig. 23.17 Schematic diagram of the intergrain space change in SnO_2 matrix after annealing: (**a**) – before and (**b**) after annealing (Adapted with permission from Korotcenkov and Han 2009, Copyright 2009 Elsevier)

More visually, one can observe in AFM images of the SnO_2 films doped by copper a disappearance of the fine-dispersed phase after thermal treatment. The diffuse image of grains' edges in metal oxide with high contents of fine-dispersed phase appears due to low resolution of the SEM technique, which does not allow discrimination of the small grains located on the surface of basic oxide's crystallites. Data on the grain size changes during the annealing process of undoped and Cu-doped SnO_2 films are presented in Fig. 23.16. Those data were obtained using XRD measurements.

Because fine amorphous-like phase fills intercrystallite space, the coalescence of this fine phase with bigger crystallites increases the area of contact between intercrystallites, creating conditions at which conductive channel between crystallites is not being overlapped under any conditions. This situation is shown in Fig. 23.17. The last one should be accompanied by a decrease of sensor response. As the study conducted by Korotcenkov and Han (2009) has shown, the second oxide phase, presenting in metal oxide matrix, does not hinder this process (see Fig. 23.17), because crystallites' intergrowth takes place through the fine-dispersed phase of base oxide. It should be noted that Varela et al. (1990) reached a similar conclusion. They stated that very wide particle size distribution, i.e., the presence of both big and small particles, leads to grain growth and agglomerate densification first.

One can also find in Chaps. 2 and 10 (Vol. 1) and Chap. 14 (Vol. 2) additional information related to various additives influence on structural and gas-sensing properties of metal oxides.

References

Alfredsson M, Cora F, Dobson DP, Davy J, Brodholt JP, Parker SC, Price GD (2007) Dopant control over the crystal morphology of ceramic materials. Surf Sci 601:4793–4800

Aragón FH, Coaquira JAH, Candela DS, Baggio Saitovitch E, Hidalgo P, Gouvêa D, Morais PC (2010) Structural and hyperfine properties of Cr-doped SnO_2 nanoparticles. J Phys Conf Ser 217:012079

Azam A, Ahmed AS, Chaman M, Naqvi AH (2010) Investigation of electrical properties of Mn doped tin oxide nanoparticles using impedance spectroscopy. J Appl Phys 108:094329

Behr G, Fliegel W (1995) Electrical properties and improvement of the gas sensitivity in multiple-doped SnO_2. Sens Actuators B 26:33–37

Bueno PR, Pianaro SA, Pereira EC, Longo E, Varela JA (1998) Investigation of the electrical properties of SnO_2 varistor system using impedance spectroscopy. J Appl Phys 84:3700–3705

Cabot A, Arbiol J, Morante JR, Weimar U, Bârsan N, Göpel W (2000) Analysis of the noble metal catalytic additives introduced by impregnation of as obtained SnO_2 sol–gel nanocrystals for gas sensors. Sens Actuators B 70:87–100

Carreno NLV, Maciel AP, Leite ER, Lisboa-Filho PN, Longo E, Valentino A, Probst LED, Paiva-Santos CO, Schreiner WH (2002) The influence of cations segregation on the methanol decomposition on nanostructured SnO_2. Sens Actuators B 86:185–192

Choi S-D, Lee D-D (2001) CH_4 sensing characteristics of K-, Ca-, Mg impregnated SnO_2 sensors. Sens Actuators B 77:335–338

Comini E (2006) Metal oxide nano-crystals for gas sensing. Anal Chim Acta 568(1–2):28–40

De Fresart E, Darville J, Gilles JM (1982) Influence of the surface reconstruction of the work function and surface conductance of (110) SnO_2. Appl Surf Sci 11/12:637–651

Dieguez A, Vila A, Cabot A, Romano-Rodriguez A, Morante JR, Kappler J, Barsan N, Weimar U, Gopel W (2000) Influence on the gas sensor performances of the metal chemical states introduced by impregnation of calcinated SnO_2 sol–gel nanocrystals. Sens Actuators B 68:94–99

Fang LM, Zu XT, Li ZJ, Zhu S, Liu CM, Zhou WL, Wang LM (2008a) Synthesis and characteristics of Fe^{3+}-doped SnO_2 nanoparticles *via* sol–gel-calcination or sol–gel-hydrothermal route. J Alloys Compd 454:261–267

Fang LM, Zu XT, Li ZJ, Zhu S, Liu CM, Wang LM, Gao F (2008b) Microstructure and luminescence properties of Co-doped SnO_2 nanoparticles synthesized by hydrothermal method. J Mater Sci Mater Electron 19:868–874

Ferroni M, Carotta MC, Guidi V, Martinelli G, Ronconi F, Sacerdoti M, Traversa E (2001) Preparation and characterization of nanosized titania sensing film. Sens Actuators B 77:163–166

Fisher S, Shierbaum K-D, Gopel W (1996) Submonolayer Pt on TiO_2 (110) surfaces: electronic and geometric effects. Sens Actuators B 31:13–18

Fliegel W, Behr G, Werner J, Krabbes G (1994) Preparation, development of microstructure, electrical and gas-sensitive properties of pure and doped SnO_2 powders. Sens Actuators B 19:474–478

Fonash SJ (1981) Solar cell device physics. Academic, New York

Gao X, Wachs IE (2000) Investigation of surface structures of supported vanadium oxide catalysts by UV-vis-NIR diffuse reflectance spectroscopy. J Phys Chem B 104:1261–1268

Gas'kov A, Rumyantseva M (2009) Metal oxide nanocomposites: Synthesis and characterization in relation with gas sensing phenomena. In: Baraton MI (ed) *Sensors* for Environment, Health and Security. Springer Science + Business Media B.V., Dordrecht, the Netherlands, pp 3–29

Gutierrer FJ, Ares L, Robla JI, Horillo MC, Sayago I, Getino JM, de Agapito JA (1993) NO_x tin dioxide sensors activities, as a function of doped materials and temperature. Sens Actuators B 15–16:354–356

Henshaw GS, Ridley R, Williams DE (1996) Room-temperature response of platinised tin dioxide gas-sensitive resistors. J Chem Soc Faraday Trans 92:3411–3417

Hernandez-Ramirez F, Prades JD, Tarancon A, Barth S, Casals O, Jimenez-Diaz R, Pellicer R, Rodriguez J, Juli MA, Romano-Rodriguez A, Morante JR, Mathur S, Helwig A, Spannhake J, Mueller G (2007a) Portable microsensors based on individual SnO_2 nanowires. Nanotechnology 18:495501

Hernandez-Ramirez F, Tarancon A, Casals O, Arbiol J, Romano-Rodriguez A, Morante JR (2007b) High response and stability in CO and humidity measures using a single SnO_2 nanowire. Sens Actuators B 121:3–17

Ioannides T, Verykios XE (1996) Charge transfer in metal catalysts supported on doped TiO_2: a theoretical approach based on metal–semiconductor contact theory. J Catal 161:560–569

Ionescu R, Vancu A (1994) Time-dependence of the conductance of SnO_2:Pt:Sb in atmospheres containing oxygen, carbon monoxide and water vapour. I. Non-oscillatory behavior. Appl Surf Sci 74:207–212

Ionescu R, Vancu A, Tomescu A (1994) Time-dependence of the conductance of SnO_2:Pt:Sb in atmospheres containing oxygen, carbon monoxide and water vapour. II. Oscillatory behavior. Appl Surf Sci 74:213–219

Ivanovskaya MI, Kotsikaua DA, Taurino A, Siciliano P (2007) Structural distinctions of Fe_2O_3–In_2O_3 composites obtained by various sol–gel procedures, and their gas-sensing features. Sens Actuators B 124:133–142

Jimenez VM, Espinos JP, Caballero A, Contreras L, Fernandez A, Justo A, Gonzalez-Elope AR (1999) SnO_2 thin films prepared by ion beam induced CVD: preparation and characterization by X-ray absorption spectroscopy. Thin Solid Films 353:113–123

Kaji T, Oono H, Nakahara T, Yamazoe N, Seiyama T (1980) Fixation of palladium(II) and copper(II) complexes on the surface of stannic oxide (SnO_2) and their catalytic activity for propylene oxidation. J Chem Soc Jpn 7:1088–1093

Kao KC, Hwang W (1981) Electrical transport in solids. Pergamon Press, Oxford

Kawamura F, Yasui I, Sunagawa I (2001a) Effects of supersaturation and impurity on step advancement on TiO_2 (110) faces grown from high-temperature solution. J Cryst Growth 233:517–522

Kawamura F, Yasui I, Kamei M, Sunagawa I (2001b) Habit modifications of SnO_2 crystals in SnO_2-Cu_2O flux system in the presence of trivalent impurity cations. J Am Ceram Soc 84(6):1341–1346

Klein A, Sauberlich F, Spath B, Schulmeyer T, Kraft D (2007) Non-stoichiometry and electronic properties of interfaces. J Mater Sci 42:1890–1900

Kohl D (1990) The role of noble metals in the chemistry of solid state gas sensors. Sens Actuators B 1:158–165

Korotcenkov G (2005) Gas response control through structural and chemical modifications of metal oxide films: state of the art and approaches. Sens Actuators B 107:209–232

Korotcenkov G (2007) Practical aspects in design of one-electrode semiconductor gas sensors: status report. Sens Actuators B 121:664–678

Korotcenkov G, Cho BK (2009) Thin film SnO_2-based gas sensors: film thickness influence. Sens Actuators B 142:321–330

Korotcenkov G, Cho BK (2011) Instability of metal oxide-based conductometric gas sensors and approaches to stability improvement. Sens Actuators B 156:527–538

Korotcenkov G, Cho BK (2012) The role of grain size on the thermal instability of nanostructured metal oxides used in gas sensor applications and approaches for grain size stabilization. Prog Cryst Growth Charact Mater 58(4):167–208

Korotcenkov G, Han SD (2009) (Cu, Fe, Co and Ni)-doped SnO_2 films deposited by spray pyrolysis: doping influence on thermal stability of SnO_2 film structure. Mater Chem Phys 113:756–763

Korotcenkov G, Boris I, Brinzari V, Luchkovsky Y, Karkotsky G, Golovanov V, Cornet A, Rossinyol E, Rodriguez J, Cirera A (2004) Gas sensing characteristics of one-electrode gas sensors on the base of doped In_2O_3 ceramics. Sens Actuators B 103:13–22

Korotcenkov G, Nazarov M, Zamoryanskaya MV, Ivanov M, Cirera A, Shimanoe K (2006) Cathodoluminescence study of SnO_2 powders aimed for gas sensor applications. J Mater Sci Eng B 130(1–3):200–205

Korotcenkov G, Brinzari V, Boris I (2008) (Cu, Fe, Co or Ni)-doped SnO_2 films deposited by spray pyrolysis: doping influence on film morphology. J Mater Sci 43(8):2761–2770

Korotcenkov G, Cho BK, Nazarov M, Noh D-Y, Kolesnikova E (2010) Cathodoluminescence studies of undoped and (Cu, Fe, and Co)-doped SnO_2 films deposited by spray pyrolysis deposition. Curr Appl Phys 10:1123–1131

Leite R, Maciel AP, Weber IT, Lisbon-Filho PN, Longo E, Paiva-Santos CO, Andrade AVC, Pakoscimas CA, Maniette Y, Schreiner WH (2002) Development of metal oxide nanoparticles with high stability against particle growth using a metastable solid solution. Adv Mater 14(12):905–908

Liu L, Guo C, Li S, Wang L, Dong Q, Li W (2010) Improved H_2 sensing properties of Co-doped SnO_2 nanofibers. Sens Actuators B 150:806–810

Maciel AP, Lisboa-Filho PN, Leite ER, Paiva-Santos CO, Schreiner WH, Maniette Y, Longo E (2003) Microstructural and morphological analysis of pure and Ce-doped tin dioxide nanoparticles. J Eur Ceram Soc 23:707–713

Madler L, Sahm T, Gurlo A, Grunwaldt J-D, Barsan N, Weimar U, Pratsinis SE (2006) Sensing low concentrations of CO using flame-spray-made Pt/SnO_2 nanoparticles. J Nanopart Res 8:783–796

Madou MJ, Morrison SR (1989) Chemical sensing with solid state devices. Academic, Boston

Matko I, Gaidi M, Hazemann JL, Chenevier B, Labeau M (1999) Electrical properties under polluting gas (CO) of Pt- and Pd-doped polycrystalline SnO_2 thin films: analysis of the metal aggregate size effect. Sens Actuators B 59:210–215

Matko I, Gaidi M, Chenevier B, Charai A, Saikaly W, Labeau M (2002) Pt doping of SnO_2 thin films: a transmission electron microscopy analysis of the porosity evolution. J Electrochem Soc 149(8):H153–H158

Matsushima S, Maekawa T, Tamaki J, Miura N, Yamazoe N (1992) New methods for supporting palladium on a tin oxide gas sensor. Sens. Actuators B 9:71–78

Matsuura Y, Takahata K (1991) Stabilization of SnO_2 sintered gas sensors. Sens Actuators B 5:205–209

Miao H, Ding C, Luo H (2003) Antimony-doped tin dioxide nanometer powders prepared by the hydrothermal method. Microelectron Eng 66:142–146

Miller TA, Bakrania SD, Perez C, Wooldridge MS (2006) Nanostructured tin dioxide materials for gas sensor applications. In: Geckeler KE, Rosenberg E (eds) Functional nanomaterials. American Scientific Publishers, Stevenson Ranch, CA, pp 1–24

Min B-K, Choi S-D (2004) Role of CaO as crystallite growth inhibitor in SnO_2. Sens Actuators B 99:288–296

Nakao Y (1995) Noble metal solid sols in poly(methyl) methacrylate. J Colloid Interface Sci 171:386–391

Nayral C, Viala E, Colliere V, Fau P, Senocq F, Maisonnat A, Chaudret B (2000) Synthesis and use of a novel SnO_2 nanomaterial for gas sensing. Appl Surf Sci 164:219–226

Norris JOW (1987) The role of precious metal catalysts. In: Moseley PT, Tofield BC (eds) Solid state gas sensors. Adam Hinger, Bristol, UK, pp 124–138

Nowotny J (1988) Surface segregation of defects in oxide ceramic materials. Solid State Ionics 28–30:1235–1243

Nowotny J (1991) Interface defect chemistry of oxide ceramic materials. Unresolved problems. Solid State Ionics 49:119–128

Oswald S, Behr G, Dobler D, Werner J, Wetzig K, Arabczyk W (2004) Specific properties of fine SnO_2 powders connected with surface segregation. Anal Bioanal Chem 378:411–415

Papadopoulos CA, Avaritsiotis JN (1995) A model for the gas sensing properties of tin oxide thin films with surface catalysts. Sens Actuators B 28:201–210

Pavelko RG, Vasil'ev AA, Sevast'yanov VG, Gispert-Guirado F, Vilanova X, Kuznetsov NT (2009a) Studies of thermal stability of nanocrystalline SnO_2, ZrO_2 and SiC for semiconductor and thermocatalytic gas sensors. Russ J Electrochem 45(4):470–475

Pavelko RG, Vasil'ev AA, Llobet E, Vilanova X, Barrabés N, Medina F, Sevastyanov VG (2009b) Comparative study of nanocrystalline SnO_2 materials for gas sensor application: thermal stability and catalytic activity. Sens Actuators B 137:637–643

Ramgir NS, Hwang YK, Jhung SH, Mulla IS, Chang JS (2006) Effect of Pt concentration on the physicochemical properties and CO sensing activity of mesostructured SnO_2. Sens Actuators B 114:275–282

Rastomjee CS, Dale RS, Schaffer RJ, Jones FH, Egdell RG, Georgiadis GC, Lee MJ, Tate TJ, Cao LL (1996) An investigation of doping of SnO_2 by ion implantation and application of ion-implanted films as gas sensors. Thin Solid Films 279:98–105

Rosenfeld D, Sanjines R, Schreiner WH, Levy F (1993) Gas sensitive and selective SnO_2 thin polycrystalline films doped by ion implantation. Sens Actuators B 15–16:406–412

Sayago I, Gutierrer J, Ares L, Robla JI, Horrillo MC, Getino J, Rino J, Agapito JA (1995) Long-term reliability of sensors for detection of nitrogen oxides. Sens Actuators B 26:56–58

Sears WM, Love DA (1993) Measurements of the electrical mobility of silver over a hot tin oxide surface. Phys Rev B 47:12972–12975

Sinha AK, Wayne DM, Essex R (1992) Flux growth of pure and doped zircons. J Cryst Growth 125:431–439

Smatko V, Golovanov V, Liu CC, Kiv A, Fuks D, Donchev I, Ivanovskaya M (2010) Structural stability of In_2O_3 films as sensor materials. J Mater Sci Mater Electron 21(4):360–363

Sukharev V, Cavicchi R, Semancik S (1993) Solid state sensors based on semiconducting metal oxides with surface-dispersed metal additives: modeling of the mechanism of response formation. In: Proceedings of the NIST workshop on gas sensors: strategies for future technologies, Gaithersburg, MD, p 88

Sulz G, Kuhner G, Reiter H, Uptmoor G, Schweizer W, Low H, Lacher M, Steiner K (1993) Ni, In and Sb implanted Pt and V catalysed thin-film SnO_2 gas sensors. Sens Actuators B 15–16:390–395

Taurino AM, Epifani M, Toccoli T, Iannotta S, Siciliano P (2003) Innovative aspects in thin film technologies for nanostructured materials in gas sensor devices. Thin Solid Films 436:52–63

Tian ZM, Yuan SL, He JH, Li P, Zhang SQ, Wang CH, Wang YQ, Yin SY, Liu L (2008) Structure and magnetic properties in Mn doped SnO_2 nanoparticles synthesized by chemical co-precipitation method. J Alloys Compd 466:26–30

Tofield BC (1988) Tin dioxide gas sensors. Part 2: The role of surface additives. J Chem Soc Faraday Trans 84:441–457

Tricoli A, Graf M, Pratsinis SE (2008) Optimal doping for enhanced SnO_2 sensitivity and thermal stability. Adv Funct Mater 18:1969–1976

Varela JA, Whittemore OJ, Longo E (1990) Pore size evolution during sintering of ceramic oxides. Ceram Int 16(3):177–189

Vlachos DS, Papadopoulos CA, Avaritsiotis JN (1996) On the electronic interaction between additives and semiconducting oxide gas sensors. Appl Phys Lett 69:650–652

Wang C-T, Lai D-L, Chen M-T (2010) Surface and catalytic properties of doped tin oxide nanoparticles. Appl Surf Sci 257:127–131

Wang C-T, Chen M-T, Lai D-L (2011a) Vanadium-tin oxide nanoparticles with gas-sensing and catalytic activity. J Am Ceram Soc 94(12):4471–4477

Wang C-T, Lai D-L, Chen M-T (2011b) Surface characterization and reactivity of vanadium-tin oxide nanoparticles. Appl Surf Sci 257:5109–5114

Weber LT, Andrade R, Leite ER, Longo E (2001) A study of the $SnO_2{\cdot}Nb_2O_5$ system for an ethanol vapour sensor: a correlation between microstructure and sensor performance. Sens Actuators B 72:180–183

Wynblatt P, Rohrer GS, Papillon F (2003) Grain boundary segregation in oxide ceramics. J Eur Ceram Soc 23:2841–2848

Xu C, Tamaki J, Miura N, Yamazoe N (1992) Stabilization of SnO_2 ultrafine particles by additives. J Mater Sci 27(4):963–971

Yamaura H, Moriya K, Miura N, Yamazoe N (2000) Mechanism of sensitivity promotion in CO sensor using indium oxide and cobalt oxide. Sens Actuators B 65:39–41

Yamazoe N (1991) New approaches for improving semiconductor gas sensors. Sens Actuators B 5:7–19

Yamazoe N, Kurokawa Y, Seiyama T (1983) Effects of additives on semiconductor gas sensors. Sens Actuators 4:283–289

Yang H, Han S, Wang L, Kim IJ, Son YM (1998) Preparation and characterization of indium-doped tin dioxide nanocrystalline powders. Mater Chem Phys 56:153–156

Yang H, Jin W, Wang L (2003) Synthesis and characterization of V_2O_5-doped SnO_2 nanocrystallites for oxygen-sensing properties. Mater Lett 57:3686–3689

Yuasa M, Masaki T, Kida T, Shimanoe K, Yamazoe N (2009) Nano-sized PdO loaded SnO_2 nanoparticles by reverse micelle method for highly sensitive CO gas sensor. Sens Actuators B 136:99–104

Zhang J, Colbow K (1997) Surface silver clusters as oxidation catalysts on semiconductor gas sensors. Sens Actuators B 40:47–52

Zhang G, Liu M (2000) Effect of particle size and dopant on properties of SnO_2-based gas sensors. Sens Actuators B 69:144–152

Zhang J, Liu D, Colbow K (1993) Growth of fractal clusters on thin solid films and scaling of the active zone. Phys Rev B 48:9130–9133

Chapter 24
Bulk and Structure Modification of Polymers

24.1 Modifiers of Polymer Structure

Most polymeric materials are not a single polymer but they contain chemicals that modify some physical and/or chemical behavior. These additives are generally added to modify properties, assist in processing, and introduce new properties to a material (Carraher 2008). Some of these additives are present in minute amounts and others are major amounts of the overall composition.

24.1.1 Solvents (Porogens)

It is known that gas penetrability and specific surface area are two of the most important parameters of polymers intended for application in gas sensors. This means that polymers synthesized should be porous. Experiments have shown that this parameter of polymers is controlled mainly by the solvent (Pichon and Chapuis-Hugon 2008). The solvent serves to bring all the components in the polymerization—the template, the functional monomer(s), the cross-linker, and the initiator—into one phase, and therefore just the physical and chemical characteristics of the solvent determine both the accuracy of the assembly between the template and the monomer and the creation of the pores in polymers (Cormack and Elorza 2004). For this reason, the solvent is commonly referred to as the "porogen." When polymerization occurs, solvent molecules occupy space in the polymer network and create the pores required to allow the diffusion of the template out of the network, and its subsequent diffusion back into the polymer, during recognition. This means that the porosity of the polymers can be determined by the overall concentration of monomers and cross-linkers in the solution. A favorable solvent for molecular imprinting polymers (MIPs) will create well-developed pores within the network and increase the total pore volume. However, a large amount of solvent can ultimately lead to the formation of microspheres and nanospheres instead of a large and stable cross-linked network (Bergmann and Peppas 2008).

Aside from its dual role as a solvent and a pore-forming agent, the solvent in a covalent polymerization must also be judiciously chosen so that it simultaneously maximizes the likelihood of complex formation between the template and the functional monomers. More specifically, the use of a highly thermodynamic solvent tends to result in polymers with well-developed pore structures and high specific surface areas, while the use of a thermodynamically poor solvent leads to polymers with poorly developed pore structures and low specific surface areas. Historically, chloroform has been used as a highly thermodynamic solvent; other solvents that have been investigated include dimethyl

G. Korotcenkov, *Handbook of Gas Sensor Materials*, Integrated Analytical Systems,
DOI 10.1007/978-1-4614-7388-6_24, © Springer Science+Business Media New York 2014

Table 24.1 Solvents typically used for preparing polymer films

Solvent	Polymer
Toluene/ethanol	Ethyl cellulose (EC); poly(vinyl chloride-*co*-isobutyl vinyl ether) (PVC-iBVE); poly(4-*tert*-butyl styrene) (PTBS); polystyrene (PS); poly(vinyl methyl ketone) (PVMK)
Tetrahydrofuran	Poly(tetrafluor ethylene-*co* vinylidenfluorid-*co*-propylene) (PFE-VFP); poly(4-vinyl phenol) (PVPh); poly(vinyl chloride) (PVC)
Chloroform	Poly(styrene-*co*-acrylonitrile) (PSAN); cellulose acetate (CAc); polysulfone (PSu); poly(bisphenol A carbonate) (PC); poly(methyl methacrylate) (PMMA)
Dimethylformamide	Poly(acrylonitrile) (PAN)

sulfoxide, toluene, carbon tetrachloride, *n*-hexane, and benzene (Bergmann and Peppas 2008). Several examples of solvents typically used for preparing polymer films are listed in Table 24.1. It was established that chloroform is a really good solvent for the polymer, while other solvents, for example, carbon tetrachloride and *n*-hexane, are pore solvents (Odian 2004). The inherent viscosity of the polymer increases as the reaction mixture contains a larger proportion of chloroform.

A solvent must be chosen that will not interfere with the template–monomer complex. If the solvent does indeed create an interaction, it could inhibit the formation of the imprinted sites. As a result, many imprinting systems shun polar solvents and instead utilize nonpolar solvents in order to maximize attraction of the template by the functional monomers (Bergmann and Peppas 2008). In this context, water is an especially poor solvent choice for MIPs because of its highly polar nature. Water is both a hydrogen-bond donator and a hydrogen acceptor. Thus, many hydrogen bonds that are formed during covalent imprinting can be destroyed by the sheer amount of water molecules present. In addition, many cross-linking agents that are soluble in water have little structural integrity, thus limiting these materials in extraction applications, such as high-pressure liquid chromatography. However, if hydrophobic forces are being used to drive the complexation, then water could well be the solvent of choice. In this case, polar solvents are preferred because their use helps to stabilize hydrogen bonds. Supercritical CO_2 has been proposed as an alternative porogen for MIP production (Alexander et al. 2006).

Finally, it is necessary to know that the better the reaction medium as a solvent for the polymer, the longer the polymer stays in solution and the larger the polymer molecular weight. With a solvent medium that is a poor solvent for polymer, the molecular weight is limited by precipitation. In addition to the effect of a solvent on the course of a polymerization, the solvent is a poor or good solvent for the polymer; solvents affect polymerization rates and molecular weights due to preferential salvation or other specific interactions with either the reactants or transition state of the reaction or both.

24.1.2 Cross-Linkers

In polymers, the cross-linker fulfils three major functions (Cormack and Elorza 2004). First, the cross-linker is important in controlling the morphology of the polymer matrix, whether it is a gel type, macroporous, or a microgel powder. In particular, from a polymerization point of view, high cross-link ratios are generally preferred to access permanently porous (macroporous) materials. Second, it serves to stabilize the binding sites in imprinted polymers. In other words, cross-links can act to lock in "memory" preventing free-chain movement. It is known that most successful MIP networks involve

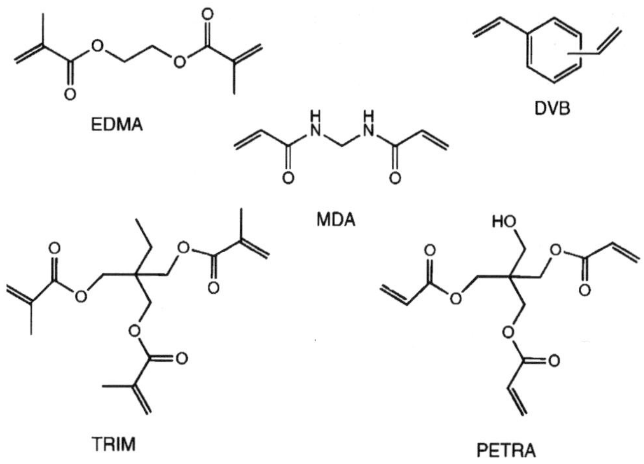

Fig. 24.1 Structure of cross-linkers used for imprinting

hydrogel components, such as polyethylene glycol (PEG) or hydroxyethylmethacrylate (HEMA), that absorb large amounts of water, causing them to swell exponentially in volume. In an MIP system this can lead to further swelling in the template cavities (Pichon and Chapuis-Hugon 2008). Finally, it imparts adequate mechanical stability to the polymer matrix. Cross-linking can also be effected through application of heat, mechanically, through exposure to ionizing and nonionizing (such as microwave) radiation, through exposure to active chemical agents, or through any combination of these. Chemical cross-linking generally renders the material insoluble.

A number of known cross-linkers are commercially available, a few of which are capable of simultaneously complexing with a template and thus acting as functional monomers (see Fig. 24.1). Ethylene glycol dimethacrylate (EDMA) is the most commonly used cross-linker for methacrylate-based systems, primarily because it provides materials with mechanical and thermal stability, good wettability in most rebinding media, and rapid mass transfer with strong recognition properties (Alexander et al. 2006; Sellergren et al. 2009). With the exception of trimethacrylate monomers, such as trimethylolpropane trimethacrylate (TRIM) (Kempe and Mosbach 1995), no other cross-linking monomers provide similar recognition properties for such a large variety of target templates. Divinylbenzene (DVB) has proved to be a superior matrix monomer for some templates, but it is most commonly used in combination with other polymerization formats, such as emulsion, precipitation, or suspension. Polar protic cross-linking monomers, such as methylenediacrylamide (MDA) (Hart and Shea 2001) and pentaerythritoltrimethacrylate (PETRA) (Manesiotis et al. 2005), have been useful for imprinting and applications in more polar solvents.

However it should be noted that the cross-linking agent and its concentration must be very carefully chosen. For example, according to Pichon and Chapuis-Hugon (2008), the choice of cross-linker is important to the imprinting binding sites obtained, since the binding capacity of the polymer increases in relation to the degree of cross-linking. The nature of the interactions developed by the cross-linker must be taken into consideration as well. In addition, the mole ratio of functional monomers to the cross-linker must also be taken into consideration so that the functional cavities are sufficiently spaced to allow the individual binding pockets to swell (Pichon and Chapuis-Hugon 2008). If there is too little cross-linking, the template cavities will be too close to each other, thus creating a larger, less recognitive pore. However, the amount of cross-linking must not be so high as to limit diffusion of the template into the network. Polymers with cross-link ratios in excess of 80 % are often the norm. For the same reason that the reactivity ratios of functional monomers need to be matched in a cocktail polymerization to ensure smooth incorporation of the comonomers, the reactivity ratio of the cross-linker should ideally also be matched to that of the functional monomer(s).

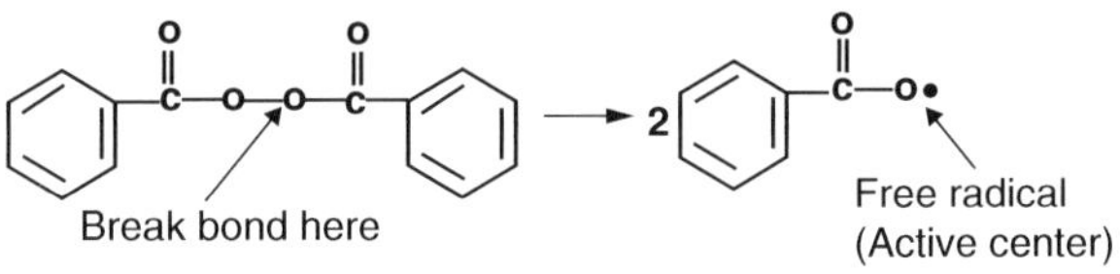

Fig. 24.2 Diagram illustrated the forming of free radical

Break bond here

Free radical (Active center)

We also need to take into account that in some gas sensor applications, where the swelling effect plays a major role, the use of cross-linkers can be limited. Cross-linking increases the strength of the cross-linked material but decreases its flexibility and increases its brittleness. Most chemical cross-linking is not easily reversible.

The application of cross-linker can produce other effects as well. For example, Matsuguchi et al. (2003) analyzed the influence of cross-linker on the characteristics of QCM-based SO_2 sensors and found that sensors with cross-linked structure had lower sorption ability but faster sorption/desorption rates. The latter is known to be very important for sensors designed for in situ measurements. In the case of poly(styrene-*co*-chloromethyl styrene), the use of the cross-linked structure clearly increased the sensitivity to NO_2 of piezoelectric devices (Matsuguchi et al. 2005).

However, this effect was observed only at a certain concentration of cross-linker. Larger concentrations of cross-linkers did not increase sensitivity.

Chen et al. (2006) have shown that cross-linking agent isophorone diisocyanate (IPDI) added to the carbon black (CB)/waterborne polyurethane (WPU) composite latexes can both weaken the unwanted negative vapor coefficient (NVC) effect and improve reproducibility of CB/WPU composites acting as gas-sensing materials in the environment of organic solvent vapors. Moreover, the maximum magnitude of response and response rate of the composites are significantly increased after cross-linking treatment as well. Chen et al. (2006) believe that the mechanism responsible for the sensing behavior of the composites remains unchanged even though the matrix polymer has been cross-linked, while the interaction between CB particles and the matrix polymer is enhanced due to the appearance of the cross-linking structure. As a result, the movement of the fillers in the swollen matrix is localized, which ensures the reversible breakdown and establishment of the conductive networks throughout the composites.

24.1.3 Initiators

Functional initiators find application in conventional radical polymerization for the synthesis of various polymers. The initiator functionality can be considered by reason of the presence of functional end groups, such as hydroxyl and carboxyl, or azo and perester bonds, which undergo dissociation to the alkyl, alkoxy, or acyloxy radicals under the influence of temperature or irradiation and initiate the polymerization (Moad and Solomon 1995). A *free radical* is simply a molecule with an unpaired electron. The tendency for this free radical to gain an additional electron in order to form a pair makes it highly reactive, so that it breaks the bond on another molecule by stealing an electron. As a result, the formation of two additional molecules with an unpaired electron (which are other free radicals) takes place. Free radicals are often created by the division of a molecule (known as an *initiator*) into two fragments along a single bond. The diagram in Fig. 24.2 shows the formation of a radical from its initiator, in this case benzoyl peroxide.

During the activation of the initiator consisting of a heterocyclic, aryl substituted, or aryl-ring fused sulfonium salt, a carbon–sulfur bond is broken via a ring-opening reaction, leading to formation of a sulfide and a carbocation (carbenium ion) within the same molecule. The functionality of initiators is utilized in various ways to achieve a particular purpose. In particular, thermal decomposition of azo initiators such as azo-bis-isobutyronitrile (AIBN) is the most commonly used source of free radicals in the formation of both DVB- and (meth)acrylate-based MIPs. The photochemical decomposition of this compound allows MIPs to be prepared at a low temperature and with a resulting increase in separation efficiency of the polymers (Alexander et al. 2006). Azo-dialkyl peroxides, azo-diacyl peroxides, azo-peresters, azo-hydroperoxides, etc., are other examples of initiators used for radical polymerization (Pabin-Szafko et al. 2005). The first three groups of azo-peroxy compounds can play a role of bifunctional initiators in generation of block copolymers. They can also play a role of traditional initiator in radical polymerization of just one type of monomers. Azo-diacyl peroxides and azo-peresters were tested as initiators in styrene and acrylamide polymerization processes and in preparation of block copolymers from vinyl and acrylic monomers (Czech et al. 2008).

It should be noted that, since the activation does not lead to fragmentation of the initiator molecule into smaller molecules, no molecular sulfur-containing decomposition products form that would otherwise evaporate or migrate from the polymer causing bad smells.

24.1.4 *Plasticizers*

A plasticizer is a material incorporated into a plastic to increase its workability and flexibility. The addition of a plasticizer may lower the melt viscosity, elastic modulus, and glass transition temperature (T_g). Plasticization can occur through addition of an external chemical agent or may be incorporated within the polymer itself (Carraher 2008). Internal plasticization can be produced through copolymerization, giving a more flexible polymer backbone, or by grafting another polymer onto a given polymer backbone. Thus, poly(vinylchloride-*co*-vinyl acetate) is internally plasticized because of the increased flexibility brought about by the change in structure of the polymer chain. The presence of bulky groups on the polymer chain increases segmental motion and placement of such groups through grafting which acts as an internal plasticizer. Internal plasticization achieves its end goal at least in part by discouraging association between polymer chains. External plasticization is achieved through incorporation of a plasticizing agent into a polymer through mixing and/or heating.

Plasticizers used should be relatively nonvolatile, nonmobile, inert, inexpensive, nontoxic, and compatible with the system to be plasticized. They can be divided based on their solvating power and compatibility (Carraher 2008). Primary plasticizers are used as either the sole plasticizer or the major plasticizer with the effect of being compatible with some solvating nature. Secondary plasticizers are materials that are generally blended with a primary plasticizer to improve some performance such as flame or mildew resistance or to reduce cost. The division between primary and secondary plasticizers is at times arbitrary.

The three main chemical groups of plasticizers are phthalate esters, trimellitate esters, and adipate esters. Most plasticizers are classified as general-purpose, performance, or specialty plasticizers (Carraher 2008). General-purpose plasticizers are those that offer good performance inexpensively. Most plasticizers belong to this group. Performance plasticizers offer added performance over general-purpose plasticizers, generally with added cost. Performance plasticizers include fast-solvating materials such as butyl benzyl phthalate and dihexyl phthalate, low-temperature plasticizers such as di-*n*-undecylphthalate and di-2-ethylhexyladipate (DOA), and the so-called permanent plasticizers such as tri-2-ethyl hexyl trimellitate (TOTM), triisononyl trimellitate, and diisodecyl phthalate. The plasticizers tributyl phosphate (TBP), tris(2-ethylhexyl)phosphate (TOP), 2-(octyloxy)benzonitrile (OBN), and 2-nitrophenyl octyl ether (NPOE) are also used to plasticize the polymers intended for gas sensor fabrication (Apostolidis 2004). Specialty plasticizers include materials that provide important properties such as reduced migration, improved stress–strain behavior, flame resistance, and increased stabilization. In all cases, performance is varied through the introduction of different alcohols into the final plasticizer product. There is a balance between compatibility and migration. Generally, the larger the ester group the less the migration, up to a point where compatibility becomes a problem and a limiting factor.

Regarding plasticizer influence on gas-sensing parameters of polymer-based gas sensors, we can say that this effect depends on many factors, including the technological route used, the type of polymer, and the target gas. For example, Apostolidis (2004) found that although the plasticizers did not coercively increase sensitivity of polystyrene-based optical sensors, they enhanced performance due to decrease of response time. As mentioned before, the permeability P of a gas into a polymer is linked to the diffusion coefficient D and the solubility S of a gas in a particular polymer. Thus, increasing the plasticizer content in a polymer matrix usually leads to an increase of the diffusion coefficient of the gas into a particular polymer. As a result, the response time of the sensor material is affected and the response is more rapid than that of a non-plasticized matrix. A positive effect of plasticizer influence was also observed for PVC-based ammonia sensors. The sensitivity of PVC was increased dramatically upon plastification. The addition of plasticizers, e.g., TBP or TOP, enabled an enhancement of oxygen sensitivity of the PTBS-based materials as well. The same effect of sensitivity increase was observed in optical carbon dioxide sensors. However, it should be noted that in other combinations we can have the opposite effect. The resulting matrix, for example, may offer lower permeability than the non-plasticized due to overcompensation of this effect by a reduced solubility of the target gas in the softened matrix. Thus, the observed decrease of sensitivity can be attributed to a decrease of solubility of the target gas with increasing plasticizer content. In particular, it was found that OBN and NPOE at concentrations of 10 wt% in the PTBS and PS matrix caused a decrease in sensitivity of the oxygen sensors. In addition Apostolidis (2004) found that the introduction of plasticizers in polymer matrix increased temperature dependence of CO_2 sensor parameter.

24.2 Approaches to Polymer Functionalizing

As a rule, most polymeric materials have a hydrophobic, chemically inert surface; untreated nonpolar polymer surfaces often have adverse problems in adhesion, coating, painting, coloring, lamination, packaging, colloid stabilization, etc. To solve these problems, an enormous number of basic and applied pieces of research in various fields using different innovative techniques including chemical and physical processes have been devoted to the functionalizing of polymeric materials (Garbassi et al. 1994; Chan 1994; Hoffman 1996; Uyama et al. 1998; Kang and Zhang 2000; Kato et al. 2003).

24.2.1 Polymer Doping

Research carried out has shown that many different approaches can be used for polymer functioning (Skotheim 1986; Nalwa 1997). Doping is the most common process for conducting polymer modification. Examples of doping influence on sensor response of polymer-based gas sensors are shown in Fig. 24.3. The concept of doping is the unique, central, underlying, and unifying theme which distinguishes conducting polymers from all other types of polymers (Chiang et al. 1977; MacDiarmid 2001). During the doping process an organic polymer, either an insulator or semiconductor having low conductivity, typically in the range from 10^{-10} to 10^{-5} S/cm, can be converted to a polymer which is in the "metallic" conducting region (~1–10^4 S/cm). The controlled addition of known, usually small (≤10 %) nonstoichiometric quantities of chemical species results in dramatic changes in the electronic, electrical, magnetic, optical, and structural properties of the polymer.

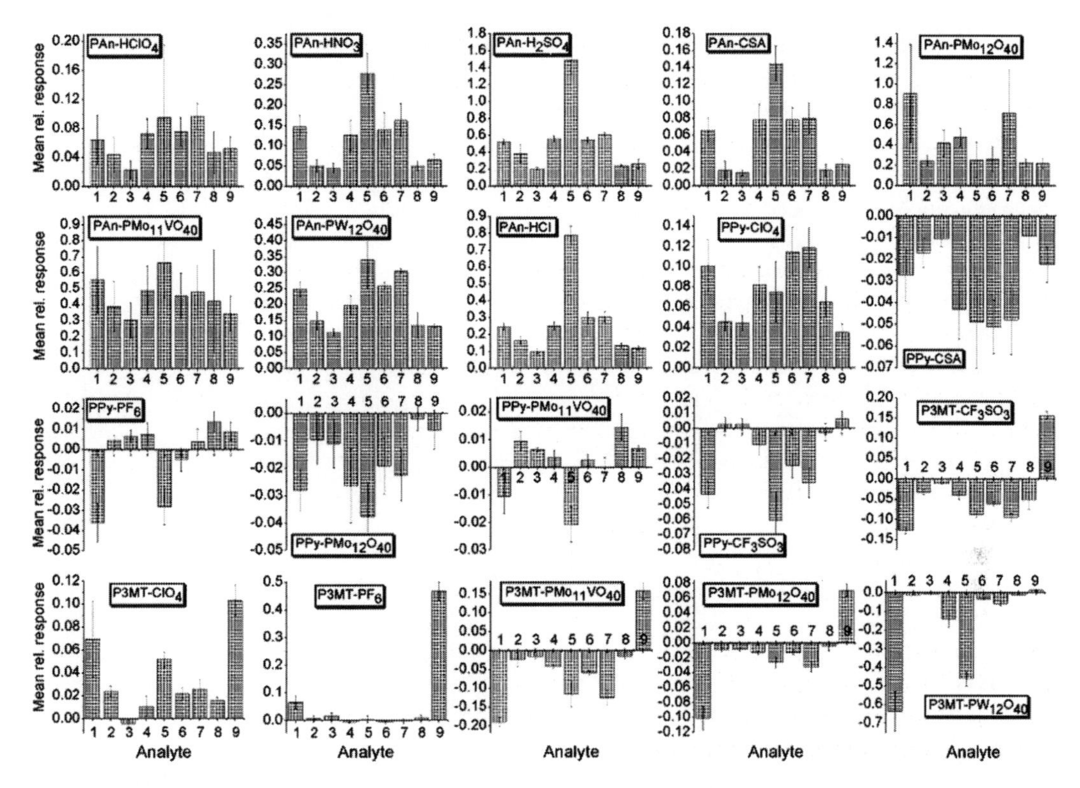

Fig. 24.3 Mean values and standard deviations of responses of 20 investigated polymer-based sensors to saturated vapors of 9 used analytes: (1) acetone, (2) benzene, (3) xylene, (4) amyl alcohol, (5) ethyl alcohol, (6) isobutyl alcohol, (7) isopropyl alcohol, (8) toluene, (9) chloroform. The polymer and doping additives indicate for every sensor (Reprinted with permission from Kukla et al. 2009, Copyright 2009 Elsevier)

In the doped state the backbone of a conducting polymer consists of a delocalized π-system. In the undoped state the polymer may have a conjugated backbone such as in *trans*-$(CH)_x$ which is retained in a modified form after doping, or it may have a non-conjugated backbone, as in polyaniline (leucoemeraldine base form), which becomes truly conjugated only after *p*-doping, or a non-conjugated structure as in the emeraldine base form of polyaniline which becomes conjugated only after protonic acid doping. Doping is reversible to produce the original polymer with little or no degradation of the polymer backbone. Both doping and undoping processes, involving dopant counterions, which stabilize the doped state, may be carried out chemically or electrochemically (Kanatzidis 1990). Transitory doping by methods which introduce no dopant ions is also known (Ziemelis et al. 1991).

In general, doping includes the following processes:

1. *Redox doping* (*ion doping*)
   All conductive polymers can undergo either *p*- or *n*-redox doping by chemical and/or electrochemical processes. Dopants used for selected polymers are listed in Table 24.2. The *p*-doping, i.e., partial oxidation of the π-backbone of an organic polymer, was first discovered by treating *trans*-$(CH)_x$ with an oxidizing agent such as iodine. The *n*-doping, i.e., partial reduction of the backbone π-system of an organic polymer, was also discovered using *trans*-$(CH)_x$ by treating it with a reducing agent (Chiang et al. 1977). These doping processes change the number of electrons associated with the polymer backbone; hence the conductivity of the polymer is also changed. For example, undoped polyacetylene is silvery, insoluble, and intractable, with a

Table 24.2 Conductivities of conductive polymers with selected dopants

Polymer	Doping materials	Conductivity (S/cm)
Polyacetylene	I_2, Br_2, Li, Na, AsF_5	10^4
Polypyrrole	BF_4^-, ClO_4^-, tosylate	$500–7.5\times10^3$
Polythiophene	BF_4^-, ClO_4^-, tosylate, $FeCl_4^-$	10^3
Poly(3-alkylthiophene)	BF_4^-, ClO_4^-, $FeCl_4^-$	$10^3–10^4$
Polyphenylenesulfide	AsF_5	500
Polyphenylene-vinylene	AsF_5	10^4
Polythienylene-vinylene	AsF_5	2.7×10^3
Polyphenylene	AsF_5, Li, K	10^3
Polyisothi-anaphthene	BF_4^-, ClO_4^-	50
Polyazulene	BF_4^-, ClO_4^-	1
Polyfuran	BF_4^-, ClO_4^-	100
Polyaniline	HCl	200

Source: Reprinted with permission from Kumar and Sharma (1998). Copyright 1998 Elsevier

conductivity similar to that of semiconductors. However, when it was weakly oxidized by compounds such as iodine, it turned a golden color and its conductivity increased to about 10^4 S/m. The same situation takes place with other conjugated polymers.

Doping of polymers with the aim of electronic device fabrication has been an important research area since the 1980s. Research has shown that typical oxidizing dopants, which can be used for polymer modification, include chlorine, bromine, arsenic pentachloride, iron(III) chloride, and $NOPF_6$, in addition to iodine. Many experiments have been done using halogen dopants (Cl, Br, I) because polymer doping with these elements can be very easily achieved by simple chemical processes. A typical reductive dopant is liquid sodium amalgam or preferably sodium naphthalide. The main criterion for application of the above-mentioned dopants is their ability to oxidize or reduce the polymer without lowering its stability or whether or not they are capable of initiating side reactions that inhibit the polymers' ability to conduct electricity. An example of the latter is the doping of a conjugated polymer with bromine and chlorine. It has been shown that if the high-electron affinity of Cl and Br allows one to obtain a stable charge transfer complex (CT-complex) between these halogens and polymers, in return there is often a progressive attack of the polymer backbone by the halogen (Safoula et al. 1999, 2001). For example, bromine is too powerful an oxidant and adds across the double bonds to form sp3 carbons. As a result the chlorine and bromine doping induce partial degradation of the polymer. The same problem may also occur with $NOPF_6$ if left too long. Therefore, it appears that iodine doping, even if it is less stable, is more promising because it does not induce any polymer degradation. After iodine doping, only CT-complex formation takes place (Napo et al. 1999; Safoula et al. 2001).

2. *Photo and charge-injection doping*

   When ICPs are exposed to radiation whose energy is greater than the bandgap energy of the polymer, electrons hop across the gap and the polymer undergoes photo doping. However, we need to take into account that charge carriers disappear rapidly due to recombination of electrons and holes when irradiation is discontinued. Charge-injection doping is most conveniently carried out using a metal/insulator/semiconductor configuration involving a metal and a conducting polymer separated by a thin layer of a high-dielectric strength insulator.

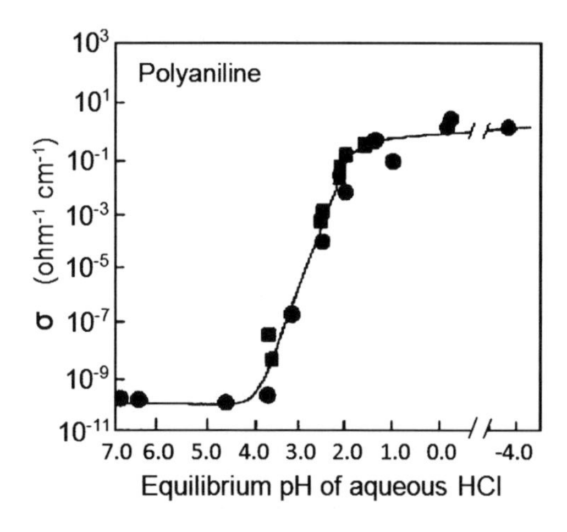

Fig. 24.4 Conductivity of emeraldine base as a function of pH of HCl dopant solution as it undergoes protonic acid doping (Reprinted with permission from MacDiarmid 2001, Copyright 2001 Elsevier)

3. *Non-redox doping*
   In this process, the number of electrons associated with the polymer backbone does not change during the doping process, but instead the energy levels of these electrons are rearranged during doping. An example of this occurs when an ICP such as emeraldine, base form of polyaniline, becomes highly conductive by immersing the polymer in an acid. Emeraldine base interacts with aqueous protonic acids to produce the protonated emeraldine which is highly conductive (Chiang and Macdiarmid 1986; Macdiarmid et al. 1987). This process can increase the conductivity of the polymer by several orders of magnitude. An example of such influence is shown in Fig. 24.4.

24.2.2 Polymer Grafting

Grafting is another approach used for polymer functioning (Uchida and Ikada 1996; Ranby 1999; Kang and Zhang 2000; Zhao and Brittain 2000). Among the surface modification techniques developed to date, surface grafting has emerged as a simple, useful, and versatile approach to improve surface properties of polymers for many applications. According to Gopal et al. (2007), the grafting has the following advantages: (1) the ability to modify the polymer surface to have distinct properties through the choice of different monomers; (2) the controllable introduction of graft chains with a high density and exact localization to the surface, without affecting the bulk properties; and (3) long-term chemical stability, which is assured by covalent attachment of graft chains (Gopal et al. 2007). The latter factor contrasts with physically coated polymer chains that can in principle be removed rather easily. Experiment has shown that surface grafting provides versatile techniques for introducing functional groups such as amine, imine, hydroxyl, carboxylic acid, sulfonate, and epoxide onto a broad range of conventional polymeric substrates, most of which have a nonpolar, less reactive surface (see Table 24.3). Hydroxyl, amine, carboxyl, and sulfone groups are known as hydrophilic functional groups (Van der Bruggen 2009). Usually functional groups are localized on the side chains. A typical polymer consists of a backbone or main chain, which is a long, repeating chain of smaller units called monomers, and side chains. A side chain is simply a relatively short branch of the polymer molecule, usually several atoms or groups of atoms, that are connected to the polymer backbone. There may be a few or many. Sometimes even the branches (side chains) have branches (side chains). The presence of these side chains can affect the physical properties of a polymer. For example, high-density polyethylene, with its near absence of side chains, is harder and more abrasion resistant and will withstand

Table 24.3 Functional groups used for functionalizing of polymer used in gas sensors

Backbone	Side chains
PPy	S: alkyl; S: alkoxy; S: hydroxyalkyl; S: carboxyalkyl; S: alkyl sulfonic acid; S: amine; S: ester group
PAni	S: alkoxy; S: sulfonic acid; S: phenyl; S: boronate
PTh	S: alkyl; S: alkoxy; S: ester group; S: alkthio; S: carboxyl alkyl
PEdot	S: alkoxy; S: ether group
PA	S: amine
PEB	S: alkoxy

Source: Reprinted from Bai and Shi (2007). Published by MDPI
PEB poly(diethylyl benzene)

higher temperatures, compared to low-density polyethylene that has numerous molecular branches or side chains. The functional groups introduced with the help of side chains can be utilized to react further with small or large molecules through covalent or non-covalent linkage. Functionalization is achieved by either direct grafting of functional monomer or postderivatization of graft chains.

The introduction of grafts to the backbones of conducting polymers has two effects (Bai and Shi 2007). First, most of the side chains are able to increase the solubility of conducting polymers. Because of this they can be processed into the sensing film by LB technology, spin-coating, ink-printing, or other solution-assistant method. Second, some functional chains can adjust the physical and chemical properties of conducting polymers, including the space between molecules (Li et al. 2006) or dipole moments (Torsi et al. 2003), or bring additional interactions with analytes, which may enhance the response, shorten the response time, or produce new sensitivity to other gases (Ruangchuay et al. 2004).

The techniques to initiate grafting are (1) chemical, (2) photochemical and/or via high-energy radiation, (3) the use of a plasma, and (4) enzymatic (Nady et al. 2001; Kato et al. 2003; Van der Bruggen 2009). The choice of a specific grafting technique depends on the chemical structure of the polymer and the desired characteristics after surface modification. For example, modification of polymer surfaces can be rapidly and cleanly achieved by plasma treatment due to the possibility of the formation of various active species on the surface of polyethylene (PE), polypropylene (PP), polytetrafluoroethylene (PTFE), etc. By variation of plasma treatment parameters, surfaces with different properties can be obtained. Possible gases include CF_4, Ar, O_2, H_2, He, Ne, N_2, and CO_2, in addition to H_2O. The surface is bombarded with ionized plasma components to generate radical sites. Bonds that can be attacked by radicals are C–C, C–H, and C–S bonds, with exclusion of the aromatic C–H and C–C bonds. This is similar to photodegradation. The generated radicals can subsequently react with gas molecules (depending on the plasma), schematically shown for Ar in Fig. 24.5. Remaining radical sites bind with oxygen or nitrogen after contact with the air.

Van der Bruggen (2009), analyzing chemical modification of polymer membranes, have noted that CO_2-plasma treatment leads to the incorporation of oxygen in the membrane surface in the form of carbonyl, acid, and ester groups, yielding an increase in hydrophilicity. Modification of polymer surfaces with CO_2 plasmas in general leads to surface oxidation and the formation of hydrophilic surfaces. A fast reaction was observed; the treated membranes had a better fouling resistance. Continued plasma treatment, however, resulted in membrane degradation. H_2O plasma treatment also leads to the incorporation of oxygen containing functional groups on the surface. O_2 plasmas have a similar effect, with reported functional groups mainly being hydroxyl, carbonyl, and carboxyl groups. Nitrogen-containing plasma systems, on the other hand, yield amine, imine, amide, and nitrile

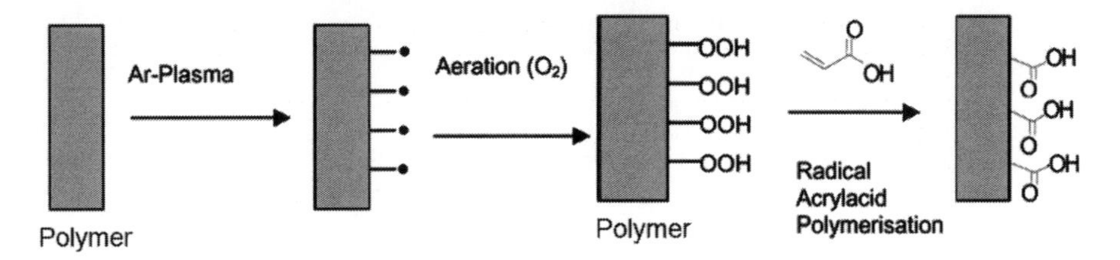

Fig. 24.5 Schematic representation of polyacrylic acid (PAAc) graft modification by Ar plasma treatment (Reprinted with permission from Tyszler et al. 2006, Copyright 2006 Elsevier)

functional groups on the membrane surface. Through post reactions after contact with air, oxygen compounds can also be present.

However, plasma treatment is generally slow and expensive in fiber applications. Plasma treatment also results in deposition of a macromolecular structure, graft polymerization, etching, roughening, and cross-linking. Besides surface modifications, plasma treatment under certain conditions could favor an increase in film crystallinity and even new crystallization occurring with a concentration gradient depending on the film thickness. The effects of plasma treatment on the surface of polymeric materials were discussed in detail, comparing the energies of plasma with those of the chemical bonds in polymers (Nakamatsu and Delgado-Aparicio 1997). Chemical modification makes it possible to introduce various functional groups on the polymer surface using various reactions such as sulfonation, chloromethylation, aminomethylation, and lithiation (Guiver and Apsimon 1988; Michael et al. 1989; Breitbach et al. 1991). The main challenge for modification by chemical treatment of commercial polymer membranes is that the modification agent may partly block the pores of the membranes. Even if the modified membranes are less prone to fouling, the total flux after modification is generally smaller than before modification (Anke Nabe et al. 1997).

We would like to note that the possibility of keeping the bulk properties of polymers without changes during their modification is a good feature of the grafting. The performance of polymeric materials in many applications, especially in gas sensors, relies largely upon the combination of bulk (e.g., mechanical) properties in combination with the properties of their surfaces. However, polymers very often do not possess the surface properties needed for these applications. Vice versa, those polymers that have good surface properties frequently do not possess the mechanical properties that are critical for their successful application. Due to this dilemma, (surface) modification of polymers without changing the bulk properties has been a topical aim in research for many years, mostly because surface modification provides a potentially easier route than, e.g., polymer blending, to obtain new polymer properties important for gas sensor design (Kato et al. 2003; Gopal et al. 2007).

24.2.3 Role of Polymer Functionalization in the Gas-Sensing Effect

It must be stated that the above-mentioned studies related to polymer surface modification have, in the main, not been focused on the design of gas-sensing material. As a result, we did not find any comprehensive analysis and summarizing of doping or surface modification influences on operating characteristics of polymer-based gas sensors. We have only individual papers related to this topic of research, which indicate that conducting polymers doped with different ions may give distinct responses to a specific analyte (Bai and Shi 2007). In addition these papers testify that the great importance for gas-sensing characteristics have the nature of dopants (Brie et al. 1996; Van and Potje-Kamloth 2001), the molecular sizes of dopants (de Souza et al. 2001), and the doping levels

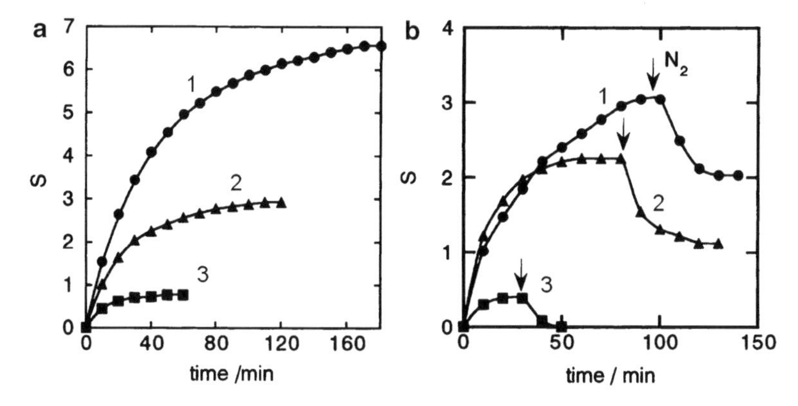

Fig. 24.6 Response characteristics of the sensors using amino-functional copolymers (CMS mole fraction = 0.210 (**a**) and 0.365 (**b**)) measured for 50 ppm of SO_2 at 30 °C (**a**) and 50 °C(**b**); (1) DMEDA, (2) DMPDA, and (3) DPEDA (Reprinted with permission from Matsuguchi et al. 2001, Copyright 2001 Elsevier)

(Kawai et al. 1998; Nicho et al. 2001; Ruangchuay et al. 2004). In particular, Fig. 24.3 illustrates the influence of doping (grafting) on gas-sensing characteristics of polymer-based conductometric gas sensors designed by Kukla et al. (2009). It is seen that the doping (grafting) is a really powerful instrument for influencing the operating parameters of gas sensors. For example, Chabukswar et al. (2001) have shown that PAni doped with small inorganic ions showed an increased resistance to ammonia, while acrylic acid-doped PAni exhibited the opposite response. Jain et al. (2003) have found that camphorsulfonic acid (CSA)-doped PAni shows the best response comparing with those doped with diphenyl phosphate (DPPH) and maleic acid (Mac) when detecting water vapor. Hong et al. (2004) reported that strong acid dopants resulted in better reversibility of a PAni-based chemiresistor although a worse response. Xu et al. (2006) compared PAni films doped with ClO_4^- and dodecyl benzene sulphate and found that the latter PAni had a higher response to ammonia gas. A great difference in sensing performances between Cl^-, SO_4^{2-}, and NO_3^- doped PPy composites was also observed by Guernion et al. (2002). Matsuguchi et al. (2001) have shown that using different kinds of diamine compounds, *N,N*-dimethylethylenediamine (DMEDA), *N,N*-dimethylpropanediamine (DMPDA), and *N,N*-dimethyl-*p*-phenylenediamine (DPEDA), for polymer functionalizing, one can control parameters of the QCM-type SO_2 styrene (St)-*co*-chloromethyl styrene (CMS) copolymer-based gas sensors. They found that the response time was the shortest for a sensor using the DPEDA functional copolymer, though the sensitivity was the smallest at the same measuring temperature. These research results are shown in Fig. 24.6. Table 24.4, presented in a review by Bai and Shi (2007), summarizes modified polymers designed for gas sensor applications.

Analyzing the absence of any summarizing in the field of doping and surface modification influence on operating characteristics of polymer-based gas sensors, one can assume it is quite possible that this situation takes place because results obtained are specific for every experiment. It was established that consequences of surface modification are strongly dependent on the polymer used and the modification method selected for surface functioning. For example, Brie et al. (1996) studied PPy-based NH_3 sensors doped by ClO_4^- and *p*-toluenesulfonate (TsO^-) and found that high conductivity of doped PPy and high relative response ($\Delta R/R_0$) resulted in a low initial resistance R_0. However, Anitha and Subramanian (2003) reported an opposite trend in PAni-based chemiresistor during organic solvents detection; the relative change in conductance decreased with the increase of original conductance.

We also assume that addition surface modification and doping strongly affect temporal stability of functionalized polymer surface (see Chap. 19 (Vol. 2)). For example, de Leeuw et al. (1997) reported that many polymer-based devices were not designed due to the lack of environmentally stable *n*-type-doped

Table 24.4 Modified conducting polymers designed for gas sensing

Backbone	Side chain
Polypyrrole (PPy)	S: alkyl; S: alkoxy; S: hydroxyalkyl; S: carboxyalkyl
	S: alkyl sulfonic acid; S: amine; S: ester group
	G: to PVA; C: with thiophene
Polyaniline (PAni)	S: alkoxy; S: sulfonic acid; S: phenyl
	S: boronate; G: to SWNT
Polythiophene (PTh)	S: alkyl; S: alkoxy; S: ester group
	S: alkthio; S: carboxyl alkyl
	C: poly(3-octylthiophene-*co*-thienylethanol)
	C: with PS; C: with PMA; C: with PBA
Poly (3,4-thylenedioxythiophene) (PEdot)	S: alkoxy; S: ether group
Polyacetylene (PA)	S: amine
Poly(diethylyl benzene) (PEB)	S: alkoxy

Source: Data from Bai and Shi (2007)
S side chain, *C* copolymer, *G* graft

conducting polymers. Research has shown that some heavily *p*-type-doped polymers are stable under ambient conditions, e.g., polypyrrole and polyaniline. Conjugated materials that are stable as an undoped or very slightly doped *p*-type-doped material are also known, e.g., polythiophene. To date, no *n*-type-doped conjugated polymers having similar stability are known. The difficulty in arriving at such a polymer is related to the well-known instability of organic anions, especially of carbanions. Typically, these ions are easily oxidized in contact with air or water. This means that design and synthesis of an *n*-type-conducting polymer with high stability is not trivial. As we know, temporal stability is one of the most important problems in the field of gas sensor design. In addition, we need to take into account that, in general, the doping of polymers is a reversible process, and therefore, for charge stabilization on the polymer backbone, counter "dopant" ions should be introduced in a polymer during all chemical and electrochemical *p*- and *n*-doping processes (MacDiarmid 2001). Taking into account the present situation, we will limit our discussion of approaches which can be used for functionalizing of gas-sensing polymer by information presented earlier.

We need to recognize that the above-mentioned conclusion about the absence of any summarizing of doping and surface functionalizing influence on properties of gas-sensing polymers is not related to polymer-based membranes intended for electrochemical gas sensors. In membrane manufacturing, surface functionalization of preformed membranes has already become a key technology (Nady et al. 2001). Usually the aims of surface modification of a membrane are the following: (1) minimization of undesired interactions (adsorption or adhesion, or in more general terms membrane fouling) that reduce the performance as described previously and (2) improvement of the selectivity or even the formation of entirely novel separation functions (Ulbricht 2006).

An overall comparison between the different methods used for surface modification of polymer membranes is presented in Table 24.5, which was prepared by Nady et al. (2001). Please note that it is not always straightforward to interpret and compare results, because many parameters may be influenced simultaneously by one modification method. Thus, Table 24.5 gives a general impression only.

According to Nady et al. (2001), all the surface modification methods mentioned earlier allow modification without affecting the bulk properties too much when appropriate conditions are selected; mostly the flux is similar to the base membrane or slightly lower. Complete and seemingly permanent hydrophilic modification of poly(arylsulfone) membranes is achieved by blending and photoinduced grafting. A few studies showed that chemical redox initiation grafting could be successfully applied to PES ultrafiltration membranes (Belfer et al. 2000: Reddy et al. 2005). However, chemical treatment

Table 24.5 Advantages and disadvantages of modification methods applied to polymer membranes

Modification method	Flux after modification	Simplicity/ versatility	Reproducibility	Environmental aspects	Cost-effectiveness
Coating	L	H	H	H	H
Blending	H	E	H	H	H
Composite	H	H	L	H	H
Chemical	L	H	L	L	H
Grafting initiated by					
Chemical	H	H	H	L	H
Photochemical	H	H	H	H	H
Radiation	H	L	L	L	L
Plasma	H	L	L	H	L
Combined methods	L	L	H	L	L

Source: Reprinted with permission from Nady et al. (2001). Copyright 2001 Elsevier
L low, *H* high, *E* excellent

usually employs harsh treatment; it may often lead to undesirable surface changes and contamination and may not be the best choice in environmental terms. Plasma treatment is probably one of the most versatile poly(arylsulfone) membrane surface treatment techniques. For example, simple inert gas, nitrogen, or oxygen plasmas have been used to increase the surface hydrophilicity of membranes, and water plasma treatments have successfully rendered asymmetric PSF membranes permanently hydrophilic. However, high costs and technical complexity remain drawbacks for large-scale use. Combination of two or three modification techniques is complex in terms of cost-effectiveness and environmental drawbacks but could lead to multifunctional membranes that are of great interest for "membranes of the future." Such membranes may need more functions than "only" providing a selective barrier with high performance (flux and stability). To be complete, it should be noted that all mentioned methods influence membrane smoothness/roughness (Rana and Matsuura 2010).

Of course, the pretreatment of polymer-sensing film may also affect the performance of the gas sensors. For example, Jun et al. (2003) have found that soaking PPy in methanol solution during sensor preparing can shorten the response time to methanol vapors, which could be interpreted as the removal of excess counterions. The pretreatment of the electrodes' surface is sometimes able to optimize the contact resistance as well as the sensitivity. On a rough Au electrode surface, the electrodeposited PPy adherence is much better than on a smooth Au electrode (Cui and Martin 2003). By chemical modification of the surface, the PAni could be grafted onto Si substrate with good adherence (Chen et al. 2001). These technologies are useful in the fabrication of electrochemical sensors. Heating, as shown by Geng et al. (2006), is another way to change the sensitivity of sensors.

References

Alexander C, Andersson HS, Andersson LI, Ansell RJ, Kirsch N, Nicholls IA, O'Mahony J, Whitcombe MJ (2006) Molecular imprinting science and technology: a survey of the literature for the years up to and including 2003. J Mol Recognit 19:106–180

Anitha G, Subramanian E (2003) Dopant induced specificity in sensor behaviour of conducting polyaniline materials with organic solvents. Sens Actuators B Chem 92:49–59

Anke Nabe A, Staude E, Belfort G (1997) Surface modification of polysulfone ultrafiltration membranes and fouling by BSA solutions. J Membr Sci 133:57–72

Apostolidis A (2004) Combinatorial approach for development of optical gas sensors: concept and application of high-throughput experimentation. PhD thesis, University of Regensburg, Regensburg
Bai H, Shi G (2007) Gas sensors based on conducting polymers. Sensors 7:267–307
Belfer S, Fainchtain R, Purinson Y, Kedem O (2000) Surface characterization by FTIR-ATR spectroscopy of polyethersulfone membranes - unmodified, modified and protein fouled. J Membr Sci 172:113–124
Bergmann NM, Peppas NA (2008) Molecularly imprinted polymers with specific recognition for macromolecules and proteins. Prog Polym Sci 33:271–288
Breitbach L, Hinke E, Staude E (1991) Heterogeneous functionalizing of polysulfone membranes. Angew Makromol Chem 184:183–196
Brie M, Turcu R, Neamtu C, Pruneanu S (1996) The effect of initial conductivity and doping anions on gas sensitivity of conducting polypyrrole films to NH_3. Sens Actuators B Chem 37:119–122
Carraher CE Jr (2008) Polymer chemistry. Taylor and Francis Group, Boca Raton, FL
Chabukswar VV, Pethkar S, Athawale AA (2001) Acrylic acid doped polyaniline as an ammonia sensor. Sens Actuators B Chem 77:657–663
Chan C-M (1994) Polymer surface modification and characterization. Hanser, Munich
Chen YJ, Kang ET, Neoh KG, Tan KL (2001) Oxidative graft polymerization of aniline on modified Si(100) surface. Macromolecules 34:3133–3141
Chen SG, Hu JW, Zhang MQ, Rong MZ, Zheng Q (2006) Improvement of gas sensing performance of carbon black/waterborne polyurethane composites: effect of crosslinking treatment. Sens Actuators B Chem 113:361–369
Chiang JC, Macdiarmid AG (1986) 'Polyaniline': protonic acid doping of the emeraldine from to the metallic regime. Synth Met 13:193–205
Chiang CK, Fincher CR Jr, Park YW, Heeger AJ, Shirakawa H, Louis EJ, MacDiarmid AG (1977) Electrical conductivity in doped polyacetylene. Phys Rev Lett 39:1098–1101
Cormack PAG, Elorza AZ (2004) Molecularly imprinted polymers: synthesis and characterisation. J Chromatogr B 804:173–182
Cui XY, Martin DC (2003) Fuzzy gold electrodes for lowering impedance and improving adhesion with electrodeposited conducting polymer films. Sens Actuators A Phys 103:384–394
Czech Z, Butwin A, Herko E, Hefczyc B, Zawadiak J (2008) Novel azo-peresters radical initiators used for the synthesis of acrylic pressure-sensitive adhesives. eXPRESS Polym Lett 2(4):277–283
De Leeuw DM, Simenon MMJ, Brown AR, Einerhand REE (1997) Stability of *n*-type doped conducting polymers and consequences for polymeric microelectronic devices. Synth Met 87:53–59
De Souza JEG, dos Santos FL, Barros-Neto B, dos Santos CG, de Melo CP (2001) Polypyrrole thin films gas sensors. Synth Met 119:383–384
Garbassi F, Morra M, Occhiello E (1994) Polymer surfaces. Wiley, Chichester
Geng L, Wang SR, Zhao YQ, Li P, Zhang SM, Huang WP, Wu SH (2006) Study of the primary sensitivity of polypyrrole/r-Fe_2O_3 to toxic gases. Mater Chem Phys 99:15–19
Gopal R, Zuwei M, Kaur S, Ramakrishna S (2007) Surface modification and application of functionalized polymer nanofibers. In: Mansoori GA, George TF, Assoufid L, Zhang G (eds) Molecular building blocks for nanotechnology: from diamondoids to nanoscale materials and applications. Springer, Berlin, pp 72–91
Guernion N, Costello BPJD, Ratcliffe NM (2002) The synthesis of 3-octadecyl- and 3-docosylpyrrole, their polymerisation and incorporation into novel composite gas sensitive resistors. Synth Met 128:139–147
Guiver MD, Apsimon JW (1988) The modification of polysulfone by metalation. J Polym Sci C Polym Lett 26:123–127
Hart BR, Shea KJ (2001) Synthetic peptide receptors: molecularly imprinted polymers for the recognition of peptides using peptide-metal interactions. J Am Chem Soc 123:2072–2073
Hoffman AS (1996) Surface modification of polymers: physical, chemical, mechanical and biological methods. Macromol Symp 101:443–454
Hong KH, Oh KW, Kang TJ (2004) Polyaniline-nylon 6 composite fabric for ammonia gas sensor. J Appl Polym Sci 92:37–42
Jain S, Chakane S, Samui AB, Krishnamurthy VN, Bhoraskar SV (2003) Humidity sensing with weak acid-doped polyaniline and its composites. Sens Actuators B Chem 96:124–129
Jun HK, Hoh YS, Lee BS, Lee ST, Lim JO, Lee DD, Huh JS (2003) Electrical properties of polypyrrole gas sensors fabricated under various pretreatment conditions. Sens Actuators B Chem 96:576–581
Kanatzidis MG (1990) Polymeric electrical conductors. Chem Eng News 1990(December):36–54
Kang ET, Zhang Y (2000) Surface modification of fluoropolymers via molecular design. Adv Mater 12:1481–1494
Kato K, Uchida E, Kang E-T, Uyama Y, Ikada Y (2003) Polymer surface with graft chains. Prog Polym Sci 28:209–259
Kawai T, Kojima S, Tanaka F, Yoshino K (1998) Electrical property of poly(3-octyloxythiophene) and its gas sensor application. Jpn J Appl Phys Pt 1 37:6237–6241

Kempe M, Mosbach K (1995) Receptor binding mimetics: a novel molecularly imprinted polymer receptor. Tetrahedron Lett 36:3563–3566

Kukla AL, Pavluchenko AS, Shirshov YM, Konoshchuk NV, Posudievsky OY (2009) Application of sensor arrays based on thin films of conducting polymers for chemical recognition of volatile organic solvents. Sens Actuators B Chem 135:541–551

Kumar D, Sharma RC (1998) Advances in conductive polymers. Eur Polym J 34(8):1053–1060

Li B, Sauve G, Iovu MC, Jeffries-El M, Zhang R, Cooper J, Santhanam S, Schultz L, Revelli JC, Kusne AG, Kowalewski T, Snyder JL, Weiss LE, Fedder GK, McCullough RD, Lambeth DN (2006) Volatile organic compound detection using nanostructured copolymers. Nano Lett 6:1598–1602

MacDiarmid AG (2001) "Synthetic metals": a novel role for organic polymers. Curr Appl Phys 1:269–279

Macdiarmid AG, Chiang JC, Richter AF, Epstein AJ (1987) Polyaniline: a new concept in conducting polymers. Synth Met 18:285–290

Manesiotis P, Hall AJ, Courtois J, Irgum K, Sellergren B (2005) An artificial riboflavin receptor prepared by a template analogue imprinting strategy. Angew Chem Int Ed 44:3902–3906

Matsuguchi M, Tarnai K, Sakai Y (2001) SO_2 gas sensors using polymers with different amino groups. Sens Actuators B Chem 77:363–367

Matsuguchi M, Sakurada K, Sakai Y (2003) Effect of crosslinked structure on SO_2 gas sorption properties in amino-functional copolymers. J Appl Polymer Sci 88:2982–2987

Matsuguchi M, Kadowaki Y, Tanaka M (2005) A QCM-based NO_2 gas detector using morpholine-functional cross-linked copolymer coatings. Sens Actuators B Chem 108:572–575

Michael D, Guiver MD, Kutowy O, Apsimon JW (1989) Functional group polysulphones by bromination–metalation. Polymer 30:1137–1142

Moad G, Solomon DH (1995) The chemistry of free radical polymerization. Elsevier Science, Oxford

Nady N, Franssen MCR, Zuilhof H, Mohy Eldin MS, Boom R, Schroën K (2001) Modification methods for poly(arylsulfone) membranes: a mini-review focusing on surface modification. Desalination 275:1–9

Nakamatsu KJ, Delgado-Aparicio VLF (1997) Modificacion de superficies de polymeros con plasma. Rev Plasticos Mod 74:262–268

Nalwa HS (1997) Handbook of organic conductive materials and polymers. Wiley, New York, NY

Napo K, Safoula G, Bernede JC, D'Almeida K, Touirhi S, Alimi K, Barreau A (1999) Influence of the iodine doping process on the properties of organic and inorganic polymer thin films. Polym Degrad Stab 66:257–262

Nicho ME, Trejo M, Garcia-Valenzuela A, Saniger JM, Palacios J, Hu H (2001) Polyaniline composite coatings interrogated by a nulling optical-transmittance bridge for sensing low concentrations of ammonia gas. Sens Actuators B Chem 76:18–24

Odian G (2004) Principles of polymerization. Wiley Interscience, Hoboken, NJ

Pabin-Szafko B, Wisniewska E, Szafko J (2005) Functional azo-initiators—synthesis and molecular characteristics. Polimery 50:271–278

Pichon V, Chapuis-Hugon F (2008) Role of molecularly imprinted polymers for selective determination of environmental pollutants—a review. Anal Chim Acta 622:48–61

Rana D, Matsuura T (2010) Surface modifications for antifouling membranes. Chem Rev 110:2448–2471

Ranby B (1999) Surface modification and lamination of polymers by photografting. Int J Adh Adhes 19:337–343

Reddy AVR, Trivedi JJ, Devmurari CV, Mohan DJ, Singh P, Rao AP, Joshi SV, Ghosh PK (2005) Fouling resistant membranes in desalination and water recovery. Desalination 183:301–306

Ruangchuay L, Sirivat A, Schwank J (2004) Electrical conductivity response of polypyrrole to acetone vapor: effect of dopant anions and interaction mechanisms. Synth Met 140:15–21

Safoula G, Touihri S, Bernede JC, Jamali M, Rabiller C, Molinie P, Napo K (1999) Properties of the complex salt obtained by doping the poly(N-vinylcarbazole) with bromine. Polymer 40:531–539

Safoula G, Napo K, Bernede JC, Touihri S, Alimi K (2001) Electrical conductivity of halogen doped poly-(N-vinylcarbazole) thin films. Eur Polym J 37:843–849

Sellergren B, Schillinger E, Lanza F (2009) Experimental combinatorial methods in molecular imprinting. In: Potyrailo RA, Mirsky VM (eds) Combinatorial methods for chemical and biological sensors. Springer Science+Business Media, New York, NY, pp 173–200

Skotheim TA (1986) Handbook of conducting polymers, vols 1 and 2. Marcel Dekker, New York, NY

Torsi L, Tanese MC, Cioffi N, Gallazzi MC, Sabbatini L, Zambonin PG, Raos G, Meille SV, Giangregorio MM (2003) Side-chain role in chemically sensing conducting polymer field effect transistors. J Phys Chem B 107:7589–7594

Tyszler D, Zytner RG, Batsch A, Brugger A, Geissler S, Zhou H, Klee D, Melin T (2006) Reduced fouling tendencies of ultrafiltration membranes in wastewater treatment by plasma modification. Desalination 189:119–129

Uchida E, Ikada Y (1996) Surface modification of polymers by UV-induced graft polymerization. Curr Trends Polym Sci 1:135–146

Ulbricht U (2006) Advanced functional polymer membranes. Polymer 47:2217–2262

Uyama Y, Kato K, Ikada Y (1998) Surface modification of polymers by grafting. Adv Polym Sci 137:1–39

Van der Bruggen B (2009) Chemical modification of polyethersulfone nanofiltration membranes: a review. J Appl Polymer Sci 114:630–642

Van CN, Potje-Kamloth K (2001) Electrical and NO_x gas sensing properties of metallophthalocyanine-doped polypyrrole/silicon heterojunctions. Thin Solid Films 392:113–121

Xu K, Zhu LH, Li J, Tang HQ (2006) Effects of dopants on percolation behaviors and gas sensing characteristics of polyaniline film. Electrochim Acta 52:723–727

Zhao B, Brittain WJ (2000) Polymer brushes: surface-immobilized macromolecules. Prog Polym Sci 25:677–710

Ziemelis KE, Hussain AT, Bradley DDC, Friend RH, Rilhe J, Wegner G (1991) Optical spectroscopy of field-induced charge in poly(3-hexyl thienylene) metal-insulator-semiconductor structures: evidence for polarons. Phys Rev Lett 66:2231–2234

Chapter 25
Surface Functionalizing of Carbon-Based Gas-Sensing Materials

25.1 Surface Functionalizing of Carbon Nanotubes and Other Carbon-Based Nanomaterials

As indicated in Chap. 1 (Vol. 2), in general, CNTs do not have sensing response to all gases and vapors but only those with high adsorption energy or that can interact with them. Pristine CNT-based gas sensors are currently limited to sensing gases such as NH_3, NO_2, SO_2, O_2, and NO. Gas molecules such as toxic gases (CO) and water, however, cannot be detected since they do not react with (adsorb on) the surface of pure carbon SWCNTs (Terrones et al. 2004). For H_2 detection, bare CNT does not exhibit appreciable sensitivity as well. Experimental and theoretical studies established that the sensitivity of CNT sensors can be improved by doping the CNTs (Zhang and Zhang 2009). It was found that if the surface of the tube is doped with a donor or an acceptor, drastic changes in the electronic properties are observed. The N-doped nanotubes, for example, exhibit a higher reactivity toward reactants when compared to un-doped tubes due to the introduction of nitrogen species and the structural irregularity of carbon hexagonal rings (see Fig. 25.1). The N substitution reactions are also able to create radicals over nanotube surfaces, which can react with suitable reactants. As a result of the binding of the molecules to the doped locations—because of the presence of holes (B-doped tubes) or donors (N-doped tubes), the surfaces became more reactive and sensitive to surrounding gas. Using first-principle calculations, Peng and Cho (2003) demonstrated that B- or N-doped SWCNT-based sensors can detect CO and water molecules, and, more importantly, the response of these sensors can be controlled by adjusting the doping level of heteroatoms in a nanotube.

It is also important to point out that, in spite of high sensitivity, CNT-based sensors are nonselective, which limits their use for sensing purposes in real samples. This means that mechanisms need to be developed to increase the selectivity of the detectors. As shown before, surface functionalizing is the most effective methods suitable for these purposes.

At present there are two main approaches for the surface functionalizing of CNTs: covalent functionalization and non-covalent functionalization, depending on the types of linkages of the functional entities onto the nanotubes (Zhang et al. 2008). Several mechanisms of covalent and non-covalent modification of CNTs are illustrated in Fig. 25.2. It should be noted that altering the nanotube surface, besides influencing gas-sensing properties, strongly affects solubility properties, which can affect the ease of fabrication of CNT sensors. The reviews by Hirsh (2002), Balasubramanian and Burghard (2005), Shen et al. (2008), and Zhang and Zhang (2009) discuss many of the functionalizations that have commonly been used. Currently, most covalently functionalized CNTs are based on esterification or amidation of carboxylic acid groups that are introduced on defect sites of the CNTs during acid treatment. Experiments have shown that the CNT ends and sidewall "defect" have greater electrochemical

G. Korotcenkov, *Handbook of Gas Sensor Materials*, Integrated Analytical Systems,
DOI 10.1007/978-1-4614-7388-6_25,

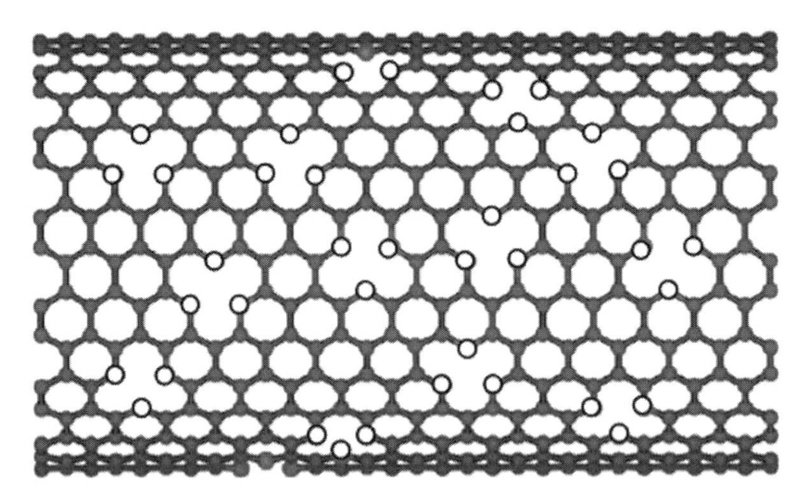

Fig. 25.1 Nitrogen-doped carbon nanotube scheme: molecular model of CN_x nanotubes containing pyridine-like and highly coordinated N atoms (Reprinted with permission from Terrones et al. 2002, Copyright 2002 Springer)

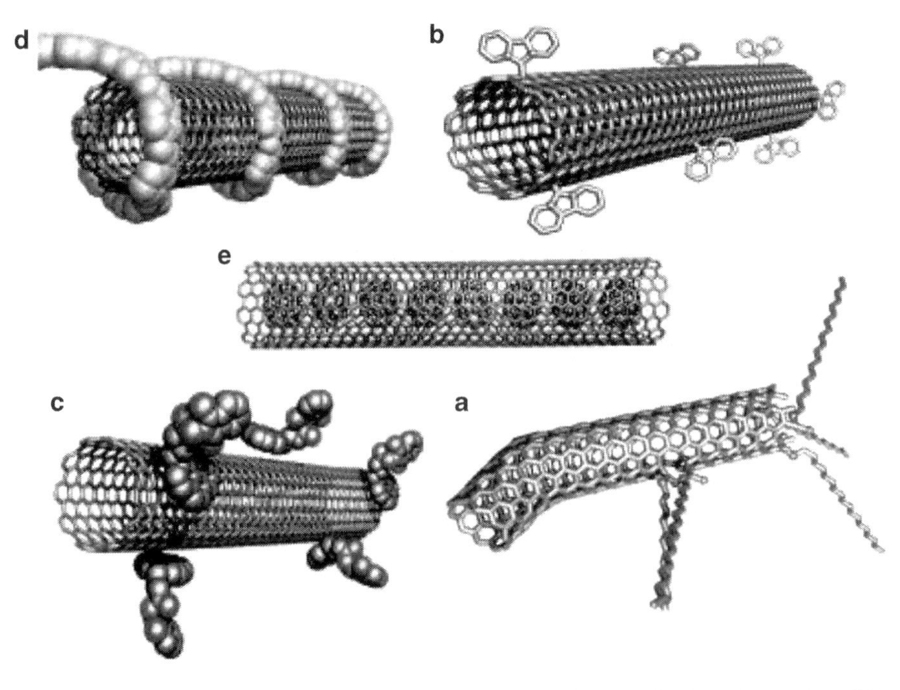

Fig. 25.2 Functionalization possibilities for SWNTs: (**a**) defect-group functionalization, (**b**) covalent sidewall functionalization, (**c**) non-covalent exohedral functionalization with surfactants, (**d**) non-covalent exohedral functionalization with polymers, and (**e**) endohedral functionalization with, for example, C_{60}. For methods **b**–**e**, the tubes are drawn in idealized fashion, but defects are found in real situations (Adapted with permission from Hirsh 2002, Copyright 2002 Wiley)

activity (Banks and Compton 2005; Dumitrescu et al. 2009). Introducing defect sites along the sidewall of the CNTs can be carried out during the purification (oxidation) process as well (Zhang and Zhang 2009). Experiment has shown that carbon nanotubes could possess structural defects such as: (1) five- or seven-membered rings within the carbon network; (2) sp^3-hybridized defects, with R=H and OH

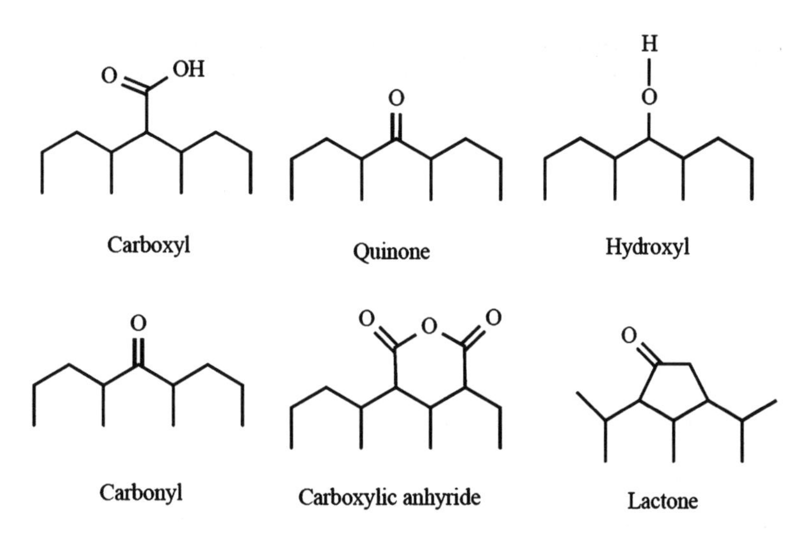

Fig. 25.3 Simplified schematic of some acidic surface groups bonded to aromatic rings on activated carbon

groups; (3) –COOH groups introduced by nanotube damage under oxidative conditions; and (4) open ends terminated with –COOH groups, or even other terminal groups, such as $-NO_2$, –OH, –H, and =O (Hirsh 2002).

It is important that the oxidation in the gas or liquid phase could be used to increase the concentration of surface oxygen groups, while heating under an inert atmosphere might be used to remove selectively some of these functions. It was shown that gas phase oxidation of the carbon mainly increased the concentration of hydroxyl and carbonyl surface groups, while oxidations in the liquid phase increased the concentration of carboxylic acids (Figueiredo et al. 1999). Carboxyl, carbonyl, phenol, quinone, and lactone groups (Yang 2003) on carbon surfaces are shown in Fig. 25.3.

Analysis of published results testifies that the treatment with strong acids, such as HNO_3, H_2SO_4, and HCl, and oxidizing agents, such as $KMnO_4/H_2SO_4$, $K_2Cr_2O_7/H_2SO_4$, and OsO_4, is one of the main steps in CNT functionalizing. This treatment, usually used after CNTs synthesis, removes the end caps and may shorten the length of the CNTs. Acid treatment also adds oxide groups, primarily carboxylic acids, to the tube ends and defect sites. Further chemical reactions can be performed at these oxide groups to functionalize with groups such as amides, thiols, or others. As an example, one can consider polyaniline which, after exposure to HCl, shows a rapid drop in resistance within a short period of time. According to Huang et al. (2003) and Virji et al. (2004), this doping of polyaniline is achieved by protonation of the imine nitrogens by HCl. The charge created along the backbone by this protonation was counterbalanced by the resulting negatively charged chloride counterions. The change in conductivity was brought about by the formation of polarons (radical cations) that travel along the polymer backbone. The mechanism for base dedoping of polyaniline is different than that for acid doping and may account for the slower response times and smaller resistance changes.

Some modifiers typically used for CNT functionalizing are listed in Table 25.1. It should be noted that the approaches used for CNT functioning are similar to those used for modification of polymer, activated carbon, fullerenes, and graphene surfaces.

Most of the previous reports with reference to gas sensors are based on utilization of the carboxylic acid (–COOH) group, which provides reactive sites for interacting with different gases. For example, Fu et al. (2008) recently demonstrated experimentally that sensors made of carboxylated SWCNTs were sensitive to CO, with a lower detection limit of 1 ppm, whereas pristine SWCNTs did not respond.

Table 25.1 Modifiers used for surface functionalization of CNTs

Surface modifiers	Examples
Functional groups	Carboxylic acid (–COOH) group; 3-aminopropyltriethoxysilane; amides, etc.
Polymers	PEI; Nafion; PABS; PPy; PDPA; PMMA; Poly(3-methylthiophene); polystyrene; PEG; POAS; poly(vinyl acetate); polyisoprene, etc.
Metals	Pd; Pt; Ag; Au; Sn; Rh; Al
Metal complex	Eu^{3+}-containing dendrimer complex
Metal oxides	SnO_2; WO_3; TiO_2

Source: Data from Jimenez-Cadena et al. (2007) and Zhang and Zhang (2009)
PABS poly-(*m*-aminobenzene sulfonic acid), *PEG* poly(ethylene glycol), *PEI* polyethyleneimine, *PDPA* polydiphenylamine, *PMMA* poly (methyl methacrylate), *POAS* poly(*o*-anisidine), *PPy* polypyrrole

The authors exploited the different responses of carboxylated and pristine SWCNTs to differentiate between CO, NO, and NO_2. The COOH functionality is crucial to CO sensing, and the CO molecules can be absorbed on carboxylic acid functionalities through weak hydrogen bonding. However, it should be noted that the influence of CNTs surface functionalizing by other functional groups is analyzed as well. Tran et al. (2008) investigated the effect of $-NH_2$-functionalized SiO_2 surface on the gas-sensing properties of SWCNTs modified with 3-aminopropyltriethoxysilane (APTES). Tran et al. (2008) have found that the relative resistance change of the SWCNTs to NO_2 already at ppm level in the case of the APTES-treated surface was twice as large compared to the case without surface treatment under the same conditions. Tran et al. (2008) believe that the amine groups in the APTES monolayer, electron donating in nature, played a role of charge transfer to the semiconducting SWCNTs, and hence the amount of electrons transferred from SWCNTs to NO_2 molecules increased. The recovery time of the gas sensor during indicated treatment was not changed.

Non-covalent functionalization is mainly based on supramolecular complexation using various adsorptive and wrapping forces, such as van der Waals and π-stacking interactions, without destruction of the physical properties of CNTs (Zhao and Stoddart 2009). It was established that functionalized CNT sensors often offer a higher sensitivity and a better selectivity compared with pristine CNT sensors (Qi et al. 2003; Valentini et al. 2003, 2004a, b; Zhang et al. 2008; Zhang and Zhang 2009). This means that doping and surface functionalization of CNTs may broaden the application range of CNT-based gas sensors. Several examples of gas sensors based on modified CNTs are presented in Table 25.2.

As can be seen from the data presented in Tables 25.1 and 25.2, surface modification by metal nanoparticles, oxides, and polymers is also a promising approach to optimizing gas-sensing characteristics of CNT-based devices. It must be said that clusters of noble metals or metal oxides immobilized on carbon-based materials are quite common in modern chemistry, especially as hydrogenation catalysts. Methods such as wet or dry impregnation, deposition–precipitation, deposition–reduction, or ion-exchange protocols are routinely applied for these purposes (Toebes et al. 2001). Most of them rely on treatment with aqueous solutions of suitable precursors, such as metal salts. For example, in Table 25.3, a survey of electrodeposition parameters used for decorating CNTs with different metals is provided. In contrast to platinum, palladium, and nickel, the electrodeposition of silver and gold usually requires stabilizers in order to obtain regular particles (Balasubramanian and Burghard 2008). The size and distribution of the nanoparticles can in general be controlled by the magnitude of the applied potential and the concentration of the metal salt in solution.

It was found that the above-mentioned dopants attached to CNTs change the electronic properties of CNTs and can therefore be used for surface functionalization of CNTs (Wang et al. 2007; Zhang and Zhang 2009). The analysis of operating characteristics of CNT-based gas sensors with functionalized

Table 25.2 Gas sensors based on modified carbon nanotubes

Analyte	CNT material	Detection limit
NO_2	Metal-decorated CNTs	100 ppb
	Metal oxide (WO_3)-decorated CNTs	500 ppb
	Metal oxide (SnO_2)-decorated CNTs	~Several ppm
	Polymer-coated (PEI; PABS) CNTs	100 ppt–20 ppb
NO	Metal-decorated (Pt; Pd; Au; Ag) CNTs	10 ppm
	Metal oxide (SnO_2)-decorated CNTs	2 ppm
	Polymer-coated (PEI) CNTs	5 ppb
NH_3	Metal-decorated CNTs	5 ppm
	Metal oxide (WO_3, SnO_2)-decorated CNTs	5–10 ppm
	Polymer-coated (PANI; PABS) CNTs	20 ppb–5 ppm
	Atomically doped CNTs ca.	1 %
H_2	Pd-decorated CNTs	10 ppm
	Pt-decorated CNTs	0.4 %
	Cryogenically cooled CNT optical probe	4 %
CH_4	Pd-decorated CNTs	6 ppm
SO_2	Polymer-coated (PEI) CNTs	N/A
CO	Metal-decorated (Pt; Rh) CNTs	<2,000 ppm
	Metal oxide-decorated CNTs	10 ppm
	Radially deformed CNTs	N/A
	Atomically doped CNTs	N/A
	Polymer-coated (PANI) CNTs	100 ppm
H_2S	Metal oxide-decorated (Pd) CNTs	50 ppm
O_2	Eu^{3+}-containing dendrimer complex	<1 %
CO_2	Polymer-coated (PEI/starch) CNTs	500 ppm
EtOH	Metal oxide-decorated (SnO_2) CNTs	10 ppm
VOCs	Polymer-functionalized poly(vinyl acetate) CNTs	4 ppm
	Metal oxide (TiO_2) decorated CNTs	N/A

Source: Reprinted with permission from Kauffman and Star (2008). Copyright 2008 Wiley

Table 25.3 Representative examples of various parameters used for decorating SWCNTs with metal nanoparticles

Nanoparticle	Metal salt	Stabilizer	Supporting electrolyte	Solvent
Pd	$Na_2(PdCl_4)$	–	$LiClO_4$	Ethanol
Pt	$H_2(PtCl_6)$	–	$LiClO_4/HClO_4$	Ethanol
Au	$KAuCl_4$	(Poly)vinylpyrrolidone	$LiClO_4$	Water
Ag	AgCN	$K_4P_2O_7$	KCN	Water
	$AgNO_3$	–	KNO_3	Water
Ni	$NiSO_4$	–	Na_2SO_4	Water

Source: Reprinted with permission from Balasubramanian and Burghard (2008). Copyright 2008 The Royal Society of Chemistry

surface carried out by Jimenez-Cadena et al. (2007) has shown that surface modification can be accompanied by improvement of both sensitivity and selectivity during detection of NO_2, SO_2, H_2, NO, NH_3, CO, acetone, ethanol, methane, chloroform, and gasoline. A tenfold increase in sensor response can be achieved. For example, Qi et al. (2003) showed that non-covalent drop coating of polyethyleneimine (PEI) and Nafion (a polymeric perfluorinated sulfonic acid ionomer) onto SWCNTs FETs resulted in gas sensors with improved response and selectivity for NO_2 and NH_3 (Qi et al. 2003). They found that the PEI functionalization changed the SWCNTs from *p*-type to *n*-type semiconductors, and the

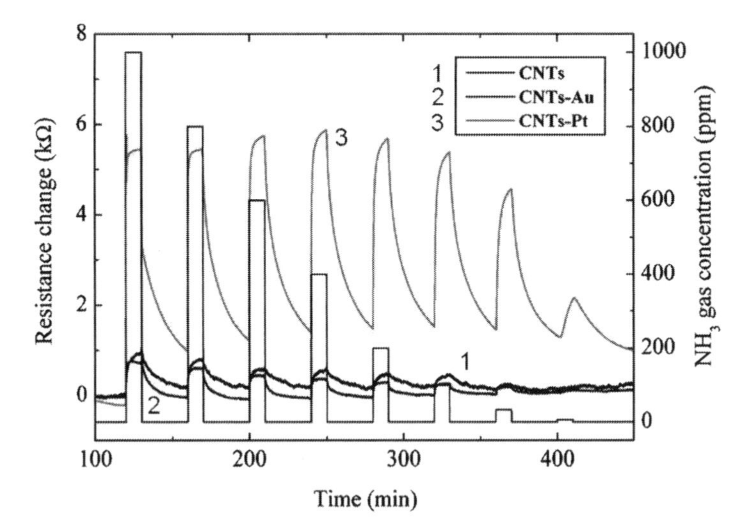

Fig. 25.4 Time response of the electrical resistance change for a chemiresistor based on unfunctionalized MWCNT films and Pt- and Au-modified MWCNT films toward NH_3 at working temperature of 150 °C. The MWCNT film thickness is 300 nm and the nominal thicknesses of Pt and Au catalysts are 3 and 5 nm, respectively (Reprinted with permission from Penza et al. 2007, Copyright American Institute of Physics)

sensors based on PEI-modified SWCNTs were able to detect less than 1 ppb NO_2 with a response time of 1–2 min (defined as the time for 80 % conductance change to take place) while being insensitive to NH_3. In contrast to PEI-coated sensors, Nafion coated SWCNTs were insensitive to NO_2 but exhibited a good sensitivity toward NH_3. Penza et al. (2007) have found that surface modification by Pt clusters strongly increase sensitivity to NH_3 (see Fig. 25.4).

The shortening of response and recovery times can also take place after surface functioning. This effect, in particular, was observed in H_2 SWCNT-based sensors after modification by Pd nanoparticles (Kong et al. 2001). According to Kong et al. (2001), the dissociation of adsorbed H_2 molecules on the surface of Pd nanoparticles causes electron transfer from Pd to SWCNT and reduces the hole carriers in the *p*-type SWCNT, and hence causes a decrease in conductance. The process is reversible as dissolved atomic hydrogen in Pd can combine with O_2 in air to form OH which will further combine with atomic hydrogen to form water and then leave the Pd-SWCNT system, thus recovering the sensor's initial conductance.

It should be noted that surface functionalizing also helps to resolve problems such as disentanglement of bundles, separation–purification, and dispersion–solubilization of CNTs, which are very important for preparation of CNT-based sensors and nanocomposites. It is known that synthetic chemistry primarily takes place in solvents. Thus, disentanglement and uniform dispersions of CNTs in several solvents have to be carried out in order to proceed with chemical reactions. Dispersion becomes difficult because CNTs are extremely resistant to wetting and are difficult to separate due to strong van der Waals interactions. Nevertheless, by using ultrasonic procedure in acids it is possible to wet carbon nanotubes. Intercalated molecules play the role of disrupting and compensate for the loss of van der Waals attractions between carbon nanotubes. In general it was observed that ionic, covalent, and non-covalent functionalization and polymer wrapping procedures could be effective for achievement of uniform carbon nanotube dispersion.

One can find in review papers prepared by Hirsh (2002), Liu (2005), Jimenez-Cadena et al. (2007), Kauffman and Star (2008), Zhang and Zhang (2009), Sun et al. (2011), and Schaetz et al. (2012) more detailed analyses of CNT surface functionalizing.

In closing it should be noted that carbon black, fullerenes, and graphene can be functionalized using the methods discussed above (Tsubokawa 1992; Shen et al. 2008; Marques et al. 2011; Schaetz et al. 2012). Moreover, the most indicated methods such as surface modifications by oxidation, grafting of polymers, treatment with surfactant, plasma treatment, and spattering were first designed

Table 25.4 The related chemical groups change at different plasma treatment conditions

Plasma gaseous	Increased chemical groups	Decreased chemical groups	Reference
O_2	–C–OOH, C=O	–C–OH, C–O–C	Domingo-Garcia et al. (2000)
N_2	–C–OH, C–O–C–, O=C–O, pyridine and quaternary nitrogen	–C=O (aromatic ring)	Brüser et al. (2004)
NH_3	N–H		Boudou et al. (2000)
CO_2	–C–OOH, C=O		Pai et al. (2006)
H_2O	–C–OOH, C=O		Pai et al. (2006)

mainly for CB (Tsubokawa 1992). Different plasma treatment and changes of related chemical functional groups on the surface of carbon were listed in Table 25.4. By using these treatments, the dispersibility of CB in solvents and compatibility in polymer matrices were markedly improved. In addition, surface functionalizing of CB enabled various functions to be given to carbon black such as photosensitivity, bioactivity, crosslinking ability, and amphiphilic properties. For example, the vapor-grown carbon fibers were modified using NH_3, O_2, CO_2, H_2O, and HCOOH plasma gases to increase the wettability, and the results show that the oxidation strength was $O_2>CO_2>H_2O>$HCOOH (Brüser et al. 2004). The functionalization of fullerenes also has positive effects. For example, sensitivity of fullerene sensors toward polar vapors (e.g., ethanol and water) was enhanced by >50-fold through the deposition of a metal-fullerene hybrid film containing both C_{60} and aluminum due to higher surface areas and possibly metal–fullerene bonding (Grynko et al. 2009). UV exposure further enhanced the sensitivity of both pristine and C_{60}–Al hybrid films through the introduction of reactive sites on the C_{60} surface (Grynko et al. 2009). Sensitivity of C_{60}-based QCM and SAW sensors toward polar and non-polar vapors such as volatile organic alcohols, aldehydes, and acids was enhanced by derivatizing the C_{60} with supramolecular host compounds such as crown ethers and cryptands (Shih et al. 2001; Lin and Shih 2003). The proposed mechanism of this sensitivity enhancement involved a combination of enhanced chelation by the cryptand/crown ether and enhanced reactivity of the C_{60} at the cryptand/crown ether binding site (Shih et al. 2001; Lin and Shih 2003). Another approach to generating supramolecular host compounds for vapor sensing with fullerenes involved liquid crystals (Dickert et al. 1997) where rigid linear (thermotropic liquid crystals) and globular (fullerenes) compounds formed a 1:1 stoichiometry sensing film, disturbing the close packing of both species and, thus, forming cavities and diffusion channels.

25.2 The Role of Defects in Graphene Functionalizing

Although many outstanding properties of graphene are due to the inherently low concentration of defects, nanoengineering of graphene-based devices, for example, gas sensors, requires the introduction of structural defects or impurities that allow achieving the desired functionality of material used (Banhart et al. 2011). For example, it was established that the as-synthesized GO transistor showed small response to gases such as NO_2 (Fowler et al. 2009). In contrast, the RGO, which has larger numbers of defects generated during the high-temperature reduction process (Jung et al. 2008), was highly responsive to NO_2. This means that the sensitivity of RGO-based sensors is really strongly dependent on the defect structure of graphene and this parameter of graphene-based sensors can be modified by controlling the reduction treatment process (Robinson et al. 2008; Jung et al. 2008.). In addition, high-temperature (~1000 °C)-reduced GO shows faster chemoresistive response than the chemical and low-temperature reduction upon exposure to water vapor. In another study, Sundaram et al. (2008) have found that graphene surface chemically modified by Pd nanoparticles shows improved sensitivity toward certain analytes such as H_2, which cannot be directly detected with unmodified material.

It is intuitively clear that defects associated with dangling bonds should enhance the reactivity of graphene. Indeed, numerous simulations (Boukhvalov and Katsnelson 2008; Cantele et al. 2009) indicate that hydroxyl, carboxyl, or other groups can easily be connected to vacancy-type defects. It was established that the behavior of intrinsic zero- or one-dimensional defects in graphene such as vacancies or line defects is governed by the reconstruction of the graphenic lattice around defects, which is a unique property among all known materials. The same is true for graphene edges that are normally saturated with hydrogen. Simulations also show that reconstructed defects without dangling bonds such as Stone–Wales (SW) defects or reconstructed vacancies locally change the density of π-electrons (Boukhvalov and Katsnelson 2008; Peng and Ahuja 2008) and may also increase the local reactivity (Duplock et al. 2004). The Stone–Wales (SW) defect is the simplest example of defects in graphene, which does not involve any removed or added atoms. Four hexagons are transformed into two pentagons and two heptagons by rotating one of the C–C bonds by 90° (Stone and Wales 1986). One can find in the review prepared by Banhart et al. (2011) detailed analyses of structural defects in graphene. Indeed, experiments provide evidence that the trapping of metal atoms occurs in reconstructed vacancies (Cretu et al. 2010).

Thus, the controlled creation of defects with a high spatial selectivity can be used for the local functionalization of graphene samples, i.e., for the creation of graphene-based sensors with the designed properties by various applications. Numerous theoretical reports on the sensing properties of doped graphene confirmed the above-mentioned conclusions (Hwang et al. 2007; Ao et al. 2008; Dan et al. 2009; Zhang et al. 2009). For example, it was established that, by introducing substituent impurities into graphene through chemical doping, the local electronic structures around the dopants could be modified. In particular, as confirmed by first-principle studies (Ao et al. 2008, 2010), incorporating Al into graphene causes distortion to the electron density distribution around the dopant. In this case, C atoms surrounding the Al dopant attract electrons due to their high electron affinity, whereas on the Al dopant, a decrease in electron density can be observed. The charge redistribution makes the Al an active site for CO adsorption. This charge redistribution effect can be confirmed with advanced field theoretical methods (Peres et al. 2006, 2009).

It was found that the doping of graphene by foreign species can be done in several ways. An injection of charge into the electron system of graphene can be achieved from metal contacts (Giovanetti et al. 2008) or by attaching organic molecules on a perfect graphene layer (Wehling et al. 2008). However, the adsorption of organic species is weak and desorption starts at temperatures below 100 °C (Coletti et al. 2010). Nevertheless, the localization of dopants in graphene could be achieved on defects where an increased reactivity prevails. Foreign atoms on substitutional sites appear to be unfavorable because the strong scattering of the conduction electrons at such atoms might cause the electronic properties of graphene to deteriorate (Castro Neto et al. 2009). Reconstructed defects, however, could be more attractive because the coherence of the graphene lattice is preserved and the foreign dopant atoms are firmly attached (Cretu et al. 2010). Doping by substitutional impurities is quite straightforward if B atoms (*p*-doping) or N atoms (*n*-doping) are used. As shown before, the same approach is normally used for functionalizing CNTs.

Experiments have shown that structural defects in graphene, which can be used for graphene functionalizing, can be generated by irradiation with electrons or ions due to the ballistic ejection of carbon atoms (Banhart 1999; Compagnini et al. 2009; Krasheninnikov and Nordlund 2010). The atom can be sputtered away from graphene or get adsorbed on the sheet and migrate on its surface as an adatom. The reactions of carbon atoms in a graphene layer with other species can lead to the loss of atoms and hence to defects. However, the high inertness of graphene (apart from edge positions that are highly reactive) only allows a very limited number of possible reactions at room temperature.

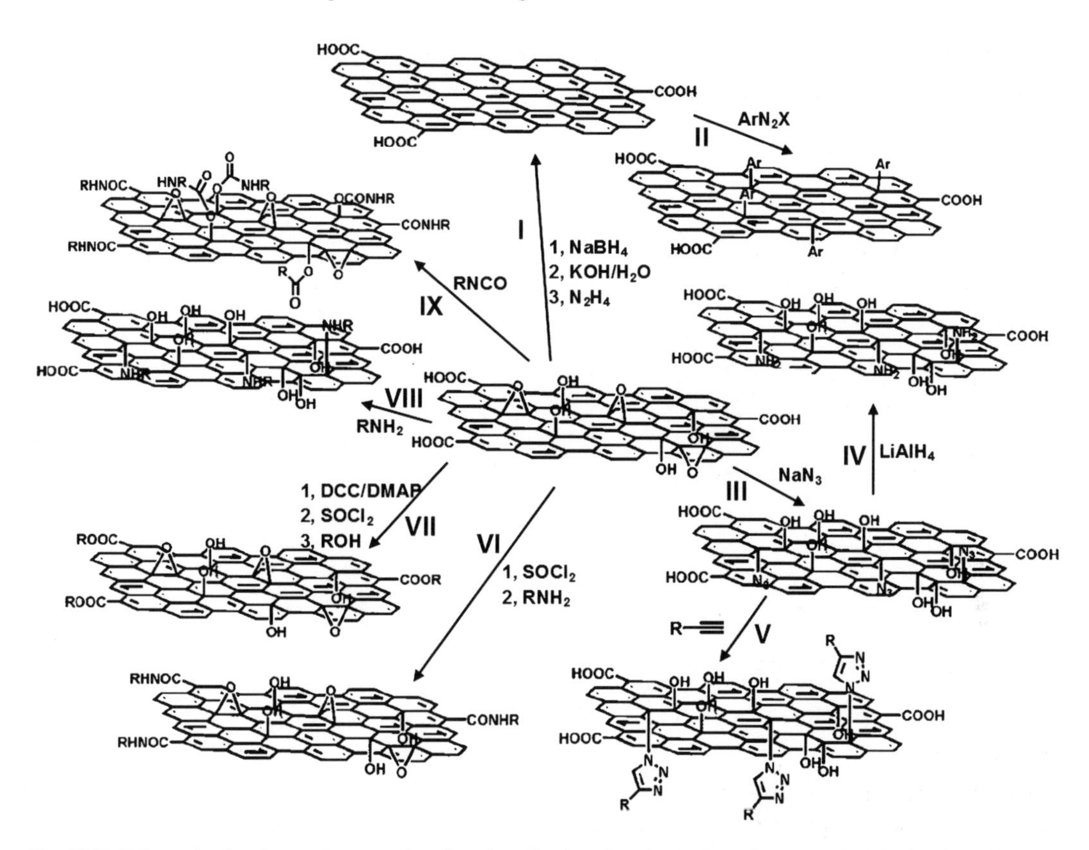

Fig. 25.5 Schematic showing various covalent functionalization chemistry of graphene or GO. *I*: Reduction of GO into graphene by various approaches. *II*: Covalent surface functionalization of reduced graphene via diazonium reaction. *III*: Functionalization of GO by the reaction between GO and sodium azide. *IV*: Reduction of azide functionalized GO (azide–GO) with $LiAlH_4$ resulting in the amino-functionalized GO. *V*: Functionalization of azide–GO through click chemistry (R–ChCH/$CuSO_4$). *VI*: Modification of GO with long alkyl chains by the acylation reaction between the carboxyl acid groups of GO and alkylamine (after $SOCl_2$ activation of the COOH groups); *VII*. Esterification of GO by DCC chemistry or the acylation reaction between the carboxyl acid groups of GO and ROH alkylamine (after $SOCl_2$ activation of the COOH groups). *VIII*: Nucleophilic ring-opening reaction between the epoxy groups of GO and the amine groups of an amine-terminated organic molecular. *IX*: Treatment of GO with organic isocyanates leading to the derivatization of both the edge carboxyl and surface hydroxyl functional groups via formation of amides or carbamate esters (RNCO) (Adapted with permission from Loh et al. 2010, Copyright 2010. Royal Society of Chemistry)

The oxidation, for example, in an oxidizing acid (HNO_3 or H_2SO_4), is the most common approach acceptable for graphene functionalizing. In such a treatment it is possible to attach oxygen and hydroxyl (OH) or carboxyl (COOH) groups to graphene. When graphene is covered more or less uniformly with hydroxyl or carboxyl groups, the material is called graphene oxide, which is essentially a highly defective graphene sheet functionalized with oxygen groups (Bagri et al. 2010). It is important that graphite oxide (GO) with a wide range of oxygen functional groups both on the basal planes and at the edges of graphene oxide sheets becomes readily exfoliated in water. The above-mentioned reactive oxygen functional groups of graphene oxide can then be chemically modified to produce homogeneous colloidal suspensions in various solvents and to influence the properties of graphene-based materials (Park et al. 2009). For these purposes, various methods of covalent and non-covalent functionalization can be applied (see Fig. 25.5). It should be noted that these methods are similar to the approaches designed for CNTs functionalizing.

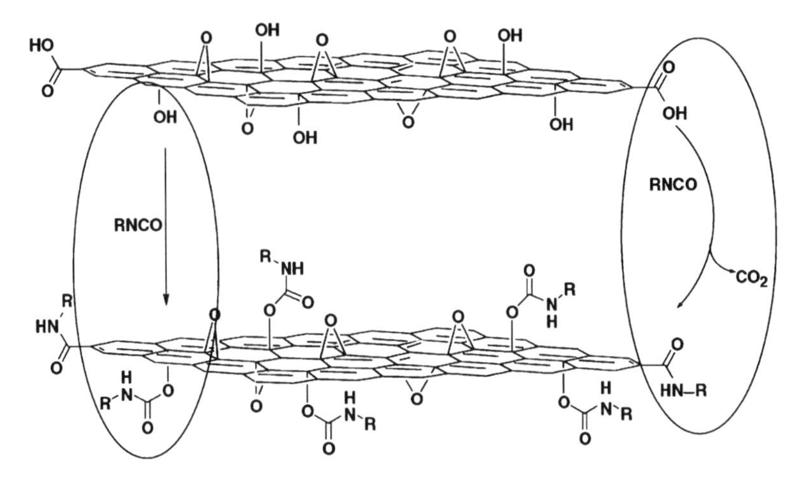

Fig. 25.6 Proposed reactions during the isocyanate treatment of GO where organic isocyanates react with the hydroxyl (*left oval*) and carboxyl groups (*right oval*) of graphene oxide sheets to form carbamate and amide functionalities, respectively (Reprinted with permission from Stankovich et al. 2006a, Copyright 2006 Elsevier)

Treatment of isocyanates reduced the hydrophilicity of graphene oxide by forming amide and carbamate esters from the carboxyl and hydroxyl groups of graphene oxide, respectively. As a result, such isocyanate-derivatized graphite oxides no longer exfoliate in water but readily form stable dispersions in polar aprotic solvents (such as *N,N*-dimethylformamide (DMF)), consisting of completely exfoliated, functionalized individual graphene oxide sheets with thickness ~1 nm (Stankovich et al. 2006a). The proposed reaction is shown in Fig. 25.6. This dispersion also facilitated the intimate mixing of the graphene oxide sheets with matrix polymers, providing a novel synthesis route to make graphene–polymer nanocomposites (Stankovich et al. 2006b). In particular, using the indicated graphene treatment, the polystyrene–graphene composites with percolation threshold near 0.1 vol.% were synthesized (Stankovich et al. 2006b). Such a low percolation threshold for a three-dimensional isotropic case is evidently due to the extremely high aspect ratio of the graphene sheets and their excellent homogeneous dispersion in these composites.

In order to use carboxylic acid groups on graphene oxide to anchor other molecules, the carboxylic acid groups have been activated by thionyl chloride ($SOCl_2$), 1-ethyl-3-(3-dimethylaminopropyl)-carbodiimide (EDC), and *N,N*-dicyclohexylcarbodiimide (DCC or 2-(7-aza-1*H*-benzotriazole-1-yl)-1,1,3,3-tetramethyluronium hexafluorophosphate (HATU)) (Singh et al. 2011). The subsequent addition of nucleophilic species, such as amines or alcohols, produced covalently attached functional groups on graphene oxide via the formation of amides or esters. The attachment of hydrophobic long, aliphatic amine groups on hydrophilic graphene oxide improved the dispersability of modified graphene oxide in organic solvents (Niyogi et al. 2006). Subrahmanyam et al. (2008) have shown that soluble graphene layers in solvents such as CCl_4, CH_2Cl_2, and THF can also be generated by the covalent attachment of alkyl chains to graphene layers by the reduction of graphite fluoride with alkyl lithium reagents.

In order to enhance the solubility of graphene oxide nanosheets in water, the graphene oxide nanosheets were functionalized with allylamine (Wang et al. 2009). Si and Samulski (2008) introduced a small number of *p*-phenyl-SO_3H groups into the graphene oxide before it was fully reduced and the resulting graphene remained soluble in water and did not aggregate.

The amine groups and hydroxyl groups on the basal plane of graphene oxide have also been used to attach polymers through either grafting-onto or grafting-from approaches. To grow a polymer from graphene oxide, an atom transfer radical polymerization (ATRP) initiator (i.e., a-bromoisobutyrylbromide)

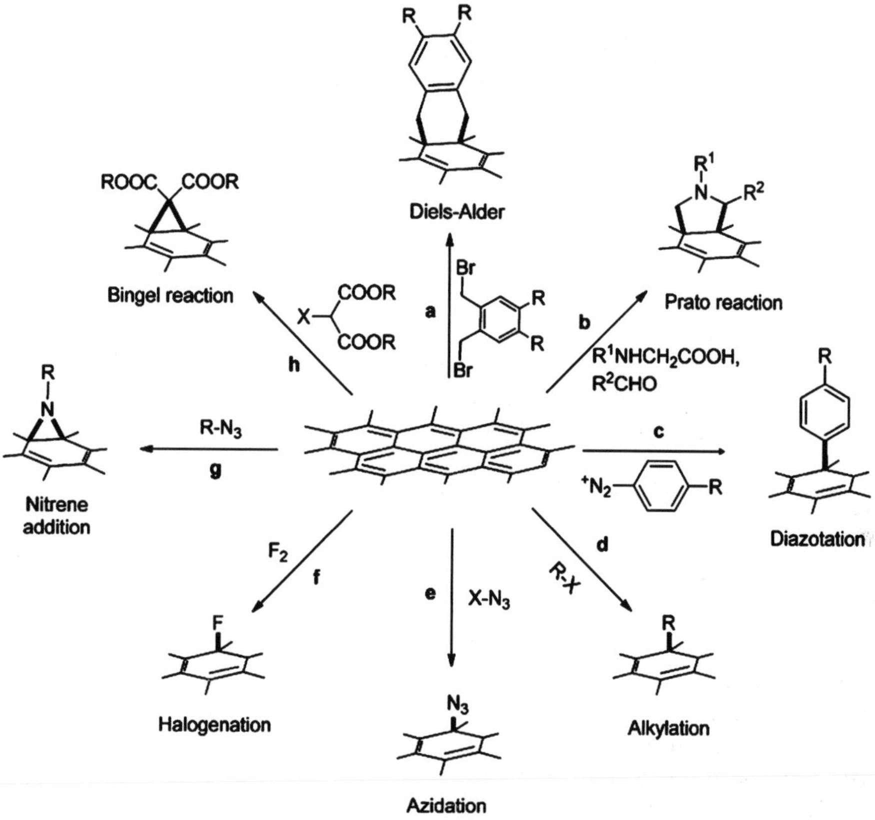

Fig. 25.7 Covalent functionalization of sp^2-hybridized carbons: (**a**) Diels–Alder reaction; (**b**) Prato reaction; (**c**) diazonium chemistry; (**d**) alkylation of graphene oxide/activated graphenes; (**e**) azidation; (**f**) halogenations; (**g**) nitrene addition; and (**h**) Bingel reaction (Reprinted with permission from Schaetz et al. 2012, Copyright 2012 American Chemical Society)

was attached to graphene surfaces (Fang et al. 2009). The following living polymerization produced graphene oxide with polymers that enhanced the compatibility of solvents and other polymer matrices.

Besides the carboxylic acid groups, the epoxy groups on graphene oxide can be used to attach different functional groups through a ring-opening reaction. The preferred mechanism for this type of reaction involves a nucleophilic attack of amines on α-carbon. Various amine ending chemicals such as octadecylamine (Wang et al. 2008), an ionic liquid 1-(3-aminopropyl)-3-methylimidazolium bromide (Yang et al. 2009) with an amine end group, and APTES have reacted with epoxy groups. This surface modification is of particular interest when attempting to disperse the GO in polar solvents (Yang et al. 2009).

Some of the most important routes of graphene surface functionalizing, which are not based on surface-bound oxygen and carbonyl moieties, are depicted in Fig. 25.7. It is important that most sp^2-hybridized carbon scaffolds (including carbon nanotubes, fullerenes, and graphitic carbon shells) are amenable to these reactions.

The non-covalent functionalization of graphene oxide utilizes the weak interactions (i.e., π–π interaction, Van der Waals interaction, and electrostatic interaction) between the graphene oxide and target molecules. Electrostatic interaction takes place mainly due to the fact that its surface is negatively charged due to the presence of oxygen functional groups. The sp^2 network on graphene oxide provides π–π interactions with conjugated polymers and aromatic compounds that can stabilize reduced graphene oxide that resulted from chemical reduction and produce functional composite graphene–polymer materials, which can include poly(sodium 4-styrenesulfonate) (PSS), sulfonated polyaniline, poly(3-hexylthiophene) (P3HT), conjugated polyelectrolyte, tetrasulfonate salt of copper phthalocyanine (TSCuPc), porphyrin, and cellulose derivatives (Singh et al. 2011). Aromatic molecules have large aromatic planes and can anchor onto the reduced graphene oxide surface without disturbing its electronic conjugation. This type of functionalization has some advantages in certain areas, such as chemical sensors and biomedical materials.

Plasma treatment and adsorption of atomic hydrogen on a graphene surface followed by its self-organization and hydrogen island formation (Balog et al. 2010) can also be referred to in the context of graphene treatment by chemical methods.

It is clear that the high stability of defects in graphene in its different configurations is of particular importance for devices intended for the gas sensor market. The study of the influence of intrinsic defects on the electronic properties of graphene is still in its infancy, and experimental data relating defect concentration with the changes in electronic and optical characteristics are urgently needed (Banhart et al. 2011). On the other hand, it is clear that extrinsic defects such as foreign atoms on different positions have a strong influence on the electron–electron interaction and thus charge distribution and the band structure of graphene.

References

Ao Z, Yang J, Li S, Jiang Q (2008) Enhancement of CO detection in Al doped graphene. Chem Phys Lett 461:276–279

Ao Z, Li S, Jiang Q (2010) Correlation of the applied electrical field and CO adsorption/desorption behavior on Al-doped graphene. Solid State Commun 150:680–683

Bagri A, Mattevi C, Acik M, Chabal YJ, Chhowalla M, Shenoy VB (2010) Structural evolution during the reduction of chemically derived graphene oxide. Nat Chem 2:581–587

Balasubramanian K, Burghard M (2005) Chemically functionalized carbon nanotubes. Small 1:180–192

Balasubramanian K, Burghard M (2008) Electrochemically functionalized carbon nanotubes for device applications. J Mater Chem 18:3071–3083

Balog R, Jørgensen B, Nilsson L, Andersen M, Rienks E, Bianchi M, Fanetti M, Lægsgaard E, Baraldi A, Lizzit L (2010) Bandgap opening in graphene induced by patterned hydrogen adsorption. Nat Mater 9:315–319

Banhart F (1999) Irradiation effects in carbon nanostructures. Rep Prog Phys 62:1181–1221

Banhart F, Kotakoski J, Krasheninnikov AV (2011) Structural defects in graphene. ACS Nano 5(1):26–41

Banks CE, Compton RG (2005) Exploring the electrocatalytic sites of carbon nanotubes for NADH detection: an edge plane pyrolytic graphite electrode study. Analyst 130:1232–1239

Boudou JP, Martinez-Alonzo A, Tascon JMD (2000) Introduction of acidic groups at the surface of activated carbon by microwave induced oxygen plasma at low pressure. Carbon 38:1021–1029

Boukhvalov DW, Katsnelson MI (2008) Chemical functionalization of graphene with defects. Nano Lett 8:4373–4379

Brüser V, Heintze M, Brandl W, Marginean G, Bubert H (2004) Surface modification of carbon nanofibres in low temperature plasmas. Diamond Relat Mater 13:1177–1181

Cantele G, Lee Y-S, Ninno D, Marzari N (2009) Spin channels in functionalized graphene nanoribbons. Nano Lett 9:3425–3429

Castro Neto AH, Guinea F, Peres NMR, Novoselov KS, Geim AK (2009) The electronic properties of graphene. Rev Mod Phys 81:109–162

Coletti C, Riedl C, Lee DS, Krauss B, Patthey L, von Klitzing K, Smet JH, Starke U (2010) Charge neutrality and band-gap tuning of epitaxial graphene on SiC by molecular doping. Phys Rev B 81:235401

Compagnini G, Giannazzo F, Sonde S, Raineri V, Rimini E (2009) Ion irradiation and defect formation in single layer graphene. Carbon 47:3201–3207

Cretu O, Krasheninnikov AV, Rodriguez-Manzo JA, Sun L, Nieminen R, Banhart F (2010) Migration and localization of metal atoms on graphene. Phys Rev Lett 105:196102

Dan Y, Lu Y, Kybert NJ, Luo Z, Johnson ATC (2009) Intrinsic response of graphene vapor sensors. Nano Lett 9:1472–1475

Dickert FL, Zenkel ME, Bulst W-E, Fischerauer G, Knauer U (1997) Fullerene/liquid crystal mixtures as QMB- and SAW-coatings – detection of diesel- and solvent-vapours. Fresenius J Anal Chem 357:27–31
Domingo-Garcia M, Lopez-Garzon FJ, Perez-Mendoza M (2000) Effect of some oxidation treatments on the textural characteristics and surface chemical nature of an activated carbon. J Colloid Interface Sci 222:233–240
Dumitrescu I, Unwin PR, Macpherson JV (2009) Electrochemistry at carbon nanotubes: perspective and issues. Chem Commun (Camb) (45):6886–6901
Duplock EJ, Scheffler M, Lindan PJD (2004) Hallmark of perfect graphene. Phys Rev Lett 92:225502
Fang M, Wang K, Lu H, Yang Y, Nutt S (2009) Covalent polymer functionalization of graphene nanosheets and mechanical properties of composites. J Mater Chem 19:7098–7105
Figueiredo JL, Pereira MFR, Freitas MMA, Orfao JJM (1999) Modification of the surface chemistry of activated carbons. Carbon 37:1379–1389
Fowler JD, Allen MJ, Tung VC, Yang Y, Kaner RB, Weiller BH (2009) Practical chemical sensors from chemically derived graphene. ACS Nano 3:301–306
Fu D, Lim H, Shi Y, Dong X, Mhaisalkar SG, Chen Y, Moochhala S, Li L-J (2008) Differentiation of gas molecules using flexible and all-carbon nanotube devices. J Phys Chem 112(3):650–653
Giovanetti G, Khomyakov PA, Brocks G, Karpan VM, van den Brink J, Kelly PJ (2008) Doping graphene with metal contacts. Phys Rev Lett 101:026803
Grynko D, Burlachenko J, Kukla O, Kruglenko I, Belyaev O (2009) Fullerene and fullerene-aluminum nanostructured films as sensitive layers for gas sensors. Semicond Phys Quantum Electron Optoelectron 12:287–289
Hirsh A (2002) Functionalization of single-walled carbon nanotubes. Angew Chem Int Ed 41(11):1853–1859
Huang JX, Virji S, Weiller BH, Kaner RB (2003) Polyaniline nanofibers: facile synthesis and chemical sensors. J Am Chem Soc 125:314–315
Hwang EH, Adam S, Sarma SD (2007) Transport in chemically doped graphene in the presence of adsorbed molecules. Phys Rev B 76:195421
Jimenez-Cadena G, Riu J, Rius FX (2007) Gas sensors based on nanostructured materials. Analyst 132:1083–1099
Jung I, Dikin D, Park S, Cai W, Mielke SL, Ruoff RS (2008) Effect of water vapor on electrical properties of individual reduced graphene oxide sheets. J Phys Chem C 112:20264
Kauffman DR, Star A (2008) Carbon nanotube gas and vapor sensors. Angew Chem Int Ed 47:6550–6570
Kong J, Chapline MG, Dai HJ (2001) Functionalized carbon nanotubes for molecular hydrogen sensors. Adv Mater 13(18):1384–1386
Krasheninnikov AV, Nordlund K (2010) Ion and electron irradiation-induced effects in nanostructured materials. J Appl Phys 107:071301
Lin H-B, Shih J-S (2003) Fullerene C_{60}-cryptand coated surface acoustic wave quartz crystal sensor for organic vapors. Sens Actuators B 92:243–254
Liu P (2005) Modifications of carbon nanotubes with polymers. Eur Polym J 41:2693–2703
Loh KP, Bao QL, Ang PK, Yang JX (2010) The chemistry of graphene. J Mater Chem 20(12):2277–2289
Marques PAAP, Gonçalves G, Cruz S, Almeida N, Singh MK, Grácio J, Sousa ACM (2011) Functionalized graphene nanocomposites (ch. 11). In: Hashim A (ed) Advances in nanocomposite technology. InTech, pp 247–272
Niyogi S, Bekyarova E, Itkis ME, McWilliams JL, Hamon MA, Haddon RC (2006) Solution properties of graphite and graphene. J Am Chem Soc 128(24):7720–7721
Pai YH, Ke JH, Huang HF, Lee CM, Zen JM, Shieu FS (2006) CF_4 plasma treatment for preparing gas diffusion layers in membrane electrode assemblies. J Power Sources 161:275–281
Park S, Dikin DA, Nguyen SBT, Ruoff RS (2009) Graphene oxide sheets chemically cross-linked by polyallylamine. J Phys Chem C 113(36):15801–15804
Peng X, Ahuja R (2008) Symmetry breaking induced bandgap in epitaxial graphene layers on SiC. Nano Lett 8:4464–4468
Peng S, Cho K (2003) Ab initio study of doped carbon nanotube sensors. Nano Lett 3(4):513–517
Penza M, Cassano G, Rossi R, Alvisi M, Rizzo A, Signore MA, Dikonimos T, Serra E, Giorgi R (2007) Enhancement of sensitivity in gas chemiresistors based on carbon nanotube surface functionalized with noble metal (Au, Pt) nanoclusters. Appl Phys Lett 90:173123
Peres N, Guinea F, Castro NA (2006) Electronic properties of disordered two-dimensional carbon. Phys Rev B 73:125411
Peres N, Tsai S-W, Santos J, Ribeiro R (2009) Scanning tunneling microscopy currents on locally disordered graphene. Phys Rev B 78–79:155442
Qi P, Vermesh O, Grecu M, Javey A, Wang Q, Dai H (2003) Toward large arrays of multiplex functionalized carbon nanotube sensors for highly sensitive and selective molecular detection. Nano Lett 3:347–351
Robinson JT, Perkins FK, Snow ES, Wei Z, Sheehan PE (2008) Reduced graphene oxide molecular sensors. Nano Lett 8:3137–3140
Schaetz A, Zeltner M, Stark WJ (2012) Carbon modifications and surfaces for catalytic organic transformations. ACS Catal 2:1267–1284

Shen W, Li Z, Yiong LY (2008) Surface chemical functional groups modification of porous carbon. Recent Patents Chem Eng 1:27–40
Shih J-S, Mao Y-C, Sung M-F, Gau G-J, Chiou C-S (2001) Piezoelectric crystal membrane chemical sensors based on fullerene C_{60}. Sens Actuators B 76:347–353
Si YC, Samulski ET (2008) Synthesis of water soluble graphene. Nano Lett 8:1679–1682
Singh V, Joung D, Zhai L, Das S, Khondaker SI, Seal S (2011) Graphene based materials: past, present and future. Prog Mater Sci 56:1178–1271
Stankovich S, Piner RD, Nguyen ST, Ruoff RS (2006a) Synthesis and exfoliation of isocyanate-treated graphene oxide nanoplatelets. Carbon 44:3342–3347
Stankovich S, Dikin DA, Dommett GHB, Kohlhaas KM, Zimney EJ, Stach EA, Piner RD, Nguyen ST, Ruoff RS (2006b) Graphene-based composite materials. Nature 442:282–286
Stone AJ, Wales DJ (1986) Theoretical studies of icosahedral C_{60} and some related species. Chem Phys Lett 128:501–503
Subrahmanyam KS, Vivekchand SRC, Govindaraj A, Rao CNR (2008) A study of graphenes prepared by different methods: characterization, properties and solubilization. J Mater Chem 18:1517–1523
Sun J-T, Hong C-Y, Pan C-Y (2011) Surface modification of carbon nanotubes with dendrimers or hyperbranched polymers. Polym Chem 2:998–1007
Sundaram RS, Navarro CG, Balasubramaniam K, Burghard M, Kern K (2008) Electrochemical modification of graphene. Adv Mater 20:3050–3053
Terrones M, Ajayan PM, Banhart F, Blase X, Carroll DL, Charlier JC, Czerw R, Foley B, Grobert N, Kamalakaran R, Kohler-Redlich P, Ruhle M, Seeger T, Terrones H (2002) N-doping and coalescence of carbon nanotubes: synthesis and electronic properties. Appl Phys A 74:355–361
Terrones M, Jorio A, Endo M, Rao AM, Kim Y, Hayashi T, Terrones H, Charlier JC, Dresselhaus G, Dresselhaus MS (2004) New direction in nanotube science. Mater Today 30–45
Toebes ML, van Dillen JA, de Jong KP (2001) Synthesis of supported palladium catalysts. J Mol Catal A 173:75–98
Tran TH, Lee J-W, Lee K, Lee YD, Ju B-K (2008) The gas sensing properties of single-walled carbon nanotubes deposited on an aminosilane monolayer. Sens Actuators B 129:67–71
Tsubokawa N (1992) Functionalization of carbon black by surface grafting of polymers. Prog Polym Sci 17:417–470
Valentini L, Cantalini C, Lozzi L, Armentano I, Kenny JM, Santucci S (2003) Reversible oxidation effects on carbon nanotubes thin films for gas sensing applications. Mater Sci Eng C 23:523–529
Valentini L, Cantalini C, Armentano I, Kenny JM, Lozzi L, Santucci S (2004a) Highly sensitive and selective sensors based on carbon nanotubes thin films for molecular detection. Diam Relat Mater 13:1301–1305
Valentini L, Bavastrello V, Stura E, Armentano I, Nicolini C, Kenny JM (2004b) Sensors for inorganic vapor detection based on carbon nanotubes and poly(*o*-anisidine) nanocomposite material. Chem Phys Lett 383(5–6):617–622
Virji S, Huang JX, Kaner RB, Weiller BH (2004) Polyaniline nanofiber gas sensors: examination of response mechanisms. Nano Lett 4:491–496
Wang R, Zhang D, Sun W, Han Z, Liu C (2007) A novel aluminum-doped carbon nanotubes sensor for carbon monoxide. J Mol Struct 806(1–3):93–97
Wang S, Chia PJ, Chua LL, Zhao LH, Png RQ, Sivaramakrishnan S, Zhou M, Goh RGS, Friend RH, Wee ATS, Ho PKH (2008) Band-like transport in surface-functionalized highly solution-processable graphene nanosheets. Adv Mater 20(18):3440–3446
Wang GX, Wang B, Park J, Yang J, Shen XP, Yao J (2009) Synthesis of enhanced hydrophilic and hydrophobic graphene oxide nanosheets by a solvothermal method. Carbon 47:68–72
Wehling TO, Novoselov KS, Morozov SV, Vdovin EE, Katsnelson MI, Geim AK, Lichtenstein AI (2008) Molecular doping of graphene. Nano Lett 8:173–177
Yang RT (2003) Adsorption. Wiley, Hoboken, NJ
Yang H, Shan C, Li F, Han D, Zhang Q, Niu L (2009) Covalent functionalization of polydisperse chemically-converted graphene sheets with amine-terminated ionic liquid. Chem Commun 2009(26):3880–3882
Zhang W-D, Zhang W-H (2009) Carbon nanotubes as active components for gas sensors. J Sensor 2009:160698
Zhang T, Mubeen S, Myung NV, Deshusses MA (2008) Recent progress in carbon nanotube-based gas sensors. Nanotechnology 19:332001
Zhang YH, Chen YB, Zhou KG, Liu CH, Zeng J, Zhang HL, Peng Y (2009) Improving gas sensing properties of graphene by introducing dopants and defects: a first-principles study. Nanotechnology 20:185504
Zhao YL, Stoddart JF (2009) Noncovalent functionalization of single-walled carbon nanotubes. Acc Chem Res 42:1161–1171

Chapter 26
Structure and Surface Modification of Porous Silicon

26.1 Structure and Morphology Control of Porous Silicon

We need to recognize that morphology control and structure reproducibility present great problems with porous Si, especially mesoporous and nanoporous silicon applied in gas sensors. It was established that there are too many factors which influence PSi morphology. PSi porosity and morphology depend on electrolyte composition, current density, lighting, magnetic field, ultrasonic agitation and etching time during anodization, pH and temperature of solution used, orientation and doping of wafer, surface patterning, etc. (Korotcenkov and Cho 2010a). Several examples of the indicated factors that influence PSi porosity and pore size are shown in Fig. 26.1.

Moreover, we should note that to date no complete understanding of the mechanisms that determine PSi morphology exists because the Si porosification is a very complicated process which, as indicated before, depends on the great number of parameters involved. Nevertheless, some general trends can be derived for different types of starting Si substrates. Table 26.1 shows the summarized information concerning those peculiarities. More detailed analysis of the influence of electrochemical etching conditions on porous semiconductor parameters such as the pore diameter, the pore depth and the pore growth direction, one can find in the following sections of present chapter, as well as in several published articles (Bomchil et al. 1989; Lehmann and Gruning 1997; Korotcenkov and Cho 2010a).

In was also found that porous layers formed can have a great variation in microstructure. One can see in Fig. 26.2 how big the variation could be in microstructure of PSi layers formed under different conditions. Schematically possible structures of porous Si are shown in Fig. 26.3.

According to Zhang (2005), microstructures of PSi layer can have the following peculiarities:

1. Pores can be either straight with smooth walls (Fig. 26.3Aa) or can be branched (Fig. 26.3Ab–Af). Research has shown that straight large pores with smooth walls can be formed by backside illumination of *n*-Si of the (100) orientation (Lehmann 1993).
2. PSi can have a surface transition layer. It was found that the pores at the surface usually have smaller diameter than those in the bulk of PSi (Fig. 26.3Bb) (Zhang 1991; Smith and Collins 1992). Such an increase in pore diameter from the surface to bulk is due to the transition from pore initiation to steady growth. The thickness of this transition layer is related to the size of pores; the smaller the pores, the thinner the surface transition layer.
3. PSi can have a layer structure (Fig. 26.3Bc). Two-layer PSi, with a micro-PSi layer on the top of a macro-PSi layer is formed only under certain conditions. For *p*-Si, a two-layer structure was observed only on lowly doped substrates. For moderately or highly doped *p*-Si or for *n*-Si in the dark, the formation of a two-layer PSi has not been observed. For *n*-Si, the formation of a two-layer

G. Korotcenkov, *Handbook of Gas Sensor Materials,* Integrated Analytical Systems,
DOI 10.1007/978-1-4614-7388-6_26, © Springer Science+Business Media New York 2014

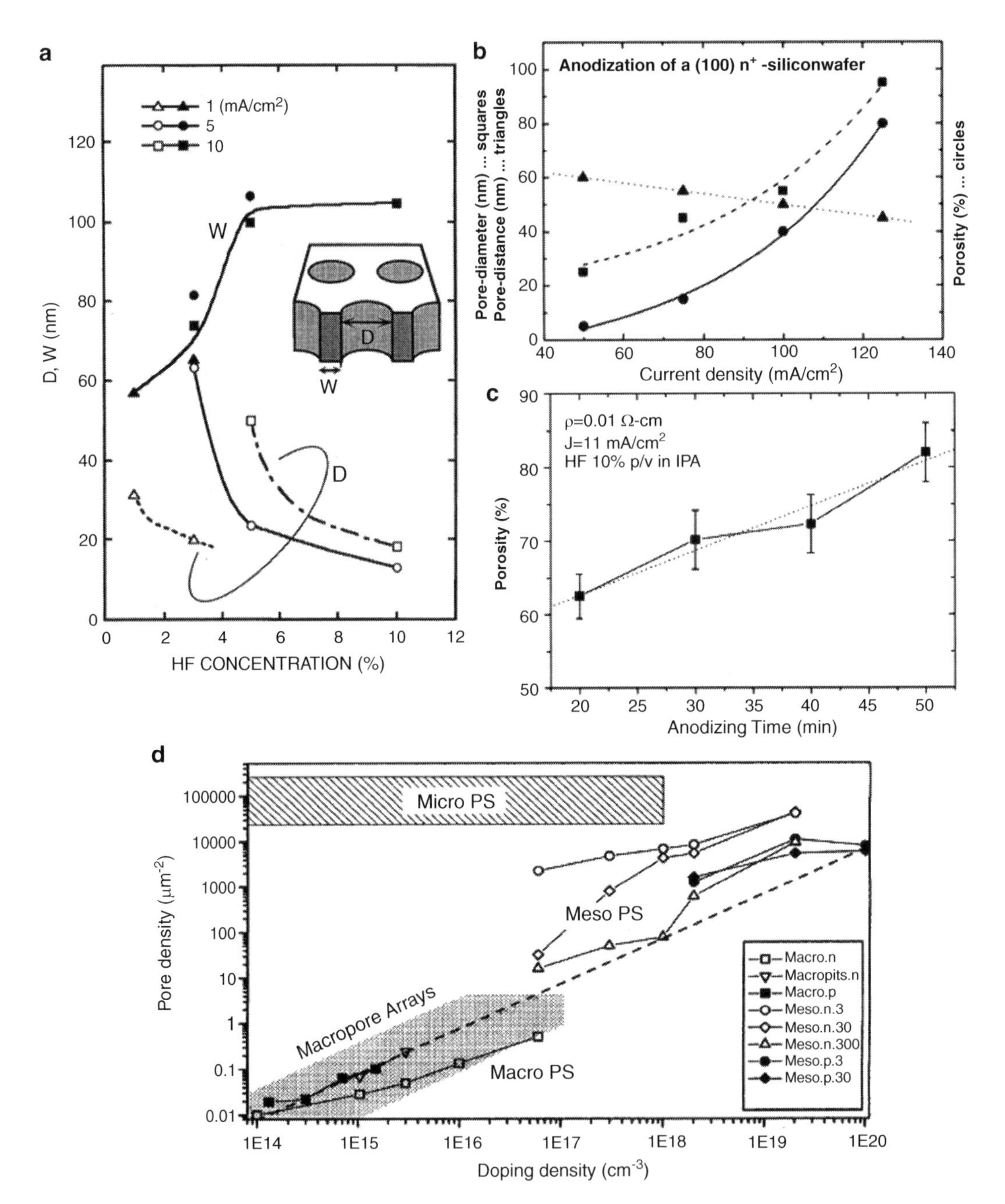

Fig. 26.1 (**a**) The pore diameter, *D*, and the width of Si wall, *W*, as a function of HF concentration for several samples with various anodization current densities (Reprinted with permission from Nakagawa et al. 1998, Copyright 1998 The Japan Society of Applied Physics). (**b**) Relation between current density and pore diameter and porosity as well during anodization of a (100) $n+$Si wafer in a 10 wt% aqueous HF solution (Reprinted with permission from Rumpf et al. 2009, Copyright 2009: Wiley). (**c**) Porosity as a function of the Si anodization time ($\rho \sim 0.01$□ cm; I ~ 11 mA/ᶜm2) (Reprinted with permission from Torres et al. 2008, Copyright 2008 Elsevier). (**d**) Pore density vs. silicon electrode doping density for PSi layers of different size regimes ($I \sim 3$–300 mA/cm^2). The *dashed line* shows the pore density of a triangular pore pattern with a pore pitch equal to double the space charge region (SCR) width at 3 V. Note that only macropores on *n*-type substrates may show a pore spacing significantly exceeding the SCR width. The regime of stable macropore array formation is indicated by a *dot pattern* (Reprinted with permission from Lehmann et al. 2000, Copyright 2000 Elsevier)

Table 26.1 Effect of anodization parameters on PSi formation

An increase of	Parameters		
	Porosity	Etching rate	Critical current
HF concentration	Decreases	Decreases	Increases
Current density	Increases	Increases	–
Anodization time	Increases	Almost constant	–
Temperature	–	–	Increases
Wafer doping (*p*-type)	Decrease	Increases	Increases
Wafer doping (*n*-type)	Increases	Increases	

Source: Reprinted with permission from Bisi et al. (2000). Copyright 2000 Elsevier

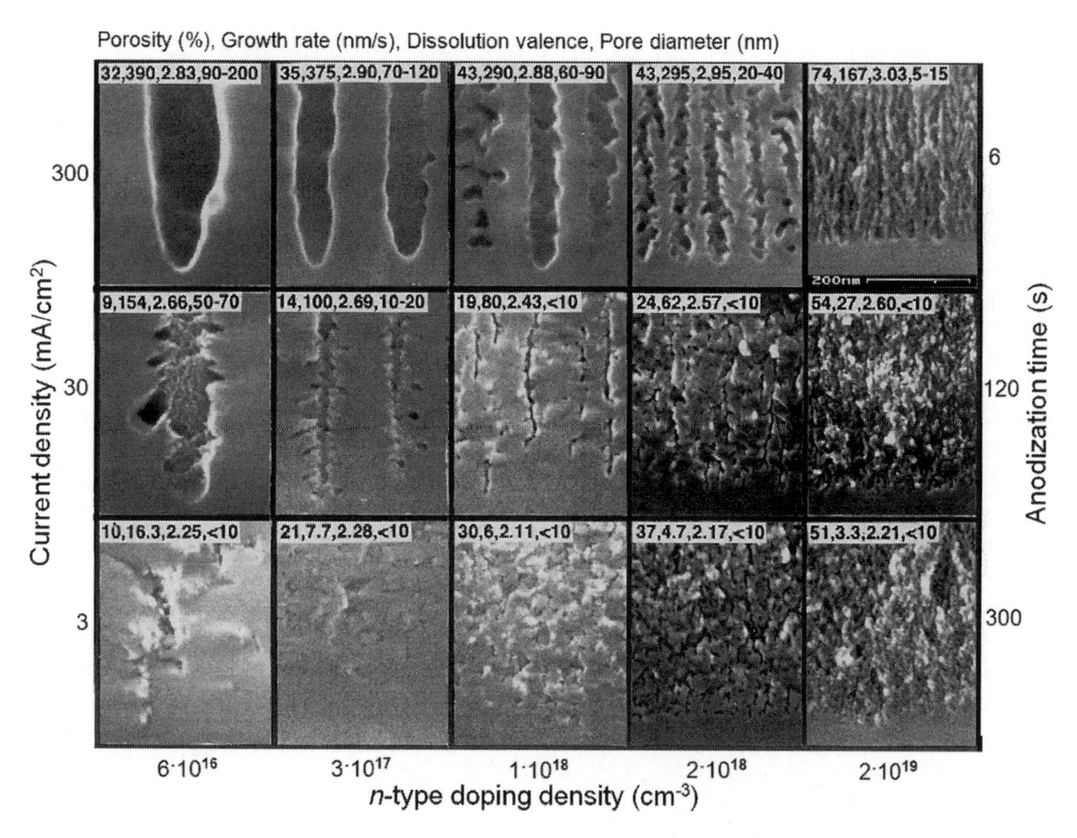

Fig. 26.2 Scanning electron micrographs of the interface between bulk and porous silicon for *n*-type-doped (100) silicon electrodes anodized galvanostatically in ethanoic HF (Reprinted with permission from Lehmann et al. 2000, Copyright 2000 Elsevier)

PSi is associated with front illumination, although it can also be formed with back illumination (Smith and Collins 1992; Lehmann 1993; Levy-Clement et al. 1994). For the micro-PSi layer formed on front illuminated *n*-Si, the pore diameter is less than 2 nm and the thickness of PSi changes with illumination intensity and the amount of charge passed.

4. Individual pores, depending on formation conditions, may propagate straight in the preferred direction with very little branching (Fig. 26.3Bb, Bc) or with the formation of numerous side or branched pores (Fig. 26.3Ad–Af). The orientation of primary pores is in general in the <100> direction for all the PSi formed on all types of (100) substrates (Smith and Collins 1992; Jager et al. 2000). In general, the conditions that favor the formation of small pores also favor branching. The branched pores can

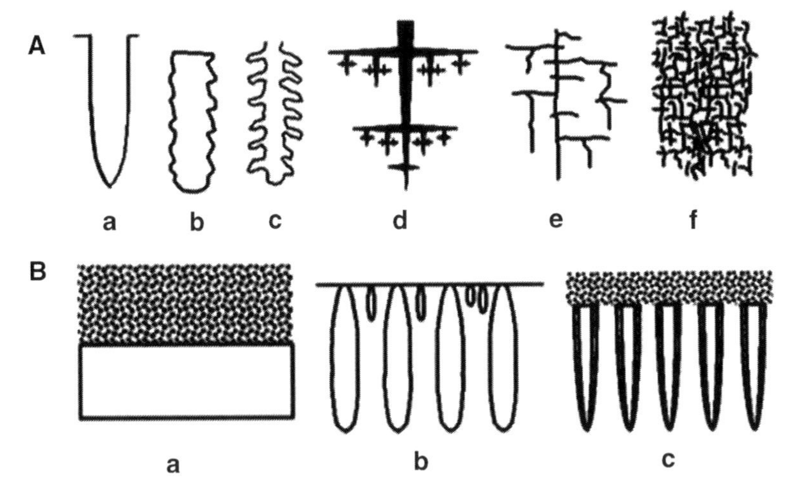

Fig. 26.3 Variants of microstructure of the PSi layers fabricated using different parameters of anodization: (**A**) (variants of branching): (**a**) (smooth pore wall); (**b**) (branches shorter than diameter); (**c**) (second level branches only); (**d**) (dendritic branches); (**e**) (main pores with second and third level branches); (**f**) (dense, random, and short branched); (**B**) (depth variation of PSi layer): (**a**) (single layer of microporous Si); (**b**) (single layer of macroporous Si with smaller pores near the surface); (**c**) a layer of micropores Si on the top of macropores Si (Idea from Zhang 2005)

have second level, third level, or further levels of branches. Branched and hierarchical pore structure has been found to form on all types of substrates. Branched pores are generally smaller than the primary pores (Smith and Collins 1992; Jager et al. 2000). The degree of branching and interpore connection depends strongly on doping concentration. For heavily doped materials, pores are generally branched (Beale et al. 1985). However, the most highly connected branching structure was found in the PSi layers formed in lowly doped *p*-Si and in microporous layers in *n*-Si formed under illumination (Smith and Collins 1992). On the other hand, well-separated and straight pores with smooth walls were generally found in PSi formed in moderately or lowly doped *n*-Si in the dark. Also, on *n*-Si, smooth and straight pores without branching can be obtained under back illumination using a surface patterned substrate. The macropores formed on lowly doped *p*-Si generally have no side pores longer than the pore diameter (Fig. 26.3Ab) (Lehmann and Ronnebeck 1999).

5. Pore diameters of the PSi layer formed under a given set of conditions can have different distributions. Normal, log-normal, bimodal, fractal, and nonuniform distributions have been found for PSi (Yaron et al. 1993; Osaka et al. 1997). For the PSi formed on heavily doped silicon, the pore diameters have a narrower distribution at a lower current at a given HF concentration and the distribution is narrower at a lower HF concentration at a given current density. Bimodal distributions of the pore diameters are generally associated with two-layer PSi formed on lowly doped *p*-Si and illuminated *n*-Si. Pores with multiple distributions have been observed for the PSi that has a surface micropore layer and smaller branched pores in addition to the main pores (Binder et al. 1996). Illumination during formation of porous layers on *p*-Si can also affect the distribution of pore diameters. It was found that illumination increases the amount of smaller nanocrystals, while it reduces the amount of larger crystals (Thonisson et al. 1996). It was also shown that for the PSi formed under illumination, the relative number of small crystals tends to increase with reduction of light wavelength (Thonisson et al. 1996).

The above-mentioned analysis indicates that operating characteristics of gas sensors based on porous Si also will be strongly dependent on both the process of electrochemical etching and the parameters of the Si wafers used. For example, the optimization of the sensor performance needs a better control of the thickness and porosity of the PS membrane and its wetting ability. Salgado et al. (2006)

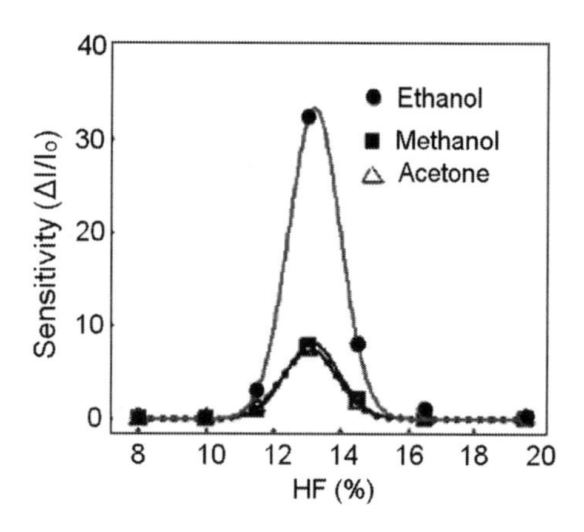

Fig. 26.4 Sensitivity to ethanol, methanol, and acetone (I/I_0) of PSi-based conductometric sensors vs. HF concentration in solution for electrochemical etching (Reprinted with permission from Irajizad et al. 2004, Copyright 2004 Elsevier)

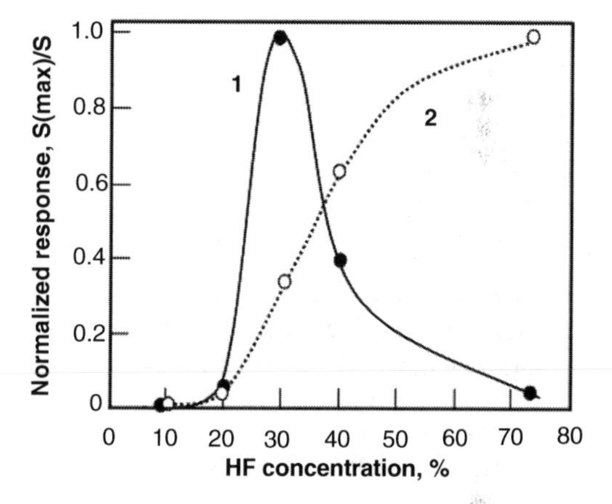

Fig. 26.5 The influence of HF concentration in electrolytes for porosification on the maximum sensitivity of (1) polysilicon and (2) SiC (d–400 to 500 nm) to humidity; the sensitivity was measured for a relative humidity (RH) change from 10 to 90 % RH (Adapted with permission from Connolly et al. 2002, Copyright 2002 Elsevier)

concluded that a very thin PSi layer is not enough to detect some appreciable resistance changes, and for thick PS layers the diffusion time of the molecules plays an important role in producing fast responses. Figure 26.4 shows how strong this influence can be. It was found that maximum sensitivity corresponds to PSi with maximum porosity (Irajizad et al. 2004).

The same result, concerning the strong sensitivity dependence on porous microstructure as well as on the thickness of the porous layer, was obtained by Pancheri et al. (2003, 2004). They observed a large microstructural transformation in PSi even for an HF concentration change between 13 and 15 %. Pancheri et al. (2003, 2004) assumed that the observed different sensitivity most likely originated in the degree of interconnection between the conducting microchannels. In the structures with lower degree of interconnection, high resistance paths, which are highly sensitive to gas, dominate the overall resistance, as in a series arrangement. On the other hand, in structures with higher interconnections, low resistance and insensitive paths frequently allow local bypasses across the sensitive paths (parallel arrangement). Thus, in the latter case, the sensitivity is dramatically inhibited.

All the above-mentioned results indicate that, for attainment of the essential parameters of chemical sensors, it is necessary to use porous layers with optimal thickness and porosity, which in most cases are being established experimentally. For example, Connolly et al. (2002) found that if the maximum sensitivity of porous silicon and polysilicon to humidity was achieved using 30 % HF, the best result for humidity sensitivity of porous SiC was obtained using 73 % HF (see Fig. 26.5). The layer microstructure in terms of pore shape, pore size, and pore distribution is very important for the capillary condensation mechanisms. This means that all technological parameters of PSi forming should have very strong control. As a result, we cannot expect that PSi-based gas sensors will have good reproducibility of their parameters.

Table 26.2 Surface treatments applied for optimization of PSi-based sensors

Surface treatment	Type of sensor	Effect
Cu S (chemical deposition)	Conduct.	Sensitivity to NH_3
Cu/Pd (magnetron sputtering)	Conduct.	H_2 sensitivity
Au (electroless)	Conduct.	Increased sensitivity to NH_3
Sn (electroless)	Conduct.	Increased sensitivity to NO, CO
Au (8–40 nm, DC sputtering)	Conduct.	Increased sensitivity to NO_2
Pd (electron beam evaporation)	Conduct.	Increased H_2 sensitivity Improved response time
Surface texturing	Conduct.	Increased sensitivity to EtOH
Burn-in process (55 °C)	Capacitance RH sensor	Increased sensitivity to RH Decreased response time
Carbonization	Capacitance RH sensor	Improved stability Decreased hysteresis
CH*x* group (thermal process)	Heterostruct.	O_2 sensors
Congo Red molecules (impregnation)	Optical	Appearance sensitivity to HCl vapors
Silanation	Optical	Stable hydrophilic surface
Cu (chemical deposition)	Optical	Increased sensitivity to methanol
Pd (immersion, clusters $\approx$ 50–100 nm)	Optical	Increased sensitivity to H_2
3-Amino-1-propanol	Optical	Appearance sensitivity to CO_2
Co phthalocyanine (immersion)	PL	Improved stability
Methyl 10-undeacetonate (derivatization)	PL	Improved stability
Grignard reagents (RT, $t=2$ h) (derivatization)	PL	Improved stability
TEOS/ethanol/water solution (RT, $t=3$–232 h)	PL	Increase of integral PL intensity
10 % H_2SeO_3 acid (RT, $t=30$–150 s)	PL	Increase of PL intensity
Pd (evaporation, $d \approx 20$ nm)	CPD	Increased sensitivity to H_2

Source: Reprinted with permission from Korotcenkov and Cho 2010b. Copyright 2010 Taylor and Francis

26.1.1 *Surface Modification of Porous Semiconductors to Improve Gas-Sensing Characteristics*

As shown in the previous section, to achieve the essential parameters of gas sensors, it is necessary to use porous layers with optimal thickness and porosity. However, it should not be forgotten that the surface chemistry of the inner walls of the pores controls the adsorption of gases as well as the capillarity condensation. Therefore, in designing sensors based on porous materials, the opportunity to control these processes, using various treatments for surface functionalizing and stabilization, should not be ignored. As demonstrated earlier for metal oxide-based sensors, such an approach makes it possible to optimize better the parameters of gas sensors. The results of numerous research projects, which can be found in Table 26.2, have shown that such an approach for the design of gas sensors based on porous semiconductors is effective as well.

For example, Sharma et al. (2007b) demonstrated that texturizing a silicon surface before porosification is a simple and effective way to form highly porous, highly luminescent, thick films of PSi with reduced stress, improved stability, and superior mechanical properties. Good results may also be obtained by partial oxidation of porous silicon. This approach was discussed in Chap. 22 (Vol. 2). Fürjes et al. (2003) found that, after partial oxidation, the sensitivity of both the surface

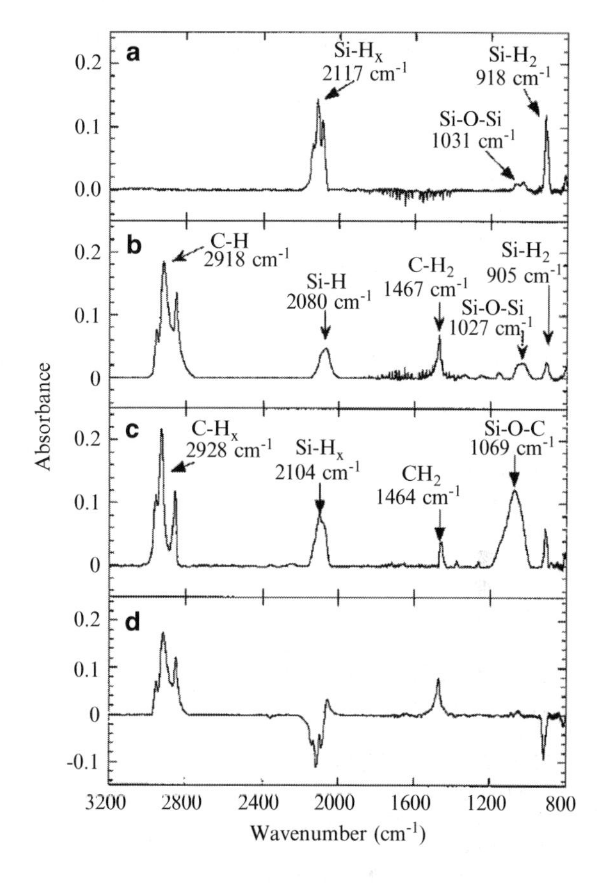

Fig. 26.6 Diffuse reflectance infrared Fourier transform spectra for (**a**) freshly prepared PSi before functionalization and PSi derivatized with (**b**) 1-decene and (**c**) decylaldehyde and (**d**) a difference DRIFTS spectrum of (**b**)–(**a**) (Reprinted with permission from Boukherroub et al. 2001, Copyright 2001 American Chemical Society)

potential and the electroconductivity of the PSi to a change of H_2S concentration increased considerably. Fürjes et al. (2003) supposed that the positive influence of PSi oxidation was connected with a decrease of density of local electron states at the PSi–SiO_2 interface. Unfortunately, thermally oxidized PSi is not completely stable, especially under high humidity (Bjorkqvist et al. 2003). It has also been established that the surface chemistry can be changed by silanization (Stewart and Buriak 1998), carbonization (Salonen et al. 2000; Bjorkqvist et al. 2004a, b), or functionalization by the covalent binding of functional groups (O'Halloran et al. 1997; Buriak and Allen 1998a; Mahmoudi et al. 2007b).

Most of the above-mentioned approaches are based on PSi surface modification with Si–C bonds. Finally we obtain derivatized PSi surface, in which the Si–H*x* bonds were replaced by Si–C bonds due to hydrosilylation of alkenes or alkynes on the PSi surface (Salonen and Lehto 2008). The change of FTIR spectra of porous silicon after derivatization is shown in Fig. 26.6. The absorption intensity of $\nu_{\text{Si–Hx}}$ and δ_{Si-H_2} decreases substantially after the reaction, indicating that most of the hydrogen has reacted with the unsaturated C=C and C=O double bonds (Fig. 26.6d). This result is consistent with a hydrosilylation reaction that preferentially consumes the more reactive SiH_3 and SiH_2 species (negative bands for the SiH*x* stretch modes around 2,117 cm^{-1} and the Si–H_2 scissors mode at 915 cm^{-1}).

The group of Buriak (Stewart and Buriak 1998, 2001; Buriak and Allen 1998b; Buriak 1999; Robins et al. 1999; Buriak et al. 1999; Holland et al. 1999) introduced three different approaches to obtain the chemically derivatized PSi surface: Lewis acid-mediated hydrosilylation, white light-promoted hydrosilylation, and cathodic electrografting. The highest treatment efficiency (the proportion of replaced Si–H*x*) of 28 % was obtained with one pentene using Lewis acid-mediated hydrosilylation (Stewart et al. 2000). Boukherroub et al. (2001) extended the usable techniques for hydrosilylation by introducing a simple thermally promoted approach for hydrosilylation. Later on, they used microwaves to improve the treatment efficiency (Boukherroub et al. (2003) and also produced a hydrophilic derivatized PSi surface with undecylenic acid Boukherroub et al. (2002)). Due to the simplicity, thermal hydrosilylation is an interesting choice, considering the biomedical applications. At the beginning, the treatment efficiency remained quite low (20–30 %), but further improvements in the treatments have uplifted the efficiency to 80 % (Lees et al. 2003). The hydrosilylation, its chemical and biological applications, and also some other approaches to PSi stabilization have been reviewed in detail elsewhere (Song and Sailor 1999; Salonen and Lehto 2008). Instead of the organic liquids used in hydrosilylation, the use of gaseous hydrocarbons could be considered due to their small size and rapid diffusion into the pores. Indeed, the poor treatment efficiency due to the low substitution levels generally obtained with the long organic molecules used in hydrosilylation may be avoided by using small gas molecules, such as acetylene or acetone vapor (Salonen et al. 2000, 2002, 2004; Orlov and Skvortsov 2001; Lakshmikumar and Singh 2002, 2003).

Long-term stability studies of differently stabilized porous silicon samples have shown that thermal carbonization of PSi is an even more efficient method of stabilization than thermal oxidation (Bjorkqvist et al. 2003). Thermal carbonization of PSi has been studied since the year 2000 (Salonen et al. 2000, 2002, 2004; Tuura et al. 2008). This technique uses an interesting property of acetylene molecules. Adsorbed acetylene molecules stick so strongly on the Si surface at room temperature that they remain on the surface up to dissociation at temperatures above 400 °C. At the same time, hydrogen from the surface termination of the as-anodized PSi desorbs and the carbon atoms bind to the silicon atoms resulting in the carbonized PSi surface. Due to the above-mentioned properties and the rapid and easy diffusion of the relatively small acetylene molecules, complete carbonization of the surface can be achieved. It was established that a thermally carbonized porous silicon (TC-PSi) surface is at least as stable in a humid atmosphere as a thermally oxidized PSi surface. The thermally carbonized surface has been found to be very stable in chemically harsh environments and even in HF and KOH solutions. Moreover, the sensitivity of TC-PSi is presumably better due to its larger specific surface area. Mahmoudi et al. (2007a) have shown that plasma in a methane/argon mixture can also be used for deposition of a hydrocarbon layer. A longtime stabilization of the PL intensity of CH_x/PSi has been confirmed by PL measurements at intervals of 1 month in aging time.

Deposition of polymer layers on the porous structure has similar effects and changes the stability and sensitivity of PSi-based devices as well (Vrkoslav et al. 2005; Xia et al. 2005; Benilov et al. 2007). For example, Xia et al. (2005) established that PDMS (polydimethylsiloxane) monolayers provide good protection and some characteristic improvement for PL of PSi. The measurements have shown that the PDMS monolayer provided a strong armature to PSi under a variety of stringent conditions such as in the base solution. Vrkoslav et al. (2005) also showed that impregnation of porous silicon with cobalt phthalocyanine ($Co^{II}Pc$) is an effective way to improve the stability of the photoluminescence quenching response.

The effect of PSi surface modification by $Co^{II}Pc$ is clearly shown in Fig. 26.7. For $Co^{II}Pc$-modified PSi films, the shortening of the PL quenching time and prolongation of the PL recovery time for a homological set of linear alcohols was also observed. According to results obtained by Vrkoslav et al. (2005), the protection of the PSi surface with $Co^{II}Pc$ results in a substantial increase of resistance against slow ambient-temperature oxidation. Regarding the accelerating of PL quenching in modified PSi, Vrkoslav et al. (2005) believed that the reduction of pore size and the increase of PSi surface polarity due to $Co^{II}Pc$ impregnation were the main reasons for these effects. Hedrich et al. (2000) established that methyl 10-undeacetonate derivatization of the PSi surface also increased the PL time stability of PSi-based sensors. However, it should be remembered that the better stability of the sensor parameters is paid for by a decrease of sensor response. For example, PSi surface stabilization by $Co^{II}Pc$ was accompanied by a decrease of sensitivity of a factor of 1.5–2 (Vrkoslav et al. 2005). Kim and Laibinis (1998) established that methyl 10-undeacetonate derivatization of the PSi surface also increased the PL time stability of PSi-based sensors.

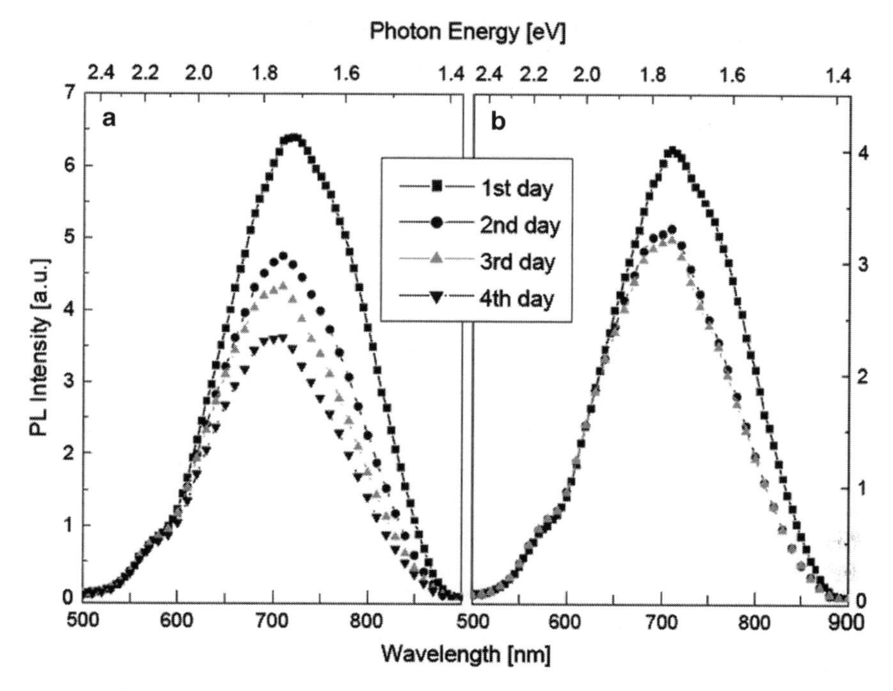

Fig. 26.7 Evolution of PL spectra for (**a**) unmodified PSi and (**b**) $Co^{II}Pc$-impregnated PSi during long-term measurements of PL quenching of *n*-alcohols (Reprinted with permission from Vrkoslav et al. 2005, Copyright 2005 Elsevier)

Very interesting results were reported by Rocchia et al. (2003). Experiments carried out by the authors of this chapter showed that, with a surface modification by 3-amino-1-propanol, one can fabricate PSi-based conductometric devices that are sensitive to CO_2. Unfortunately, the chemical nature of the surface species both before and after the binding of CO_2 is not clear at this time. In addition, the detection limit given in this work is still far from market requirements, but the reversibility and low cost of this system represents a starting point for future development of PSi-based CO_2 gas sensors.

It has been shown that the improvement of sensitivity and selectivity can be achieved by deposition of palladium (Polishchuk et al. 1998) and copper (Arwin et al. 2003) on the PSi layers. Arwin et al. (2003) showed that sensitivity to methanol increased by a factor of 2. Similar experiments with ethanol and propanol did not show an increase in sensitivity. For PSi-based Pd-catalyzed H_2 sensors, it has been established that (1) the distribution of Pd over the porous skeleton plays a significant role dictating the dynamics of the sensor (Rahimi et al. 2005) and (2) the response time and stability of the sensor are influenced by the complex Pd–H interactions (McLellan 1997), which in turn depend critically on the concentration of binding sites (pores). Mlcak et al. (1994) established that Pd-doped PSi films show enhanced catalytic activity to hydrogen oxidation into water, which implies an increase of the reaction rate. It was also found that the catalytic activity of Pd-doped porous silicon at 160 °C is significantly higher than that of a planar surface covered with Pd. Rahimi et al. (2005) observed that sensors made with porous silicon and palladium nanoparticles demonstrated a significant decrease in resistivity with respect to time when exposed to hydrogen. The Pd nanoparticles also decreased the adsorption and desorption times, which increased the sensitivity, response, and regeneration times of the sensor. The optimizing effect of PSi coating by Pd clusters was observed for optical H_2 sensors as well (Lin et al. 2004). H_2 gas at a concentration as low as 0.17 vol.% can be detected in a few seconds by monitoring either the change in optical thickness or the change in reflected light intensity obtained from the interferometric reflectance spectrum.

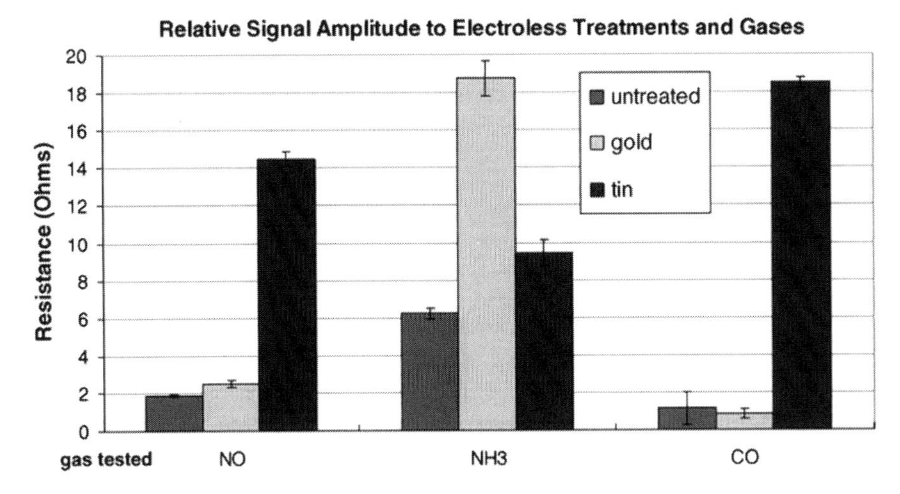

Fig. 26.8 Comparison of response for sensors that are untreated, treated with electroless gold, or treated with electroless tin and tested with 30 repeat pulses of 20 ppm NO_x, NH_3, or CO. Their average impedance change is given (Reprinted with permission from Lewis et al. 2005, Copyright 2005 Elsevier)

Baratto et al. (1999) found that gold deposited by sputtering catalyzed the porous silicon response to NO_2. The response to interfering gases such as CO, CH_4, ethanol, and methanol was negligible. Further, the gold penetrated into the pores instead of forming a continuous layer. Surface modification with gold also promoted an increase of sensor response to NH_3 and SO_2, whereas surface modification with tin was more effective in improving sensitivity to NO and CO (see Fig. 26.8) (Lewis et al. 2005). Here the electroless process was used for surface modification of the porous silicon. Modified sensors could detect a number of analytes at very low concentrations, including CO (<5 ppm), NO_x (<1 ppm), SO_2 (<1 ppm), and NH_3 (500 ppb).

There are also many methods used for biofunctionalization of PSi surfaces. However, we will not discuss them. One can find this information, for example, in review papers prepared by Salonen and Lehto (2008).

References

Arwin H, Wang G, Jansson R (2003) Gas sensing based on ellipsometric measurement on porous silicon. Phys Status Solidi A 197:518–522

Baratto C, Sberveglieri G, Comini E, Faglia G, La Ferrara V, Lancellotti L, Di Francia G, Quercia L, Guidi V, Boscarino D, Rigato V (1999) Gold catalysed porous silicon sensor for nitrogen dioxide. In: CD proceeding of the 13th European conference on solid-state transducers, EUROSENSORS XIII, The Hague, the Netherlands, 12–15 September 1999, pp 105–108, Abstract 4P10

Beale MIJ, Chew NG, Uren MJ, Cullis AG, Benjamin JD (1985) Microstructure and formation mechanism of porous silicon. Appl Phys Lett 46(1):86–88

Benilov A, Gavrilchenko I, Benilova I, Skryshevsky V, Cabrera M (2007) Influence of pH solution on photoluminescence of porous silicon. Sens Actuators A 137:345–349

Binder M, Edelmann T, Metzger TH, Mauckner G, Goerigk G, Peisl J (1996) Bimodal size distribution in p^- porous silicon studied by small angle X-ray scattering. Thin Solid Films 276:65–68

Bisi O, Ossicini S, Pavesi L (2000) Porous silicon: a quantum sponge structure for silicon based optoelectronics. Surf Sci Rep 38:1–126

Bjorkqvist M, Salonen J, Paski J, Laine E (2003) Comparison of stabilizing treatments on porous silicon for sensor applications. Phys Status Solidi (a) 197:374–377

Bjorkqvist M, Salonen J, Paski J, Laine E (2004a) Characterization of thermally carbonized porous silicon humidity sensor. Sens Actuators A 112:244–247

Bjorkqvist M, Salonen J, Laine E (2004b) Humidity behavior of thermally carbonized porous silicon. Appl Surf Sci 222:269–274

Bomchil G, Halimaoui A, Herino R (1989) Porous silicon: the material and its applications in silicon-on-insulator technologies. Appl Surf Sci 41/42:604–613

Boukherroub R, Morin S, Wayner DDM, Bensebaa F, Sproule GI, Baribeau JM, Lockwood DJ (2001) Ideal passivation of luminescent porous silicon by thermal, noncatalytic reaction with alkenes and aldehydes. Chem Mater 13: 2002–2011

Boukherroub R, Wojtyk JTC, Wayner DDM, Lockwood DJ (2002) Thermal hydrosilylation of undecylenic acid with porous silicon. J Electrochem Soc 149:H59–H63

Boukherroub R, Petit A, Loupy A, Chazalviel JN, Ozanam F (2003) Microwave-assisted chemical functionalization of hydrogen-terminated porous silicon surfaces. J Phys Chem B 107:13459

Buriak JM (1999) Silicon-carbon bonds on porous silicon surfaces. Adv Mater 11:265–267

Buriak JM, Allen MJ (1998a) Lewis acid mediated functionalization of porous silicon with substituted alkenes and alkynes. J Am Chem Soc 120:1339–1340

Buriak JM, Allen MJ (1998b) Photoluminescence of porous silicon surfaces stabilized through Lewis acid mediated hydrosilylation. J Luminesc 80:29–35

Buriak JM, Stewart MP, Geders TW, Allen MJ, Choi HC, Smith J, Raftery D, Canham LT (1999) Lewis acid mediated hydrosilylation on porous silicon surfaces. J Am Chem Soc 121:11491–11502

Connolly EJ, O'Halloran GM, Pham HTM, Sarro PM, French PJ (2002) Comparison of porous silicon, porous polysilicon and porous silicon carbide as materials for humidity sensing applications. Sens Actuators B 99:25–30

Fürjes P, Kovács A, Cs D, Ádám M, Müller B, Mescheder U (2003) Porous silicon-based humidity sensor with interdigital electrodes and internal heaters. Sens Actuators B 95:140–144

Hedrich F, Billat S, Lang W (2000) Structuring of membrane sensors using sacrificial porous silicon. Sens Actuators A 84:315–323

Holland JM, Stewart MP, Allen MJ, Buriak JM (1999) Metal mediated reactions on porous silicon surfaces. J Solid State Chem 147:251–258

Irajizad A, Rahimi F, Chavoshi M, Ahadian MM (2004) Characterization of porous poly-silicon as a gas sensor. Sens Actuators B 100:341–346

Jager C, Finkenberger B, Jager W, Christophersen M, Carstensen J, Foll H (2000) Transmission electron microscopy investigations of the formation of macropores in *n*- and *p*-Si(001)/(111). Mater Sci Eng B 69–70:199–204

Kim NY, Laibinis PE (1998) Derivatization of porous silicon by Grignard reagents at room temperature. J Am Chem Soc 120:4516–4517

Korotcenkov G, Cho BK (2010a) Silicon porosification: state of the art. Crit Rev Solid State Mater Sci 35(3):153–260

Korotcenkov G, Cho BK (2010b) Porous semiconductors: advanced material for gas sensor applications. Crit Rev Solid State Mater Sci 35(1):1–37

Lakshmikumar ST, Singh PK (2002) Formation of carbonized porous silicon surfaces by thermal and optically induced reaction with acetylene. J Appl Phys 92:3413–3415

Lakshmikumar ST, Singh PK (2003) Stabilization of porous silicon surface by low-temperature photo-assisted reaction with acetylene. Curr Appl Phys 3:185–189

Lees IN, Lin H, Canaria CA, Gurtner C, Sailor MJ, Miskelly GM (2003) Chemical stability of porous silicon surfaces electrochemically modified with functional alkyl species. Langmuir 19:9812–9817

Lehmann V (1993) The physics of macropore formation in low doped *n*-type silicon. J Electrochem Soc 140(10): 2836–2843

Lehmann V, Gruning U (1997) The limits of macropore array fabrication. Thin Solid Films 297:13–17

Lehmann V, Ronnebeck S (1999) The physics of macropore formation in low-doped p-type silicon. J Electrochem Soc 146:2968–2975

Lehmann V, Stengl R, Luigart A (2000) On the morphology and the electrochemical formation mechanism of mesoporous silicon. Mater Sci Eng B 69–70:11–22

Levy-Clement C, Lagoubi A, Tomkiewicz M (1994) Morphology of porous n-type silicon obtained by photoelectrochemical etching I. Correlations with material and etching parameters. J Electrochem Soc 141:958–967

Lewis SE, DeBoer JR, Gole JL, Hesketh PJ (2005) Sensitive, selective, and analytical improvements to a porous silicon gas sensor. Sens Actuators B 110:54–65

Lin H, Gao T, Fantini J, Sailor MJ (2004) A porous silicon-palladium composite film for optical interferometric sensing of hydrogen. Langmuir 20:5104–5108

Mahmoudi BE, Gabouze N, Haddadi M, Mahmoudi BR, Cheraga H, Beldjilali K, Dahmane D (2007a) The effect of annealing on the sensing properties of porous silicon gas sensor: use of screen-printed contacts. Sens Actuators B 123:680–684

Mahmoudi BE, Gabouze N, Guerbous L, Haddadi M, Beldjilali K (2007b) Long-time stabilization of porous silicon photoluminescence by surface modification. J Lumin 127:534–540

McLellan RB (1997) The kinetic and thermodynamic effects of vacancy interstitial interactions in Pd–H solutions. Acta Mater 45:1995–2000

Mlcak R, Tuller HL, Greiff P, Sohn J, Niles L (1994) Photoassisted electrochemical micromachining of silicon in HF electrolytes. Sens Actuators A 40:49–55
Nakagawa T, Sigiyama H, Koshida N (1998) Fabrication of periodic Si nanostructure by controlled anodization. Jpn J Appl Phys 37:7186–7189
O'Halloran GM, Kuhl M, Trimp PJ, French PJ (1997) The effect of additives on the adsorption properties of porous silicon. Sens Actuators A 61:415–420
Orlov AM, Skvortsov AA (2001) Effects of high-temperature carbonizing and plasma processing on the optical properties of porous silicon. Inorg Mater 37:761–766
Osaka T, Ogasawara K, Nakahara S (1997) Classification of the pore structure of n-type silicon and its microstructure. J Electrochem Soc 144:3226–3237
Pancheri L, Oton CJ, Gaburro Z, Soncini G, Pavesi L (2003) Very sensitive porous silicon NO_2 sensor. Sens Actuators B 89:237–239
Pancheri L, Oton J, Gaburro Z, Soncini G, Pavesi L (2004) Improved reversibility in aged porous silicon NO_2 sensors. Sens Actuators B 97:45–48
Polishchuk V, Souteyrand E, Martin JR, Strikha VI, Skryshevsky VA (1998) A study of hydrogen detection with palladium modified porous silicon. Anal Chim Acta 375:205–210
Rahimi F, Irajizad A, Razi F (2005) Characterization of porous polysilicon impregnated with Pd as a hydrogen sensor. J Phys D: Appl Phys 38:36–40
Robins EG, Stewart MP, Buriak JM (1999) Anodic and cathodic electrografting of alkynes on porous silicon. Chem Commun 1999:2479–2480
Rocchia M, Garrone E, Geobaldo F, Boarino L, Sailor MJ (2003) Sensing CO_2 in a chemically modified porous silicon film. Phys Status Solidi (a) 197:365–369
Rumpf K, Granitzer P, Poelt P, Krenn H (2009) Transition metals specifically electrodeposited into porous silicon. Phys Status Solidi (c) 6(7):1592–1595
Salgado GG, Becerril TD, Santiesteban HJ, Andres ER (2006) Porous silicon organic vapor sensor. Opt Mater 29:51–55
Salonen J, Lehto V-P (2008) Fabrication and chemical surface modification of mesoporous silicon for biomedical applications. Chem Eng J 137:162–172
Salonen J, Lehto V-P, Björkqvist M, Laine E, Niinistö L (2000) Studies of thermally-carbonized porous silicon surfaces. Phys Status Solidi (a) 182:123–126
Salonen J, Laine E, Niinisto L (2002) Thermal carbonization of porous silicon surface by acetylene. J Appl Phys 91:456–461
Salonen J, Bjorkqvist M, Laine E, Niinisto L (2004) Stabilization of porous silicon surface by thermal decomposition of acetylene. Appl Surf Sci 225:389–394
Sharma SN, Bhagavannarayana G, Kumar U, Debnath R, Mohan SC (2007) Role of surface texturization on the gas-sensing properties of nanostructured porous silicon films. Physica E 36:65–72
Smith RL, Collins SD (1992) Porous silicon formation mechanisms. J Appl Phys 71(8):R1–R22
Song JH, Sailor MJ (1999) Chemical modification of crystalline porous silicon surfaces. Comments Inorg Chem 21:69–84
Stewart MP, Buriak JM (1998) Photopatterned hydrosilylation on porous silicon. Angew Chem Int Ed 37:3257–3260
Stewart MP, Buriak JM (2001) Exciton-mediated hydrosilylation on photoluminescent nanocrystalline silicon. J Am Chem Soc 123:7821–7830
Stewart MP, Robins EG, Geders TW, Allen MJ, Choi HC, Buriak JM (2000) Three methods for stabilization and functionalization of porous silicon surfaces via hydrosilylation and electrografting reactions. Phys Status Solidi (a) 182:109–115
Thonisson M, Berger MG, Arens-Fisher R, Gluck O, Kruger M, Luth H (1996) Illumination-assisted formation of porous silicon. Thin Solid Films 276:21–24
Torres J, Martinez HM, Alfonso JE, Lopez LD (2008) Optoelectronic study in porous silicon thin films. Microelectronics J 39(3–4):482–484
Tuura J, Bjorkqvist M, Salonen J, Lehto V-P (2008) Electrically isolated thermally carbonized porous silicon layer for humidity sensing purposes. Sens Actuators B 131:627–632
Vrkoslav V, Jelınek I, Matocha M, Kral V, Dian J (2005) Photoluminescence from porous silicon impregnated with cobalt phthalocyanine. Mater Sci Eng C 25:645–649
Xia B, Xiao S-J, Wang J, Guo D-J (2005) Stability improvement of porous silicon surface structures by grafting polydimethylsiloxane polymer monolayers. Thin Solid Films 474:306–309
Yaron AA, Bastide S, Maurice JL, Clement CL (1993) Morphology of porous n-type silicon obtained by photoelectrochemical etching. 2. Study of the tangled Si wires in the nanoporous layers. J Lumin 57:67–71
Zhang XG (1991) Mechanism of pore formation on *n*-type silicon. J Electrochem Soc 138:3750–3756
Zhang GX (2005) Porous silicon: morphology and formation mechanisms. In: Vayenas C (ed) Modern aspects of electrochemistry, vol 39. Springer, New York, p 65

Part VI
Technology and Sensing Material Selection

Chapter 27
Technological Limitations in Sensing Material Applications

Good technological effectiveness and processability, i.e., the ability to produce, under control and with reproducibility, powders, films, and ceramics with the required structural and morphological properties, is also an important criterion in selecting a material for a gas sensor (Barsan et al. 1999). Both complicated techniques and the absence of a technological base for mass production can also considerably limit the application of a particular sensing material. This confirms again that, for practical use, considerations of stability, reliability, and technological effectiveness are determinative.

Today, we do not have any technical problems in fabricating binary oxides, solid electrolytes, and standard semiconductors with specified electrophysical, chemical, and structural properties. In the literature, one can find a great number of works devoted to the elaboration of various techniques for deposition of sensing materials (Nenov and Yordanov 1996; Will et al. 2000; Van Tassel and Randall 2006; Viswanathan et al. 2006; Vayssieres 2007; Milchev 2008; Tiemann 2008). These methods were analyzed in detail by Korotcenkov (2010). However, for more complicated oxides (for example, binary oxides modified by various additives), there are still many problems to be resolved when attempting deposition of these materials.

We should note that the processability is an important factor for polymer materials as well. It is known that conjugated polymers may be made by a variety of techniques, including cationic, anionic, radical chain growth, coordination polymerization, step-growth polymerization, or electrochemical polymerization. Electrochemical polymerization occurs by suitable monomers which are electrochemically oxidized to create an active monomeric and dimeric species which react to form a conjugated polymer backbone. The main problem with electrically conductive plastic stems from the very property that gives it its conductivity, namely, the conjugated backbone. This causes many such polymers to be intractable, insoluble films or powders that cannot melt. There are two main strategies for overcoming these problems: either to modify the polymer so that it may be more easily processed or to manufacture the polymer in its desired shape and form. There are, at this time, four main methods used to achieve these aims (Pratt 2003). The first method is to manufacture a malleable polymer that can be easily converted into a conjugated polymer. This is done when the initial polymer is in the desired form and then, after conversion, is treated so that it becomes a conductor. The treatment used is most often thermal treatment. The precursor polymer used is often made to produce highly aligned polymer chain, which are retained upon conversion. The second method is the synthesis of copolymers or derivatives of a parent conjugated polymer with more desirable properties. This method is the more traditional one for making improvements to a polymer. What is done is to try to modify the structure of the polymer to increase its processability without compromising its conductivity or its optical properties. All attempts to do this on polyacetylene have failed as they always significantly reduced its conductivity. However, such attempts on polythiophenes (PTs) and polypyrroles proved more fruitful. The hydrogen on carbon 3 on the thiophene or the pyrrole ring was replaced with an alkyl group with at least four carbon atoms

G. Korotcenkov, *Handbook of Gas Sensor Materials,* Integrated Analytical Systems, DOI 10.1007/978-1-4614-7388-6_27, © Springer Science+Business Media New York 2014

Table 27.1 Stability and processing attributes of some conducting polymers

Polymer	Conductivity (W^{-1}·cm^{-1})	Stability (doped state)	Processing possibilities
Polyacetylene (PAc)	10^3–10^5	Poor	Limited
Polyphenylene (PPE)	10^3	Poor	Limited
Polyphenylene sulfide (PPS)	10^2	Poor	Excellent
Poly(*p*-phenylene vinylene) (PPV)	10^3	Poor	Limited
Polypyrroles (PPy)	10^2	Good	Good
Polythiophenes (PTs)	10^2	Good	Excellent
Polyaniline (PANI)	10	Good	Good

Source: Data extracted from Pratt (2003)

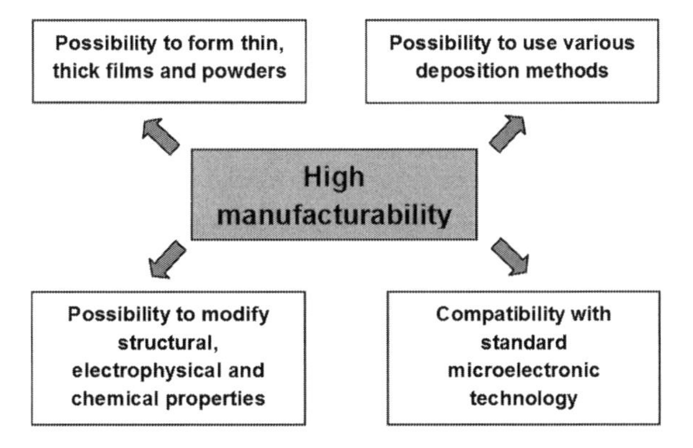

Fig. 27.1 Factors characterizing high manufacturability of sensing materials

in it. One or more CH-groups in the thiophene or the pyrrole rings were replaced with an alkyl groups. The number of carbon atoms in the alkyl group may vary from 1 to 20 carbon atoms. The resulting polymer, when doped, has a comparable conductivity to its parent polymer whilst be able to melt and it is soluble. A water soluble version of these polymers has been produced by placing a carboxylic acid group or a sulfonic acid group on the alkyl chains. If sulfonic acid groups are used along with built-in ionizable groups, then such a system can maintain charge neutrality in its oxidized state, and so they effectively dope themselves. Such polymers are referred to as "self-doped" polymers. The third method is to grow the polymer in its desired shape and form. An insulating polymer impregnated with a catalyst system is fabricated in its desired form. This is then exposed to the monomer, usually a gas or a vapor. The monomer then polymerizes on the surface of the insulating plastic producing a thin film or a fiber. This is then doped in the usual manner. The final method is the use of the Langmuir–Blodgett trough to manipulate the surface-active molecules into a highly ordered thin film whose structure and thickness are controllable at the molecular layer. We can see that there are numerous approaches to achieve the required properties of polymeric materials, and therefore the optimal polymer should give maximum technological possibilities for resolving this task. The processing possibilities for several most used polymers are presented in Table 27.1.

An important aspect of the technological effectiveness of the material is also its potential to be adapted for modern microelectronic techniques (see Fig. 27.1). This aspect is especially evident in the area of microminiaturization of high-temperature solid electrolyte gas sensors. In this case, the difficulties concern agglomeration of a very fine metallic electrode structure on the oxide surface at high temperature and film cracks due to thermal expansion mismatch between the thin film and the substrate. Thermal stress results for heating from ambient temperature to operating temperature for films deposited at ambient temperature or for cooling to ambient temperature for films deposited or

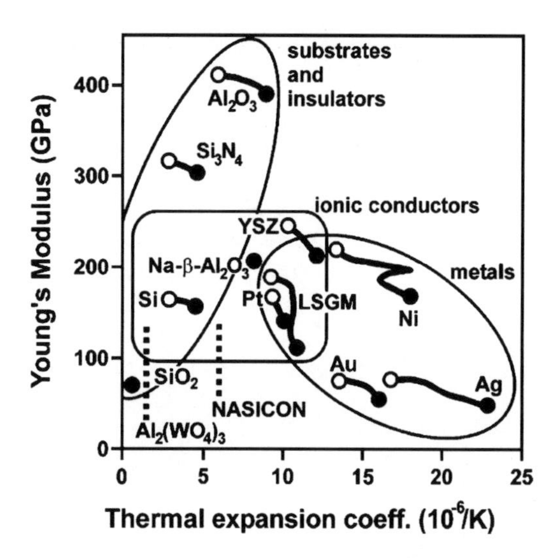

Fig. 27.2 Young's modulus and linear thermal expansion coefficients (relative to 300 K) between 300 K (*open circles*) and 1,000 K (*closed circles*) of electrode metals, ionic conductors, and substrate/insulator materials for solid-state ionic micro gas sensors (polycrystalline materials). Young's modulus of NASICON and $Al_2(WO_3)_4$ were not available (Reprinted with permission from Dubbe 2003, Copyright 2003 Elsevier)

recrystallized at higher temperatures. Therefore, when choosing a material for a planar microminiature solid electrolyte-based gas sensor, one must consider these possibilities. For integration of solid-state ionic devices on silicon, the chemical compatibility of all applied materials with silicon and silicon processing has to be considered as well. Chemical compatibility factors may lead to the exclusion of materials with otherwise promising functionality.

A detailed analysis of compatibility of various materials used for solid-state ionic gas sensors fabrication with silicon technology was carried out by Dubbe (2003). He compared materials such as the solid Na^+ ion conductors NASICON and β-alumina, the Al^{3+} and Sc^{3+} ion conductors $(Al,Sc)_2((Mo,W)O_4)_3$, and the O^{2-} ion conductors LSGM ($La_{0.9}S_{0.1}Ga_{0.8}Mg_{0.2}O_{3-\delta}$), ZrO_2, CeO_2, and HfO_2 with Ca, Y, or Mg dopants. It was established that the thermal expansion coefficient of silicon is much lower than most electrode and electrolyte materials (see Fig. 27.2). Only $Al_2(WO_4)_3$ is an exception. Composites containing this solid electrolyte could be tailored to match the thermal expansivity of Si. Si_3N_4 and SiO_2 are common insulator materials or membrane materials for microhotplates but have thermal expansion coefficients even lower than silicon. Al_2O_3 has the highest thermal expansion coefficient among the substrate and insulator materials in Fig. 27.2, comparable with NASICON and β-alumina. Sapphire or polycrystalline Al_2O_3 was often used as substrate for YSZ thin-film devices (Kondo et al. 1993; Saji et al. 1993). The thermal expansivity of Pt is not much different from that of YSZ and LSGM solid electrolytes. Due to the high thermal expansion coefficients of Ag, Au, and Ni, these metals could be applied on Si or Al_2O_3 substrates or on solid electrolytes only in the form of very thin films.

Thermodynamic stability of the interface with silicon is also important for correct selection of sensor material. According to Hubbard and Schlom (1996), for MgO and ZrO_2, thermodynamic stability of their interfaces with silicon was predicted. For HfO_2, Al_2O_3, CaO, and Y_2O_3, experimental evidence showed that these oxides also form stable interfaces with silicon. The CeO_2/Si interface is thermodynamically unstable, in accordance with experimental observations of formation of a Ce_2O_3 interlayer at the CeO_2/Si interface. Present analysis shows that, for the majority of metal oxides used, the interface with Si is stable. However, ZrO_2 or YSZ thin films are permeable for oxygen at elevated temperatures, so that oxygen diffuses to the silicon substrate and a SiO_2 layer is formed at the interface (Jia et al. 1995). It should be noted that this problem exists for all metal oxides used for gas sensor design.

Interdiffusion at the interface is another parameter which depends on the nature of the material used and controls stability of devices. Interdiffusion of Zr and Si at the YSZ/Si interface was investigated by Prusseit et al. (1992), showing very slow diffusion of Zr into Si. However, we cannot say the same regarding materials containing alkali ions. Deterioration of dielectric thin films is known to arise from alkali ions, which lead to increased leakage currents in SiO_2 gate dielectrics (Chang and Sze 1996). This could rule out the application of any alkali-ion conductor in silicon devices, since alkali ions are also fast diffusors and they can segregate from Si into SiO_2. Imanaka et al. (2000) and Adachi et al. (2001) have shown that Al^{3+} and Sc^{3+} ion conductors such as $(Al,Sc)_2((Mo,W)O_4)_3$ are promising alternative cation ion conductors, which can replace alkali-ion conductors in Sibased devices. In such MOS-integrated circuits aluminum can be used as the interconnect material. Alternatively, the MOS circuits could be protected with an alkali-ion impermeable Si_3N_4 coating (Hull 1999).

It is clear that all materials introduced into a silicon microdevice should be inert to the succeeding processing steps. The oxides YSZ, CeO_2, and HfO_2 are corrosion resistant to both diluted acids and alkali solutions, including KOH or TMAH (tetramethylammonium hydroxide) solutions for anisotropic etching of silicon (Sze 1994). However, NASICON and β-alumina are sensitive to aqueous solutions.

Thus, according to Dubbe (2003), solid electrolytes such as $(Al,Sc)_2((Mo,W)O_4)_3$ are most appropriate for microelectronic designs. These electrolytes have low thermal expansion coefficients and good chemical compatibility with silicon technology. The most common material for macroscopic and thick-film solid-state ionic gas sensors, yttria-stabilized zirconia (YSZ), is also well suited for integration on silicon devices. At the same time, however, the integration of alkali-ion conductors with silicon technology is problematic because of possible degradation of the silicon devices due to contamination by alkali ions (Dubbe 2003).

We need also to take into account that some materials are incompatible with technological methods used for microelectronic designs (Brinker and Scherer 1989; Randhaw 1991; Hitchman and Jensen 1993; Bunshah 1994; Hecht et al. 1994; Glocker and Shah 1995; Arthur 2002; Choy 2003; Christen and Eres 2008; Jaworek and Sobczyk 2008). For example, during polymer sputtering using electron-beam techniques, chemical decomposition of polymers is possible, which naturally limits the application of such materials.

Implementation of 1D structures in devices intended for broad application presents several serious problems as well. Technological approaches used in manufacturing sensors based on individual 1D structures differ fundamentally from the methods used in standard silicon technology. As a result, while working with individual 1D structures, we face, as discussed earlier, great difficulties with their manipulation and formation of ohmic contacts. Therefore, new approaches, new technological decisions, as well as time will be required to solve these problems. Taking into account the above-mentioned discussions related to the manufacturability of different sensing materials, one can suppose that nanobelts (nanoribbons) probably could be the most demanding one-dimensional structure to exploit for gas-sensing applications (Korotcenkov 2008). Nanobelts are thin and plain belt-type structures with rectangular cross section. At present, nanobelts have been obtained using various methods for nearly all oxides used in gas sensors (Dai et al. 2003; Wang 2004). Typical nanobelts have widths of 20–300 nm and lengths from several micrometers to hundreds or even some thousands of micrometers. The typical width-to-thickness ratio for nanobelts ranges from 5 to 10. For comparison, for nanowires (or nanorods), this ratio equals 2–5 (Dai et al. 2003). Nanobelts do not have the mechanical strength of nanotubes. However, they have structural homogeneity and crystallographic perfection. It is well known that crystallographic defects may destroy quantum-size effects. Because of the zero defects of nanobelts, structural defects will not be a problem as observed for nanowire-type structures. It is necessary to emphasize that suitable geometry, high homogeneity of the structure, and long length are important advantages of nanobelts for mass manufacturing of gas sensors. Besides, nanobelts have flexible structures, and, therefore, they could be curved up to 180° without being damaged. It is known that nanotubes do not have such properties. This fact gives additional advantages to nanobelts for sensor designs.

Separation and selection of 1D nanostructures are also a great problem (Belin and Epron 2005). This problem is especially hard to resolve for carbon nanotubes, because during the process of separation we need to take into account both diameter and the type of conductivity of nanotubes. It is known that carbon nanotubes, which are typically grown in bundles, can have both semiconducting and metallic type of conductivity. Therefore, during the last decade many attempts were made to design an acceptable method for carbon nanotube separation (Collins et al. 2001; Chattopadhyay et al. 2003; Krupke et al. 2003; Zheng et al. 2003).

For example, Collins et al. (2001) proposed a method based on the selective destruction of metallic nanotubes, which could be realized with bursts of electricity. The ropes containing both semiconducting and metallic tubes are deposited on a flat surface of silicon oxide. Then electrodes on the top of the ropes are fabricated by lithography technique. A voltage is applied to the tubes through the electrodes. The metallic CNTs are destroyed by current-induced oxidation, and only the semiconducting tubes remain (carbon nanotubes transistors. http://www.research.ibm.com). However, this technique is only useful for transistor geometries and cannot be extended to bulk separation.

Krupke et al. (2003) have shown that the bulk separation of various CNT types could be carried out using alternating current dielectrophoresis. It was established that, when submitted to an electric field, the SWNTs have an induced dipole moment which can be used to move these tubes selectively. Of course, the bundles must be separated previously in individual tubes using an ultrasonication treatment of the nanotubes with surfactants such as the SDS (sodium dodecyl sulfate). The dielectrophoresis is then applied to this solution. The metallic tubes are deposited onto the electrodes, while the semiconducting ones remain in suspension.

Chattopadhyay et al. (2003) established that CNT types can also be separated using a precipitation route. Chattopadhyay et al. (2003) found that the SWNTs with octadecylamine (ODA) in tetrahydrofuran (THF) formed a stable suspension. Therefore the authors assume that the stability is achieved by the physisorption and the organization of the ODA molecules along the SWNT sidewalls. Whereas the metallic tubes properties are insensitive, the electrical properties of semiconducting nanotubes are modified by the adsorption of alkylamines (Kong and Dai 2001). Thus, the ODA physisorption on the SWNTs might improve the stability of the semiconducting tubes in the suspension instead of the metallic ones. In this case, there is a separation between ODA-stabilized semiconducting and metallic tubes.

Zheng et al. (2003) found that nanotube separation using the DNA-assisted method is acceptable as well. Experiment has shown that the nucleic acid polymers form a hybrid material with CNTs. The phosphate groups (from DNA) on a DNA–CNT hybrid provide a negative surface charge density which is dependent on the DNA sequence and the electronic property of the tube. In contrast, the metallic tubes in DNA–CNT hybrid contributes to a weak surface charge density. Finally, the ion-exchange liquid chromatography with DNA–CNT hybrid suspensions is used by the authors in order to separate the semiconductor and metallic nanotubes.

However, we need to recognize that the separation of nanotubes with semiconductor and metallic conductivity is only the first step in sensor fabrication. Moreover, the methods mentioned above do not resolve the problems of separation and selection of the fibers with specific diameter. The manipulation of individual nanotubes at the stage of sensor fabrication, especially taking into account mass production, also requires novel technological decisions and approaches.

References

Adachi G, Imanaka N, Tamura S (2001) Rare earth ion conduction in solids. J Alloys Compd 323–324:534–539

Arthur JA (2002) Molecular beam epitaxy. Surf Sci 500:189–217

Barsan N, Schweizer-Berberich M, Gopel W (1999) Fundamental and practical aspects in the design of nanoscaled SnO_2 gas sensors. A status report. Fresenius J Anal Chem 365:287–304

Belin T, Epron F (2005) Characterization methods of carbon nanotubes: a review. Mater Sci Eng B 119:105–118

Brinker CJ, Scherer GW (1989) Sol-gel science: the physics and chemistry of sol-gel processing. Academic, New York

Bunshah RF (ed) (1994) Handbook of deposition technologies for films and coatings, 2nd edn. Noyes, Park Ridge, NJ

Chang CY, Sze SM (eds) (1996) ULSI technology. McGraw-Hill, New York

Chattopadhyay D, Galeska I, Papadimitrakopoulos F (2003) A route for bulk separation of semiconducting from metallic single wall carbon nanotubes. J Am Chem Soc 125:3370–3375

Choy KL (2003) Chemical vapour deposition of coatings. Prog Mater Sci 48(2):57–170

Christen HM, Eres G (2008) Recent advances in pulsed-laser deposition of complex oxides. J Phys Condens Matter 20:264005

Collins PG, Arnold MS, Avouris P (2001) Engineering carbon nanotubes and nanotube circuits using electrical breakdown. Science 292:706–709

Dai ZR, Pan ZW, Wang ZL (2003) Novel nanostructures of functional oxides synthesized by thermal evaporation. Adv Funct Mater 13(1):9–24

Dubbe A (2003) Fundamentals of solid state ionic micro gas sensors. Sens Actuators B 88:138–148

Glocker DA, Shah I (eds) (1995) Handbook of thin film process technology. Institute of Physics, Bristol, UK

Hecht G, Richter F, Hahn J (eds) (1994) Thin films. DGM Informationgesellschaft, Oberursel, Germany

Hitchman ML, Jensen KF (1993) CVD: principles and applications. Academic, San Diego, CA

Hubbard KJ, Schlom DG (1996) Thermodynamic stability of binary oxides in contact with silicon. J Mater Res 11:2757–2776

Hull R (ed) (1999) Properties of crystalline silicon. INSPEC, London

Imanaka N, Kobayashi Y, Tamura S, Adachi G (2000) Trivalent ion conducting solid electrolytes. Solid State Ionics 136(137):319–324

Jaworek A, Sobczyk AT (2008) Electrospraying route to nanotechnology: an overview. J Electrostat 66:197–219

Jia QX, Wu XD, Zhou DS, Foltyn SR, Tiwari P, Peterson D, Mitchell TE (1995) Deposition of epitaxial yttria-stabilized zirconia on single-crystal Si and subsequent growth of an amorphous SiO_2 interlayer. Philos Mag Lett 72:385–391

Kondo H, Saji K, Takahashi H, Takeuchi M (1993) Thin-film air fuel ratio sensor. Sens Actuators B 13(14):49–52

Kong J, Dai H (2001) Full and modulated chemical gating of individual carbon nanotubes by organic amine compounds. J Phys Chem B 105:2890–2893

Korotcenkov G (2008) The role of morphology and crystallographic structure of metal oxides in response of conductometric-type gas sensors. Mater Sci Eng R 61:1–39

Korotcenkov G (ed) (2010) Chemical sensors: fundamentals of sensing materials. Vol. 1. General approaches. Momentum, New York

Krupke R, Hennrich F, Lohneysen H, Kappes MM (2003) Separation of metallic from semiconducting single-walled carbon nanotubes. Science 301:344–347

Milchev A (2008) Electrocrystallization: nucleation and growth of nano-clusters on solid surfaces. Russ J Electrochem 44:619–645

Nenov TG, Yordanov SP (1996) Ceramic sensors: technology and applications. Technomic, Basel

Pratt CM (2003) Effects of metal ions on the synthesis and properties of conducting polymers. PhD thesis, Kingston University, London, UK

Prusseit W, Corsepius S, Zwerger M, Berberich P, Kinder H, Eibl O, Jaekel C, Breuer U, Kurz H (1992) Epitaxial $YBa_2Cu_3O_{7-}\delta$ films on silicon using YSZ/Y_2O_3 buffer layers. Physica C 201:249–256

Randhaw H (1991) Review of plasma-assisted deposition processes. Thin Solid Films 196:329–349

Saji K, Kondo H, Takahashi H, Futata H, Angata K, Suzuki T (1993) Development of a thin-film oxygen sensor for combustion control of gas appliances. Sens Actuators B 13(14):695–696

Sze SM (ed) (1994) Semiconductor sensors. Wiley, New York

Tiemann M (2008) Repeated templating. Chem Mater 20:961–971

Van Tassel JJ, Randall CA (2006) Mechanisms of electrophoretic deposition. Key Eng Mater 314:167–174

Vayssieres L (2007) An aqueous solution approach to advanced metal oxide arrays on substrates. Appl Phys A 89:1–8

Viswanathan V, Laha T, Balani K, Agarwal A, Seal S (2006) Challenges and advances in nanocomposite processing techniques. Mater Sci Eng R 54:121–285

Wang ZL (2004) Functional oxide nanobelts: materials, properties and potential applications in nanosystems and biotechnology. Annu Rev Phys Chem 55:159–196

Will J, Mitterdorfer A, Kleinlogel C, Perednis D, Gauckler LJ (2000) Fabrication of thin electrolytes for second-generation solid oxide fuel cells. Solid State Ionics 131:79–96

Zheng M, Jagota A, Semke ED, Diner BA, McClean RS, Lustig SR, Richardson RE, Tassi NG (2003) DNA-assisted dispersion and separation of carbon nanotubes. Nat Mater 2:338–342

Chapter 28
Technologies Suitable for Gas Sensor Fabrication

The production of high-quality materials suitable for use in gas sensors is one of the most important tasks of modern materials science. As shown in previous chapters, materials used in gas sensors need to fulfill a range of requirements related to the crystallographic structure, chemical composition, electrophysical properties, catalytic activity, and so on. These materials also show a great deal of variation. Materials for gas sensors can come in a variety of forms, including films, ceramics, or powders. Their structure may be amorphous, glassy, nanocrystalline, polycrystalline, single crystalline, or epitaxial. They may be either dense or porous. These materials may be elementary substances, complex compounds, or composites. Polymers, metals, dielectrics, and semiconductors can also be used as materials for chemical sensors. They may be either organic or inorganic in nature.

This vast amount of variation indicates that it is impossible to produce such a wide range of materials using just one method. The possible differences in the physical–chemical properties of the materials are too great; so too are the resulting differences in the conditions required for the synthesis and deposition of these materials. Therefore, for preparing gas sensor materials with required properties we have to use various methods (see Fig. 28.1). These techniques differ in deposition rates, substrate temperature during deposition, precursor materials, the necessary equipment, expenditure, and the quality of the resulting films. A short account of these methods is presented in Table 28.1. One can find more detailed analysis of these methods in a vast array of quality reviews devoted to the subject (Brinker and Scherer 1990; Randhaw 1991; Hitchman and Jensen 1993; Hecht et al. 1994; Bunshah 1994; Brinker et al. 1996; Glocker and Shah 1995; Nenov and Yordanov 1996; Huczko 2000; Simon et al. 2001; Arthur 2002; Willmott 2004; Tay et al. 2006; Vahlas et al. 2006; Van Tassel and Randall 2006; Vayssieres 2007; Viswanathan et al. 2006; Christen and Eres 2008; Jaworek and Sobczyk 2008a, b; Milchev 2008; Tiemann 2008; Korotcenkov and Cho 2010; Sahner and Tuller 2010).

It is clear that the selection of a method acceptable for sensor material synthesis or deposition during sensor design and fabrication is a complicated task, and we need to analyze many different factors, including the type of technology which will be used for sensor fabrication: ceramic, thick-film, or thin-film technologies. Of course, every technology has advantages and disadvantages. Tables 28.2 and 28.3 present a comparison of several of these methods. Limitations of technology based on the use of 1D nanomaterials (nanowires, nanotubes, etc.) were discussed in Chap. 27 (Vol. 2).

G. Korotcenkov, *Handbook of Gas Sensor Materials,* Integrated Analytical Systems,
DOI 10.1007/978-1-4614-7388-6_28, © Springer Science+Business Media New York 2014

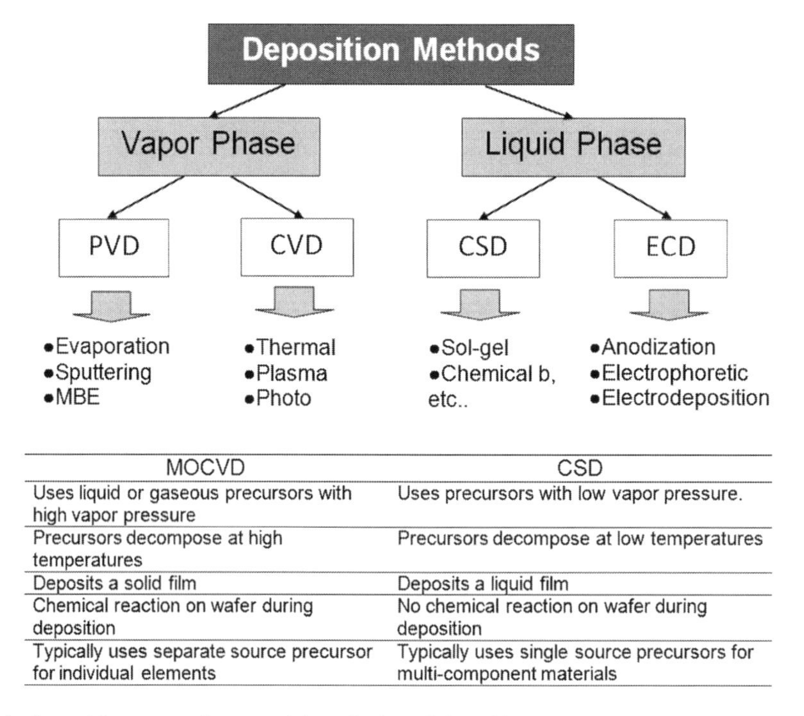

MOCVD	CSD
Uses liquid or gaseous precursors with high vapor pressure	Uses precursors with low vapor pressure.
Precursors decompose at high temperatures	Precursors decompose at low temperatures
Deposits a solid film	Deposits a liquid film
Chemical reaction on wafer during deposition	No chemical reaction on wafer during deposition
Typically uses separate source precursor for individual elements	Typically uses single source precursors for multi-component materials

Fig. 28.1 Methods used for gas-sensing material synthesis and deposition

28.1 Ceramic Technology

Ceramic elements have been investigated since the 1950s and are still employed as gas chemiresistors (Vandrish 1996). The simplest ceramic sensor design is a sintered block arrangement. The block is prepared by sintering metal oxide powders, with the electrode wires embedded in the block. The techniques used to produce ceramic gas sensors are similar to those used to fabricate thermistors, resistors, and electrodes from ceramics. The regimes of annealing and high-temperature processing are optimized for each material to provide either dense ceramics for solid electrolyte electrochemical gas sensors or porous ceramics for other types of gas sensors. As indicated before, a porous structure is necessary to increase both the surface-to-volume ratio and the gas penetrability of a gas-sensing matrix. The final shape of the ceramic element may be a sphere, tablet, or cylinder, which contains a heater and electrical contact wires.

Metal oxide powders for ceramic sensors can be synthesized using various methods such as sol–gel, flame pyrolysis, and hydrothermal synthesis. Many of these technologies offer some specific advantages for fabricating nanoparticles, such as simplicity, flexibility, low cost, ease of use on large substrates, and the ability to modify the composition by the addition of dopants and modifiers. These methods are well described in reviews and books (Comini et al. 2009; Korotcenkov and Cho 2010). Modification of the gas-sensing properties of these ceramics is possible by changing the parameters of synthesis and sintering (Song and Park 1994). By varying synthesis conditions, the characteristic size of the crystallites in the ceramic samples might be reduced down to 4 nm (Yu et al. 1997), which promotes high gas sensitivity (see Chap. 2 (Vol. 1)). Ball milling is an efficient operation which can also be used to adjust the parameters of ceramic sensors (Choi et al. 2005). Powders obtained using chemical methods usually have low gas permittivity due to strong agglomeration; ball milling destroys dense agglomerates and makes the gas-sensing "3D" matrix more homogeneous (Yamazoe and Miura 1992).

Table 28.1 Short description of the methods used for gas sensor fabrication

Method	Description
Physical vapor deposition (PVD)	PVD covers a number of deposition technologies in which material is released from a source and transferred to the substrate. The two most important technologies are evaporation and sputtering
Vacuum evaporation	*Vacuum evaporation* (including sublimation) is a PVD process where material from a thermal vaporization source reaches the substrate without collision with gas molecules in the space between the source and substrate. The vacuum is required to allow the molecules to evaporate freely in the chamber and then subsequently condense on all surfaces
	There are two popular evaporation technologies, which are e-beam evaporation and resistive evaporation, each referring to the heating method. In e-beam evaporation, an electron beam is aimed at the source material causing local heating and evaporation. In resistive evaporation, usually a tungsten boat, containing the source material, is heated electrically with a high current to make the material evaporate
Molecular beam epitaxy	*Epitaxial techniques* are techniques of arranging atoms in single-crystal fashion on crystalline substrates so that the lattice of the newly grown film duplicates that of the substrate. If the film is of the same material as the substrate, the process is called homoepitaxy, epitaxy, or simply epi. In molecular beam epitaxy (MBE), the heated single-crystal sample is placed in an ultrahigh vacuum (10^{-10} Torr) in the path of streams of atoms from heated cells that contain the materials of interest. These atomic streams impinge on the surface, creating layers whose structure is controlled by the crystal structure of the surface, the thermodynamics of the constituents, and the sample temperature. The deposition rate of MBE is very low
Sputtering	*Sputter deposition* is a method of depositing thin films by sputtering material from a "target." Sputtering is a term used to describe the mechanism in which atoms are ejected from the surface of a material when that surface is stuck by sufficiency energetic ions generated by low-pressure gas plasma, usually an argon plasma. The sputtering takes place at a much lower temperature than evaporation. Sputtered atoms ejected into the gas phase in vapor form are not in their thermodynamic equilibrium state and tend to condense on all surfaces in the vacuum chamber including the substrate
	There are several modifications of sputtering. However, the basic principle of sputtering is the same for all sputtering technologies. The differences typically relate to the manner in which the ion bombardment of the target is realized
Reactive sputtering	*Reactive sputtering* is a modification of the conventional sputtering process. During this process, reactive gas is introduced into the sputtering chamber in addition to the argon plasma. As a result the compound, which is formed by the elements of that gas combining with the sputter material, can be deposited
Laser ablation	*Laser ablation* is the process of removing material from a solid surface by irradiating it with a laser beam. The ablation process takes place in a vacuum chamber—either in vacuum or in the presence of some background gas. In the case of oxide films, oxygen is the most common background gas
	At low laser flux, the material is heated by the absorbed laser energy and evaporates or sublimates. At high laser flux, the material is typically converted to a plasma. As a result, a supersonic jet of particles (plume) with composition similar to the composition of the material is ejected normal to the target surface. The plume, similar to the rocket exhaust, expands away from the target with a strong forward-directed velocity distribution of the different particles. The ablated species condense on the substrate placed opposite to the target
Chemical vapor deposition (CVD)	*Chemical vapor deposition* refers to the formation of a nonvolatile solid material from the reaction of chemical reactants, called precursors, being in vapor phase in the right constituents. A reaction chamber is used for this process, into which the reactant gases are introduced to decompose and react with the substrate to form thin film or powders
	There are several main classification schemes for chemical vapor deposition processes. These include classification by the pressure (atmospheric, low-pressure, or ultrahigh vacuum), characteristics of the vapor (aerosol or direct liquid injection), or plasma processing type (microwave plasma-assisted deposition, plasma-enhanced deposition, remote plasma-enhanced deposition)

(continued)

Table 28.1 (continued)

Method	Description
Rheotaxial growth with following thermal oxidation (RGTO)	*RGTO* technique includes the following two steps: (1) the growth of metal in high vacuum onto a flat substrate kept at a temperature higher than melting temperature of metal used (usually Sn or In) and (2) slow oxidation of the deposited film during an annealing cycle at a maximum temperature of 793 K; this process ensures a complete transformation of the metallic Sn or In into SnO_2 or In_2O_3
Sol–gel method	The *sol–gel process* is a wet-chemical technique used for the fabrication of both glassy and ceramic materials. In this process, the sol (or solution) evolves gradually toward the formation of a gel-like network containing both a liquid phase and a solid phase. In other words the sol–gel process is the formation of an oxide network through polycondensation reactions of a molecular precursor in a liquid. A "sol" is a stable dispersion of colloidal particles or polymers in a solvent. These particles may be amorphous or crystalline. A "gel" consists of a three-dimensional continuous network, which encloses a liquid phase. In a colloidal gel, the network is built from agglomeration of colloidal particles
	Typical precursors are metal alkoxides and metal chlorides, which undergo hydrolysis and polycondensation reactions to form a colloid. The basic structure or morphology of the solid phase can range anywhere from discrete colloidal particles to continuous chain-like polymer networks
Electrospinning	*Electrospinning* is a simple method for generating nanofibers made of polymers, ceramics, and composites. In the electrospinning process, a high voltage is used to create an electrically charged jet of solution, mainly polymer based, which dries or solidifies to leave a very fine (typically on the micro- or nanoscale) fiber. One electrode is placed into the spinning solution and the other attached to a collector. An electric field is applied to the end of a capillary tube that contains the liquid (polymer) fluid held by its surface tension. When a sufficiently high voltage is applied to a liquid droplet, the body of the liquid becomes charged, and electrostatic repulsion counteracts the surface tension and the droplet is stretched; at a critical point, a stream of liquid erupts from the surface. The process of electrospinning does not require the use of coagulation chemistry or high temperatures to produce solid threads from solution. However, for preparing metal oxide fibers, the following annealing is required. Electrospinning from molten precursors is also practiced; this method ensures that no solvent can be carried over into the final product
	Successful electrospinning requires the use of an appropriate solvent and polymer system to prepare solutions exhibiting the desired viscoelastic behavior. The traditional setup for electrospinning works well for most conventional polymers, but it cannot be easily applied to polymers with limited solubilities (e.g., conjugated polymers) or low molecular weights
Chemical deposition	Unit species of material to be deposited is applied in liquid/solution form
	Substrates act as a physical support and no reaction. Deposition carried out at lower temperatures (<100 °C) typically atmospheric pressures
Chemical solution deposition (CSD)	The *chemical solution deposition* method used for preparation of oxide films comprises the deposition of a liquid sol on a substrate and the conversion of gel films to ceramic films via heat treatment
Liquid-phase deposition (LPD)	*Liquid-phase deposition* is a method for the non-electrochemical production of polycrystalline ceramic films at low temperatures. along with other aqueous solution methods [chemical bath deposition (CBD), successive ion layer adsorption and reaction (SILAR), and electroless deposition (ED) with catalyst] has been developed as a potential substitute for vapor-phase and chemical-precursor systems. The method involves immersion of a substrate in an aqueous solution containing a precursor species (commonly a fluoro-anion) which hydrolyzes slowly to produce a supersaturated solution of the desired oxide, which then precipitates preferentially on the substrate surface, producing a conformal coating

(continued)

Table 28.1 (continued)

Method	Description
Electrochemical deposition (ECD) (electroplating)	The *electrodeposition* process, which is typically restricted to electrically conductive materials and is carried out in a liquid solution of ions (electrolyte), is well suited to make films of metals such as copper, gold, and nickel. The films can be made in any thickness from <0.1 to >100 μm. Other materials including metal oxides can be deposited as well. There are basically two technologies for plating: *electroplating* and *electroless* plating
	During *electroplating*, when an electrical potential is applied between a conducting area on the substrate and a counter electrode (usually platinum) in the liquid, a chemical redox process takes place resulting in the formation of a layer of material on the substrate and usually some gas generation at the counter electrode. This method, although more complicated, allows for more operator control
	In the *electroless* plating process, a more complex chemical solution is used, in which deposition happens spontaneously on any surface which forms a sufficiently high electrochemical potential with the solution. The deposition is from a solution containing a metal salt and a reducing agent as well as various other additives such as stabilizers, surfactants, etc. This process is desirable since it does not require conductive substrates, any external electrical potential, and contact to the substrate during processing. Unfortunately, it is also more difficult to control with regard to film thickness and uniformity
Electrophoretic deposition (EPD)	*Electrophoretic deposition* is a particulate-forming process. It is a high-level efficient process for production of films or coatings on electrically conducting objects from colloidal suspensions. It begins with a dispersed powder material in a solvent and uses an electric field to move the powder particles into a desired arrangement on an electrode surface. Deposition on the electrode occurs via particle coagulation. The technique allows depositing thin and thick films and the shaping of bulk objects with metallic, polymeric, or ceramic particles
	There are four defining characteristics of EPD: (1) it begins with particles which are well dispersed and able to move independently in solvent suspension; (2) the particles have a surface charge due to electrochemical equilibrium with the solvent; where the particles would normally be electrically neutral, a compound might be bonded to them to give them an electrical charge in suspension; (3) there is electrophoretic motion of the particles in the bulk of the suspension; and (4) a rigid (finite shear strength) deposition of the particles is formed on the deposition electrode
Template-based synthesis	*Template-based synthesis* involves the fabrication of the desired material within the pores or channels of a nanoporous template. A template may be defined as a central structure within which a network forms in such a way that removal of the template creates a filled cavity with morphological and/or stereochemical features related to those of the template. Track-etch membranes, porous alumina, and other nanoporous structures have been characterized as templates. Electrochemical and electroless depositions, chemical polymerization, sol–gel deposition, and chemical vapor deposition have been presented as major template synthetic strategies. *Template-based synthesis* can be used to prepare nanostructures of conductive polymers, metals, metal oxides, semiconductors, carbons, and other solid matter
	If the templates that are used have cylindrical pores of uniform diameter, monodisperse nanocylinders of the desired material are obtained within the voids of the template material. Depending on the operating parameters, these nanocylinders may be solid (a nanorod) or hollow (a nanotubule). The nanostructures can remain inside the pores of the templates or they can be freed and collected as an ensemble of free nanoparticles. Alternatively, they can protrude from the surface like the bristles of a brush

(continued)

Table 28.1 (continued)

Method	Description
Casting	*Casting* is a simple technology, which can be used for a variety of materials (mostly polymers). In this process, the material to be deposited is dissolved in liquid form in a solvent. The material can be applied to the substrate by spraying or spinning. Once the solvent is evaporated, a thin film of the material remains on the substrate. The thicknesses that can be cast on a substrate range all the way from a single monolayer of molecules (adhesion promotion) to tens of micrometers. The control on film thickness depends on exact conditions, but can be sustained within ±10 % over a wide range. Delamination and cracking can occur if the liquid film is too thick. This method gives a more uniform and a more reproducible membrane than dip coating
Spin coating	In the *spin-coating* process, the substrate spins around an axis which should be perpendicular to the coating area. The quality of the coating depends on the rotation velocity, rheological parameters of the coating liquid, and surrounding atmosphere. The coating thickness varies between several hundreds of nanometers and up to 10 μm. Desired thickness is obtained by precursor dilution, spin speed, and number of layers. Equipment similar to that of spin-coat tracks used for photoresist deposition
Spray coating	Precursor is atomized to form a fine aerosol which is then deposited on a slowly rotating wafer. Deposition enhanced by electrostatic charging of aerosol. Desired thickness is controlled by adjusting deposition time and number of layers. The coating step is suitable for establishing an in-line process
Dip coating	*Dip-coating* techniques can be described as a process where the substrate to be coated is immersed in a liquid and then withdrawn with a well-defined withdrawal speed under controlled temperature and atmospheric conditions. The coating thickness is mainly defined by the withdrawal speed, by the solid content, and the viscosity of the liquid
	The applied coating may remain wet for several minutes until the solvent evaporates. This process can be accelerated by heated drying. In addition, the coating may be exposed to various thermal, UV, or IR treatments for stabilization
Langmuir–Blodgett film	In the *Langmuir–Blodgett* (LB) process, a monolayer of film-forming molecules (stearic acid is often used as a model molecule) on an aqueous surface is compressed into a compact floating film and transferred to a solid substrate by passing the substrate through the water surface
Soft lithography	The term "*soft lithography*," a low-cost alternative to traditional photolithography, describes a patterning technology that allows the shaping of colloidal suspensions on a micrometer scale. The core element of the process is an elastomeric mold of the desired micropattern, which is prepared by polymerizing an appropriate organic precursor around a positive master mold. In most cases, poly dimethylsiloxane (PDMS) is the polymer of choice
Micro contact printing	In *micro contact printing*, the patterned PDMS mold is used as a stamp to transfer an appropriate organic solution onto the substrate to be patterned. Hydrophobic and hydrophilic regions are created on the substrate, which is then dipped into a colloidal suspension of the coating material. During the dip-coating process, the suspension wets the substrate selectively, and the desired micropattern is replicated with a resolution of 5 μm, well below the typical resolution limit of screen-printed patterns
Slip casting	*Slip casting* is a technique in which a suspension (slip) is poured into a porous mold (generally made of plaster). The mold's pores absorb the liquid, and particles are compacted on the mold walls by capillary forces, i.e., solidify, producing parts of uniform thickness. Once dried to the leather-hard stage, the molds are opened and the cast object removed to dry completely before firing
Tape casting	*Tape casting* is a technique for continuous production of ceramic or other tapes according to the "doctor blade principle." In this process, a suspension of ceramic, metal, or polymer particles in an organic solvent or water, mixed together with strengthening plasticizers and/or binders, can be used. The slip is cast onto a precisely machined stone plate, on which the carrier film is moved smoothly and without perturbations. By means of the doctor blade, the slurry is spread homogeneously on the surface of the tape. Drying and firing are final stages of the actual tape forming

(continued)

Table 28.1 (continued)

Method	Description
Screen printing	*Screen printing* is a printing technique that uses a woven mesh to support an ink-blocking stencil. The paste (ink) used is a mixture of the material of interest, an organic binder, and a solvent. The attached stencil forms open areas of mesh that transfer ink or other printable materials which can be pressed through the mesh as a sharp-edged image onto a substrate. A roller or squeegee is moved across the screen stencil, forcing or pumping ink past the threads of the woven mesh in the open areas. After printing, the wet films are allowed to settle for 15–30 min to flatten the surface and are dried. This removes the solvents from the paste. Subsequent firing burns off the organic binder, metallic particles are reduced or oxidized, and glass particles are sintered. It can be used to print on a wide variety of substrates, including paper, paperboard, plastics, glass, metals, fabrics, and many other materials
Inkjet printing	*Inkjet technologies*, which are based on the 2D printer technique of using a jet to deposit tiny drops of ink onto substrate, are perfectly suited to controllably dispense small and precise amounts of "liquid" to precise locations. The available inkjet technologies include (1) continuous inkjet, (2) drop-on-demand inkjet, (3) thermal inkjet, and (4) piezo inkjet The "liquid" materials can encompass low- to high-viscosity fluids, colloidal suspensions, frits, metallic suspensions, and almost any other material that can be dispersed in a liquid carrier material. The carrier material can be aqueous- or nonaqueous-based solvent material. When printed, liquid drops of these materials instantly cool and solidify to form a layer of the part. Usually inkjet printing is accompanied by thermal treatment

Table 28.2 Advantages and disadvantages of synthesis and deposition methods usually used during gas sensor fabrication

Advantages	Disadvantages
Vacuum evaporation	
• High-purity films can be deposited from high-purity source material	• Large-volume vacuum chambers are generally required to keep an appreciable distance between the hot source and the substrate
• Source of material to be vaporized may be a solid in any form and purity	• Inconstancy of evaporation rates during the deposition process
• The line-of-sight trajectory and limited-area sources allow the use of masks to define areas of deposition on the substrate and shutters between the source and substrate to prevent deposition when not desired	• A considerable difference in composition between the evaporated and deposited materials. Therefore, many compounds and alloy compositions, due to stoichiometry problem, can only be deposited with difficulty. The same problems take place for materials that have high melting temperature and low saturated vapor pressure
• Deposition rate monitoring and control are relatively easy	• Line-of-sight trajectories and limited-area sources result in poor film-thickness uniformity and in poor surface coverage on complex surfaces
• It is the least expensive of the PVD processes	• Possible contamination from the evaporator
	• The need to periodically load the evaporator
Sputtering	
• Low-defect-density films of almost all materials used in gas sensors, including high-melting-point materials can be grown on unheated substrates	• Difficulties in plasma stabilization, particularly at low pressure and in large areas
• The opportunity to synthesize compounds and alloys that cannot be obtained by thermal evaporation of materials in a high vacuum	• The rate's dependence on electrical power

(continued)

Table 28.2 (continued)

Advantages	Disadvantages
• The high adhesion of a film	• A large particle energy resulting in surface damage due to surface bombardment. This problem is particularly important for the deposition of polymer materials and films, which cannot be thermally processed for recovery to the initial state
• A high coefficient of use of the sputtered material	
• Becoming homogeneous through thickness coverings	
• The opportunity to create apparatus and production lines for continuous operation	
RGTO technique	
• The RGTO technique allows using of any metal to form a metal oxide film	• The RGTO technology has very low reproducibility
• The coverings may have a very high porosity, i.e., high surface area	• The parameters of the films formed are highly dependent on the mode of deposition, particularly its thickness
• The two-stage technology simplifies the technology used to form the required topology of the metal oxide film with high chemical stability. It is well known that metals are not as chemically stable as metal oxides	• The technological cycle has an unacceptably long duration
Laser ablation	
• PLD permits precise control of deposition of multilayer structures in situ in one technological step	• Low productivity of the method
• Any metal, ceramic, alloy, or intermetallic compound, as well as fully reacted metals, can be deposited on virtually any substrate, including plastic, paper, metals, and ceramics	• Relatively high investment costs
• Films can be deposited at either low (including room temperature) or high temperature	• The composition and thickness depend on too many deposition conditions, such as wavelength, energy and shape of the laser pulse, focusing geometry, process atmosphere, and substrate temperature
• The method is very convenient for the quick preparation and study of new sensing materials	• Difficulties in the deposition of thick layers
• Process is compatible with oxygen and other reactive gases	• Difficulties in attaining the necessary stoichiometry of materials containing volatile components
• PLD technology creates the possibility for controlled deposition of ultrathin coverings, which is very attractive for surface modification of the materials for gas sensors	• Difficulties in scaling up to large wafers
	• Due to repeated interaction of the laser beam with the target, structural changes occur on the surface with craters forming. Therefore, the composition and properties of the deposited material will depend on the duration of the deposition process
Chemical vapor deposition	
• CVD has the ability to coat complex shapes internally and externally because it is a non-line-of-sight process with strong throwing power	• Up-front capital costs can be high, with complex handling, safety, and automatic systems

(continued)

Table 28.2 (continued)

Advantages	Disadvantages
• CVD can produce uniform films with strong reproducibility and adhesion at reasonably high deposition rates	• High-temperature requirements may limit substrate choices
• CVD can produce multilayered coatings and coatings for a variety of metals, alloys, and compounds	• Some substrates can be attacked by the coating gases
• CVD provides the ability to control crystal structure, surface morphology, and the orientation of the products by controlling the process parameters	• Poor adhesion or lack of metallurgical bonding is possible
• Coatings are dense, and their purity can be controlled	• Masking portions that are not to be coated can be difficult
	• The difficulty with stoichiometry control of multicomponent materials is possible
Deposition from aerosol phase	
• The required equipment for aerosol deposition is simple and safe, and the process is straightforward, rapid, reliable, and inexpensive	• Fairly low reproducibility and the homogeneity of the film-thickness distribution over the substrate's area, especially if the substrate used is not flat
• The deposition of films does not require a vacuum at any stage. Deposition can be carried out at atmospheric pressure	• Other technological problems are associated with the plotting of sensitive material on small areas
• It is possible to use simple, less expensive, and less toxic precursors, which do not possess the high pressure of vapor under saturation. This creates the possibility to apply some salts and metal–organic compounds that cannot be used with standard methods such as CVD and MOCVD	• Low effectiveness in the use of the precursor during the film deposition process
• The deposition rate, the thickness, and the composition of the films can be easily controlled. It offers an extremely simple way to dope films with virtually any element in any proportion, merely by adding it in some form to the spray solution	
Sol–gel technique	
• Sol–gel is low-cost wet-chemical technology, which offers the possibility to prepare solids with predetermined structure, including thick porous ceramics needed for gas sensors, by varying the experimental conditions	• Weak bonding and as a result low adhesion and low wear resistance. Therefore, the sol–gel technique is very substrate dependent, and the thermal expansion mismatch prevents its wide application
• Multicomponent compounds and doped materials may be prepared with a controlled stoichiometry by mixing sols of different compounds and using multiple different dopants	• Large shrinkage during drying processing, which is accompanied by cracking effect, increased with film thickness increasing
• There is the possibility of independent control over porosity, crystal structure, and grain size	• The presence of residual hydroxyl and residual carbon
• This method can easily shape materials into complex geometries in a gel state. So, materials in different shapes as films, fibers, powders, and bulk could be obtained	• High cost of raw materials
• Very small quantities of raw material can be involved and hence the cost of metal–organic precursors is not a consideration	• Long processing time
• The possibility to synthesise ceramic material at a temperature close to room temperature opens the opportunity of incorporating volatile components or soft dopants, such as fluorescent dye molecules and organic chromophores, in synthesized matrix	

Table 28.3 Summary of the advantages and limitations of the novel deposition methods with respect to potential applications in metal oxide gas sensing

Method	Advantages	Limitations
Soft lithography	Feature resolutions down to 1 μm. Simple and versatile patterning of molds	Drying conditions need to be carefully controlled. Complex optimization of the precursor systems required
Direct writing		
MAPLE	Direct patterning with feature resolution down to 10 μm. Suitable for a broad variety of materials, including nanosized precursors	Limited to 2D, low aspect ratios
Inkjet	Versatile direct patterning tool. Potential for use in high-throughput material discovery	Limited to 2D. Ink optimization is complex
3D printing	Versatile tool to access. Line widths <1 μm can be achieved	3D patterns. Ink optimization is complex. Not yet available for many materials systems
In situ spraying	Simple, inexpensive multilayer deposition. Homogeneous films	Limited to films (no direct patterning)
Structure replication	Synthesis of high surface area materials	For many metal oxides: two-step process (template of silica matrix + nanocasting) required
Nanofiber growth	Very high surface-to-volume ratio	Single nanofibers: limited reproducibility. Difficult to assemble into real-world devices
Nanocarving	Well-defined nanostructured devices with high surface area	Available only for a limited number of materials
Electric field-assisted methods		
Electrospinning	Simple setup operating under ambient atmosphere. Unoriented, porous fiber mats with high surface areas	Without additional treatment: poor adhesion to substrate
Electrospraying	Simple setup operating under ambient atmosphere. Preparation of films with nanoscaled features	Deposition parameters need to be carefully controlled

Source: Reprinted with permission from Sahner and Tuller (2010), Copyright 2010 Springer

Despite complex geometry and chemical structure, the ceramic sensors have rather high gas sensitivity and long-term stability of functional properties. The high sensitivity seems to be a result of the small size of crystallites that constitute the powders and high porosity of gas-sensing matrix. The long-term stability seems to be a result of thoroughly developed aging procedures. An additional advantage of the ceramics is their rather low production cost, even in small quantities.

The major disadvantages are the difficulty of reliably producing sensors with the same parameters on a large scale, large thermal sluggishness, insufficient mechanical durability, and little compatibility with most microelectronic devices fabricated using planar technologies. In addition, conventional ceramic substrates are fairly large in size and, in the case of metal oxide conductometric gas sensors, require heaters which consume power of the order of 0.5–1.5 W per sensor element to achieve operating temperatures of about 400 °C. Such power consumption is a serious drawback to the use of MOX gas sensors in bus-connected sensor networks or multisensor arrays.

28.2 Planar Sensors

Planar constructions of gas sensors are generally considered today to be better than ceramics for the development of gas-sensing devices. These structures can be fabricated by a number of microelectronic protocols as "single-sided" or "double-sided" designs. The choice depends mostly on how to

Table 28.4 Conventional techniques for fabricating planar gas sensors

Thick-film technologies	Thin-film technologies
Sol–gel	Chemical vapor deposition (thermal, plasma, laser induced)
Flame pyrolysis	Sputtering (reactive, cathode)
Precipitation	Evaporation (reactive, thermal, arc, laser)
Screen printing	Spray pyrolysis
Dip coating	Molecular beam epitaxy
Drop coating	Electroplating
Spin coating	Ion plating
	Photolithography
	Etching

deposit major functional elements of the sensor, the heater, and the gas-sensitive layer in the case of chemiresistors. These elements may be deposited on the same (front) side of the substrate ("single-sided" design) or on different sides of the substrate ("double-sided" design) (Oyabu 1982; Schierbaum et al. 1990). The double-sided construction requires employing two-sided photolithography or other deposition protocol and makes it difficult to wire the sensor at the housing. However, this type of design allows smaller chip size and decreased power loss. Another advantage of this construction is the option to apply different materials when forming the heater and contact electrodes.

Planar gas sensors can be fabricated via thin-film or thick-film technologies. These technologies involve various protocols (see Table 28.4), which result in different structural and functional properties of the sensors. However, the general architecture of thin-film and thick-film sensors is similar. Planar sensors are very robust; the temperature homogeneity over the sensing layer is good. However, power consumption of conductometric metal oxide gas sensors fabricated on the standard platform is still quite high, about 0.2–1.0 W at operating temperatures of 300–500 °C, depending on the substrate dimensions. In the case of microfabricated sensors, which used hotplate platform, the power consumption can be reduced up to several mW (see Chap. 7 (Vol. 1)).

Thick-film and thin-film planar gas sensors are generally favored compared to sintered ceramic-based sensors because of their lower power consumption and sometimes better sensor performance in terms of response time and gas sensitivity. However, as for the ceramic elements, the gas sensitivity is also highly dependent on the film porosity, film thickness, operating temperature, presence of additives, and crystallite size (Korotcenkov 2005, 2008). Thick-film and thin-film technologies are frequently combined to produce planar sensors, especially in laboratory investigations. For example, in conductometric gas sensors, the gas-sensing polycrystalline layers may be prepared via thin-film technologies (say, magnetron or cathode sputtering), while the heaters are formed with thick-film technology.

28.3 Thick-Film Technology

28.3.1 General Description

The fabrication processes, utilizing thick-film technologies, have been developed over 40 years and are rather specific to the particular devices (Janata 1989; Madou and Morrison 1989; Moseley et al. 1991). Thick-film technologies include such methods as spray deposition, dip coating, drop coating, centrifugal coating, conformal coverage with thermoplastic transfer molding, and screen printing (Heule and Gauckler 2001; Lee et al. 2007b). A short description of these methods is presented in Table 28.1. However, screen printing is the most widely used technology (White and Turner 1997), utilizing a fine sieve pressed tightly onto the substrate and further covered by a print (see Chap. 8 (Vol. 1)).

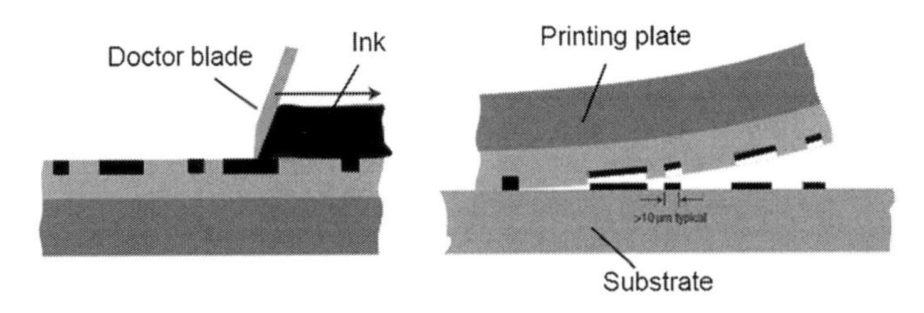

Fig. 28.2 Conventional gravure printing technology. The doctor blade fills the recessed features of a printing plate with ink. Then particles dispersed in the ink are transferred from the plate onto the substrate (Idea from Kraus et al. 2007)

The lowest resolution limit of screen printing is typically about 50 μm. The thickness of the layers fabricated by thick-film technologies is commonly in the range 2–300 μm. In many ways, the microstructure of the thick-film compares with that of ceramics and is a function of grain sintering conditions. As a rule, films prepared by thick-film technology are porous. Reproducibility of sensors fabricated by thick-film technology is better than for ceramic elements.

Screen-printing pastes contain four basic components: the functional substance, binders, resins, and organic solvents (see Chap. 8 (Vol. 1)). To remove organic solvents, the layers are dried with an infrared belt drier or a conventional oven at temperatures of around 150 °C. After drying, the adhesion of the layers to the substrate is enhanced. In general, further annealing of the thick films is performed at high temperatures, in the range 300–1,200 °C. During this step, frequently called firing, the glass component melts, the fine powders sinter, and the overall layer, a solid composite material, is attached to the substrate. The thickness of the fabricated layer depends on the paste viscosity and the size of the apertures in the mesh (Madou and Morrison 1989; Comini et al. 2009). The thermal treatment has to be gradual to minimize the temperature-induced stresses which appear under sudden heating and cooling (Moseley and Tofield 1987; Sberveglieri 1992). These steps can be repeated using materials appropriate for fabricating specific areas of the device.

It is worth noting that various laboratories are continuing intensive research toward finding new methods to reduce the characteristic size of sensors fabricated by thick-film technologies. One method which was developed recently is to cast a suspension into appropriate photoresist structures (Schoenholzer et al. 2000). Another method, which is more cost efficient, uses a soft lithography of liquid materials (Xia and Whitesides 1998). The use of liquid inorganic precursor polymers has already been demonstrated for the fabrication of microstructured ceramics (Yang et al. 2001) and other hierarchically ordered oxides (Yang et al. 1998). In the soft lithography process, the micropatterns are transferred by casting a silicone rubber, poly(dimethylsiloxane) (PDMS), against the master structure. The PDMS is then peeled off, cut, and used as a mold that forms microcapillaries on a substrate which can be filled with a liquid. This technique is referred to as micromolding in capillaries (MIMIC). The most striking result of using PDMS as the mold material is that this elastomeric material readily establishes a reversible conformal contact on the molecular level to a variety of substrates, thus sealing the capillaries optimally. Additionally, the master structures may be reused many times to cast PDMS molds. It is important that the MIMIC technique does not require clean-room conditions. Experiments carried out by Heule et al. (2001) have shown that one can use a colloidal dispersion of ceramic powders with solid contents of 0.1–40 vol.% to form the film. These methods enabled the fabrication of microstructured ceramic lines with a spatial resolution of 10 μm which can be integrated into a miniaturized MOX-based gas sensor (Heule and Gauckler 2001).

Gravure printing is also a very promising method for depositing colloidal nanoparticle-based layers with high resolution (Fig. 28.2). A detailed description of this method is given elsewhere (Kraus et al. 2007), where it was shown that this approach allows the possibility of achieving resolution in the submicrometer range.

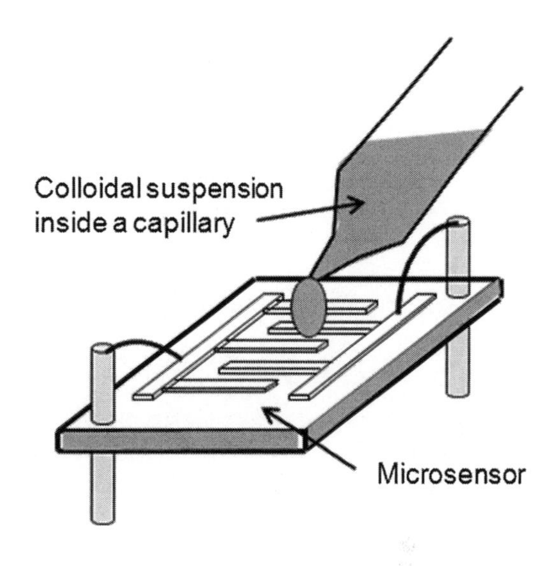

Fig. 28.3 Deposition of MOX sol onto a microheater using the drop dispenser technique (Idea from Spannhake et al. 2009)

However, in many cases, especially during micromachining technology when the prefabricated microhotplates are used, lithographic processing and screen printing of the sensing layers are not possible (Spannhake et al. 2009). The sensing layers, therefore, need to be deposited over shadow masks, which should allow placing the layers exclusively onto predefined spots between the electrodes. This requires only moderate precision, so the masks can be made of mechanically stiff ceramics which are conventionally produced in a MEMS foundry in the form of micromachined silicon wafers. Another promising method for locally depositing a gas-sensitive layer is based on the use of drop coating (Fig. 28.3). This method allows placing small volumes of gas-sensitive material as a sol on top of the hotplate followed by thermal annealing to form a gel (Vincenzi et al. 2001; Guidi et al. 2002; Francioso et al. 2006; Epifani et al. 2007). Because of the good thermal insulation, the center regions of the hotplates can be heated to temperatures of 300–700 °C, necessary to transform the colloidal suspension into a solid film.

Direct writing of powders via inkjets is another promising technology for local deposition of MOXs on microhotplates. This method offers the possibility of combining the advantages of thick films and micromachining. A general review of the state-of-the-art in inkjet printing of various materials can be found elsewhere (Calvert 2001; Zhao et al. 2002; Bietsch et al. 2004). The scheme of the method is drawn in Fig. 28.4. Inkjet printing is a noncontact technique which does not require any masks for design and repeated production of microscale patterns. Usually inkjet printing involves the drop formation of polymer or metal oxide-laden inks through the use of a piezoelectric dispensing printhead. The printhead consists of a glass capillary surrounded by a piezoelectric material. Through an applied voltage, the piezoelectric material provides compression on the glass capillary, and droplets of picoliter volume can be produced depending on the orifice of the glass capillary. There are several properties that must be controlled during inkjet printing including, but not limited to, solid loading, density, viscosity, surface tension, and evaporation rate of the ink as well as printing settings of applied voltage, vacuum level, and orifice diameter. These properties must be regulated to produce consistent drops of a specific volume which corresponds to the amount of material deposited on the miniaturized sensor substrate. The physical size of the microhotplate also dictates the positioning of the drops produced by the inkjet printer, which must be precise and reproducible. In contrast to parallel processes such as photolithography, inkjet printing makes it possible to fabricate each sample individually, although this contradicts the requirements of mass-scale production. Still, modern computer-driven setups could help to build such individual devices with the required automation and

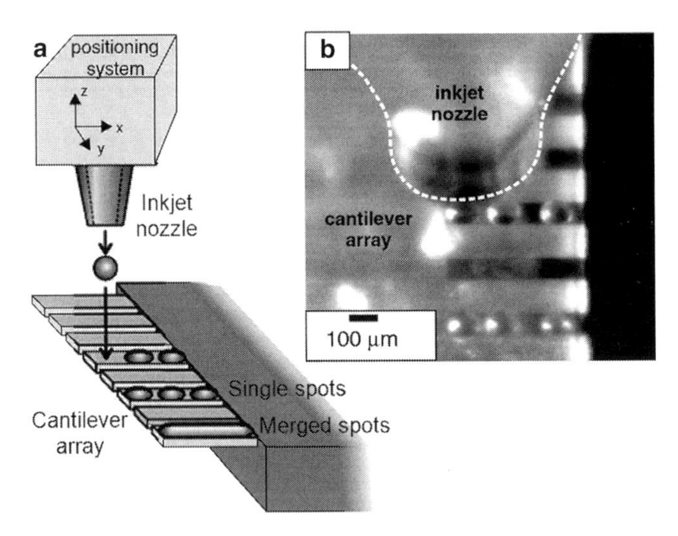

Fig. 28.4 Inkjet printing of individual droplets onto a cantilever array (**a**) as a scheme and (**b**) as seen by a video camera. A positioning system allows accurate placement of single droplets onto selected cantilevers. When deposited with a small pitch, the droplets merge into a continuous layer covering the entire cantilever length. For demonstration, three droplets of water are deposited onto selected cantilevers. Owing to the oblique view of the camera, only the central cantilever is in focus (Reprinted with permission from Bietsch et al. 2004, Copyright 2004 IOP)

accuracy. Moreover, this method enables one to fabricate high-quality patterns on a big variety of flexible paper- or polymer-based substrates and is suitable for the inexpensive production of sensor modules for microelectromechanical systems. Such inkjet technology has been utilized to print aqueous 2D MOX suspensions with a size of ~37 μm (Windle and Derby 1999), which is sufficient to design sensors using micromachining (Liu 1995).

It should be said that at present this method is used mostly for deposition of polymer-based materials. However, sensors with acceptable parameters fabricated using inkjet printing of carbon black composites (De Girolamo Del Mauro et al. 2011), carbon nanotubes (Kim et al. 2009), graphene (Dua et al. 2010), and metal oxide gas-sensing layers, such as $Cr_{2-x}TixO_{3+z}$ (Peter et al. 2011), WO_3 (Kukkola et al. 2012), SnO_2 (Lee et al. 2007a), and In_2O_3 (Pashchanka et al. 2012) are presented. These layers were incorporated in resistive (Kim et al. 2009; Crowley et al. 2010), cantilever (Bietsch et al. 2004), optical (O'Toole et al. 2009), and FET-based (Maklin et al. 2008) gas sensors. Examples of different types of inkjet-printed resistive gas sensors are listed in Table 28.5.

The performance of gas sensors fabricated by thick-film technology depends critically on the applied materials, primarily on the pastes employed to make the gas-sensing layer and electrodes. The contact electrodes, heater, and sensing layer are formed by sequential deposition of the corresponding pastes on the substrate with subsequent annealing. Therefore, the substances used to fabricate, for example, conductometric gas sensors can be either fine metal powders to prepare electrodes, heaters, and temperature sensors or fine-grained MOX powders and polymers to prepare gas-sensitive layers. In choosing these materials, important considerations are good adhesion, similar coefficients of expansion to prevent stress-related damage during sudden heating and cooling, the ability to retain their characteristics throughout the fabrication process, easy availability, and low cost, among others. Furthermore, the paste and composite materials used in the sensor device have to provide adequate and reliable electrical contacts for solid–solid interfaces.

Table 28.5 Comparison of different types of inkjet-printed resistive gas sensors

Active material	Analyte (sensitivity,□ R/R)	Lowest conc. measure	References
PANI	NH_3 (~0.32 % per ppm)	2.5 ppm	Crowley et al. (2008)
	NO_x, TMA, TEA, H_2S (~0.2–2.57 % per ppm)	100 ppm	
PANI ($CuCl_2$ doped)	H_2S (~16 % per ppm)	2.5 ppm	Crowley et al. (2010)
Polypyrrole	CH_3OH, C_2H_5OH, C_3H_7OH, $CHCl_3$ and C_6H_6 (~0.006–0.018 % per ppm)	5,000 ppm	Mabrook et al. (2006b)
UV-curable polymers	H_2O vapor (exponential decrease of R with relative humidity)		Cho et al. (2008)
Thiophene-based polymers	CH_3COCH_3, CH_2Cl_2, $C_6H_5CH_3$, and c-C_6H_{12} (~3–10 % per ppm)	10–170 ppm	Li et al. (2007)
PEDOT:PSS	CH_3OH and C_2H_5OH ((~3–7)×10^{-4} per ppm)	2,850 ppm	Mabrook et al. (2006a)
Polystyrene/carbon black	CH_3COCH_3 (~2×10^{-4} % per ppm)	600 ppm	De Girolamo Del Mauro et al. (2011)
PMAS–SWCNTs	CH_3OH, C_2H_5OH, 2-C_3H_7OH, C_4H_9OH, $CHCl_3$ and $C_6H_5CH_3$ (56.5 %, 31.3 %, 10.9 %, 6.7 %, 0 % and 0 %, respectively)	3 %	Small and Panhuis (2007)
MWCNTs	CH_3OH, C_2H_5OH, 2-C_3H_7OH, NH_3, H_2O (~50 %, <10 %, <10 %, ~150 %, and ~100 %, respectively)	Sat. vapors	Kordas et al. (2006)
SWCNTs	CH_3OH (~0.02 % per ppm)	300 ppm	Mabrook et al. (2009)
SWCNTs	NO_2 (~10 % per ppm)	50 ppb	Kim et al. (2009)
SWCNTs	NH_3 (~2.5×10^{-3} % per ppm)	4 %	Yang et al. (2009)
Reduced graphene oxide	NO_2 (~0.6 % per ppm)	0.5 ppm	Dua et al. (2010)
	Cl_2 (~0.13 % per ppm)	6 ppm	
	NH_3, CH_3OH, C_2H_5OH and CH_2Cl_2 (~22 %, ~10 %, ~9 % and ~6 %, respectively)	Sat. vapors	
WO_3:Ag	NO (~10 % per ppm)	Sub-ppm	Kukkola et al. (2012)
WO_3:Pd, Pt	H_2 (10 % (Pd) and ~10^3 % (Pt) per ppm)		
In_2O_3	CO (~40 % per ppm)	~1 ppm	Pashchanka et al. (2012)
	H_2 (~6–10 % per ppm)		

PABS polyaminobenzene sulfonic acid; *SWCNT* single-walled carbon nanotube; *PANI* polyaniline; *CNT* carbon nanotube; *PMAS* poly(2-methoxyaniline-5-sulfonic acid); *MWCNT* multiwalled carbon nanotube; *PEDOT* poly(3,4-ethylene dioxythiophene); *PSS* polystyrene sulfonated acid; *TMA* trimethylamine; *TEA* triethylamine
Source: Data from Kukkola et al. (2012), etc.

28.3.2 Powder Technology

Nanoscaled powders are the main materials used to prepare ceramics and thick films for a variety of applications during gas sensor design and fabrication. It was found that the various synthesis techniques used for producing nanoscaled powders can generally be divided into three major types: assembly of clusters/nanoparticles produced by (1) wet-chemical routes, (2) gas-phase synthesis, and (3) electrolytic deposition.

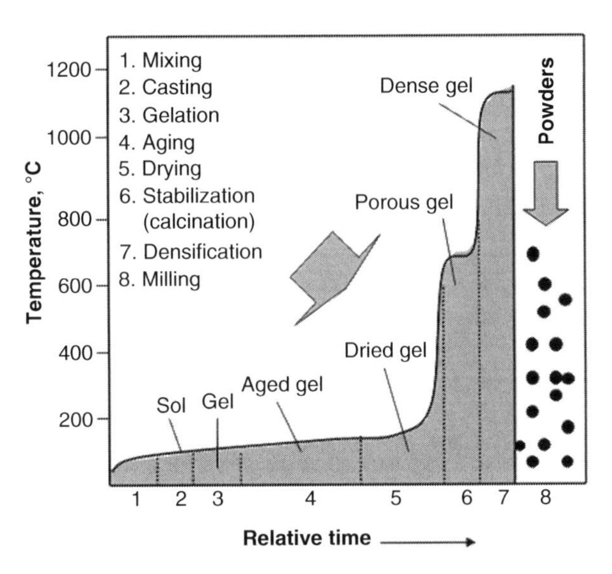

Fig. 28.5 Typical stages of the sol–gel route

28.3.2.1 Sol–Gel Process

The challenge in synthesizing nanostructure powders using wet-chemical routes is to control and engineer the physical properties (size, shape, composition, etc.) of the starting powders, since these affect the properties of the final products. Many advanced processing techniques, such as sol–gel synthesis, spray pyrolysis, and precipitation, have been used to produce fine powders (Hench and Ulrich 1984; Lavernia and Wu 1996; Niesen and De Guire 2001; Vahlas et al. 2006). Sol–gel synthesis is the most used method. It has been established that, through this process, homogeneous inorganic oxide materials with desirable properties of hardness, optical transparency, chemical durability, tailored porosity, and thermal resistance can be produced at room temperatures. In a typical sol–gel process, the precursor is subjected to a series of hydrolysis and polymerization reactions to form a colloidal suspension called a sol (see Fig. 28.5). A *sol* is a dispersion of solid particles (~0.01–1 μm) in a liquid in which only Brownian motion suspends the particles. Further heat treatment leads to the transformation of gel to xerogels, powders, or ceramic materials in various forms. In a *gel*, liquid and solid are dispersed within each other, presenting a solid network containing liquid components. With further drying and heat treatment, the gel is converted into dense ceramic or glass particles. If the liquid in a wet gel is removed under a supercritical condition, a highly porous and extremely low-density material called an aerogel is obtained. Details of the sol–gel process are discussed more extensively in several excellent review articles (Livage et al. 1988; Brinker and Scherer 1990; Minh and Takahashi 1995; Brinker et al. 1996; Bagwell and Messing 1996; Livage 1997; Narendar and Messing 1997; Troczynski and Yang 2001; Olding et al. 2001). Synthesis using sol–gel processing techniques is generally based on the hydrolytic polycondensation of metal alkoxides, which corresponds to the nucleophilic attack of H_2O on the metal, producing metal hydroxide groups. In the following steps, the metal hydroxide groups react with either the alkoxide (heterocondensation) or the hydroxide groups (homocondensation), giving rise to M–O–M bridges (Zeigler and Fearon 1990; Brinker and Scherer 1990).

During sol–gel formation, the viscosity of the solution gradually increases, so the gel may exhibit spontaneous shrinkage, called syneresis, or aging. Depending on the sol–gel processing conditions, such as the Me/H_2O ratio, the type and concentration of the catalyst, alkoxide precursors, and temperature, gel formation can take place on a time scale ranging from a number of seconds to a number of months. The network is initially supple, allowing for further condensations, and bond formation induces construction of the network and expulsion of liquid from the pores. Gels which have been

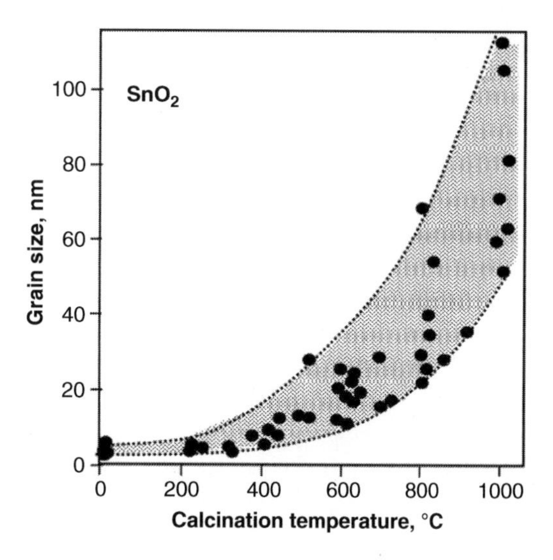

Fig. 28.6 Influence of calcination temperature on size of SnO_2 powders prepared by the sol–gel method

dried by evaporation under normal conditions are called xerogels, and they are interesting materials because of their porosity and high surface area, making them useful as catalyst substrates, sensing materials, and filters. They are also useful in the preparation of dense ceramics by sintering (densification driven by interfacial energy) (Brinker et al. 1984).

The characteristics and properties of a sol–gel inorganic network are related to a number of factors that affect the rate of hydrolysis and condensation reactions: the pH level, the temperature and time of the reaction, the reagent concentrations, the nature and concentration of the catalyst, the H_2O/M molar ratio (R), the aging temperature, and the drying time. Among these factors, the pH level, the nature and concentration of the catalyst, the H_2O/M molar ratio (R), and the temperature have been identified as being the most important (Brinker and Scherer 1990). However, it has been shown that, regardless of the pH level, hydrolysis occurs through nucleophilic attack of the oxygen contained in water on the M atom, as proven by the reaction of isotopically labeled water with tetraethyl orthosilicate (TEOS), which produces only unlabeled alcohol in both acid- and base-catalyzed systems (Bradley et al. 1978; Livage et al. 1988). Although hydrolysis can occur without the addition of an external catalyst, it is more rapid and complete when a catalyst is employed. Mineral acids (HCl) and ammonia are used most often; however, other catalysts may be used as well, including acetic acid, KOH, amines, KF, and HF. Under basic conditions, the hydrolysis reaction is first order in the base concentration. With weaker bases, such as ammonium hydroxide and pyridine, measurable speeds of reaction are produced only if large concentrations are present. Compared to acidic conditions, base hydrolysis kinetics is more strongly affected by the nature of the solvent. It was established that an increased value of R (H_2O/M molar ratio) is expected to promote the hydrolysis reaction.

Research has shown that stabilized colloidal metal hydroxides, which are synthesized during the sol–gel process, are very far from thermodynamic equilibrium. All of these products are very unstable at temperatures over 150 °C and do not have a crystalline structure. Furthermore, depending on the formation procedure, the material also presents traces of other chemical species, such as chlorides or traces of organic precursors. Therefore, thermal treatment (calcination) is required to transfer these hydroxides into their stoichiometric oxide form. This process is accompanied by the growth of grains, which depends on the temperature of calcinations (see Fig. 28.6 and see also Chap. 20 (Vol. 2)). The modes of thermal treatment commonly used in the calcination of some oxide phases are presented in Table 28.6 (Tahar et al. 1997; Chung et al. 1998; Sangaletti et al. 1999; Li et al. 1999; Nayral et al. 2000; Ivanovskaya 2000; Llobet et al. 2000; Yue and Gao 2000; Leite et al. 2001; Zakrzewska 2001; Kaya et al. 2002). The fundamentals of the processes taking place in nanoceramics during sintering are given by Lu (2008).

Table 28.6 Some metal oxides prepared by the sol–gel method and their parameters

Metal oxide	Precursor, temperature of hydrolysis	Temperature and time of calcination	Grain size
SnO_2	$[Sn(NMe_2)_2]_2$, 135 °C	600 °C, 6 h	10–20 nm
SnO_2	$SnCl_4$, 150 °C	700 °C, 1 h	~10 nm
In_2O_3	$InCl_3$, 110 °C	600 °C, 1 h	20–27 nm
In_2O_3	$In(NO_3)_3$, 60 °C	500–800 °C, 1 h	18–50 nm
In_2O_3	$In(OAc)_3$, 3 days	400–700 °C, 30 min	23–37 nm
WO_3	$W(OC_2H_5)_6$, RT, 1 h	400–700 °C	
WO_3	$(NH_4)_{10}W_{12}O_{41}$ $5H_2O$	300–1,000 °C, 3–5 h	
TiO_2	$Ti(OC_2H_5)_4$, RT, 1 h	350–500 °C, 4 h	10–15 nm
TiO_2	$TiCl_4$	750 °C, 2 h	Whiskers
Ta_2O_5	$TaCl_5$, 24 h	600–700 °C, 4 h	~50 nm
Fe_2O_3	$Fe(OC_2H_4OCH_3)_3$, 90 °C	400 °C, 2 h	~5–30 nm
ZrO_2	Zr(acac), 220 °C	600 °C, 4 h	~20–60 nm
TiO_2–WO_3	$Ti(OC_3H_7)_4 + H_2WO_4$	700–900 °C	
TiO_2–SnO_2	$Ti(OC_3H_7)_4 + SnCl_2$	600–700 °C	10–50 nm
$BaSnO_3$	$Ba(OH)_2 + K_2SnO_3$ $3H_2O$	1,000 °C	200 nm

Thus, by controlling the parameters of hydrolysis, polymerization, and condensation, it is possible to vary over wide ranges the structure and properties of the sol–gel-derived inorganic network. At the same time, such strong dependence of the produced materials on the parameters of the sol–gel process shows the need to follow the synthesis and annealing conditions strictly to obtain reproducible powders and ceramics with the required properties. It should be noted that the sol–gel process can be used for preparing organic/inorganic networks as well.

28.3.2.2 Gas-Phase Synthesis

However, we have to recognize that preparing powders using wet-chemical routes is a complicated process. From an industrial point of view, cost-effective and less complicated synthesis techniques are required for large-scale applications. Therefore, the gas-phase synthesis methods have been developed quite extensively over the last few years. In all of these techniques, the primary products are nanosized powders, which are subsequently transformed by various consolidation techniques to materials in either bulk form or coatings, with or without porosity. Some of the most widely used techniques are summarized schematically in Fig. 28.7. The powder synthesis processes shown are gas condensation (GPC), chemical vapor condensation (CVC), microwave plasma (MPP), and low-pressure combustion flame synthesis (CFS). These processes can all be interchanged and used in similar chamber designs. The mechanisms involved in the process of particle formation for two sources, thermal evaporation and electrospraying, are presented schematically in Fig. 28.8 (Nakaso et al. 2003).

A range of collection devices has been used to separate nanoparticles from the gas (Choy 2003; Vahlas et al. 2006). Traditionally, a rotating cylindrical device cooled with liquid nitrogen has been employed for particle collection. The nanoparticles are subsequently removed from the surface of the cylinder with a scraper, usually in the form of a metallic plate. However, the simplest method of collecting nanopowders is to use a mechanical filter with a small pore size. Nanoparticles ranging from 2 to 50 nm in size may be extracted from the gas flow by thermophoretic forces from an applied permanent temperature gradient and then deposited loosely on the surface of the collection device as a powder of low density with no agglomeration.

For some applications, especially solid-state gas sensor design, it is necessary to use nanoscaled powders with sizes smaller than 5–10 nm (Nayral et al. 2000; Leite et al. 2001). In many cases, however,

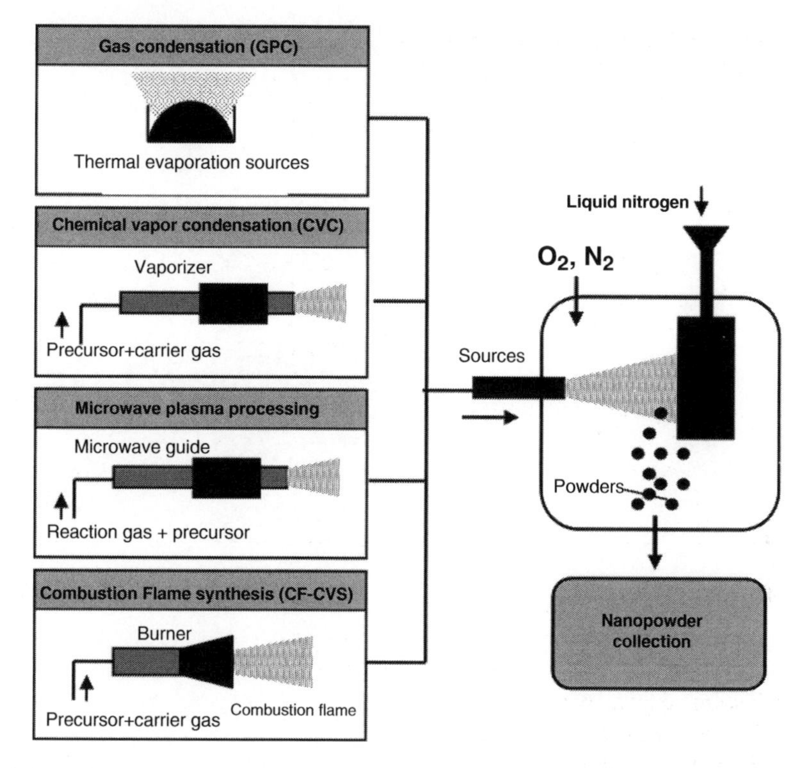

Fig. 28.7 Techniques for gas-phase synthesis of nanocrystalline powders (Adapted with permission from Hahn 1997, Copyright 1997 Elsevier)

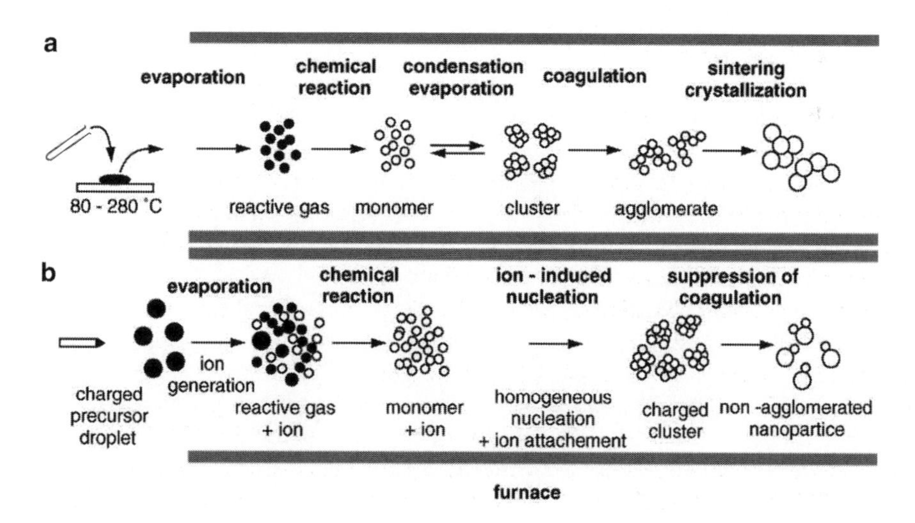

Fig. 28.8 Mechanisms in the process of particle formation in (**a**) homogeneous thermal CVD and (**b**) electrospray (ES)-CVD (Reprinted from Vahlas et al. 2006, Copyright 2006 Elsevier)

the deposited powders are agglomerated and are large in size. Mechanical milling is an effective method of resolving this problem (Hadjipanayis and Siegel 1994; Koch 1997). Mechanical attrition, or ball milling, which induces heavy cyclic deformation in powders, is a technique that produces nanostructures not by cluster assembly but by the structural decomposition of coarser-grained structures as the result of severe plastic deformation. This has become a popular method for producing nanocrystalline

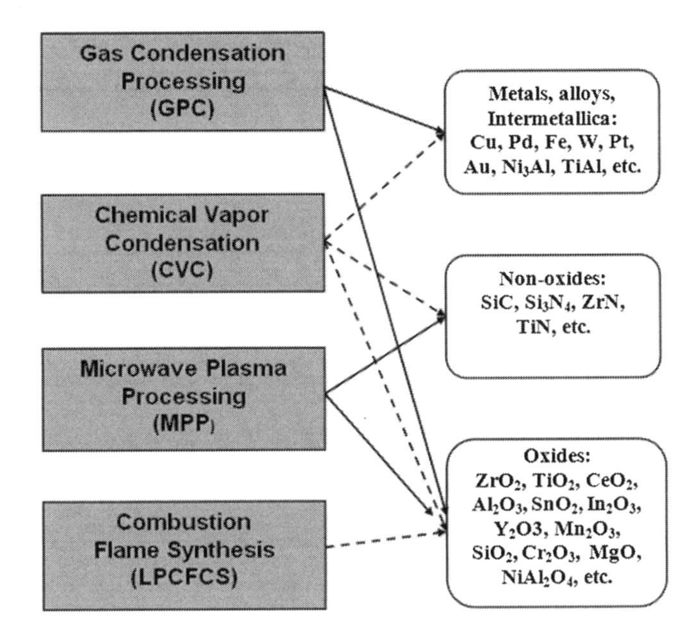

Fig. 28.9 Typical nanocrystalline materials synthesized using various powder technology methods

materials because of its simplicity, the relatively inexpensive equipment (on the laboratory scale) needed, and the applicability to essentially all classes of materials. The most commonly cited advantage is the possibility for easily scaling up material to tonnage quantities for various applications.

Figure 28.9 summarizes the capabilities of the various gas-phase synthesis methods that have been described. It is evident that all materials can be prepared by means of gas-phase synthesis in a nanocrystalline microstructure. For each case, it is necessary to determine which technique is most appropriate in terms of cleanliness of the powder surfaces, degree of agglomeration, particle size and distribution, phases, and quantities. In addition to single-phase materials, some of the techniques are also capable of synthesizing metal/metal, metal/ceramic, and ceramic/ceramic composites, as well as coated nanoparticles, potentially leading to interesting applications in the near future.

28.3.3 Advantages and Disadvantages of Thick-Film Technology

Barsan and Weimar (2001) considered thick-film technology an excellent technique for fabricating many types of gas sensors because it allows one to make the highly porous sensitive layers of nanostructured films. In general, thick-film technology has several significant advantages for sensor development:

1. Flexibility to develop various sensor constructions
2. Extensive choice of materials
3. Easy integration with electronic circuits
4. Low cost even in low-volume production
5. The possibility to use automatic fabrication processes
6. Compatibility with other (micro)electronic devices

In addition, the composition of thick films can be functionalized via incorporation of numerous catalytic promoter impurities, and they can be used to make mesoporous structures with high specific area (Moseley and Tofield 1987; Madou and Morrison 1989; Ihokura and Watson 1994; Morrison 1994; Barsan et al. 1999; Marek et al. 2003; Graf et al. 2004).

Compared to ceramic elements, thick films are more mechanically reliable and can be mounted in standard housings for integrated micromechanical schemes. Compared to thin films, thick films are less sensitive to the quality of the substrates. The parameters of thick film-based sensors are more reliable than those of ceramic elements, and, in general, the gas sensitivity is higher. The operating regime of these sensors can be maintained to keep constant either the working current or the working voltage. If this regime is chosen at the maximum temperature dependence of gas sensitivity, these sensors are easily employed as gas alarms.

Although thick-film technology is an attractive option for the fabrication of sensors, it does have some drawbacks:

1. The different materials employed are not very compatible and depend strongly on the manufacturing process.
2. These processes require complicated curing cycles under high temperatures carried out over long time frames.
3. The thick-film sensors are still not very reproducible in series.
4. The resolution of this technology is limited to a possible minimum line width of 100 μm (Sinner-Hettenbach 2000).
5. The surface of screen-printed layers is rather rough. In addition, thick-film heaters are not always very reproducible, with variations in their temperature coefficient of resistance and nominal resistance.

Some of the mentioned disadvantages of thick films and ceramic elements, such as reliability of sensor characteristics under production in series, long times to stabilize the functional properties, and rather high power consumption of sensors fabricated on conventional platform, are substantially improved by the application of thin-film microelectronic technologies.

28.4 Thin-Film Technology

Compared to technologies employed to fabricate thick films and ceramics, the development of thin-film sensor elements is based on well-managed deposition processes (Wu et al. 1993; Liu 1995). Gas-sensitive films are produced via all the available thin-film technologies, among which one can note thermal evaporation and sputtering (Stryhal et al. 2002; Saadeddin et al. 2007), laser ablation (Phillips et al. 1996), spray pyrolysis (Tiburcio-Silver and Sanchez-Juarez 2004), chemical vapor deposition (Heilig et al. 1999; Choy 2000), atomic-layer deposition (ALD) (Takada 2001), and rheotaxial growth and thermal oxidation (RGTO) (Sberveglieri et al. 1992). One can find a short description of these methods in Table 28.1. Each method has advantages and disadvantages. For example, spray pyrolysis and chemical vapor deposition seem to be the cost-effective techniques which are attractive for the production of inexpensive sensors (Sberveglieri et al. 1993; Labeau et al. 1993; Rumyantseva et al. 1996; Brousse and Schleich 1996; Olvera et al. 1996). Spray pyrolysis is quite flexible in terms of materials and structures that can be used to design gas sensors and to deposit composite materials (Korotcenkov et al. 2001b). ALD allows one to deposit highly homogeneous thin films with excellent coverage and thickness control. Films grown by ALD are generally dense, pinhole-free, and extremely conformal to the underlying substrate. Furthermore, film-thickness control at the monolayer level can be readily achieved by simply counting the number of ALD cycles (Göpel and Reinhardt 1996). The RGTO method allows working with metallic films at the stage of sensing-layer shaping. Reactive sputtering yields high reproducibility (Lalauze et al. 1991) and long-term stability of film functional properties (Sayago et al. 1995a, b). Magnetron sputtering allows one to vary the crystalline structure of the film from amorphous to single crystal (epitaxial) just by changing the substrate temperature and the rate of deposition (LeGore et al. 1997; Kissin et al. 1999a), whereas the film stoichiometry is

Table 28.7 Comparison between chemical vapor deposition and physical vapor deposition coating techniques

Chemical vapor deposition (CVD)	Physical vapor deposition (PVD)
Sophisticated reactor and/or vacuum system	*Sophisticated reactor and vacuum system*
Simpler deposition rigs with no vacuum system has been adopted in variants of CVD, AACVD, ESAVD, FACVD, and CCVD	
Expensive techniques for LPCVD, PECVD, PACVD, MOCVD, EVD, ALE, UHVCVD	*Expensive techniques*
Relatively low-cost techniques for AACVD and FACVD	
Non-line-of-sight process	*Line-of-sight process*
Therefore, it can coat complex-shaped components and deposit coating with good conformal coverage	Therefore, it has difficulty coating complex-shaped components and producing conformal coverage
Tend to use volatile/toxic chemical precursors	*Tend to use expensive sintered solid targets/sources*
Less volatile/more environmentally friendly precursors have been adopted in variants of CVD such as AACVD, ESAVD, and CCVD	Potential difficulties in large-area deposition and varying the composition or stoichiometry of the deposits
Multisource precursors tend to produce nonstoichiometric films	*Both single and multiple targets* do not guarantee the deposition of stoichiometric films because different elements will evaporate or sputter at different rates, except with the laser ablation method
Single-source precursors (AACVD, PICVD) have overcome such problems	
High deposition temperatures in conventional CVD	*Low to medium deposition temperatures*
Low to medium deposition temperatures can be achieved using variants of CVD such as PECVD, PACVD, MOCVD, AACVD, ESAVD	

CCVD combustion chemical vapor deposition; *MOCVD* metal–organic-assisted CVD; *PECVD* plasma-enhanced CVD; *FACVD* flame-assisted CVD; *AACVD* aerosol-assisted CVD; *ESAVD* electrostatic-atomization CVD; *LPCVD* low-pressure CVD; *APCVD* atmospheric-pressure CVD; *PACVD* photo-assisted CVD; *TACVD* thermal-activated CVD; *EVD* electrochemical vapor deposition; *RTCVD* rapid thermal CVD; *UHVCVD* ultrahigh-vacuum CVD; *ALE* atomic-layer epitaxy; *PICVD* pulsed-injection CVD
Source: Adapted with permission from Choy (2003), Copyright 2003 Elsevier

driven by the amount of oxygen in the vacuum chamber (Williams and Coles 1993; 1995; DiGiulio et al. 1996; Miccoci et al. 1996; Kissin et al. 1999b). The minimum crystallite size in the gas-sensitive polycrystalline MOX thin films is now down to 3 nm (Barbi et al. 1995). According to Demarne and Grisel (1993), DC magnetron sputtering allows one to use lower temperatures than radio-frequency sputtering, a definite advantage for Si substrates. A more detailed comparison of selected deposition methods is presented in Tables 28.2 and 28.7. Table 28.8 compares two main deposition methods of thin-film technology such as PVD and CVD, frequently used in the design of gas sensors for film deposition. The same comparison can be made for CVD and CSD methods (see Table 28.8).

Sensors prepared using thin-film technologies have gas-sensitive layers of typical thickness up to 1 μm. The thin-film technologies allow one to control reliably the physical properties, such as the thickness, morphology, microstructure, and stoichiometry of the gas-sensitive layers which is extremely important because these parameters govern the overall sensor performance (Korotcenkov 2008). Therefore, these deposition techniques are of great interest to both research laboratories and manufacturers. The conventional planar thin-film technologies employ masks made by photolithography with a resolution of 1–10 μm, which may be brought down to 10–100 nm by using ion-beam or electron-beam lithography. These techniques can be used to fabricate gas sensors of a few millimeters in size. Another option is to form multisensor arrays on a single substrate, which extends the long-term performance of the sensor as well as allowing selective analysis of gases using the "electronic nose" concept (Gardner et al. 1995; Althainz et al. 1996).

Table 28.8 Comparative characterization of CVD and CSD methods

	MOCVD	CSD
Merits	• Control of microstructure and hence properties	• Simple, inexpensive means of film deposition
	• Compatible with MEMS technology	• Low-temperature process
	• Mature systems have good thickness control and repeatability	• Rapid means for studying film–substrate interaction, grain growth behavior, effect of dopants, etc.
	• Excellent conformal coating and thickness scaling	• Precise stoichiometry control
	• In situ deposition	• Deposition of multicomponent materials without any problems
Limitations	• More sophisticated and expensive than CSD technique	• Conformal coating and thickness scaling
	• Deposition of multicomponent materials requires strong control	• Crack formation and delamination of material during drying process
	• Not very flexible to variation of deposition parameters	• Additional thermal treatment is necessary
	• Long cycle of development	• Deposition of continuous films with nm thickness

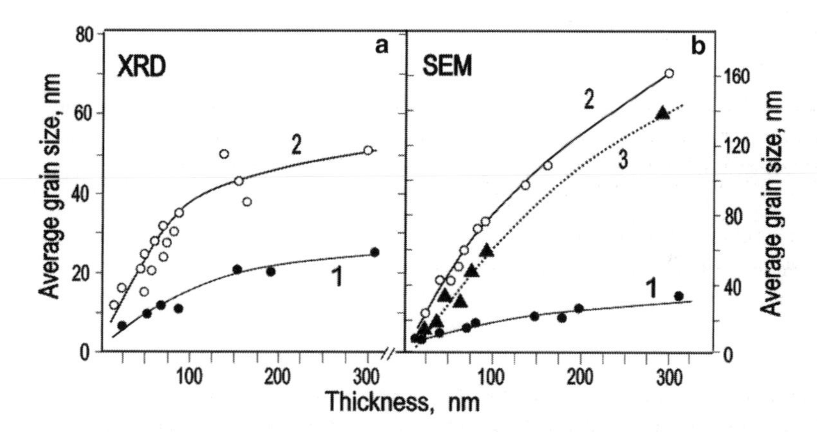

Fig. 28.10 Grain-size dependence on thickness of films prepared by pyrolysis at different temperatures, calculated using (**a**) XRD data and (**b**) SEM images: (*1*) T_{pyr} = 350–375 °C; (*2*) T_{pyr} = 450–475 °C; (*3*) T_{pyr} = 510–535 °C (Reprinted with permission from Korotcenkov et al. 2005b, Copyright 2005 Elsevier)

Regarding morphology of the film deposited using thin-film technology, one can say that the morphology of MOX films deposited using thin-film technologies is much more diverse than that of films made using thick-film technologies. For example, as has been established, "thin" and "thick" film gas sensors have completely different dependencies of the grain size on the film thickness. In both ceramic and thick-film gas sensors, the size of the metal oxide grains does not depend on the thickness of the sensing layer. In materials prepared using thick-film and ceramic technologies, the method of powder preparation (precursor material, aging time, pH, etc.) and the sintering temperature are the main parameters that control the grain size (Risti et al. 2002; Vuong et al. 2004; Amjoud et al. 2005). In contrast, for thin-film sensors, which were fabricated using metal oxide deposition at temperatures higher than 200–300 °C, the grain size of the metal oxides is usually determined directly by the thickness of the deposited film (Korotcenkov et al. 2005a, b). Increasing the film thickness leads to larger grain size (see Fig. 28.10). The morphology of films deposited using thin-film technologies depends on a number of factors. For example, it was found that the grain size of SnO_2 and In_2O_3 films deposited by spray pyrolysis depends on such deposition parameters as

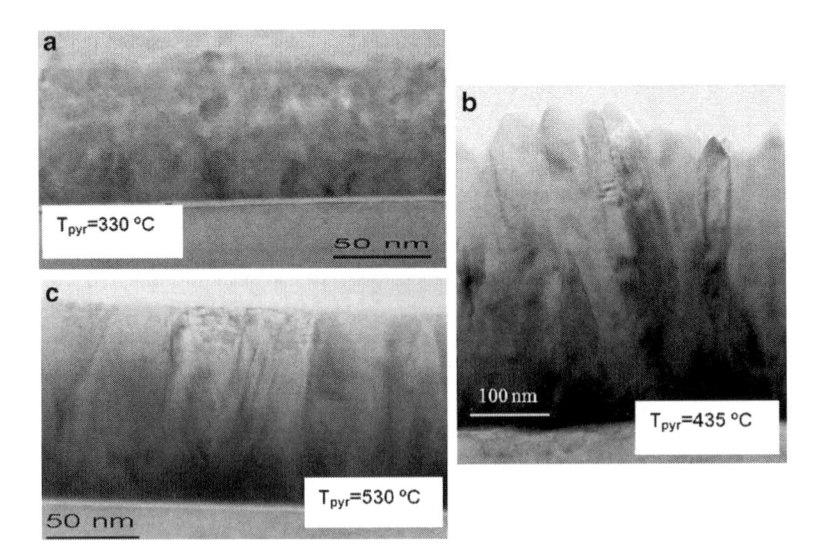

Fig. 28.11 TEM cross-section micrographs of SnO_2 films deposited by spray pyrolysis: (**a**) grained structure, T_{pyr} = 330–350 °C, d = 70–80 nm; (**b**, **c**) columnar structure, (**b**) T_{pyr} = 475 °C, d = 300 nm, (**c**) T_{pyr} = 510 °C, d = 75–100 nm (Reprinted with permission from Korotcenkov et al. 2005b, Copyright 2005 Elsevier)

the pyrolysis temperature, the film thickness, the distance between the atomizer and the substrate, and the properties of the precursor employed (Palatnik et al. 1972; Brinzari et al. 2002; Korotcenkov et al. 2001a, 2005a, b). Such distinctions make it impossible to apply regularities established for sensors fabricated by thick-film technology to those fabricated by thin-film technology, making the present research necessary.

The same effect is observed in other MOXs (Bender et al. 2002; Kiriakidis et al. 2007) made by various methods such as magnetron sputtering (Bender et al. 2002; Suchea et al. 2006; Kiriakidis et al. 2007), pulsed laser deposition (PLD) (Dolbec et al. 2003), and plasma-enhanced chemical vapor deposition (PECVD) (Huang et al. 2006). With these techniques the grain size also depends on the total gas pressure and the oxygen partial pressure in the deposition chamber. For example, it has been found that changing the oxygen partial pressure from 10 to 200 mTorr during SnO_2 PLD leads to an increase of the SnO_2 grain size from 3 nm to 10 nm (Dolbec et al. 2003).

The two-dimensionality of the grains which compose the oxide films is another feature of thin-film morphology which should be considered in analyzing the influence of grain size on the gas-sensing effect. In many cases, MOX films deposited by standard thin-film methods have a columnar structure (see Fig. 28.11) (Korotcenkov et al. 2004, 2005a, b). This means that the in-plane size of grains may differ significantly from the grain size measured in the growth direction. Ordinary analytical methods employed to analyze film structures and grain size, such as X-ray diffraction (XRD), scanning electron microscopy (SEM), and atomic force microscopy (AFM), provide mainly in-plane grain sizes. It is worth noting that the two-dimensionality effects are stronger in thicker films. The grained structure (see Fig. 28.11a) that is typical for MOX using ceramic or thick-film technology is usually observed only in thin films deposited at low temperatures (Korotcenkov et al. 2005b). With spray pyrolysis, for example, the grained structure of SnO_2 films was observed only when the deposition temperatures did not exceed 400 °C.

Detailed study has also shown that, in spite of a considerably smaller thickness, metal oxide films prepared using thin-film technologies do not have the porosity found in sensors fabricated by "thick" or "ceramic" technologies. Typical SEM images of the intergrain boundary in SnO_2 films deposited at temperatures higher than 350 °C are shown in Fig. 28.12. It is seen that the necks between grains are absent. As shown before, SnO_2 films deposited at temperatures higher than 350 °C have a columnar structure in which grains can grow through the entire film thickness (see Fig. 28.12). This means that films deposited by thin-film technology have a larger contact area between crystallites.

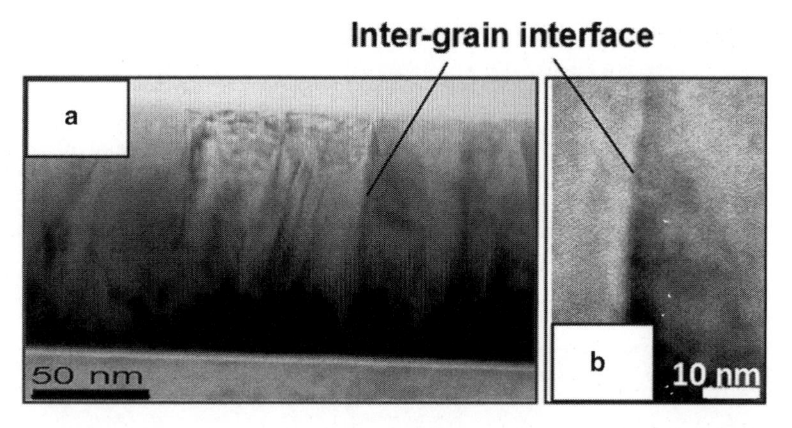

Fig. 28.12 (**a**), (**b**) SEM images of the SnO_2 films deposited at $T_{pyr}=520$ °C obtained at different magnifications ($d=120$ nm)

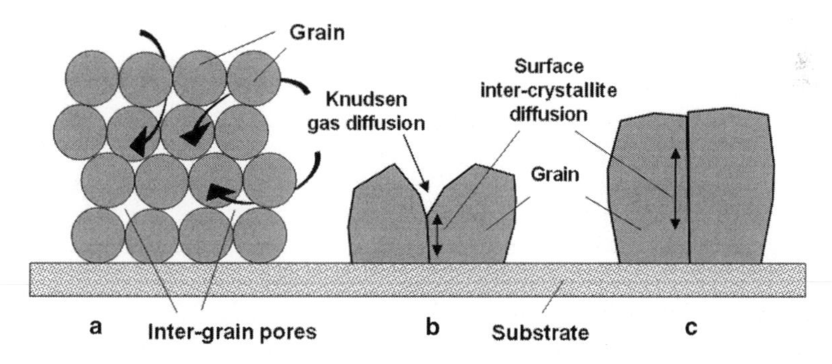

Fig. 28.13 Models, illustrating diffusion processes dominating in films formed using (**a**) thick-film and (**b, c**) thin-film technologies

Moreover, the comparison of the images of films having different thicknesses shows that the area of indicated contacts increases considerably with the film growth. Korotcenkov and co-workers (Korotcenkov et al. 2007a, b; Korotcenkov and Cho 2009) have shown that kinetics of sensor response in thin-film devices is being controlled by adsorption/desorption processes and the surface diffusion in inter-crystalline space. In the frame of this model, the time required for the diffusion of oxygen or oxygen vacancies into inter-crystalline space should increase along with the film-thickness growth. The specific character of such a film structure is shown in Fig. 28.13. However, it should be noted that the above-mentioned processes start to be dominating in kinetics of gas-sensing effect only when film thickness exceeds 60–80 nm. Experiments have shown that in films with thicknesses smaller than 60–80 nm there is no diffusion limitation in kinetics of sensor response (Korotcenkov and Cho 2009). In this case, sensitivity is mainly determined by the efficiency of the surface reactions.

Thus, a short survey of technologies suitable for gas sensor fabrication shows that we do not have ideal technology and we cannot name the best method of sensing-layer deposition. The selection of the method optimal for each specific engineering application should take into account the properties of the deposited material, the construction of the sensor design, and the possible consequences for the sensor's parameters during the application of the method selected. Will et al. (2000) considered and analyzed current methods and their possibilities for forming thin-film solid electrolytes. The results of that comparison are presented in Table 28.9.

Table 28.9 Comparison of methods for producing thin and dense solid electrolytes

	Film characteristics		Process features	
Technique	Structure	Deposition rate or thickness	Cost	Characteristics and limitations
Vapor phase				
Thermal spray technologies		100–500 μm h^{-1}		High deposition rates, various compositions, possible thick and porous coatings, high temperatures necessary
EVD	Columnar	3–50 μm h^{-1}	Expensive	High reaction temperatures necessary, corrosive gases equipment and processing costs
CVD	Columnar	1–10 μm h^{-1}	Expensive	Various precursor materials possible, high reaction equipment temperatures necessary, corrosive gases
PVD (RF and magnetron) sputtering	Amorphous to polycrystalline	0.25–2.5 μm h^{-1}	Expensive	Tailor-made films, dense and crack-free, low film equipment deposition temperatures, multipurpose technique, relatively low deposition rate
Laser ablation	Amorphous to polycrystalline		Expensive	Intermediate deposition temperatures, difficult equipment upscaling, time sharing of laser, relatively low deposition rate
Spray pyrolysis	Amorphous to polycrystalline	5–60 μm h^{-1}	Economical	Robust technology, upscaling possible, easy control of polycrystalline parameters, corrosive salts, post-thermal treatment usually necessary
Liquid phase				
Sol–gel, liquid-precursor route processes	Polycrystalline	0.5–1 μm for each coating	Economical	Various precursors possible, very thin films, low temperature sintering, coating and drying/heating have to be repeated 5–10 times, crack formation during drying, many process parameters
Solid phase				
Tape casting	Polycrystalline	25–200 μm		Robust technology, upscaling possible, crack formation
Slip casting and slurry coating	Polycrystalline	25–200 μm	Economical	Robust technology, crack formation, slow
Tape calendering	Polycrystalline	5–200 μm		Upscaling possible, co-calendering possible
EPD	Polycrystalline	1–200 μm		Short formation time, little restriction to shape of substrate, suitable for mass production, high deposition rates, inhomogeneous thickness
Transfer printing	Polycrystalline	5–100 μm	Economical	Robust technology, rough substrate surfaces possible, adhesion on smooth substrates difficult
Screen printing	Polycrystalline	10–100 μm	Economical	Robust technology, upscaling possible, crack formation

Source: Adapted with permission from Will et al. (2000), Copyright 2000 Elsevier

Based on the analysis carried out in the present section, it can be concluded that only simple stoichiometric compounds can be deposited using standard CVD and PVD methods, as each component has to be evaporated at a different temperature due to their different vapor pressures. Moreover, the CVD process also applies very toxic precursors (Choy 2000, 2003). The constituents have to be deposited from independently controlled sources, adding complexity to the system. At the same time, PVD is a line-of-sight process, meaning that it has difficulty in coating complex-shaped components.

Other methods also have limitations. Magnetron sputtering of multicomponent materials requires expensive sintered solid targets and produces surface damage. Ceramic powder methods and spray pyrolysis methods, on the other hand, have the potential to be good candidates for complicated stoichiometric compositions or for mixtures of materials. However, sol–gel-coated films tend to crack and have a thickness limitation for each layer, meaning that the process needs to be repeated to obtain the required thickness. Spray pyrolysis has limitations in repeatability. The investment cost for CVD and PVD apparatus is high compared to the droplet and powder techniques, whereas the setup for spray pyrolysis is inexpensive and simple. Standard CVD methods, however, have the advantage of being able to coat large areas uniformly, and they feature easily controlled deposition rates and film thicknesses. On the other hand, all liquid-precursor methods, such as sol–gel and slurry coating, are time-, labor-, and energy-intensive, because coating and drying/sintering have to be repeated in order to avoid crack formation.

According to consideration of the general requirements for sensor technology, one can expect the ideal method in any application to meet the following criteria:

- Compatibility with the process of manufacture of the chemical sensor
- No impairment of, or effect on, the properties of the bulk materials used in the device
- Ability to deposit the required type of material with the required thickness and structure
- Improvement in the quality of the designed sensor
- Ability to coat the engineering components uniformly with respect to both size and shape
- Cost-effectiveness in terms of the cost of the substrate, depositing material, and coating technique
- Ecologically clean and safe for attending personnel

It is obvious that there is currently no perfect method that meets all requirements. Therefore, in practice it is always necessary to search for the best compromise and choose the method which meets as many requirements as possible. At the same time, however, these requirements may change considerably during the device's elaboration phase or during industrial development. Properties that may be preferred during elaboration, such as multifunctionality, an opportunity to vary the parameters of the technological process and the deposited material, and the speed of reorganizing the technological process, differ greatly from the properties necessary during industrial fabrication, such as compatibility with basic technological processes, reproducibility, productiveness, and cost. Such differences in requirements are inevitable, and it is necessary to take them into account during research targeted toward the elaboration of devices designed for application in the gas sensor market.

28.5 Polymer Technology

There are two main options for incorporating polymers into gas sensors (Gardner and Bartlett 1995; Kumar and Sharma 1998; Harsanyi 1995, 2000):

1. Preprocessed polymer films are synthesized and shaped by extrusion, stretched into sheet forms, and covered by metal film separately from the sensor structures. They can then be attached (typically by gluing) to inorganic sensor surfaces.
2. Polymerization occurs directly on the sensor surfaces. The synthesis and shaping process occur on the sensor surface.

Of course, the second method is more progressive and is therefore the more commonly used technique for the fabrication of thin polymer layers used in the majority of gas sensors.

Table 28.10 Synthesis techniques for some conductive polymers used in chemical sensor fabrication

Polymer	Method
Polyacetylene	Chemical polymerization
Polythiophene	Chemical polymerization
	Electrochemical polymerization
Polyaniline	Chemical polymerization
	Electrochemical polymerization
Polyisoprene	Inclusion polymerization
Polybutadiene	Inclusion polymerization
Polysiloxane	Pyrolysis
Poly(2,3-dimethyl-butadiene)	Inclusion polymerization
Polypyrrole	Chemical polymerization
	Electrochemical polymerization
	Photochemical polymerization
Poly(*p*-phenylene-terephthalamide)	Electrochemical polymerization
Polypyrrole–polyamide composites	Electrochemical polymerization
PVC	Chemical polymerization
Polystyrene	Concentrated emulsion polymerization
Tetraphenylporphyrin	Vacuum polymerization
Poly(*p*-phenylene)	Chemical polymerization
Poly(α-naphthylamine)	Electrochemical polymerization
3-Octylthiophene-3-methylthiophene	Electrochemical polymerization
Poly(1,4-phenylene)	Electrochemical polymerization

28.5.1 Methods of Polymer Synthesis

There is no single method for synthesizing polymers targeted for gas sensors. Most polymers, except ionomeric polymers, can be synthesized using standard methods of polymerization, both conventional and specific routes, including the polycondensation process and metal-catalyzed polymerization techniques. Thus, polymers may be synthesized by any of the following techniques:

- Chemical or radical polymerization
- Electrochemical polymerization
- Photochemical polymerization
- Metathesis polymerization
- Concentrated emulsion polymerization
- Inclusion polymerization
- Solid-state polymerization
- Plasma polymerization
- Pyrolysis
- Soluble-precursor polymer preparation

These methods are described in detail in various comprehensive reviews (Malkin and Siling 1991; Skolheim 1986; Kumar and Sharma 1998; Gurunathan et al. 1999; Malinauskas 2001; Reisinger and Hillmyer 2002). A summary of the reported literature highlighting the polymerization techniques of some widely used conductive polymers is presented in Table 28.10.

Among all the techniques listed above, chemical polymerization is the most used for preparing large amounts of conductive polymers, since it is performed without electrodes (Malinauskas 2001). Chemical polymerization (oxidative coupling) is followed by the oxidation of monomers to a cation radical and their coupling to form dications. The repetition of this process generates a polymer, and

Table 28.11 Electrochemical data for some heterocyclic and aromatic monomers used for electrochemical polymerization

Monomer	Oxidation potential (V)
Pyrrole	0.6–0.8
Bipyrrole	0.55
Terpyrrole	0.26
Thiophene	2.07
Bithiophene	1.31
Terthiophene	1.05
Azulene	0.91
Pyrene	1.30
Carbazole	1.82
Fluorene	1.62
Fluoranthene	1.83
Aniline	0.71

all classes of conjugated polymers can be synthesized using this technique. Chemical polymerization is conducted with relatively strong chemical oxidants, such as ammonium peroxydisulfate (APS), ferric ions, permanganate or bichromate anions, or hydrogen peroxide. The reaction is controlled by the concentration and oxidizing power of the oxidant. Chemical polymerization occurs in the bulk of the solution, and the resulting polymers precipitate as insoluble solids. The polymer formed by chemical synthesis is generally a black powder.

Another chemical method used in the preparation of polymers is chemical vapor deposition. This method uses reagents in the gaseous phase, with the polymer being formed on a substrate present in the reagent vapor. An alternative is to have only the monomer in gaseous form, with the oxidant being present in liquid form on the surface of the substrate to be coated.

Electrochemical polymerization is usually carried out on the working electrode in a single-compartment cell with a three-electrode configuration, including a working electrode (generally Pt, but may vary according to the final requirements), a reference electrode (saturated calomel electrode), and a secondary electrode (Pt, Ni, or C). Generally, organic solvents are used in electrochemical polymerization, but aqueous solutions have been employed as well. Electrochemical polymerization can be carried out potentiometrically using a suitable power supply (potentiogalvanostat). Generally, potentiostatic conditions are recommended to obtain thin films, while galvanostatic conditions are recommended to obtain thick films. The advantage of this method is that precise flow control and rate of film deposition can be maintained by varying the potential/current conditions of the working electrode in the system.

The voltage potential is the most important parameter in controlling the polymerization process. For electrochemical oxidation, a certain electropolymerization potential (EP) must be applied to the solution for the monomer to be oxidized. Table 28.10 gives the peak oxidation potentials for some of the aromatic compounds that can produce conducting polymers using the electrochemical technique (Miller 1982). Table 28.11 shows that the electrochemically polymerizable monomers reported to date have relatively lower anodic oxidation potential peaks, which are at oxidation potential smaller than 2.1 V. No polymerization occurs below the electropolymerization potential. Above the EP, the rate of polymerization increases with the potential, which may be affected by a number of parameters, including monomer concentration, electrolyte concentration, and the nature of the electrode. For example, the EP for pyrrole is generally between 0.6 and 0.8 V. During pyrrole polymerization, lower current densities lead to the formation of a more crystalline polymer with fewer cross-linkages, while higher current densities produce rougher films.

Using this technique, a variety of conductive polymers has been generated, such as polypyrrole, polythiophene, polyaniline, polyphenylene oxide pyrrole, and polyaniline/polymeric acid composite.

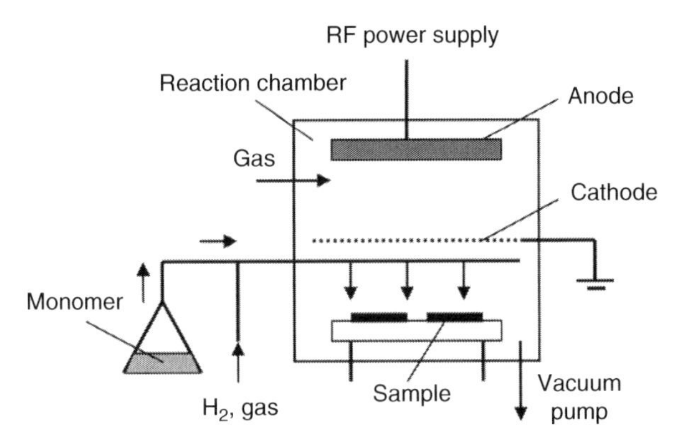

Fig. 28.14 Configuration of device for polymer film deposition by reactive sputtering

The degree of polymer doping during electrochemical polymerization depends on the dopant concentration, the amount of charge passed, and the voltage applied.

Photochemical polymerization takes place in the presence of UV irradiation. This technique utilizes photons to initiate a polymerization reaction in the presence of photosensitizers. Pyrrole has recently been photopolymerized using a ruthenium(II) complex as the photosensitizer. Under photoirradiation, Ru(II) is oxidized to Ru(III), and the polymerization is initiated by a one-electron-transfer oxidation process. Polypyrrole (Ppy) films can be obtained through the photosensitized polymerization of pyrrole using a copper complex as the photosensitizer. Photopolymerization of benzo(C)thiophene has been carried out in acetonitrile using CCl_4 and tetrabutylammoniumbromide. Photosensitive polymers have a unique advantage in processing, as polymerization and shaping can occur simultaneously with UV illumination. Photosensitive polymers can be applied and patterned with the same technology used by photoresists.

Plasma polymerization is a technique that is used in the preparation of ultrathin uniform layers (5–10 nm) that adhere strongly to an appropriate substrate. An electric glow discharge is used to create low-temperature "cold" plasma. A schematic diagram of the apparatus used for plasma polymerization is provided in Fig. 28.14. The device used for plasma polymerization is commonly constructed with an anode, a mesh cathode, and a monomer supply ring. The stage is cooled by water because the probability that the monomer radicals will stick to the wafer decreases at high temperature, and the plasma-polymerization rate decreases along with the temperature. The monomer liquid is cooled at 0 °C to maintain the vapor pressure. Argon gas is supplied to the chamber through the holes in the anode and is ionized between the anode and the mesh cathode. The advantage of this technique is that it eliminates a number of steps needed in the conventional coating process (Favia and De Agostino 1998).

Metathesis polymerization is unique, differing from all other polymerizations in that all the double bonds in the monomer remain in the polymer. It is a natural outgrowth of Ziegler–Natta polymerization in that the catalysts used are similar and often identical. This usually involves a transition metal compound plus an organometallic alkylating agent. Metathesis polymerization is further divided into three classes: ring-opening metathesis of cycloolefins (ROMP); metathesis of alkynes, acyclic or cyclic; and metathesis of diolefins. By far the greatest amount of work has been done on ROMP.

Pyrolysis is probably one of the oldest approaches utilized to synthesize conductive polymers by heating the polymer to form extended aromatic structures, thus eliminating heteroatoms. The product of polymer hydrolysis can be a film, a powder, or a fiber, depending on the form and nature of the standing polymer and the pyrolysis condition.

Nevertheless, conductive polymers have also been synthesized using other techniques, such as chain polymerization, step-growth polymerization, chemical vapor deposition, solid-state polymerization, soluble-precursor polymer preparation, and concentrated emulsion polymerization, to name just a few. Most of these techniques, however, are time consuming and involve the use of costly chemicals.

28.5.2 Fabrication of Polymer Films

Thin polymer films can be deposited using a variety of techniques that vary in complexity and applicability (Gardner and Bartlett 1995; Matsumoto et al. 1998; Pique et al. 2003; Chrisey and Hubler 1994). The choice of deposition technique depends on the physicochemical properties of the material, the film quality requirements, and the substrate being coated. The simplest methods involve the application of a liquid-solution polymer in a volatile solvent, including aerosol, dipping, spin coating, and Langmuir–Blodgett (LB) dip coating (Harsanyi 1995).

The Langmuir–Blodgett technique is based on the transfer of an insoluble polymer monolayer on a substrate with a hydrophilic surface as it is raised from the liquid covered by this polymer monolayer (Gaines 1966; Osada and DeRossi 2000). It is also possible to dip a substrate with a hydrophobic surface in water covered by a polymer monolayer and then slowly draw it back out. The possibility of depositing ordered films with known and controlled thickness (in the range ±2.5 nm) is the main advantage of the Langmuir–Blodgett technique. In principle, the LB technique can be used to prepare mono- and multimolecular layers and architectures with high perfection, different layer symmetries, and molecular orientations. This method, however, has very low technological effectiveness, making its application in real chemical sensor fabrication processes unlikely. The number of polymers which can be used for preparing polymer films with the LB method is also very limited.

Part of polymers formed by chemical polymerization can deposit spontaneously on the surface of various materials immersed in the polymerization solution. The distribution of the resulting polymers between the precipitated and deposited forms depends on many variables and varies within a broad range. To coat materials with a polymer layer, it is desirable to shift this distribution toward the surface-deposited form, whereas bulk polymerization should be diminished as much as possible. This can usually be achieved by choosing appropriate reaction conditions, such as the concentration of the solution components, the concentration ratio of oxidant to monomer, the reaction temperature, and an appropriate treatment of the surface of the material to be coated by conducting polymers. Although a bulk polymerization cannot be completely suppressed, a reasonably high yield of surface-deposited polymers can be achieved by adjusting the reaction conditions (Malinauskas 2001).

Screen printing is a widely used technology in processing polymer composite materials available in paste form. The dipping method and spin-coating technique depend on the solubility of the conductive polymers, previously synthesized by the chemical polymerization technique. In this case, the surface to be coated is enriched either with a monomer or an oxidizing agent and is then treated with a solution of either oxidizer or monomer, respectively. A major advantage of this process is that the polymerization occurs almost exclusively at the surface; no bulk polymerization takes place in the solution. For some polymer materials, the surface can be enriched with a monomer by its sorption from the solution. Enrichment of the surface by an oxidizer can be achieved either by using an ion-exchange mechanism or by deposition of an insoluble layer of oxidizer. The disadvantage of this process is that it is limited by materials that can be covered or enriched with a layer of either monomer or oxidizer in a separate stage, preceding the surface polymerization.

Several techniques are applicable to bulk polymer materials, such as vacuum deposition technologies. Vacuum deposition processes may be used to obtain thin polymer films that have high density, thermal stability, and insolubility in organic solvents, acids, and alkalis. These polymer deposition techniques involve in situ polymerization on a substrate surface affected by various factors. The layers can be deposited on any substrates that cannot be damaged by the vacuum processes. The following are all examples of vacuum deposition processes (Skolheim 1986; Harsanyi 2000):

- Vacuum pyrolysis consisting of sublimation, a pyrolysis, and a deposition-polymerization process
- Vacuum polymerization stimulated by electron bombardment
- Vacuum polymerization initiated by UV irradiation

- Vacuum evaporation using either a resistance-heated solid polymer source or, more effectively, an electron beam
- Radio-frequency sputtering of a polymer target in a plasma composed of polymer fragments with argon added to the plasma
- Plasma or glow discharge polymerization of monomer gases or vapors

Most of the currently available techniques, however, are not generic enough to be capable of simultaneously depositing polymer thin films without affecting their chemical integrity and physico-chemical properties while also producing thin, uniform, and solvent-free coatings in a discrete or continuous fashion. Furthermore, most of these techniques are not appropriate for the fabrication of multilayers, since they rely on the application of a solvent solution containing the material of interest, which may dissolve any previously deposited layers. Therefore, intensive research has been carried out over the last few years attempting to perfect the methods for the deposition of polymer films.

The most successful approach so far has been a modification of the PLD method, described earlier. Previous work with UV PLD showed the ability of this technique to deposit thin films of various types of polymer materials. However, it has also been established that the PLD of polymers in a standard variant is limited to a small class of materials.

The recently developed matrix-assisted pulsed laser evaporation (MAPLE) method considerably extends the possibilities of PLD for the deposition of polymeric materials. MAPLE differs from PLD in the way the target is prepared and in the laser energy regime under which the laser–material interactions at the target take place (Pique et al. 2003). The MAPLE deposition process has been used successfully to deposit various types of polymer and organic materials, including chemoselective polymers. MAPLE offers significant advantages for the fabrication of chemical sensors, since it allows for the deposition of solvent-free chemoselective polymers on a variety of substrate surfaces. For example, using MAPLE, highly uniform films of siloxane fluoroalcohol (SXFA) have been deposited on the surface of surface acoustic wave (SAW) resonators. The performance of these MAPLE-coated sensors was comparable to that of spray-coated SAW devices.

It should be noted that in the CVD method with laser activation, the method of cooling the substrate during polymer deposition is also effective. This cooling method creates conditions that hamper the polymer's degradation and allows for a considerable increase in the deposition rate. For example, by irradiating a cooled substrate with an excimer laser in an organic gas environment, polymethylmethacrylate (PMMA) films have been selectively deposited with a high deposition rate (Takashima et al. 1994).

It should be stressed that the appearance of effective dry methods for polymer deposition is an important achievement, because the standard wet processes do not promote device integration into modern semiconductor processing (Harsanyi 2000).

28.6 Deposition on Fibers

28.6.1 Specifics of Film Deposition on Fibers

In principle, it is possible to use any of the previously mentioned methods for deposition of sensitive materials on fibers. It is only necessary to take into account the following factors:

- Deposition takes place on a small area, and, as a result, the process has very low effectiveness of use of the deposited material.
- All sides of the fiber need to be covered uniformly. An SEM image of a fiber with a covering deposited using the PVD method without rotation is shown in Fig. 28.15.
- The fiber may be too long and have poor thermal contact with the substrate's holder.

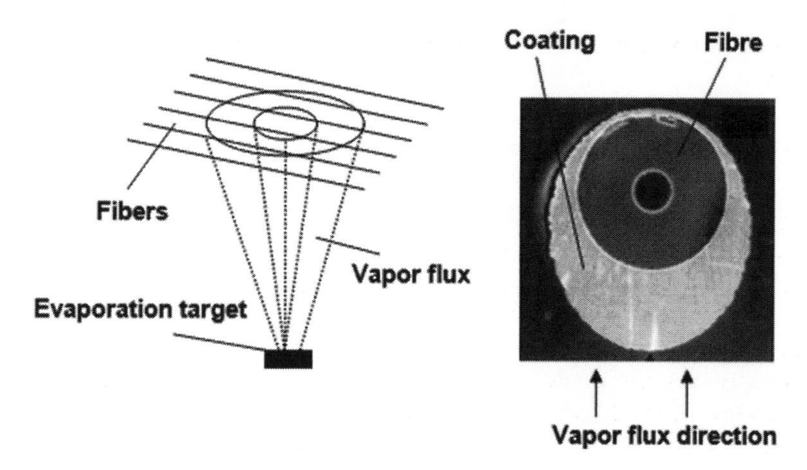

Fig. 28.15 Schematic diagrams and SEM image for fiber coating by thermal evaporation

- The fiber may be polymeric, which may impose a series of limitations on the thermal parameters of deposition.
- The fiber has lowered mechanical durability, while the covering needs to have high adhesion because the fiber may bend.

28.6.2 Coating Design and Tooling

Coating methods for fibers do not require specific changes in the deposition process. Therefore, for sensing the material deposition on the fiber, the same modes may be used on the lane surfaces as are used during deposition. The details of the deposition on the fiber become apparent only when specific constructions of the substrate's holders and heaters are used, which fix the fiber's position, rotation, uniform heating, and displacement (Kashima et al. 1991; Kaneko and Nittono 1997).

At present there are no fixed rules for elaborating these units. Every designer resolves this problem individually, taking into account the deposition method being used and available equipment. For example, the fiber may be pulled through a reactor with hot walls in the process of CVD (Choy 2000, 2003). Alternatively, the fiber may be heated to the reaction temperature by an electrical current applied through a mercury electrode. In one conventional technique, the fiber was heated in a wave-guide-type microwave applicator. A hot-wall CVD fiber-coating system capable of coating a fiber in a continuous manner is depicted in Fig. 28.16. This system may operate at atmospheric pressure and has separate furnaces for desizing the fiber prior to deposition and for thermal treatment of the coating after deposition. A supply spool and motorized take-up spool complete the system.

Of course, all of these approaches have specific advantages and disadvantages. For example, the disadvantages of the direct heating technique are that it is limited to electrically conductive fibers, each reaction chamber can accommodate only one fiber at a time, and the mercury used for electrical contact with the fiber has a highly toxic vapor. The microwave technique is not limited to electrically conductive fibers and does not involve mercury. However, it is limited to one fiber per applicator, and the waveguide applicator is intrinsically energy inefficient because its proper operation depends on the absorption of a substantial portion of the incident microwave power in a dummy load. It should be kept in mind, however, that the effectiveness of the microwave-cavity applicator has improved considerably. The fibers can be electrically conductive or nonconductive, so there is no need for mercury, and microwave energy is being utilized more efficiently than in the older waveguide/dummy-load

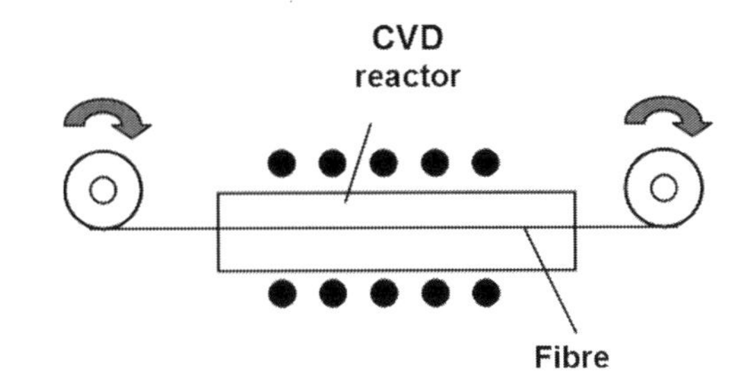

Fig. 28.16 Apparatus for CVD on fiber

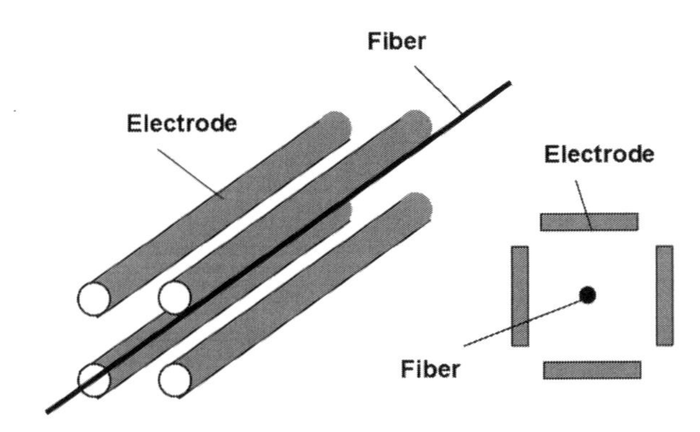

Fig. 28.17 Apparatus with multielectrode configuration (Adapted from Kashima et al. 1991. Copyright 1991 Elsevier)

microwave applicators. An effective way of improving the uniformity of a fiber's covering using the PDV method is through simultaneous use of several deposition sources located around the fiber (Kashima et al. 1991), as shown in Fig. 28.17.

Another area that is critical to the successful production of fiber coatings is the tool used to hold the fibers in the coating chamber. Because most fibers must be coiled and packed into the tool, it is important that the fibers remain unstressed. This requirement may necessitate tooling in a standard coating chamber that reduces the distance from the source to the substrate. Another approach is to have tooling in the form of a drum. Even in such cases, however, the number of fibers that can be coated is limited by the amount of epoxy in the fiber bundles. Even cured epoxy will outgas in a vacuum chamber, with most of the gas being water vapor absorbed from ambient air. Such outgassing of the epoxy affects the adhesion of the coating to the fiber, the packing density of the coating, and the refractive index of the deposited films.

During deposition of sensing material on polymer fibers using magnetron sputtering, it is necessary to take into account that, depending on the evaporation conditions, PVD-coated fibers can present quite different surface properties (Dietzel et al. 2000). Deposition of reactive metals induces both textural and chemical changes on the fiber surface. Nitrogen and carbon react with the polymer surface and form new structures as a result of an etching process. A higher concentration of reactive gas leads to a more drastic recombination of the polymer surface and results in a fibroid fiber structure. Dietzel et al. (2000) established that different noble and reactive gases produce different layer adhesion. Thus, nitrogen and acetylene plasmas lead to better layer adhesion than does argon plasma. Higher N_2 concentrations produce superior layer adhesion, and pure Zr-based layers adhere better than Ti-based layers.

Temperature control of the fiber during coating is another parameter that must be monitored. Research has shown that the refractive indices of a "cold"-coated film are usually less than that of a hot film, even with the use of ion-assisted deposition. In many cases they are also nonuniform. If the fibers have epoxy or plastic jackets, as is usually the case, they cannot be exposed to high temperatures. The temperature controller of the coating system is usually programmed to a temperature between 35 °C and 90 °C; however, the heat from the electron-beam gun or ion gun will raise the

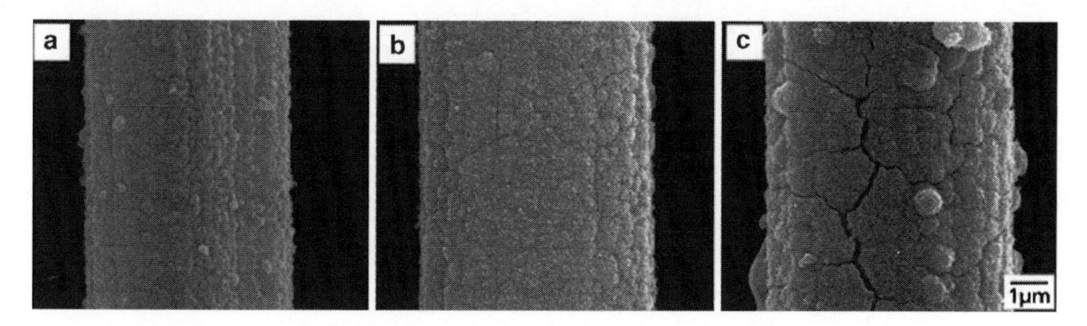

Fig. 28.18 SEM images of carbon fiber after (**a**) 2-h, (**b**) 4-h, and (**c**) 6-h exposure to a TiO_2 LPD solution (Reprinted with permission from Kern and Gadow 2004, Copyright 2004 Elsevier)

temperature of the fiber during coating. If the temperature of the fiber exceeds a safe limit, the ion or electron gun must be shut off to allow the fiber to cool down.

To resolve this problem and establish a successful fiber-coating system, the coating chamber is set up with either a broad-band, high-current, low-voltage ion gun or a suitable plasma system to densify the coating. This is known as an ion-assisted process and can produce dense films on cold substrates, such as fibers.

In addition to densifying the coating, an ion gun may also be used to clean the surface prior to coating, thus improving adhesion of the coating to the fiber tips. For this to be possible, the ion gun must be capable of handling oxygen gases, because an oxide coating is the most robust that can be applied to a fiber. However, in spite of the intensive development of dry deposition methods for various films on a fiber, the dip-coating method continues to be the most widely used (Kaneko and Nittono 1997). So far, compared to wet deposition methods, this method is a more commercially attractive alternative to costly vapor deposition technologies (Kern and Gadow 2002; Herbig and Loebmann 2004). The process offers several unique properties, including the ability to use a wide range of compositions and the ability to apply coatings to complex substrates, both of which make this process advantageous for coating fibers with sensing materials. For example, Herbig and Loebmann (2004) found that by using a dip-coating procedure the sol–gel technique could be used to apply a coating to individual fibers. SEM measurements confirmed that an effective coating had been applied to the fibers (see Fig. 28.18). Kern and Gadow (2004) further showed that the continuous liquid-phase coating method is technically and economically feasible for the production of coated carbon fibers used in the manufacture of various composite materials.

Some fibers can be coated with polymer layers by immersing them in an electrolyte, where electrochemical polymerization takes place, or by vapor-phase treatment of oxidant-containing carriers with the monomers (Malinauskas 2001). In this latter case, fibrous materials are charged by sorption with $FeCl_3$, provided from aqueous or ethereal solution, and then treated with a pyrrole vapor, either in vacuum or from its solution in toluene.

References

Althainz P, Goschnick J, Ehrmann S, Ache HJ (1996) Multisensor microsystem for contaminants in air. Sens Actuators B Chem 33:72–76

Amjoud M, Rhouta B, Alimoussa A, Hajji L, Mezzane D, Ahamdane H (2005) Effect of pH adjustment in sol-gel synthesis route on grain size of tin dioxide intended for gas sensors application. Phys Chem News 22:120–124

Arthur JA (2002) Molecular beam epitaxy. Surf Sci 500:189–217

Bagwell RB, Messing GL (1996) Critical factors in the production of sol–gel derived porous alumina. Key Eng Mater 115:45–64

Barbi GB, Santos JP, Serrini P, Gibson PN, Horrillo MC, Manes L (1995) Ultrafine grain-size tin-oxide films for carbon monoxide monitoring in urban environments. Sens Actuators B Chem 25:559–563

Barsan N, Schweizer-Berberich M, Gopel W (1999) Fundamental and practical aspects in the design of nanoscaled SnO_2 gas sensors. A status report. Fresenius J Anal Chem 365:287–304

Barsan N, Weimar U (2001) Conduction model of metal oxide gas sensors. J Electroceram 7:143–167

Bender M, Gagaoudakis E, Douloufakis E, Natsakou E, Katsarakis N, Cimalla V, Kiriakidis G, Fortunato E, Nunes P, Marques A, Martins R (2002) Production and characterization of zinc oxide thin films for room temperature ozone sensing. Thin Solid Films 418:45–50

Bietsch A, Zhang J, Hegner M, Lang HP, Gerber C (2004) Rapid functionalization of cantilever array sensors by inkjet printing. Nanotechnology 15:873–880

Bradley DC, Mehrotra RC, Gaur DP (1978) Metal alkoxides. Academic, London

Brinker CJ, Clark DE, Ullrich DR (eds) (1984) Better ceramics through chemistry. Elsevier, New York, NY

Brinker CJ, Scherer GW (1990) Sol-gel science: the physics and chemistry of sol-gel processing. Academic, San Diego, CA

Brinker CJ, Hurd AJ, Schunk PR, Ashely CS, Cairncross RA, Samuel J, Chen KS, Scotto C, Schwartz RA (1996) Sol-gel derived ceramic films—fundamentals and applications. In: Stern K (ed) Metallurgical and ceramic protective coatings. Chapman & Hall, London, pp 112–151

Brinzari V, Korotcenkov G, Schwank J, Lantto V, Saukko S, Golovanov V (2002) Morphological rank of nano-scale tin dioxide films deposited by spray pyrolysis from $SnC_{14}{\cdot}5H_2O$ water solution. Thin Solid Films 408:51–58

Brousse T, Schleich DM (1996) Sprayed and thermally evaporated SnO_2 thin films for ethanol sensors. Sens Actuators B Chem 31:77–79

Bunshah RF (ed) (1994) Handbook of deposition technologies for film and coatings: science, technology and applications. Noyes, Park Ridge, NJ

Calvert P (2001) Inkjet printing for materials and devices. Chem Mater 13:3299–3305

Cho N, Lim T, Jeon Y, Gong M (2008) Inkjet printing of polymeric resistance humidity sensor using UV-curable electrolyte inks. Macromol Res 16:149–154

Choi U-S, Shimanoe K, Yamazoe N (2005) Influences of ball-milling time on gas-sensing properties of Co_3O_4–SnO_2 composites. Sens Actuators B Chem 107:516–522

Choy KL (2000) Vapour processing of nanostructured materials. In: Nelwa HS (ed) Handbook of nanostructured materials and nanotechnology, vol 1. Academic, San Diego, CA, pp 533–577

Choy KL (2003) Chemical vapour deposition of coatings. Prog Mater Sci 48(2):57–170

Chrisey D, Hubler G (eds) (1994) Pulsed laser deposition of thin films. Wiley, New York, NY

Christen HM, Eres G (2008) Recent advances in pulsed-laser deposition of complex oxides. J Phys Condens Matter 20:264005

Chung WK, Sakai G, Shimanoe K, Miura N, Lee DD, Yamazoe N (1998) Preparation of indium oxide thin film by spin-coating method and its gas-sensing properties. Sens Actuators B Chem 46:139–145

Comini E, Faglia G, Sberveglieri G (2009) Electrical-based gas sensing. In: Comini E, Faglia G, Sberveglieri G (eds) Solid state gas sensing. Springer, New York, NY, pp 47–107

Crowley K, Morrin A, Hernandez A, O'Malley E, Whitten PG, Wallace GG, Smyth MR, Killard AJ (2008) Fabrication of an ammonia gas sensor using inkjet-printed polyaniline nanoparticles. Talanta 77:710–717

Crowley K, Morrin A, Shepherd RL, in het Panhuis M, Wallace GG, Smyth MR, Killard AJ (2010) Fabrication of polyaniline-based gas sensors using piezoelectric inkjet and screen printing for the detection of hydrogen sulfide. IEEE Sens J 10:1419–1426

De Girolamo Del Mauro A, Grimaldi IA, Loffredo F, Massera E, Polichetti T, Villani F, Di Francia G (2011) Geometry of the inkjet-printed sensing layer for a better volatile organic compound sensor response. J Appl Polym Sci 122:3644–3650

Demarne V, Grisel A (1993) A new SnO_2 low temperature deposition technique for integrated gas sensors. Sens Actuators B Chem 15–16:63–67

Dietzel Y, Przyborowski W, Nocke G, Offermann P, Hollstein F, Meinhardt J (2000) Investigation of PVD arc coatings on polyamide fabrics. Surf Coat Technol 135:75–81

DiGiulio M, Serra A, Tepore A, Rella R, Siciliano P, Mirenghi L (1996) Influence of the deposition parameters on the physical properties of tin oxide thin films. Mater Sci Forum 203:143–148

Dolbec R, El Khakani MA, Serventi AM, Saint-Jacques RG (2003) Influence of the nanostructural characteristics on the gas sensing properties of pulsed laser deposited tin oxide thin films. Sens Actuators B Chem 93:566–571

Dua V, Surwade S, Ammu S, Agnihotra S, Jain S, Roberts K, Park S, Ruoff R, Manohar S (2010) All-organic vapor sensor using inkjet-printed reduced graphene oxide. Angew Chem Int Ed Engl 49:2154–2157

Epifani M, Francioso L, Siciliano P, Helwig A, Mueller G, Dı az R, Arbiol J, Morante JR (2007) SnO_2 thin films from metalorganic precursors: synthesis, characterization, microelectronic processing and gas-sensing properties. Sens Actuators B Chem 124:217–226

Favia P, De Agostino R (1998) Plasma treatments and plasma deposition of polymers for biomedical applications. Surf Coat Technol 98:1102–1106

Francioso L, Russo M, Taurino AM, Siciliano P (2006) Micrometric patterning process of sol–gel SnO_2, In_2O_3 and WO_3 thin film for gas sensing applications: towards silicon technology integration. Sens Actuators B Chem 119:159–166

Gaines GL Jr (1966) Insoluble monolayers at liquid-gas interfaces. Wiley-Interscience, New York, NY

Gardner JW, Bartlett PN (1995) Application of conducting polymer technology in microsystems. Sens Actuators A Phys 51:57–66

Gardner JW, Pike A, de Rooij NF, Koudelka-Hep M, Clerc PA, Hierlemann A, Gopel W (1995) Integrated array sensor for detecting organic solvents. Sens Actuators B Chem 26–27:135–139

Glocker DA, Shah I (eds) (1995) Handbook of thin film process technology. Institute of Physics Pubiishing, Bristol

Göpel W, Reinhardt G (1996) Metal oxide sensors: new devices through tailoring interfaces on the atomic scale. In: Baltes H, Göpel W, Hesse J (eds) Sensors update, Sensor technology—applications markets, vol 1. VCH, Weinheim, pp 49–120

Graf M, Barrettino D, Zimmermann M, Hierlemann A, Baltes H, Hahn S, Barsan N, Weimar U (2004) CMOS monolithic metal-oxide sensor system comprising a micro hotplate and associated circuitry. IEEE Sens J 4(1):9–16

Guidi V, Butturi MA, Carotta MC, Cavicchi B, Ferroni M, Malagu C, Martinelli G, Vincenzi D, Sacerdoti M, Zen M (2002) Gas sensing through thick film technology. Sens Actuators B Chem 84:72–77

Gurunathan K, Murugan AV, Marimuthu R, Mulik UP, Amalnerkar DP (1999) Electrochemically synthesised conducting polymeric materials for applications towards technology in electronics, optoelectronics and energy storage devices. Mater Chem Phys 61:173–191

Hadjipanayis GC, Siegel RW (eds) (1994) Nanophase materials. Kluwer, Dordrecht

Hahn H (1997) Gas phase synthesis of nanocrystalline materials. NanoStruct Mater 9:3–12

Harsanyi G (1995) Polymeric sensing films: new horizons in sensorics? Sens Actuators A Phys 46–47:85–88

Harsanyi G (2000) Sensors in biomedical applications: fundamentals, technology and applications. Technomic, Basel

Hecht G, Richter F, Hahn J (eds) (1994) Thin films. DGM Informationgesellschaft, Oberursel

Heilig A, Bârsan N, Weimar U, Gopel W (1999) Selectivity enhancement of SnO_2 gas sensors: simultaneous monitoring of resistances and temperatures. Sens Actuators B Chem 58:302–309

Hench LL, Ulrich DR (eds) (1984) Ultrastructure processing of ceramics, glasses, and composites. Wiley, New York, NY

Herbig B, Loebmann P (2004) TiO_2 photocatalysts deposited on fiber substrates by liquid phase deposition. J Photochem Photobiol A Chem 163:359–365

Heule M, Gauckler LJ (2001) Gas sensors fabricated from ceramic suspensions by micromolding in capillaries. Adv Mater 13:1790–1793

Heule M, Meier L, Gauckler LJ (2001) Micropatterning of ceramics on substrates towards gas sensing applications. Mater Res Soc Symp Proc 657:EE9.4

Hitchman ML, Jensen KF (1993) CVD: principles and applications. Academic, San Diego, CA

Huang H, Tan OK, Lee YC, Tse MS (2006) Preparation and characterization of nanocrystalline SnO_2 thin films by PECVD. J Cryst Growth 288:70–74

Huczko A (2000) Template-based synthesis of nanomaterials. Appl Phys A 70:365–376

Ihokura K, Watson J (1994) The stannic oxide Gas sensor: principle and application. CRC, Boca Raton, FL

Ivanovskaya M (2000) Ceramic and film metal oxide sensors obtained by sol-gel method: structural features and gas-sensitive properties. Electron Technol 33(1/2):108–112

Janata J (1989) Principle of chemical sensors. Plenum, New York, NY

Jaworek A, Sobczyk AT (2008a) Electrospraying route to nanotechnology: an overview. J Electrostatics 66:197–219

Jaworek A, Sobczyk AT (2008b) Electrospraying route to nanotechnology: an overview. J Electrostatics 66:197–219

Kaneko T, Nittono O (1997) Improved design of inverted magnetrons used for deposition of thin films on wires. Surf Coat Technol 90:268–274

Kashima T, Matsuda Y, Fujiyama H (1991) Development of the quadrupole plasma chemical vapour deposition method for low temperature, high speed coating on an optical fibre. Mater Sci Eng A 139:79–84

Kaya C, He JY, Gu X, Butler EG (2002) Nanostructured ceramic powders by hydrothermal synthesis and their applications. Microporous Mesoporous Mater 54:37–49

Kern F, Gadow R (2002) Liquid phase coating process for protective ceramic layers on carbon fibers. Surf Coat Technol 151–152:418–423

Kern F, Gadow R (2004) Deposition of ceramic layers on carbon fibers by continuous liquid phase coating. Surf Coat Technol 180–181:533–537

Kim J, Yun J, Song J, Han C (2009) The spontaneous metal-sitting structure on carbon nanotube arrays positioned by inkjet printing for wafer-scale production of high sensitive gas sensor units. Sens Actuators B Chem 135:587–591

Kiriakidis G, Suchea M, Christoulakis S, Horvath P, Kitsopoulos T, Stoemenos J (2007) Structural characterization of ZnO thin films deposited by dc magnetron sputtering. Thin Solid Films 515:8577–8581

Kissin VV, Voroshilov SA, Sysoev VV (1999a) Oxygen flow effect on gas sensitivity properties of tin oxide film prepared by r.f. sputtering. Sens Actuators B Chem 55:55–59

Kissin VV, Voroshilov SA, Sysoev VV (1999b) A comparative study of SnO_2 and SnO_2:Cu thin films for gas sensor applications. Thin Solid Films 348:307–314

Koch CC (1997) Synthesis of nanostructured materials by mechanical milling: problems and opportunities. Nanostruct Mater 9:13–22

Kordas K, Mustonen T, Toth G, Jantunen H, Lajunen M, Soldano C, Talapatra S, Kar S, Vajtai R, Ajayan PM (2006) Inkjet printing of electrically conductive patterns of carbon nanotubes. Small 2:1021–1025

Korotcenkov G, Brinzari V, DiBattista M, Schwank J, Vasiliev A (2001a) Peculiarities of SnO_2 thin film deposition by spray pyrolysis for gas sensor application. Sens Actuators B Chem 77:244–252

Korotcenkov G, Brinzari V, Schwank J, Cerneavschi A (2001b) Possibilities of aerosol technology for deposition of SnO_2-based films with improved gas sensing characteristics. J Mater Sci Eng C 19:73–77

Korotcenkov G, Cerneavschi A, Brinzari V, Vasiliev A, Cornet A, Morante JR, Cabot A, Arbiol J (2004) In_2O_3 films deposited by spray pyrolysis as a material for ozone gas sensors. Sens Actuators B Chem 99:304–310

Korotcenkov G (2005) Gas response control through structural and chemical modifications of metal oxide films: state of the art and approaches. Sens Actuators B Chem 107:209–232

Korotcenkov G, Brinzari V, Ivanov M, Cerneavschi A, Rodriguez J, Cirera A, Cornet A, Morante JR (2005a) Structural stability of In_2O_3 films deposited by spray pyrolysis during thermal annealing. Thin Solid Films 479:38–51

Korotcenkov G, Cornet A, Rossinyol E, Arbiol J, Brinzari V, Blinov Y (2005b) Faceting characterization of SnO_2 nanocrystals deposited by spray pyrolysis from $SnCl_4$-$5H_2O$ water solution. Thin Solid Films 471:310–319

Korotcenkov G, Ivanov M, Blinov I, Stetter JR (2007a) Kinetics of In_2O_3-based thin film gas sensor response: the role of "redox" and adsorption/desorption processes in gas sensing effects. Thin Solid Films 515(7–8):3987–3996

Korotcenkov G, Brinzari V, Stetter JR, Blinov I, Blaja V (2007b) The nature of processes controlling the kinetics of indium oxide-based thin film gas sensor response. Sens Actuators B Chem 128:51–63

Korotcenkov G (2008) The role of morphology and crystallographic structure of metal oxides in response of conductometric-type gas sensors. Mater Sci Eng R 61:1–39

Korotcenkov G, Cho BK (2009) Thin film SnO_2-based gas sensors: film thickness influence. Sens Actuators B Chem 142:321–330

Korotcenkov G, Cho BK (2010) Methods of sensing materials synthesis and deposition. In: Korotcenkov G (ed) Chemical sensors: fundamentals of sensing materials, vol 1, General approaches. Momentum, New York, NY, pp 214–303

Kraus T, Malaquin L, Schmid H, Riess W, Spencer ND, Wolf H (2007) Nanoparticle printing with single-particle resolution. Nature Nanotechnol 2:570–576

Kukkola J, Mohl M, Leino A-R, Toth G, Wu M-C, Shchukarev A, Popov A, Mikkola J-P, Lauri J, Riihimaki M, Lappalainen J, Jantunena H, Kordas K (2012) Inkjet-printed gas sensors: metal decorated WO_3 nanoparticles and their gas sensing properties. J Mater Chem 22:17878–17886

Kumar D, Sharma RC (1998) Advances in conductive polymers. Eur Polym J 34(8):1053–1060

Labeau M, Gautheron B, Delabouglise G, Pena J, Ragel V, Varela A, Roman J, Martinez J, Gonzalez-Calbet JM, Regi-Vallet M (1993) Synthesis, structure and gas sensitivity properties of pure and doped SnO_2. Sens Actuators B Chem 15–16:379–389

Lalauze R, Breuli P, Pijolat C (1991) Thin films for gas sensors. Sens Actuators B Chem 3:175–182

Lavernia E, Wu Y (1996) Spray atomization and deposition. Wiley, Chichester

Lee D-H, Chang Y-J, Stickle W, Chang C-H (2007a) Functional porous tin oxide thin films fabricated by inkjet printing process. Electrochem Solid State Lett 10(11):K51–K54

Lee S, Lee G-G, Kim J, Kang S-JL (2007b) A novel process for fabrication of SnO_2-based thick film gas sensors. Sens Actuators B Chem 123:331–335

LeGore LJ, Greenwood OD, Paulus JW, Frankel DJ, Lad RJ (1997) Controlled growth of WO_3 films. J Vac Sci Technol A 15:1223–1227

Leite ER, Cerri JA, Longo E, Valera JA, Paskocima CA (2001) Sintering of ultrafine undoped SnO_2 powders. J Eur Ceram Soc 21:669–675

Li GL, Wang GH, Hong JM (1999) Morphologies of rutile form TiO_2 twins crystals. J Mater Sci Lett 18:1243–1246

Li B, Santhanam S, Schultz L, Jeffries-EL M, Iovu MC, Sauve G, Cooper J, Zhang R, Revelli JC, Kusne AG, Snyder JL, Kowalewski T, Weiss LE, McCullough RD, Fedder GK, Lambeth DN (2007) Inkjet printed chemical sensor array based on polythiophene conductive polymers. Sens Actuators B Chem 123:651–660

Liu C (1995) Development of chemical sensors using microfabrication and micromachining techniques. Mater Chem Phys 42:87–90

Livage J (1997) Sol-gel processes. Solid State Mater Sci 2:132–136

Livage J, Henry M, Sanchez C (1988) Sol-gel chemistry of transition metal oxides. Prog Solid State Chem 18:259–342

Llobet E, Molas G, Molinas P, Calderer J, Vilanova X, Brezmes J, Sueiras JE, Correig X (2000) Fabrication of highly selective tungsten oxide ammonia sensors. J Electrochem Soc 147(3):776–779

Lu K (2008) Sintering of nanoceramics. Int Mater Rev 53:21–38

Mabrook MF, Pearson C, Petty MC (2006a) Inkjet-printed polymer films for the detection of organic vapors. IEEE Sens J 6:1435–1444

Mabrook MF, Pearson C, Petty MC (2006b) Inkjet-printed polypyrrole thin films for vapour sensing. Sens Actuators B Chem 115:547–551

Mabrook MF, Pearson C, Jombert AS, Zeze DA, Petty MC (2009) The morphology, electrical conductivity and vapour sensing ability of inkjet-printed thin films of single-wall carbon nanotubes. Carbon 47:752–757

Madou MJ, Morrison SR (1989) Chemical sensing with solid state devices. Academic, Boston

Maklin J, Mustonen T, Halonen N, Toth G, Kordas K, Vahakangas J, Moilanen H, Kukovecz A, Konya Z, Haspel H, Gingl Z, Heszler P, Vajtai R, Ajayan PM (2008) Inkjet printed resistive and chemical-FET carbon nanotube gas sensors. Phys Status Solidi B 245:2335–2338

Malinauskas A (2001) Chemical deposition of conducting polymers. Polymer 42:3957–3972

Malkin AY, Siling MI (1991) Scientific principles of present-day and future technologies of synthesis and processing polycondensation polymers. Rev Polymer Sci 33:2135–2160

Marek J, Trah H-P, Suzuki Y, Yokomori I (eds) (2003) Sensors for automotive technology. VCH, Weinheim

Matsumoto Y, Yoshida K, Ishida M (1998) A novel deposition technique for fluorocarbon films and its applications for bulk- and surface-micromachined devices. Sens Actuators A Phys 66:308–314

Miccoci G, Serra A, Siciliano P, Tepore A, Ali-Adib Z (1996) CO sensing characteristics of reactively sputtered SnO_2 thin films prepared under different oxygen partial pressure values. Vacuum 47:1175–1177

Miller JS (ed) (1982) Catalysis and electrocatalysis. Am. Chem. Soc. Symp. Ser. vol 192. American Chemical Society, Washington, DC

Milchev A (2008) Electrocrystallization: Nucleation and growth of nano-clusters on solid surfaces. Russ J Electrochem 44:619–645

Minh NQ, Takahashi T (1995) Science and technology of ceramic fuel cells. Elsevier, Amsterdam

Morrison SR (1994) Chemical sensors. In: Sze SM (ed) Semiconductor sensors. Wiley, New York, NY, pp 404–408

Moseley PT, Tofield BC (eds) (1987) Solid state gas sensors. Adam Hilger, Bristol

Moseley PT, Norris JOW, Williams DE (1991) Techniques and mechanisms in gas sensing. Adam Hilger, Bristol

Nakaso K, Han B, Ahn KH, Choi M, Okuyama K (2003) Synthesis of non-agglomerated nanoparticles by an electrospray assisted chemical vapor deposition (ES-CVD) method. J Aerosol Sci 34:869–881

Narendar Y, Messing GL (1997) Mechanisms of phase separation in gel-based synthesis of multicomponent metal oxides. Catal Today 35:247–268

Nayral C, Viala E, Fau P, Senocq F, Jumas JC, Maisonnat A, Chaudret B (2000) Synthesis of tin and tin oxide nanoparticles of low size dispersity for application in gas sensing. Chem Eur J 6:4082–4090

Nenov TG, Yordanov SP (1996) Ceramic sensors: technology and applications. Technomic, Basel

Niesen TP, De Guire MR (2001) Review: deposition of ceramic thin films at low temperatures from aqueous solutions. J Electroceram 6:169–207

Olding T, Sayer M, Barrow D (2001) Ceramic sol-gel composite coatings for electrical insulation. Thin Solid Films 398–399:581–586

Olvera ML, Maldonaldo A, Asomoza R (1996) Characterization of a thin film tin oxide gas sensor deposited by chemical spraying. AIP Conf Proc 378:376–381

Osada Y, DeRossi DE (eds) (2000) Polymer sensors and actuators. Springer, Berlin

O'Toole M, Shepherd R, Wallace GG, Diamond D (2009) Inkjet printed LED based pH chemical sensor for gas sensing. Anal Chim Acta 652(1–2):308–314

Oyabu T (1982) Sensing characteristic of SnO_2 thin film gas sensors. J Appl Phys 53:2785–2787

Palatnik LS, Fuks MI, Kosevich VM (1972) Mechanism of formation and substructure of condensed films. Science, Moscow (in Russian)

Pashchanka M, Gurlo A, Prasad RM, Nicoloso N, Riedel R, Schneider JJ (2012) Inkjet printed In_2O_3 and In_2O_3/CNT hybrid microstructures for future gas sensing application. In: Proceedings of the 14th international meeting on chemical sensors, IMCS 2012, Nuremberg, 20–23 May, 791–794

Peter C, Kneer J, Wöllenstein J (2011) Inkjet printing of titanium doped chromium oxide for gas sensing application. Sens Lett 9(2):807–811

Phillips HM, Li Y, Bi X, Zhang B (1996) Reactive pulsed laser deposition and laser induced crystallization of SnO_2 transparent conducting thin films. Appl Phys A 63:347–351

Pique A, Auyeung RCY, Stepnowsk JL, Weir DW, Arnold CB, McGill RA, Chrisey DB (2003) Laser processing of polymer thin films for chemical sensor applications. Surf Coat Technol 163–164:293–299

Randhaw H (1991) Review of plasma-assisted deposition processes. Thin Solid Films 196:329–349

Risti M, Ivanda M, Popovi S, Musi S (2002) Dependence of nanocrystalline SnO_2 particle size on synthesis route. J Non-Crystal Solids 303:270–280

Reisinger JJ, Hillmyer MA (2002) Synthesis of fluorinated polymers by chemical modification. Prog Polym Sci 27:971–1005

Rumyantseva MN, Labeau M, Senateur JP, Delabouglise G, Boulova MN, Gaskov AM (1996) Influence of copper on sensor properties of tin dioxide films in H_2S. Mater Sci Eng B 41:228–234
Saadeddin I, Pecquenard B, Manaud JP, Decourt R, Abrugère C, Buffeteau T, Campet G (2007) Synthesis and characterization of single- and co-doped SnO_2 thin films for optoelectronic applications. Appl Surf Sci 253: 5240–5249
Sahner K, Tuller HL (2010) Novel deposition techniques for metal oxide: prospects for gas sensing. J Electroceram 24:177–199
Sangaletti L, Depero LE, Allieri B, Pioselli F, Angelucci R, Poggi A, Tagliani T, Nicoletti S (1999) Microstructural development in pure and V-doped SnO_2 nanopowders. J Eur Ceram Soc 19:2073–2077
Sayago I, Gutierrer FJ, Ares L, Robla JI, Horrillo MC, Getino J, Rino J, Agapito JA (1995a) The effect of additives in tin oxide on the sensitivity and selectivity to NO*x* and CO. Sens Actuators B Chem 26:19–23
Sayago I, Gutierrer J, Ares L, Robla JI, Horrillo MC, Getino J, Rino J, Agapito JA (1995b) Long-term reliability of sensors for detection of nitrogen oxides. Sens Actuators B Chem 26:56–58
Sberveglieri G (ed) (1992) Gas sensors. Kluwer, Dordrecht
Sberveglieri G, Faglia G, Groppelli S, Nelli P, Taroni A (1992) A novel PVD technique for the preparation of SnO_2 thin-films as C_2H_5OH sensors. Sens Actuators B Chem 7:721–726
Sberveglieri G, Nelli P, Benussi GP, Depero LE, Zocchi M, Rossetto G, Zanella P (1993) Enhanced response to methane for SnO_2 thin films prepared with the CVD technique. Sens Actuators B Chem 15–16:334–337
Schierbaum KD, Vaihinger S, Gopel W (1990) Prototype structure for systematic investigations of thin-film gas sensors. Sens Actuators B Chem 1:171–175
Schoenholzer U, Hummel R, Gauckler LJ (2000) Microfabrication of ceramics by filling of photoresist molds. Adv Mater 12:1261–1263
Simon I, Barsan N, Bauer M, Weimar U (2001) Micromachined metal oxide gas sensors: opportunities to improve sensor performance. Sens Actuators B Chem 73:1–26
Sinner-Hettenbach M (2000) SnO_2 (110) and nano-SnO_2: characterization by surface analytical techniques. Ph.D. thesis, University of Tübingen
Skolheim TA (ed) (1986) Handbook of conducting polymers. Marcel Dekker, New York, NY
Small WR, Panhuis M (2007) Inkjet printing of transparent, electrically conducting single-walled carbon-nanotube composites. Small 3:1500–1503
Song K-H, Park SJ (1994) Factors determining the carbon monoxide sensing properties of tin oxide thick films calcined at different temperatures. J Am Ceram Soc 77:2935–2939
Spannhake J, Helwig A, Schulz O, Muller G (2009) Micro-fabrication of gas sensors. In: Comini E, Faglia G, Sberveglieri G (eds) Solid state gas sensing. Springer, Berlin, pp 1–46
Stryhal Z, Pavlik J, Novak S, Mackova A, Perina V, Veltruska K (2002) Investigations of SnO_2 thin films prepared by plasma oxidation. Vacuum 67:665–671
Suchea M, Katsarakis N, Christoulakis S, Nikolopoulou S, Kiriakidis G (2006) Low temperature indium oxide gas sensors. Sens Actuators B Chem 118:135–141
Tahar RBH, Ban T, Ohya Y, Takahashi Y (1997) Optical, structural and electrical properties of indium oxide thin films prepared by the sol-gel method. J Appl Phys 82:865–870
Takada T (2001) A temperature drop on exposure to reducing gases for various metal oxide thin films. Sens Actuators B Chem 77:307–311
Takashima K, Minami K, Esashib M, Nishizawa J (1994) Laser projection CVD using the low temperature condensation method. Appl Surf Sci 79/81:366–374
Tay BK, Zhao ZW, Chua DHC (2006) Review of metal oxide films deposited by filtered cathodic vacuum arc technique. Mater Sci Eng R 52:1–48
Tiburcio-Silver A, Sanchez-Juarez A (2004) Regeneration processes study on spray-pyrolyzed SnO_2 thin films exposed to CO-loaded air. Sens Actuators B Chem 102:174–177
Tiemann M (2008) Repeated templating. Chem Mater 20:961–971
Troczynski T, Yang Q (2001) Process for making chemically bonded sol-gel ceramics. US Patent 6,284,682, May 2001
Vahlas C, Caussat B, Serp P, Angelopoulos GN (2006) Principles and applications of CVD powder technology. Mater Sci Eng R 53:1–72
Vandrish G (1996) Ceramic applications in gas and humidity sensors. Key Eng Mater 122–124:185–224
Van Tassel JJ, Randall CA (2006) Mechanism of electrophoretic deposition. In: Boccaccini AR, Van der Biest O, Clasen R (eds) Electrophoretic deposition: fundamentals and applications. Trans Tech Publications, Zurich
Vayssieres L (2007) An aqueous solution approach to advanced metal oxide arrays on substrates. Appl Phys A 89:1–8
Vincenzi D, Butturi MA, Stefancich M, Malagu C, Guidi V, Carotta MC, Martinelli G, Guarnieri V, Brida S, Margesin B, Giacomozzi F, Zen M, Vasiliev AA, Pisliakov AV (2001) Low-power thick-film gas sensor obtained by a combination of screen printing and micromachining techniques. Thin Solid Films 391:288–292
Viswanathan V, Laha T, Balani K, Agarwal A, Seal S (2006) Challenges and advances in nanocomposite processing techniques. Mater Sci Eng R 54:121–285

Vuong DD, Sakai G, Shimanoe K, Yamazoe N (2004) Preparation of grain size-controlled tin oxide sols by hydrothermal treatment for thin film sensor application. Sens Actuators B Chem 103:386–391

White NM, Turner JD (1997) Thick-film sensors: past, present and future. Meas Sci Technol 8:1–20

Will J, Mitterdorfer A, Kleinlogel C, Perednis D, Gauckler LJ (2000) Fabrication of thin electrolytes for second-generation solid oxide fuel cells. Solid State Ion 131:79–96

Williams G, Coles GSV (1993) NO*x* response of tin dioxide based gas sensors. Sens Actuators B Chem 15–16:349–353

Williams G, Coles GSV (1995) The influence of deposition parameters on the performance of tin dioxide NO_2 sensors prepared by radio-frequency magnetron sputtering. Sens Actuators B Chem 25:469–473

Willmott PR (2004) Deposition of complex multielemental thin films. Prog Surf Sci 76:163–217

Windle J, Derby B (1999) Ink jet printing of PZT aqueous ceramic suspensions. J Mater Sci Lett 18:87–90

Wu Q, Lee K-M, Lin C-C (1993) Development of chemical sensors using microfabrication and micromachining techniques. Sens Actuators B Chem 13–14:1–6

Yang P, Deng T, Zhao D, Feng P, Pine D, Chmelka BF, Whitesides GM, Stucky GD (1998) Hierarchically ordered oxides. Science 282:2244–2246

Yang H, Deschatelets P, Brittain ST, Whitesides GM (2001) Fabrication of high performance ceramic microstructures from a polymeric precursor using soft lithography. Adv Mater 13:54–58

Yang L, Zhang R, Staiculescu D, Wong CP, Tentzeris MM (2009) A novel conformal RFID-enabled module utilizing inkjet printed antennas and carbon nanotubes for gas detection applications. IEEE Antennas Wireless Propagat Lett 8:653–656

Yamazoe N, Miura N (1992) Some basic aspects of semiconductor gas sensors. In: Yamauchi S (ed) Chemical sensor technology, vol 4. Elsevier, Amsterdam

Yu KN, Xiong X, Liu Y, Xiong C (1997) Microstructural change of nano-SnO_2 grain assemblages with the annealing temperature. Phys Rev B 55:2666–2671

Yue Y, Gao Z (2000) Synthesis of mesoporous TiO_2 with crystalline frame work. Chem Commun 1755–1756

Xia Y, Whitesides GM (1998) Soft lithography. Angew Chem Int Ed 37:550–575

Zakrzewska K (2001) Mixed oxides as gas sensors. Thin Solid Films 391:229–238

Zeigler JM, Fearon FWG (eds) (1990) Silicon based polymer science: a comprehensive resource. ACS advances in chemistry series No. 224. American Chemical Society, Washington, DC

Zhao X, Evans JRG, Edirisinghe MJ (2002) Direct ink-jet printing of vertical walls. J Am Ceram Soc 85:2113–2115

Chapter 29
Outlook: Sensing Material Selection Guide

As our review shows, we do not have an ideal sensing material. All materials have both advantages and shortcomings. A detailed comparison of the parameters of various sensing materials is presented in Tables 29.1 and 29.2. It is seen that some sensing materials have poor selectivity, some are highly sensitive to humidity, some are stable only at low temperatures, some degrade while interacting with ozone, some require high temperatures for operating, and so on. Therefore, in choosing a sensing material for a particular application, the selected material should capitalize on its advantages, while its shortcomings should minimally influence the characteristics of the final device. For example, organic semiconductors do not interact as strongly with oxygen as inorganic semiconductors (Sadaoka 1992). Moreover, polymers can have a huge variety of properties, and they can be easily modified to obtain the required selectivity during interaction with analytes at low operating temperatures (Walton 1990; Harsanyi 1994, 2000). However, polymers have worse thermal and long-term stability of parameters than some other materials (Harsanyi 1994, 2000) (see Table 29.3). Other disadvantages of polymers are high sensitivity to water vapor and low sensitivity to hydrocarbons and other hydrophobic molecules. For instance, the humidity responses of conductive polymer sensors are sometimes so high that the small signals produced by important volatiles are lost, leading to a lack of discrimination (Clements et al. 1998).

It should be noted that the absence of an ideal sensor material is quite natural. Gas sensors are being used to measure a very wide variety of analytes in an equally large number of environments. Just for that reason there are different requirements for materials intended for use in different sensors. Optical gas sensors require materials with either transparency or intensity of luminescence, depending on the surroundings. Conductometric gas sensors require materials with maximum efficiency of chemical reaction upon a change in resistance. Mass-sensitive sensors require materials with large and reversible adsorption. The choice of a suitable material for humidity sensors should be based on good sensitivity over the entire range of humidity and temperature, on low hysteresis and high stability of parameters over time and thermal cycling, and on exposure to the various chemicals presented in the environment (Kulwicki 1991). In other words, the choice of material for a particular gas sensor is always a compromise decision, demanding consideration of sometimes contradictory requirements. For example, high basicity of oxides is advantageous for the formation of protonic charge carriers in proton conductors. On the other hand, basic oxides are expected to react easily with acidic or even amphoteric gases such as SO_3, CO_2, or H_2O to form sulfates, carbonates, or hydroxides (Kreuer 1997, 2003). A comparative analysis of sensing materials acceptable for application in solid-state gas sensors is presented in Table 29.1.

As we have emphasized throughout, sensing materials for different applications require different properties, which may be important only for a specific type of sensor (Korotcenkov 2005). Every new

G. Korotcenkov, *Handbook of Gas Sensor Materials,* Integrated Analytical Systems,
DOI 10.1007/978-1-4614-7388-6_29, © Springer Science+Business Media New York 2014

Table 29.1 Advantages and disadvantages of sensing materials acceptable for application in chemical sensors

Sensing material	Main advantages	Main disadvantages	Type of sensors optimal for material application
Metal oxides (poly- or nanocrystalline)	Good chemical and thermal stability; high sensitivity; sensitivity to a wide range of gases; long lifetime; good potential for microminiaturization; wide range of operating temperatures; easily functionalized; low-cost technology	Poor selectivity; long-term drift; poisoning	Sensors for tough conditions; conductometric gas sensors; support for catalytic gas sensors; electronic nose; electrodes for high-temperature electrochemical sensors; pH sensors; humidity sensors; optical sensors; capacitance sensors
Metal oxide 1D structures (individual structures)	Chemical stability; thermal stability; high sensitivity (can detect single molecules); low thermal mass allows rapid heating with low power consumption	Low selectivity; long-term drift; expensive technology (problems with separation, selection, manipulation, and electrical contacts); poor reproducibility	Conductometric sensors; FET sensors; sensors for microenvironments
Metal oxide 1D structures (arrays)	Chemical stability; thermal stability; high sensitivity	Low selectivity; long-term drift	Conductometric gas sensors; sensors for tough conditions; pH sensors
Carbon 1D structures (individual structures)	High surface area; RT operation; can be functionalized for chemical specificity (can detect single molecules); low thermal mass allows rapid heating with low power consumption	Individual CNTs have a broad range of possible conductance values; low selectivity; expensive technology (problems with separation, selection, manipulation, and electrical contacts); poor reproducibility; necessity of functionalizing; insufficiently good stability	Conductometric gas sensors; sensors for microenvironments; biosensors
Carbon 1D structures (arrays)	High surface area; RT operation	Low selectivity; necessity of functionalizing; insufficiently good stability	Conductometric gas sensors; membranes; biosensors; separation
Standard semiconductors (Si, GaAs, etc.)	Compatibility with standard microelectronic technology; good parameter control; good potential for miniaturization; well-established manufacturing methods including micromachining and patterning; RT operation	Poor selectivity; poor stability; drift; low sensitivity	Low-temperature gas Schottky barriers and FET gas sensors; CHEMFET sensors
Wide-band-gap semiconductors (GaN, SiC, etc.)	Compatibility with standard microelectronic technology; good potential for microminiaturization; good chemical and thermal stability; can be operated at high temperatures and in harsh environments	Poor selectivity; complex technology; low sensitivity	Specific applications; Schottky barriers and FET gas sensors; electronic nose

Porous silicon	High surface area; control of many parameters can be used for sensor design; RT operation	Poor stability; drift; poor reproducibility; complicated surface chemistry	Specific low-temperature sensors; biosensors; optical sensors; fabrication hotplate platforms
Bulk metals (films)	Compatibility with standard microelectronic technology; good potential for microminiaturization	Limited number of metals acceptable for application (expensive noble metals required); poor stability of ultrathin films at high temperatures; drift; limited number of analyte that can be tested	Electrodes; catalysts; filters for all types of sensors; surface plasmon resonance sensors; H_2 sensors; work function sensors
Metal nanoparticles	High catalytic activity; good chemical selectivity for specific analytes	Poor reproducibility; poor stability; drift; poisoning	Low-temperature sensors; surface plasmon resonance sensors; pellistors; biosensors
Solid electrolytes	High stability; parameters readily predictable; acceptable selectivity for specific analytes	Operate at high temperatures; limited number of analytes that can be tested; relatively high cost	High-temperature electrochemical gas sensors; sensors for tough conditions; electronic nose
Ionic liquids		Instability	RT electrochemical sensors
Zeolites	Small pore diameter; shape selectivity; good possibility for controlling pore parameters; high stability	Speed of response; unstable in hydrothermal environments	Membranes; filters for all types of sensors; SAW sensors; cantilever-based sensors
Polymers	Good selectivity; sensitivity to a wide range of analytes; RT operation; easily functionalized	Poor chemical and thermal stability; may be chemically incompatible with certain chemical environments; long-term drift; insufficient lifetime; long response time; high sensitivity to humidity; poor sensitivity to hydrocarbons and other hydrophobic molecules	Low-temperature sensors; SAW sensors; biosensors; humidity sensors; conductometric gas sensors; membranes; electrochemical sensors; cantilever-based sensors; optical and fiber-optic sensors; electronic noise
Biological objects	High selectivity; RT operation	Poor stability; short lifetime; low technological effectiveness; hard to control; problematic in large-scale production	Biosensors; pathogen detection; toxicity indicators
Calixarenes	Good sensitivity; easy to synthesize and functionalize; high selectivity	Relatively soluble in aqueous solutions	Electrochemical sensors; separation; mass-sensitive sensors
Composites	Many possibilities for control of sensing properties	Control of composition during deposition is more difficult; difficult to fabricate sensors reproducibly with consistent characteristics	Same as for basic material
Metal oxide-based composites	Same as for nano- and polycrystalline metal oxides	Same as for nano- and polycrystalline metal oxides	Catalytic filters; specific conductometric gas sensors
Carbon-based composites	Same as for carbon 1D structures (arrays)	Same as for carbon 1D structures (arrays)	Same as for carbon 1D structures (arrays)
Polymer-based composites	Same as for polymers	Same as for polymers	Same as for polymers

Table 29.2 Advantages and disadvantages of various materials for solid-state conductometric gas sensor applications

	Material					
Property	Polymers	Metals	Solid electrolytes	Ionic salts	Metal oxides	Covalent semiconductors
Stability	Very poor	Good	Good	Poor	Very good	Poor
Operating temperature (°C)	RT–200	RT–500	200–900	RT–100	RT–800	RT–400
Sensitivity	Good	Average	Good	Good	Good	Average
Selectivity	Very good	Average	Very good for separate analyte	Good	Poor	Poor
Manufacturability	Average	Very good	Good	Average	Very good	Average

Source: Reprinted with permission from Wilson et al. (2001). Copyright 2001 IEEE
RT room temperature

Table 29.3 Summary of advantages and disadvantages of polymer-based sensors

Sensor application	Advantages compared to inorganic based sensors	Disadvantages
Surface acoustic wave sensor	High sensitivity; shock resistance	Limited temperature range
Humidity sensors	High sensitivity; possibility of integration	Stability problems
Gas sensors	High selectivity; high sensitivity; RT operation	Long-term drift
Ion-selective sensors	High selectivity; wide choice of ionophores	Short lifetime

Source: Data from Harsanyi (1994)

Table 29.4 Parameters that should be optimized and controlled for successful development of inorganic, organic, and biological sensing materials

Group of sensing materials	Type of sensing material	Optimized and controlled material parameters
Inorganic	Catalytic metals	Surface additives; porosity; layered structure; grain size; alloying; deposition method
	Metal oxide materials	Base single or mixed metal oxides; deposition method and conditions of base; metal oxide(s); annealing method and conditions; dopant(s); doping method and conditions; purity of materials
	Plasmonic nanostructures and nanoparticles	Substrate type; nanoparticle material; nanoparticle shape, size, morphology; nanoparticle arrangement; surface functionality
	Plasmonic nanoparticles in polymers	Size of nanoparticle; strength of polymer/particle interaction; polymer grafting density; polymer chain length; binding constant; pH influence; redox state; selectivity; toxicity; poisoning agents
Organic	Indicators	Analyte-responsive reagent; polymer matrix; analyte-specific ligand; plasticizer; other agents (stabilizing phase transfer, etc.); common solvent
	Polymeric compositions	Polymerization conditions; types of heterocycles; additive(s); side groups; dopant; oxidation state
	Conjugated polymers	Electrode material; thickness; morphology
	Molecularly imprinted polymers	Functional monomer(s); template concentration; cross-linker; porogen; monomer(s)/template ratio; physical conditions during polymerization
Biological	Surface-immobilized bioreceptors	Immobilization technique; receptor–surface spacer; receptor–receptor spacer
	Matrix-immobilized bioreceptors	Immobilization technique; receptor density; matrix hydrophilicity; matrix charge; matrix chemical composition; matrix thickness

Source: Reprinted with permission from Potyrailo and Mirsky (2009). Copyright 2009 Springer

application creates its own requirements for sensing materials. As a result, in the literature one can find a lot of materials that are being tested for the purpose of evaluating their potential for application to the design of gas sensors. As we have shown in the present volumes, this process of synthesizing and studying new sensing materials is in progress and involves research into new types of technologies and materials. Fullerenes, carbon and metal oxide nanotubes, nanowires, metal, and semiconductor nanoparticles, which have been discussed in detail in the present volumes, include such materials. We do not know yet where a qualitative leap to a world of nanosized ranked structures might eventually lead. However, the last results obtained in this field are encouraging (Chao and Shih 1998; Kong et al. 2000; Varghese et al. 2001; Baena et al. 2002; Cantalini et al. 2003; Valentini et al. 2003, 2004; Wang 2003, 2004; Kolmakov and Moskovits 2004; Penza et al. 2004; Gao and Wang 2005; Zhang et al. 2004, 2005; Baratto et al. 2005; Yu et al. 2005; Ramgir et al. 2005; Comini 2006; Sberveglieri et al. 2007; Sysoev et al. 2010).

In closing we would like to note that, as follows from the present volumes, the gas-sensing effect is a very complicated phenomenon which is dependent on many and various factors. As a result, for designing optimal material for gas sensors we have to optimize many parameters of the gas-sensing matrix. In particular, Table 29.4 provides a summary of parameters that should be optimized and controlled for successful development of inorganic, organic, and biological sensing materials. Inorganic sensing materials include catalytic metals for field-effect devices, metal oxides for conductometric and cataluminescent sensors, plasmonic, and semiconductor nanocrystal materials. Organic sensing materials include indicator dyes (free, polymer immobilized, and surface confined), polymeric compositions, homo- and copolymers, conjugated polymers, and molecularly imprinted polymers. Biological materials include surface- and matrix-immobilized bioreceptors. One can see that every type of gas-sensing material requires a specific approach to this process.

References

Baena JR, Gallego M, Valcarcel M (2002) Fullerenes in the analytical sciences. Trends Anal Chem 21(3):187–198

Baratto C, Comini E, Faglia G, Sberveglieri G, Zha M, Zappettini A (2005) Metal oxide nanocrystals for gas sensing. Sens Actuators B 109:2–6

Cantalini C, Valentini L, Lozzi L, Armentano I, Kenny JM, Santucci S (2003) NO_2 gas sensitivity of carbon nanotubes obtained by plasma enhanced chemical vapor deposition. Sens Actuators B 93:333–337

Chao YC, Shih JS (1998) Adsorption study of organic molecules on fullerene with piezoelectric crystal detection system. Anal Chim Acta 374:39–46

Clements J, Boden N, Gibson TD, Chandler RC, Hulbert JN, Ruck-Keene EA (1998) Novel, self-organizing materials for use in gas sensor arrays: beating the humidity problem. Sens Actuators B 47:37–42

Comini E (2006) Metal oxide nano-crystals for gas sensing. Anal Chim Acta 568:28–40

Gao PX, Wang ZL (2005) Nanoarchitectures of semiconducting and piezoelectric zinc oxide. J Appl Phys 97:044304

Harsanyi G (1994) Polymer films in sensor applications. Technomic, Lancaster, PA

Harsanyi G (2000) Polymer films in sensor applications: a review of present uses and future possibilities. Sens Rev 20(2):98–105

Kolmakov A, Moskovits M (2004) Chemical sensing and catalysis by one-dimensional metal oxide nanostructures. Annu Rev Mater Res 34:151–180

Kong J, Franklin NR, Zhou C, Chapline MG, Peng S, Cho K, Dai H (2000) Nanotube molecular wires as chemical sensors. Science 287:622–625

Korotcenkov G (2005) Gas response control through structural and chemical modification of metal oxides: state of the art and approaches. Sens Actuators B 107:209–232

Kreuer KD (1997) On the development of proton conducting materials for technological applications. Solid State Ionics 97:1–15

Kreuer KD (2003) Proton-conducting oxides. Annu Rev Mater Res 33:333–359

Kulwicki BM (1991) Humidity sensors. J Am Ceram Soc 74:697–708

Penza M, Antolini F, Vittori-Antisari M (2004) Carbon nanotubes as SAW chemical sensors materials. Sens Actuators B 100:47–59

Potyrailo RA, Mirsky VM (2009) Introduction to combinatorial methods for chemical and biological sensors. In: Potyrailo RA, Mirsky VM (eds) Combinatorial methods for chemical and biological sensors. Springer, New York, pp 3–24

Ramgir NS, Mulla IS, Vijayamohanan KP (2005) A room temperature nitric oxide sensor actualized from Ru-doped SnO_2 nanowires. Sens Actuators B 107:708–715

Sadaoka Y (1992) Organic semiconductor gas sensors. In: Sberveglieri G (ed) Gas sensors. Kluwer Academic, Dordrecht, pp 187–218

Sberveglieri G, Baratto C, Comini E, Faglia G, Ferroni M, Ponzoni A, Vomiero A (2007) Synthesis and characterization of semiconducting nanowires for gas sensing. Sens Actuators B 121:208–213

Sysoev VV, Strelcov E, Sommer M, Bruns M, Kiselev I, Habicht W, Kar S, Gregoratti L, Kiskinova M, Kolmakov A (2010) Single-nanobelt electronic nose: engineering and tests of the simplest analytical element. ACS Nano 4(8):4487–4494

Valentini L, Cantalini C, Lozzi L, Armentano I, Kenny JM, Santucci S (2003) Reversible oxidation effects on carbon nanotubes thin films for gas sensing applications. Mater Sci Eng C 23:523–529

Valentini L, Cantalini C, Armentano I, Kenny JM, Lozzi L, Santucci S (2004) Highly sensitive and selective sensors based on carbon nanotubes thin films for molecular detection. Diamond Relat Mater 13:1301–1305

Varghese OK, Kichambre PD, Gong D, Ong KG, Dickey EC, Grimes CA (2001) Gas sensing characteristics of multi-wall carbon nanotubes. Sens Actuators B 81:32–41

Walton DJ (1990) Electrically conducting polymers. Mater Design 11:142–152

Wang ZL (2003) Nanobelts, nanowires, and nanodiskettes of semiconducting oxides—from materials to nanodevices. Adv Mater 15:432–436

Wang ZL (2004) Functional oxide nanobelts: materials, properties and potential applications in nanosystems and biotechnology. Annu Rev Phys Chem 55:159–196

Wilson DM, Hoyt S, Janata J, Booksh K, Obando L (2001) Chemical sensors for portable, handheld field instruments. IEEE Sensors J 1:256–274

Yu C, Hao Q, Saha S, Shi L, Kong X, Wang ZL (2005) Integration of metal oxide nanobelts with microsystems for nerve agent detection. Appl Phys Lett 86:063101

Zhang Y, Ago H, Liu J, Yumura M, Uchida K, Ohshima S, Iijima S, Zhu J, Zhang X (2004) The synthesis of In, In_2O_3 nanowires and In_2O_3 nanoparticles with shape-controlled. J Cryst Growth 264:363–368

Zhang Y, Kolmakov A, Libach Y, Moskovits M (2005) Electronic control of chemistry and catalysis at the surface of an individual tin oxide nanowire. J Phys Chem B 109:1923–1929

About the Author

Ghenadii Korotcenkov has more than 40-year experience as a teacher and scientific researcher. He received his Ph.D. in Physics and Technology of Semiconductor Materials and Devices in 1976, and his Habilitate Degree (Dr. Sci.) in Physics and Mathematics of Semiconductors and Dielectrics in 1990. For many years he led the scientific Gas Sensor Group and managed various national and international scientific and engineering projects carried out in Laboratory of Micro- and Optoelectronics, Technical University of Moldova. Since 2008, Korotcenkov has been a research Professor in Gwangju Institute of Science and Technology, Republic of Korea.

Korotcenkov's research results are well known in the study of Schottky barriers, MOS structures, native oxides, and photoreceivers based on III–V compounds. His current research interests, starting from 1995, include material sciences and surface science, focused on metal oxides and solid state gas sensor design.

Korotcenkov is the author or editor of 16 books and special issues, 12 invited review papers, 19 book chapters, and more than 190 peer-reviewed articles. He is a holder of 18 patents. His research activities have been honored by Award of the Supreme Council of Science and Advanced Technology of the Republic of Moldova (2004), The Prize of the Presidents of Ukrainian, Belarus and Moldovan Academies of Sciences (2003), Senior Research Excellence Award of Technical University of Moldova (2001, 2003, 2005), Fellowship from International Research Exchange Board (1998), National Youth Prize of the Republic of Moldova (1980), among others.

G. Korotcenkov, *Handbook of Gas Sensor Materials,* Integrated Analytical Systems,
DOI 10.1007/978-1-4614-7388-6, © Springer Science+Business Media New York 2014

Index

G. Korotcenkov, *Handbook of Gas Sensor Materials,* Integrated Analytical Systems, DOI 10.1007/978-1-4614-7388-6,

D

N

O

Printed in Great Britain
by Amazon